Geophysical Monograph 78

Climate Change in Continental Isotopic Records

P. K. Swart
K. C. Lohmann
J. McKenzie
S. Savin

Editors

American Geophysical Union

Published under the aegis of the AGU Books Board.

Library of Congress Cataloging-in-Publication Data

Climate change in continental isotopic records / P.K. Swart . . . [et al.], editors.
    p.   cm. — (Geophysical monograph : 78)
    Includes bibliographical references.
    ISBN 0-87590-037-2
    1. Climate changes.   2. Paleoclimatology.   3. Isotope geology.
4. Radioactive tracers.   I. Swart, P. K. (Peter K.)   II. Series.
QC981.8.C5C5195   1993
551.6—dc20   93-37709
                                                          CIP

ISSN: 0065-8448

ISBN: 0-87590-037-2 ✓

♾ This book is printed on acid-free paper.

Printed in the United States of America.

CONTENTS

CONTENTS (continued)

PREFACE

This volume is designed to provide paleoclimatologists with information on past climate from sources other than the deep sea. For climate modelers aware of the fact that current global climate models do not always accurately predict the temperatures of continental interiors, the analyses presented here may offer much needed constraints on their models. Because analysis of the oxygen and carbon isotopic composition of benthic and planktonic foraminifera has provided the basis for most of the work done on climatic variation, our interpretation has been biased by records from the marine environment and does not provide an adequate indication of the temperature of continental interiors. Our eventual aim is to stimulate geologists, geochemists, and others to collect data on continental isotopic indicators of climate (CIICs) throughout the world, perhaps concentrating on certain geological intervals. In fact, a global continental climate program, styled after an earlier program which studied climate in the marine realm (CLIMAP), has been proposed. With such data it may eventually be possible to construct paleo maps of surface temperature and/or the oxygen isotopic composition of precipitation during specific time periods. One suitable time period, for example, would be the Cretaceous. In this case further information may be gleaned on controversies such as whether the polar regions were in fact warm at this time or whether temperature gradients similar to today's were present.

The concept of continental isotopic indicators of climate was the theme of an AGU Chapman Conference attended by over 100 scientists in Jackson Hole, Wyoming between June 10 and 14, 1991. Oral and poster presentations explored potential applications of CIICs to past record of climate change. A representative selection of the papers presented at the meeting form the basis for this volume, which is divided into four sections, dealing with hydrology, lacustrine environments, regoliths, and organic materials.

The volume starts with the hydrological cycle, as it is the basis for all interpretation involving so-called CIICs.

In fact, without question, the most used diagram in the CIIC meeting was the contour map of the world showing the oxygen isotopic composition of global precipitation. This figure is also reproduced numerous times within this volume. In spite of its widespread use, the data on which this map is based are sparse. Further data are presented in this volume along with models which attempt to understand local and global variations in the oxygen and hydrogen isotopic composition of precipitation. However, further work is undoubtedly necessary on this aspect.

The second section deals with the continental analogues of the deep sea, the lacustrine environments. Of all the so-called CIICs, lake sediments have perhaps the greatest potential to give us climatic records similar to those obtained from the Deep Sea and Ocean Drilling programs. However, there are more than subtle differences between lakes and oceans, some of which are not completely understood. This volume devotes nine papers to this subject; some interpret results directly from lacustrine sediments, and others attempt to understand more fully the mechanisms that might lead to the observed patterns of isotopes.

The next section investigates the isotopic composition of what we call regolith. This category includes the formation and alteration of soil carbonates (calcretes), clay minerals, oxides and glasses. The premise here is that these minerals are influenced by the isotopic composition of the water and the temperature at which alteration takes place. By analyzing the chemical composition, one may be able to indirectly predict paleotemperatures and isotopic compositions.

The final section covers the isotopic composition of organic material, which can occur in a variety of environments, both terrestrial and aquatic. The isotopes that can be potentially analyzed in organic material include those of carbon, hydrogen, and oxygen, but substantial yet interesting problems of interpretation still remain.

The editors thank contributors to this volume and all the participants of the meeting. We see this volume as a beginning, not an end. The meeting was supported with help from the American Geophysical Union, the National Science Foundation, and the Swiss ProClim program. In addition, we particularly wish to thank Avis Miller, who helped extensively with preparation of this volume and the meeting organization. Finally, we wish to thank all the reviewers, who performed an outstanding job.

Peter K. Swart
Editor

Introduction

PETER K. SWART

Rosenstiel School of Marine and Atmospheric Sciences, Miami, Florida

JUDITH A. MCKENZIE

Geological Institute, ETH-Zentrum, Switzerland

KYGER C LOHMANN

Department of Geology, University of Michigan, Ann Arbor, Michigan

Much of our knowledge about the variations in past climate has been gained in the past 30 years through analysis of the oxygen and carbon isotopic composition of benthic and planktic foraminifera. Because these indicators of climate variations are restricted to the marine record, our information is naturally biassed towards variations in climate as experienced by marine organisms. While there is no doubt that the climatic variations recorded in marine environments do reflect global changes in general, these indicators do not reflect the climate of continental interiors.

Among the numerous techniques used in the study of the climate of continents are direct means, such as a synthesis of fauna and flora information, as well as indirect chemical measurements. This volume concentrates on continental isotopic indicators of climate (CIICs). Examples of such indicators are the hydrogen and oxygen isotopic ratios of cellulose, ice cores, and clays; the oxygen and carbon isotopic compositions of carbonates (cements, altered whole rock and calcretes), cave deposits, and lake sediments; and the oxygen isotopic composition of phosphorites. Papers included in this volume explore some of these indicators.

Most materials formed or influenced by sub-aerial processes involve precipitation from an aqueous phase originating either directly from, or modified by, meteoric processes. Through these mechanisms, materials acquire oxygen and hydrogen isotopic signatures indicative of both the tempera-

ture of formation and the fluid from which they form. A major influence on the hydrogen and oxygen isotopic composition of natural waters is exerted by the atmospheric part of the global water cycle. During the process of evaporation, water vapor becomes preferentially enriched in the lighter isotopes of H and O, while the residual fluid becomes isotopically heavier. Similarly during precipitation, rain and snow are enriched in the heavier isotopes, leaving atmospheric moisture depleted of these isotopes. Although there are many exceptions and variations, the general trend of atmospheric precipitation is one in which precipitation in equatorial regions is relatively heavy in oxygen and hydrogen isotopes, becoming progressively lighter towards the poles (see Rozanski et al., this volume). This worldwide trend is principally a consequence of the fact that most of the evaporation occurs in the sub-tropics and then is transported from tropical regions towards the higher latitudes. Along this path, water is lost as precipitation, and the residual water vapor becomes isotopically depleted according to processes approximating a Rayleigh distillation model. An oxygen vs. hydrogen isotopic plot of global precipitation falls on a straight line known as the Meteoric Water Line (MWL). The slope of this line represents the difference in the magnitude of the equilibrium fractionation factor (α) of hydrogen and oxygen during evaporation. Studies of fossil materials including Precambrian systems support the notion that ancient waters may follow a relationship similar to the MWL. However, while application of the MWL back through time should be treated with caution, at the moment there are no compelling arguments that the systematics of ancient meteor-

Climate Change in Continental Isotopic Records
Geophysical Monograph 78

ic waters were radically different from those of the modern day. Variations in the O and H isotopic composition of rainfall also correlate particularly well with the average monthly surface temperature, and therefore isotopic composition of meteoric water can be used as a proxy indicator of temperature.

In summary, the oxygen and hydrogen isotopic composition of meteoric water is recorded by a combination of sub-aerial diagenetic processes as well as by biological processes. Examination of the geographical distribution of these patterns should reveal changes in the patterns of climate. At the present day, strong latitudinal gradients exist in the oxygen and hydrogen isotopic composition of meteoric waters which can be correlated with temperature. In previous geological time periods in which climates were more equable, such gradients may not have been present to the same extent.

As an introduction, the following paragraphs present examples of continental isotopic indicators of climate. Papers within this volume which deal with these topics in more detail are referenced.

Cements and Diagenetically Transformed Carbonate Sediments

Groundwaters generally integrate the isotopic composition of the local rainfall. Exceptions to this may occur in situations where standing water is allowed to evaporate and hence become isotopically enriched in ^{18}O and D. In young carbonate terrains, waters soon become saturated with respect to calcium carbonate, dissolving metastable precursors and converting the local rock to cemented limestone. The isotopic composition of the formed limestone reflects both the temperature of precipitation and the isotopic composition of local fluids. The alteration of sediments in freshwater lenses can be established by examining the paragenetic sequence of the rock. A characteristic C and O isotopic pattern is found in such rocks as water-dominated systems shift the oxygen isotopic composition of the sediment away from the original O isotopic value towards one reflecting that of the meteoric fluid, but with a wider range of carbon isotopic values (see Figure 1). In such a system the O isotopic composition is water dominated, but the C is dictated by the isotopic composition of the precursor and secondary organic input. When preserved in carbonate precipitates, this trend of variable $\delta^{13}C$ and invariant $\delta^{18}O$ (meteoric calcite line, MCL) identifies a characteristic $\delta^{18}O$-carbonate value for the freshwater phreatic portion of each diagenetic system. Because the composition of meteoric calcite will vary geographically, its isotopic composition will reflect variations in climate. The identification of a MCL requires variation in the degree of rock-water interaction, either spatially or temporally. Therefore, in mono-mineralic calcite terrains, the isotopic values of meteoric calcites will not record a high

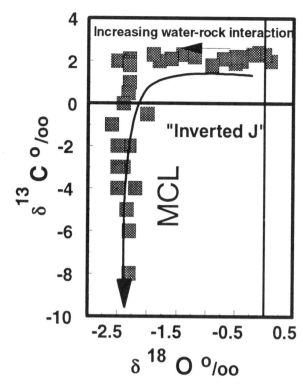

Figure 1: Idealized plot of variations in $\delta^{18}O$ and $\delta^{13}C$ of meteoric vadose and phreatic carbonates (inverted "J" curve).

degree of water-rock interaction. The application of this principle to the interpretation of paleoclimates is shown in this volume by papers from Smith and Dorobek and Rossinsky and Swart.

Calcretes

Calcretes can be defined as products of terrestrial processes within the weathering zone in which calcium carbonate has accumulated and replaced the existing soil. While the carbon isotopic composition of this carbonate can be extremely variable depending on the contribution of isotopically depleted CO_2 from vegetative processes, the oxygen isotopic composition generally is related to the temperature of precipitation and isotopic composition of local meteoric fluid. Care must be taken, however, to recognize calcretes influenced by evaporative water. Changes in the O isotopic composition of calcretes from the Plio-Pleistocene of East Africa have been used to estimate changes in climate. Several papers in this volume address calcretes, including a discussion of the fundamentals by Cerling and Quade and a discussion of their application to Holocene climatic problems by Kelly et al.

Lake Deposits

Lake waters can exhibit a wide range of isotopic compositions depending upon the isotopic composition of the rainfall in the catchment area and its seasonality, the amount and temperature of evaporation, and the relative humidity. An excellent record of past climate change can be secured from the study of discrete components of lake sediments. For example, the oxygen isotopic composition of authigenic carbonates and diatoms can be used to obtain a surface water signal of changes in temperature and meteoric conditions, while the composition of calcareous bottom dwellers, such as ostracods, can be used to filter out temperature effects, assuming that the bottom water temperatures remained constant. The occurrence of carbon and oxygen isotopic covariance in sedimented authigenic carbonates, or lack thereof, can provide climatically important information on the hydrologic state of a lake, such as whether it is an open or closed system, as well as the residence time of the water in the lake. In addition, the $\delta^{13}C$ of the autochthonous (amorphous) organic matter in conjunction with organic geochemical studies can be related to climatically induced changes of a lake's trophic state. These techniques have been extensively applied to sediments from lakes with waters that span the entire salinity range from fresh to hypersaline. The advantages of isotopic studies of climatic change on lake sediments are as follows: (1) the reservoirs are always small compared with the oceans, and the geochemical perturbations are large, providing a clear signal; (2) the response time is immediate; (3) lake sediments are frequently varved or laminated, so that high-resolution signals are possible; (4) sedimentation rates are high, so that high-resolution stratigraphy is possible; (5) lakes are ubiquitous through the geological record and are present at all latitudes. Numerous papers discuss climatic implications of lacustrine deposits both on the recent timescale (as in the papers by McKenzie and Hollander and by Chivas and coworkers), and towards progressively older periods (such as the work by Casanova and Hillaire-Marcel, Curtis and Hodell, Dettman and Lohmann, Palacios-Fest et al., and Patterson et al.).

Clays

Clays and related minerals formed as weathering products approach isotopic equilibrium with the environment in which they are formed. Oxygen and hydrogen isotopic compositions of clay minerals generally fall on a line which parallels the MWL offset by the appropriate equilibrium factor. This approach has been used to examine surface and supergene clays hydrated from volcanic glasses in the Tertiary of continental North America. These data reveal a systematic shift of the δD contours, interpreted as indicating more temperate climatic conditions during this time interval. This volume presents details of the principles involved in the application of near-surface alteration products as paleoclimate indicators (Yapp, Bird et al., Arehart and O'Neil).

Organic Material

The stable O and H isotopic composition of organic material can be used in several ways to reconstruct paleoclimate. Although tree rings are perhaps the most obvious example of this approach, relevant papers in this volume will not focus on climate variations revealed by modern trees in the 100 to 1000 year time scale. However, tree cellulose is often found in older geological deposits.

The oxygen and hydrogen isotopic composition of cellulose in plant material can be directly related to the isotopic composition of the water and temperature. These data suggest that δD values of meteoric water during this time period were considerably heavier than at the present time. There appears, however, to be separate relationships between H, O and water composition and temperature depending upon the species of plant being examined, and therefore care must be taken to develop relationships using a single species.

Oxygen and hydrogen isotopes of cellulose can be combined through a two-stage model involving evapotranspiration and equilibrium isotopic exchange. This later approach has been used to predict humidity based on oxygen and hydrogen isotopic analyses of cellulose, without consideration of factors such as temperature, $\delta^{18}O$ of atmospheric vapor and leaf-boundary dynamics. Three excellent papers (by Edwards; Leavitt; and Jirikowic et al.) detail the types of environmental information that can be obtained from the isotopic analysis of organic material.

Other Climatic Indicators

Although this volume is by no means a comprehensive review of all possible types of CIICs (other examples are phosphatic material in fish and mammal bones, cave deposits, and radiogenic isotopes), it serves as a starting point for further study in the field.

Isotopic Patterns in Modern Global Precipitation

Kazimierz Rozanski, Luis Araguás-Araguás, and Roberto Gonfiantini

International Atomic Energy Agency, P.O. Box 100, 1400-Vienna, Austria

The International Atomic Energy Agency (IAEA), in cooperation with the World Meteorological Organization (WMO), has been conducting a world-wide survey of hydrogen ($^2H/^1H$) and oxygen ($^{18}O/^{16}O$) isotope composition of monthly precipitation since 1961. At present, 72 IAEA/WMO network stations are in operation. Another 82 stations belonging to national organizations continue to send their results to the IAEA for publication. The paper focuses on basic features of spatial and temporal distribution of deuterium and ^{18}O in global precipitation, as derived from the IAEA/WMO isotope database. The internal structure and basic characteristics of this database are discussed in some detail. The existing phenomenological relationships between observed stable isotope composition of precipitation and various climate-related parameters such as local surface air temperature and amount of precipitation are reviewed and critically assessed. Attempts are presented towards revealing interannual fluctuations in the accumulated isotope records and relating them to changes of precipitation amount and the surface air temperature over the past 30 years.

INTRODUCTION

Decisive improvements in mass-spectrometry techniques shortly after World War II [*Nier, 1947; Nier et al., 1947; McKinney et al., 1950*] enabled precise measurements of the natural abundances of oxygen-18 and deuterium in meteoric waters. First publications on this subject appeared within a few years [*Dansgaard, 1953, 1954; Epstein and Mayeda, 1953; Friedman, 1953*]. The first attempt to summarize the available information on the isotopic composition of freshwaters worldwide (including precipitation) was published by Craig in 1961 [*Craig, 1961*].

In the same year the International Atomic Energy Agency (IAEA), in cooperation with the World Meteorological Organization (WMO), initiated a world-wide survey of the isotope composition of monthly precipitation. The programme was launched with the primary objective of collecting systematic data on isotope content of precipitation on a global scale (deuterium, oxygen-18 and tritium), characterizing their spatial and temporal variability and, consequently, providing basic isotope data for the use of environmental isotopes in hydrogeological investigations. It appeared soon that the collected data are also very useful in other water-related fields such as oceanography, hydrometeorology and climatology.

The operation of the network started with more than 100 meteorological stations in 65 countries and territories, collecting monthly composite precipitation samples for isotope analyses. The number of stations varied and reached a maximum of about 220

stations in 1963-64. After a revision in 1977, operation of some stations was stopped. During the past decade, the number of IAEA/WMO network stations was oscillating around eighty. Approximately the same number of stations belonging to national organizations continued to send their results to the IAEA for publication. Since 1990, approximately 15 new stations have been incorporated in the network, mainly representing regions with poor data coverage. Detailed technical procedures for the collection and shipment of samples, to be followed by the stations, and a standardized data reporting format were introduced from the beginning of operation of the network. The isotope analyses are performed in the IAEA laboratory and in laboratories of cooperating institutions in Member States. In addition to the isotope data, certain meteorological variables are also recorded (type and amount of precipitation, surface air temperature, vapor pressure) and reported to the IAEA.

The isotope and meteorological data are published regularly by the IAEA in the form of data books [*IAEA*, 1969, 1970, 1971, 1973, 1975, 1979, 1983, 1986, 1990] and are also available on magnetic tape or on floppy disks. Information on isotopic composition of precipitation includes the tritium content (reported as $^3H/^1H$ ratios, expressed in Tritium Units) and the stable isotope ratios ($^2H/^1H$ and $^{18}O/^{16}O$), all with their analytical errors as stated by the laboratories performing the analyses. Basic statistical treatment of the data accumulated till 1978 is available as a separate volume [*IAEA*, 1981]. A revised and extended version of this treatment, covering the data gathered up to the end of 1987, was published by the IAEA in 1992 [*IAEA*, 1992].

The basic characteristics of the network, as at the end of 1987, are summarized in Figures 1 to 3. The total number of stations ever operated is 379. Both tritium and stable isotopes are available for approximately 64 percent of the stations. Of the 251 stations for which the stable isotope data are available, 219 have an oxygen-18 record

Climate Change in Continental Isotopic Records
Geophysical Monograph 78

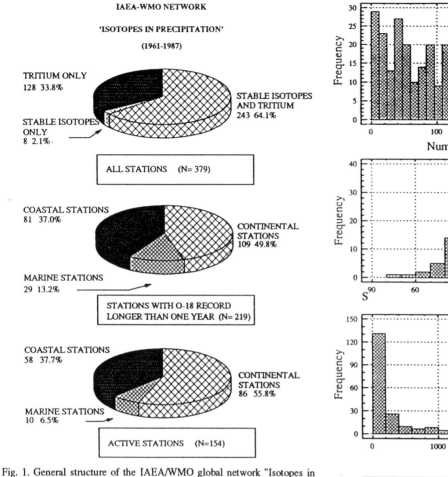

Fig. 1. General structure of the IAEA/WMO global network "Isotopes in Precipitation".

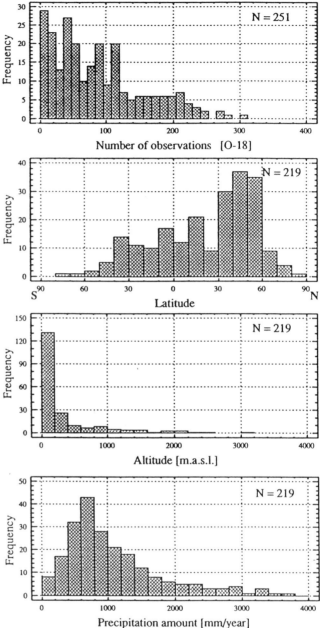

Fig. 2. (a) Frequency distribution histogram of the length of ^{18}O records available for the IAEA/WMO stations, as at the end of 1987; (b) Latitudinal distribution of the network stations for which a minimum of one year of ^{18}O record is available; (c) Frequency distribution histogram of the network stations with respect to altitude of their location; (d) Frequency distribution histogram of the network stations with respect to the amount of annual precipitation.

longer than one complete year (at least 12 monthly ^{18}O values). Only these stations were considered in this study. Although only one year of ^{18}O record may not always properly reflect average conditions at a given station, we have decided to include in the discussion also these relatively short records in order to improve spatial coverage with the data. Geographical distribution of the network stations is rather inhomogeneous (Figure 3); more than 50 percent are located in mid- northern latitudes, between 30°N and 60°N (Figure 2). Only 16 out of 219 stations are located in high latitudes (< 60°). The low-altitude stations (up to 200 m a.s.l.) constitute about 60 percent of the selected stations. Mountainous regions (above 1000 m a.s.l.) are represented by 24 stations. The length of the available oxygen-18 record also varies considerably (Figure 2). For 64 out of 219 stations, the record is shorter than 3 years. There are 32 stations for which at least 20 years of ^{18}O record is available. The Vienna station has the longest ^{18}O record (348 monthly data, 31 years of operation).

To date, approximately 180,000 isotope and meteorological values have been accumulated in the IAEA database. Critical analysis of the quality of stored information has recently been carried out, resulting in removing inconsistent data, obvious outliers, typing errors, etc. Altogether 530 stable isotope, 28 tritium and 325 meteorological data

have been removed in this way from the database. Similar analysis carried out recently for several southern hemisphere stations revealed that a substantial part of the stable isotope record for Rarotonga, Kaitaia and Invercargill are unreliable, most probably due to evaporation of precipitation samples after collection [*Taylor*,

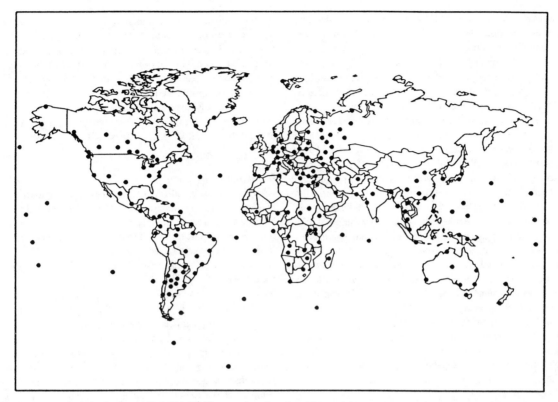

Fig. 3. Geographical distribution of the IAEA/WMO network stations for which a minimum of one complete year of stable isotope record is available.

1990]. These erroneous results were also removed from the database. Although some errors may still be present in the record, we believe that the overall quality of the database is now substantially improved. The revised statistical treatment mentioned above has been based on this cleaned database.

The stable isotope ratios of water, $^2H/^1H$ and $^{18}O/^{16}O$, are expressed by convention as parts per thousand deviation relative to the standard V-SMOW (Vienna Standard Mean Ocean Water). Delta notation, δ^2H or $\delta^{18}O$, commonly used to report the measured isotope variations, is defined by the following equation:

$$\delta = \left(\frac{R_{SAMPLE}}{R_{STANDARD}} - 1 \right) \times 1000 \quad [^o/_{oo}] \qquad (1)$$

where R_{SAMPLE} and $R_{STANDARD}$ stand for the isotope ratios $^2H/^1H$ or $^{18}O/^{16}O$ of the sample and the standard, respectively.

GLOBAL DISTRIBUTION PATTERNS

The early data gathered by the IAEA/WMO network were reviewed by Dansgaard [*Dansgaard*, 1964]. He gave a salient phenomenological evaluation of these data, relating the observed distribution of heavy isotopes to a number of environmental parameters characterizing the given sampling site, such as latitude, altitude, distance to the coast, amount of precipitation, surface air temperature. The reviews which followed [*Yurtsever and Gat*, 1981; *Gonfiantini*, 1982, *Gonfiantini*, 1985] essentially confirmed the early

findings of Dansgaard. It was soon realized that the above mentioned empirical relationships (so called "effects") can in fact be considered as a measure of the average degree of rain-out of moisture from the given air mass, on the way from the source region to the site of precipitation.

The theoretical approaches to explain the observed isotope variations in meteoric waters evolved from the "isolated air mass" models based on Rayleigh condensation with immediate removal of precipitation or with a part of the condensate being kept in the cloud during the rain-out process [*Dansgaard*, 1964; *Friedman et al.*, 1964; *Craig and Gordon*, 1965; *Gat and Dansgaard*, 1972; *Taylor*, 1972; *Merlivat and Jouzel*, 1979; *Siegenthaler and Matter*, 1983; *Van der Straaten and Mook*, 1983; *Covey and Haagenson*, 1984; *Jouzel and Merlivat*, 1984; *Gedzelman*, 1988; *Johnsen et al.*, 1989]. The Rayleigh condensation concept has also been used in connection with classical equations of water vapor transport in the atmosphere, to model regional isotope fields [*Eriksson*, 1965; *Rozanski et al.*, 1982; *Sonntag et al.*, 1983; *Fisher and Alt*, 1985; *Fisher*, 1990; *Gedzelman and Lawrence*, 1990; *Ingraham and Craig*, this volume]. The global-scale modelling of deuterium and oxygen-18 distribution patterns in precipitation was recently carried out with the aid of general circulation models (GCMs) of the atmosphere [*Joussaume et al.*, 1984a,b; *Jouzel at al.*, 1987a; *Jouzel et al.*, 1991). It appears that major characteristics of the global isotope fields are properly reproduced by the currently available GCMs.

A considerable amount of theoretical and applied work carried out during the past three decades resulted in a fairly good understanding of the processes controlling the isotopic evolution of atmospheric waters at different levels of the water cycle. It became apparent that, like many other atmospheric properties, the isotopic composition of atmospheric water vapor and, consequently, of precipitation exhibits a broad spectrum of temporal variations with characteristic periods ranging from minutes up to hundreds or thousands of years.

Isotope studies of individual storm events, initiated by *Dansgaard* [1953] and continued by many other authors [e.g. *Gambell and Friedman*, 1965; *Miyake et al.*, 1968; *Ambach et al.*, 1975; *Gedzelman and Lawrence*, 1982; *Rindsberger et al.*, 1990], revealed that stable isotope composition of successive portions of precipitation collected during single rain event may vary dramatically. *Rindsberger et al.* [1990] in the regional study of individual rain events in Israel found that quite often the pattern of evolution of $\delta^{18}O$ in time is V- or W-shaped: a sharp decrease of the δ values was usually observed with a minimum value sometime in the middle of the shower. The most depleted isotope values corresponded usually to the period of most intense rain. This pattern was attributed to the passage of the frontal system(s). The range of in-storm $\delta^{18}O$ variations was quite large, reaching in some cases 10 to 12 per mil. Similar results were obtained by other authors [*Gambell and Friedman*, 1965; *Miyake et al.*, 1968; *E.M. Adar* - unpublished data, 1992]. The general conclusion of these studies was that the isotopic composition of precipitation from a given storm depends strongly on meteorological history of the air in which the precipitation is produced and through which it falls.

Precipitation samples collected on the per-event basis also reveal a strong linkage between their isotope signature and the storm's path, structure and evolution [*Rindsberger et al.*, 1983; *Gedzelman and Lawrence*, 1990; *Nativ and Riggio*, 1990]. For instance, detailed analysis of meteorological processes producing precipitation over northeastern United States led Gedzelman and co-workers [*Gedzelman and Lawrence*, 1982; *Lawrence et al.*, 1982] to the conclusion that the further south the storm track and the warm front, the lower δD value of precipitation. They also demonstrated that convective clouds produce precipitation with higher δ values than stratiform clouds with the same base and top.

A distinction should be made between liquid (rain) and solid (snow, hail) precipitation. Whereas isotopic composition of snow or hail collected at the ground reflects in-cloud conditions, raindrops undergo evaporation and isotopic exchange with atmospheric vapor on their fall down to the surface. Theoretical estimates [*Miyake et al.*, 1968; *Jouzel*, 1986], laboratory experiments [*Steward*, 1975] as well as isotope analyses of parallel samples of water vapor and rain collected at the ground [*Schoch-Fischer et al.*, 1984; *Jacob and Sonntag*, 1991] indicate that isotope exchange of raindrops with surrounding moisture is efficient enough to bring isotopic composition of rain close to the equilibrium value at the ground-level temperature. Substantial evaporative enrichment of raindrops beneath the cloud is mainly observed for light rains falling in relatively dry atmosphere.

The information available to date about deuterium and oxygen-18 content of global precipitation will be reviewed and discussed below in some detail. The review is based on the IAEA/WMO database (1961-1987, stations with at least a one-year record of oxygen-18). In some cases we use in the discussion more recent data sets as well

as other published data, not included in the network. Regional maps of ^{18}O content in precipitation are shown in Figures 4 to 9. For each station two values are reported: long-term arithmetic and long-term weighted mean $\delta^{18}O$ (weighing by the amount of precipitation).

One should keep in mind that the IAEA/WMO database, represented by monthly composite samples of precipitation, provides only statistically averaged, episodic information about isotopic composition of atmospheric water vapor, resulting from successive precipitation events, each usually with highly variable isotope characteristics.

Latitudinal distribution

The annual mean values of ^{18}O content in precipitation are plotted in Figure 10a as a function of latitude. The available data from high-latitude regions (mainly Greenland and Antarctica) are also schematically indicated on the figure. They represent in most cases freshly collected snow [*Lorius and Merlivat*, 1977; *Peel et al.*, 1988; *Johnsen et al.*, 1989; *Petit et al.*, 1991; *A. D. Fisher* - unpublished data, 1991].

The distribution shown in Figure 10a can be qualitatively understood in view of the fact that the major global source of water vapor is the tropical ocean; approximately 65 percent of the global evaporation flux over the oceans originates between 30°S and 30°N [*Peixoto and Oort*, 1983]. Poleward transport of this vapor is connected with gradual rain-out and the resulting reduction of total precipitable water in the atmosphere. This process can be modelled using the above mentioned Rayleigh approach. In fact, *Fisher* [1990] was able to reproduce fairly well the $\delta^{18}O$ distribution shown in Figure 10a using a zonally-averaged global isotope model with distributed water vapor sources. The remarkable spread of the data shown in Figure 10a has several sources: (1) precipitation collected at mid-latitude continental stations is in general more depleted in heavy isotopes than at coastal or marine stations located at the same latitude (see discussion below). This effect is especially obvious in the northern hemisphere; (2) a strong vertical gradient in the isotope content of atmospheric moisture [*Taylor*, 1972; *Ehhalt*, 1974; *Rozanski and Sonntag*, 1982] results in an apparent tendency to observe precipitation more depleted in heavy isotopes with increasing elevation of the sampling site. Perhaps the most characteristic example of this effect is the Izobamba station located in the Andes close to the equator (elevation 3058 m a.s.l.), which exhibits the isotope composition of precipitation typical for mid-latitude regions; (3) some tropical maritime stations tend to show excessive depletion in the heavy isotope content when compared with coastal or continental stations located within the same latitude belt, due to pronounced amount effect; (4) regional temperature anomalies may result in the average isotopic composition of precipitation falling off the general latitudinal trend. For instance, some coastal stations in the North Atlantic region (Reykjavik, Lista, Isfjord, Valentia) are characterized by relatively enriched $\delta^{18}O$ values, probably due to the positive temperature anomaly caused by the presence of the Gulf Stream. A substantial contribution of local vapor sources for some of the stations in the region, as suggested by *Johnsen et al.* [1989], may also contribute to the observed effect.

The role of temperature in establishing the latitudinal gradient in the heavy isotope composition of precipitation is illustrated in Figure 10b. It shows the 3-month averages of $\delta^{18}O$ (winter and summer

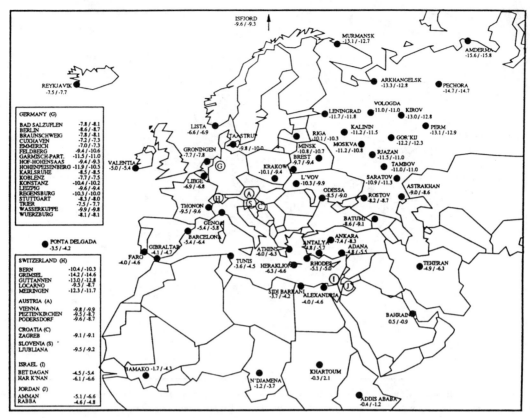

Fig. 4. Geographical distribution of the IAEA/WMO network stations in Europe. The stations are identified by name. For each station the long-term arithmetic mean (first number) and the weighted mean (second number) $\delta^{18}O$ values are indicated.

months, respectively) for continental and coastal stations, as a function of latitude. Reduced temperatures of the lower atmosphere over mid and high latitudes during winter months result in a substantial reduction of the total precipitable water in the atmosphere. For instance, for the latitude band of 40°N to 50°N the zonally-averaged total precipitable water drops from 25 kg m^{-2} in summer (JJA) to 12 kg m^{-2} during winter (DJF), with the values close to equator remaining essentially constant throughout the whole year [*Peixoto and Oort*, 1983]. This leads to a larger effective rain-out of the air masses resulting in a greater depletion of the heavy isotope content of precipitation during winter. As can be seen in Figure 10b, seasonal differences of the isotope signal are more pronounced in the northern hemisphere, reflecting the role of continents in establishing seasonal thermal gradients.

Continentality

The regional maps shown in Figures 4 to 9 reveal a great variety of distribution patterns of $\delta^{18}O$ over the continents. Detailed discussion of this distribution is beyond the scope of this paper. Here, we will focus only on the major features. In general, the distribution of $\delta^{18}O$ mimics the topography of the continents; mountainous chains are marked by more negative $\delta^{18}O$ values. This feature is called altitude effect. Also, there is an apparent tendency to observe more negative $\delta^{18}O$ values in precipitation with increasing distance from the coast. This feature is known as a "continental effect".

Basically the same mechanism is responsible for both effects: gradual removal of moisture from air masses which move inland or are orographically uplifted, coupled with preferential removal of the heavy isotopes during condensation process. This implies that the ocean is a major source of water vapor over the continents. Such an assumption might, however, not always be fulfilled, e.g. in the vicinity of large continental water bodies or in the interior of continents, where the re-evaporated moisture plays an important role in the atmospheric water balance [*Salati et al.*, 1979; *Sonntag et al.*, 1983; *Ingraham and Taylor*, 1986].

Systematic studies of isotopic composition of near-ground atmospheric water vapor carried out in Europe [*Schoch-Fischer et al.*, 1984; *Rozanski*, 1986; *Jacob and Sonntag*, 1991] and in the northeast United States [*White and Gedzelman*, 1984] revealed a strong positive correlation between δD of the vapor and the specific humidity of air. *Sonntag et al.* [1983] demonstrated that, under certain assumptions, the monthly and seasonal mean values of the specific humidity measured at ground level can be converted to the total precipitable water, using only ground-level characteristics of the atmosphere. Further, they showed that the ratio of total precipitable water measured over the continent to that over the vapor source region (monthly or seasonal averages) is a good measure of the average degree of rain-out of the air masses moving inland, and, consequently, can be used to calculate the average isotopic depletion of the vapor

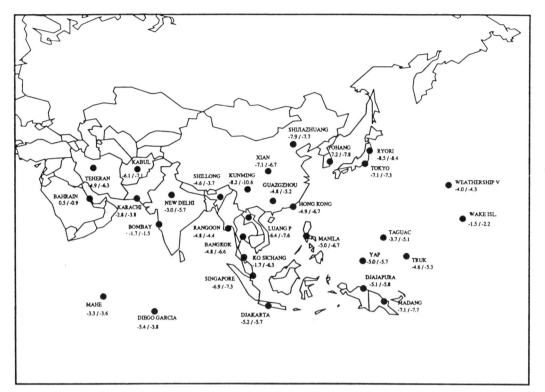

Fig. 5. Geographical distribution of the IAEA/WMO network stations in Asia and Far East. The stations are identified by name. For each station the long-term arithmetic mean (first number) and the weighted mean (second number) $\delta^{18}O$ values are indicated.

and precipitation over the continents.

Perhaps the most characteristic example of the continental effect is supplied by the available data for the European continent (Figure 11a). The seasonal and annual mean $\delta^{18}O$ values for selected European stations are plotted as a function of the distance from the Atlantic coast. The Atlantic coast was chosen due to predominant westerly circulation over western and central Europe [*Korzun*, 1974; *Peixoto and Oort*, 1983]. Two stations representing major source regions for the vapor entering the European continent are also shown in Figure 11a (Weathership E and Ponta Delgada). A distinct, gradual depletion of ^{18}O content in precipitation is observed with increasing distance from the coast. The overall depletion reaches 8 per mil over a distance of about 4500 km. The extent of the ^{18}O depletion varies seasonally: in winter it is much more pronounced than in summer. This effect has been explained by seasonal differences of water balance over the continent. In summer, the plant cover returns to the atmosphere substantial amounts of water during the transpiration process. This backward flux of vapor reduces the inland gradient of total precipitable water [*Peixoto and Oort*, 1983] and the resulting effective degree of rain-out of air masses moving eastward, which in turn reduces the extent of isotopic depletion of precipitation during summer months [*Rozanski et al.*, 1982]. The gradual isotopic depletion with increasing distance from the Atlantic coast seems to persist till the Ural Mountains. Unfortunately, no reliable data are so far available for western and central Siberia.

An apparent nonlinearity of the continental effect, limited to the winter months, is also seen in Figure 11a. A substantially larger slope of the best fit line (approximately 3.8‰ per 1000 km) is observed till the longitude band of about 17°E to 20°E, as compared with the average slope further east (about 1.6‰ per 1000 km). The decrease of the slope has been explained by additional moisture supply from the south (Mediterranean Sea, Black Sea, Caspian Sea), substantially contributing to water balance over eastern Europe during winter months [*Sonntag et al.*, 1983]. In fact, *Rozanski et al.* [1982] were able to reproduce the seasonal mean δD values in European precipitation only till the above mentioned longitude band, assuming that the Atlantic Ocean is the only source of water vapor over Europe. Significant south-north component in the horizontal vapor flux over eastern Europe during winter months is visible also on the global maps of the zonal and meridional horizontal vapor flow [*Korzun*, 1974; *Kuznetsova*, 1990] and on maps of total precipitable water [*Peixoto and Oort*, 1983].

Like western and central Europe, the South American continent is under the prevailing influence of the Atlantic ocean [*Ratisbona*, 1976]. The Andes constitute in fact an effective barrier against air masses transporting moisture from the Pacific Ocean. This fact has important consequences for spatial distribution of $\delta^{18}O$ in precipitation over this continent (Figure 7). For the Amazon Basin, the continental gradient of $\delta^{18}O$ in precipitation over the distance of the first 2000 kilometers from the coast is substantially smaller when compared to Europe (Figure 11b). It reaches approximately 1.5‰ per 1000 km, compared to about 2.0‰ per 1000 km for Europe,

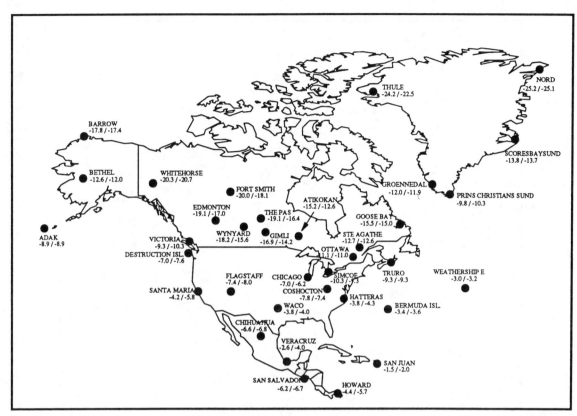

Fig. 6. Geographical distribution of the IAEA/WMO network stations in the north American continent. The stations are identified by name. For each station the long-term arithmetic mean (first number) and the weighted mean (second number) $\delta^{18}O$ values are indicated.

when long-term annual mean $\delta^{18}O$ values are considered. This reduced "continentality" of the isotope signal has been explained by intense recycling of moisture within the basin by the evapotranspiration [*Salati et al.*, 1979; *Gat and Matsui*, 1991]. Two other features of the data points presented in Figure 11b are worth being mentioned here: (1) the continental gradient of $\delta^{18}O$ in precipitation over the basin becomes very small west of Manaus; during the rainy period $\delta^{18}O$ is stabilized already at a distance of about 1000 kilometers from the coast. Such an effect can be observed in Europe only during summer, however at much greater distance from the coast (Figure 11a). This points to a high intensity of recycling of moisture by evapotranspiration in the Amazon Basin; (2) there are relatively large seasonal fluctuations of $\delta^{18}O$ in precipitation already at the entrance to the Amazon Basin (station Belém) which are then propagated westward. Origin of these variations remains unclear. *Matsui et al.* [1983] noted that very low $\delta^{18}O$ in precipitation and in atmospheric moisture at Belém are associated with the passage of the Intertropical Convergence Zone (ITCZ) over the region. The Izobamba station, shown for comparison in Figure 11b, reveals much higher isotope depletion than the stations located within the basin. It results from orographic uplift of the air masses on the eastern slopes of the Andes.

The data coverage for the North American continent (Figure 6) is not as good as for Europe. On the other hand, spatial distribution patterns of $\delta^{18}O$ and δD in precipitation over this continent are substantially more complex, reflecting seasonally varying influence of the air masses of different origin: (1) Gulf of Mexico and subtropical Atlantic in the south-east; (2) subtropical Pacific in the south-west; (3) north Pacific in the west and north-west, and (4) Arctic in the north [*Bryson and Hare*, 1974]. This results in a strong apparent linkage between isotope signature of precipitation and the air mass trajectories as reported by several authors [*Lawrence et al.*, 1982; *Fritz et al.*, 1983; *Nativ and Riggio*, 1990; *Lawrence and White*, 1991; *Friedman et al.*, 1992]. On a regional scale, a distinct continental effect recorded in surface and shallow ground waters has been observed by *Ingraham and Taylor* [1986] along a traverse through Northern California and by *Yonge et al.*, [1989] in southwestern Canada.

Isotope data for the African continent are scarce (Figure 8). Three sources of vapour can in principle contribute to precipitation collected over the continent: the Atlantic Ocean, the Indian Ocean and the Mediterranean Sea. The contribution of the Mediterranean is limited to North Africa. Precipitation patterns over Western and Central Africa are controlled by seasonal shift of the ITCZ with associated NE or SE trade winds and permanent monsoonal flow originating in the southern subtropical Atlantic [*Lacaux et al.*, 1992]. Eastern Africa receives most of its precipitation from the Indian Ocean. For the stations located between 0 and 15°N the rainy period usually lasts from June to September, whereas the south-east of Africa receives most of its precipitation between November and April. It is interesting to compare the stations lying along the trajectory of the Atlantic monsoonal flow (Kano, N'djamena,

Fig. 7. Geographical distribution of the IAEA/WMO network stations in the south American continent. The stations are identified by name. For each station the long-term arithmetic mean (first number) and the weighted mean (second number) $\delta^{18}O$ values are indicated.

Geneina, Khartoum). An apparent increase of $\delta^{18}O$ with increasing distance from the coast is observed. Surprisingly enough, the Addis Ababa station, situated at an elevation of 2360 m a.s.l., reveals the highest ^{18}O enrichment of precipitation among the stations of the region. Most of the data points cluster along the Global Meteoric Water Line (cf. Figure 27) which excludes substantial evaporative enrichment of rain collected at this station.

It has been suggested [*Sonntag et al.*,1979] that rainforest of the Congo Basin may represent an important source of moisture for the regions situated north and north-east of the basin thus leading to relatively high $\delta^{18}O$ and δD values of Addis Ababa precipitation. Because the transpiration proceeds without isotope differentiation [*Zimmermann et al.*, 1967], the moisture released in this process will be isotopically much heavier than the atmospheric water vapor of maritime origin. Consequently, rain produced from such recycled moisture can easily reach positive $\delta^{18}O$ and δD values, which often

happens at Addis Ababa. Further, more detailed studies of rain and atmospheric moisture in the region would be needed to verify this hypothesis.

Recently, *Joseph et al*, [1992] proposed an alternative explanation for the apparent east-west gradient in ^{18}O content of precipitation and shallow ground waters observed in the Sahelo-Sudanese Zone in Africa (between 10°N and 17°N), which includes the network stations discussed above. They suggest that a major supply of moisture for this region is from the Indian Ocean. The vapor is transported westward by the zonal flows of East African Jet and Tropical Easterly Jet. In their model, the rain collected at Addis Ababa during the rainy period represents in fact first condensation stage of maritime moisture brought by the Indian monsoon.

Seasonality

The regular seasonal variations of deuterium and ^{18}O content of monthly precipitation, with precipitation isotopically depleted in

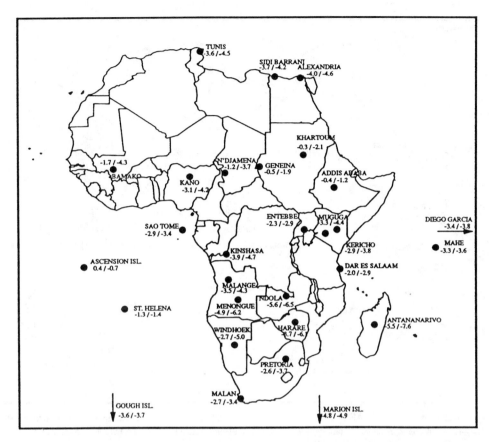

Fig. 8. Geographical distribution of the IAEA/WMO network stations in Africa. The stations are identified by name. For each station the long-term arithmetic mean (first number) and the weighted mean (second number) $\delta^{18}O$ values are indicated.

winter and enriched in summer, are common at mid- and high-latitude stations of the IAEA/WMO network. These seasonal differences are due to several factors: (1) seasonally changing temperature at mid- and high latitudes, with only minor fluctuations in the tropics. It results in seasonal variations of the total precipitable water in the atmosphere at these latitudes due to a varying degree of rain-out of the air masses transported poleward. This, in turn, via the Rayleigh mechanism, induces seasonality in the isotope signal observed in precipitation; (2) seasonally modulated evapotranspiration flux over the continents induces seasonal differences in the atmospheric water balance; (3) seasonally changing source areas of the vapor and/or different storm trajectories.

The seasonal fluctuations of $\delta^{18}O$ and δD, observed at many tropical island stations, have a different origin. In this case, the ^{18}O and deuterium content in precipitation is usually well correlated with the amount of precipitation (see discussion below). Isotopically depleted precipitation is observed during the rainy period.

Figure 12 illustrates the seasonal variability of $\delta^{18}O$ and temperature records for selected stations of the network. Groups of stations representing marine, coastal and continental environments at low, mid- and high latitudes are shown. It is obvious that the continental stations reveal much higher seasonal variations of both $\delta^{18}O$ and temperature than is the case for marine or coastal stations.

The gradual enhancement of seasonal variations of the isotope and temperature records with increasing distance from the coast is illustrated by Figure 13, showing the long-term monthly means of $\delta^{18}O$ and temperature for several European stations. The amplitude of $\delta^{18}O$ signal increases from about 2.5‰ at the coast (the Valentia station) to approximately 10‰ at the station Moskva, 3200 km inland. The above discussed substantial reduction of the continental effect during summer is also visible in Figure 13a. The large amplitude of seasonal variations of the heavy isotope composition of precipitation at more continental sites in Europe has been attributed to a combined effect of: (1) reduced continental isotope gradient in precipitation during summer due to recycling of atmospheric moisture by evapotranspiration, and (2) seasonally changing thermal gradient between the source regions (subtropical Atlantic) and the continent, leading to a larger degree of rain-out of air masses during winter months [*Rozanski et al.*, 1982].

The amount effect

The apparent correlation between the amount of monthly precipitation and its isotopic composition was first observed by *Dansgaard* [1964] and named the "amount effect". Figure 14a illustrates this relationship for tropical marine stations of the network (islands) located between 20°S and 20°N. The long-term monthly and annual means of $\delta^{18}O$ are plotted there as a function of the average

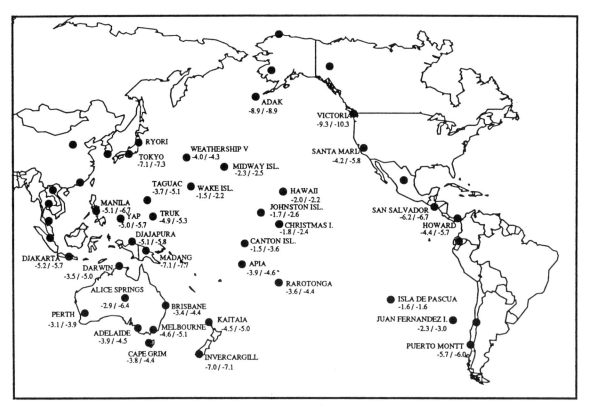

Fig. 9. Geographical distribution of the IAEA/WMO network stations in the Pacific region. The stations are identified by name. For each station the long-term arithmetic mean (first number) and the weighted mean (second number) $\delta^{18}O$ values are indicated.

monthly precipitation. The same isotope data are plotted on Figure 14b as a function of surface air temperature (monthly and annual means). Apparently, there is a strong inverse relationship between the mean monthly or annual $\delta^{18}O$ of precipitation and the amount of monthly precipitation for these stations whereas the correlation with the mean monthly (annual) surface air temperature is virtually nonexistent.

Figure 15 shows seasonal variations of $\delta^{18}O$, temperature and amount of precipitation at two tropical island stations: Apia (13.80°S, 171.7°W) and Taguac, Guam Island (13.55°N, 144.83°E). Both stations experience only minor fluctuations of temperature throughout the year. The $\delta^{18}O$ is inversely correlated with the amount of monthly precipitation: the correlation coefficient is equal to 0.55 for Apia and 0.71 for Taguac [IAEA, 1992].

The inverse relationship between precipitation $\delta^{18}O$ (δD) values and precipitation intensity in a single storm was reported by several authors [Matsuo and Friedman, 1967; Miyake et al., 1968, Mook et al., 1974; Ambach et al., 1975]. This led Yapp [1982] to the development of a cloud model simulating isotopic composition of precipitation during convective showers, in which the precipitation rate at a given height is controlled mainly by the vertical velocity of ascending air mass. He was able to reproduce monthly mean δD values of precipitation recorded at two tropical islands (Wake Island and Johnston Island), as a function of rain intensity recorded at these stations. He further showed that the amount effect may be explicable,

in part, as a consequence of the extent of rain-out process of deep convective clouds because a rough correspondence was noted between monthly mean precipitation intensity and precipitation amount at these two tropical, oceanic island sites.

Another process contributing to the amount effect is connected with isotope exchange and partial evaporation of raindrops below the cloud base. Already Dansgaard [1964] noted that during months with low precipitation, evaporative enrichment of raindrops may play an important role (low relative humidity beneath the cloud base) which is not the case during the rainy period. On the other hand, heavy showers will tend to modify the heavy isotope content of atmospheric moisture beneath the cloud towards a more negative value via the isotope exchange with falling raindrops. This, in turn, should help to preserve the in-cloud isotope signatures (low $\delta^{18}O$ and δD values) of raindrops collected at the ground.

For some tropical continental stations the apparent correlation between the heavy isotope composition and the amount of precipitation seems to be controlled not only by local rain-out processes but also by changes of the isotopic composition of the vapor in the source region. The Manaus station (3.12°S, 60.00°W) may serve as a good example of such a situation (Figure 16). Gradual decrease of $\delta^{18}O$ values observed between February and May at this station coincide with the maximum of precipitation, suggesting the typical amount effect. However, at the same time the ITCZ crosses the eastern margin of the Amazon Basin, pushing isotopically depleted

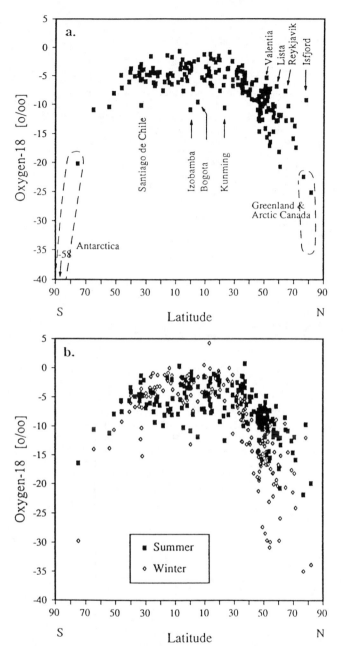

Fig. 10. (a) Long-term annual mean [18]O content in precipitation, derived from the database of the IAEA/WMO global network, plotted as a function of latitude. The $\delta^{18}O$ data for polar regions, gathered outside the network, are schematically indicated by contour lines [*Johnsen et al.*, 1989; *Lorius and Merlivat*, 1977; *Peel et al.*, 1988; *Petit et al.*, 1991; *D. A. Fisher* - unpublished data, 1991]; (b) Long-term seasonal mean [18]O content in precipitation, derived from the database of the network, plotted as a function of latitude. Summer season: June, July, August (NH) and December, January, February (SH). Winter season: December, January, February (NH) and June, July, August (SH).

moisture inland [*Matsui et al.*, 1983]. This negative isotope signal is visible also in the Izobamba station some 2000 km westward (Figure 16).

Figure 17 illustrates seasonal variations of $\delta^{18}O$, temperature and amount of monthly precipitation at two stations controlled by a monsoon climate: New Delhi (28.58°N, 77.20°E) and Hong Kong (22.32°N, 114.10°E). Contrary to tropical oceanic islands (Figure 15), the temperature at these stations reveals a distinct seasonal trend, with the maximum corresponding to monsoon period. Interestingly, the $\delta^{18}O$ is inversely correlated with monthly temperature with the slope equal to -0.42 $°/_{oo}$ per °C for Hong Kong and -0.10 $°/_{oo}$ per °C for New Delhi. This confirms a predominant role of the amount effect in establishing the observed seasonal variations of $\delta^{18}O$ in precipitation at these stations. The correlation coefficient of the monthly $\delta^{18}O$-precipitation relationship is equal to 0.45 and 0.71 for New Delhi and Hong Kong, respectively [*IAEA*, 1992].

The role of local temperature

The apparent link between local surface air temperature and the heavy isotope composition of precipitation has attracted much attention since the very beginning of isotope studies. This interest was stimulated mainly by the potential importance of stable isotopes as palaeoclimatic indicators. Numerous studies carried out during the past three decades were directed towards reconstruction of past climatic changes from records of isotopic composition of ancient precipitation preserved in various environmental archives such as glacier ice, sediments, groundwater, organic matter, and others [e.g. *Dansgaard et al.*, 1982; *Jouzel at al.*, 1987b; *Eicher and Siegenthaler*, 1976; *Winograd et al.*, 1988; *Yapp and Epstein*, 1977; *Becker et al.*, 1991; *Stute et al.*, 1992].

Dansgaard [1964] in his classical review presented the empirical relationship between the annual averages of $\delta^{18}O$ of precipitation and local surface air temperature, derived from the data gathered during the first three years of operation of the IAEA/WMO network. The relationship was developed for mid- and high northern latitude coastal stations. The slope of this relationship is 0.69$°/_{oo}$ per °C for $\delta^{18}O$ and 5.6$°/_{oo}$ per °C for δD. This relationship was frequently used in isotope-aided palaeoclimatic reconstructions. However, doubts often arose whether spatial relations between isotopic composition of precipitation and climatic variables, derived for the present-day conditions, can be used with confidence to interpret isotope records preserved in various environmental archives, as they usually reflect long-term linkage between isotopic composition of precipitation and climate on a given area.

In principle, three different types of the isotope-temperature relationship can be derived from the database of the IAEA/WMO network: (1) spatial relation between the long-term (annual) averages of $\delta^{18}O$ (δD) of precipitation and surface air temperature for different stations; (2) temporal relation between short-term (seasonal) changes of $\delta^{18}O$ (δD) and temperature for a single station or group of stations, and (3) temporal relation between long-term (interannual) changes of $\delta^{18}O$ (δD) and temperature at a given location.

Figure 18 shows the spatial relation between long-term annual arithmetic means of $\delta^{18}O$ and surface air temperature, derived for the entire set of the IAEA/WMO stations. It contains also the best fit lines of the published data for polar regions (Greenland, Antarctica). The data presented in Figure 18 confirm the temperature dependence of the isotope-temperature coefficient expected from theoretical

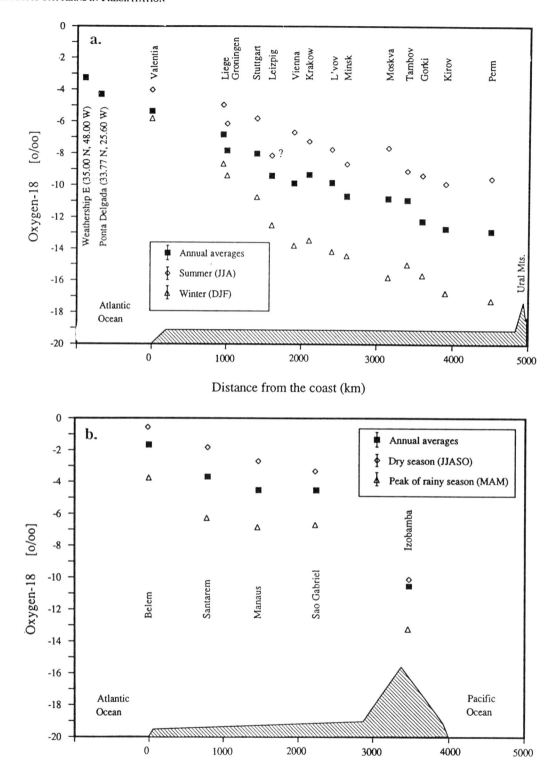

Fig. 11. (a) ^{18}O content of precipitation for selected European stations of the IAEA/WMO global network, plotted as a function of the distance from the Atlantic coast. Long-term annual, summer (June, July, August) and winter (December, January, February) mean δ^{18}O values are indicated for each station; (b) ^{18}O content of precipitation for selected stations of the network located in the Amazon Basin. The mean annual and seasonal δ^{18}O values are indicated for each station.

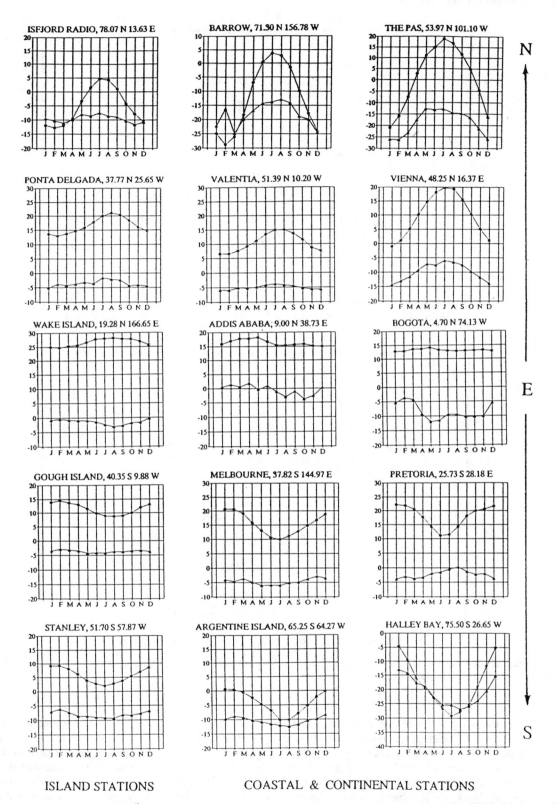

ISLAND STATIONS   COASTAL & CONTINENTAL STATIONS

Fig. 12. Seasonal variations of $\delta^{18}O$ of precipitation and surface air temperature for selected stations of the IAEA/WMO global network, representing marine, coastal and continental environment of both hemispheres. Full triangles stand for long-term monthly means of $\delta^{18}O$ whereas open squares represent long-term monthly average temperature.

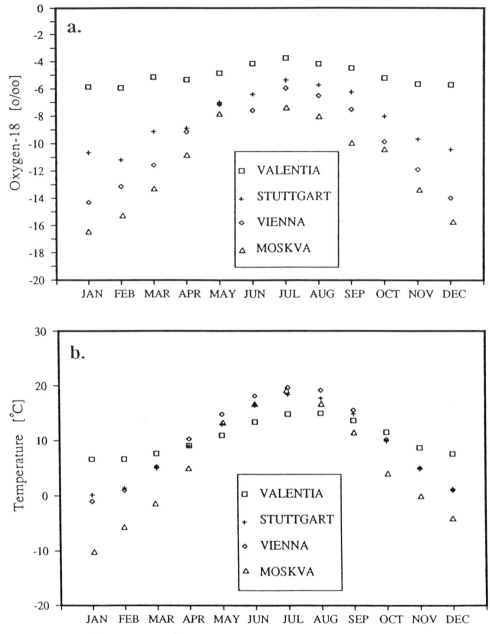

Fig. 13. Seasonal variations of $\delta^{18}O$ of precipitation (a) and surface air temperature (b), for selected European stations of the IAEA/WMO global network with increasing distance from the Atlantic coast.

considerations, with increasing slope of the $\delta^{18}O$-temperature relationship for reduced condensation temperatures. The slope of best fit line of the $\delta^{18}O$ data representing the temperature range between 0 and 20°C is equal to about 0.58°/ₒₒ per °C, which is in close agreement with theoretical predictions for this temperature range [*Van der Straaten and Mook*, 1983], based on the Rayleigh condensation model with isobaric cooling of precipitating air masses. Lack of correlation for the high-temperature range (20 to 30°C) results from dominance of the amount effect in the tropics.

At polar regions the slope of the $\delta^{18}O$-temperature relation is substantially higher: the values reported in the literature range between 0.67°/ₒₒ per °C for southern and western Greenland [*Johnsen et al.*, 1989], 0.76°/ₒₒ per °C for East Antarctica [*Lorius and Merlivat*, 1977] and about 0.90°/ₒₒ per °C for the Antarctic Peninsula [*Peel et al.*, 1988]. The slope of 0.9°/ₒₒ per °C has also been reported by *Picciotto et al.* [1960] for the coastal Antarctic station (King Baudouin Base, 70.26°S, 24.19°E), using the temperature of the cloud sheet where precipitation was formed. Substantial spread of the data

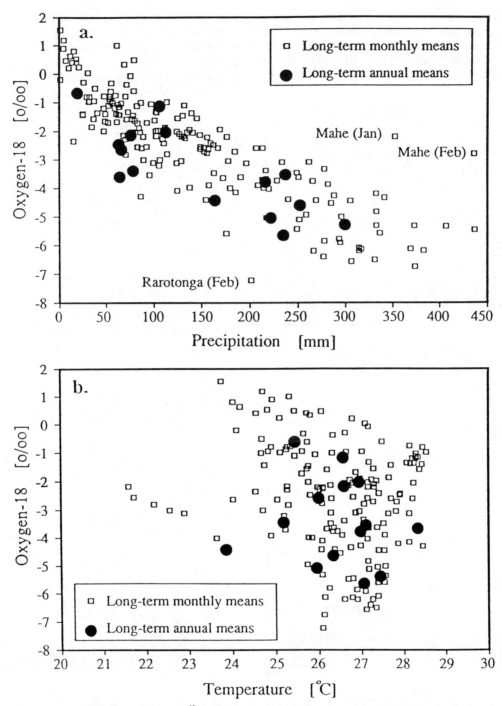

Fig. 14. (a) Long-term monthly and annual mean $\delta^{18}O$ values for tropical island stations of the IAEA/WMO global network (20°S to 20°N), plotted as a function of mean monthly precipitation. Annual mean $\delta^{18}O$ values are plotted as a function of the mean annual precipitation divided by 12; (b) Long-term monthly and annual mean $\delta^{18}O$ values for the same group of stations, plotted as a function of the mean monthly (annual) surface air temperature.

points, seen in Figure 18, confirms the fact that local surface air temperature is not always a good indicator of the average degree of rain-out of a given air mass. For instance, a comparison of the Whitehorse (60.72°N, 135.00°W) and Barrow (71.30°N, 156.70°W) stations shows that in spite of more than 10°C lower mean annual temperature at Barrow, the annual mean $\delta^{18}O$ of precipitation for this station is almost two per mil higher than at Whitehorse, with similar annual distribution of precipitation. This probably reflects a substantially smaller average degree of rain-out of air masses precipitating at Barrow, with the moisture originating in the north Pacific and in the Arctic. The Whitehorse station, although situated only about 500 km from the Pacific, is effectively shielded from its direct influence by the Saint Elias Mountains, and receives moisture

already isotopically depleted from both the south and south-east [*Holdsworth et al.*, 1991].

The seasonal relationship between $\delta^{18}O$ and temperature is summarized in Figure 19 and Table 2. Figure 19a shows relative changes of long-term monthly means of $\delta^{18}O$ and temperature for the mid-latitude IAEA/WMO network stations situated in the northern

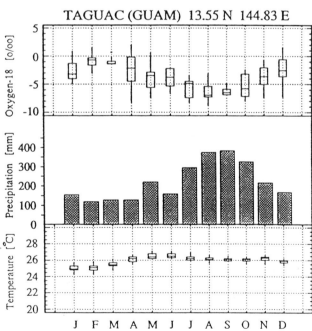

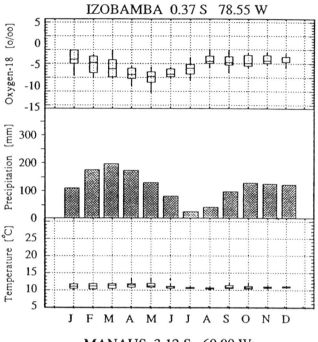

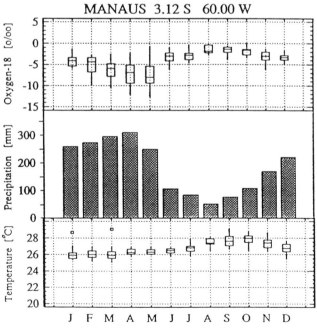

Fig. 15. Seasonal variations of $\delta^{18}O$, precipitation amount and surface air temperature for two tropical marine stations of the IAEA/WMO global network. The $\delta^{18}O$ and temperature data are presented in the form of box-and-whisker plots.

Fig. 16. Seasonal variations of $\delta^{18}O$, precipitation amount and surface air temperature for two stations of the IAEA/WMO global network located in the equatorial region of the south American continent. The $\delta^{18}O$ and temperature data are presented in the form of box-and-whisker plots.

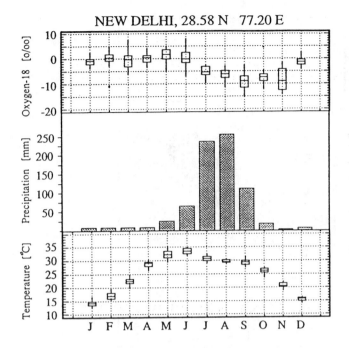

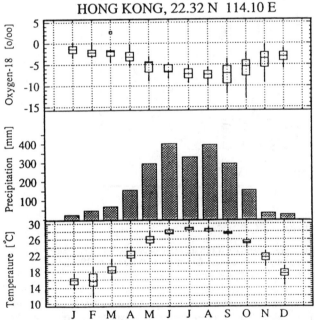

Fig. 17. Seasonal variations of $\delta^{18}O$, precipitation amount and surface air temperature for two tropical stations of the IAEA/WMO global network controlled by monsoon climate. The $\delta^{18}O$ and temperature data are presented in the form of box-and-whisker plots.

hemisphere (40°N to 60°N). The relative changes $\Delta\delta^{18}O$ and ΔT (long-term monthly minus long-term annual mean) were calculated in order to directly compare different stations. The same procedure was adopted to calculate seasonal $\Delta\delta^{18}O/\Delta P$ coefficients listed in Table 2. The slope of the best fit line is equal to 0.31°/$_{oo}$ per °C, a substantially lower value than that derived from the spatial

relationship (1). In fact, it is apparent from Figure 19 that the $\Delta\delta^{18}O$-ΔT relationship is nonlinear, with a higher slope for negative departures $\Delta\delta^{18}O$ and ΔT and only small changes of $\delta^{18}O$ for the upper end of the temperature range, in agreement with theoretical predictions. The slope of the $\Delta\delta^{18}O$-ΔT relationship, assuming linear dependence among these two variables, change from 0.66°/$_{oo}$ per °C ($r^2 = 0.47$) to 0.17°/$_{oo}$ per °C ($r^2 = 0.14$) for surface air temperatures lower than -10°C and higher than 10°C, respectively.

Very similar isotope variations have been observed in atmospheric water vapor sampled continuously at the ground level at Heidelberg, Germany (49.8°N, 8.7°E), between 1981 and 1988 [*Jacob and Sonntag*, 1991]. The authors conclude that the lack of a strong relationship between $\delta^{18}O$ and temperature in summer results mainly from an admixture of water vapor which is released by plant transpiration and has a relatively constant isotopic composition, close to weighted annual mean of precipitation. Higher slope of $\Delta\delta^{18}O$-ΔT relation for the low-temperature end is understood in view of a higher effective degree of rain-out of air masses in winter [*Rozanski et al.*, 1982] and a substantial contribution of snow, the formation of which is accompanied by an additional fractionation effect [*Jouzel and Merlivat*, 1984].

Figure 19b shows the relationship between amplitudes of seasonal changes of $\delta^{18}O$ and surface air temperature, defined as a difference between long-term averages of $\delta^{18}O$ and temperature for summer and winter months (JJA - DJF for the northern hemisphere and DJF - JJA for the southern hemisphere, respectively), for selected stations of the IAEA/WMO network. In general, more continental climate implies larger amplitude of seasonal variations of $\delta^{18}O$. The slope of the best fit line is equal to 0.40°/$_{oo}$ per °C. However, for a number of stations the ratio $\Delta\delta^{18}O/\Delta T$ is negative. This indicates that seasonal changes of $\delta^{18}O$ at these stations are controlled by factors other than local surface air temperature (for instance, by seasonal changes of storm trajectories or the amount of precipitation).

Long-term trends

The relationship between long-term changes of the heavy isotope composition of precipitation and the surface air temperature at a given location is probably the most relevant as far as palaeoclimatic applications of stable isotopes are concerned. *Siegenthaler and Matter* [1983] investigated this relationship for selected stations of mid- and high latitudes and found significant correlations for only 4 out of 14 stations. However, the length of the analyzed records was rather short at that time (between 8 and 13 years). Recently, *Lawrence and White* [1991] examined in detail the available meteorological and isotope data for several stations of the IAEA/WMO network. They concluded that correlations between climate parameters (temperature, amount of precipitation) and the isotopic composition of precipitation, derived from the interannual variations of these parameters, are usually limited to certain seasons of the year.

At present, much longer records of isotope and meteorological data are available for a number of stations of the IAEA/WMO network, reaching in several cases three decades (cf. Table 1). Therefore, it was interesting to search for long-term trends in these records and to try to identify the response of isotope composition of precipitation to the postulated global warming trend during the past decade. The selection of stations for analysis was guided by several factors: (1)

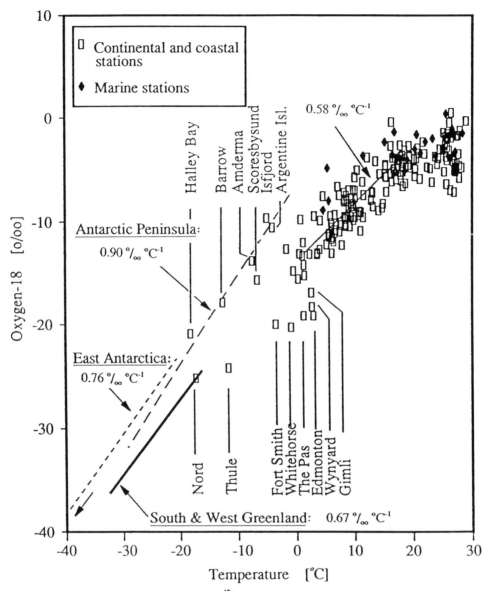

Fig. 18. Relationship between the long-term annual means of ^{18}O content in precipitation and surface air temperature for all stations of the IAEA/WMO global network. The best fit lines of the published data for southern and western Greenland [*Johnsen et al.,* 1989], East Antarctica [*Lorius and Merlivat,* 1977] and the Antarctic Peninsula [*Peel et al.,* 1988] are also shown.

length and quality of the available isotope and meteorological records; (2) continuity of the records, and (3) different climatic regimes represented by the stations. Twelve stations have been selected using the criteria outlined above. Eight of them are located in Europe: Meiringen, Guttannen, Grimsel and Bern (Switzerland), Thonon-les-Bains (France), Vienna (Austria), Groningen (The Netherlands) and Krakow (Poland). The remaining four stations are: Ottawa (Canada), Hong Kong (South East Asia), Taguac (Guam Island, West Pacific) and Argentine Island (Antarctic Peninsula).

The analysis of the available time series of monthly $\delta^{18}O$ and temperature data for the selected stations started by removing the seasonal component from the records. This was done by calculating a 12-month running average of $\delta^{18}O$ and temperature. Single-month

gaps in the records were filled by the long-term mean monthly values. The calculation was interrupted whenever more than three consecutive months were missed in the records. Then, the differences $\Delta\delta^{18}O$ and ΔT were calculated for individual stations by subtracting the running average curves from the long-term annual means. Finally, the resulting curves were smoothed by again applying a 12-month running average. Spatial averaging of ^{18}O and temperature signals (composite curves for the Swiss stations and for the whole of Europe) was done by calculating average differences $\Delta\delta^{18}O$ and ΔT (arithmetic averaging of the differences for individual stations), the procedure usually adopted for calculation of spatially averaged temperature records [*Hansen and Lebedeff,* 1987].

Figure 20 shows the trend curves of $\Delta\delta^{18}O$ and ΔT calculated for

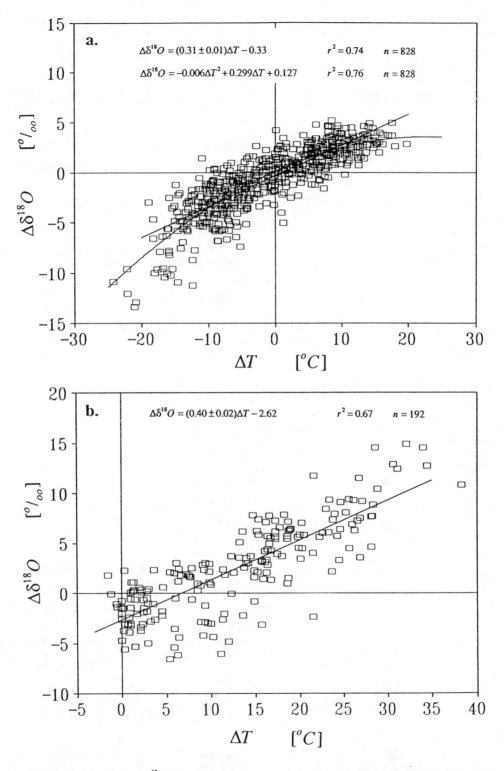

Fig. 19. (a) Seasonal correlation between $\delta^{18}O$ and surface air temperature derived for the stations of the IAEA/WMO global network situated between 40°N and 60°N. The differences between long-term monthly and long-term annual means of $\delta^{18}O$ and surface air temperature are plotted on the figure. (b) The relationship between amplitudes of seasonal changes of $\delta^{18}O$ and surface air temperature for the selected stations of the network, defined as a difference between long-term averages of $\delta^{18}O$ and temperature for summer and winter months (JJA - DJF for the northern hemisphere and DJF - JJA for the southern hemisphere, respectively).

TABLE 1. Summary of isotope and meteorological data for 219 stations of the IAEA/WMO network "Isotopes in Precipitation". The summary covers the time period 1961-1987. The stations marked by an asterisk continue operation after 1987. The isotope data are reported as per mil deviations from the standard V-SMOW. For calculation of the weighted means, weighing by amount of precipitation has been adopted. The number of available monthly data has been reported for both deuterium and ^{18}O data.

#	Station name	Latitude (deg.)	Longitude (deg.)	Altitude (m.a.s.l)	O18 Mean	O18 W.mean	O18 n	O18 Start	O18 End	D Mean	D W.mean	D n	D Start	D End	cont.	DE n	DE Mean	DE W.mean	Prec (mm)	Temp (C)	Vap.P. (mb)
1	Adak	51.88 N	176.65 W	4	-8.83	-8.89	68	62-01	73-05	-64.2	-64.4	61	62-01	73-05		60	6.5	6.4	1466	4.8	7.3
2	Adana	36.98 N	35.30 E	73	-4.70	-5.50	91	64-03	87-12	-24.3	-28.0	82	64-03	87-12	*	81	13.5	16.1	659	19.0	15.3
3	Addis Ababa	9.00 N	38.73 E	2360	-0.36	-1.31	102	61-03	87-09	8.5	1.8	95	61-03	87-09	*	93	11.6	13.2	1181	16.3	11.0
4	Adelaide	34.93 S	138.58 E	43	-4.06	-4.63	113	62-01	84-12	-20.7	-25.3	124	62-01	84-12	*	112	11.6	11.7	493	17.0	11.5
5	Alexandria	31.20 N	29.95 E	-3	-3.97	-4.60	73	61-10	73-03	-17.2	-21.9	64	61-10	73-03		63	14.6	16.4	191	20.2	16.4
6	Alice Springs	23.80 S	133.88 E	546	-2.84	-6.70	155	62-01	87-09	-14.8	-37.8	140	62-01	87-09	*	134	8.0	12.9	298	20.8	9.4
7	Amderma	69.46 N	61.41 E	53	-15.63	-15.82	36	80-04	83-12	-114.5	-114.00	27	80-04	83-12		26	15.0	14.5	456	-6.7	4.1
8	Amman	31.98 N	35.93 E	850	-5.07	-6.59	18	65-11	68-05	-20.0	-27.8	18	65-11	68-05		18	20.2	23.5	890	-	-
9	Ankara (Central)	39.95 N	32.88 E	902	-7.52	-8.34	212	63-07	87-12	-51.6	-56.7	186	64-03	87-12	*	185	9.0	10.5	401	11.7	8.3
10	Antalya	36.88 N	30.70 E	49	-4.70	-5.62	139	63-05	75-10	-24.6	-28.5	130	63-05	75-04		127	13.8	16.9	1098	18.5	14.2
11	Antananarivo	18.90 S	47.53 E	1300	-4.95	-6.41	71	61-12	77-09	-26.4	-36.0	67	62-10	77-09		66	13.9	15.4	1390	17.8	15.8
12	Apia	13.80 S	171.78 W	0	-3.75	-4.48	111	62-10	87-12	-17.8	-23.3	109	62-10	87-12	*	108	12.1	12.4	3024	26.4	28.6
13	Argentine Island	65.25 S	64.27 W	0	-10.54	-10.80	280	64-01	87-12	-84.5	-85.5	262	64-01	87-12	*	261	0.6	1.8	363	-4.0	4.1
14	Arkhangelsk	64.58 N	40.50 E	13	-13.21	-12.77	55	61-11	84-09	-99.8	-95.2	40	61-11	84-09		40	7.9	9.4	548	1.2	6.4
15	Ascension Island	7.92 S	14.42 W	15	0.29	-0.76	73	61-12	76-06	10.0	2.5	74	61-11	76-06		73	7.7	8.6	238	25.5	22.1
16	Astrakhan	46.25 N	48.03 E	-18	-8.99	-8.58	54	80-04	84-12	-68.5	-67.8	43	81-01	84-09		43	6.0	6.2	200	11.0	10.1
17	Asuncion	25.27 S	57.63 W	65	-4.45	-6.20	29	61-11	67-11	-25.7	-37.8	29	61-11	67-11		29	10.5	11.5	1417	24.1	20.4
18	Athens	37.90 N	23.73 E	27	-6.03	-6.30	55	60-11	76-01	-33.2	-34.2	50	60-11	74-03		38	13.2	13.1	395	18.0	12.5
19	Atikokan	48.75 N	91.62 W	393	-15.19	-12.61	76	75-08	82-07	-111.7	-91.5	75	75-08	82-07		75	10.3	9.7	683	1.4	-
20	Bad Salzuflen	52.10 N	8.73 E	100	-7.99	-8.22	119	78-01	87-03	-56.2	-58.1	119	78-01	87-03	*	119	7.8	7.7	859	8.8	-
21	Bahrain	26.27 N	50.62 E	2	0.57	-0.97	89	61-11	87-03	13.6	4.6	91	61-11	87-12		84	8.9	14.1	66	26.5	23.5
22	Bamako	12.63 N	8.03 W	329	-1.88	-4.44	111	62-05	79-08	-15.8	-30.7	103	62-05	79-09		103	4.3	7.0	917	27.6	18.4
23	Bangkok	13.73 N	100.50 E	2	-4.84	-6.60	185	68-07	87-11	-31.9	-43.6	157	68-08	87-11	*	152	8.6	8.9	1425	28.2	29.1
24	Barbados	13.07 N	59.48 W	50	-1.03	-1.14	186	61-02	87-12	-0.9	-4.0	188	61-02	87-12		170	7.6	8.0	1257	26.6	26.9
25	Barcelona	41.38 N	2.11 E	65	-5.23	-6.11	35	84-01	87-02	-27.7	-38.2	21	85-04	87-02		21	11.4	11.6	520	16.1	13.0
26	Barranquilla	10.88 N	74.78 W	14	-4.60	-5.09	85	71-07	87-06	-29.3	-33.9	77	72-08	87-06		75	7.2	8.0	743	27.4	28.9
27	Barrow	71.30 N	156.78 W	7	-18.54	-19.61	49	80-04	69-10	-132.9	-131.1	51	80-04	69-11		47	10.3	10.4	133	-12.7	2.9
28	Batumi	41.39 N	41.38 E	6	-8.61	-9.09	44	80-04	84-12	-54.2	-57.5	44	81-01	84-08	*	42	14.7	15.4	2612	14.8	14.0
29	Belem	1.43 S	48.48 W	24	-1.78	-2.71	233	65-03	87-09	-6.4	-13.9	201	65-03	87-02		189	8.6	9.1	2845	26.0	29.6
30	Berlin	52.47 N	13.40 E	50	-8.66	-8.67	114	78-01	87-12	-61.4	-61.6	114	78-01	87-12		114	7.9	7.9	584	8.7	9.0
31	Bermuda Island	32.37 N	64.68 W	6	-3.35	-3.53	48	62-01	66-01	-16.1	-16.7	45	62-01	65-12	*	45	10.5	11.5	1372	21.5	18.3
32	Bern	46.92 N	7.50 E	511	-10.40	-10.25	204	71-01	87-12										1090	8.5	8.3
33	Bet Dagan	32.00 N	34.82 E	30	-4.30	-5.33	137	61-01	85-04	-17.1	-21.8	99	62-01	87-12	*	95	16.8	18.4	531	19.2	15.9
34	Bethel	60.78 N	161.80 W	41	-12.52	-11.95	50	61-06	66-02	-98.7	-94.5	48	61-06	65-12		48	1.6	1.2	325	-1.9	5.0
35	Bogota	4.70 N	74.13 W	2547	-8.35	-9.50	120	71-08	87-06	-55.1	-65.5	113	72-06	87-06		112	9.8	10.6	781	13.2	12.1
36	Bombay	18.90 N	72.82 E	10	-1.72	-1.51	91	61-06	77-09	-5.1	-4.1	48	61-06	77-09		48	8.8	8.7	1984	27.4	26.9
37	Brasilia	15.85 S	47.93 W	1061	-4.07	-4.96	136	65-03	87-06	-22.2	-29.9	122	62-01	87-06	*	118	11.7	11.4	1474	20.6	16.6
38	Braunschweig	52.30 N	10.45 E	88	-7.98	-8.10	118	78-01	87-12	-57.6	-58.5	117	78-01	87-12		117	6.1	6.2	643	8.3	-
39	Brest	52.07 N	23.41 E	142	-9.71	-9.40	44	80-04	87-12	-73.4	-71.1	35	81-01	83-12		35	7.4	7.2	591	8.5	9.2
40	Brisbane	27.43 S	153.08 E	0	-3.50	-4.49	226	62-01	87-12	-13.3	-21.1	218	62-01	87-12		215	13.8	13.3	1197	21.0	16.5
41	Buenos Aires	34.58 S	58.48 W	24	-5.03	-5.01	106	61-07	85-04	-26.1	-26.0	87	61-07	84-12		87	10.7	10.4	1113	17.5	14.9
42	Canton Island	2.77 S	171.72 W	2	-1.51	-3.65	47	62-01	66-02	-6.5	-24.7	50	62-01	66-04	*	47	6.7	7.7	798	28.3	27.5
43	Cape Grim	40.68 S	144.69 E	90	-3.80	-4.41	95	79-09	87-12	-20.3	-24.3	90	79-09	87-12		90	9.8	10.6	782	12.3	11.8
44	Cayenne	4.83 N	52.37 W	8	-1.67	-2.09	91	62-05	75-08	-7.6	-10.0	60	62-05	75-08		60	8.9	10.4	3737	25.5	28.0
45	Ceara Mirim	5.80 S	35.20 W	8	-1.58	-2.30	148	61-12	82-12	-3.1	-9.1	115	61-12	82-12		113	10.3	11.0	1324	25.7	25.6
46	Chicago (Midway)	41.78 N	87.75 W	189	-6.90	-6.04	192	61-12	79-10	-50.0	-43.0	170	62-01	79-10	*	169	7.0	7.0	908	10.2	10.0
47	Chihuahua	28.63 N	106.07 W	1423	-6.41	-6.69	127	62-06	87-12	-41.7	-42.9	122	62-02	87-12		121	9.0	10.4	345	18.2	10.8
48	Christmas Island	1.98 N	157.46 W	3	-1.83	-2.43	23	81-09	64-04	-7.9	-11.4	23	81-09	64-04		23	6.9	8.1	845	-	-
49	Corrientes	24.47 S	58.83 W	55	-5.86	-6.75	33	66-01	84-12	-32.6	-37.3	13	66-01	83-02		13	10.8	11.8	1520	21.2	19.4
50	Coshocton	40.37 N	81.80 W	344	-7.90	-7.49	65	61-11	71-06	-50.7	-47.3	64	61-11	71-06		64	12.4	12.5	977	-	-
51	Cuiaba	15.60 S	56.10 W	165	-4.07	-5.33	203	61-11	87-06	-22.5	-31.9	190	61-11	87-06	*	187	9.6	10.4	1273	25.8	24.2
52	Cuxhaven	53.87 N	8.72 E	12	-7.20	-7.29	119	78-01	87-12	-51.8	-52.1	119	78-01	87-12		119	5.8	6.2	829	8.3	-
53	Dar Es Salaam	6.88 S	39.20 E	55	-2.03	-2.83	126	61-01	73-10	-7.1	-13.3	119	61-01	73-10		119	8.5	9.3	1139	25.7	25.6
54	Darwin	12.43 S	130.87 E	26	-3.49	-4.90	161	61-01	87-12	-18.7	-28.2	144	62-04	87-12	*	143	10.8	11.5	1775	27.7	24.8
55	Destruction Island	47.67 N	124.48 W	21	-6.88	-7.47	48	62-04	66-03	-49.5	-52.9	48	62-04	66-03		48	5.6	7.0	2012	-	-
56	Diego Garcia Island	7.32 S	72.40 E	1	-3.46	-4.03	182	62-01	87-12	-20.0	-23.7	170	62-01	87-12	*	159	9.1	9.6	2195	26.9	29.0
57	Djakarta	2.53 S	140.72 E	3	-4.98	-5.24	268	62-01	87-12	-31.6	-32.9	234	62-01	87-12		234	6.7	6.9	2332	27.2	29.9
58	Djajapura	6.18 S	106.83 E	8	-5.02	-5.58	168	62-01	87-12	-30.9	-34.8	169	62-01	87-12		167	9.7	9.7	1710	27.1	28.2
59	Edmonton (Industrial)	53.57 N	113.52 W	671	-19.14	-17.05	52	61-05	66-02	-149.0	-130.9	53	61-05	66-03		52	5.6	5.3	430	3.2	6.3
60	Emmerich	51.83 N	6.60 E	43	-6.98	-7.30	94	80-01	87-12	-50.0	-50.0	94	80-01	87-12		94	5.8	5.9	744	9.1	-
61	Entebbe (Airport)	0.05 N	32.45 E	1155	-2.26	-2.91	120	61-02	74-09	-5.8	-11.2	115	61-02	74-09		105	12.6	12.4	1613	21.6	20.5

TABLE 1. Continued

No.	Station name	Latitude (deg.)	Longitude (deg.)	Altitude (m.a.s.l)	O‑18 Mean	O‑18 W.mean	O‑18 Start	O‑18 End	O‑18 n	D Start	D End	D n	D W.mean	D Mean	D‑exc. n	D‑exc. Mean	D‑exc. W.mean	Prec (mm)	Temp (C)	Vap. P. (mb)
62	Faro	37.01 N	7.96 W	9	-3.93	-4.39	78-11	87-12	68	78-11	87-11	63	-25.2	-21.7	60	9.5	10.0	365	17.5	14.0
63	Feldberg/Schwarzwald	47.52 N	8.00 E	1493	-9.35	-10.57	81-01	85-08	50	81-01	85-08	50	-73.0	-65.0	50	11.5	11.5	1879	3.1	-
64	Flagstaff	35.13 N	111.67 W	2137	-7.26	-8.07	61-12	74-07	114	61-12	74-07	101	-65.9	-59.8	101	-1.6	3.5	533	7.5	5.5
65	Fort Smith	3.72 S	38.55 W	27	-1.83	-2.65	61-08	65-09	41	61-08	69-06	72	-11.8	-4.4	41	5.5	7.6	1705	26.4	27.3
66	Fortaleza	60.02 N	111.97 W	203	-19.95	-18.10	65-04	85-12	126	65-04	84-12	97	-146.4	-162.0	97	9.0	8.3	360	-3.2	5.4
67	Garmisch-Partenk.	47.48 N	11.07 E	720	-11.49	-10.95	78-01	87-12	119	78-01	87-12	119	-79.9	-84.6	119	7.4	7.8	1383	6.4	-
68	Geneina	13.48 N	22.45 E	805	-0.44	-1.92	68-06	76-09	48	68-06	76-09	36	-8.5	2.0	36	5.5	5.9	455	26.2	11.1
69	Genoa (Sestri)	44.42 N	8.85 E	2	-5.41	-5.78	61-11	87-12	216	61-11	87-12	216	-35.1	-33.1	213	9.9	10.3	1117	14.5	13.2
70	Gibraltar	36.15 N	5.35 W	5	-4.18	-4.70	61-10	87-12	145	61-10	87-12	142	-25.7	-22.6	139	10.8	11.5	747	18.2	15.5
71	Gimli	50.62 N	96.98 W	223	-16.87	-14.15	75-08	82-07	76	75-08	82-07	73	-103.7	-124.3	72	9.3	9.0	551	2.6	-
72	Goose Bay	53.32 N	60.42 W	46	-15.54	-15.08	61-05	66-02	57	61-05	69-06	85	-109.8	-112.4	54	12.9	12.3	960	0.5	5.7
73	Gor'Kij	56.13 N	43.49 E	82	-12.16	-12.27	80-05	83-12	42	80-05	83-12	33	-89.8	-90.4	32	9.8	11.5	662	5.3	7.8
74	Gough Island	40.35 S	9.88 W	54	-3.63	-3.73	61-04	87-09	236	61-04	74-09	76	-20.8	-20.3	76	7.7	8.0	3193	11.6	11.2
75	Grimsel	46.57 N	8.33 E	1950	-14.23	-14.69	70-08	87-12	209									2079	1.3	4.8
76	Groennedal	61.22 N	48.12 W	27	-11.47	-11.29	61-01	74-08	97	62-05	74-08	74	-87.7	-88.0	67	3.3	2.8	1018	1.2	5.0
77	Groningen	53.21 N	6.57 E	0	-7.70	-7.84	64-03	87-11	284	64-03	87-12	280	-51.2	-50.2	280	11.4	11.6	758	9.5	-
78	Guangzhou	23.13 N	113.32 E		-4.71	-5.18	86-04	87-11	16	86-04	87-11	16	-34.2	-30.4	16	7.3	7.0	1748	23.2	7.8
79	Guttannen	46.65 N	8.30 E	1055	-13.06	-12.87	65-07	87-12	208									1683	6.2	-
80	Halley Bay	75.50 S	26.65 W	0	-20.87	-20.67	65-07	87-12	232	65-07	87-12	213	-155.4	-158.0	212	7.1	7.5	432	-18.2	1.8
81	Har Kna'an (Tirat Yael)	32.97 N	35.50 E	964	-6.21	-6.74	64-03	87-12	77	64-03	87-12	45	-28.7	-27.8	44	21.2	23.2	734	16.5	10.6
82	Harare	17.83 S	31.02 E	1471	-3.88	-6.14	60-11	83-12	175	60-11	83-12	110	-32.4	-16.1	106	12.1	13.4	809	18.0	12.9
83	Hattems	35.27 N	77.55 W	3	-3.82	-4.29	62-02	76-12	93	62-01	76-12	93	-22.5	-19.0	93	11.6	11.9	1386	16.6	16.3
84	Hawaii	19.72 N	155.07 W	9	-2.02	-2.29	62-02	69-10	87	62-01	69-10	85	-6.4	-4.8	85	10.9	11.4	3312	23.3	21.5
85	Heraklion (Yraklion)	35.33 N	25.18 E	47	-6.29	-6.61	63-05	74-02	22	63-05	74-02	23	-35.5	-32.8	22	16.9	15.8	524	18.8	13.5
86	Hof-Hohensaas	50.32 N	11.88 E	567	-9.44	-9.30	83-01	87-12	60	83-01	87-12	60	-66.6	-67.8	60	7.8	7.7	738	6.2	8.2
87	Hohenpeissenberg	47.80 N	11.02 E	977	-11.93	-11.38	74-01	87-12	168	74-01	87-12	209	-80.2	-84.2	168	10.4	10.3	1053	6.5	9.0
88	Hong Kong	22.32 N	114.17 E	65	-5.01	-6.70	61-01	87-11	193	61-01	87-11	193	-43.2	-30.0	185	9.7	10.0	2156	23.0	22.7
89	Howard Afb	8.92 N	79.60 W	13	-4.56	-5.65	68-04	87-12	165	68-07	87-12	150	-37.0	-28.8	138	8.0	8.8	1626	27.0	28.7
90	Invercargill	46.42 S	168.32 E	2	-6.95	-7.11	77-04	87-12	113	61-07	87-12	107	-47.3	-46.7	103	9.0	9.3	1037	9.8	10.0
91	Isfjord Radio	78.07 N	13.63 E	6	-9.55	-9.17	61-07	75-05	98	61-07	75-05	75	-59.9	-63.3	75	8.2	7.7	488	-5.1	3.9
92	Isla De Pascua	27.17 S	109.43 W	41	-1.57	-1.57	65-09	75-10	26	65-09	75-10	17	-5.1	-3.6	13	12.8	13.1	989	20.6	19.2
93	Izobamba	0.37 S	78.55 W	3059	-10.68	-11.05	73-06	84-11	111	73-06	84-11	102	-78.9	-75.8	101	9.8	10.0	1287	11.1	10.5
94	Johnston Island	16.73 N	169.52 W	2	-1.72	-2.58	62-01	76-12	93	62-01	76-12	90	-13.6	-6.4	90	7.6	7.3	798	26.0	25.2
95	Juan Fernandez Island	33.62 S	78.83 W	6	-2.50	-2.99	64-01	75-11	24	64-01	75-11	21	-16.8	-13.7	21	2.1	3.8	1062	15.1	12.8
96	Kabul (Karizimir)	34.67 N	69.08 E	1860	-6.01	-7.15	62-06	87-12	93	62-01	87-12	70	-36.1	-33.2	70	14.6	17.2	327	11.5	7.1
97	Kaitaia	35.07 N	173.28 E	76	-4.54	-5.01	62-10	87-12	193	62-10	87-12	180	-28.6	-24.7	180	11.7	11.7	1382	15.4	14.3
98	Kalinin	56.54 N	35.54 E	31	-12.19	-11.57	61-05	84-07	51	61-05	84-07	43	-86.7	-93.3	42	8.8	10.0	723	5.3	8.1
99	Kano	12.05 N	8.53 E	476	-2.88	-3.81	61-07	73-09	39	61-07	73-09	33	-27.2	-19.1	33	5.6	5.2	763	26.2	16.0
100	Karachi	24.90 N	67.13 E	23	-2.96	-3.93	61-07	75-08	45	61-07	75-08	43	-23.5	-19.2	39	6.2	9.0	195	25.7	22.0
101	Karlsruhe	49.02 N	8.38 E	120	-2.90	-3.83	81-01	87-11	79	61-01	87-11	79	-59.8	-59.5	79	8.1	8.1	851	9.8	-
102	Kericho	0.37 S	35.35 E	2130	-0.44	-2.10	67-11	70-05	24	67-11	69-12	19	-16.3	-10.0	19	11.5	13.4	1037		-
103	Khartoum	15.60 N	32.55 E	382	-3.88	-4.70	62-06	78-08	40	62-06	78-08	37	-10.0	1.7	36	3.6	6.1	147	29.0	12.9
104	Kinshasa (Binza)	4.37 S	15.25 E	438	-13.01	-12.73	80-04	83-10	60	61-01	83-10	59	-24.2	-17.9	59	13.2	13.5	1285	24.3	25.0
105	Kirov	58.39 N	49.37 E	164	-10.43	-7.45	81-03	83-10	43	81-03	83-10	32	-101.3	-104.0	32	3.5	3.2	679	3.7	7.1
106	Koblenz	50.35 N	7.58 E	97	-10.18	-7.68	80-01	87-12	36	80-01	87-12	29	-54.6	-56.2	80	5.2	5.1	637	10.2	-
107	Konstanz	47.68 N	9.18 E	447	-5.75	-6.28	78-01	87-12	15	78-01	87-12	14	-74.7	-77.0	118	6.5	6.8	873	8.8	9.4
108	Ko Samui	9.28 N	100.03 E	0	-5.79	-9.35	79-02	83-11	80	79-02	83-11	80	-27.7	-24.2	28	10.6	10.8	1442	27.8	29.4
109	Ko Sichang	13.17 N	100.80 E	4	-10.06	-10.61	84-06	87-11	118	84-06	87-11	118	-38.9	-35.0	14			1239	27.8	28.0
110	Krakow	50.07 N	19.88 E	205	-10.61	-9.35	75-03	87-12	154	75-03	87-12	154	-66.5	-72.2	154	8.4	8.3	619	7.7	9.4
111	Kunming	23.02 N	102.68 E	1841	-5.99	-9.41	86-04	87-11	19	86-04	87-11	19	-71.0	-51.8	19	13.9	14.6	1115	15.5	-
112	La Suela	30.58 S	64.58 W	900	-6.83	-5.99	81-08	84-03	22	81-08	84-03	24	-61.8	-62.5	24	11.5	12.4	1122	18.3	13.8
113	Leipzig	51.35 N	12.43 E	125	-6.66	-9.41	66-01	70-05	24	66-01	70-05	51	-47.5	-47.5	50	7.2	7.3	552	9.0	9.4
114	Liege	50.70 N	5.47 E	190	-6.83	-6.89	61-02	77-08	52	61-02	77-08	109	-48.6	-46.2	109	7.3	7.5	821	7.4	9.0
115	Lista	58.10 N	6.57 E	13	-6.66	-6.89	61-02	87-12	129	61-02	87-12	48	-64.4	-64.9	48	9.3	9.5	1025	9.9	10.1
116	Ljubljana	46.04 N	14.31 E	299	-9.24	-9.28	72-09	87-12	48	72-09	87-12							1474	11.6	9.1
117	Locarno	46.17 N	8.78 E	379	-9.47	-8.64	61-03	64-11	179	61-03	64-11	26	-50.9	-42.5	26	7.9	8.4	1850		26.1
118	Luang-Prabang	19.88 N	102.13 E	305	-6.39	-7.62	80-08	83-12	45	80-08	83-12	35	-71.8	-75.8	35	8.5	9.0	1257	25.7	8.9
119	L'vov	49.49 N	23.57 E	329	-10.28	-9.86	61-02	87-08	125	61-02	82-07	105	-49.7	-44.8	102	11.3	11.7	708	7.8	29.5
120	Madang	5.22 S	145.80 E	4	-7.04	-7.70	61-02	63-08	20	61-02	63-08	20	-15.0	-14.2	20	12.8	15.6	3218	27.0	28.6
121	Mahe	4.62 S	55.45 E	1	-3.29	-3.57	61-04	75-09	127	61-04	75-09	127	-12.9	-8.9	126	13.5	14.7	2597	27.1	13.9
122	Malan (Cape Town)	33.97 S	18.60 E	44	-2.81	-3.45	69-01	83-03	86	69-01	83-04	66	-26.1	-21.1	63	12.3	12.0	513	16.3	17.5
123	Malange	9.55 S	16.37 E	1139	-3.67	-4.48	65-07	87-08	183	65-07	87-08	156	-33.1	-25.9	152	12.0	11.8	1060	21.9	29.0
124	Manaus	3.12 S	60.02 W	60	-4.55	-5.52	61-02	76-03	48	61-02	76-03	46	-44.2	-34.6	46	5.6	8.8	2211	26.8	27.3
125	Manila	14.52 N	121.00 E	14	-5.11	-6.74												1955	27.0	

TABLE 1. Continued

Station name	Latitude (deg.)	Longitude (deg.)	Altitude (m.a.s.l.)	O-18 Mean	O-18 W.mean	O-18 n	O-18 Start	O-18 End	*	D Mean	D W.mean	D n	D Start	D End	*	Excess n	Excess Mean	Excess W.mean	Prec (mm)	Temp (C)	Vap.P. (mb)
126 Maracay	10.25 N	67.65 W	442	-3.91	-4.01	64	61-05	75-08		-25.5	-27.9	48	61-05	75-08		48	5.8	4.2	806	24.5	22.5
127 Marion Island	46.88 S	37.87 E	26	-4.91	-4.93	194	61-04	87-12		-33.4	-33.5	79	61-04	76-03		79	7.0	7.0	2452	5.5	7.6
128 Meiringen	46.73 N	8.20 E	632	-12.29	-11.77	209	70-08	87-12											1278	7.9	8.9
129 Melbourne	37.82 S	144.97 E	28	-4.68	-5.13	248	81-10	84-12	*	-25.5	-28.2	211	62-01	87-12	*	200	11.3	11.5	635	15.4	11.4
130 Mendoza	32.88 S	68.85 W	827	-5.90	-5.59	27													363	17.1	11.1
131 Menongue	14.67 S	17.70 E	1348	-4.82	-6.18	59	62-01	83-03		-28.8	-38.0	46	69-01	83-03		45	9.6	11.2	895	20.4	13.4
132 Midway Is.	28.22 N	177.37 W	13	-2.33	-2.52	264	62-02	87-12	*	-8.8	-10.6	233	62-02	87-12	*	218	9.4	9.7	1090	22.3	20.6
133 Minsk	52.52 N	27.32 E	225	-10.79	-10.70	44	80-04	83-12		-77.7	-78.6	36	81-01	83-12		36	10.9	10.3	676	6.8	8.2
134 Moskva	55.75 N	37.57 E	157	-11.42	-10.99	63	70-01	79-12		-91.4	-89.1	115	81-01	79-01		63	5.4	5.4	693	5.4	7.5
135 Muzuza	1.22 S	36.63 E	2070	-3.56	-4.38	14	67-02	68-06		-13.7	-20.6	14	67-02	68-06		14			1268		
136 Murmansk	68.58 N	33.03 E	46	-13.10	-12.68	42	81-10	83-09		-97.8	-93.2	34	81-01	83-12		31	7.2	8.2	489	0.6	5.7
137 Nancunan	34.03 N	67.97 W	572	-5.64	-5.77	30													427	15.4	10.1
138 N'djamena	12.13 N	15.03 E	294	-1.51	-4.11	81	64-03	78-09		-6.2	-22.5	70	64-03	78-09		68	7.4	11.6	557	27.9	16.6
139 Ndola	13.00 S	28.65 E	1331	-6.03	-6.59	79	61-07	85-05		-35.1	-39.5	69	68-11	85-04		68	12.5	13.3	1172	20.4	14.8
140 New Delhi	28.58 N	77.20 E	212	-3.40	-5.81	216	61-01	87-09	*	-19.3	-37.2	183	61-01	87-09	*	181	6.4	8.3	774	25.1	17.4
141 Nord	81.60 N	16.67 W	35	-25.01	-24.98	93	80-04	83-12		-185.6	-185.6	63	81-01	83-12		62	11.1	11.5	164	-17.1	1.9
142 Odessa	46.48 N	30.63 E	64	-8.44	-9.09	40	70-02	83-12		-60.5	-64.5	32	67-06	83-12		31	6.9	8.2	459	10.0	10.5
143 Ottawa	45.32 N	75.67 W	114	-11.18	-10.89	202	80-04	83-12		-77.0	-74.2	188	81-01	87-12		164	11.2	11.3	881	5.8	8.3
144 Pechora	65.07 N	57.06 E	56	-14.74	-14.67	45	80-04	83-12	*	-111.2	-109.7	35	81-01	83-12	*	35	9.7	9.3	623	-0.5	5.7
145 Perm	58.01 N	56.18 E	161	-13.02	-12.90	42	62-02	76-12		-93.2	-92.1	32	62-02	76-12		32	14.9	14.0	630	2.4	7.0
146 Perth	31.95 S	115.97 E	17	-3.12	-3.87	98	65-05	71-08		-12.0	-14.9	93	66-05	71-08		93	13.0	16.2	809	18.6	13.2
147 Petzenkirchen	48.15 N	15.15 E	252	-9.45	-8.68	64	61-02	76-12		-68.5	-63.4	61	61-03	76-12		60	8.8	7.8	726	8.6	9.4
148 Podersdorf	47.85 N	16.85 E	120	-9.57	-8.73	41	61-02	68-12	*	-68.9	-63.4	41	66-05	68-12	*	24			566		
149 Pohang	36.03 N	129.38 E	6	-7.31	-7.78	110	65-03	76-12		-44.5	-51.0	118	61-03	76-12		110	12.3	10.1	1106	13.2	12.4
150 Ponta Delgada	33.77 N	25.65 W	175	-3.53	-4.20	70	65-08	74-12		-16.0	-19.9	72	65-08	74-12		70	12.4	13.7	1058	16.6	15.5
151 Porto Alegre	30.08 S	51.18 W	7	-4.63	-4.71	200	61-03	83-07	*	-24.8	-25.2	175	65-03	83-07	*	172	11.3	12.6	1305	19.6	17.8
152 Porto Velho	8.77 S	63.92 W	105	-5.74	-6.82	86	76-03	81-10		-34.4	-43.0	73	76-03	81-10		72	9.8	11.4	2210	25.3	28.1
153 Pretoria	25.73 S	28.18 E	1330	-2.57	-3.70	190	64-05	83-03		-9.2	-16.8	121	64-05	83-12		121	11.4	12.1	682	17.8	11.9
154 Prins Christians Sund	60.02 N	43.07 W	76	-9.76	-10.30	16	65-12	78-05	*	-63.5	-64.5	7	65-12	78-05	*	7	14.9		2928	1.2	8.1
155 Puerto Montt	41.47 S	72.93 W	13	-5.72	-6.07	40	76-03	75-09		-39.5	-41.9	27	76-03	75-09		27	6.4	7.7	1884	10.0	10.3
156 Rabba	31.20 N	35.75 E	970	-4.55	-4.81	12	78-01	68-02		-16.0	-16.9	12	78-01	68-02		12	22.6	23.6	700		
157 Rarotonga	21.20 S	159.80 W	6	-3.47	-4.26	85	61-04	87-12	*	-17.8	-23.7	85	61-04	87-12	*	85	10.0	10.5	1953	23.9	24.2
158 Regensburg	49.02 N	12.07 E	377	-10.29	-10.02	119	79-01	87-12		-76.0	-73.5	119	79-04	87-12		119	6.4	6.7	639	7.7	
159 Reykjavik	64.13 N	21.93 W	14	-7.49	-7.65	147	61-04	76-12		-53.7	-55.2	98	61-04	76-12		98	6.5	6.8	870	4.7	7.2
160 Rhodes (Maritsai)	36.38 N	28.10 E	42	-5.11	-4.95	23	63-05	76-03		-25.3	-23.5	18	63-05	76-03		18	17.5	19.6	647	18.6	13.9
161 Riga	56.97 N	24.07 E	3	-10.12	-10.34	55	80-04	84-12	*	-71.3	-73.3	44	81-11	84-09	*	42	10.9	11.0	646	6.5	8.4
162 Rio De Janeiro	22.90 S	43.17 W	26	-3.81	-4.54	182	61-11	87-09		-18.4	-26.2	135	61-11	87-09		132	12.3	12.0	1167	23.8	22.7
163 Rjazan	54.37 N	39.43 E	135	-11.45	-10.99	42	80-05	83-12		-87.6	-83.3	36	80-05	83-12		35	5.6	6.9	617	5.7	8.0
164 Rostov-Na-Donu	47.25 N	39.82 E	77	-8.29	-8.83	51	80-09	84-09		-64.5	-67.1	36	81-01	84-09		36	8.8	9.6	641	9.8	10.1
165 Ryori	39.02 N	141.50 E	260	-8.37	-8.13	86	79-01	86-06		-53.2	-53.0	89	79-01	86-06		86	13.5	12.1	1124	10.0	10.5
166 Salta	24.78 S	65.40 W	1187	-4.01	-4.53	19	65-02	83-12	*	-1.3	-2.9	118			*				697	17.7	14.8
167 Salvador (Ondina)	13.00 S	38.52 W	45	-1.45	-1.73	169	79-05	87-07		-1.3	-2.9	118	65-02	85-04		112	11.1	11.4	2106	25.2	25.7
168 San Andres	34.57 N	88.55 W	0	-4.40	-3.52	35													1368		
169 San Juan (Puerto Rico)	18.43 N	66.00 W	4	-1.57	-2.19	45	81-09	83-07		-3.8	-5.6	18	68-08	73-04		18	8.7	8.3	1356	27.0	25.7
170 San Luis	33.30 S	66.35 W	709	-5.17	-4.08	17	80-04	83-11	*						*				792	16.8	11.2
171 S. Petersburg	59.58 N	30.18 E	4	-11.66	-11.80	44	69-01	78-10		-87.1	-86.8	34	69-01	78-10		34	7.1	8.4	672	6.2	8.3
172 San Salvador (Ilopango)	13.70 N	89.12 W	615	-4.59	-6.67	98	78-10	87-12	*	-42.3	-45.8	74	78-10	84-07	*	73	10.1	9.5	1509	23.2	20.6
173 Santa Maria	34.90 N	120.45 W	79	-3.65	-6.01	62	75-08	82-07		-26.5	-38.7	60	75-08	82-07		60	7.3	9.4	312	13.7	11.3
174 Santiago	33.45 S	70.70 W	520	-10.26	-10.16	13	62-01	75-07		-63.5	-69.0	13	62-01	75-07		12	16.0	17.8	323	14.0	11.4
175 Santiago Del Estero	27.78 S	64.27 W	187	-6.89	-4.90	21	72-08	84-08					72-08						775	20.1	17.1
176 Sao Gabriel	0.13 S	67.08 W	87	-1.31	-4.65	100	61-11	83-01		-26.3	-28.8	101	61-11	83-01		89	9.2	9.1	2764	25.5	28.1
177 Sao Tome	0.38 N	6.72 E	8	-7.99	-3.39	123	80-12	76-12		-12.5	-17.7	94	81-01	76-12		94	8.8	9.2	939	25.2	26.2
178 Saratov	51.34 N	46.02 E	166	-12.66	-12.55	82	75-08	84-12		-72.4	-74.5	38	75-08	84-09		36	6.2	6.8	422	7.5	7.9
179 Scoresbysund	70.50 N	22.00 W	0	-13.65	-13.67	40	85-07	87-11											442	-7.3	3.2
180 Shijiazhuang	38.02 N	114.25 E	80	-7.85	-7.77	29	81-09	87-11	*	-53.1	-53.3	29	69-01	78-11	*	29	10.6	8.8	472	13.8	
181 Shillong	25.57 N	91.88 E	1598	-4.59	-3.67	31	69-01	78-10		-27.5	-28.5	30	78-10	78-10		30	12.7	13.2	2187	17.4	15.1
182 Sidi Barrani	31.63 N	25.96 E	23	-3.65	-4.18	45	78-10	87-12		-17.2	-20.4	46	75-08	87-12		45	11.5	12.7	152	19.5	12.7
183 Simcoe	42.85 N	80.27 W	240	-10.26	-9.27	81	75-08	82-07		-69.1	-62.2	78	68-06	82-07		78	11.5	11.4	941	8.6	
184 Singapore (Airport)	1.35 N	103.90 E	32	-6.89	-7.26	97	62-05	75-12	*	-43.9	-46.2	78	62-05	82-07	*	52	13.4	13.7	2164	26.3	29.2
185 St. Helena	15.97 S	5.70 W	604	-1.31	-1.42	80	72-08	75-12		3.1	2.4	52	61-11	75-12		74	13.9	14.1	893	16.9	16.6
186 Stanley	51.70 S	57.87 W	51	-7.99	-8.08	95	75-08	76-12		-57.5	-58.0	74	75-08	76-11		92	6.4	6.4	605	5.7	8.0
187 Ste. Agathe	46.05 N	74.28 W	395	-12.66	-12.55	82	75-08	82-07		-89.0	-87.8	92	61-02	82-07		80	13.5	13.6	1200	4.1	
188 Stuttgart (Cannstatt)	48.83 N	9.20 E	315	-8.01	-8.19	226	61-02	87-12		-59.9	-58.4	223	61-02	87-12		223	5.8	5.7	655	9.3	9.4
189 Taastrup	55.67 N	12.30 E	28	-9.76	-10.02	50	65-11	71-03	*	-68.5	-70.7	51	65-11	71-03	*	50	9.1	9.2	515		

TABLE 1. Continued

#	Station name	Latitude (deg.)	Longitude (deg.)	Altitude (m.a.s.l)	O-18 Mean	O-18 W.mean	O-18 n	O-18 Start	O-18 End	*	D Mean	D W.mean	D n	D Start	D End	*	Dex n	Dex Mean	Dex W.mean	Prec (mm)	Temp (C)	Vap.P. (mb)
190	Taguac	13.55 N	144.83 E	110	-3.73	-5.21	112	61-12	77-03		-20.3	-31.1	106	61-12	77-03		106	9.7	10.8	2659	26.0	28.2
191	Tambov	52.44 N	41.28 E	139	-11.05	-10.95	22	80-04	83-12		-84.9	-86.4	15	82-08	83-12		15	5.8	7.6	558	6.7	7.9
192	Teheran	35.68 N	51.32 E	1200	-5.22	-6.76	159	61-02	87-03		-30.3	-41.4	137	61-02	87-03		136	8.8	11.3	211	16.8	6.7
193	The Pas	53.97 N	101.10 W	272	-16.31	-19.47	73	75-08	82-07		-144.5	-122.8	70	75-08	82-07		70	7.8	6.8	379	1.5	-
194	Thonon-Les-Bains	46.22 N	6.28 E	385	-9.62	-9.69	196	63-06	87-01	*	-	-	2	85-01	87-03	*	1	-	8.6	1008	10.2	-
195	Thule	76.52 N	68.83 W	77	-24.14	-22.72	59	66-01	71-09	*	-184.2	-171.9	45	66-01	70-03	*	43	9.3	11.4	189	-11.6	2.3
196	Tokyo	35.68 N	139.77 E	4	-7.10	-7.30	219	61-03	79-12		-43.8	-46.4	184	61-03	79-12		183	12.7	6.8	1378	15.5	13.3
197	Trier	49.75 N	6.07 E	273	-7.48	-7.75	118	78-01	87-12	*	-53.3	-55.2	118	78-01	87-12		118	6.6	10.3	829	8.7	-
198	Truk	7.47 N	151.85 E	2	-4.94	-5.35	93	68-03	77-05		-30.1	-32.9	72	68-03	77-05		72	9.8	10.2	3582	27.5	29.9
199	Truro	45.37 N	63.27 W	40	-9.48	-9.53	91	75-08	83-12		-65.8	-66.1	91	75-08	83-12		91	10.1	10.8	1008	10.1	7.0
200	Tunis (Carthage)	36.83 N	10.23 E	4	-3.58	-4.54	145	68-04	87-11		-21.5	-26.6	129	68-04	82-08		129	7.5	-	481	18.2	15.0
201	Ushuaia	54.78 S	68.28 W	10	-10.53	-10.41	36	81-10	85-04		-80.3	-73.8	4	82-05	82-08	*	4	-	7.1	299	6.2	7.3
202	Valentia (Observatory)	51.93 N	10.25 W	9	-5.01	-5.34	209	60-03	87-12	*	-33.6	-35.7	179	60-03	87-12	*	175	6.8	7.5	1426	10.5	10.9
203	Veracruz	19.20 N	96.13 W	16	-2.86	-4.13	169	62-04	87-06		-16.1	-27.1	136	62-04	87-06		133	7.7	5.0	1518	25.4	25.9
204	Victoria	48.25 N	16.37 E	203	-9.87	-9.96	72	75-08	82-07		-70.9	-70.8	70	75-08	82-07		70	3.8	8.0	610	10.0	9.5
205	Vienna (Hohe Warte)	48.65 N	123.43 W	20	-9.32	-10.31	306	61-02	87-12	*	-71.2	-76.8	302	61-02	87-12	*	302	7.7	6.6	909	10.7	-
206	Vologda	59.17 N	39.52 E	118	-13.16	-12.75	45	80-04	83-12		-101.2	-98.9	35	81-01	83-12		35	6.4	9.4	565	3.6	7.5
207	Waco	31.62 N	97.22 W	156	-3.75	-4.04	96	62-01	76-12		-19.8	-22.9	96	62-01	76-12		96	10.2	8.0	810	19.4	16.1
208	Wake Island	19.28 N	166.65 E	3	-1.46	-2.11	151	62-02	76-12		-3.9	-10.3	132	62-02	76-12		122	8.1	10.7	908	26.6	25.7
209	Wasserkuppe Rhoen	50.50 N	9.95 E	921	-9.88	-9.78	120	78-01	87-12	*	-68.3	-67.6	120	78-01	87-12	*	120	10.8	9.5	1127	4.2	-
210	Weathership E	35.00 N	48.00 W	0	-2.90	-3.16	129	62-11	73-06		-14.9	-16.4	102	62-11	73-06		101	9.5	10.7	527	20.4	18.9
211	Weathership V	31.00 N	164.00 E	0	-3.99	-4.35	96	61-05	71-09		-22.8	-23.7	84	61-05	66-01		83	8.9	5.4	832	19.1	18.1
212	Whitehorse	60.72 N	135.07 W	702	-20.86	-20.95	54	61-04	66-01		-162.6	-162.2	54	61-04	75-04		54	4.4	12.6	269	-1.1	4.7
213	Windhoek	22.57 S	17.10 E	1728	-2.59	-5.03	135	61-04	80-12		-11.3	-24.5	91	61-04	80-12		87	10.1	6.0	385	19.3	7.6
214	Wuerzburg	49.80 N	9.90 E	259	-8.07	-8.10	120	78-01	87-12	*	-59.0	-58.9	120	78-01	87-12		120	5.6	7.9	612	8.3	-
215	Wynyard	51.77 N	104.20 W	561	-17.84	-15.55	77	75-08	82-07		-139.2	-119.3	76	75-08	82-07		72	6.4	9.4	354	3.0	-
216	Xian	34.30 N	108.93 E	397	-7.12	-6.74	19	85-09	87-10	*	-45.7	-43.0	19	85-09	87-10		19	12.4	6.0	530	13.9	-
217	Yangoon	16.77 N	96.17 E	20	-4.79	-4.34	20	61-05	63-12		-32.0	-28.9	20	61-05	63-12		20	6.3	9.5	2456	27.3	27.7
218	Yap	9.49 N	138.09 E	0	-4.98	-5.65	99	68-06	76-12		-29.1	-34.7	65	68-06	76-12		65	8.6	-	2812	27.1	29.1
219	Zagreb	45.49 N	15.59 E	165	-9.08	-9.08	94	80-01	87-12	*	-64.5	-64.3	94	80-01	87-12	*	94	8.2	8.4	840	11.0	10.4

TABLE 2. Seasonal and long-term $\delta^{18}O$-temperature and $\delta^{18}O$-precipitation coefficients derived for selected stations of the IAEA/WMO global network. See text for definition of $\Delta\delta^{18}O$, ΔT and ΔP for both seasonal and long-term relationships.

STATION	$\overline{T}$ (°C)	$\overline{P}$ (mm)	Seasonal $\Delta\delta^{18}O/\Delta T$ (°/∞ /°C)	r^2	Long-term $\Delta\delta^{18}O/\Delta T$ (°/∞ /°C)	r^2	Seasonal $\Delta\delta^{18}O/\Delta P$ (°/∞ / 10 mm)	r^2	Long-term $\Delta\delta^{18}O/\Delta P$ (°/∞ / 10 mm)	r^2
1. Vienna (1961-1990) 48.25°N, 16.30°E; 203 m.a.s.l.	9.9	610	0.39 ± 0.02	0.93	0.65 ± 0.05	0.36	1.83 ± 0.60	0.48	-0.18 ± 0.05	0.039
2. Ottawa (1970-1990) 45.32°N, 75.60°W; 114 m.a.s.l.	5.8	884	0.31 ± 0.02	0.94	0.49 ± 0.08	0.14	2.49 ± 0.88	0.44	0.11 ± 0.05	0.017
3. Argentine Island (1965-1987) 65.25°S, 64.20°W; 0 m.a.s.l.	-4.1	377	0.31 ± 0.03	0.89	0.59 ± 0.04	0.45	-0.88 ± 0.64	0.16	0.10 ± 0.06	0.015
4. Hong Kong (1961-1965) (1973-1987) 22.32°N, 114.10°E; 65 m.a.s.l	22.9	2219	-0.42 ± 0.04	0.91	0.25 ± 0.13	0.02	-0.13 ± 0.02	0.75	-0.06 ± 0.01	0.310
5. Taguac (Guam Isl.) (1961-1967) (1973-1977) 13.55 N, 144.80 E; 110 m.a.s.l.	26.0	2659	-2.43 ± 0.98	0.38	-1.28 ± 0.83	0.02	-0.19 ± 0.02	0.89	-0.17 ± 0.01	0.750
6. Apia (1962-1967) (1972-1977) 13.80 S, 171.70 W 2 m.a.s.l.	26.4	3024	-4.29 ± 0.51	0.88	-1.15 ± 0.19	0.29	-0.13 ± 0.02	0.78	0.03 ± 0.01	0.070

the Vienna station, having the longest $\delta^{18}O$ record in the network. It is clear that the smoothing does not change the character of the trend curves. The trend curves of $\Delta\delta^{18}O$ for the Swiss stations are shown in Figure 21a. They reveal a strong coherence among the four stations (correlation coefficient between each pair of stations higher than 0.8). This indicates the presence of common mechanisms controlling the long-term behaviour of ^{18}O content in precipitation in this region.

The composite trend curve of ΔT for the Swiss stations (Figure 22a) indicates a warming trend in the order of 1.5°C between 1986 and 1990, associated with an increase of $\Delta\delta^{18}O$ by about 2°/∞ above the 1970-1990 average, during the same time interval. The composite trend curve of temperature for all European stations shown in Figure 22b reveals the same trend and is in qualitative agreement with regional estimates published by *Hansen and Lebedeff* [1987], based on a much larger data set. Correlation analysis suggests that up to 45 per cent of the variations observed for trend curves of $\Delta\delta^{18}O$ representing European stations can be attributed to temperature changes, although the strength of this coupling may vary substantially from place to place, even on a regional scale [*Rozanski et al.*, 1992]. Figure 23 illustrates the correlation between $\Delta\delta^{18}O$ and ΔT for the Vienna station, derived from the trend curves shown in Figure 20. The slope of the best fit line is equal to 0.65±0.05 °/∞ per °C.

The trend curves for stations representing other regions reveal a different behaviour. The Ottawa station, although located at a similar latitude to the Vienna station, does not reveal any warming trend over the last decade. To the contrary, a general cooling is visible (Figure 22c). There is a poor correlation between $\Delta\delta^{18}O$ and ΔT for this station (see Table 2). The Hong Kong data, representing tropical, monsoon-type climate, shows relatively small fluctuations of $\Delta\delta^{18}O$ and ΔT, without apparent correlation among them (Fig. 22d). The Taguac data (tropical, oceanic climate) reveal similar behaviour.

Argentine Island, located at a high latitude in the southern hemisphere, is characterized by the strongest link between ^{18}O and temperature among the stations analyzed. The trend curves show a very distinct maximum around 1985. The $\Delta\delta^{18}O/\Delta T$ coefficient derived from the trend curves of ^{18}O and temperature amounts to 0.59±0.04 °/∞ per °C. This is in excellent agreement with an independent estimate of this coefficient based on comparison of δD variations measured in the snow core collected at Dalinger Dome (Antarctic Peninsula) and covering the time period 1953-1980, with instrumental records of temperature in this region [*Aristarain et al.*, 1986]

The trend curves of the long-term changes of precipitation amount (ΔP) at selected stations were calculated in an analogous way as for the temperature and $\delta^{18}O$. They are compared with the $\Delta\delta^{18}O$ curves in Figure 24. The slopes $\Delta\delta^{18}O/\Delta T$ and $\Delta\delta^{18}O/\Delta P$ derived from regression analysis of the trend curves are summarized in Table 2. For the analyzed continental stations (Vienna, Ottawa, Argentine

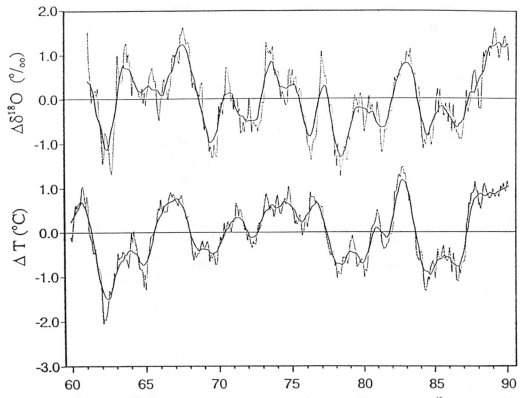

Fig. 20. Long-term trend curves of $\delta^{18}O$ and surface air temperature for the Vienna station. The trend curves $\Delta\delta^{18}O$ and ΔT were calculated by applying a 12-month running average over the monthly time series and subtracting the resulting curves from the long-term annual means of $\delta^{18}O$ and temperature (broken lines). The resulting curves were then smoothed by applying again a 12-month running average (heavy lines).

Island) there is no significant correlation between long-term changes of $\delta^{18}O$ and the amount of precipitation. Temperature appears to be a major parameter controlling long-term behaviour of isotopic composition of precipitation in a continental environment. On the contrary, Taguac and Hong Kong data reveal significant correlation between long-term changes of $\delta^{18}O$ and precipitation amount with no apparent link to the fluctuations of temperature at these stations. Exceptionally strong correlation between $\Delta\delta^{18}O$ and ΔP for Taguac station ($r^2 = 0.75$) has already been noted by *Lawrence and White* [1991].

δD - $\delta^{18}O$ relationship

A close relationship between deuterium and oxygen-18 isotopic composition of freshwaters (including precipitation) when plotted on the δD - $\delta^{18}O$ diagram, was first noted by *Friedman* [1953]. Craig, in his global survey of deuterium and oxygen-18 content of freshwaters [*Craig*, 1961], suggested the best fit line of his data points, $\delta D = 8\delta^{18}O + 10$, which was later named the Global Meteoric Water Line (GMWL). A few years later, Dansgaard [*Dansgaard*, 1964] introduced the concept of "deuterium excess", defined as $d = \delta D - 8\cdot\delta^{18}O$.

A clear distinction has to be made between the deuterium excess calculated using the above equation which defines location of individual data points on the δD - $\delta^{18}O$ plot with respect to the GMWL,

and the intercept of the best fit line of all data points available, which frequently has a different slope than the GMWL. Figure 25a shows the long-term annual mean $\delta^{18}O$ and δD values for all analyzed stations of the IAEA/WMO network, plotted on the δD - $\delta^{18}O$ diagram. The least square fit of the data points results in the following equations:

a) long-term arithmetic means:

$$\delta D_a = (8.17 \pm 0.06)\delta^{18}O + (10.35 \pm 0.65) \quad r^2 = 0.99 \quad n = 206 \quad (2)$$

b) long-term weighted means (weighing by amount of precipitation):

$$\delta D_w = (8.20 \pm 0.07)\delta^{18}O + (11.27 \pm 0.65) \quad r^2 = 0.98 \quad n = 205 \quad (3)$$

The above equations are identical, within the quoted errors, with the earlier estimates based on the IAEA database [*Dansgaard*, 1964; *Yurtsever and Gat*, 1981] and confirm that Craig's equation $\delta D = 8\cdot\delta^{18}O + 10$ is a good approximation of the locus of points representing average isotopic composition of freshwaters worldwide.

The global δD - $\delta^{18}O$ relationship is well understood. It was properly reproduced by both "isolated air mass" models based on the Rayleigh approach [*Merlivat and Jouzel*, 1979] and by general circulation models [*Joussaume et al.*, 1984a; *Jouzel et al.*, 1987a]. It

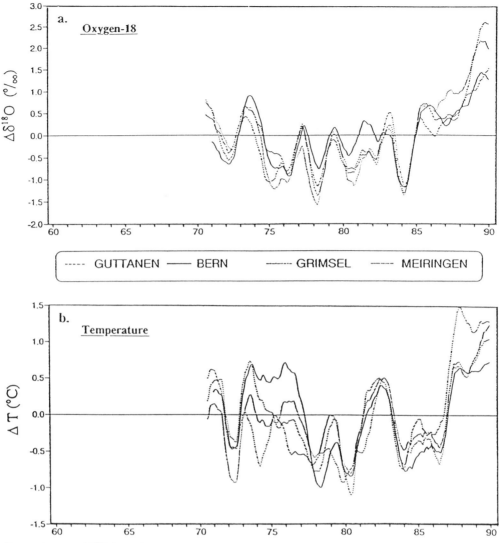

Fig. 21. The trend curves of $\delta^{18}O$ in precipitation (a) and surface air temperature (b) for four Swiss stations of the IAEA/WMO global network, derived from the available time series of monthly data (see text for details).

has been demonstrated that the value of the intercept of the global meteoric water line is controlled by the evaporation process in the major source regions of the vapor (subtropical ocean). The sea surface temperature, humidity of the air and the wind speed are the main controlling parameters. In-cloud processes in general do not modify significantly the deuterium excess value of the vapor, as long as formation of rain is considered. It has also been shown that the slope of the GMWL is controlled in the first instance by the ratio of equilibrium isotope enrichments for deuterium and oxygen-18, respectively [*Dansgaard*, 1964; *Merlivat and Jouzel*, 1979].

Comparison of the long-term annual means of $\delta^{18}O$ and δD (Figure 25a) with the long-term monthly means for the same set of stations (Figure 25b), reveals in the latter case a slightly lower slope of the δD - $\delta^{18}O$ relationship and generally higher spread of data points. This effect is more obvious when monthly data for individual stations

are plotted on the δD - $\delta^{18}O$ diagram (Figures 26 and 27). The comparison of δD - $\delta^{18}O$ plots for selected marine (Figure 26) and continental (Figure 27) stations shows that in the former case the range of variations is smaller and a scatter of the data points is usually substantially larger. The regression equations with corresponding correlation coefficients for the monthly data presented in Figures 26 and 27 are summarized in Table 3. In an extreme situation, represented by the St. Helena station (see Figure 26), the monthly data points form a tight cluster, with a very poor correlation between δD and $\delta^{18}O$. At this station essentially all precipitation probably comes from nearby sources and represents the first stage of the rain-out process. This probably accounts for the poor correlation. Most of the data points have positive δD values. This reflects a higher evaporation temperature at the vapor source, when compared to average condensation temperatures. The generally weaker

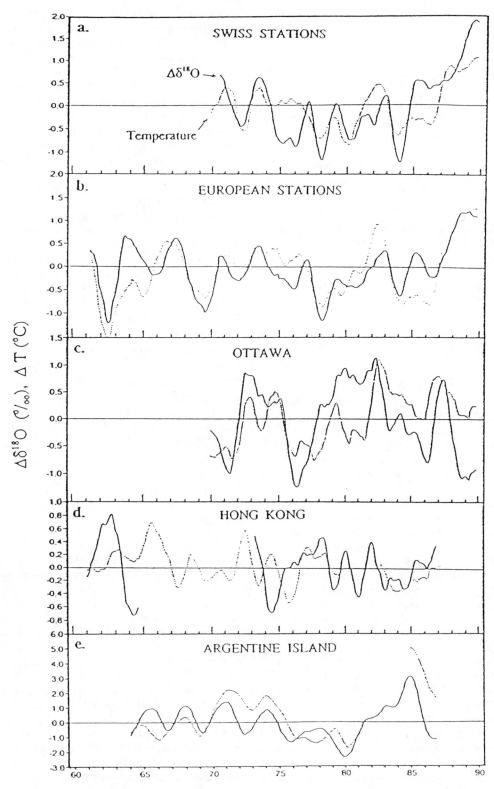

Fig. 22. The long-term trends of $\delta^{18}O$ in precipitation and surface air temperature for selected stations of the IAEA/WMO global network. (a) composite trend curves of $\Delta\delta^{18}O$ and ΔT for the four Swiss stations presented in Figure 21; (b) composite trend curves of $\Delta\delta^{18}O$ and ΔT for selected European stations; (c,d,e) trend curves of $\Delta\delta^{18}O$ and ΔT for Ottawa, Hong Kong and Argentine Island, respectively.

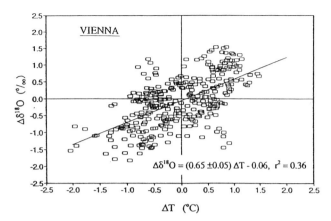

Fig. 23. Relationship between the trend curves of $\Delta\delta^{18}O$ and ΔT calculated for the Vienna station having the longest ^{18}O record among the stations of the IAEA/WMO global network.

correlation of δD and $\delta^{18}O$ values for the marine stations of the network [see *IAEA*, 1992] may reflect varying contributions of air masses with different source characteristics and a low degree of rain-out.

The imprint of local conditions at source regions of the vapor and different storm trajectories on the isotopic composition of precipitation can be seen also at a number of coastal and continental stations. For instance, the Bet Dagan station receives rains associated with Mediterranean or European air masses, each characterized by a different deuterium excess value [*Gat and Carmi*, 1987; *Rindsberger et al.*, 1983, 1990]. This results in a large spread of data points on the δD - $\delta^{18}O$ plot. A similar situation is observed for Tokyo station, which receives precipitation of Pacific origin in summer, whereas in winter the westerly circulation dominates bringing precipitation from the Japan Sea and the China Sea.

Even in cases where the stations receive precipitation essentially from one source, distinct seasonal trends can be identified on δD - $\delta^{18}O$ plots of monthly data. European stations of the network serve as an example of such a situation. The plot of the mean monthly values of the deuterium excess for the Valentia and Vienna stations (Figure 28) reveal significantly higher values of this parameter in winter than in summer. This, in turn, leads to the slope of δD - $\delta^{18}O$ relation distinctly lower than 8 when an attempt is made to fit the data with a straight line (see Table 3). It has been suggested [*Rozanski*, 1987] that this apparent seasonality of the d-excess values in Europe may be generated by two effects: (1) lower relative humidity of air (normalized to the sea surface temperature) in the source region in winter leads to enhanced kinetic fractionation during the evaporation process and, consequently, higher d-excess signature of the vapor. (2) lower relative humidity over the continent in summer facilitates partial evaporation of raindrops below the cloud base. This, in turn will lead to reduction of the deuterium excess values of summer rains. For stations with substantial contribution of snow precipitation, higher deuterium excess values during winter may result from additional kinetic fractionation during snow formation [*Jouzel and Merlivat*, 1984].

Fitting monthly isotope data of individual stations with a straight line is not always justified theoretically; the above examples have

shown that seasonally varying influence of different sources of vapor with different isotope characteristics, different storm trajectories or evaporation and isotope exchange processes below the cloud base, may often lead to a more complex relationship between δD and $\delta^{18}O$ of precipitation recorded at a given station.

Figure 29a shows latitudinal distribution of the deuterium excess values derived from the long-term annual means of δD and $\delta^{18}O$ for the selected stations of the network. Stations of the northern hemisphere reveal a relatively large spread of the d-values, without any clear trend. An exception is the group of stations located in the Mediterranean region, which are characterized by distinctly higher deuterium excess values, reflecting specific conditions of vapor formation over the Mediterranean Sea [*Gat and Carmi*, 1970; *Rindsberger et al.*, 1983, 1990]. For the southern hemisphere, a slight increase of the d-values can be noted between the equatorial region and 30°S, followed by an apparent reduction of this parameter further south, with a distinct minimum around 65°S (Argentine Island). The Perth station (31.95°S. 115.90°E) constitutes an exception to this trend. Contrary to other Australian stations of the network, in Perth, as in the Mediterranean region, winter precipitation prevails. Although these apparent variations of the deuterium excess may properly reflect different conditions at vapor source regions (relative humidity, wind speed, sea surface temperature), scarcity of data documenting variability in these conditions does not allow any firm conclusions.

The role of vapor source regions in establishing the deuterium excess values observed in precipitation is illustrated by Figure 29b showing the d-values for selected stations of the network, as a function of a fraction of winter precipitation in the total precipitation collected at the given station. For the stations with dominating winter precipitation, an apparent increase of the deuterium excess is observed, probably due to lower relative humidity (normalized to the sea surface temperature) over the ocean in winter and/or substantial contribution of snow precipitation formed under non-equilibrium conditions.

Frequency distribution of monthly data

The above discussion demonstrated that the isotopic composition of monthly composite samples of precipitation collected throughout the year is controlled by a number of different processes, both of a regional and local nature. To illustrate the relative importance of these processes in formation of the average isotopic composition of precipitation at a given location, frequency distribution histograms of monthly $\delta^{18}O$ values have been constructed for several stations representing different environments. These histograms are shown in Figure 30. Two marine stations (Weathership E and Apia) reveal a rather symmetric distribution, coupled with a relatively small range for the $\delta^{18}O$ values. This appears to be a typical $\delta^{18}O$ distribution for oceanic islands [*IAEA*, 1992]. The mid-latitude continental stations in contrast are characterized by a much broader range of $\delta^{18}O$ values and a distribution which is usually skewed towards negative values (Vienna, Ottawa). An interesting example of bimodal distribution is represented by the data from the New Delhi station. Two distinct peaks are associated with two distinct modes of precipitation regime: the more negative peak represents abundant monsoon precipitation (June-September), whereas light rains occurring between October

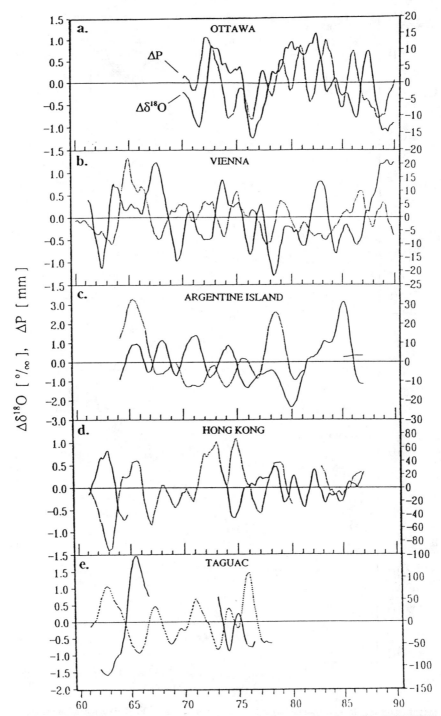

Fig. 24. The long-term trends of $\delta^{18}O$ and amount of precipitation, calculated for selected stations of the IAEA/WMO global network: Vienna (a), Ottawa (b), Argentine Island (c), Hong Kong (d) and Taguac (e). See text for details of calculation of $\Delta\delta^{18}O$ and ΔP.

and June form the second peak. Detailed analysis of isotope and precipitation data for this station suggests that isotopic composition of precipitation representing the dry period is probably modified by evaporation and isotope exchange below the cloud base [*Datta et al.*, 1991].

CONCLUDING REMARKS

The IAEA/WMO network "Isotopes in Precipitation" has accumulated during the past three decades a unique set of basic data on spatial and temporal distribution patterns of deuterium and oxygen-18 isotope composition of precipitation on the global scale.

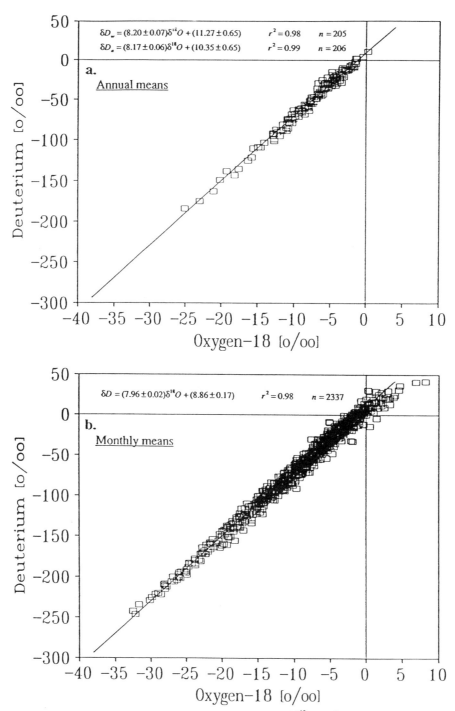

Fig. 25. Global relationship between long-term annual (a) and monthly (b) means of $\delta^{18}O$ and δD in precipitation, derived for all stations of the IAEA/WMO global network. Heavy lines indicate the position of the Global Meteoric Water Line: $\delta D = 8\cdot\delta^{18}O + 10$.

This database has been extensively used in the past in numerous applications of environmental isotopes to hydrology, oceanography and climatology. The basic design and structure of the network appears to be correct, although certain modifications would be necessary to meet current needs of new fields of applications of stable isotope data, especially in climatology and global atmospheric modelling.

The present-day global distribution patterns of deuterium and oxygen-18 in meteoric waters reveal a close relationship between some climatically relevant meteorological parameters such as surface

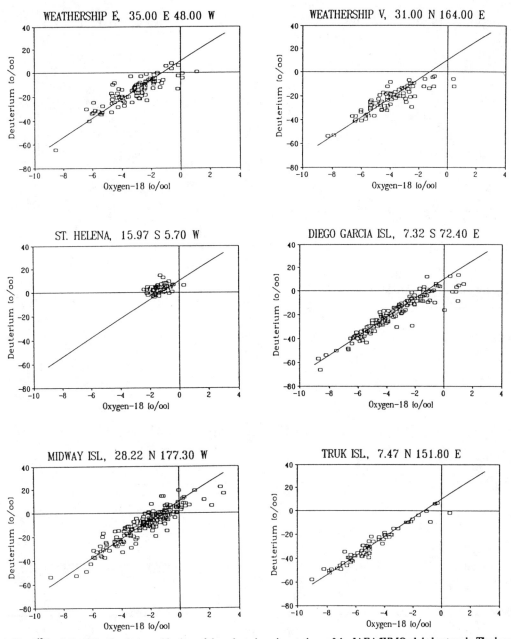

Fig. 26. δD - δ^{18}O relationships for the monthly data of the selected marine stations of the IAEA/WMO global network. The heavy line indicates the position of the Global Meteoric Water Line: δD = 8•δ^{18}O + 10.

air temperature or amount of precipitation and isotopic composition of precipitation. These apparent correlations have been applied in numerous studies to extract palaeoclimatic information from records of isotopic composition of ancient precipitation preserved in various climatic archives (polar ice cores, lake deposits, groundwater, organic matter). Quantitative interpretation of these records is, however, hampered by our limited understanding of physical processes controlling global isotope behaviour. Whereas the link between isotope signature of precipitation and climate at polar regions is at present relatively well understood, this is much less the

case for temperate and tropical areas. The IAEA/WMO network is providing ground-truth data needed to improve our knowledge of these processes.

Today, long records of isotope and meteorological data are available for a number of stations of the IAEA/WMO network, reaching in several cases three decades, a time scale comparable with climatic fluctuations. Therefore, a search for long-term trends in the isotope records is becoming meaningful. We have demonstrated that the long-term changes of ^{18}O content of precipitation in some regions closely follow long-term changes of surface air temperature and

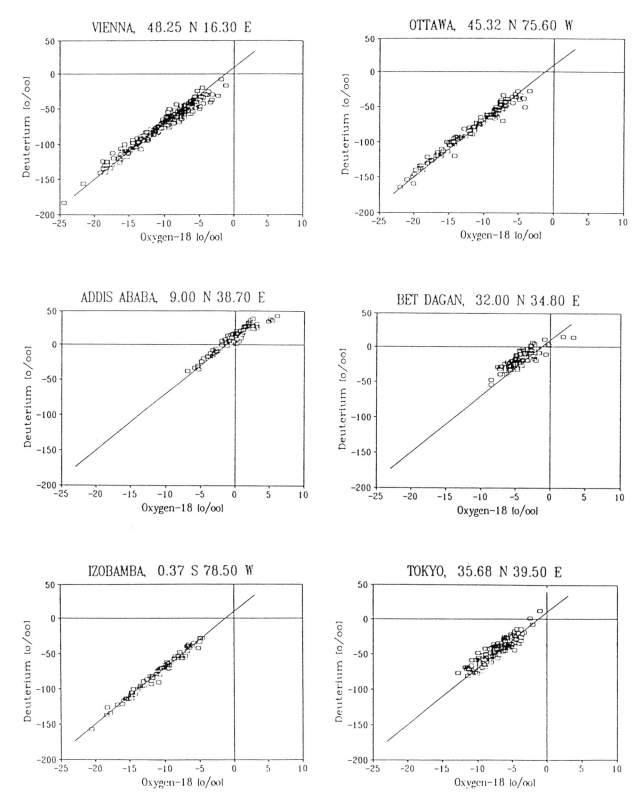

Fig. 27. δD - δ^{18}O relationships for the monthly data of selected continental stations of the IAEA/WMO global network. The heavy line indicates the position of the Global Meteoric Water Line: δD = 8·δ^{18}O + 10.

TABLE 3. Regression equations of δD and δ¹⁸O monthly data for selected stations of the IAEA/WMO global network.

STATION	Regression equation	r^2	STATION	Regression equation	r^2
Continental and coastal stations			Marine stations		
Vienna	$\delta D = 7.07\ \delta\ ^{18}O - 1.38$	0.961	Weathership E	$\delta D = 5.96\ \delta\ ^{18}O + 2.99$	0.738
Ottawa	$\delta D = 7.44\ \delta\ ^{18}O + 5.01$	0.973	Weathership V	$\delta D = 5.51\ \delta\ ^{18}O - 1.10$	0.737
Addis Ababa	$\delta D = 6.95\ \delta\ ^{18}O + 11.51$	0.918	St. Helene	$\delta D = 2.80\ \delta\ ^{18}O + 6.61$	0.158
Bet Dagan	$\delta D = 5.48\ \delta\ ^{18}O + 6.87$	0.695	Diego García Isl.	$\delta D = 6.93\ \delta\ ^{18}O + 4.66$	0.880
Izobamba	$\delta D = 8.01\ \delta\ ^{18}O + 10.09$	0.984	Midway Isl.	$\delta D = 6.80\ \delta\ ^{18}O + 6.15$	0.840
Tokyo	$\delta D = 6.87\ \delta\ ^{18}O + 4.70$	0.835	Truk Isl.	$\delta D = 7.07\ \delta\ ^{18}O + 5.05$	0.940

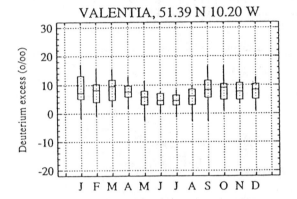

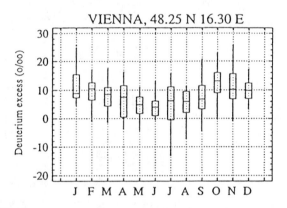

Fig. 28. Seasonal variations of the deuterium excess values in precipitation collected at two European stations of the IAEA/WMO global network: Valentia and Vienna. The data are presented in the form of box-and-whisker plots.

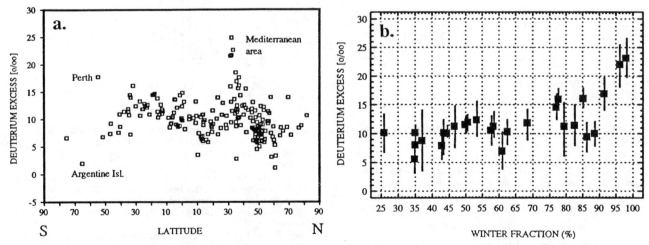

Fig. 29. (a) Latitudinal distribution of deuterium excess values calculated from the long-term annual means of δ¹⁸O and δD in precipitation, available for the stations of the IAEA/WMO global network; (b) the deuterium excess values calculated from the long-term annual means of δ¹⁸O and δD in precipitation for selected stations of the network and plotted as a function of the fraction of winter precipitation in the mean annual precipitation recorded at the given station.

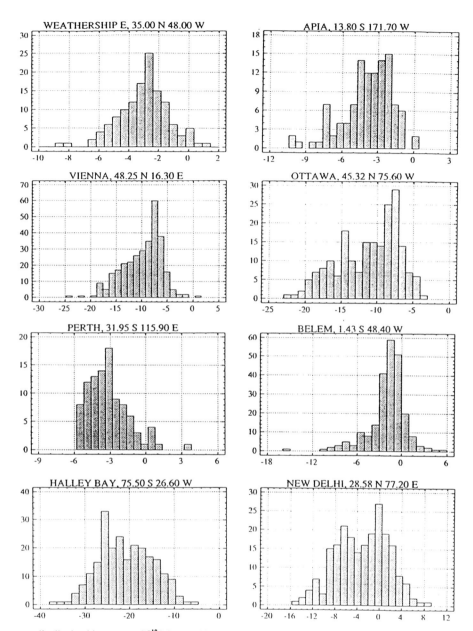

Fig. 30. Frequency distribution histograms of $\delta^{18}O$ in monthly precipitation, calculated for selected stations of the IAEA/WMO global network.

amount of precipitation, confirming the importance of stable isotopes as climatic indicators.

Acknowledgements. The continuous support and cooperation of numerous individuals and institutions in the IAEA and WMO Member States in maintaining the network activities is highly acknowledged. Thorough reviews of Carol Kendall and James R. Lawrence greatly improved the manuscript.

REFERENCES

Ambach, W., M. Elsässer, H. Moser, W. Rauert, W. Stichler, and P. Trimborn, Variationen des Gehaltes an Deuterium, Sauerstoff-18 und Tritium während einzelner Niederschläge, *Wasser und Leben*, 27, 186-192, (1975).

Aristarain, A.J., J. Jouzel, and M. Pourchet, Past Antarctic Peninsula climate (1850-1980) deduced from an ice core isotope record, *Clim. Change, 8*, 69-89, 1986.

Becker, B., B. Kromer, and P. Trimborn, A stable-isotope tree-ring timescale of the Late Glacial/Holocene boundary, *Nature, 353*, 647-649, 1991.

Bryson, R.A., and F.K. Hare, Climates of North America, in *World Survey of Climatology*, edited by H.E. Landsberg, pp. 321-356, Elsevier, Amsterdam, 1974.

Covey, C., and P.L. Haagenson, A model of oxygen isotope composition of precipitation: Implication for paleoclimate data, *J. Geophys. Res.*, 89, 4647-4655, 1984.

Craig, H., Isotopic variation in meteoric waters, *Science, 133*, 1702-1703, 1961.

Craig, H., and L. Gordon, Deuterium and oxygen 18 variation in the ocean and marine atmosphere, in *Stable Isotopes in Oceanographic Studies and Palaeotemperatures, Spoleto 1965*, edited by E. Tongiorgi, pp. 9-130, Consiglio Nazionale della Richerche, Pisa, Italy, 1965.

Dansgaard, W., The abundance of ^{18}O in atmospheric water and water vapour, *Tellus, 5*, 461-469, 1953.

Dansgaard, W., The ^{18}O abundance of fresh water, *Geochim. Cosmochim. Acta, 6*, 241-260, 1954.

Dansgaard, W., Stable isotopes in precipitation. *Tellus, 16*, 436-468, 1964.

Dansgaard, W., H. Clausen, N. Gundestrup, C. Hammer, S. Johnsen, P. Kristinsdottir, and N. Reeh, A new Greenland deep ice core, *Science, 218*, 1273-1277, 1982.

Datta, P.S., S.K. Tyagi, and H. Chandrasekharan, Factors controlling stable isotope composition of rainfall in New Delhi, India, *J. Hydrol., 128*, 223-236, 1991.

Eicher, U., and U. Siegenthaler, Palynological and oxygen isotope investigations on Late-Glacial sediment cores from Swiss lakes, *Boreas, 5*, 109-117, 1976.

Ehhalt, D.H., Vertical profiles of HTO, HDO and H_2O in the troposphere. *NCAR Technical Note, NCAR-TN/STR-100*, 133pp., 1974.

Epstein, S., and T. Mayeda, Variations of the ^{18}O content of waters from natural sources, *Geochim. Cosmochim. Acta, 4*, 213-224, 1953.

Eriksson, E., Deuterium and oxygen 18 in precipitation and other natural waters: Some theoretical considerations, *Tellus, 27*, 498-512, 1965.

Fisher, D.A., and B.T. Alt, A global oxygen isotope model, semiempirical, zonally averaged, *Ann. Glaciol., 7*, 117-124, 1985.

Fisher, D.A., A zonally-averaged stable-isotope model coupled to a regional variable-elevation stable-isotope model, *Ann. Glaciol., 14*, 65-71, 1990.

Friedman, I., Deuterium content of natural waters and other substances, *Geochim. Cosmochim. Acta, 4*, 89-103, 1953.

Friedman, I., A.C. Redfield, B. Schoen, and J. Harris, The variation of the deuterium content of natural waters in the hydrologic cycle, *Rev. Geophys., 2*, 1-124, 1964.

Friedman, I., G.I. Smith, J.D. Gleason, A. Warden, and J.M. Harris. Stable isotope composition of waters in southeastern California, 1. Modern precipitation, *J. Geophys. Res., 97*, 5795-5812, 1992.

Fritz, P., R.J. Drimmie, S.K. Frape, and K. O'Shea, The isotopic composition of precipitation and groundwater in Canada, in *Isotope Techniques in Water Resources Development*, pp. 539-550, International Atomic Energy Agency, Vienna, Austria, 1987.

Gambell, A.W., and I. Friedman, Note on the great variation of deuterium/hydrogen ratios in rainfall for a single storm event, *J. Appl. Meteorol., 4*, 533-535, 1965.

Gat, J.R., and W. Dansgaard, Stable isotope survey of the fresh water occurrences in Israel and the northern Jordan rift valley, *J. Hydrol., 16*, 177-212, 1972.

Gat, J.R., and I. Carmi, Effect of climate changes on the precipitation patterns and isotopic composition of water in a climate transition zone: Case of the Eastern Mediterranean Sea area. in *The Influence of Climatic Variability on the Hydrologic Regime and Water Resources*, IAHS Publ. No. 168, pp. 513-523, 1987.

Gat, J.R., and E. Matsui, Atmospheric water balance in the Amazon Basin: An isotopic evapotranspiration model, *J. Geophys. Res., 96*, 13,179-13,188, 1991.

Gedzelman, S.D., and J.R. Lawrence, The isotopic composition of cyclonic precipitation, *J. Appl. Meteorol., 21*, 1387-1404, 1982.

Gedzelman, S.D., Deuterium in water vapour above the atmospheric boundary layer, *Tellus, 40B*, 134-147, 1988.

Gedzelman, S.D., and J.R. Lawrence, The isotopic composition of precipitation from two extratropical cyclones, *Monthly Weather Review, 118*, 495-509, 1990.

Gonfiantini, R., On the isotopic composition of precipitation, *Rendiconti Soc. It. Miner. Petr., 38*, 1175-1187, 1982.

Gonfiantini, R., On the isotopic composition of precipitation in tropical stations, *Acta Amazonica, 15*, 121-139, 1985.

Hansen, J., and S. Lebedeff, Global trends of measured surface air temperature, *J. Geophys. Res., 92*, 13,345-13,372, 1987.

Holdsworth, G., S. Fogarasi, and H.R. Krouse, Variation of the stable isotopes of water with altitude in the Saint Elias Mountains of Canada, *J. Geophys.*

Res., 96, 7483-7494, 1991.

IAEA, *World Survey of Isotope Concentrations in Precipitation*. Technical Report Series Nos. 69, 117, 129, 147, 165, 192, 226, 264, 311, International Atomic Energy Agency, Vienna, 1969, 1970, 1971, 1973, 1975, 1979, 1983, 1986, 1990.

IAEA, *Statistical treatment of environmental isotope data in precipitation*, Technical Report Series No. 206, 256pp., International Atomic Energy Agency, Vienna, 1981.

IAEA, *Statistical treatment of data on environmental isotopes in precipitation*, Technical Report Series, No. 331, 720pp., International Atomic Energy Agency, Vienna, 1992.

Ingraham, N.L., and B.E. Taylor, Hydrogen isotope study of large-scale meteoric water transport in northern California and Nevada, *J. Hydrol., 85*, 183-197, 1986.

Ingraham, N.L., and R.C. Craig, Isotopic evidence of the terrestrial recycling of meteoric water as validated by a climate model, (this volume).

Jacob, H., and C. Sonntag, An 8-year record of the seasonal variation of ^2H and ^{18}O in atmospheric water vapour and precipitation at Heidelberg, Germany, *Tellus, 43B*, 291-300, 1991.

Johnsen, S.J., W. Dansgaard, and J.W.C. White, The origin of Arctic precipitation under present and glacial conditions, *Tellus, 41B*, 452-468, 1989.

Joseph, A., J.P. Frangi and J.F. Aranyossy, Isotope characteristics of meteoric water and groundwater in the Sahelo-Sudanese Zone, *J. Geophys. Res., 97*, 7543-7551, 1992.

Joussaume, S., J. Jouzel, and R. Sadourny, Water isotope cycles in the atmosphere: First simulation using a general circulation model, *Nature, 311*, 24-29, 1984a.

Joussaume, S., R. Sadourny, and J. Jouzel, Simulation of the HDO and $H^{18}O$ cycles in an atmospheric general circulation model, *Ann. Glaciol., 5*, 208-210, 1984b.

Jouzel, J., and L. Merlivat, Deuterium and oxygen 18 in precipitation: Modelling of the isotopic effect during snow formation, *J. Geophys. Res., 89*, 11,749-11,757, 1984.

Jouzel, J., Isotopes in cloud physics: multiphase and multistage condensation processes, in *Handbook of Environmental Isotope Geochemistry, Vol. 2 The Terrestrial Environment, B*, pp. 61-105, Elsevier, Amsterdam-Oxford-New York-Boston, 1986.

Jouzel, J., G.L. Russel, R.J. Suozzo, R.D. Koster, J.W.C. White, and W.S. Broecker, Simulations of the HDO and $H^{18}O$ atmospheric cycles using the NASA GISS general circulation model: The seasonal cycle for present-day conditions, *J. Geophys. Res., 92*, 14,739-14,760, 1987a.

Jouzel, J., C. Lorius, J. Petit, C. Genthon, N. Barkov, V. Kotlyakov, and V. Petrov, Vostok ice core: a continuous isotope temperature record over the last climatic cycle (160,000 years), *Nature, 329*, 403-408, 1987b.

Jouzel, J., R.D. Koster, R.J. Suozzo, G.L. Russel, J.W.C. White and W.S. Broecker, Simulations of the HDO and $H_2^{18}O$ atmospheric cycles using the NASA GISS general circulation model: Sensitivity experiments for present-day conditions, *J. Geophys. Res., 96*, 7495-7507, 1991.

Korzun, V.I. (Ed.) *World Water Balance and Water Resources of the Earth*. Report of the USRR Committee for the IHD, 663pp, Gidromet. Izdatel., Leningrad, 1974, (English translation: *Studies and Reports in Hydrology, No. 25*, UNESCO, Paris, 1978.)

Kuznetsova, L.P., Use of data on atmospheric moisture transport over continents and large river basins for the estimation of water balances and other purposes. *Technical Documents in Hydrology*, 105pp., UNESCO, Paris, 1990.

Lacaux, J.P., R. Delmas, G. Kouadio, B. Cros, and M.O. Andreae, Precipitation chemistry in the Mayombé forest of equatorial Africa, *J. Geophys. Res. 97*, 6195-6206, 1992.

Lawrence, J.R., S.D. Gedzelman, J.W.C. White, D. Smiley, and P. Lazov, Storm trajectories in eastern U.S.: D/H isotopic composition of precipitation, *Nature, 296*, 638-640, 1982.

Lawrence, J.R., and J.W.C. White, The elusive climate signal in the isotopic composition of precipitation, in *Stable Isotope Geochemistry: A Tribute to Samuel Epstein*, edited by H.P. Taylor, Jr., J.R. O'Neil and I.R. Kaplan, pp. 169-185, The Geochemical Society, Special Publication No. 3, 1991.

Lorius, C., and L. Merlivat. Distribution of mean surface stable isotope values in East Antarctica: observed changes with depth in the coastal area. in *Isotopes and Impurities in Snow and Ice*, IAHS Publ. No. 118, pp. 127-137, 1977.

Matsui, E., E. Salati, M.N.G. Ribeiro, M.C. Reis, A.C. Tancredi, and J.R. Gat, Precipitation in the central Amazon Basin: The isotopic composition of rain and atmospheric moisture at Belém and Manaus, *Acta Amazonica, 13*, 307-369, 1983.

Matsuo, S., and I. Friedman, Deuterium content in fractionally collected rain, *J. Geophys. Res., 72*, 6374-6376, 1967.

McKinney, C.R., J.M. McCrea, S. Epstein, H.A. Allen, and H.C. Urey, Improvements in mass spectrometers for the measurement of small differences in isotope abundance ratios, *Rev. Sci. Instrum., 21*, 724-756, 1950.

Merlivat, M., and J. Jouzel, Global climatic interpretation of the deuterium - oxygen-18 relationship for precipitation, *J. Geophys. Res., 84*, 5029-5033, 1979.

Miyake, Y., O. Matsubaya and C. Nishihara, An isotopic study on meteoric precipitation, *Pap. Meteor. Geophys., 19*, 243-266, 1968.

Mook, W.G., D.J. Groeneveld, A.E., Brown, and A.J. Van Ganswijk, Analysis of a run-off hydrograph by means of natural ^{18}O, in *Isotope Techniques in Groundwater Hydrology, Vol. I*, pp. 145-153, International Atomic Energy Agency, Vienna, 1974.

Nativ, R., and R. Riggio, Precipitation in the southern High Plains: Meteorologic and isotopic patterns, *J. Geophys. Res., 95*, 22,559-22,564, 1990.

Nier, A.O., A mass spectrometer for isotope and gas analysis, *Rev. Sci. Instrum., 18*, 398-404, 1947.

Nier, A.O., E.P. Ney, and M.G. Inghram, A null method for the comparison of two ion currents in a mass spectrometer, *Rev. Sci. Instrum., 18*, 294-301, 1947.

Peel, D.A., R. Mulvaney, and B.M. Davison. Stable-isotope/air-temperature relationship in ice cores from Dolleman Island and the Palmer Land Plateau, Antarctic Peninsula, *Ann. Glaciol., 10*, 130-136, 1988.

Peixoto, J.P., and A.H. Oort, The atmospheric branch of the hydrological cycle and climate, in *Variations in the Global Water Budget*, edited by A. Street-Perrott et al., pp. 5-65, D. Reidel Publishing Company, 1983.

Petit, J.R., J.W.C. White, N.W. Young, J. Jouzel, and Y.S. Korotkevich, Deuterium excess in recent Antarctic snow, *J. Geophys. Res., 96*, 5113-5122, 1991.

Picciotto. E., X. De Maere and I. Friedman, Isotopic composition and temperature of formation of Antarctic snows, *Nature, 187*, 857-859, 1960.

Ratisbona, L.R., The climate of Brazil, in *Climates of Central and South America*, edited by W. Schwerdtfeger, pp. 219-293, Elsevier-Science, New York, 1976.

Rindsberger, M., M. Magaritz, I. Carmi, and D. Gilad, The relation between air mass trajectories and the water isotope composition of rain in the Mediterranean Sea area, *Geophys. Res. Lett., 10*, 43-46, 1983.

Rindsberger, M., S. Jaffe, S. Rahamin, and J.R. Gat, Patterns of the isotopic composition of precipitation in time and space: data from the Israeli storm water collection program, *Tellus, 42B*, 263-271, 1990.

Rozanski, K., and C. Sonntag, Vertical distribution of deuterium in atmospheric water vapour, *Tellus, 34*, 135-141, 1982.

Rozanski, K., C. Sonntag and K.O. Münnich, Factors controlling stable isotope composition of modern European precipitation, *Tellus, 34*, 142-150, 1982.

Rozanski, K., D, ^{18}O and ^3H in atmospheric water vapour and precipitation collected at Krakow station during the time period June 1981 - Dezember 1985, *Report INT 202/I*, Institute of Physics and Nuclear Techniques, Academy of Mining and Metallurgy, Krakow, 1986.

Rozanski, K., Deuterium and oxygen-18 in atmospheric part of hydrological cycle, *Scientific Bulletins of Academy of Mining and Metallurgy, No. 1098*, 1-100pp., (in Polish), 1987.

Rozanski, K., L. Araguás-Araguás, and R. Gonfiantini, Relation between long-term trends of oxygen-18 isotope composition of precipitation and climate, Science, (in press), 1992.

Salati, E., A. Dall'Olio, E. Matsui, and J.R. Gat, Recycling of water in the Amazon Basin: An isotopic study, *Water Resour. Res., 15*, 1250-1258, 1979.

Schoch-Fischer, H., K. Rozanski, H. Jacob, C. Sonntag, J. Jouzel, G. Östlund and M.A. Geyh, Hydrometeorological factors controlling the time variation of D, ^{18}O and ^3H in atmospheric water vapour and precipitation in the northern westwind belt, in Isotope Hydrology 1983, pp. 3-31, International Atomic Energy Agency, Vienna, 1984.

Siegenthaler, U., and H. Matter, Dependence of $\delta^{18}O$ and δD in precipitation on climate, in *Paleoclimate and Paleowaters: A Collection of Environmental Isotope Studies*, pp. 37-51, International Atomic Energy Agency, Vienna, Austria, 1983.

Sonntag, C. E. Klitzsch, E.P. Löhnert, E.M. El-Shazly, K.O. Münnich, Ch. Junghans, U. Thorweihe, K. Weistroffer and F.M. Swailem, Palaeoclimatic information from deuterium and oxygen-18 in carbon-14 dated north Saharian groundwaters, in *Isotope Hydrology 1978, Vol. II*, pp. 569-581, International Atomic Energy Agency, Vienna, 1979.

Sonntag, C., K. Rozanski, K.O. Münnich, and H. Jacob, Variations of deuterium and oxygen-18 in continental precipitation and groundwater and their causes, in *Variations in the Global Water Budget*, edited by A.Street-Perrott et al., pp. 107-124, D. Reidel Publishing Company, 1983.

Steward, M.K., Stable isotope fractionation due to evaporation and isotopic exchange of falling water drops: Applications to atmospheric processes and evaporation of lakes, *J. Geophys. Res., 80*, 1133-1146, 1975.

Stute, M., P. Schlosser, J.F. Clark, and W.S. Broecker, Paleotemperatures in the southwestern United States derived from noble gases in ground water, *Science, 256*, 1000-1003, 1992.

Taylor, C.B., The vertical variations of the isotopic concentrations of tropospheric water vapour over continental Europe and their relationship to tropospheric structure, *DSIR Report INS-R-107*, 44pp., Lower Hutt, New Zealand, 1972.

Taylor, C.B., Stable isotope composition of monthly precipitation samples collected in New Zealand and Rarotonga, DSIR Physical Sciences Report No. 3, 102pp., Lower Hutt, New Zealand, 1990.

White, J.W.C., and S.D. Gedzelman, The isotopic composition of atmospheric water vapor and the concurrent meteorological conditions, *J. Geophys. Res., 89*, 4937-4939, 1984.

Winograd, I.J., B.J. Szabo, T.B. Coplen, and A.C. Riggs, A 250,000-year climatic record from Great Basin vein calcite: implications for Milankovitch theory, *Science, 242*, 1275-1280, 1988.

Van der Straaten C.M., and W.G. Mook, Stable isotope composition of precipitation and climatic variability, in *Paleoclimate and Paleowaters: A Collection of Environmental Isotope Studies*, pp. 53-64, International Atomic Energy Agency, Vienna, 1983.

Yapp, C.J., and S. Epstein, Climatic implications of D/H ratios of meteoric water over North America (9.500-22.000 B.P.) as inferred from ancient wood cellulose C-H hydrogen, *Earth Planet. Sci. Lett., 34*, 333-350, 1977.

Yapp, C.J., A model for the relationships between precipitation D/H ratios and precipitation intensity, *J. Geophys. Res., 87*, 9614-9620, 1982.

Yonge, C.J., L. Goldenberg, and H.R. Krouse, An isotope study of water bodies along a traverse of southwestern Canada, *J. Hydrol., 106*, 245-255, 1989.

Yurtsever, Y., and J.R. Gat, Atmospheric waters. in *Stable Isotope Hydrology: Deuterium and Oxygen-18 in the Water Cycle*, edited by J.R. Gat and R. Gonfiantini, Technical Report Series, No. 210, pp. 103-142, International Atomic Energy Agency, Vienna, 1981.

Zimmermann, U., D.H. Ehhalt, and K.O. Münnich. Soil water movement and evapotranspiration: Changes in the isotopic composition of water, in *Isotopes in Hydrology*, pp. 567-585, International Atomic Energy Agency, Vienna, 1967.

K. Rozanski, L. Araguás-Araguás and R. Gonfiantini, International Atomic Energy Agency, P.O. Box 100, 1400-Vienna, Austria.

Interpreting Continental Oxygen Isotope Records

P. M. GROOTES

*Quaternary Isotope Laboratory, University of Washington
Seattle, Washington 98195*

Continental isotope records cannot be interpreted as only the result of local temperature changes. They may reflect various other processes as well. An example is the tropical Quelccaya ice cap, where the oxygen isotopic composition of precipitation is determined mainly by air mass stability. When continental isotope records are compared with each other or with ocean sediment isotope records, one needs to know what processes are reflected by each of the isotope records. If a record of past temperature changes is desired, corrections should be applied to eliminate non-temperature influences such as the effect of global ice volume on the isotopic composition of ocean and atmosphere. An example is the continental isotope record from Devils Hole, Nevada, which due to its narrow range in isotope values is sensitive to such corrections. The corrected record correlates strongly with the SPECMAP deep-sea isotope record of global climate change.

INTRODUCTION

Much of our quantitative knowledge of climate fluctuations in the past has been derived from the long and detailed isotope records preserved in ice sheets and deep-sea sediments. The main paleoclimate indicator has been the abundance ratio of the oxygen isotopes of mass 18 (^{18}O) and mass 16 (^{16}O) in water and carbonates. The difference in physical properties caused by the mass difference leads to temperature-dependent isotope fractionation in phase changes and chemical reactions. This allows the isotopic abundance ratio $^{18}O/^{16}O$ to be used as a tracer for the study of climate and the hydrologic cycle, as well as of carbonate precipitation and dissolution and photosynthesis; in short: all processes that involve water and leave a record.

Comparison of different types of paleoclimatic records from ice sheets and ice caps, peat bogs, lake and ocean sediments, loess deposits, speleothems, and tree rings, from different parts of the world, provides a mosaic of local responses to global climate change and therefore a detailed picture of such change. For a comparison it is essential that the timing of the different records is accurately known, and that differences in response time for different records are considered. Correlating changes in isotopic composition between records can lead to incorrect conclusions when, for example:

Climate Change in Continental Isotopic Records
Geophysical Monograph 78

1. Factors that contribute to or dominate the isotope signal differ between records.
2. Records represent accumulation over part of the seasonal cycle, as may be the case in some ice cores and continental records, and this part differs for different records [Grootes and Steig, 1992; Paterson and Hammer, 1987].
3. The isotope records show the response of different parts of the global climate system to climate forcing, and the lag between forcing and response differs among the records; e.g. composition of precipitation from the atmosphere responds rapidly while the change in ocean composition due to ice sheet melting is quite slow.

I will discuss two continental isotope records, namely that of Quelccaya, where temperature is not the major parameter controlling isotopic composition, and that of Devils Hole, Nevada, where the timing of the records has raised serious questions about the causes of global climate change.

QUELCCAYA: A TROPICAL ICE CORE ISOTOPE RECORD

The Quelccaya ice cap, Peru, (13°56'S, 70°50'W, elevation 5670 m) is situated in the tropics in the easternmost glaciated mountain chain of the Peruvian Andes. Two cores drilled to bedrock, at about 160 m depth, during the summer of 1983, have yielded a record of environmental change over the last 1500 years [Thompson et al., 1985, 1986]. The snow deposited on the ice cap shows a strong seasonal isotope cycle with snow that is up to 20‰ richer in ^{18}O during the slightly colder winter. This seasonal isotope cycle is much larger than can be attributed to the seasonal change in average surface

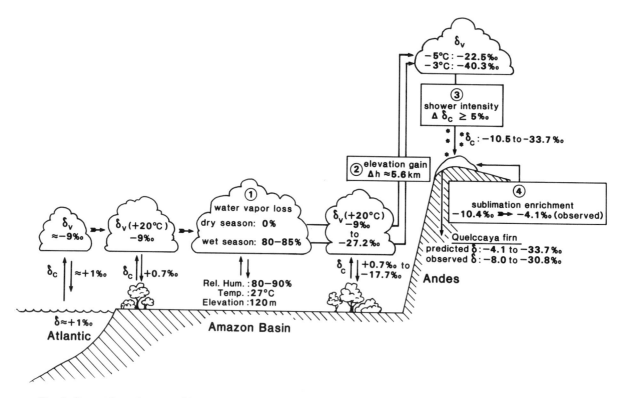

Fig. 1. Oxygen isotopic composition of atmospheric water vapor and precipitation from the tropical Atlantic Ocean across the Amazon Basin to Quelccaya. The ^{18}O depletion accompanying net water vapor loss from the air is calculated in three steps using a Rayleigh condensation equation, while 4) considers post-depositional isotope enrichment [Grootes et al., 1989]. Step 1), Water vapor depletion over the Amazon Basin varies from 0 % (dry season) to 85 % (wet season), resulting in a seasonal $\delta^{18}O$ change in precipitation of up to 18.4‰ ; Step 2), a 5.6 km increase in surface elevation from the Amazon Basin to Quelccaya results in a $\delta^{18}O$ decrease in precipitation of about 11‰; Step 3), strong convection in summer showers can increase the range of the seasonal $\delta^{18}O$ cycle by at least 5‰ (-10.5 to -33.7‰). Enrichment of $\delta^{18}O$ at the surface during the dry season may produce snow with $\delta^{18}O$ = -4.1‰. The observed range of $\delta^{18}O$ values in snow pits (-8.0 to -30.8‰) and its phase agree with the predicted range of -4.1 to -33.7‰.

temperature of about 2°C over Quelccaya [Thompson et al., 1984], and the relationship between $\delta^{18}O$ and temperature is the opposite of the well known $\delta^{18}O$-temperature correlation for middle and high latitudes [e.g. Dansgaard, 1964; Rozanski et al., 1982]. This is an extreme example of the need to include parameters other than temperature in the interpretation of a continental isotope record.

The general circulation patterns over southern tropical South America [Taljaard, 1972], as well as local observations [Thompson et al., 1984], link precipitation on Quelccaya to air masses from the Amazon Basin, both in summer and in winter. This is supported by observations, reported by Aravena et al. [1989], linking precipitation events in the Cordillera Occidental in Northern Chile (south of Quelccaya at about 20°S, 69°W) to moist air masses from the Amazon Basin. Over the tropical Amazon Basin the amount of water vapor lost from the air by rainfall is determined by the convection of the air masses. In summer and fall (November-April) humid, unstable air masses extend over the Amazon

Basin and abundant precipitation falls from convective showers.

Rainfall during the wet summer season may be two to three times greater than evapotranspiration of the rainforest [Dall'Olio et al., 1979], which leads to a significant depletion of the amount of water vapor in the air, as well as of the heavy isotopic water molecules. During the winter (June-August) a large part of the Amazon Basin, extending to the Andes, comes under the influence of dry stable air and easterly winds from the Atlantic anticyclone [Ratisbona, 1976; Taljaard, 1972]. Rainfall during this period is low and the water lost is recycled by evapotranspiration of the rain forest. Thus, there is little or no water vapor loss or isotope depletion during the dry season. A seasonal cycle in air mass stability over the almost isothermal Amazon Basin thus leads to a clear seasonal isotope cycle in precipitation in the basin and in adjacent areas like Quelccaya and the Cordillera Occidental of Northern Chile. The seasonal isotope cycle observed in the snow on Quelccaya [Thompson et al., 1984] can be produced quantitatively using a Rayleigh distillation model for isotopic fractionation during air mass movement from the Atlantic Ocean over the Amazon Basin to the Quelccaya ice cap (Fig.1), [Grootes et al., 1989].

Although there may be other ways to explain the observed isotope cycle, our model provides a simple explanation consistent with field observations. The model implies that in tropical areas and at lower latitudes, where the seasonal temperature cycle is less prominent, factors like air circulation pattern, air-mass stability, and the character of the precipitation event will influence or even determine the seasonal isotope cycle. This complicates the interpretation of low-latitude isotope records, but holds the promise of providing better insights into the changes in climate processes affecting the hydrologic cycle in the past.

Another example of an unconventional explanation for an observed tropical isotopic gradient is given by Aravena et al. [1989]. They observed in northern Chile a decrease in heavy oxygen isotopes in precipitation with increasing elevation, going from the Pampa del Tamarugal up the western slopes of the Cordillera del Medio. Field observations showed, however, that the precipitation was derived from air masses coming from the east over the mountains. Therefore the classical interpretation, that the gradient is caused by cooling of air forced to rise up the mountain, cannot be used. Aravena et al. suggest that evaporation and isotopic enrichment of raindrops falling through dry air below the clouds may lead to precipitation that is isotopically heavier as the ground surface is further below the crest of the mountains. Without direct observations of the precipitation process, the isotope record preserved on and in the ground would be interpreted quite differently.

Although the seasonal cycle in the oxygen isotope record at Quelccaya is not caused by local temperature changes, the long-term isotope changes do correlate with certain indices of global climate. Thompson et al. [1986] report a positive correlation between decadal $\delta^{18}O$ averages of two Quelccaya ice cores and departures of Northern Hemisphere decadal temperatures over the past 500 years from the 1881-1975 mean. The correlation may reflect a slightly lower mean temperature over Quelccaya or increased depletion over a slightly cooler, less evaporative Amazon Basin during the Little Ice Age. Another possible explanation is suggested by the preservation of a strong (about 6‰) seasonal isotope cycle in the ice from about AD 1850 to 1500. This implies not only colder conditions on Quelccaya as smoothing of the isotope record increases with temperature, but also a possible decrease in post-depositional enrichment during firnification. Thus, large scale or global climate changes may be recorded in low latitude and tropical ice caps.

DEVILS HOLE: A LONG CONTINENTAL RECORD OF GLACIAL-INTERGLACIAL CHANGE

Winograd et al. [1988] obtained a 250,000-year long $\delta^{18}O$ record from a calcitic vein in Devils Hole, Great Basin, Nevada

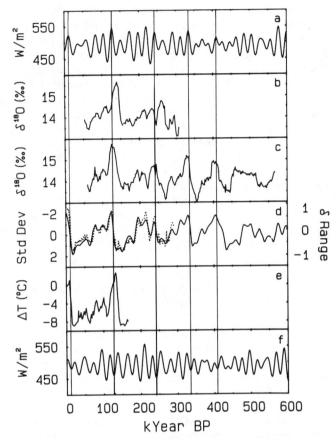

Fig. 2. Oxygen isotope records of glacial-interglacial climate change and high-latitude insolation forcing. a. June insolation at 60°N [Berger and Loutre, 1991]; b. Devils Hole oxygen isotope record DH2 [Winograd et al., 1988]; c. Devils Hole oxygen isotope record DH11 [Winograd et al., 1992]; d. solid line: SPECMAP standardized deep-sea stacked oxygen isotope record, expressed in standard deviation units on left-hand scale [Imbrie et al., 1984], dotted line: normalized high-resolution stacked deep-sea oxygen isotope record with units on right-hand scale [Martinson et al., 1987]; e. glacial-interglacial temperature cycle derived from Vostok ice core isotopes [Jouzel et al.,1987]; f. December insolation at 60°S [Berger and Loutre, 1991]. Each curve is plotted on its original time scale.

(36.4°N, 116.3°W). The deposit has been carefully analyzed and dated using both the ^{230}Th-^{234}U and the ^{234}U-^{238}U methods (Fig.2b). More recently a second core, taken from the same deposit, provided a continuous 500,000-year climatic record for the Great Basin [Winograd et al., 1992] with somewhat younger dating based on high-precision mass spectrometric techniques [Ludwig et al., 1992] (Fig. 2c).

Figure 2 compares the Devils Hole records with the standardized deep-sea records of Imbrie et al. [1984, SPECMAP] and Martinson et al. [1987], with the Vostok ice core record [Lorius et al., 1985, Jouzel et al., 1987], and with insolation forcing in the Northern and Southern Hemisphere (NH and SH) [Berger and Loutre, 1991].

Devils Hole. The Devils Hole calcite records the isotopic

composition of groundwater fed by infiltration of precipitation in the recharge area. Precipitation responds directly to temperature as is is evident from the strong seasonal $\delta^{18}O$ cycle frequently observed (International Atomic Energy Agency, 1981). Possible delayed responses to climate change may come from changes in the isotopic composition of the ocean-source-area of the water vapor caused by waxing or waning of the ice sheets and from changes in air circulation related to the presence or absence of the Laurentide ice sheet. The travel of the groundwater from the recharge area(s) to Devils Hole produces a delay that is poorly known, while along the way also mixing of infiltrating waters of different age may occur [Winograd et al., 1992].

Vostok core. The Vostok ice core is formed by snow accumulating in central East Antarctica. The precipitation will respond directly to climate change, but changes in sea ice extent around Antarctica and in the isotopic composition and surface conditions of the southern oceans will contribute a (minor) delayed component [Grootes and Stuiver, 1987].

Ocean sediment. The benthic $\delta^{18}O$ record shows the change in ocean isotopic composition caused by melting or growth of isotopically light ice sheets. This benthic $\delta^{18}O$ signal will be delayed relative to direct atmospheric records by an amount equivalent to the response time of the ice sheet. The total glacial-interglacial benthic $\delta^{18}O$ change was about 1.6‰ of which on average 1.2‰ (75%) was ice volume and the remainder deep-ocean temperature [Labeyrie et al., 1986; Chappel and Shackleton, 1986; Birchfield, 1987].

Noticeable is that Devils Hole ages in the newly determined DH11 time scale are significantly younger than in DH2. The glacial-interglacial cycles in DH11 generally agree with those observed in the deep-sea record, but differ in the duration and timing of the glacial-interglacial transitions. I will discuss the significance of the correspondences and differences between DH11 and the ocean and ice core records in Fig.2.

Timing of the Major Glacial-Interglacial Cycles

The timing of the major interglacial episodes over the last 400 ky is very similar in DH11 and SPECMAP (Figs. 2c, d) The timing of SPECMAP's many (negative $\delta^{18}O$) "warming" episodes has been adjusted ("tuned") to match over the full record the phase-relationship observed over the last 150 ky between such changes in the Indian Ocean sediment core RC11-120 [Hays et al., 1976] and NH insolation peaks (Fig. 2a). Most of the isotope increases in the independent DH11 record show the same timing. Exceptions are the insolation cycles immediately preceding a termination. The 60°N maximum is weak for these cycles and we see a large ice volume reflected in the oceans and low but rising $\delta^{18}O$ values in DH11. The comparison suggests that insolation forcing significantly influences both the Devils Hole groundwater and the deep-sea ice volume record but that at times of maximum glaciation the changes in the global climate system are determined largely by other factors, possibly ice dynamics.

Winograd et al. [1988, 1992] selected for comparison the timing of the prominent changes in long-term average

isotopic composition, associated with the glacial to interglacial transitions (Termination II to V), and the major "troughs" (actually periods of heavier isotope values in the ocean caused by storage of isotopically light water in the continental ice sheets) in the marine record at about 65 ky and 225 ky ago. These features were assumed to be contemporaneous in the different records, yet Winograd et al. [1988, 1992] note that, with the exception of the 65 ky trough, the Devils Hole Th/U ages for the onset or the midpoint of the heavy isotope increase are significantly higher than the corresponding ages on the marine time scale. Winograd et al. [1988, 1992] conclude from the Devils Hole results that the insolation cycles related to orbital parameters (Milankovitch cycles) may not be responsible for the glacial to interglacial transitions, as their forcing towards a warmer climate occurs after the warming itself is well underway. This conclusion is of major importance for the understanding of global climate change.

Duration of the Glacial to Interglacial Transitions

In DH11 the time from the onset of the $\delta^{18}O$ increase to the end of the rapid rise is about 23, 24, and 21 ky for Termination II, III, and IV respectively. It takes another 5 to 7 ky from there to the center or top of the $\delta^{18}O$ peak. The length of these intervals in the DH11 core has been determined directly by dating samples from the base and the top of the rise by high-precision Th/U mass spectrometry and must be real.

The melting of the ice sheets responsible for the decrease in deep-sea $\delta^{18}O$ is well documented for Termination I, the last glacial to interglacial transition. Sarnthein and Tiedemann [1990] report significant similarities between Terminations I through VI in a sediment core off northwest Africa. This suggests that the response time of ice sheet melting at the most recent glacial-interglacial transition, which is much better dated than the older ones, can be applied to the older transitions. Most events during the last glacial-interglacial transition have been dated using ^{14}C. The onset of a decreasing trend in $\delta^{18}O$ is dated at 15.1±.2 ky in a number of Atlantic sediment cores [Sarnthein and Tiedeman, 1990; Fairbanks, 1989, 1990; Lehman and Keigwin, 1992; Keigwin et al., 1991; Duplessy et al., 1986]. The decrease is about 70% complete by 9.1±.2 ky BP, and 90% by 6.5 ky [Fairbanks, 1989]. These are conventional radiocarbon ages with a 400-year correction for reservoir age. Recent work by Bard et al. [1990] using mass spectrometer Th/U and accelerator mass spectrometer ^{14}C measurements on coral reefs, indicates that ^{14}C ages across the last glacial-interglacial transition are from 1000 to 3000 years too young [Mazaud et al., 1991; Bard et al., 1993]. Using a calibration by Stuiver and Braziunas [1993] based on these results, we get about 11 ky, from 18 to 7 ky ago, for the last main ice sheet melting period recorded in Atlantic deep-sea sediment cores.

A continental sediment record of ice sheet and climate fluctuations from Andøya [Vorren et al., 1988] confirms the information obtained from the ocean records. The island Andøya, 69°15'N, northern Norway, is situated on the northwestern margin of the Fennoscandian ice sheet in the North Atlantic. The sedimentary record on Andøya shows a brief period of maximum ice extent 19.0-18.5 ky BP, and a marked warming starting at 16 ky BP (conventional radiocarbon ages). The Th/U corrected "calendar age" of the warming is 19.4 ky BP. This agrees well with the onset of melting documented in Atlantic Ocean sediments about 1 ky later. Both could be in response to summer insolation, which was relatively high in the Northern Hemisphere from 15 to 6 ky BP after a minimum at about 22 ky ago [Berger and Loutre, 1991]. The Andøya record only extends to 11 ky BP and therefore provides no information on the completion of deglaciation. The Andøya climate record agrees well with the latest detailed climate reconstruction from two ice cores from the summit of the Greenland ice sheet [Johnsen et al., 1992] (Grootes, unpublished data). Both records show low $\delta^{18}O$ values about 23 ky ago during the insolation minimum and large climate fluctuations between 15 and 11 ky ago.

Vostok, Antarctica, provides the only published ice core record of Termination II. Sowers et al. [1991] obtained a record of the relative timing of atmospheric warming, increase in atmospheric CO_2, and ice sheet melting for Termination II from air bubbles in the Vostok ice core. Poor core quality made it impossible to obtain the same information for Termination I. The air trapped in the bubbles shows across the glacial to interglacial transition an increase in CO_2 concentration and a decrease in $\delta^{18}O_{(air\ O_2)}$. The drop in $\delta^{18}O$ of air oxygen is due to the lowering of the oxygen isotope ratio of the water of the oceans by ice sheet melting, followed by evaporation of this lighter water and photosynthetic production of lighter atmospheric oxygen. Sowers et al. estimate that the turnover time of the atmospheric O_2 reservoir is 2 to 3 ky, so changes in the air-O_2 isotope ratio should lag those in ocean water by that amount. The onset of the glacial-interglacial change in $\delta D/\delta^{18}O$, pCO_2, and $\delta^{18}O_{atm}$ is found in the Vostok core at 1958.9±1, 1992.7±5, and 1921.1±13 m depth respectively [Sowers et al., 1991]. With the Vostok depth/age relationship of Jouzel et al. [1987] and estimates of the ice age at pore close-off from Barnola et al. [1991] these depth differences indicate that the initial temperature increase at 144.7±0.2 ky ago was followed after 0.9±0.8 ky by an increase in atmospheric CO_2 and after 7.3±1.1 ky by a decrease in atmospheric $\delta^{18}O$. With a lag between $\delta^{18}O_{atm}$ and $\delta^{18}O_{ocean}$ of 2 ky the ocean change due to ice sheet melting started 5.3±1.1 ky after Antarctic temperatures began to increase, at 139.4±1.1 ky ago. The rapid phase of ice sheet melting lasted till about 133 ky and melting was completed by 129 ky ago. The melting of the (Northern Hemisphere) ice sheets during Termination II lasted about 10 ky. This is very similar to the 11 ky obtained from radiocarbon dates of ocean and coral cores for Termination I. Although the 10 to 11 ky melting time is quite similar for terminations I and II, the timing of this melting relative to the insolation cycles is different (Fig. 2). This may be due to the different pattern of insolation changes (Fig. 2a, f) and to the fact that the CO_2 increase preceded melting by about 4 ky during Termination II while it started at

the same time for Termination I [Sowers et al., 1992]. The 10 to 11 ky for ice sheet melting is significantly shorter than the 21 to 24 ky obtained by Th/U dating for the glacial to interglacial transition in Devils Hole. If we accept both datings, the glacial to interglacial increase in $\delta^{18}O$ in groundwater recorded in Devils Hole calcite cannot be directly related to the decrease in global ice volume monitored in the deep sea, though both records are monitoring global climate change.

The Age of Termination II

Discussion of the age discrepancies between Devils Hole and SPECMAP has centered on the age of Termination II, 140±3 ky in DH11 and 128±3 in the SPECMAP record [Imbrie et al., 1984]. The age of Termination II is a cornerstone of the SPECMAP chronology because it is one of only two independently dated points in the chronology, the other one being the Brunhes/Matuyama magnetic polarity boundary. The value of 128 ky coincides with the exceptional peak in high latitude NH summer insolation at 127±1 ky [Berger and Loutre, 1991] and provides a strong argument for the current concept of climate forcing by summer insolation changes at high northern latitudes.

The Devils Hole DH11 record has been carefully dated using mass spectrometric uranium series dating. The sample material is generally of good quality [Ludwig et al., 1992]. High-precision mass spectrometer dates of 132 ky for the top of the rapid increase in $\delta^{18}O$ and 150 ky for its beginning place "Termination II", the midpoint of the rise, at 140±3 ky in DH11. In view of the extensive quality control [Ludwig et al., 1992] these dates must be considered firm.

Direct dating of Termination II in the deep-sea is limited due to a lack of good dating material, and an age of 127±6 ky [Broecker and Van Donk, 1970] is commonly used. Th/U dating of coral terraces formed during high sea level episodes in different parts of the world gave a cluster of dates mostly in the range of 120 to 130 ky. For the α-counting technique used for these dates a typical analytical error is ±10 ky [Harmon et al., 1979], and most dates were statistically indistinguishable. The conclusion was that a high sea level episode occurred about 125 ky ago, and that Termination II was 127 ky ago. Imbrie et al. [1984] started with this date and an age of 730 ky for the Brunhes/Matuyama boundary in the construction of their SPECMAP chronology. They assumed that for the climate changes observed in the ocean record the rate of response of the global climate system had at any instant been proportional to the magnitude of the insolation forcing. They obtained 128±3 ky for Termination II by allowing adjustment of the age of the stage 5/6 boundary in the later stages of their tuning. Although the tuning was to the variations in obliquity and in the precession index and therefore independent of latitude and season, the value of 127 ky for Termination II tied the SPECMAP time scale to high-latitude summer insolation in the Northern Hemisphere which peaked at 127 ky. The concept of climate forcing by Milankovitch insolation cycles, and orbital tuning of age-depth scales has been highly successful in matching observed climate fluctuations in a simple way to solar forcing, while producing reasonable accumulation rate fluctuations in the deep-sea cores [Imbrie and Imbrie, 1980; Imbrie et al., 1984; Shackleton and Pisias, 1985; Shackleton et al., 1990; Martinson et al., 1987; Hilgen, 1991].

High-precision Th/U mass spectrometer dates of high sea levels associated with the penultimate interglacial at two Bahamian coral reefs range from 132 to 120 ky (2σ of 1.5 ky) [Chen et al., 1991]. Chen et al. conclude that the samples represent a prolonged period of high sea level, starting possibly at or before 132 ky ago. High precision dates (2σ of 1.1 to 1.3 ky) between 130 and 122 ky were obtained for corals from Barbados and the New Hebrides [Edwards et al., 1987]. The rapid rise in global sea level at the end of stage 6 (Termination II) preceded the high sea level stands and thus must be 132 ky or older instead of 127 ky. The new age for Termination II, although significantly greater than the old value of 127 ky if one considers the Th/U age uncertainty of 1.5 ky (2σ), is compatible with the 6 ky uncertainty of the old value. The isotope results of SPECMAP [Imbrie et al.,1984] indicate high sea levels not earlier than 124 ky, while Martinson et al. [1987] predict these at 125,190 or later. At this point the deep-sea time scale is thus 5 to 6 ky too young if high sea level occurred after 130 ky ago, 7 to 8 ky if we accept the earlier dates up to 132 ky. It should be noted here that for the age of Termination II the oldest reliable date for a high sea level is important and not the average of all dates obtained for different phases of reef growth during the interglacial high sea level episode. Adjustment of the SPECMAP time scale by 7 to 8 ky gives 135 ky for Termination II and reduces the difference in timing between SPECMAP and DH11 to 5 ky, close to the level of uncertainty in the time scales. The major problem with the proposed adjustment is that it places high sea level before the NH insolation maximum. In a model where climate changes in a linear response to NH insolation forcing, and sea level change is caused by the growth or melting of the NH ice sheets, this is impossible. Such a model ignores, however, the direct influence of SH insolation on Antarctica and its surrounding ocean as well as the non-linearity of response created by changes in albedo and greenhouse gas concentrations and by the internal dynamics of the large ice sheets.

A detailed Southern Hemisphere perspective of events during the glacial to interglacial transition corresponding to Termination II is provided by the Vostok ice core [Sowers et al., 1991; above]. The onset of the $\delta^{18}O$ increase, i.e. of local warming, at Vostok at 144.7 ky does not fit a NH forcing. Yet it occurs about halfway an increase in December insolation at 60°S that starts at 151 ky. Thus SH forcing may have started a warming trend over Antarctica that around 139 ky ago [Sowers et al., 1991] initiated a rapid decrease in $\delta^{18}O$ in the oceans which lasted till about 133 ky ago. When SH insolation decreased after 138 ky ago, the increase in NH insolation that is held responsible for the penultimate glacial to interglacial transition was underway. Interhemispheric coupling via the North Atlantic Deep Water (NADW) and the concentration of

greenhouse gases in the atmosphere must then be held responsible for continued oceanic warming around Antarctica in spite of decreasing local insolation. Johnson and Andrews [1986] suggested that melting of substantial Antarctic ice shelves caused the early part of the oceanic change at Termination II, preceding the melting of the NH ice sheets. The debate on the Vostok- SPECMAP time scale discrepancy is ongoing. Jouzel et al. [1987] found no convincing arguments to change the Vostok chronology. Petit et al. [1990] find, however, that if dust in the Vostok core and magnetic susceptibility in core RC11-120 are used as chronostratigraphic marker, the climate records are roughly in phase at Termination II. Petit et al. conclude from coral-reef terrace dates of 126 ky for the peak of the penultimate interglacial that the deep-sea-core dating is correct, and, consequently, the Vostok time scale 10 ky too old. As discussed above, the new coral terrace dates indicate high sea levels from 130 ky onward and may be as early as 132 ky which leads to an age of about 135 ky for Termination II. This value is close to the midpoint of the rapid $\delta^{18}O$ increase from 139 to 133 ky ago in the Vostok ice core.

The conclusion from the foregoing discussion is that the timing of Termination II in the deep-sea and in the Vostok ice core is quite similar when the deep-sea time scale is adjusted based on the new mass spectrometer Th/U dating of coral terraces. The completion of the glacial-interglacial change coincides in DH11 and the other two records, but its start and, consequently, its midpoint are earlier. This discrepancy is discussed further below. It appears reasonable and necessary to include SH insolation, interhemispheric ocean/atmosphere coupling via greenhouse gases and North Atlantic Deep Water, and ice sheet dynamics in the discussion of global climate change. Doing so goes, however, at the expense of the simple and detailed time scale obtained with the current simple Milankovitch theory of climate forcing by NH summer insolation.

Glacial-Interglacial Transition at Devils Hole

The long duration of the glacial to interglacial increases in $\delta^{18}O$ in DH11 still needs to be explained. The Devils Hole calcite records the changes in isotopic composition of groundwater recharged by precipitation in the Ash Meadows groundwater basin in the southern part of the Great Basin. Such isotopic changes can reflect varying conditions (i) in the oceanic water-vapor source-area, (ii) along the air mass trajectory from the source area, and (iii) in the precipitation area. Winograd et al. [1988] discuss seven possible sources of variation, and conclude that of the local parameters only surface temperature during recharging precipitation (winter/spring) is important. Two changes in the ocean source-area can be quantified. These are the change in the isotopic composition of the oceans caused by the sequestering of isotopically light water in the ice sheets, and the effect of a change in kinetic isotope fractionation and sea surface temperature in the water-vapor source-region over the ocean. For the ice volume correction I chose the stacked benthic oxygen isotope record of Pisias et al. [1984] tabulated with their high-resolution chronology by Martinson et al. [1987]. The tabulated $\delta^{18}O$ values have been "normalized", and range from -1 to +1. The glacial to interglacial change in benthic foram $\delta^{18}O$ values is generally about 1.6‰, of which about 0.4‰ is attributed to a change in water temperature [Labeyrie et al., 1986; Chappell and Shackleton, 1986; Birchfield, 1987]. No detailed independent information on the relative timing of the changes in deep-sea temperature and global ice volume is available, so I treated the record as if at any time 3/4 of the change observed in benthic $\delta^{18}O$ is caused by ice volume change, and 1/4 by temperature. The $\delta^{18}O$ corrections can then be obtained from the tabulated normalized $\delta^{18}O$ values. To obtain a 1.2‰ range for the ice volume correction the tabulated values (range -1 to +1) were multiplied by 0.6‰. This is to be added to the Devils Hole $\delta^{18}O$ values. The time scale used is that of SPECMAP increased by 5 ky at 40 ky, linearly increasing to 7 ky at stage 6 at about 135 ky. From 135 ky on (revised time scale) the offset was kept constant in view of the close correlation between SPECMAP and DH11. The resulting ocean composition correction curve is plotted in Fig. 3b.

During periods of maximum glaciation (deep-sea isotope stages 2,4,6,...) increased zonal temperature gradients cause stronger winds. This is documented by increased dust concentrations in ice cores [e.g. Thompson and Mosley-Thompson, 1981; Petit et al., 1981]. Jouzel et al. [1982] interpreted a change in deuterium excess "d", observed in the Dome C isotope record, as indicative of reduced kinetic fractionation of water vapor across the ocean-atmosphere boundary layer caused by turbulent mixing brought about by the higher windspeeds over the ocean during periods of maximum glaciation. Turbulent mixing may increase the $\delta^{18}O$ value of atmospheric water vapor in the source-area by 0.7‰ [Grootes and Stuiver,1987]. Cooling of the ocean surface in the water-vapor source-area during periods of maximum glaciation by an estimated 1°C leads to larger fractionation between water and vapor, and to vapor that is about 0.1‰ lighter. The net result is water vapor in the source-area that is about 0.6‰ heavier during periods of maximum glaciation. For the moment I assume this correction applies to periods of maximum glaciation as indicated by isotope stages 2,4,6,.. in the deep-sea record and that turbulent mixing "switches on/off" rapidly (Fig. 3a). A more detailed record of this correction may be obtainable from the dust and deuterium excess records of ice cores.

The effect of the corrections on the Devils Hole continental isotope record can be seen in Fig. 3a,b,c. The corrected curve is by no means final, since the corrections still have considerable uncertainty in their timing. Important is that, with an observed glacial-interglacial range of $\delta^{18}O$ values of only about 2.5‰, the Devils Hole isotope record is more sensitive to corrections than the polar ice core records, where the glacial-interglacial change is at least 5‰. The main effect of the corrections for ice volume and windspeed is an increase in the amplitude of the $\delta^{18}O$ fluctuations.

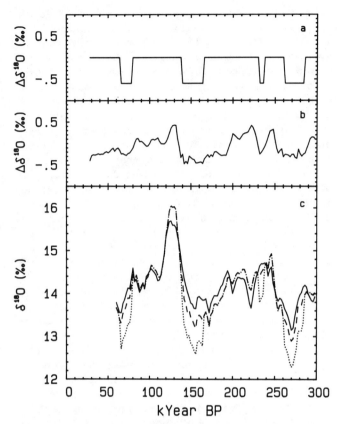

Fig. 3. Source-area oxygen isotope corrections and the Devils Hole continental oxygen isotope record DH11. a. Change in $\delta^{18}O$ of atmospheric water vapor in the ocean source region related to absence of kinetic fractionation during cold periods (-0.7‰), and colder surface ocean (-1°C gives +0.1‰). It is assumed that these conditions occur only during cold episodes and that they can change very rapidly, effectively stepwise. b. Change in ocean isotopic composition caused by changes in global ice volume. Time scales of both a. and b. are based on the high-resolution Martinson et al. [1987] deep-sea timescale. A correction ranging linearly from +5 ky at 40 ky ago to +7 ky for 135 ky (corrected) was used for both a. and b. Beyond 135 ky the correction is kept constant at 7 ky. c. The $\delta^{18}O$ correction values from 3.a,b are added to the Devils Hole DH11 record; solid line is the original DH11 record, dashed the DH11 record after correction for the effects of global ice volume change, and dotted the result after correction for both ice volume and kinetic fractionation-surface ocean temperature.

To evaluate other factors related to conditions in the oceanic water-vapor source-area (i) and along the air mass trajectory (ii), we need to know the airstream pattern of the area. The southern Great Basin is presently year-round under the influence of westerly airflow from a subtropical Pacific anticyclone [Bryson and Hare, 1974]. Precipitation in winter comes from cyclonic disturbances, in summer from thunderstorms in invading tropical air from the Gulf of Mexico.

The increase in $\delta^{18}O$ in DH11 started as early as 156 ky ago, well before Vostok or SPECMAP, and continued more or less uniformly up till 133 ky ago [Winograd et al.,1992]. If we accept the apparently consistent conclusions from the Vostok ice core record and the ocean record, then DH11 shows a 1‰ increase in $\delta^{18}O$ between 156 and 140 ky ago while global ice volume was at its maximum and NH insolation was low and for most of the time decreasing towards a minimum. The apparent contradiction between increasing $\delta^{18}O$ and a glacial maximum with decreasing insolation may be reconcilable. Reduced kinetic fractionation across a turbulent atmosphere-ocean boundary layer and a cooler vapor source-area may generate about half the increase. A reduced temperature difference between a cooler ocean surface and the Great Basin precipitation temperature could explain the remainder. The effect of a 1°C-cooling of the ocean surface will be the same as that of a 1°C-warming of the Great Basin, namely an increase of $\delta^{18}O$ in precipitation in the Great Basin by 0.6‰ [Rozanski et al., 1992]. No detailed sea surface data are available for Termination II. CLIMAP data for the Last Glacial Maximum, 18 ky ago, show for February sea surface temperatures between 0 and 2°C colder than today in the 30° to 40° latitude band in the eastern Pacific upstream from Devils Hole [CLIMAP project members, 1981]. A southward shift of the colder northern Pacific westerlies, maybe in part due to the presence of the Cordilleran and Laurentide ice sheets or to sea ice, may also have brought water vapor from a colder ocean surface to the Great Basin. At the same time winter temperatures in central North America may have been milder due to the blocking of Arctic air by the Laurentide ice sheet [Bryson and Hare, 1974] and this effect may have extended to the Great Basin. This blocking may have offset the effects of reduced insolation. The DH11 $\delta^{18}O$ increase may thus reflect reduced kinetic fractionation under full glacial conditions and an initial cooling of the ocean source-area relative to the Great Basin due to lower sea surface temperatures, and/or a southward shift of circulation patterns, and/or blocking of cold arctic air in winter, followed by global glacial to interglacial warming between 140 and 133 ky ago and return of interglacial air circulation and kinetic fractionation. The magnitude of the glacial to interglacial warming at Devils Hole is obscured because its effect on the Devils Hole isotope record was counteracted by the return of the ocean source temperature and kinetic fractionation to interglacial values and by the dilution of the ocean with isotopically light meltwater (a change of 1.2‰). Instead of the 3.5°C calculated from the observed 2.1‰ increase in $\delta^{18}O$ and the 0.6‰/°C gradient [Rozanski et al.,1992], one obtains a local increase in winter/spring temperature of about 7°C for a $\delta^{18}O$ increase of 4.2‰ to produce the 2‰ increase observed. This temperature increase contradicts the earlier requirement for little or no change in temperature in the Great Basin while the ocean surface cooled. This can still be true if the interglacial to glacial cooling in the Great Basin occurred during the earlier phases of the glacial. The temperature change is much greater than the about 2°C derived for low-latitude ocean surfaces but close to the range of 4.2 to 6.5°C deduced from the change in the elevation of the 0°C isotherm during the peak of the last glacial [Broecker and Denton, 1989] and the 5°C derived from noble

gas studies in the Carrizo aquifer, Texas [Stute et al., 1992]. Although the explanation presented above uses corrections consistent with available evidence it postulates their exact timing and it is not neccessarily correct. Other ways to produce the same observed isotope curve may exist. The discussion demonstrates, however, that the rather small glacial-interglacial differences in isotopic composition observed in lower latitude records [another example is the last glacial-interglacial transition in the isotope record of the Dunde ice cap, China, Thompson et al., 1989] can be significantly influenced by factors other than local temperature.

CONCLUSIONS

Continental isotope records of climate change frequently show only a narrow range of isotope values. This increases the importance of isotopic changes in the water vapor in the oceanic source-area, and of ocean surface temperature changes relative to temperature changes in the precipitation area. Where possible, source-area corrections should be applied before the interpretation of continental isotope records, even if one does not plan to translate the observed isotope changes into temperature changes. Interpretation of lower latitude continental isotope records in terms of local temperature changes requires great care and independent corroborating evidence because, as demonstrated for Quelccaya, temperature may not be the main factor controlling the isotopic composition of the precipitation.

The discrepancies between the new Devils Hole record DH11 and the deep-sea chronology present interesting interpretation problems. The Devils Hole record shows, just like the deep-sea, evidence of the importance of the precessional insolation cycles for climate forcing. Adjustment of the age of Termination II by 7 ky based on recent high-precision coral dates brings agreement between the deep-sea and Vostok chronologies and with the later part of the DH11 glacial to interglacial transition. However, adjusting the timing of Termination II implies the abandoning of high-latitude NH summer insolation as the sole forcing function for climate change. The evidence from the records suggests the need for a forcing that includes both SH and NH insolation, interhemispheric coupling, non-linear response due to feedback from albedo and greenhouse gases, and the effects of ice sheet dynamics on climate response. The long duration of the glacial to interglacial transitions in the DH11 record needs to be understood before the more secure dating of the continental Devils Hole record DH11 can be fully used to further refine our understanding of global climate change caused by orbitally determined (Milankovitch) insolation changes.

Continental isotope records are valuable because they provide information on the regional expression of changes in global climate in the past and they may offer a better opportunity than the deep-sea to obtain accurate dates for such climate changes. Their full potential can, however, only be realized when we take care to understand all the factors determining the local changes in the isotope record and when the correlation with other isotope records is either based on accurate unambiguous dating or on a detailed understanding of the events to be correlated in both records.

Acknowledgments. This work was supported by NSF grants DPP-8822073 and DPP-8915924, Division of Polar Programs. Discussions with Ike Winograd, Minze Stuiver, and Eric Steig and comments of a reviewer contributed to the development of this paper.

REFERENCES

Aravena, R., H. Pena, A. Grilli, O. Suzuki, and M. Mordeckai, Isotopic variation of rains and origin of the air masses in the Chilean Altiplano, in *Isotope Hydrology Investigations in Latin America, Proceedings Regional Seminar for Latin America on the Use of Isotope Techniques in Hydrology*, Salazar, Mexico, Sept.28-Oct.2, pp. 129-142, IAEA, 1989.

Bard, E., B. Hamelin, R.G. Fairbanks, and A. Zindler, Calibration of the ^{14}C timescale over the past 30,000 years using mass spectrometric U-Th ages from Barbados corals, *Nature*, *345*, 405-410, 1990

Bard, E., M. Arnold, R.G. Fairbanks, and B. Hamelin, $230Th/234U$ and ^{14}C ages obtained by mass spectrometry on corals, *Radiocarbon*, 1993, in press.

Barnola, J.M., P. Pimienta, D. Raynaud, and Y.S. Korotkevich, CO_2-climate relationship as deduced from the Vostok ice core: A reexamination based on new measurements and on a reevaluation of the air dating, *Tellus B*, *43*, 83-90, 1991.

Berger, A. and M.F. Loutre, Insolation values for the climate of the last 10 million years, *Quat. Sci. Rev.*, *10*, 297-317, 1991.

Birchfield, G.E., Changes in deep-ocean water $\delta^{18}O$ and temperature from the last glacial maximum to the present, *Paleoceanography*, *2*, 431-442, 1987.

Broecker, W.S. and G.H. Denton, The role of ocean-atmosphere reorganizations in glacial cycles, *Geochim. Cosmochim. Acta 53*, 2465-2501, 1989.

Broecker, W.S. and J. Van Donk, Insolation changes, ice volumes, and the O^{18} record in deep-sea cores, *Rev. Geophys. Space Phys.*, *8*, 169-198, 1970.

Bryson, R.A. and F.K. Hare, The climates of North America, in *World Survey of Climatology*, vol. 11, *Climates of North America*, edited by R.A. Bryson and F.K. Hare, pp. 1-47, Elsevier-Science, New York, 1974.

Chappell, J. and N.J. Shackleton, Oxygen isotopes and sea level, *Nature*, *324*, 137-140, 1986.

Chen, J.H., H.A. Curran, B. White, and G.J. Wasserburg, Precise chronology of the last interglacial period: 234U-230Th data from fossil coral reefs in the Bahamas, *G.S.A. Bull.*, *103*, 82-97, 1991.

CLIMAP Project Members, Seasonal reconstructions of the Earth's surface at the last glacial maximum. *Geological Society of America Map and Chart Series 36*, 1981.

Dall'Olio, A., E. Salati, C.T. Azevedo, and E. Matsui, Modelo de fracionamento isotopico da agua na bacia Amazonica: Primeira aproximacao, *Acta Amazonica*, *9*, 675-687, 1979.

Dansgaard, W., Stable isotopes in precipitation, *Tellus*, *16*, 436-468, 1964.

Duplessy, J-C, M. Arnold, P. Maurice, E. Bard, J. Dupratt, and J. Moyes, Direct dating of the oxygen-isotope record of the last deglaciation by ^{14}C accelerator mass spectrometry, *Nature*, *320*, 350-352, 1986.

Edwards, R.L., J.H. Chen, T.-L. Ku, and G.J. Wasserburg, Precise timing of the last interglacial period from mass spectrometric determination of Thorium-230 in corals, *Science*, *236*, 1547-1553, 1987.

Fairbanks, R.G., A 17,000-year glacio-eustatic sea level record:influence of glacial melting rates on the Younger Dryas event and deep-ocean circulation, *Nature*, *342*, 637-642, 1989.

Fairbanks, R.G., The age and origin of the "Younger Dryas climate event" in Greenland ice cores, *Paleoceanography*, *5*, 937-948, 1990.

Grootes, P.M. and E.J. Steig, Uncertainties in the isotope-temperature relationship in ice cores, *EOS, 73, Spring suppl.*, 107, 1992.

Grootes, P.M., and M. Stuiver, Ice sheet elevation changes from isotope profiles, *Proc. Symp. "The Physical Basis of Ice Sheet Modeling"*, Vancouver, 1987, edited by E.D. Waddington and J.S. Walder, *IAHS publ. no.170*, 269-281, 1987.

Grootes, P.M., M. Stuiver, L.G. Thompson, and E. Mosley-Thompson, Oxygen isotope changes in tropical ice, Quelccaya, Peru, *J. Geophys. Res.*, *94*, 1187-1194, 1989.

Harmon, R.S., T.-L. Ku, R.K. Matthews, and P.L. Smart, Limits of U-series analyses: Phase I results of the Uranium-Series Intercomparison Project, *Geology*, *7*, 405-409, 1979.

Hays, J.D., J. Imbrie, and N.J. Shackleton, Variations in the Earth's orbit: Pacemaker of the Ice Ages, *Science*, *194*, 1121-1132, 1976.

Hilgen, F.J., Astronomical calibration of Gauss to Matuyama sapropels in the Mediterranean and implication for the geomagnetic polarity time scale, *Earth Plan. Sci. Lett.*, *104*, 226-244, 1991.

Imbrie, J. and J.Z. Imbrie, Modelling the climatic response to orbital variations, *Science*, *207*, 943-954, 1980.

Imbrie, J., J.D. Hays, D.G. Martinson, A. McIntyre, A.C. Mix, J.J. Morley, N.G. Pisias, W. Prell, and N.J. Shackleton, The orbital theory of Pleistocene climate: support from a revised chronology of the marine ^{18}O record, in *Milankovitch and Climate*, edited by A. Berger, J. Imbrie, J. Hays, G. Kukla and B. Saltzman, pp.269-305, D. Reidel, Hingham, Mass, 1984.

International Atomic Energy Agency, Statistical treatment of environmental isotope data in precipitation, *Tech. Rep. Ser.*, *206*, pp. 255, IAEA, Vienna, Austria, 1981.

Johnsen, S.J., H.B. Clausen, W. Dansgaard, K. Fuhrer, N. Gundestrup, C.U. Hammer, P. Iversen, J. Jouzel, B. Stauffer, and J.P. Steffensen, Irregular glacial interstadials recorded in a new Greenland ice core, *Nature, 359*, 311-313, 1992.

Johnson, R.G. and Andrews, J.T., Glacial terminations in the oxygen isotope record of deep sea cores: Hypothesis of massive Antarctic ice-shelf destruction, *Palaeogeogr. Palaeoclimatol. Palaeoecol.*, *53*, 107-138, 1986.

Jouzel, J., C. Lorius, J.R. Petit, C. Genthon, N.I. Barkov, V.M. Kotlyakov, and V.M. Petrov, Vostok ice core: a continuous isotope temperature record over the last climatic cycle (160,000 years), *Nature, 329*, 403-407, 1987.

Jouzel, J., L. Merlivat, and C. Lorius, Deuterium excess in an East Antarctic ice core suggests higher relative humidity at the oceanic surface during the last glacial maximum, *Nature, 299*, 688-691, 1982.

Keigwin, L.D., G.A. Jones, and S.J. Lehman, Deglacial meltwater discharge, North Atlantic deep circulation, and abrupt climate change, *J. Geophys. Res.*, *96*, 16, 811-16, 826, 1991.

Labeyrie, L.D., J.J. Pichon, M. Labracherie, R. Ippolito, J. Duprat, and J.C. Duplessy, Melting history of Antarctica during the past 60,000 years, *Nature, 322*, 701-706, 1986.

Lehman, S.J. and L.D. Keigwin, Sudden changes in North Atlantic circulation during the last deglaciation, *Nature, 356*, 757-762, 1992.

Lorius, C., J. Jouzel, C. Ritz, L. Merlivat, N.I. Barkov, Y.S. Korotkevich, and V.M. Kotlyakov, A 150,000-year climatic record from Antarctic ice, *Nature, 316*, 591-596, 1985.

Ludwig, K.R., K.R. Simmons, B.J. Szabo, I.J. Winograd, J.M. Landwehr, A.C. Riggs, and R.J. Hoffman, Mass-spectrometric 230Th-234U-238U dating of the Devils Hole calcite vein, *Science, 258*, 284-287, 1992.

Martinson, D.G., N.G. Pisias, J.D. Hays, J. Imbrie, T.C. Moore Jr., and N.J. Shackleton, Age dating and the orbital theory of the Ice Ages: Development of a high-resolution 0 to 300,000-year chronostratigraphy, *Quat. Res.*, *27*, 1-29, 1987.

Mazaud, A., C. Laj, E. Bard, M. Arnold, and E. Tric, Geomagnetic field control of ^{14}C production over the last 80 ky: implications for the radiocarbon time-scale, *Geophys. Res. Lett.*, *18*, 1885-1888, 1991.

Paterson, W.S.B. and C.U. Hammer, Ice core and other glaciological data, in *The Geology of North America, vol. K-3, North America and adjacent oceans during the last deglaciation*, edited by W.F.Ruddiman and H.E.Wright, Jr, pp. 91-109, Geological Society of North America, Boulder, 1987.

Petit, J.R., M. Briat, and A. Royer, Ice age aerosol content from East Antarctic ice core samples and past wind strength, *Nature*, *293*, 391-394, 1981.

Petit, J.R., L. Mounier, J. Jouzel, Y.S. Korotkevich, V.I. Kotlyakov, and C. Lorius, Paleoclimatological and chronological implications of the Vostok core dust record, *Nature, 343*, 56-58, 1990.

Pisias, N.G., D.G. Martinson, T.C. Moore Jr., N.J. Shackleton, W. Prell, J. Hays, and G. Boden, High resolution stratigraphic correlation of benthic oxygen isotope records spanning the last 300,000 years, *Marine Geology, 56*, 119-136, 1984.

Ratisbona, L.R., The climate of Brazil, in *World Survey of Climatology*, vol.12, *Climates of Central and South America*, edited by W. Schwerdtfeger, pp.219-293, Elsevier-Science,New York, 1976.

Rozanski, K, L. Araguas-Araguas, and R. Gonfiantini, Relation between long-term trends of oxygen-18 isotope composition of precipitation and climate, *Science, 258*, 981-985, 1992.

Rozanski, K., C. Sonntag, and K.O. Münnich, Factors controlling stable isotope composition of European precipitation, *Tellus, 34*, 142-150, 1982.

Sarnthein, M. and R. Tiedemann, Younger Dryas-style cooling events at glacial terminations I-VI at ODP site 658: Associated benthic δ^{13}C anomalies constrain meltwater hypothesis, *Paleoceanography, 5*, 1041-1055, 1990.

Shackleton, N.J., A. Berger, and W.R. Peltier, An alternative astronomical calibration of the Lower Pleistocene time scale based on ODP Site 677, *Trans. R. Soc. Edinburgh; Earth Sci, 81*, 251-261, 1990.

Shackleton, N.J. and N.G. Pisias, Atmospheric carbon dioxide, orbital forcing, and climate, in *The Carbon Cycle and Atmospheric CO$_2$: Natural variations Archean to Present, Geophysical Monograph 32*,edited by E.T. Sundquist and W.S. Broecker, pp.303-317, American Geophysical Union, Washington, D.C., 1985.

Sowers, T., M. Bender, and J.M. Barnola, Records of Antarctic temperature, atmospheric CO$_2$ and ice volume during the last deglaciation, *EOS, 73, Fall suppl.*, 260, 1992.

Sowers, T., M. Bender, D. Raynaud, Y.S. Korotkevich, and J. Orchardo, The δ^{18}O of atmospheric O$_2$ from air inclusions in the Vostok ice core: Timing of CO$_2$ and ice volume changes during the penultimate deglaciation, *Paleoceanography, 6*, 679-696, 1991.

Stuiver, M. and T.F. Braziunas, Modeling, atmospheric ^{14}C influences, and radiocarbon ages of marine samples back to 10,000 BC, *Radiocarbon*, 1993, in press.

Stute, M., P. Schlosser, J.F. Clark, and W.S. Broecker, Paleotemperatures in the southwestern United States derived from noble gases in ground water, *Science, 256*, 1000-1003, 1992.

Taljaard, J.J., Synoptic meteorology of the southern hemisphere, in *Meteorology of the Southern Hemisphere*, edited by C.W.Newton, *Meteorol. Monogr. 13*, 139-213, 1972.

Thompson, L.G., E. Mosley-Thompson, M.E. Davis, J.F. Bolzan, J. Dai, T. Yao, N. Gundestrup, X. Wu, L. Klein, Z. Xie, Holocene-Late Pleistocene climatic ice core records from Quinghai-Tibetan Plateau, *Science, 246*, 474-477, 1989.

Thompson, L.G., E. Mosley-Thompson, P.M. Grootes, M. Pourchet, and S. Hastenrath, Tropical glaciers: Potential for ice core paleoclimatic reconstructions, *J. Geophys. Res., 89*, 4638-4646, 1984

Thompson, L., E. Mosley-Thompson, J.F. Bolzan, and B.R. Koci, A 1500-year record of tropical precipitation in ice cores from the Quelccaya Ice Cap, Peru, *Science, 229*, 971-973, 1985.

Thompson, L.G., E. Mosley-Thompson, W. Dansgaard, and P.M. Grootes, The Little Ice Age as recorded in the stratigraphy of the tropical Quelccaya ice cap, *Science, 234*, 361-364, 1986.

Thompson, L.G. and E. Mosley-Thompson, Microparticle concentration variations linked with climatic change: evidence from polar ice cores, *Science, 212*, 812-815, 1981.

Vorren, T.O., K-D. Vorren, T. Alm, S. Gulliksen, and R. Lovlie, The last deglaciation (20,000 to 11,000 B.P.) on Andøya, northern Norway, *Boreas, 17*, 41-77, 1988.

Winograd, I.J., T.B. Coplen, J.M. Landwehr, A.C. Riggs, K.R. Ludwig, B.J. Szabo, P.T. Kolesar, and K.M. Revesz, Continuous 500,000-year climate record from vein calcite in Devils Hole, Nevada, *Science, 258*, 255-260, 1992.

Winograd, I.J., B.J. Szabo, T.B. Coplen, and A.C. Riggs, A 250,000-year climatic record from Great Basin vein calcite: Implications for Milankovitch theory, *Science, 242*, 1275-1280, 1988.

P.M. Grootes, Quaternary Isotope Laboratory, University of Washington, Seattle, Washington 98195.

Constraining Estimates of Evapotranspiration with Hydrogen Isotopes in a Seasonal Orographic Model

Neil L. Ingraham

Water Resources Center, Desert Research Institute, Las Vegas, Nevada 89132-0040

Richard G. Craig

Department of Geology, Kent State University, Kent, Ohio 44242-0001

Precipitation and groundwater hydrogen isotope data are used to constrain a climate–isotope model that solves δD at a 10 km spacing along a 400 km traverse at 41° 30'N across northern California. Parameters of the isotope model are δD and evapotranspiration (ET) rate at the coast, advection velocity, and ET rate at 400 km inland. A good fit of the model to the data along most of the traverse is obtained when the δD of groundwater at the coast was set at –90 per mil, the advection velocity is 3.4 m/sec, and the evapotranspiration rate is 20%. This renders a poor fit for the first 60 km of the traverse. A lower advection velocity (1.0 m/sec) accommodates the observed –43 per mil value at the coast and provides acceptable fit to the marked decrease in δD along the first 60 km of the traverse.

Introduction

Geographic variations in stable isotopic ratios of meteoric water resulting from meteoric processes have recently been used to characterize large–scale hydrologic systems. The nature and degree of depletion of the heavier isotopes by rain–out [the 'Continental Effect' of Friedman et al., 1964] can be interpreted in terms of the vertical fluxes (precipitation and evapotranspiration or ET) of meteoric water.

Rozanski et al. [1982] modeled the stable isotopic composition of atmospheric water vapor and precipitation in central Europe with respect to distance from the Atlantic Ocean, and included the effects of winter precipitation and summer ET. They documented decreases in δD of 3.3 per mil/100 km in winter precipitation and 1.3 per mil/100km in summer precipitation. Rozanski et al. [1982] concluded that variations in local precipitation were primarily controlled by regional–scale precipitation/ET events upwind, modified only slightly by local temperature fluctuations.

Salati et al. [1979] observed a depletion in δ¹⁸O of 0.075 per mil/100 km in precipitation in the Amazon Basin and attributed it to a large contribution of moisture recycled by ET. Gat and Matsui [1991] have recently attempted to determine the ratio of evaporation to transpiration in the Amazon Basin by using a steady–flux model incorporating deuterium excess values (defined as $d = \delta D - 8\,\delta^{18}O$). Ingraham and Taylor [1986, 1991] observed marked depletion in meteoric water along traverses from the Pacific Ocean across California. The slopes of the depletion trends were interpreted as being controlled by the recycling of meteoric water by evaporation and transpiration. Their transects extended into the Great Basin where, due to the hydrologically closed nature of the region, no further depletion in the hydrogen isotopic ratios of meteoric water were observed. The lack of large isotopic depletion in groundwaters with increasing distance was explained as representing a closed hydrologic system [Ingraham and Taylor, 1986, 1991]. However, the lack of geographic depletion in precipitation may be obtained in an open system where ET recycling serves to maintain downwind precipitation as observed by Salati et al. [1979] for the Amazon Basin.

Not all researchers have observed the effects of ET on the stable isotopic ratios of inland meteoric waters. Smith et al. [1979] compared the isotopic composition of precipitation in southwestern California and western Nevada to a model of isotopic fractionation in a pseudo–adiabatically rising air mass. They were able to obtain good results without including the upwind recycling of evapotranspired water. However, their model neglected possible precipitation over the ocean prior to the air mass reaching the coast. Yonge et al. [1989] modeled the stable isotopic ratios of surface waters across western Canada to the Great Divide (over 500 km distance inland) in terms of a single–stage Rayleigh distillation curve, concluding that

Climate Change in Continental Isotopic Records
Geophysical Monograph 78

precipitation west of the divide did not appear to be greatly affected by ET.

In contrast to the results of Yonge et al. [1989], we hypothesize that ET recycling is an important component of the hydrologic cycle of northern California and the Great Basin. Understanding this component is important for current hydrologic studies as well as for climate change prognostics in which the affects of varying evaporation and transpiration rates and modification of vegetation patterns or infiltration rates could alter the contribution from recycling. The research presented here is a continuation of that reported by Craig et al. [1992], who modeled the hydrogen isotope ratios of meteoric water along a transect across northern California. Rain-out and recycling were modeled along an air mass flowline using seasonal average temperature and total precipitation in an area dominated by orographic precipitation. The amount of ET recycling was held constant along the transect, and no acceptable fit to the observed isotopic composition of precipitation we obtained when the isotopic effects of ET recycling were ignored. Here we extend the model by including: (1) linear variations in ET rate along the transect studied and (2) fractionation during the evaporation of surface water. This extended model was applied to a transect across northern California to provide an understanding of the importance of the terrestrial recycling of water.

PROCEDURES

The traverse, 10 km wide, at approximately 41°30'N, the latitude of Traverse I of Ingraham and Taylor [1991], was aligned parallel to the dominant wet-season 800 mb windfield to represent the maximum changes in isotopic composition of water vapor as it is advected from west to east. The traverse includes several physiographic and hydrologic provinces in northern California, and allows calculation of water vapor advection into the Great Basin in northern Nevada. Shallow groundwaters were considered to be the type of sample that best-represented the average isotopic composition of precipitation, being the least sensitive to the extreme yearly, seasonal and in-storm isotopic variability [such as found by Smith et al., 1979; Ingraham and Taylor, 1991], and yet reflecting the current climate and hydrologic regime.

Thirty precipitation samples and 44 surface and shallow groundwater samples were collected along the traverse to augment data from the literature, primarily Ingraham and Taylor [1986, 1991]. More than 65 samples resulted, providing hydrogen isotope data at approximately 10 km spacing, matching that of the climate model. Rain was collected for stable isotopic analysis using 1 liter glass collection tanks each with a 2 cm (~0.75 inch) pvc pipe constriction. The collectors contained 100 ml of mineral oil to control evaporation, and were buried in the ground for stability as well as security. During the winter rainy season between mid-December 1990 and late March 1991, precipitation was collected from the coast to a point approximately 100 km inland. All water samples were sealed in 125 ml (4 oz) poly-sealed glass bottles. The hydrogen isotope ratios were determined by reducing a 5 µl aliquot of the sample to hydrogen gas using zinc as a reducing agent [Kendall and Coplen, 1985]. The hydrogen gas was then analyzed in a

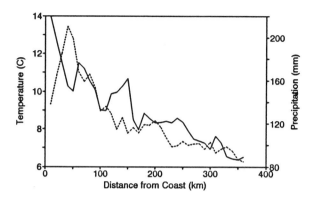

Fig. 1 Climate model solutions of wet-season average temperature (solid) and total wet-season precipitation (dashed) along the traverse studied.

Nuclide 3-60 double collector mass spectrometer. All data are reported in the standard 'δ' notation as a per mil variation from SMOW; the reproducibility of the δD values is ±1.0 per mil (1σ).

Model Description

Our method of isotopic modeling follows closely that of Rozanski et al. [1982]; however, we differ in our use of a grid spacing almost two orders of magnitude more fine. In addition, Rozanski et al. [1982] were able to use 10-year averages of observational data to define the climate of their large grid boxes. Such data are not available at the scale needed to represent orographic influences in western North America. To represent these orographic influences we use mean monthly estimates of temperature and precipitation (Figure 1) derived from the climate model calibrated with five year averages of instrumental records.

The climate model employed [Craig and Stamm, 1989; Stamm and Craig, 1989] considers mass balance, wind trajectory, physiographic controls, and radiation balance when determining the climate at a specific location. The equations are derived from a canonical analysis of the instrumental record and can produce solution of precipitation and temperature on a 10 km horizontal spacing based on a digital elevation model at that scale. This model differs from more conventional, primitive-equation Eulerian models in that it is statistically-based and computes independent variables in a semi-Lagrangian mode. By adjusting the isotope model to duplicate observed stable isotopic ratios in precipitation and groundwaters, we are able to study their variations along the atmospheric flowpath at a level of detail where the isotopic variation within a grid cell is negligible.

Parameters of the Isotope Model

Four parameters can be specified in the isotope model, (1) the δD of precipitation at the coast, (2) the advection velocity (which remains constant throughout the traverse), (3) the rate of ET at the coast and, (4) the rate of ET at 400 km inland, the end of the traverse.

The choice of the advection velocity determines the amount of water vapor crossing the coast. Greater advection velocities increase the absolute amount of water vapor (and thus precipitable water) crossing the coast. A large quantity of precipitable water would decrease the distillation fraction and result in less depletion in the δD of the modeled precipitation and shallow groundwater would then be more enriched than observed.

ET in the model is expressed as a fraction of precipitation and varies linearly along the traverse between the coast and 400 km inland. The evapotranspired vapor is returned to the atmosphere and mixed by mass balance with the water vapor already present. ET is an important factor for not only the water balance along the traverse but for the isotopic balance as well. All ET in the model is treated as evaporation and is accompanied by fractionation [Majzoub, 1971], determined by Rayleigh distillation at surface temperature. Thus, the model does not differentiate between evaporation and transpiration, which effectively occurs without fractionation [Wershaw et al., 1966; Zeigler et al., 1976]. However the inclusion of fractionation during evaporation [here and Gat and Matsui, 1991], albeit without a kinetic effect, represents an expansion of earlier models [Craig et al., 1992; and Rozanski et al., 1982] in which ET return was not fractionated. Without fractionation during transpiration, the returning vapor has the same δD as the source precipitation, producing smaller isotopic changes in atmospheric vapor downwind than if all ET is evaporation. Our treatment of ET introduces a bias towards excessive evolution of the atmospheric vapor that is greatest in heavily vegetated areas such as the western portion of the traverse. We expect that this bias is less than would result if fractionation during evaporation were ignored.

The climate model does not contain a storage term to allow residual water to accumulate for ET in subsequent time periods. Since water cannot be stored, the only source of ET is the precipitation in the current modeled season. The amount of water available for ET could either be larger or smaller than actual. However, most probably, the amount of water available in each modeled cell will be less than actual in the absence of a storage term in the model. Thus ET determined by the model will tend to be smaller than actual; but fractionation during ET will be excessive. We expect to introduce a storage term in subsequent versions of the model.

Model Mechanics

The model is adjusted to consider the wet–season, defined as November through March, when 50% to 80% of precipitation falls, as shown in Figure 2. We considered instantaneous precipitable water content (10.6 mm) to the 325 mb level at the coast. The average annual value in the U.S. is 17 mm [Court, 1974, p. 206] which corresponds closely to the northern California coastal average. Assuming an equivalent correspondence of monthly values, we calculate the total wet–season water content, Q, from the following relation:

$$Q = P\ V\ T\ W \qquad (1)$$

where:

P = average instantaneous wet–season precipitable
water content (m),
V = advection velocity (m/sec),
T = time (1.296×10^7 sec),
W = cell width (10,000 m).

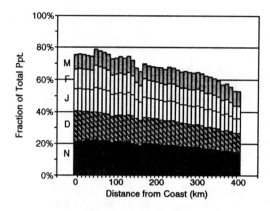

Fig. 2 Modeled monthly contribution to total annual precipitation along the traverse for the five month wet–season.

Precipitation is not uniformly distributed throughout the wet–season (Figure 2). Since the model uses average precipitable water during this period, the amount of water available in the wettest months (November and December, Figure 2) is underestimated. Thus, distillation would be greater in the model than observed rendering a δD more depleted than observed, during this wetter period. The amount of precipitable water available in the dryer months of the wet–season (January and February) is overestimated, yielding less distillation and more enriched isotopic compositions than observed during this period.

Advection velocity estimates are based on the data of Schutz and Gates [1971] at the 800 mb level during the wet–season for their cell centered on 122.5°W and 40°N. The average velocity from December through February is reported to be 3.73 m/sec, whereas that from June through August is reported to be 2.7 m/sec. The average 800 mb velocity for the wet–season is expected to fall between these values. If most wet–season precipitation condenses at or below the 800 mb level the average transport velocity will be less than the 800 mb velocity.

An average wet–season temperature was used in the model for evaporation, and condensation temperatures for precipitation were approximated by extrapolation from the surface to 2 km with an environmental lapse rate of 6.5°C/1000m [Rogers and Yau, 1989]. Modeled fractionation factors will be too large in months with actual temperatures warmer than the average wet–season temperature. Thus, for these months, the model precipitation will tend to be more depleted than observed precipitation. Since precipitation is not uniformly distributed over the wet–season, this can produce a (slight) bias in the results, in addition, the average wet–season surface temperatures everywhere exceed 0°C; thus, a provision for fractionation of solid precipitation was not made in the model.

The cumulative effect of the biases described above is difficult to integrate during the five month period and throughout the 400 km transect. However, since evaporation during raindrop fall is not represented in the model, we constrained δD of the modeled precipitation to be depleted relative to the observed precipitation data. With respect to the groundwater, the existence of a systematic bias in the model is not obvious. In the absence of a

TABLE 1. The distance from the coast (in km), occurrence (of the surface and shallow groundwater), location, and hydrogen isotopic composition of the samples collected for this study.

No.	km	Surface and shallow groundwater Occurrence	Location	δD	No.	km	Precipitation Location	δD
1	1	Creek	41–43–30/124–08–15	−44	1	0	41–28–30/124–03–30	−33
2	1	Creek	42–16–45/124–29–00	−42	2	7.5	41–30–45/123–58–45	−44
3	1	Creek	42–09–45/124–21–15	−42	3	0	41–35–45/124–05–45	−37
4	20	Spring	42–11–30/124–07–45	−47	4	1	41–43–30/124–08–15	−38
5	26	Spring	41–42–15/123–56–15	−55	5	0	41–59–00/124–12–45	−38
6	34	Creek	41–42–45/123–51–00	−54	6	1	42–06–30/124–20–15	−37
7	30	Spring	41–51–30/123–53–45	−54	7	0	42–16–45/124–29–00	−37
8	34	Spring	41–52–30/123–50–45	−56	9	20	42–11–30/124–07–45	−44
9	28	Spring	41–57–30/123–45–15	−62	10	16	41–48–30/124–03–00	−41
10	28	Well	41–58–30/123–45–15	−60	11	31	41–52–30/123–50–45	−48
11	58	Spring	42–14–00/123–42–00	−69	13	52	42–04–45/123–42–00	−59
12	64	Creek	42–16–45/123–37–15	−70	14	81	42–05–45/123–24–45	−73
13	70	Creek	42–09–45/123–31–15	−70	15	72	42–10–00/123–30–30	−71.5
14	58	Spring	42–02–30/123–33–00	−74	16	61	42–12–00/123–39–00	−65
15	63	Spring	41–56–00/123–29–45	−83	18	59	42–13–45/123–42–30	−62
16	61	Spring	41–31–30/123–31–30	−89	19	67	42–16–30/123–37–00	−67
17	62	Creek	41–36–15/123–31–00	−73	20	72	41–49–30/123–22–45	−75
18	63	Creek	41–36–45/123–29–45	−70	21	66	41–53–30/123–26–00	−79
19	64	Creek	41–37–15/123–29–00	−73	22	62	41–55–30/123–28–30	−70
20	68	Creek	41–42–30/123–26–15	−75	24	56	42–03–45/123–38–00	−62
21	80	Creek	41–50–45/123–18–00	−81	25	21	41–49–30/123–58–15	−47
22	91	Creek	41–47–30/123–09–45	−88	26	30	41–49–15/123–49–30	−49
23	103	Creek	41–47–00/123–01–30	−94	27	15	41–42–30/123–56–30	−44
24	114	Creek	41–50–30/122–53–15	−94	28	18	41–42–15/123–56–15	−48
25	126	Creek	41–51–45/122–45–00	−96	31	25	41–42–30/123–52–00	−48
26	137	Creek	41–50–15/122–37–30	−97	32	26	41–42–45/123–51–00	−49
27	126	Well	41–37–00/122–23–30	−113	33	59	41–42–00/123–26–45	−68
28	172	Well	41–38–00/122–11–45	−104	34	47	41–34–00/123–31–45	−59
29	196	Well	41–58–15/121–55–15	−106	35	48	41–31–30/123–31–00	−62
30	200	Well	42–00–15/121–52–00	−105	36	78	41–52–15/123–18–00	−71
31	226	Well	41–59–45/121–33–15	−92				
32	231	Well	41–58–30/121–29–45	−91				
33	255	Well	41–36–30/121–13–00	−105				
34	332	Well	41–33–30/120–18–00	−103				
35	363	Well	41–36–15/119–55–30	−107				
36	368	Well	41–34–45/119–52–30	−118				
37	386	Creek	41–35–15/119–39–15	−121				
38	410	Spring	41–45–45/119–22–00	−120				
39	437	Spring	41–52–30/119–02–45	−123				
40	473	Well	41–55–15/118–37–00	−128				
41	543	Well	40–57–45/117–47–15	−121				
42	490	Well	40–53–15/118–24–45	−134				
43	432	HS	40–46–15/119–06–45	−127				
44	410	HS	40–40–30/119–22–00	−103				

HS = hot spring (90°C)

Precipitation gages 8, 12, 17, 23, 29, and 30 could not be relocated at the end of the sampling season.

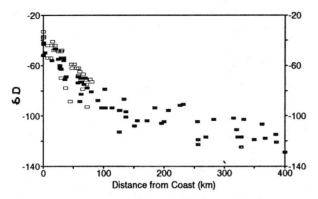

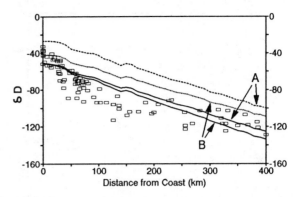

Fig. 3 Observed hydrogen isotope data of precipitation and groundwater used for this research. Some of the data were taken from the literature (primarily Ingraham and Taylor, [1991]).

Fig. 5 (A) (–50, 2.5, 10%, 20%) and (B) (–50, 2.5, 20%, 20%) showing the effect of increasing ET recycling at the coast. Both of these solutions give reasonable results for the first 60 km of the traverse, but not for the remaining portion. The model cannot accommodate the majority of the traverse with an input value of –50 per mil for precipitation at the coast.

clear indication otherwise, we assume the model is an unbiased estimator of the groundwater data. Thus, the model parameters will be adjusted so that δD of the modeled groundwater is an unbiased estimator of the average of the actual data.

Results

The δD of the precipitation and groundwater samples collected for this study are reported in Table 1, and as augmented with data from Ingraham and Taylor [1986, 1991] are shown in Figure 3. The δD values of precipitation do not mimic those of the surface and shallow groundwater data. The amount of precipitation collected was less than instrumental averages during the two different collection years [1986 and 1991]. Most

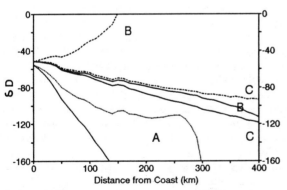

Fig. 4 (A) Model solution (–50, 1.0, 0%, 100%) in which the low advection velocity does not allow enough precipitable water to cross the coast to maintain precipitation throughout the traverse. (B) Solution (–50, 3.0, 0%, 100%) with a larger advection velocity allows enough precipitable water to cross the coast; however the divergence between the precipitation and groundwater curves indicates that the overall amount of ET recycling is too large. (C) Model solution (–50, 3.0, 0%, 20%), in which the rate of ET at 400 km inland is decreased to 20%, yields a better fit. However, the precipitation and groundwater curves do not diverge enough along the western portion of the traverse implying the need for increased evapotranspirational recycling at the coast.

of the precipitation for this research (collected in 1991) fell during March, under warmer conditions than average for the wet–season. Thus the stable isotopic compositions of the precipitation are not expected to be congruent to those of the surface and shallow groundwater.

Four surface and shallow groundwater samples collected at the coast were used to determine the initial δD of groundwater for the model. The samples range in δD from –42 to –44 per mil, and initially, the value of –50 per mil was chosen as the model's coastal precipitation value. Both the precipitation and shallow groundwater data sets show a marked decrease in δD along the first 60 km of the traverse (Figure 3). This decrease is exemplified in the δD of groundwater which decreases from about –43 per mil at the coast to almost –85 per mil at 60 km inland, for an average depletion of about 70 per mil/100km. From 60 km to 400 km inland, δD of groundwater shows less depletion with distance, varying about 12 per mil/100km.

Figures 4 through 6 illustrate the process of adjusting the model to the actual precipitation and groundwater data. The first set of Figures, 4a through 4c, demonstrate the importance of the advection velocity term. Figure 4a shows the results of using a δD of precipitation at the coast of –50 per mil, an advection velocity of 1 m/sec., ET rate at the coast of 0% and at 400 km inland of 100% (–50, 1.0, 0%, 100%). The low advection velocity does not allow enough precipitable water to cross the coast to maintain precipitation throughout the transect. Figure 4b (–50, 3.0, 0%, 100%) with a larger advection velocity allows enough precipitable water to cross the coast; however, the divergence between the precipitation and groundwater curves indicates that the overall amount of ET recycling is too large. Figure 4c (–50, 3.0, 0%, 20%) yields a better fit although the precipitation and groundwater curves do not diverge enough along the western portion of the traverse and thus require an increase in ET rate at the coast. This model setting indicates that a velocity on the order of 3.0 m/sec. and ET rate on the order of 20% produce a useful representation.

Figures 5a (–50, 2.5, 10%, 20%) and 5b (–50, 2.5, 20%, 20%)

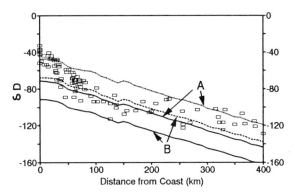

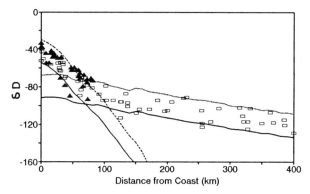

Fig. 6 (A) (–70, 2.5, 20%, 20%) and (B) (–90, 2.5, 20%, 20%) represent a reasonable divergence between the precipitation and surface water data and better fit the data between 60 km and 400 km inland. The input value of –90 per mil at the coast produces a more reasonable fit than –70 per mil, however it appears that a larger velocity would be more appropriate.

Fig. 7 Model solutions (–90, 3.4, 20%, 20%), and (–50, 1.0, 20%, 20%), the first produces a reasonable fit between 60 and 400 km inland. The second accommodates the marked decrease in δD along the first 60 km of the traverse. These results suggest that two distinct precipitation regimes exist along the traverse.

show the effect of non–zero ET rate at the coast. Both of these solutions give reasonable results for the first 60 km of the traverse, but not for the remaining portion. Setting the ET rate to 20% at the coast, as shown in Figure 5b, fits the spread between the precipitation and surface water data better than a rate of 10% as shown in Figure 5a. However, neither setting of the model accommodates the majority of the traverse with an input value of –50 per mil for groundwater at the coast. The only parameter remaining unmodified is the hydrogen isotopic composition of groundwater at the coast. And we now vary that parameter to further improve the model fit.

The initial value of δD at the coast was changed to –70 per mil as shown in Figure 6a (–70, 2.5, 20%, 20%) and –90 per mil in Figure 6b (–90, 2.5, 20%, 20%). Both of these settings produce reasonable spreads between the precipitation and surface water data and better fit the data between 60 km and 400 km inland. The input value of –90 per mil at the coast produces a more reasonable fit than –70 per mil, although it appears that a larger advection velocity is then required. A higher advection velocity will produce less–depleted precipitation since the fraction of vapor taken by precipitation will be less, as shown in Figure 4.

The final solution is shown in Figure 7. After a series of experiments (not illustrated) with various advection velocities and settings otherwise similar to those used in Figure 6, we conclude that 3.4 m/sec yields the best overall fit. The δD of the modeled vapor at the coast, using this setting is about –180 per mil and at 400 km is –220 per mil. These data compare well to those of Ehhalt [1974] who reported δD values at several levels in the atmosphere for six stations. We chose all of his wet–season values below 2 km for Santa Barbara and Death Valley, California, the nearest comparable stations. The mean (± σ) δD values are 181 (±57) and 171 (±16) per mil respectively. Figure 7 (–50, 0.8, 20%, 20%) also shows results that accommodate the marked decrease in δD along the first 60 km of the traverse. This setting primarily differs by a lower advection velocity which decreases the precipitable water, resulting in increased distillation and isotopic fractionation, and appears to explain the

first 60 km of the traverse. These results suggest that two settings of the model are required to explain the entire traverse; the anomaly appears to be the marked decrease in the isotopic composition of groundwater and precipitation observed along the first 60 km of the traverse.

Discussion

Model calculations of rain–out and recycling along the air mass trajectory, computed at a 10 km grid spacing, provide a useful representation of the geographic variations of the hydrogen isotope compositions of inland meteoric waters. These results are not practical without spatially detailed estimates of monthly temperature and precipitation for the entire wet–season.

The hydrogen isotopic composition of modeled precipitation and groundwater cannot fit the observed data along the entire traverse with any single setting of the model. The best fit for the majority of the traverse renders a poor fit for the first 60 km. There may also be fine structure to the observed isotope data which is not explained by the climate mode. However, the best fit of the model (–90, 3.4, 20%, 20%) to most of the observed data could be obtained with adjustment of the rate of ET as well as advection velocity. We believe that the model presented provides insights to the hydrologic and isotopic balance as well as to the importance of ET recycling.

The model indicates that simple Rayleigh distillation of the eastward moving air mass results in excessive depletion of δD of inland meteoric waters (Figure 4) when the effects of ET are ignored (Figure 6). The observed geographic variation in the δD of groundwater can be explained when the effects of ET are combined with those of wintertime (wet–season) precipitation. Thus, ET, or partial system closure, must be occurring in the wintertime to some degree. This observation supports the interpretation made by Ingraham and Taylor [1991] that the traverses with linear δD vs. distance trends represent partially closed systems in which precipitation is partitioned between runoff and ET. However, without detailed sampling of an individual storm along its trajectory, it is unclear whether system closure occurs during storms or between storms.

The Coastal Anomaly

The δD value of –90 per mil for groundwater at the coast, used as the best fit setting (–90, 3.4, 20%, 20%) for most of the traverse, produces an unacceptable fit along the coastal portion between 0 km and 60 km inland. A setting of the model which accommodates the marked decrease in δD along the first 60 km of the traverse (–50, 1.0, 20%, 20%) does not produce a reasonable fit for the remainder of the traverse. The discrepancy of some 40 per mil in the δD value of groundwater at the coast between the best–fit setting and observed values is, as of yet, unexplained. The discrepancy may be due to the representation of atmospheric transport or ET recycling as discussed next.

The model assumes a constant velocity, both horizontally and vertically. However, windflow is more laminar over the ocean and frictional drag of topography and vegetation probably reduces the advection velocity at the coast. Pressure–driven divergence, together with spatial variations in environmental lapse rates, could raise the condensation level inland. In these cases, the controlling advection velocity would change and so would the computed supply of precipitable water. Ultimately this would control the distillation effect during precipitation.

ET is taken as a percentage of precipitation varying linearly over the traverse. The actual pattern of variation is more complex. In addition, transpiration is not considered separately in the model, and there may be commensurately more transpiration at the coast relative to the rest of the traverse. We believe that this effect is of secondary importance compared to the effects of changing advection velocity discussed above.

Conclusions

We have used a climate model to duplicate observed hydrogen isotopic compositions of meteoric water on a traverse across northern California. A good fit of the model to most of the data was obtained when the δD of groundwater at the coast was set at –90 per mil, the advection velocity was 3.4 m/sec., and ET was geographically constant at 20%. This setting rendered a good fit for the majority of the traverse, but a poor fit for the first 60 km. In addition, no solution with an acceptable fit was obtained when the stable isotopic effects of terrestrially evapotranspired and recycled water were ignored.

ET recycling, or partial system closure, appears to be occurring along the traverse as proposed by Ingraham and Taylor [1991]. However, without detailed sampling of an individual storm, along its trajectory, it is unclear whether system closure occurs during the storm or between storms.

Acknowledgements. This work was funded by the Nevada Agency for Nuclear Projects/Nuclear Waste Project Office under U.S. Department of Energy grant number DE–FG08–85NV10461. The opinions expressed in this paper do not necessarily represent those of the State of Nevada or the U.S. Department of Energy. The stable isotope analyses were performed by C. Shadel (Desert Research Institute, Las Vegas, Nevada). Climate boundary conditions for our calculations were kindly provided by John Stamm (Princeton University, Department of Civil Engineering and Operations Research).

REFERENCES

Court, A., The Climate of the Conterminous United States. In: R.A. Bryson and F.K. Hare, (Eds.), *World Survey of Climatology, Vol. II. Climates of North America.*, 1974.

Craig, R.G., and J.F. Stamm, *A Canonical Regression Model of Climates in the Southwest U.S.*, Desert Research Institute, Reno, NV, 71 p., 1989.

Craig, R.G., N.L. Ingraham, and J.F. Stamm, High–resolution Computation of Isotopic Processes in Northern California Using a Local Climate Model in *Proceedings of the Eight PACLIM Conference, Technical Report 31 of the Interagency Ecological Studies Program,* edited by K.T. Redmond, pp. 191–203, 1992.

Ehhalt, D.H., Vertical Profiles of HTO, HDO, and H2O in the Troposphere. *NCAR Technical Note NCAR–TN/STR–100,* 1974.

Friedman, I., A.C. Redfield, B. Shoem, and J. Harris, The variations of the deuterium content of natural waters in the hydrologic cycle, *Rev. Geophysics, 2,* 177–224, 1964.

Gat, J.R. and E. Matsui, Atmospheric Water Balance in the Amazon Basin: An Isotopic Evaporation Model. *Jour. Geophys. Res., 96,* 13179–13188, 1991.

Ingraham, N.L. and B.E. Taylor, Hydrogen isotope study of large–scale meteoric water transport in Northern California and Nevada, *Jour. of Hydrology, 85,* 183–197, 1986.

Ingraham, N.L. and B.E. Taylor, Light stable isotope systematics of large–scale hydrologic regimes in California and Nevada, *Water Resources Research, 27,* 77–90, 1991.

Kendall, C. and T. Coplen, Multi sample conversion of water to hydrogen by zinc for stable isotope determination. *Anal Chem., 57,* 1437–1440, 1985.

Majzoub, M. Fractionnement en oxygéne–18 et en deutérium entre l'eau et sa vapeur: *Jour. Chim. Phys., 68,* p. 1423–1436, 1971.

Rogers, R.R. and M.K. Yau, A Short Course in Cloud Physics, 3rd ed., *Pergamon Press, Oxford, 203* p., 1989.

Rozanski, K.., C. Sonntag, and K.O. Munnich, Factors controlling stable isotope composition of European precipitation, *Tellus, 34,* 142–150, 1982.

Salati, E., A. Dall'Olio, E. Matsui, and J.R. Gat, Recycling of water in the Amazon basin: an isotopic study, *Water Resources Research, 15,* 1250–1258, 1979.

Schutz, C. and W.L. Gates., Global Climate Data for Surface, 800 mb, 400 mb: January, Advanced Research Projects Agency, *R–915–ARPA* Order No. 189–1, Rand Corp., Santa Monica, CA, 1971.

Smith, G.I., I. Friedman, H. Klieforth, and K. Hardcastle, Areal distribution of deuterium in eastern California precipitation, 1968–1969, *Jour. of Applied Meteorology, 18,* 172–188, 1979.

Stamm, J.F. and R.G. Craig, A Climate Model for the Southwestern U.S. Suitable for Paleoclimatic Reconstructions (abstract), *Geological Society of America, Abstracts with Programs, 21–6,* 1989.

Wershaw, R.L., I. Friedman, S.J. Heller and P.A. Frank, Hydrogen isotopic fractionation of water passing through trees: In: G.D. Hobson and G.C. Speers (Eds.), *Advances in organic Geochemistry. Proc. of the Third International Congress,* Pergamon, Oxford, pp. 55–67, 1966.

Yonge, C.J., L. Goldberg, and H.R. Krouse, An isotopic study of water bodies along a traverse of Southwestern Canada, *Jour. of Hydrology, 106,* 245–255, 1989.

Zeigler, H., C.B. Osborn, W. Stichler and P. Trimborn, Hydrogen isotope discrimination in higher plants: correlation with photosynthetic pathways and environment, *Planta, 128,* 85–92, 1976.

N. Ingraham, Water Resources Center, Desert Research Institute, P. O. Box 19040, Las Vegas, Nevada 89132–0040.

R. Craig, Department of Geology, Kent State University, Kent, Ohio 44242–0001.

Isotopic Indicators of Climate in Ice Cores, Wind River Range, Wyoming

DAVID L. NAFTZ[1]

U.S. Geological Survey, Cheyenne, Wyoming

ROBERT L. MICHEL

U.S. Geological Survey, Reston, Virginia

KIRK A. MILLER

U.S. Geological Survey, Cheyenne, Wyoming

Snow and ice samples were collected from Gannett, Upper Fremont, and Knife Point Glaciers in the Wind River Range of Wyoming during 1989-90 and analyzed to evaluate the quality of climatic record present in the Wind River Range Glaciers. Seasonal $\delta^{18}O$ signals are present in snowpits from Upper Fremont and Knife Point Glaciers and varied by as much as 14°/oo annually. Based on the $\delta^{18}O$, chloride, and sulfate data in snow-pit samples, meltwater percolation does shift the isotopic and anion signals from winter values; however, the seasonal $\delta^{18}O$ trends within the snow were preserved through the initial summer melt cycle. Tritium concentration profiles from two ice cores (GAN-2 and GAN-4) collected from Gannett Glacier ranged from 0 to about 400 TU. In the GAN-2 core, no tritium is found below the main peak at the 5.8-m depth, indicating no intrusion of post-1953 water below this depth. Actual inventories of tritium at the site are much smaller than calculated inputs, with 270 TU-meters remaining at GAN-2 and 510 TU-meters at GAN-4. The tritium results indicate that about 94 percent of the precipitation deposited in the last 35 years has been lost. Seasonal isotopic signals are not preserved in core GAN-2; however, ice-core samples at depth appear to contain $\delta^{18}O$ values that are in agreement with calculated mean isotopic values of precipitation being deposited at the site.

INTRODUCTION

Continuous ice cores from the ice sheets of Greenland and Antarctica have provided important, long-term climatic information [Wagenbach, 1989]. Dansgaard et al. [1985] and Dansgaard and Oeschger [1989] have utilized oxygen isotope measurements in ice cores from Dye 3 and Camp Century sites in Greenland to reconstruct surface temperatures over the last 100,000 years or more. Similarly, Lorius et al. [1985] have used oxygen isotope measurements from the Vostok area in Antarctica to reconstruct estimates of surface temperatures over the past 150,000 years.

While it has been documented that paleoclimatic information can be recovered from polar sites, questions remain about the reliability of paleoclimatic information from non-polar, high-altitude locations. A'rnason [1981] suggested that under favorable conditions, temperate glaciers may be useful in providing short-term (about 2,000 years) climatic records. High-altitude sites on glaciers and ice caps from nonpolar locations in Kenya, Peru, China, Canada, and Switzerland have been sampled for paleoclimatic information. Because of the nonpolar locations of these sites, alteration of the snow and ice by seasonal meltwater can result in changes in the isotopic composition of the ice, thus biasing the historical climate record. For example, Thompson and Hastenrath [1981] and Thompson [1981] found a distinct smoothing of the $\delta^{18}O$ patterns with depth in two shallow ice cores collected from Lewis Glacier in Kenya that they attributed to meltwater percolation processes. Long-term climatic records have been reconstructed using oxygen isotopes from ice-core samples collected at high-altitude sites in Peru and China [Thompson et al., 1984, 1986, 1988a, 1988b, and 1989].

[1]Now at: U.S. Geological Survey, Salt Lake City, Utah

Climate Change in Continental Isotopic Records
Geophysical Monograph 78

A 103 m-snow/ice core from Mt. Logan, Canada has provided a 300-year climatic record [Holdsworth and Peak, 1985; Wagenbach, 1989]. The smoothed pattern of δ^{18}O variations in ice cores from the Colle Gnifetti site in Switzerland showed good correlation to summer air temperature [Wagenbach, 1989].

Until recently, the possible existence of historical climatic records in ice cores from the midlatitude glaciers in the Rocky Mountains of the continental United States have not been evaluated. In the continental United States, the Wind River Range in central Wyoming (Figure 1) has the largest total area of glaciers in the Rocky Mountains [Denton, 1975]. This large area of glaciers with ice-coring sites at altitudes higher than 4,000 m

above sea level makes this glacial system the best suited potential area for recovering a reliable ice-core record of climate variability for the Rocky Mountains within the continental United States. Results from a reconnaissance study of the Wind River Range Glaciers [Naftz et al., 1991a and 1991b] suggest that although meltwater alteration of the ice has occurred, historical climate records may be extractable from continuous, deep ice cores from glaciers in the range. The purpose of this paper is to determine the degree and magnitude of postdepositional alteration of δ^{18}O values in snow and ice-core samples and to evaluate the potential quality of the climatic record in the Wind River Range Glaciers.

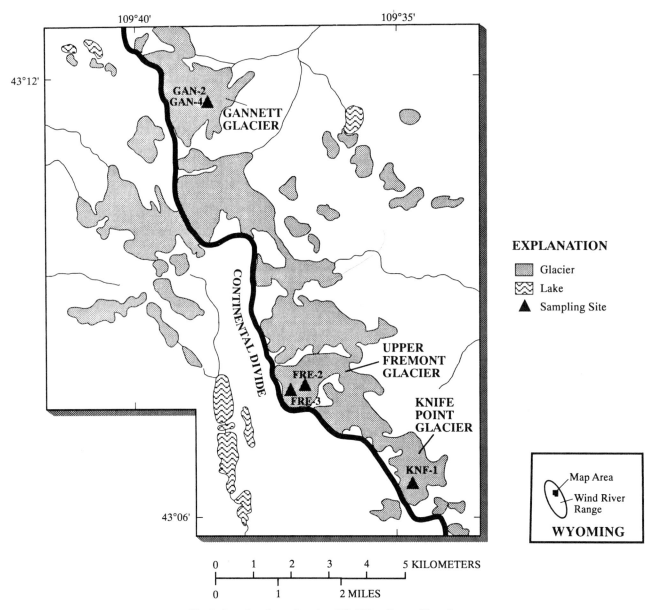

Fig. 1. Location of sampling sites, Wind River Range, Wyoming.

RESEARCH PLAN

Meltwater percolation and ablation in mid-latitude glaciers can modify significantly the $\delta^{18}O$ value and other chemical signals in ice-core and snow samples. Refreezing of the meltwater and rainwater percolating into underlying snow can dampen the seasonal isotopic signatures and cause a shift to less negative $\delta^{18}O$ values [Sharp et al., 1960; A'rnason, 1981].

The quality of the climatic record from three glaciers in the Wind River Range was evaluated based on snow and ice-core samples (Fig. 1). In 1990, three snowpits (FRE-2, FRE-3, and KNF-1) were dug and sampled on Upper Fremont and Knife Point Glaciers (Fig. 1). Snowpit altitudes ranged from 3,659 to 4,024 m above sea level. In 1989, two ice cores, designated GAN-2 and GAN-4, were obtained from Gannett Glacier at an altitude of 3,630 m above sea level. Core depths on Gannett Glacier were 9.8 m (GAN-2) and 13.8 m (GAN-4) below the glacier surface.

Snow pits were sampled, once during April 1990 and again in July 1990, to evaluate the consequences of the summer melt cycle on the snow chemistry before ice transformation. Profiles of $\delta^{18}O$ value were constructed from the snow pits to determine the seasonal signatures and to evaluate the effect of ice-melt processes by comparison with calculated values. Changes in the major-ion concentrations between the two sampling periods were also used to evaluate the effect of ice-melt processes.

Two ice cores were also sampled and chemically analyzed to evaluate longer-term effects of ice-melt processes on the climatic record. Tritium profiles from the ice cores were used to infer time lines in the ice. Tritium was also used to evaluate snow and ice loss from the site over the past 35 years by comparison to deposition patterns derived from precipitation networks. Calculated $\delta^{18}O$ values for precipitation falling at the sites were compared to measured values from the two ice cores to evaluate the effects of long-term ice-melt processes.

SAMPLING AND ANALYTICAL METHODS

Snow samples obtained from snowpits were collected from the pit face using non-powdered, latex gloves and an aluminum coring device. Samples were sealed in plastic bottles and transported frozen to the laboratory. Samples analyzed for specific conductance and major anions were placed in pre-rinsed plastic bags and also transported frozen to the laboratory. Snow density was measured using a hand-held snow density gauge. Snow temperature was measured using a digital thermometer with a J-type thermocouple.

Ice cores were obtained utilizing the Polar Ice Coring Office (PICO) hand-powered drilling system. After removal of the ice cores from the core barrel, they were placed in plastic bags, sealed, put in core tubes, and placed in snow vaults. Ice cores were transported from the field site on horseback using insulated coolers packed with dry ice.

In the laboratory, the ice cores were cut and prepared in a walk-in freezer. After cutting, each ice sample was completely scraped with a stainless steel microtome knife to remove surface contamination. Ten cm sections of the core were usually combined for analysis. To minimize sample contamination

during processing, ice and snow samples were placed in prerinsed and covered plastic containers and allowed to melt until a liquid film formed on the entire surface. The initial liquid melt was then used to rinse the outside of the samples and discarded. The remaining sample was allowed to melt and the meltwater was filtered (0.45 μm), bottled and preserved.

Concentrations of the major anions were determined by ion exchange chromatography [Fishman and Friedman, 1985] at the U.S. Geological Survey Laboratory in Doraville, Georgia. Tritium concentrations were determined by electrolytic enrichment/liquid scintillation counting [Thatcher et al., 1977] at the U.S. Geological Survey Tritium Laboratory in Reston, Virginia. The $\delta^{18}O$ isotopic values were determined in the U.S. Geological Survey Isotope Fractionation Project Laboratory in Reston, Virginia. The $\delta^{18}O$ isotopic value of the water samples was determined using a modification of the method developed by Epstein and Mayeda [1953]. $\delta^{18}O$ results are reported relative to Vienna Standard Mean Ocean Water (V-SMOW) in the °/oo notation. The two sigma uncertainty for $\delta^{18}O$ values is 0 ± 0.1°/oo.

Tritium samples were counted relative to a dilution of National Bureau of Standards sample standard #4926 (0.878×10^4 disintegrations per second on September 3, 1961). All samples were counted directly in Packard Tri-carb Scintillation systems as a mixture of 50 percent sample water and 50 percent picoflour, a scintillation cocktail made specifically for low-level tritium analyses. Sample blanks obtained from a deep well at Dulles International Airport, Virginia (tritium concentrations found to be less than 0.3 tritium units (TU)) were included with every sample set.

The standard reference materials used during analyses of the chloride and sulfate were U.S. Geological Survey standard reference water samples (SRWS) P-12 and P-13, both snowmelt standards. The mean concentration values for the undiluted reference materials were within one standard deviation of the accepted mean. Ten frozen method blanks were analyzed with the snow samples from Fremont Glacier; significant amounts of chloride or sulfate above the detection limit of 0.01 mg/L during sample processing were not present.

RESULTS AND DISCUSSION

Isotopic and major-ion variations in snow pits

Snowpits FRE-2 and FRE-3 (Fig. 1) were sampled in April 1990 and again in July 1990. Snowpit KNF-1 was sampled in April 1990. Temperature, density, $\delta^{18}O$ value, and chloride and sulfate concentration profiles were determined.

Snow temperature increased from less than 0°C during the April sampling to isothermal at 0°C during the July sampling period in pits FRE-2 and FRE-3 (Figs. 2 and 3). The density of the snowpack also increased from April to July (Figs. 2 and 3). As the snowpack becomes isothermal at 0°C, the potential exists for preferential leaching of solutes from the snow and subsequent modification of the original chemical signal during any melt cycles that may occur. Williams and Melack [1991] have found that melt-freeze cycles in the snowpack increase the concentration of solutes in meltwater.

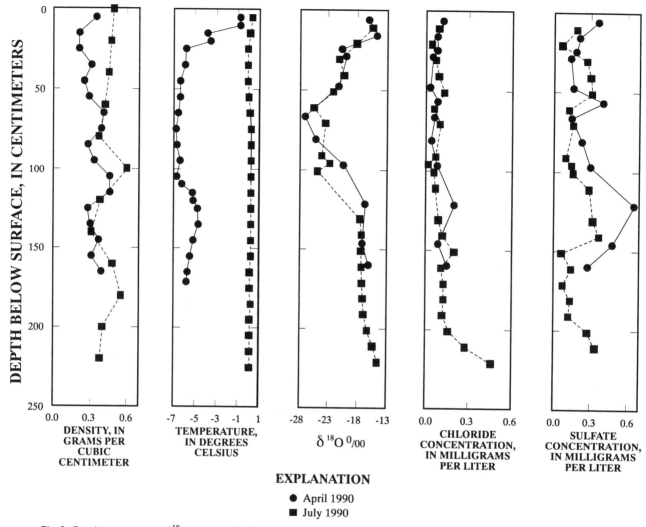

Fig. 2. Density, temperature, $\delta^{18}O$ values, and chloride and sulfate concentrations in samples from snowpit FRE-2, Upper Fremont Glacier, Wyoming.

Seasonal $\delta^{18}O$ signals appear to be present in all three snowpits (Figs. 2, 3, and 4). Based on the snowpit data, $\delta^{18}O$ values varied as much as 14°/oo annually. Near-surface samples from the snowpits (generally less than 50 cm below the surface) showed an enriched $\delta^{18}O$ signal, probably characteristic of spring and early summer precipitation. With depth, the isotopic signal becomes lighter and reaches a minimum at about 70 cm at site FRE-2, 135 cm at site FRE-3, and 170 cm at site KNF-1 (Figs. 2, 3, and 4). This minimum is probably representative of mid-winter snowfall and ranges from -28.55 at site FRE-3 to -23.00 at site KNF-1 (Figs. 3 and 4). The heavier mid-winter isotopic values observed from the KNF-1 site is probably reflecting the 365 m lower elevation of the KNF-1 site relative to the elevation of the FRE-3 site. After reaching a minimum, the $\delta^{18}O$ values begin to increase with depth at all three sites (Figs. 2, 3, and 4). This increase possibly represents late autumn and early winter snowfall.

The large seasonal variations observed in the snowpit data are consistent with the large temperature ranges observed in the measured air temperature at Upper Fremont Glacier. For the period from July 10, 1990, to July 28, 1991, the hourly mean air temperature ranged from -35.9°C to 13.3°C, a difference of 49.2°C. Large seasonal ranges in $\delta^{18}O$ values (14°/oo) have also been measured in precipitation falling on the mid-latitude glaciers in the Alps [A'rnason, 1981].

The $\delta^{18}O$ value in samples from snowpits FRE-2 and FRE-3 is slightly heavier in July compared to April. Median $\delta^{18}O$ values in samples from snowpits FRE-2 and FRE-3 show a 2.55 and 3.45°/oo shift to heavier isotopic compositions in the July samples compared to the April samples (Table 1). The observed shift to heavier isotopic values is probably the result of meltwater processes that preferentially remove the lighter oxygen isotopes, resulting in a shift to heavier isotopic values in the remaining snow. The shift to heavier isotopic values from

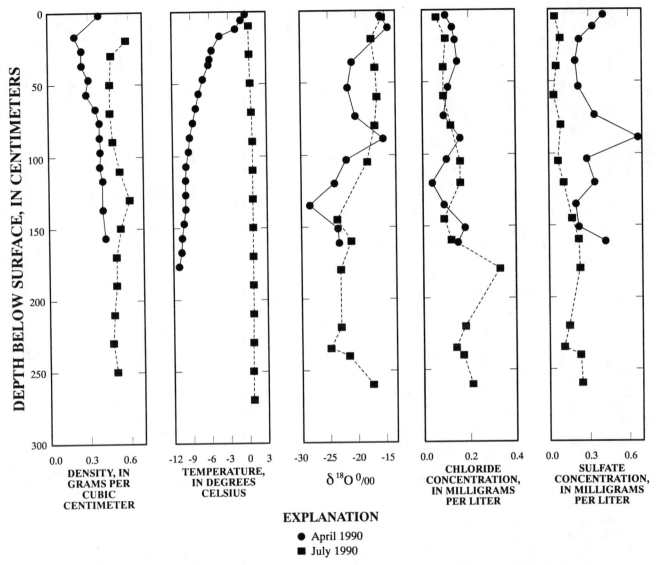

Fig. 3. Density, temperature, $\delta^{18}O$ values, and chloride and sulfate concentrations in samples from snowpit FRE-3, Upper Fremont Glacier, Wyoming.

EXPLANATION

● April 1990
■ July 1990

the April to July sampling period could also be caused by the accumulation of isotopically heavier snow during the late spring and early summer. Differentiation between the two processes is not possible with available data; however, the observed shift appears to be minor relative to the large isotopic variation in seasonal precipitation observed in the snowpit data (Figs. 2, 3, and 4). Continued meltwater leaching during firn-to-ice transitions could result in additional shifts to heavier isotopic values at depth.

The $\delta^{18}O$ values in Fremont Glacier snow samples were compared to the altitude-corrected mean $\delta^{18}O$ value determined from 11 surface-water samples collected from the Encampment River, Wyoming during 1985-87 [Tyler B. Coplen, U.S. Geological Survey, Reston, VA, writ. commun., 1991]. This site is in southern Wyoming, about 354 km southeast of Gannett

Glacier, and has a mean drainage basin altitude of 2,957 m above sea level [Lowham, 1988]. The mean $\delta^{18}O$ ratio of surface water within the Encampment River Basin was -17.8°/oo [Tyler B. Coplen, U.S. Geological Survey, Reston, VA, written commun., 1991].

In order to compare the Encampment River and ice-core isotopic data, the isotopic composition of the surface water within the Encampment River Basin was assumed to represent the isotopic composition of precipitation in the basin and has not been biased relative to precipitation values from evaporation and selective recharge or runoff. Using the range of global fractionation trends for $\delta^{18}O$ with altitude of 1.5 to 5.0°/oo/1,000 m [Yurtsever and Gat, 1981], the mean $\delta^{18}O$ value at the altitudes of sample sites FRE-2 (4,000 m), FRE-3 (4,024 m), and KNF-1 (3,659 m) were calculated (Table 1). The trends of $\delta^{18}O$ values

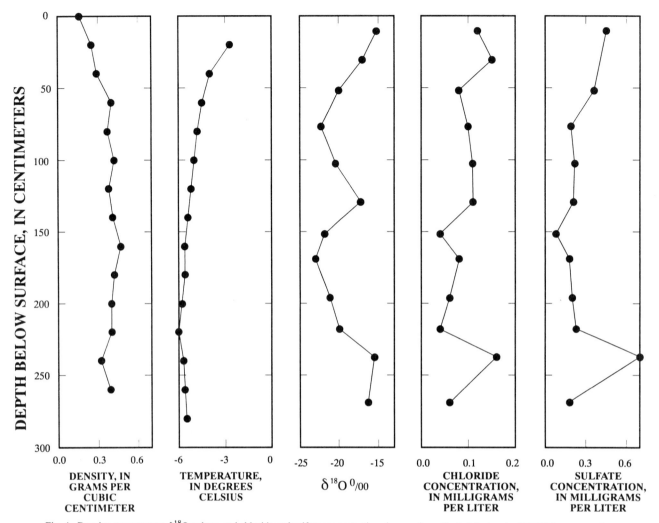

Fig. 4. Density, temperature, $\delta^{18}O$ values, and chloride and sulfate concentrations in samples collected during April 1990 from snowpit KNF-1, Knife Point Glacier, Wyoming.

with altitude calculated by Yurtsever and Gat [1981], were derived mainly from coastal data with the precipitation in the form of rain; however, these trends also agree with the altitude gradient of $5.0°/oo/1000$ m calculated by Sharp et al. [1960] with snow samples from the Blue Glacier in western Washington.

During April 1990 the median $\delta^{18}O$ ratio from all three sample sites fell within the calculated range of $\delta^{18}O$ values (Table 1). The median $\delta^{18}O$ ratio during July 1990 from sample sites FRE-2 and FRE-3 were more positive than the calculated range of $\delta^{18}O$ values by 1.91 and $1.25°/oo$ (Table 1). This observed shift to heavier isotopic values during July 1990 could suggest meltwater leaching of the snowpack during the summer melt cycle as well as accumulation of isotopically heavier snow between the April and July sampling periods. The smaller shift observed at site FRE-3 (Table 1) suggests that meltwater modification of the isotopic signature may be minimized at increased altitudes on Upper Fremont Glacier because of the decreased energy available for melting.

Another method was utilized to determine the most probable

isotopic composition of precipitation falling on Upper Fremont Glacier for comparison to the snow-pit data. Yurtsever and Gat [1981] have developed a series of regression equations that estimate the mean $\delta^{18}O$ value as a function of mean air temperature, precipitation, latitude, and altitude at a specific site. These equations are based on data collected from 91 network stations world-wide. On a global scale, variations in the mean isotopic composition at the network stations are mainly due to the temperature effect; however, on a regional scale the amount of precipitation can also control the isotopic composition [Yurtsever and Gat, 1981]. The equation used for the Wind River Range Glaciers included precipitation because of the regional effect of precipitation amount [Yurtsever and Gat, 1981]:

$$\delta^{18}O = -11.78 + (0.418*T) - (0.0084*P), \qquad (1)$$

where $\delta^{18}O$ is the mean value in $°/oo$, T is the mean air temperature in $°C$, and P is the mean precipitation in mm.

Temperature and precipitation values used in equation 1 were

TABLE 1. Median $\delta^{18}O$ Values in Snow-pit Samples from Sites Fre-2, Fre-3, and Knf-1 during April and July 1990
Compared to Calculated $\delta^{18}O$ Values

[$\delta^{18}O$ %oo relative to V-SMOW; N, number of samples; --, no data]

Sample site	Median $\delta^{18}O$				Calculated range of $\delta^{18}O$ values using Encampment River data	Calculated range of $\delta^{18}O$ values using Encampment River data
	April	N	July	N		
FRE-2	-20.00	11	-17.45	20	-19.36 to -23.02	-21.2
FRE-3	-21.60	11	-18.15	13	-19.40 to -23.14	-21.3
KNF-1	-19.95	12	--	--	-18.85 to -21.30	-19.8

estimated for the Upper Fremont Glacier site. Because of the remoteness of the Wind River Range Glaciers, long-term and site-specific temperature data are not available; however, a weather station was installed adjacent to Upper Fremont Glacier at an altitude of 3,960 m above sea level in July 1990. Air temperature was measured every minute and the hourly mean value was recorded. Annual mean air temperature at the site for the period July 11, 1990, to July 10, 1991, was -6.9°C. Annual mean temperatures used in equation 1 for sites FRE-2, FRE-3, and KNF-1 were calculated using the weather station data adjacent to Upper Fremont Glacier. The weather station data were altitude corrected for each site using a dry adiabatic lapse rate of 1°C/100 m. Mean annual precipitation for Upper Fremont Glacier is about 762 mm [Lowham, 1988].

The calculated mean $\delta^{18}O$ and the median $\delta^{18}O$ values in the three snowpits during April 1990 correlate quite well (Table 1). During July 1990, the median $\delta^{18}O$ values in snowpits FRE-2 and FRE-3 are heavier relative to the calculated $\delta^{18}O$ values (Table 1). This increase of the $\delta^{18}O$ value in the snow samples relative to the calculated values is probably the result of meltwater percolation in the snowpack during the summer melt season.

Median chloride and sulfate concentrations in April 1990 and July 1990 were also calculated to evaluate meltwater modification of snow chemistry (Table 2). The median chloride concentration at FRE-2 and FRE-3 showed a slight increase (0.01 - 0.02 mg/L) during July compared to April (Table 2). The reason for the increased chloride concentration in July is not apparent; however, the limited chloride leaching from the July snowpit data suggests a preservation of the chloride signal through a partial summer melt cycle. Naftz et al. [1991a] have found a similar preservation of chloride concentrations through

annual summer melt cycles in ice samples from Knife Point Glacier relative to chloride concentration in wet deposition.

Median sulfate concentrations show a significant decrease in July compared to April (Table 2). The median sulfate concentration decreased 0.10 mg/L at FRE-2 and 0.19 mg/L at FRE-3 from April to July 1990 (Table 2). The decrease in sulfate concentration during the summer melt period is consistent with preferential leaching of solutes from the snow. Although the absolute amount of sulfate decreased from April to July, the July sulfate profile at FRE-2 still shows a good correlation with the sulfate profile in April (Fig. 2). Naftz et al. [1991a] also found a significant (r = 0.95) correlation between mean annual sulfate concentration in ice samples from Knife Point Glacier compared to annual-weighted sulfate concentrations in wet deposition; however, the ice samples were also significantly depleted in sulfate relative to the wet deposition samples collected at significantly lower altitudes.

Based on recent monitoring data, strong orographic effects have been observed in the chemical composition of Wind River Range precipitation [A.F. Galbraith, U.S. Forest Service, Jackson, Wyo., written commun., 1991]. Sulfate has decreased in concentration with increasing altitude; therefore, the decreased sulfate concentration observed in the ice samples may partly be caused by orographic effects as opposed to meltwater leaching processes.

Based on the $\delta^{18}O$, chloride, and sulfate data in snow samples from Upper Fremont Glacier, meltwater percolation does shift the isotopic and sulfate signals; however, the seasonal $\delta^{18}O$ trends within the snowpack seem to be preserved through at least one summer melt cycle.

Isotopic variations in ice cores

The concentration of tritium in precipitation during the period of above-ground nuclear bomb testing varies by more than three orders of magnitude. Production of tritium by the testing of fusion weapons began in 1953 and reached a peak during the large nuclear bomb tests of 1962-63 [Carter and Moghissi, 1977] when concentrations attained several thousand TU. The rapid changes in concentration provide a unique time marker for snow and ice deposition [Koide et al., 1979; 1982]. As an isotope of hydrogen and a part of the water molecule, tritium is not subject to problems that affect other tracers, but follows the pathway of water movement in the environment. Age estimates and deposition rates can be obtained by comparing tritium

TABLE 2. Median Chloride and Sulfate Concentrations in Snowpit Samples from Sites Fre-2, Fre-3, and Knf-1 during April and July 1990 [Chloride and sulfate concentrations in milligrams per liter; N, number of samples; --, no data

Sample site	Chloride				Sulfate			
	April	N	July	N	April	N	July	N
FRE-2	0.08	12	0.10	21	0.26	12	0.16	21
FRE-3	0.12	12	0.13	14	0.30	12	0.11	14
KNF-1	0.09	12	--	--	0.22	12	--	--

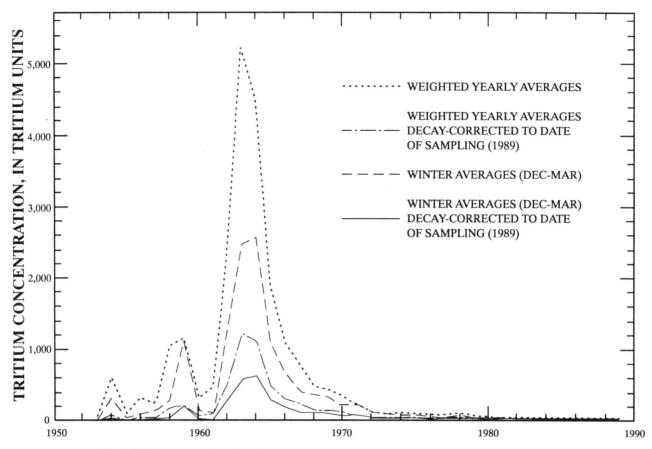

Fig. 5. Tritium concentrations in precipitation, Gannett Glacier, Wyoming (calculated from Michel [1989]).

concentrations in the ice cores to deposition patterns derived from precipitation networks. Tritium profiles in undisturbed ice cores show concentration spikes during the above ground nuclear-bomb-testing years, and little or no tritium for ice in the deeper (pre-bomb) portion of the core.

Tritium concentrations in precipitation are also influenced by seasonal and geographical effects. Due to stratospheric mixing, concentrations are larger in spring than winter, and concentrations increase with distance from the coast and with altitude [Michel, 1989]. Figure 5 shows the estimated tritium concentrations in precipitation for the winter months (December-March) and the full year [based on Michel, 1989], both at the time of deposition and decay corrected to 1989, for the latitude and longitude of Gannett Glacier. The winter average lies well below the yearly average for most years due to seasonal variations. The winter average will be used in this discussion since most snow deposition in the Wind River Range occurs during the winter.

The average winter (age-corrected to 1989) tritium concentrations resulting from snow deposition during the peak years (1963-64) would be about 650 TU in 1989 (Fig. 5). Age-corrected tritium concentrations decreased rapidly to less than 200 TU by 1966 and about 100 TU by 1969. Another significant

peak occurs in 1959 and ice from this year would have an age-corrected tritium concentration of about 200 TU in 1989.

Profiles of tritium concentrations versus depth for GAN-2 and GAN-4 ice cores are shown in Figure 6. Two distinct peaks were observed in the GAN-2 core with the largest tritium concentration being approximately 200 TU at 5.8 m below the surface. This is lower than the tritium concentration of 650 TU predicted by the decay-corrected winter deposition curve. The smaller concentration may be the result of post-depositional alteration of the tritium peak or the original concentrations may have been smaller than predicted. Tritium concentrations in the ice below the main peak are at or below the detection limit (8 TU), indicating very little post-1960's water has penetrated this zone.

In the GAN-4 core, there are several spikes throughout the core with the largest concentrations at a depth of 13 m. This pattern is suggestive of a core where intrusion of younger snow and ice into older and deeper snow and ice layers has occurred. The likelihood of this possibility is reinforced by the observation of physical differences in the ice in the deep part of the core where tritium concentrations are large. The intrusion of clear ice between layers of bubbly ice is observed at several locations within the core and coincide with tritium spikes in concentration. Because of logistical constraints, the GAN-4 core was drilled

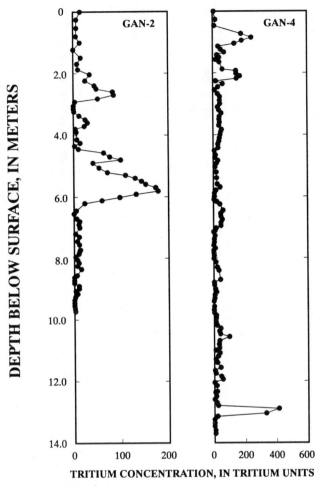

DEPTH BELOW SURFACE, IN METERS

TRITIUM CONCENTRATION, IN TRITIUM UNITS

Fig. 6. Tritium-concentration profiles in ice cores GAN-2 and GAN-4, Gannett Glacier, Wyoming.

tritium at GAN-4 is due to the movement of younger snow and water into older and deeper layers, where it could not be lost as easily by postdepositional processes. The tritium data indicate about 94 percent of the precipitation deposited in the last 35 years has been lost by runoff, ablation, evaporation, or wind redeposition.

Based on the significant loss in tritium concentration observed in the Gannett Glacier ice samples, meltwater percolation could have a major effect on the $\delta^{18}O$ isotopic values in the ice cores. In order to evaluate the extent of meltwater percolation on the Gannett Glacier ice cores, profiles of $\delta^{18}O$ values in cores GAN-2 and GAN-4 were constructed (Fig. 7). The $\delta^{18}O$ values in the ice cores were compared to the altitude-corrected mean $\delta^{18}O$ value determined from 11 surface-water samples collected from the Encampment River Basin site, as discussed previously.

The $\delta^{18}O$ values in samples from the GAN-2 core generally fall within the calculated range of $\delta^{18}O$ values in precipitation (Fig. 7). Small-scale seasonal isotopic variations are not visible in the GAN-2 core (Fig. 7). The mean $\delta^{18}O$ value for the samples collected from the GAN-2 core is -19.9°/oo (n = 27), which is within the calculated range of the isotopic composition of precipitation at the 3,600 m altitude on Gannett Glacier. Based on the $\delta^{18}O$ values in the GAN-2 profile, it appears that even with a significant tritium loss in the core, a realistic isotopic signature may be preserved within parts of the Wind River Range Glaciers.

Comparison of the $\delta^{18}O$ profile from core GAN-4 to the calculated range in $\delta^{18}O$ values showed significant meltwater leaching of the ice (Fig. 7). The mean $\delta^{18}O$ value for the samples collected from the GAN-4 core is -18.6°/oo (n=63). Hence, most $\delta^{18}O$ values in the GAN-4 profile are heavier than the M = -18.8°/oo upper range calculated for Gannett Glacier precipitation.

Comparison of the GAN-2 and GAN-4 specific conductance and isotope data show a similar feature at approximately the same depth in each core. In the GAN-2 core at 5.7 m below the surface there is a large conductance peak (larger than 10 μs/cm), which coincides with light $\delta^{18}O$ values (less than -22.0°/oo) (Fig. 7). A similar feature is shown in the GAN-4 profile at 5.2 m below the surface with a large conductance peak (larger than 10 μS/cm) and a light (less than -22.0°/oo) $\delta^{18}O$ value (Fig. 7). In the GAN-2 core this feature is about 0.1 m above the 1963-64 tritium maximum, equating to an approximate age of 1965. According to data compiled by Thompson and Love [1988], the mean summer temperature during 1965 in Pinedale, Wyoming (located about 40 km southwest of the study area at an altitude of 2190 m above sea level) was the lowest mean summer temperature recorded during 1915-86. The low summer temperatures during this period may have minimized meltwater leaching of the snow deposited during the 1964-65 season, thereby preserving the large specific conductance. The preservation of this feature in both cores since 1965 suggest that anomalous climatic signals may be preserved in the ice for longer than 30 years since the time of deposition.

Equation 1 (Yurtsever and Gat, 1981) was also used to further evaluate the effects of meltwater percolation on the isotopic

below a steep ice fall region of Gannett Glacier that contained numerous crevasses. The distribution of numerous tritium peaks throughout the core may be due to the closing of crevasses with younger snow as the glacier moved down gradient, causing the occurrence of post-bomb tritium signals adjacent to older ice. Based on the tritium profile in GAN-4, any trends observed in the other chemical constituents, such as stable isotopes and major ions, must be interpreted with extreme care. The water with a tritium concentration of about 400 TU in the GAN-4 core can only have come from the 1963-65 period as no other period could produce such large tritium concentrations.

Total deposition of tritium at the site from 1953-83 was about 17,300 TU-meters. This deposition is much more than in the surrounding areas due to the high precipitation rates at the site [Lowham, 1988]. By 1989, much of the original tritium would have been lost by decay and, if no snow or ice was lost by runoff, ablation, evaporation, or wind redeposition, about 4,300 TU-meters of tritium would remain. Actual inventories at the site are much lower, with 270 TU-meters (6 percent) at GAN-2 and 510 TU-meters (12 percent) at GAN-4. The higher retention of

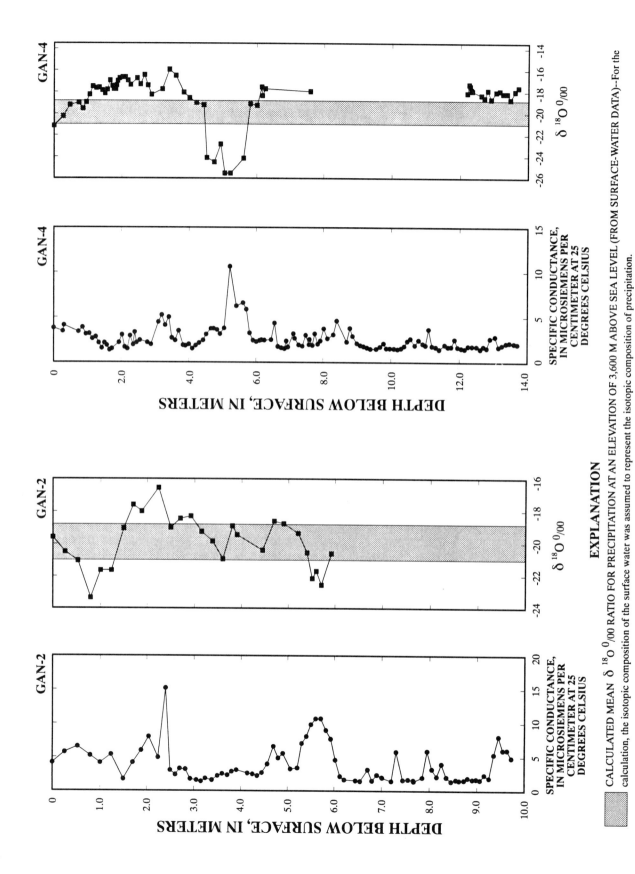

Fig. 7. Specific conductance and δ¹⁸O values from ice cores GAN-2 and GAN-4, Gannett Glacier, Wyoming

values in the Gannett Glacier ice cores. Annual mean air temperature for the drill site on Gannett Glacier was estimated to be -3.3°C. Mean annual precipitation for the Gannett Glacier site was estimated to be 762 mm [Lowham, 1988]. Using these temperature and precipitation values in equation 1, an annual mean $\delta^{18}O$ value of -19.6°/oo was calculated for the Gannett Glacier site. This annual mean isotopic value compares favorably with the range of -18.8 to -21.0°/oo, calculated using the Encampment River, Wyoming, data. The mean $\delta^{18}O$ ratio of -19.9°/oo (n = 27) in the upper 6.0 m of the GAN-2 ice core is very close to the calculated annual mean isotopic value of -19.6°/oo using equation 1. In addition, the mean $\delta^{18}O$ value from the GAN-2 ice core is similar to the range in median $\delta^{18}O$ values of -21.60 to -19.95°/oo in snow samples collected from Upper Fremont and Knife Point Glaciers before meltwater leaching (Table 1).

Although the mean $\delta^{18}O$ value of the GAN-2 ice core is similar to the calculated isotopic values for precipitation falling at the site, this does not agree with the observed 2.55 to 3.45°/oo enrichment of snow samples from Upper Fremont Glacier during one annual melt cycle (Table 1). Based on the results of the snow and ice-core samples, it appears that high resolution climatic information cannot be obtained from the ice cores; however, lower resolution climatic information may be obtained for deep ice cores.

The potential climatic record in the Wind River Range Glaciers may be similar to records present in warm glaciers in the Eurasian Arctic as observed by Ferronsky et al. [1991]. In these glaciers, meltwater-caused changes in the mean and seasonal $\delta^{18}O$ values were observed; however, long-term, lower resolution trends appeared to be preserved and useful in providing paleoclimatic information in geographical diverse areas [Ferronsky et al., 1991].

SUMMARY

Snow and ice samples were collected from Gannett, Upper Fremont, and Knife Point Glaciers in the Wind River Range of Wyoming. Changes in the physical and chemical properties of the snow on Upper Fremont Glacier during 1990 were observed to evaluate the extent and severity of meltwater modification of the snowpack chemistry. Snow temperature increased from less than 0°C during April to isothermal at 0°C during July. Seasonal $\delta^{18}O$ signals are present in both snowpits from Upper Fremont Glacier and varied by as much as 14°/oo annually. The large seasonal variations observed in the snowpit data are consistent with the large air temperature ranges of -35.9°C to 13.3°C measured hourly at Upper Fremont Glacier during 1990-91. Median $\delta^{18}O$ values in samples from both snowpits showed a 2.55 and 3.45°/oo increase in the July samples compared to the April samples. The observed shift to heavier isotopic values is probably the result of meltwater processes that preferentially removed the lighter oxygen isotopes. Median $\delta^{18}O$ values in snow samples collected during April agreed with the calculated values. Median values from both pits sampled during July increased relative to the calculated values. Based on the $\delta^{18}O$, chloride, and sulfate data in snow samples from Upper Fremont Glacier, meltwater percolation does shift the isotopic and anion

signals from winter values; however, the seasonal $\delta^{18}O$ trends within the snowpack seem to be preserved at least through the first summer melt cycle.

Ice cores from Gannett Glacier were collected from depths of 9.8 m (GAN-2) and 13.8 m (GAN-4) below the glacier surface. Tritium concentration profiles for the cores ranged from 0 to about 400 TU. In the GAN-2 core, no tritium is found below the main peak at the 5.8-m depth, indicating no intrusion of post-1953 water below this depth. In the GAN-4 core, there are several peaks throughout the core with the largest concentrations at 13 m. The distribution of numerous tritium peaks throughout the core may be due to the closing of crevasses with younger snow as the glacier moved down gradient, causing the occurrence of post-bomb tritium signals adjacent to older ice. By 1989, much of the original tritium deposited on Gannett Glacier would have been lost by decay and if no snow or ice was lost by runoff, ablation, evaporation, or wind redeposition, about 4,300 TU-meters of tritium would remain. Actual inventories of tritium at the site are much smaller, with 270 TU-meters remaining at GAN-2 and 510 TU-meters at GAN-4. The results indicate that about 94 percent of the precipitation deposited in the last 35 years has been lost.

The mean $\delta^{18}O$ value for the samples collected from the GAN-2 core was -19.9 (n = 27), which falls within the calculated range of the isotopic composition of precipitation on Gannett Glacier. Although seasonal isotopic signals are not preserved in core GAN-2, ice samples at depth appear to contain $\delta^{18}O$ values that are in agreement with calculated mean isotopic values for precipitation occurring at the site. Comparison of the $\delta^{18}O$ profile from core GAN-4 to the calculated range in $\delta^{18}O$ values (-18.8 to -21.0°/oo) for precipitation falling at the site, showed an increase relative to the upper range calculated for Gannett Glacier precipitation. Based on the results of the snow and ice-core samples it appears that high resolution climatic information cannot be obtained from ice cores in the Wind River Range; however, lower resolution climatic information may be available.

Acknowledgements. Funding for this study was provided by the Shoshone and Arapaho Tribes, Wyoming Water Development Commission, and Wyoming State Engineer. Equipment and advice provided by L.G. Thompson of the Byrd Polar Research Center were appreciated. Logistical support provided by the National Outdoor Leadership School and Hawkins and Powers Aviation during the field work was outstanding. Critical reviews by C. Drummond, K.C. Lohmann, and C. Kendall improved the manuscript significantly.

REFERENCES

A'rnason B., Ice and snow hydrology, in *Stable Isotope Hydrology: Deuterium and Oxygen-18 in the Water Cycle, IAEA Tech. Rep. Ser. 210,* edited by J.R. Gat and G. Gonfiantini, 143-175, 1981.

Carter, M.W., and A.A. Moghissi, Three Decades of Nuclear Testing, *Health Physics,* 33, 55-71, 1977.

Dansgaard, W., H.B. Clausen, N. Gundestrup, S.J. Johnsen, and C. Rygner, Dating and climatic interpretation of two deep Greenland ice cores, in *Greenland Ice Core: Geophysics, Geochemistry, and the*

Environment, AGU Monograph 33, edited by C.C. Langway,   Jr., H. Oeschger, and W. Dansgaard, 71-76, American Geophysical Union, Washington, D.C., 1985.

Dansgaard, W. and H. Oeschger, Past environmental records from the Artic, in *The Environmental Record in Glaciers and Ice Sheets,* edited by H. Oeschger and C.C. Langway, Jr., 287-318, John Wiley, New York, 1989.

Denton, G.H., Glaciers of the American Rocky Mountains, in *Mountain Glaciers of the Northern Hemisphere,* vol. 1, edited by W.O. Field, pp. 509-602, Cold Regions Research and Engineering Laboratory, Hanover, N.H., 1975.

Epstein, S. and T. Mayeda, Variation of the O-18 content of waters from natural sources, *Geochim. Cosmochim. Acta.,* 4, 213-224, 1953.

Ferronsky, V.I., A.D. Esikov, L.S. Vlasova, Yu.A. Karpychev, V.A. Polyakov, R.A. Vaikmae, and S.V. Ferronsky, Isotope studies of groundwater, ice cores, organic material, lake sediments, and permafrost ice-veins of Holocene-Pleistocene time over the USSR Continental Area (abs.), in *Proceedings of the AGU Chapman Conference on Continental Isotopic Indicators of Climate, Jackson, Wyoming,* 1991.

Fishman, M.J., and L.C. Friedman, Methods for determination of inorganic substances in water and fluvial sediments, *Tech. of Wat.-Res. Invest. of the U.S. Geol. Surv.,* book 5, Chapter A1,   672-688, 1985.

Holdsworth, G., and E. Peak, Acid content of snow from a mid-troposphere sampling site on Mount Logan, Yukon Territory, Canada, *Ann. Glac.,* 7, 153-160, 1985.

Koide, M., R. Michel, E.D. Goldberg, M.M. Herron, and C.C. Langway, Depositional history of artificial radionuclides in the Ross Ice Shelf, Antarctica, *Earth and Planetary Science Letters,* 44, 205-223, 1979.

Koide, M., R. Michel, E.D. Goldberg, M.M. Herron and C.C. Langway, Characterization of radioactive fallout from pre and post moratorium tests to polar glaciers, *Nature,* 296, 544-547,   1982.

Lorius, C., J. Jouzel, C. Ritz, L. Merlivat, N.I. Barkov, Y.S. Korotkevich, and V.M. Kotlyakov, A 150,000 year climatic record from Antarctic ice, *Nature,* 316, 591-596, 1985.

Lowham, H.W., Streamflows in Wyoming, *U.S. Geol. Surv. Wat.- Res. Invest. Rept. 88-4045,* 1988.

Michel, R.L., Tritium deposition over the continental United States, 1953-1983, in *International Association of Hydrological Sciences, Publication no. 179, Atmospheric Deposition,* edited by J. W. Delleur, pp. 109-115, Oxfordshire, United Kingdom, 1989.

Naftz, D.L., J.A. Rice, and J.R. Ranville, Glacial ice composition: A potential long-term record of the chemistry of atmospheric deposition, Wind River Range, Wyoming, *Water Resour. Res.,* 27, 1231-1238, 1991a.

Naftz, D.L., K.A. Miller, and R.B. See, Using glacial ice cores from Wyoming as long-term records of atmospheric deposition quality and climate change--A progress report, *Hydata News and Views,* 10, 23-23, July 1991b.

Sharp, R.P., S. Epstein, and I. Vidziunas, Oxygen-isotope ratios in the Blue Glacier, Olympic Mountains, Washington, U.S.A., *Journ. Geophys. Res.,* 65, 4043-4060, 1960.

Thompson, C., and C.M. Love, Reconnaissance survey: Trace metals concentration in Wind River glaciers, Wyoming Water Research Center Unnumbered Report, 36 pp., 1988.

Thompson, L.G., Ice core studies from Mt. Kenya, Africa, and their relationship to other tropical ice core studies, in *Proceedings of the Canberra Symposium, IAHS Publication No. 131,* 55-62, 1981.

Thompson, L.G., and S.L. Hastenrath, Climatic ice core studies at Lewis Glacier, Mount Kenya, *Zeitschrift fur Gletscherkunde und Glazialgeologie,* 17, 115-123, 1981.

Thompson, L.G., E. Mosley-Thompson, and B.M. Arnao, El Nino-Southern Oscillation events recorded in the stratigraphy of the tropical Quelccaya Ice Cap, Peru, *Science,* 226, 50-53, 1984.

Thompson, L.G., E. Mosley-Thompson, W. Dansgaard, and P.M. Grootes, The Little Ice Age as recorded in the stratigraphy of the tropical Quelccaya Ice Cap, *Science,* 234, 361-364, 1986.

Thompson, L.G., M.E. Davis, E. Mosley-Thompson, and K-b. Liu, Pre-Incan agricultural activity recorded in dust layers in two tropical ice cores, *Nature,* 336, 763-765, 1988a.

Thompson, L.G., E. Mosley-Thompson, X. Wu, and Z. Xie, Wisconsin/ Wurm glacial stage ice in the subtropical Dunde Ice Cap, China, *GeoJournal,* 17, 517-523, 1988b.

Thatcher, L.L., V.J. Janzer, and K.W. Edwards, Methods for determination of radioactive substances in water and fluvial sediments, *Tech. of Wat.-Res. Invest. of the U.S. Geol. Surv.* book 5, Chapter A5, 79-81, 1977.

Thompson, L.G., E. Mosley-Thompson, M.E. Davis, J.F. Bolzan, J. Dai, T. Yao, N. Gundestrup, X. Wu, L. Klein, and Z. Xie, Holocene-Late Pleistocene climatic ice core records from Qinghai-Tibetan Plateau, *Science,* 246, 474-477, 1989.

Wagenbach, D., Environmental records in alpine glaciers, in *The Environmental Record in Glaciers and Ice Sheets,* edited by H. Oeschger and C.C. Langway, Jr., 69-83, John Wiley, New York, 1989.

Williams, M.W. and J.M. Melack, Solute chemistry of snowmelt and runoff in an alpine basin, Sierra Nevada, *Water Resour. Res.,* 27, 1575-1588, 1991.

Yurtsever, Y. and J.R. Gat, Atmospheric waters, in *Stable Isotope Hydrology: Deuterium and Oxygen-18 in the Water Cycle, IAEA Tech. Rep. Ser. 210,* edited by J.R. Gat and G. Gonfiantini, 103-142, 1981.

David L. Naftz, U.S. Geological Survey, 1745 West 1700 South, Salt Lake City, UT 84104

Robert L. Michel, U.S. Geological Survey, 12201 Sunrise Valley Drive, MS 432, National Center, Reston, VA 22092

Kirk A. Miller, U.S. Geological Survey, 2617 E. Lincolnway, Suite B, Cheyenne, WY 82001

Influence of Climate on the Formation and Isotopic Composition of Calcretes

VICTOR ROSSINSKY, JR.[1] AND PETER K. SWART

Marine Geology and Geophysics, Rosenstiel School of Marine & Atmospheric Sciences,
University of Miami, 4600 Rickenbacker Causeway, Miami FL 33149

The control exerted by climate on calcrete formation throughout the Caribbean region is expressed as a change in the oxygen isotopic composition of calcretes along the climatic gradient between south Florida and the Turks and Caicos Islands, British West Indies. Along this gradient, an oxygen isotopic trend is documented as an increase in $\delta^{18}O$ from south Florida to the Turks and Caicos. This is attributed mainly to higher temperatures and faster evaporation rates in the Turks and Caicos Islands. Isotopic compositions of meteoric waters have not been measured directly, but may play a secondary role in the observed isotopic trend in calcretes.

INTRODUCTION

Numerous past studies have focused on the influence of climate on different aspects of calcrete formation. In a most general sense, formation of calcretes depends on the presence of a seasonal climate which alternates between wet and dry conditions [Goudie, 1973; Harrison, 1977; Boucot *et al.*, 1982]. Morphologies and physical make-up of calcretes have been studied along climatic gradients (mostly rainfall and evaporation), and differences documented [Harrison, 1977; Semeniuk and Searle, 1985; Semeniuk, 1986] in terms of degree of development and completeness (presence of diagnostic horizons) of the calcrete profile [Semeniuk and Searle, 1985].

Rainfall is one of the most commonly cited climatic parameters which governs calcrete formation [Harrison, 1977; Semeniuk and Searle, 1985], however, calcretes have been reported from areas with rainfall as little as 300 mm/year [Goudie, 1973]. Furthermore, soil scientists believe that $CaCO_3$ precipitation is an ongoing process in desert soils [Schlesinger, 1985]. On the other hand, Semeniuk and Searle (1985) suggested, based on field observations, that calcrete development in southwestern Australia is restricted to areas with rainfall exceeding 800 mm/year. This discrepancy can be partly explained by invoking evaporation, which is another climatic parameter instrumental in calcrete formation. A balance between rainfall and evaporation is crucial for calcretes to form so that absolute amounts of rainfall *per se* are not as important as the interplay between the two [Harrison, 1977].

In the last ten years or so, numerous studies have related carbon and oxygen isotopic signatures of calcretes to climatic conditions [Cerling *et al.*, 1977; Talma and Netterberg, 1983; Cerling, 1984; Cerling and Hay, 1986; Amundson *et al.*, 1988]. A good correlation exists between the type of flora (C-3 or C-4) present during calcrete formation and the $\delta^{13}C$ of calcrete [Cerling, 1984]. A correlation between oxygen isotopic signatures of calcretes and meteoric waters and evaporation has been shown to exist to a certain degree [Salomons *et al.*, 1978; Cerling, 1984].

Precipitation of calcretes are controlled three processes: 1) degassing of CO_2 from precipitating waters, 2) evaporation, and 3) evapotranspiration [Salomons and Mook, 1976; Salomons *et al.*, 1978; Cerling, 1984]. The original carbon isotopic composition of the fluid is determined by the relative CO_2 contributions of various components which are: 1) atmospheric (-6 °/oo) [Cerling and Hay, 1986], 2) dissolved carbon from preexisting limestone, 3) the ratio of C-3 (-24 to -34 °/oo PDB) [Smith and Epstein, 1971] and C-4 plants (-6 to -19 °/oo) [Salomons *et al.*, 1978], and 4) plant root respiration rates [Cerling, 1984]. During inorganic precipitation, degassing of CO_2 commonly only affects the $\delta^{13}C$ signature because $^{12}CO_2$ is preferentially lost during molecular diffusion [Cerling, 1984]. As CO_2 equilibrates rapidly with water relative to the rate of calcium carbonate precipitation, CO_2 degassing does not affect the oxygen isotopic composition of calcretes which primarily depends on the $\delta^{18}O$ of the precipitating fluid and its temperature. Therefore, 3 main variables affect the ultimate $\delta^{18}O$ of calcretes: 1) the original $\delta^{18}O$ of meteoric waters (rainfall, groundwater), 2) $\delta^{18}O$ enrichment of waters during evaporation prior to calcium carbonate precipitation, and 3) temperature. The major problems in interpreting oxygen isotopic compositions of calcretes can be summed up by the lack of knowledge of 1) rainfall isotopic composition, 2) amount of evaporation prior to calcrete precipitation and 3) process of calcium carbonate precipitation (CO_2 degassing or H_2O evaporation).

The purpose of this study is to (1) document isotopic variations in

---

[1]CRB Geological and Environmental Services, Inc., 2600 Douglas Road, Suite 602, Coral Gables FL 33134.

Climate Change in Continental Isotopic Records
Geophysical Monograph 78

calcretes from different settings, (2) investigate any correlations with climatic gradients, and (3) identify specific climatic parameters which can be singled out as the main causal variables for $\delta^{18}O$ variability.

SETTINGS

Pedogenic surficial calcretes were studied from six localities spanning across a gradient from south Florida to the Turks and Caicos Islands. Below are brief descriptions of the individual localities:

Caicos

The Caicos Platform comprises the southernmost platform of the Bahamian chain of platforms (Fig. 1). The northern margin is rimmed by a continuous chain of Pleistocene islands [Wanless et al., 1989]. The western and eastern margins are rimmed by fewer islands. The platform itself measures about 70 km by 100 km and is, thus, relatively small compared to Bahamian standards. Based on morphostratigraphic principles [Garret and Gould, 1984], the islands on Caicos Platform record several stages of island growth during the Pleistocene by ridge accretion which continued during the Holocene

[Wanless and Rossinsky, 1986; Wanless et al. 1989]. The southern portions of the northward facing islands consist of tidal flats [Wanless et al., 1989].

Calcretes occurring on topographic highs are petrographically and geochemically different from those occurring in topographic lows [Rossinsky and Wanless, 1992]. In both cases, the parent limestone is primarily an oolitic grainstone with minor amounts of skeletal allochems.

Bahamas

Great Exuma. Great Exuma island is located at the southwestern extent of Exuma Sound. The island is mainly comprised of eolian ridges which dip bankward in a westerly direction [Ball, 1967].

Gun Key. Gun Key is located just south of Bimini at the western margin of Great Bahama Bank (Fig. 1). The island consists of sets of ridges comprised of oolitic and skeletal grainstones. These ridges seem to be of eolian origin, based on the presence of *Cerion sp.*, gastropod molds, and detrital gypsum grains within ridge sequences. Calcretes occur on the Pleistocene limestone surface as discontinuous patches because of dissolution and karst development.

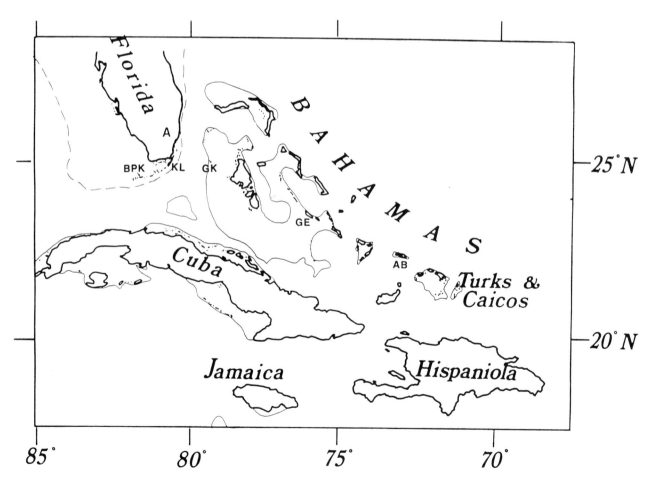

Figure 1: Map of south Florida and the Bahamas showing the location of the areas studied. Areas studied are south Florida, and the more arid Turks and Caicos Islands with stations of intermediate climate, Key Largo, Big Pine Key, Gun Key, Bahamas, Great Exuma, Bahamas, Bahamas, and Abraham Bay, Bahamas.

South Florida. South Florida is readily divisible into two separate climatological areas based on annual precipitation; the wetter mainland and the drier Florida Keys [Schomer and Drew, 1982]. Peninsular south Florida is a partly inundated carbonate platform.

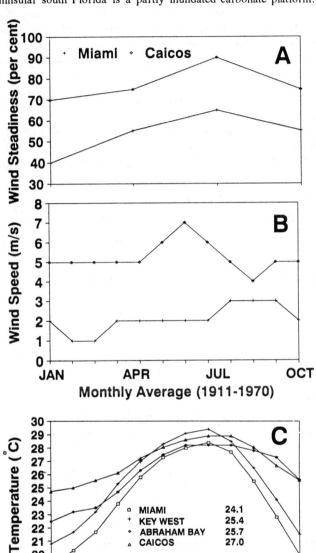

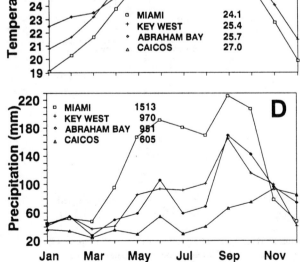

Continental influences are minor and consist of terrigenous influx of quartz sand via longshore drift [Wanless, 1976], and minor clays. The south Florida carbonate platform is separated from the Bahamas by the Straits of Florida. Calcretes from the mainland of south Florida were sampled from the Anastasia formation which occurs approximately 100 km north of Miami (Fig. 1).

The Florida Keys are an arcuate chain of islands at the southern tip of south Florida extending from Soldier Key to Key West, a distance of approximately 150 miles. They are divided into the upper keys which consist of the coralline Key Largo Limestone and the lower keys whose bedrock is comprised of the oolitic Miami limestone [Coniglio and Harrison, 1983]. The calcretes investigated in this study have been collected from Key Largo (upper keys) and Big Pine Key (lower keys) (Fig. 1).

CHARACTERIZATION OF CLIMATE

In order to characterize various climatic parameters along the Florida - Caicos climatic gradient, four stations have been selected for which climatic records are available. They are: 1) Miami (25° 48' N; 80° 16' W), 2) Key West (24° 33' N; 81° 45' W), 3) Abraham Bay (22° 22' N; 75° 28' W), and 4) Grand Turk (21° 26' N; 71° 08' W) (Fig. 1). Precipitation and rainfall have been compiled for the above stations over a period of 10 years from 1951 to 1960 (U.S. Dept. of Commerce, World Weather Records 1951-1960). Temperature and rainfall data for Grand Turk were compiled from World Wide Airfield Summaries conducted by the Naval Weather Service, a defunct government agency. The duration of the Airfield Summaries spanned 11 years, however, the exact years of the compilation are not known. Compilations from Grand Turk and the other three stations, thus, probably overlap to some degree and are comparable. No other climatic records from the Turks and Caicos Islands (except 1951 and 1952) seem to be published.

Winds, wind steadiness, specific humidity, and cloudiness data were taken from Hastenrath and Lamb (1961). These records span the period from 1911 to 1970. A comparison of those parameters is only made between Florida and Turks and Caicos Islands.

CLIMATIC PARAMETERS

Four aspects of climate, wind speed, wind steadiness, temperature, and precipitation are summarized in Figure 2 A-D. The data is presented as monthly averages over a 10 year period. Miami and Caicos are the only two stations taken into account for the first four parameters. For temperature and precipitation, Abraham Bay (Mayaguana Island, Bahamas) and Key West are considered in

Figure 2: (A) Plot of average monthly wind speed in m/sec for Miami and Caicos compiled between 1911 and 1970 [Hastenrath and Lamb, 1977]. (B) Plot of average quarterly wind steadiness in per cent for Miami and Caicos compiled between 1911 and 1970 [Hastenrath and Lamb, 1977]. (C) Plot of monthly average temperature in degrees Celsius compiled between 1951 and 1960 for Miami, Key West, Abraham Bay, and Caicos. Total averages for that time period are given in parentheses below station names (World Weather Records). (D) Plot of monthly average precipitation in mm compiled between 1951 and 1960 for Miami, Key West, Abraham Bay, and Caicos. Total averages for that time period are given in the upper left hand corner (World Weather Records).

addition to Miami and Caicos. Little variation exists in total cloudiness between Miami and Caicos averaging approximately 5 tenths for both locations. Specific humidity is recorded during January, April, July, and October with slightly higher humidities occurring in Caicos, except for July.

The climatic gradient between Caicos and south Florida manifests itself most dramatically in terms of wind regime, temperature, and precipitation (Figs. 2A, B, C, and D). These factors result in a higher evaporation rate in Caicos compared to south Florida. Wind speed, on a monthly basis (Fig. 2A), and wind steadiness (Fig. 2B), on a quarterly basis, are significantly higher in Caicos than in Miami. Monthly temperatures are shown in Figure 2C for all four stations. During the summer months there are minor differences, while during the winter months Caicos and Abraham Bay have significantly higher temperatures than south Florida. Key West has consistently higher

temperatures than Miami all year. Extreme differences can be seen in precipitation among all four stations (Fig. 2D). By far the highest precipitation occurs in Miami except for November and December. Lowest rainfall occurs in Caicos; 605 mm/year, which amounts to 40 % of Miami's average annual rainfall (1512.9 mm/year). Key West and Abraham Bay are intermediate at 970.4 and 951.0 mm/year, respectively. During the time that these measurements were taken, south Florida as well as the Caribbean Sea were affected by tropical cyclones below normal frequency [Dunn et al., 1967].

METHODS

Isotopic Analysis

Carbon and oxygen isotopic analyses were measured on CO_2 extracted from powdered calcrete samples after digestion in H_3PO_4 at $90^\circ C$ [Swart et al., 1991]. All samples were analyzed using a

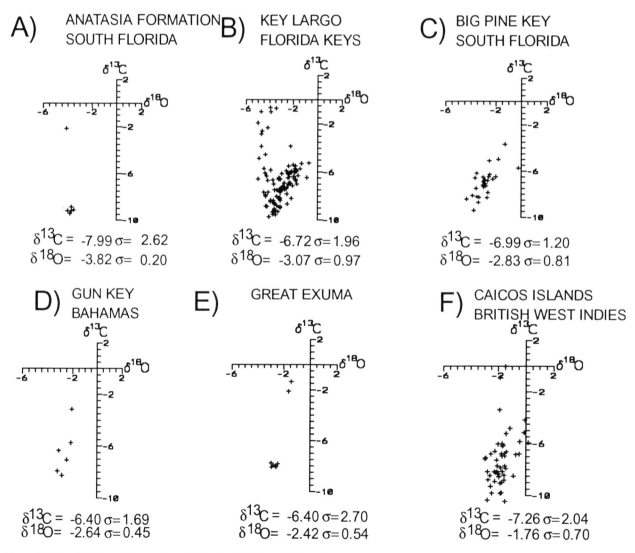

Figure 3: Plots of carbon and oxygen isotopic data for (A) Anastasia Formation, south Florida, (B) Key Largo, Florida Keys, (C) Big Pine Key, south Florida (D) Gun Key, Bahamas, (E) Great Exuma, Bahamas, and (F) Caicos. The $\delta^{13}C$ and $\delta^{18}O$ means and standard deviations are given below each graph.

Finnigan-MAT 251 mass spectrometer at the Stable Isotope Laboratory of the University of Miami. All data are corrected for conventional contributions to mass 45 and 46 following the procedure of Craig (1957), modified for a triple collector instrument and are reported relative to PDB. The external reproducibility determined by the analysis of 20 replicate samples of laboratory standard is 0.03 °/oo for oxygen and 0.02 per mil for carbon.

RESULTS

Carbon and Oxygen Isotopic Signatures in Calcretes

Approximately 400 calcrete samples were analyzed for their stable carbon and oxygen isotopic compositions. Crossplots of oxygen and carbon isotopic compositions of calcretes from individual stations are shown in Figure 3 A-F. In general, calcretes from all locations exhibit a wide range of $\delta^{13}C$ and a narrower range of $\delta^{18}O$. The standard deviation of $\delta^{18}O$ from different sites, however, varies significantly. The average calcrete composition from the Anastasia Formation (Fig. 3A) is most depleted in $\delta^{13}C$ by at least 1 °/oo compared to all other stations (except Caicos). Oxygen isotopic compositions of calcretes from the Anastasia formation show the greatest depletion in $\delta^{18}O$, a mean value of -3.82 °/oo, compared to calcretes from Caicos at the other end point of the climatic spectrum, with an average $\delta^{18}O$ composition of -1.76.

Calcretes from Key Largo, which occur predominantly in coralline limestone of the upper Florida Keys (Fig. 3B), exhibit large variations in both carbon ($\delta^{13}C$ = -6.72 °/oo) and oxygen ($\delta^{18}O$ = -3.07; σ = 0.97) isotopes. These values are similar to those obtained from Big Pine Key calcretes ($\delta^{13}C$ = -6.99; σ = 1.20 and $\delta^{18}O$ = -2.83; σ = 0.81) occurring on oolitic limestone (Fig. 3C)

Isotopic data from Gun Key (Fig. 3D) are similar to those of Great Exuma (Fig. 3E). Both sites have calcretes with a mean $\delta^{13}C$ of -6.40 °/oo although the standard deviation is considerably greater at

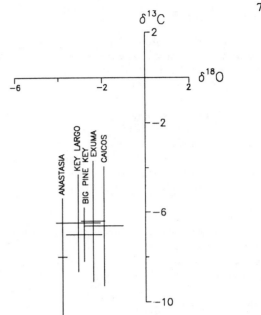

Figure 5: Plot of mean $\delta^{13}C$ and $\delta^{18}O$ and standard deviations for calcretes from Anastasia formation, Big Pine Key, Caicos, Key Largo, and Great Exuma.

Great Exuma (σ = 2.70°/oo) than at Gun Key (σ = 1.69°/oo). Oxygen isotopic data from these two sites are also similar ($\delta^{18}O$ = -2.64°/oo, σ = 0.45°/oo and $\delta^{18}O$ = -2.42°/oo, σ = 0.54°/oo) for Gun Key and Great Exuma, respectively. The calcretes from Caicos are the least depleted in $\delta^{18}O$ ($\delta^{18}O$ -1.76°/oo; σ = 0.70) (Fig. 3F). Valley calcretes and ridge calcretes [Rossinsky and Wanless, 1992] have been plotted separately in Figure 4. Whereas $\delta^{13}C$ is similar, $\delta^{18}O$ shows a significant difference of 0.86 °/oo in mean $\delta^{18}O$-values between the two calcrete types.

Mean carbon and oxygen isotopic values for all stations with their respective standard deviations are shown in Figure 5. The average carbon and oxygen isotopic signatures for all sites have been plotted with respect to the climatic gradient which exists between Caicos and

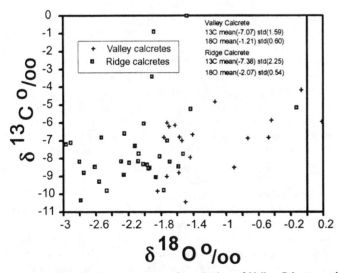

Figure 4: Plot of carbon and oxygen isotopic data of Valley Calcretes and Ridge Calcretes on Caicos islands. $\delta^{13}C$ and $\delta^{18}O$ means and standard deviations for both calcrete types are given in the upper right hand corner of the graph.

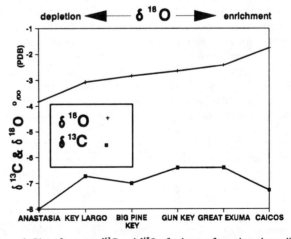

Figure 6: Plot of average $\delta^{13}C$ and $\delta^{18}O$ of calcretes from sites along climatic gradient between Caicos and the Anastasia formation in south Florida. A general depletion in $\delta^{18}O$ occurs along this climatic gradient from Caicos to south Florida and agrees generally with compositional trends observed in meteoric waters from this area. $\delta^{13}C$ shows no trend between Caicos and south Florida.

south Florida (Fig. 6). Although there is no correlation between the carbon isotopic signature and geographic location, the $\delta^{18}O$ shows a variation of approximately 2 °/oo between Caicos and south Florida.

DISCUSSION

Origins of Isotopic Variations

The absolute and relative variation in the oxygen isotopic composition from Caicos to S. Florida is a function of at least four parameters: 1) Mechanism of precipitation, 2) composition of precipitating fluids (evaporative waters), 3) temperature of precipitating fluids, 4) composition of rainfall.

Mechanism of Precipitation

The mechanism of precipitation (CO_2 loss, evapotranspiration, evaporation) will determine whether the calcretes are $\delta^{18}O$ enriched, or if they reflect values of the local meteoric waters. If calcretes precipitate as a result of evaporation rather than CO_2-degassing or evapotranspiration, the precipitating fluids will become enriched in ^{18}O through fractionation. If calcretes precipitate as a result of CO_2 degassing or evapotranspiration, processes which aren't accompanied by oxygen fractionation [Cerling, 1984], $\delta^{18}O$ of the waters will remain relatively constant, and compositional differences may be attributable to variations in rainfall $\delta^{18}O$, temperature, or the composition of the precipitating fluids.

Composition of Precipitating Fluids

Evaporation can be clearly demonstrated to take place during or prior to calcrete precipitation in the Caicos Islands. Much of the $\delta^{18}O$ variation displayed in the Caicos calcretes can be explained by differences between evaporative processes which take place in fluids precipitating calcretes in topographically low versus topographically high areas. Calcretes forming in these varied topographic settings have been termed valley calcretes and ridge calcretes, respectively [Rossinsky and Wanless, 1992]. Figure 4 shows that valley calcretes are distinguishable from ridge calcretes based on their $\delta^{18}O$ signature. The impermeable nature of ridge surface calcretes inhibits meteoric waters from penetrating into the underlying, porous grainstone, enhancing rapid surface runoff. On ridges, therefore, calcretes precipitate from fluids which evaporate significantly less than in valleys. Meteoric waters tend to accumulate in the valleys where they are subject to evaporation and, thus, become enriched in ^{18}O. Consequently, valley calcretes are 0.86 per mil heavier than ridge calcretes. This enrichment, however, can also be the result of a temperature difference of 3.5 °C between the precipitating fluids (Fig. 7). Although temperature differences between ridge and valley waters are to be expected, a difference of 3.5 °C in vadose water seems excessive. Therefore, evaporation must at least in part be responsible for the isotopic difference of 0.85 per mil between fluids precipitating valley and ridge calcretes.

Evaporation may be variable in response to local phenomena as is the case in Caicos where valley calcretes are enriched in $\delta^{18}O$ by 0.86 per mil relative to ridge calcretes (Fig. 4). However, even within valley calcretes and ridge calcretes the variability is considerable ($\sigma = 0.6$ and 0.54 respectively). This variability is probably a result of variations in the oxygen composition of rainfall. Variability of individual rainfall compositions are probably faithfully retained in the vadose zone, whereas calcretes precipitating in

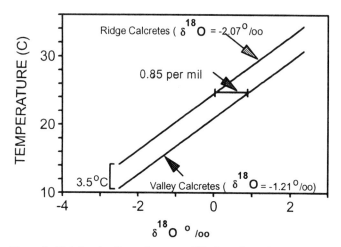

Figure 7: Plot showing lines of constant $\delta^{18}O$ for valley calcretes and ridge calcretes precipitated at different temperatures and/or oxygen isotopic composition of precipitating fluids. Average compositional differences between valley calcretes and ridge calcretes can be accounted for by a temperature difference of 3.5 degrees Celsius or 0.85 per mil difference in the oxygen isotopic composition of the precipitating fluids. The observed difference between valley calcretes and ridge calcretes is probably a combination of the two.

phreatic waters will have a more average composition. The $\delta^{18}O$ variability in calcretes between Key Largo and Big Pine Key (0.24 per mil) is for the most part a result of rain variability as local differences in evaporation are unlikely to be significant in the absence of topography. As calcretes probably precipitate in response to individual rainfall events, the entire calcrete deposit should reflect an average oxygen isotopic composition of the precipitating fluids.

Temperature

The oxygen isotopic composition of precipitated calcium carbonate is temperature dependent. Average temperatures at the different sites (Fig. 2E) are well constrained, and it is reasonable to assume that temperatures of soil and vadose waters are comparable to those of the atmosphere. However, it is not known whether the annual average temperature can be assumed to be representative of periods of calcrete precipitation. If calcretes precipitate as a result of CO_2 partial pressure, precipitation may occur preferentially during winter months and at night, when plant respiration rates and temperature are at their lowest. Calcrete precipitation favored during annual temperature maxima or minima can significantly influence the oxygen isotopic composition of the precipitated phase.

Oxygen Isotopic Compositions of Meteoric Waters

Mean annual $\delta^{18}O$ values of precipitation worldwide have been compiled by Yurtsever and Gat (1981) (Fig. 8). As a result of station sparsity on the globe, many values have to be extrapolated. In this study, average annual rainfall for south Florida and the northern Bahamas has been assumed to be -2.7 °/oo SMOW [Swart *et al.*, 1989], -2.5 °/oo SMOW for the southern Bahamas, and -2.3 °/oo SMOW for Caicos. Salomons *et al.* (1978) argued that the $\delta^{18}O$ composition of the infiltrated waters from which calcretes precipitate depends on the average precipitation. However, in south Florida,

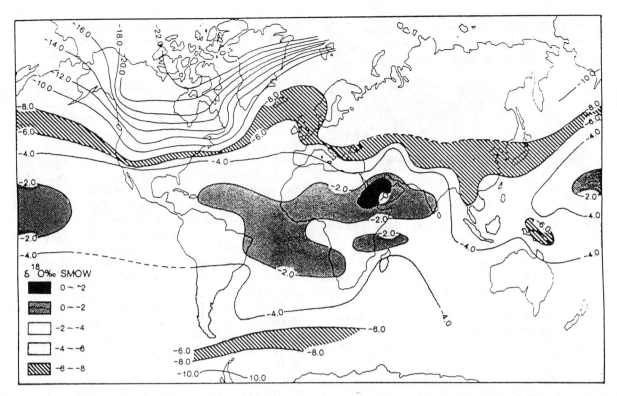

$\delta^{18}O‰$ SMOW

■	0 ~ +2
▨	0 ~ -2
□	-2 ~ -4
▤	-4 ~ -6
▨	-6 ~ -8

Figure 8: Map showing distribution of mean $\delta^{18}O$ of precipitation (based on stations having at least 2 years of record) [from Yurtsever and Gat, 1981].

groundwater has been measured to be as much as 3 °/oo heavier than rainfall [Meyers, 1990], depending on the infiltration rate of rainwater into the ground.

Meteoric water can also vary in its isotopic composition depending on its origin. In south Florida, precipitation can be derived locally from the ocean or the Everglades. Rain can also be derived from larger storm systems when south Florida happens to lie in their migratory path. The oxygen isotopic composition in all of the above cases varies significantly. Rain water in Miami has been shown to be of a highly variable nature (Swart, unpublished data). At any one locality, however, it can be assumed that waters in the phreatic zone have a fairly constant oxygen isotopic (not necessarily the same as an individual rainfall) composition which represents the average meteoric composition. In contrast, waters in the vadose zone can be highly variable, reflecting compositions of individual rainfalls. Vadose water isotopic composition can also be expected to evolve rapidly if conditions are favorable for evaporation. As calcretes precipitate in the vadose zone, they are apt to incorporate the oxygen isotopic variability of individual rainfalls reflected in the vadose waters, which may help explain the observed variability at any one locality.

INTER-STATION VARIABILITY

A gradual but consistent depletion in the average $\delta^{18}O$ of calcretes occurs from Caicos to south Florida. Average oxygen isotopic compositions of calcretes have been plotted as solid triangles in Figure 9. Plotting almost parallel above, are the calculated average

oxygen isotopic compositions of the precipitating fluids (solid circles). These have been calculated using known documented annual average temperatures and average calcrete compositions from corresponding locations. Thus, the vertical offset between the calculated values of the precipitating fluids and the measured composition of the calcretes in Figure 9 is a function of temperature.

A third line, plotted as solid squares (Fig. 9), defines the estimated average $\delta^{18}O$ of rain at each location. These estimates are based on extrapolations from Yurtsever's and Gat's (1981) compilations on average world-wide meteoric water compositions. The line defining the estimated average composition of rain can be seen to converge sharply with the line defining the calculated average of the precipitating fluids towards south Florida (Fig. 9). The oxygen isotopic difference between these two lines represents the amount of compositional change which rain water has to undergo to reach the isotopic composition of the precipitating fluids. The degree of compositional change can be seen to diminish from Caicos to south Florida and suggests that evaporation plays a primary role in the evolution of the precipitating fluids. At each station, the composition rain evolves by evaporation along the cross-hatched pattern (Fig. 9) and becomes enriched in heavy oxygen. Higher evaporation, necessary to achieve a larger compositional change in Caicos, is consistent with the observed climatic data.

TIMING OF CALCRETE FORMATION

Age constraints on calcretes in a climatic framework, are among the most fundamental prerequisites for a successful study. Cerling

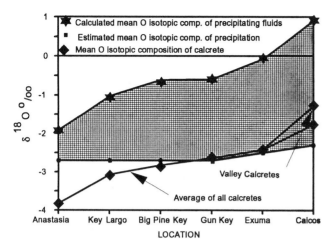

Figure 9: Plot of $\delta^{18}O$ values for average calcrete, estimated average of rainfall, and calculated average of precipitating fluids along the climatic gradient between Caicos and south Florida. The latter has been calculated based on temperature records and measured $\delta^{18}O$ in calcretes. As the Calculated Average $\delta^{18}O$ of Precipitating Fluids and the Average Calcrete $\delta^{18}O$ are nearly parallel, temperature cannot account for the $\delta^{18}O$ gradient measured in calcretes. Evaporation must be responsible for the most part in the observed ^{18}O enrichment towards Caicos. The estimated averages of $\delta^{18}O$ chosen for rainfall along indicated stations of the climatic gradient between south Florida and Caicos are reasonable in that they allow for more evaporation (cross hatched field) in Caicos than in south Florida. Differences in evaporation between the two is strongly supported by climatological data such as temperature and wind speed and steadiness.

(1984) cites the work of Salomons *et al.* (1978) as being only partially successful in showing a correlation between calcrete $\delta^{18}O$ and meteoric water $\delta^{18}O$ because of a lack of age constraints on calcretes. He argued that present-day isotopic compositions of meteoric water may not reflect past compositions from which the calcretes may have precipitated. The time in question spanned 1 million years. In another study, Cerling and Hay (1986) have shown that at Olduvai near the African East Rift Valley for any given temperature of formation, the total range in calculated $\delta^{18}O$ (SMOW) values has had a range of 8 °/oo over the last 2 million years. In this study, the time span in which calcretes may have formed is much narrower. Within this time frame it is assumed that meteoric waters did not vary to a great extent. Rossinsky and Wanless (1992) and Rossinsky *et al.* (1992) documented that the Caicos calcretes are young, perhaps forming at the present time. Comparable field criteria in the Florida keys are also present in mottled calcrete which exhibit unequivocal present-day vegetative associations. Comparisons between isotopic signatures in Caicos calcretes and south Florida calcretes only then have value if the calcretes in question are of the same ages.

Robbin and Stipp (1979) have obtained radiocarbon dates from laminated calcretes on Key Largo and Big Pine Key and concluded that their formation occurred during the last 5000 to 6000 years. The youngest dates obtained were 400 ± 70 and 270 ± 80 ybp for Key Largo and Big Pine Key respectively. This would indicate that modern Caicos calcretes and south Florida calcretes are

contemporaneous. Any geochemical differences would therefore be related, at least in part, to climatic differences.

SUGGESTIONS FOR FUTURE ISOTOPIC WORK IN RELATION TO CLIMATE

The above is a simplistic model which relies only on two more or less well known parameters - temperature and calcrete $\delta^{18}O$. Three parameters need further study to be able to relate diagenetic products to waters of $\delta^{18}O$ and climatic conditions in general. These are: 1) the actual temperature during calcrete precipitation (as discussed above diurnal or even seasonal preference for calcrete precipitation can change oxygen compositions by several per mil); 2) amount of evaporation at individual sites; and 3) a demonstrable mechanism for calcrete precipitation (evaporation or CO_2-degassing). With these parameters constrained, the average meteoric water composition can be determined quite easily from the isotopic composition of the precipitating fluid by back calculating evaporation.

In the rock record these parameters are more difficult to determine. However, petrographic, mineralogic, and paleontologic (especially palynology) evidence may be used to quantify evaporation, temperature, and perhaps even mechanisms of precipitation. Of great interest would be the establishment of meteoric water compositions during cycles of glaciation in the Pleistocene, a time period for which the quantification of the above mentioned critical parameters may be possible.

The primary and foremost objective, however, should be to test out this model by measuring meteoric waters directly. Because calcretes are demonstrably forming in Caicos today, this would be the good place to start acquiring quantitative measurements on critical parameters outlined above and testing results of the model against actual meteoric water oxygen compositions.

CONCLUSIONS

Isotopic differences between calcretes from the two climatic endpoints (Caicos and south Florida) in this study can be attributed to differences in climatic conditions. This conclusion is tentative, and more work is needed for additional supporting evidence.

A gradual enrichment in ^{18}O along the climatic gradient between south Florida and Caicos is likely to be related to warmer and drier climatic conditions and thus higher evaporation rates in the Turks and Caicos Islands. Differences in the isotopic composition of meteoric waters are not established but probably play a secondary role.

Variations in climatic parameters considered of primary importance in determining the oxygen isotopic composition of precipitated calcretes can occur over short periods of time. To establish paleoclimatic gradients based on the oxygen composition of calcretes, care must be taken to compare signatures in calcretes of same ages.

Topography can play a key role in the oxygen isotopic signature of calcretes. On Caicos, valley calcretes have been shown to be enriched in ^{18}O relative to ridge calcretes due to differing evaporation regimes within the two settings. When comparing calcretes from different climatic settings with regard to their oxygen isotopic composition, great care must be taken to compare calcretes from the same topographic setting.

Although the carbon isotopic signatures in calcretes have been shown to be climate-dependent based on the distribution of C-3 versus C-4 plants [Cerling, 1984], the following discussion will be restricted mainly to $\delta^{18}O$. Too little is known about the photosynthetic pathways of indigenous flora at the different sites as well as rates of respiration and organic decomposition, in order to relate carbon isotopic signatures to climate in a sensible way.

ACKNOWLEDGEMENTS

The authors wish to thank H. Wanless for helpful discussions. Members of the Stable Isotope Laboratory are thanked for help with analyses. Analyses were partially supported by UNICAL, EPR, GCAGS, and AAPG. The purchase of the stable isotope mass spectrometer has been supported by NSF grants to PKS EAR-9018882, EAR-8618727, and EAR-8417424.

REFERENCES

Amundson, R.G., Chadwick, O.A., Sowers, J.M., and Doner, H.E., 1988, Relationship between climate and vegetation and the stable carbon isotope chemistry of soils in the eastern Mojave Desert, Nevada, *Quaternary Research*, 29, 245-254, 1988.

Ball, M.M., Carbonate sand bodies of Florida and the Bahamas, *Jour. Sed. Petrology*, 37, 556-591, 1967.

Boucot, A.J., Gray, J., Fang, R-S., Yang, X-C., Li, Z-P., and Zhang, N., Devonian calcrete from China: its significance as the first Devonian calcrete from Asia, *Can. Jour. Earth Sci.*, 19, 1532-1534, 1982.

Cerling, T.E., The stable isotopic composition of modern soil carbonate and its relationship to climate, *Earth Planet. Sci. Lett.*, 71, 229-240, 1984.

Cerling, T.E., and Hay, An isotopic study of paleosol carbonates from Olduvai Gorge, *Quaternary Research*, 25, 63-78, 1986.

Cerling, T.E., Hay, R.L., and O'Neil, J.R., Isotopic evidence for dramatic climatic changes in East Africa during the Pleistocene, *Nature*, 267, 137-138, 1977.

Coniglio, M., and Harrison, R.,S., Holocene and Pleistocene caliche from big Pine Key, *Florida, Bull. Can. Petrol. Geol.*, 31, 3-31, 1983.

Craig, H., Isotopic standards for carbon and oxygen and correction factors for mass-spectrometric analysis of carbon dioxide, *Geochim. Cosmochim. Acta*, 12, 133-149, 1957.

Dunn, G.E., and Staff, Florida Hurricanes, ESSA Technical Memorandum WBTM SR-38 Fort Worth, Texas, 1967.

Garrett, P., and Gould, S.J., Geology of New Providence Island, Bahamas, *Geol. Soc. America Bull.*, 95, 209-220, 1984.

Goudie, A.S., *Duricrusts in tropical landscapes*, Oxford, Clarendon Press, 1973.

Harrison, R.S., Caliche profiles: indicators of near surface subaerial diagenesis, Barbados, West Indies, *Bull. Can. Petrol. Geol.*, 25, 123-396, 1977.

Hastenrath, S., and Lamb, P.J., *Climatic Atlas of the Tropical Atlantic and Eastern Pacific Ocean*, University of Wisconsin Press, 1961.

Meyers, J., *Stable isotope hydrology and diagenesis in the surficial aquifer system, southern Florida Everglades*, (Unpublished M.Sc. thesis), University of Miami, 1990.

Robbin, D.M., and Stipp, J.J., Depositional rate of laminated soilstone crusts, Florida Keys, *Jour. Sed. Petrology*, v. 49, 1979, p. 175-180.

Rossinsky, V., JR., and Wanless, H.R., Topographic and vegetative controls on calcrete formation, Turks and Caicos Islands, British West Indies, *Jour. Sed. Petrology*, 62, 84-98, 1992.

Rossinsky, V., JR., Wanless, H.R., and Swart, P.K., Penetrative calcretes and their stratigraphic implications, *Geology*, 20, 331-334, 1992.

Salomons, W., and Mook, W.G, Isotope geochemistry of carbonate dissolution and reprecipitation in soils, *Soil Science*, 122, 15-24, 1976.

Salomons, W., Goudie, A., and Mook, W.G., Isotopic composition of calcrete deposits from Europe, Africa and India, *Earth Surface Processes*, 3, 43-57, 1978.

Schlesinger, W.H., The formation of caliche in soils of the Mojave Desert, California, *Geochim. Cosmochim. Acta*, 49, 57-66, 1985.

Schomer, N.S., and Drew, R.D., An ecological characterization of the lower Everglades, Florida Bay and the Florida keys, U.S. Fish and Wildlife Service, Off. Biological Services, Washington, D.C., FWS/OBS-82/58.1, 1982

Semeniuk, V., Calcrete breccia floatstone in Holocene sand developed by storm-uprooted trees, *Sedimentary Geol.* 48, 183-192, 1986.

Semeniuk, V., Searle, D.J., Distribution of calcrete in Holocene coastal sands in relationship to climate, Southeastern Australia, *Jour. Sed. Petrology*, 55, 86-95, 1985.

Smith, B.N., and Epstein, S., Two categories of $^{13}C/^{12}C$ ratios for higher plants, *Plant Physiol.*, 47, 380-384, 1971.

Swart, P.K., Sternberg, L., Steinen, R, and Harrison, S.A., Controls of the oxygen and hydrogen isotopic composition of the waters of Florida Bay, USA, Chem. Geol. (Isotope Geoscience Section), 79, 113-123, 1989.

Swart, P.K., Burns, S., and Leder, J., Fractionation of O isotopes during the reaction of calcite with phosphoric acid as a function of temperature and technique, Chem. Geol. (Isotope Geoscience Sec.), 86, 89-96, 1991

Talma, A.S., and Netterberg, F., Stable isotope abundances in calcretes, in *Residual Deposits: Surface Related Weathering Processes and Materials*, edited by Wilson, R.C.L., Oxford, Blackwell Scientific Publ., 1983.

U. S. Department of Commerce, *World Weather Records, 1951-1960*, 1, National Climate Center, Ashville, N.C.

Wanless, H.R., Geologic setting and recent sediments of the Biscayne Bay region, Florida, *University of Miami Sea Grant*, Spec. Report No. 5, 1-31, 1976.

Wanless, H.R., and Rossinsky, V.Jr., Coastal accretion on leeward margins of carbonate platforms, Turks and Caicos Islands, British West Indies (abs.), *American Association of Petroleum Geologists Bulletin*, 70, 660, 1986.

Wanless, H.R., Tedesco, L.P., Rossinsky, V.JR., and Dravis, J.J., Field Trip T374: Carbonate Environments and Sequences of Caicos Platform, 28th Int'l. Geol. Congress, Amer. Geophys. Union, Washington, D.C., 1989.

Yurtsever, Y., and Gat, J.R., Atmospheric Waters, *in* Gat, J.R., and Gonfiantini, R., eds., Stable Isotope Hydrology, Deuterium and Oxygen - 18 in the Water Cycle: Vienna, Austria, *IAEA Technical Reports*, Ser. No. 210, 103-142, 1981.

Victor Rossinsky and Peter K. Swart, Marine Geology and Geophysics, Rosenstiel School of Marine and Atmospheric Sciences, 4600 Rickenbacker Causeway, Miami FL 33149.

Early Mississippian Climate Change: Isotopic Evidence from Meteoric Calcite

Tad M. Smith[1] and Steven L. Dorobek

Department of Geology, Texas A&M University, College Station, TX 77083

The Mississippian Mission Canyon Formation of central to southwestern Montana was deposited under semi-arid to arid climatic conditions during Osagean to early Meramecian time. Following deposition (middle Meramecian time), a pronounced climatic shift to more humid conditions occurred. Lithologic evidence for this climatic change includes extensive, post-depositional karst fabrics near the top of the Mission Canyon Formation. In addition, a record of this climatic change also may be preserved in the stable isotopic composition of early meteoric calcite cement.

The earliest calcite cement formed when the Mission Canyon platform was subaerially exposed intermittently during Osagean time. Oxygen isotopic data from these early calcite cements and from detrital carbonate components in host limestone indicate that Osagean meteoric water may have had $\delta^{18}O$ values as low as -6.0‰ SMOW.

Oxygen isotopic signatures for later calcite cement in biomoldic porosity and for calcite components in solution collapse breccias suggest Meramecian meteoric water had $\delta^{18}O$ values as low as -12.00‰. However, burial diagenesis at elevated temperatures also may have altered the isotopic composition of some Mission Canyon calcite.

Estimates for the $\delta^{18}O$ values of Osagean and Meramecian meteoric water suggest a dramatic change from arid climatic conditions during Osagean deposition to a cooler and more humid Meramecian climate. Rain-out effects during middle Meramecian time may also have contributed to the extreme depletion of ^{18}O in Meramecian precipitation.

INTRODUCTION

Paleoclimatic studies most frequently rely upon lithologic and paleontologic data for reconstruction of ancient atmospheric circulation patterns and documentation of temporal changes in climate over a given region (e.g. Parrish, 1982; Van der Zwan et al., 1985; Witzke, 1990). For example, coal and evaporites typically are used as indicators of humid and arid conditions, respectively. However, these rock types provide only a general record of climatic change over relatively long periods of geologic time (Witzke, 1990). A more detailed record of paleoclimatic conditions for a specific region might be determined by examining temporal changes in the style of meteoric diagenesis within carbonate rocks and in the geochemical "signature" of associated diagenetic mineral phases (James and Choquette, 1984; Wright, 1990; Vanstone, 1991). In carbonate rocks, stable isotopic compositions of shallow burial, meteoric calcite cements can be used to delimit sequence-specific "meteoric calcite lines" (i.e., invariant $\delta^{18}O$ and variable $\delta^{13}C$ values; Lohmann, 1988). Calcite which is used to define a meteoric calcite line for a specific carbonate sequence is inferred to

have formed in isotopic equilibrium with local meteoric waters and thus constrain the isotopic composition of local meteoric waters during meteoric diagenesis (provided diagenetic temperatures can be independently constrained). Therefore, temporal changes in the $\delta^{18}O$ values of distinct generations of meteoric calcite cement from a specific locality might provide a more detailed record of paleoclimatic change than lithologic data alone.

Stratigraphic and paleontologic data from lower Carboniferous rocks in southern Great Britain indicate that local climatic conditions were semi-arid from late Tournasian (Osagean) to mid-Visean time (Meramecian; Van der Zwan, 1985; Wright, 1990). In mid-Visean time, however, many carbonate sequences from this region were subaerially exposed and karstified, indicating a climatic shift to more humid conditions (Wright, 1988, 1990). A record of this shift also is preserved lithologically and geochemically in paleosols from siliciclastic sequences across the region (Wright, 1990; Vanstone, 1991).

A similar climatic shift is recorded in age-equivalent strata of the Mississippian Mission Canyon Formation (Osagean to early Meramecian) in central to southwestern Montana. Deposition of the Mission Canyon Formation occurred under semi-arid to arid conditions. However, post-Mission Canyon climatic conditions apparently were more humid, as suggested by the occurrence of a regional karst surface at the top of the Mission Canyon Formation. The shift to more humid conditions also may be recorded in the $\delta^{18}O$ values from two different meteoric calcite cement generations. The isotopic shift and lithologic evidence suggest a major change in

[1]Present Address: Amoco Production Company, 501 WestLake Park Blvd., Houston, TX 77253-3092.

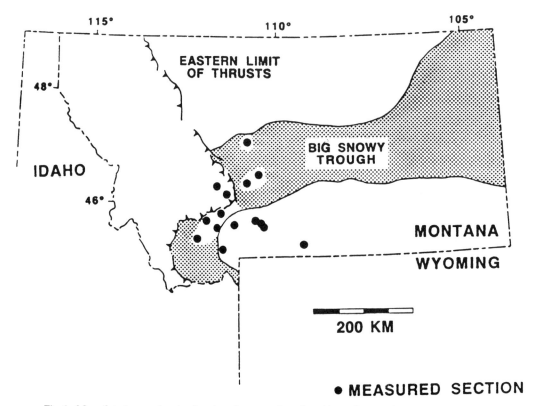

Fig. 1. Map of study area showing location of measured stratigraphic sections used in this study.

paleoclimate occurred across the Mission Canyon platform, and possibly globally, during Early Mississippian time.

GEOLOGIC AND STRATIGRAPHIC SETTING

The Mission Canyon Formation (Osagean to Meramecian) crops out in central to southwestern Montana (Fig. 1) and consists mostly of interbedded limestones and dolostones. Within the study area, the Mission Canyon Formation conformably overlies the Lower Mississippian Lodgepole Formation (Kinderhookian to Osagean) and is overlain unconformably by siliciclastics and carbonates of the Upper Mississippian Big Snowy Group (late Meramecian to early Chesterian). Regional stratigraphic relationships across the study area are shown in Figure 2. North American series/stage names are used in this paper when referring to the Mission Canyon Formation and other age-equivalent North American sequences, whereas European nomenclature is used for age-equivalent rocks from southern Great Britain (Fig. 3).

In general, the Mission Canyon Formation was deposited on a westward dipping, shallow water carbonate ramp that extended across most of Montana during Osagean and Meramecian time (Reid, 1991; Reid and Dorobek, 1991). However, deposition on the Mission Canyon ramp was affected by development of the Big Snowy trough, an elongate, east-west trending depocenter in central Montana (Fig. 1). The Big Snowy trough subsided more rapidly than the rest of the Mission Canyon ramp and influenced development of depositional facies tracts (Smith and Gilmour, 1979; Dorobek et al., 1991).

A minimum of six third- to fourth-order scale depositional sequences (or "parasequence sets") are recognized within the Mission Canyon Formation across the study area (Reid, 1991; Reid and Dorobek, 1991). The upper parts of these shallow water platform parasequence sets are characterized by completely dolomitized cryptalgalaminite facies, fenestral and pisolitic limestones, or laterally continuous solution-collapse breccias. Boundaries between parasequence sets can be traced across the entire Mission Canyon ramp and into age-equivalent slope and basinal strata in western Montana and east-central Idaho (Reid, 1991; Reid and Dorobek, 1991).

During deposition, the Mission Canyon platform was located within 2° to 5° of the paleoequator (Sandberg et al., 1983; Scotese and McKerrow, 1990). Although the Mission Canyon platform was located within equatorial latitudes, semi-arid to arid climatic conditions prevailed (Bambach et al., 1980). These depositional conditions are indicated by: 1) the widespread occurrence of completely dolomitized cryptalgalaminites; 2) dolomudstones with mudcracks, tepee structures; and evaporite pseudomorphs; 3) laterally continuous solution-collapse breccias which probably formed by evaporite dissolution during post-depositional subaerial exposure and meteoric diagenesis (Sando, 1976); and 4) caliche profiles at the tops of some third- and fourth-order depositional sequences (Reid, 1991). In addition, similar lithologic fabrics and the occurrence of bedded evaporites in correlative strata from the Williston Basin (Reid and Dorobek, 1991) suggest that semi-arid to arid conditions affected most of Montana and North Dakota during Mission

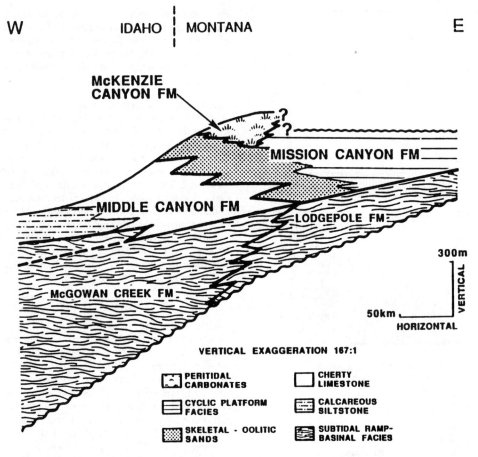

W   IDAHO | MONTANA   E

Fig. 2. Regional stratigraphic chart for the study area. From Reid (1991) and Reid and Dorobek (1991).

Canyon deposition. In addition, much of the dolomite from the Mission Canyon Formation has positive $\delta^{18}O$ values, indicative of arid conditions during dolomitization (Smith, 1991).

Middle Meramecian subaerial exposure of the Mission Canyon platform resulted in the formation of a regional unconformity and karst surface across most of Montana (Middleton, 1961; Roberts, 1966; Sando, 1988). Solution-collapse breccias associated with the karst surface probably also developed during this time (Smith and Gilmour, 1979; Sando, 1988). Estimated maximum duration of subaerial exposure of the Mission Canyon platform during Meramecian time is 13 to 15 million years across central Montana (Sando, 1988).

METHODS

Calcite cements from the Mission Canyon Formation were examined using transmitted light and cathodoluminescence petrography. Rock billets were stained with an Alizarin red/potassium ferricyanide solution (Dickson, 1965) prior to epoxying to glass thin section slides. This technique allows differentiation of calcite and dolomite and recognition of compositional zonation in transmitted light, as well as microprobe analysis and cathodoluminescence petrography of the same samples without interference from the stain.

Calcite samples for mass spectrometric analyses were collected under a binocular microscope by cleaving individual cements and grains from polished thin sections using a dental pick. All samples that were analyzed for their stable isotopic compositions are nonluminescent. Samples averaging about 30 μg were transferred immediately to reaction vessels and reacted 20 to 30 minutes in anhydrous phosphoric acid at 50°C. CO_2 was collected in an extraction line coupled directly to the inlet of a Finnigan MAT 251 isotope-ratio mass spectrometer. Isotopic enrichments were determined relative to an internal laboratory reference gas which was calibrated to PDB using the NBS 20 standard. Isotopic compositions were then converted to PDB and corrected for ^{17}O according to the procedure of Craig (1957). Precision was monitored by daily analysis of an internal laboratory reference standard and NBS 20. The standard deviation from twenty analyses of NBS 20 and six analyses of the internal reference standard is approximately 0.3‰ for $\delta^{18}O$ and 0.2‰ for $\delta^{13}C$.

CALCITE CEMENT PETROGRAPHY

Several generations of calcite cements occur in the Mission Canyon Formation. Volumetrically minor marine cement

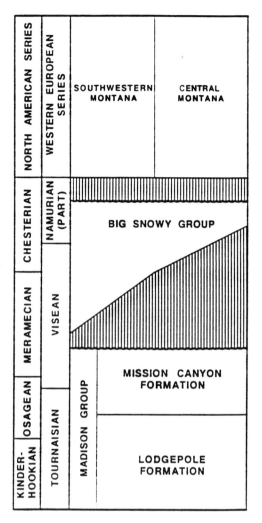

Fig. 3. Chart showing correlation between American and European stage names for the Mission Canyon Formation. Modified from Smith and Gilmour (1979) and Reid (1991).

partially occludes some fenestral porosity and will not be discussed in this paper. The most common calcite cement fills fenestral and intergranular porosity in limestone and dolomitic limestone and is nonluminescent. A second generation of petrographically similar, but isotopically distinct, nonluminescent calcite cement locally fills biomolds and dissolution porosity within solution collapse breccias. This cement is less abundant than the fenestral and intergranular calcite cement.

Fenestral and Intergranular Cement

Calcite cement that fills fenestral and intergranular porosity in limestones and dolomitic limestones is the most common cement in the Mission Canyon Formation. In transmitted light, this cement is medium to coarse crystalline equant spar and inclusion-free (Fig. 4). On artificially stained thin sections, rare purple (ferroan) zones occur locally. Under cathodoluminescence, fenestral and intergranular calcite cements generally are homogeneously nonluminescent.

Fenestral and intergranular calcite cement locally postdates minor isopachous marine cement. In addition, near the tops of some third- and fourth-order sequence boundaries this cement also postdates minor meniscus and pendant cements. Fenestral and intergranular calcite cement predates deep burial compaction and stylolitization.

Breccia and Biomoldic Cement

A second generation of petrographically similar calcite cement locally fills biomolds and dissolution porosity in solution-collapse breccias, although it is much less common than fenestral and intergranular calcite cement. In transmitted light, this cement is equant, medium to coarse crystalline, nonzoned, and has variable amounts of fluid inclusions (Fig. 5a). Under cathodoluminescence, breccia and biomoldic cements are nonluminescent (Fig. 5b), although crystal terminations locally may have dull to moderately bright zonation. Calcite cement in solution-collapse breccias locally has a bladed morphology with scalenohedral crystal terminations (Fig. 5c). Cathodoluminescent-zoned breccia and biomoldic cements were not included in this study.

TIMING OF CALCITE CEMENTATION

Calcite cement in fenestral and intergranular porosity locally replaces early dolomite, indicating that calcite cementation post-dated syndepositional dolomitization (Fig. 6a; Smith, 1991). However, fenestral and intergranular cement predates extensive mechanical compaction and stylolitization (Fig. 4). Fenestral and intergranular cement crystals also are truncated at clast margins within some of the solution-collapse breccias (Fig. 6b), indicating that these cements pre-date Meramecian subaerial exposure and karstification. These petrographic relationships constrain precipitation of fenestral and intergranular calcite cement to Osagean and early Meramecian time.

Calcite cement which fills breccia and biomoldic porosity locally is crosscut by later calcite-filled fractures (Fig. 5b), which in turn are crosscut by stylolites. These relationships indicate that breccia and biomoldic calcite cements probably formed during or after Meramecian karstification and before deep burial diagenesis and/or regional Laramide(?) deformation in the Late Cretaceous to early Tertiary.

GEOCHEMISTRY

$\delta^{18}O$ values from Osagean calcite cements that fill intergranular and fenestral porosity are –8.1 to –2.6‰; $\delta^{13}C$ values are +1.1 to +2.7‰ (Fig. 7a). $\delta^{18}O$ and $\delta^{13}C$ values from nonluminescent skeletal grains and micrite within these layers also generally are within this range of isotopic values (Fig. 7a). $\delta^{18}O$ values from Meramecian calcite cements that fill breccia and biomoldic porosity, nonluminescent breccia clasts, and dull to nonluminescent lime mud matrix from the solution-collapse breccias range from –13.8 to –8.2‰; $\delta^{13}C$ values are –1.9 to +2.2‰ (Fig. 7b).

Fe, Mn, Sr, and Na contents are below microprobe detection limits for both generations of calcite cement, although crystal terminations in some of the zoned biomoldic cement locally contain up to 2400 ppm Fe. Mg concentrations for both generations of calcite cement range from 500 to 3000 ppm.

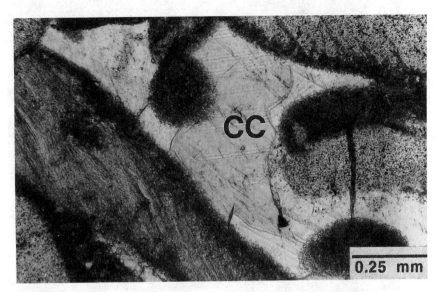

Fig. 4. Intergranular calcite cement. Intergranular and fenestral calcite cement is nonluminescent.

DISCUSSION

Geographic and stratigraphic variation in climatically sensitive rock types (e.g., coal and evaporites) provides only a general record of changing climatic conditions across a region over relatively long periods of geologic time (Witzke, 1990). More precise climatic interpretations can be made by examining temporal changes in the fabrics and geochemistry of meteoric diagenetic phases within a single lithology (Wright, 1990; Vanstone, 1991). In carbonate rocks, caliches generally are indicative of diagenesis under semi-arid conditions, whereas karst horizons suggest humid climatic conditions (James and Choquette, 1984; Wright, 1990). In addition, oxygen isotopic compositions of distinct diagenetic carbonate cements might provide another indication of climate change, provided independent estimates of diagenetic temperatures can be made.

Oxygen Isotopic Composition of Calcite Cements as Indicators of Climatic Change

Factors which control the isotopic composition of meteoric precipitation are discussed in Dansgaard (1964), Anderson and Arthur (1983), and Lawrence and White (1991). These factors include temperature, humidity, and the amount of local rainfall (Anderson and Arthur, 1983; and Lawrence and White, 1991). For a given region, a decrease in ambient air temperature or an increase in relative humidity over time should result in decreased $\delta^{18}O$ values for local meteoric waters. Depletion of ^{18}O occurs because there is less evaporation during meteoric precipitation (Anderson and Arthur, 1983; Lawrence and White, 1991). Meteoric precipitation will become even more ^{18}O-depleted if storm systems track across large areas of continental landmass or over mountain belts (Anderson and Arthur, 1983). $\delta^{18}O$ values of calcite cements that formed in near-surface, meteoric-phreatic aquifers should also reflect these temporal changes in $\delta^{18}O$ values of meteoric water.

The shift in $\delta^{18}O$ values between the Osagean and Meramecian calcite may partly reflect a local climatic change from relatively arid conditions during deposition to more humid conditions during post-Mission Canyon karstification. However, the very low $\delta^{18}O$ values for most of the Meramecian calcite indicate that this change also may partly reflect diagenesis during progressive burial at elevated temperatures. Both possibilities are discussed in this paper.

Osagean calcite. Osagean calcite cements that fill intergranular and fenestral porosity, and metastable calcite components from these same layers (echinoderm fragments and lime mud), have $\delta^{18}O$ values between -8.1 to -2.6‰ (Fig. 7a). The wide range in $\delta^{18}O$ values from these cements may reflect some combination of: 1) variable water-rock ratios during cementation, or 2) variable $\delta^{18}O$ values of Osagean meteoric waters due to fluctuations in relative humidity each time the platform was subaerially exposed. Calcite cements with more negative $\delta^{18}O$ values may reflect precipitation under more humid conditions. Meteoric recharge probably occurred during several short-term subaerial exposure events (10^4–10^5 yr) when the Mission Canyon platform intermittently aggraded to sea level following deposition of individual third- and fourth-order parasequence sets (Reid and Dorobek, 1991).

Isotopic data from Osagean meteoric calcites can be used to infer the oxygen isotopic composition of meteoric pore waters using the approach outlined in Lohmann (1988). The most ^{18}O-depleted meteoric calcite in a rock sequence typically also show the greatest variation in $\delta^{13}C$. Isotopic data from diagenetic calcites in a rock sequence that has been subjected to meteoric diagenesis often has an "inverted J-shape" on a carbon-oxygen isotopic cross-plot. Extreme carbon isotopic variation probably reflects the contribution of variable amounts of soil-derived organic carbon to

82

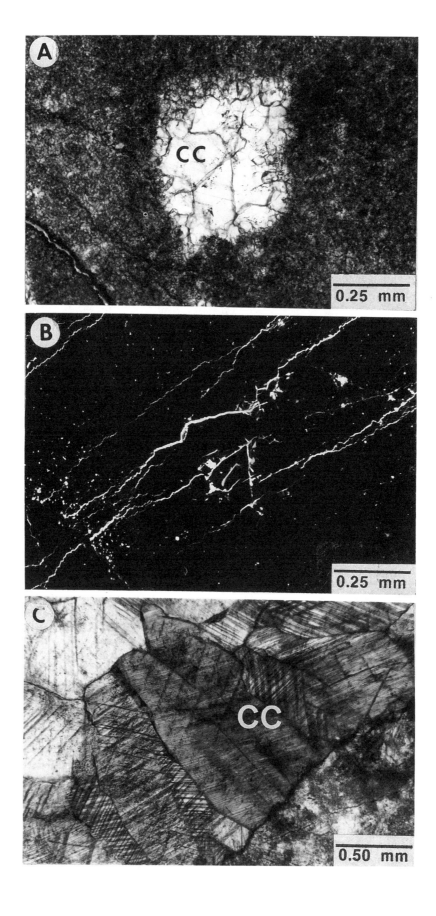

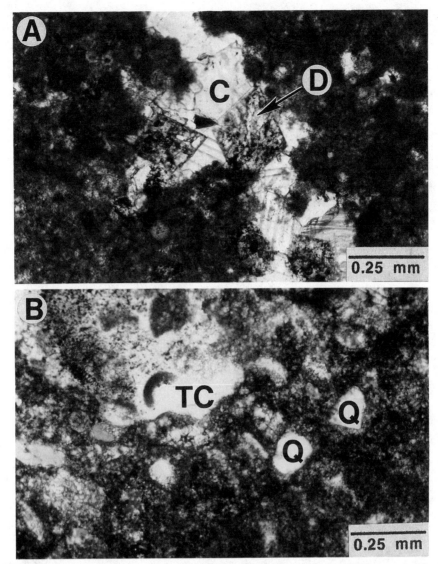

Fig. 6. A. Transmitted light photomicrograph of calcite-replaced dolomite ("dedolomite" fabric). Calcite cement (C) which replaces early-formed dolomite (D) is nonluminescent. B. Limestone clast in a solution-collapse breccia. Note truncated calcite cement (TC) at clast margin and quartz sand grains (Q) scattered throughout the breccia matrix. The truncated calcite cement is nonluminescent.

meteoric pore waters (Allan and Matthews, 1982), while the most negative and invariant $\delta^{18}O$ values record the highest water-rock ratios. Thus, meteoric calcites with the most negative $\delta^{18}O$ values (i.e., "meteoric calcite value"; sensu Lohmann, 1988) can be used to estimate the isotopic composition of meteoric pore waters.

The most negative $\delta^{18}O$ values from Osagean meteoric calcites are approximately -8‰ (Fig. 7a). However, using these data to define a unique Osagean "meteoric calcite value" is problematic. Osagean calcites with the most negative $\delta^{18}O$ values do not have corresponding negative $\delta^{13}C$ values (Fig. 7a). Positive $\delta^{13}C$ values may reflect insignificant diagenesis of organic matter during precipitation of Osagean meteoric calcite. Consequently, there would have been little

Fig. 5. A. Transmitted light photomicrograph of calcite cement (CC) which fills porosity in solution-collapse breccias. This cement petrographically is similar to calcite cement which occludes intergranular and fenestral porosity, but volumetrically is much less abundant. B. Cathodoluminescence photomicrograph of the same field of view as Figure 5A. Brightly luminescent microfractures crosscut the nonluminescent calcite cement. The brightly luminescent microfractures are crosscut by stylolites. This indicates that precipitation of the nonluminescent calcite cement in the breccia porosity predates deeper burial diagenesis. C. Bladed calcite cement (CC) partially filling breccia porosity. Under cathodoluminescence this cement is nonluminescent. Note breccia clast in lower right corner of photomicrograph.

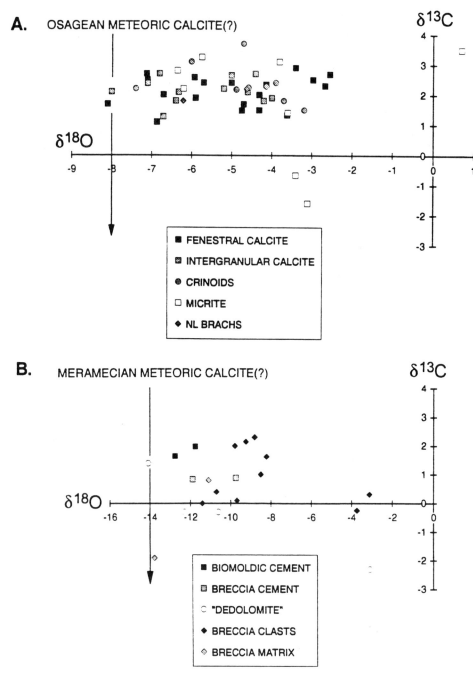

Fig. 7. Stable isotopic data from nonluminescent calcite components. Cement precipitation and grain recrystallization is interpreted to have occurred during Osagean and Meramecian meteoric diagenesis. A. Data from Osagean calcite cements and grains. These values are used to infer an Osagean meteoric calcite value of approximately –8 to –7‰. B. Data from nonluminescent Meramecian calcite cements and grains. These data are used to infer a Meramecian meteoric calcite value of approximately –11 to –12‰.

contribution of light organic carbon to meteoric pore waters. Well-developed soil profiles, other than minor caliche pisolite facies, are conspicuously absent from the Mission Canyon Formation. Poorly developed soil profiles with low organic carbon content are consistent with the generally arid climatic conditions during Osagean time. Alternatively, the uppermost parts of individual third- and fourth-order parasequence sets may have been eroded during subaerial exposure or subsequent marine transgression. Calcite with the most negative $\delta^{13}C$ values may have been in this potentially missing part of the stratigraphic section. Thus, it is not clear if the most negative $\delta^{18}O$ values from preserved

Osagean meteoric calcites record the highest water-rock ratios. Using the most negative $\delta^{18}O$ values may define an Osagean "meteoric calcite value" (sensu Lohmann, 1988) that is more positive than the "true" Osagean meteoric calcite value.

A meteoric calcite $\delta^{18}O$ value of -8‰ is lower than predicted values for calcite precipitated in isotopic equilibrium with modern, near-equatorial meteoric water (~-6‰; Anderson and Arthur, 1983; Gregory et al., 1989). However, climatic patterns for the lower Carboniferous are poorly understood (Bambach et al., 1980; Witzke, 1990), and may have been different than modern climatic patterns due to significant differences in the configuration of the continents (Scotese and McKerrow, 1990; Witzke, 1990). Therefore, assuming that -8‰ is a reasonable estimate for Osagean meteoric calcite from the Mission Canyon Formation, meteoric waters responsible for diagenesis may have had $\delta^{18}O$ values as low as -6‰ SMOW (Anderson and Arthur, 1983).

Meramecian calcite. Meramecian calcite cements in breccia and biomoldic porosity, nonluminescent breccia clasts, and micritic breccia matrix have $\delta^{18}O$ values from -13.8 to -8.2‰ (Fig. 7b). As with the Osagean calcite, the range in $\delta^{18}O$ values for the Meramecian calcite may be due to fluctuations in the oxygen isotopic composition of Meramecian meteoric

waters during karstification or to diagenesis under variable water-rock ratio conditions. The $\delta^{18}O$ values are used here to infer a local Meramecian meteoric calcite $\delta^{18}O$ value of approximately -14‰ (Fig. 7b). Therefore, Meramecian meteoric waters may have had $\delta^{18}O$ values as low as -12‰ (Anderson and Arthur, 1983). Meteoric recharge probably occurred along the regional unconformity and karst surface on top of the Mission Canyon platform.

The difference in $\delta^{18}O$ values between the Osagean and Meramecian calcites can be explained in terms of climatic changes across the Mission Canyon platform between Osagean and Meramecian time. The more negative $\delta^{18}O$ values of Meramecian calcites compared to the Osagean calcites may reflect the change from semi-arid to more humid climatic conditions between Osagean and Meramecian time. The extensive karst associated with the post-Mission Canyon unconformity is consistent with an increase in relative humidity and greater meteoric precipitation in middle Meramecian time. $\delta^{18}O$ values of Meramecian meteoric waters may have become more negative partly due to this local increase in relative humidity (Anderson and Arthur, 1983). Lower $\delta^{18}O$ values for the Meramecian components also may be partially due to lower atmospheric temperatures during Meramecian times. Lower atmospheric temperatures

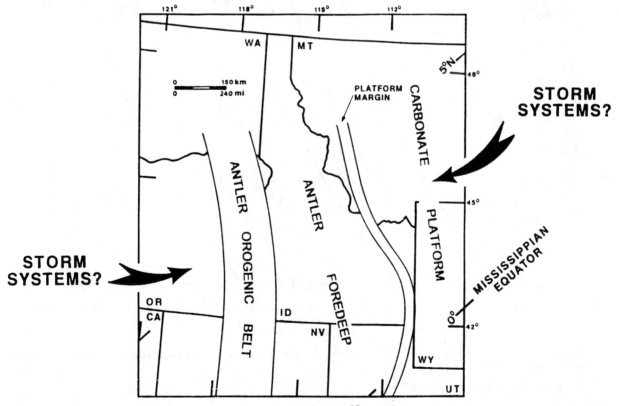

Fig. 8. Schematic diagram showing climatic patterns that may have produced ^{18}O-depleted Mississippian meteoric waters (modified from Reid, 1991). Storm systems approaching the Mission Canyon platform from the west would become depleted in [18]O due to orographic effects associated with the Antler highlands. Storm systems approaching from the east or northeast would become depleted in [18]O due to rain-out effects associated with cloud movement over a large continental land mass.

might be expected if subaerial exposure of the Mission Canyon platform was due to a glacio-eustatic fall in sea level (Ross and Ross, 1987; Reid and Dorobek, 1991).

Alternative Interpretations

Rain-out and orographic effects. Meramecian meteoric water also may have become very ^{18}O -depleted due to orographic effects associated with the Antler highlands located to the west of the Mission Canyon platform, or to rain-out effects associated with cloud movement across broad areas of exposed continental landmass. As an air mass moves over mountains and large areas of continental landmass, orographic uplift and Rayleigh distillation effects cause rainfall to become progressively depleted in ^{18}O (Anderson and Arthur, 1983). In order for Meramecian meteoric water to become depleted in ^{18}O by rain-out effects, storm systems must have moved from west to east, across the Antler highlands and foredeep before reaching the subaerially exposed Mission Canyon platform (Fig. 8). This is consistent with paleoclimatic models for the early Carboniferous (Parrish, 1982). However, other studies of the

lower Mississippian suggest that storm systems may have approached the Mission Canyon platform from the east and northeast (Fig. 8; Marsaglia and Klein, 1983). Large regions of the interior of the North American continent were subaerially exposed at this time (Sloss, 1988). Consequently, meteoric waters may have become ^{18}O -depleted due to rain-out effects associated with cloud movement across these broad areas of exposed continental landmass.

Deeper burial diagenesis. The very negative $\delta^{18}O$ values of much of the Mission Canyon calcite may be best explained by initial precipitation or later diagenetic alteration at elevated temperatures during deeper burial. Laramide deformation during Late Cretaceous to early Tertiary time may have provided an additional opportunity for isotopic exchange with warm orogenic fluids. However, these models for explaining the very negative $\delta^{18}O$ values also are problematic because of petrographic and geologic constraints.

If the Osagean and Meramecian meteoric calcites precipitated

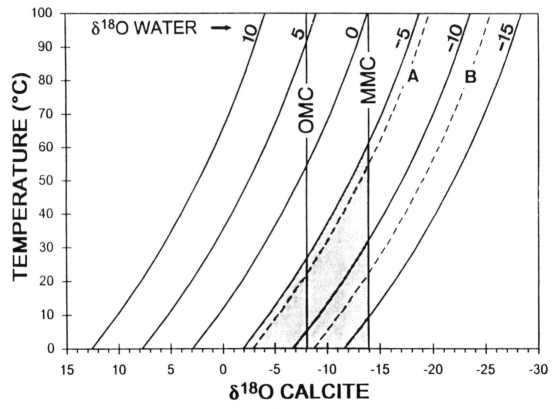

Fig. 9 Diagram showing the relationships between temperature (°C) and the isotopic composition of calcite and water. Curves were calculated using the equation of O'Neil et al. (1969):

$$1000 \ln\alpha_{cc-w} = 2.78(10^6 T_K^{-2}) - 3.39$$

OMC = Osagean meteoric calcite and MMC = Meramecian meteoric calcite. Dashed line A represents the inferred isotopic composition of Osagean meteoric water (-6‰ SMOW). Dashed line B represents the inferred isotopic composition of Meramecian meteoric water (-12‰ SMOW). Shaded area shows range of water compositions and temperatures necessary to produce Meramecian calcite with $\delta^{18}O$ values as low as -14‰ (assuming diagenesis at temperatures less than ~60°C) from solutions with similar oxygen isotopic compositions, a temperature increase of 25° to 30°C would be necessary to produce the -6‰ difference between these two generations of calcite cement (Fig. 9). This would correspond to about 1 km of burial (assuming an average geothermal gradient of 30°C/km). At these burial depths, pore waters probably would have been reducing and Mn^{2+} and/or Fe^{2+} would have been incorporated into the calcite cements (Scholle and Halley, 1985). The presence of iron-stained solution-collapse breccias in the Mission Canyon Formation and siliciclastic facies in the overlying Big Snowy Group suggest that the absence of Mn and Fe in burial pore fluids is not a viable explanation for the low Mn and Fe concentrations in Mission Canyon cements. Also, calcite cements that form in deeper burial diagenetic environments typically are luminescent (Scholle and Halley, 1985; Dorobek, 1987), while the Mission Canyon calcite cements are nonluminescent. The nonluminescent character and very low Mn and Fe contents are consistent with diagenesis in oxidizing, meteoric waters.

GLOBAL EVIDENCE FOR MISSISSIPPIAN CLIMATE CHANGE

Studies of lower Carboniferous rocks from southern Great Britain suggest that climatic conditions mostly were arid to semi-arid during late Tournasian (Osagean) through mid-Visean time (Meramecian; Wright, 1990). This is indicated by the widespread occurrence of calcretes in Tournasian and lower Visean carbonate rocks, and the presence of calcrete-bearing paleosols in age-equivalent floodplain deposits (Van der Zwan et al., 1985; Wright, 1990; Vanstone, 1991). During this time, southern Great Britain was located within 5 to 10 degrees of the paleoequator (Scotese and McKerrow, 1990).

Evaporites and desiccation features in carbonate rocks from the age-equivalent Mission Canyon Formation of western Montana and the Williston Basin of North Dakota also indicate that semi-arid to arid climatic conditions prevailed across the northwestern United States during the same time interval. In addition, much of the dolomite in the Mission Canyon Formation has high $\delta^{18}O$ values, which are indicative of arid conditions during dolomitization ($\delta^{18}O$ values as positive as +7.5‰ ; Smith, 1991). The predominance of this dolomite in restricted, evaporite-bearing peritidal facies suggests that it precipitated in hypersaline, sabkha-type environments when the Mission Canyon platform intermittently aggraded to sea level (Smith, 1991).

A shift to more humid climatic conditions in southern Great Britain is inferred to have occurred in mid-Visean time (Wright, 1988, 1990; Wright et al., 1991; Vanstone, 1991). This climatic change resulted in karstification of numerous carbonate sequences (Wright, 1988, 1990) and a change from calcrete-bearing vertisols (semi-arid climates) to dominantly leached vertisols and ferric podzols (temperate to humid climates) in siliciclastic sequences (Wright, 1990; Wright et al., 1991; Vanstone, 1991). The timing of this climatic change coincides with development of the middle Meramecian paleokarst which overlies the Mission Canyon Formation (Sando, 1988).

Although the causative mechanism(s) for the apparent Meramecian (mid-Visean) climatic shift are poorly known (Wright, 1990), the age-equivalence of numerous subaerial exposure horizons in carbonate rocks from Great Britain and across North America indicate that this climatic shift may have been due to eustatic sea level fall during middle Meramecian time (Reid, 1991; Reid and Dorobek, 1991). This eustatic sea level fall may possibly have been due to the onset of continental glaciation during the middle to late Carboniferous (Dickens, 1985; Veevers and Powell, 1987). Further work on other age-equivalent strata is warranted to develop more refined climatic models for the early Carboniferous.

CONCLUSIONS

Lithologic and stable isotopic data from the Mississippian Mission Canyon Formation indicate that local paleoclimate shifted from semi-arid conditions during Osagean to early Meramecian deposition to humid conditions after deposition (middle Meramecian time). Calcite cements which precipitated during meteoric diagenesis indicate that local Osagean meteoric waters may have had $\delta^{18}O$ values as low as -6‰ SMOW, whereas Meramecian meteoric waters may have had $\delta^{18}O$ values as low as -12‰ SMOW. This shift to more negative $\delta^{18}O$ values is consistent with a local increase in relative humidity, lower atmospheric temperatures, and with cloud movement over large areas of exposed continental landmass following a significant fall in Mississippian sea level. However, the very negative $\delta^{18}O$ values for much of the Meramecian calcite may also partly be due to later diagenesis at elevated temperatures.

The shift to more humid conditions during the middle Meramecian also is recorded in age-equivalent carbonates and siliciclastics from southern Great Britain. This indicates that the middle Meramecian (or mid-Visean) climatic change may have been global in extent. Although causative mechanisms event over widely separated geographic localities indicates that it may at least partly be due to the onset of middle to late Carboniferous continental glaciation.

Acknowledgments. This project was supported by U.S. Department of Energy Grant DE-FG05-87ER13767, Basic Energy Sciences to S.L. Dorobek. Such support does not constitute an endorsement by DOE of the views expressed in this paper. Acknowledgment also is made to the Donors of the Petroleum Research Fund, administered by the American Chemical Society, for partial support of this research (Grant #19519-G2 to S.L. Dorobek). Additional support was provided by the Geological Society of America Grants-in-Aid to the senior author. We would like to thank Tom Anderson, Steve Burns, Ethan Grossman, John Morse, and Steve Reid for their critical reviews of earlier versions of this manuscript.

REFERENCES

Anderson, T.F., and Arthur, M.A., Stable isotopes of oxygen and carbon and their application to sedimentologic and paleoenvironmental problems, in *Stable Isotopes in Sedimentary Geology*, SEPM Short Course No. 10, edited by T.F. Anderson, and M.A. Arthur, 1-1 to 1-151, 1983.

Bambach, R.K., Scotese, C.R., Ziegler, A.M., Before Pangea: The geographies of the Paleozoic World, *American Scientist, 68*, 26-38, 1980.

Craig, H., Isotopic standards for carbon and oxygen correction factors for mass spectrometric analysis of carbon dioxide, *Geochimica et Cosmochimica Acta, 12*, 133-149, 1957.

Dansgaard, W., Stable isotopes in precipitation, *Tellus*, 16, 436-468, 1964.

Dickins, J.M., Late Paleozoic glaciation, *Journal of Australian Geology and Geophysics*, 9, 163-169, 1983.

Dickson, J.A.D., A modified staining technique for carbonates in thin section, *Nature*, 205, 587, 1965.

Dorobek, S.L., Petrology, geochemistry, and origin of burial diagenetic facies, Siluro-Devonian Helderberg Group (carbonate rocks), Central Appalachians, *Bulletin of the American Association of Petroleum Geologists*, 71, 492-514, 1987.

Dorobek, S.L., Reid, S.K., Elrick, M., Bond, G.C., and Kominz, M.A., Subsidence across the Antler foreland of Montana and Idaho: Tectonic versus eustatic effects, in *Sedimentary Modeling: Computer Simulation and Methods for Improved Parameter Definition*, edited by E. Franseen, and L. Watney, Kansas Geological Society Special Volume 233, 231-251, 1991.

Gregory, R.T., Douthitt, C.B., Duddy, I.R., Rich, P.V., and Rich, T.H., Oxygen isotopic composition of carbonate concretions from the lower Cretaceous of Victoria, Australia: implications for the evolution of meteoric waters on the Australian continent in a paleopolar environment, *Earth and Planetary Science Letters*, 92, 27-42, 1989.

James, N.P., and Choquette, P.W., Diagenesis 9. Limestones - The meteoric diagenetic environment, *Geoscience Canada*, 11, 161-194, 1984.

Lawrence, J.R., and White, J.W.C., The elusive climate signal in the isotopic composition of precipitation, in *Stable Isotope Geochemistry: A Tribute to Sam Epstein*, edited by H.P Taylor, J.R. O'Neil, Jr., and I.R. Kaplan, The Geochemical Society, Spec. Pub. 3, 169-185, 1991.

Lohmann, K.C., Geochemical patterns of meteoric diagenetic systems and their application to studies of paleokarst, in *Paleokarst*, edited by N.P. James and P.W. Choquette, Springer-Verlag, New York, 58-80, 1988.

Marsaglia, K.M., and Klein, G.deV., The paleogeography of Paleozoic and Mesozoic storm depositional systems, *Journal of Geology*, 91, 117-142, 1983.

Middleton, G.V., Evaporite solution breccias from the Mississippian of southwest Montana, *Journal of Sedimentary Petrology*, 31, 189-195, 1961.

Parrish, J.T., Upwelling and petroleum source beds, with reference to the Paleozoic, *Bulletin of the American Association of Petroleum Geologists*, 66, 750-774, 1982.

Reid, S.K., Evolution of the Lower Mississippian Mission Canyon platform and distal Antler foredeep, Montana and Idaho, [unpublished Ph.D. dissertation] Texas A&M University, College Station, 115, 1991.

Reid, S.K., and Dorobek, S.L., Controls on development of third- and fourth-order depositional sequences in the Lower Mississippian Mission Canyon Formation and stratigraphic equivalents, Idaho and Montana, in *Paleozoic Paleogeography of the Western United States-II*, edited by J.D. Cooper and C.H. Stevens, Pacific Section SEPM, 67, 527-541, 1991.

Roberts, A.E., Stratigraphy of Madison Group near Livingston, Montana, and discussion of karst and solution-breccia features, *U.S. Geological Survey Prof. Paper*, 526-B, B1-B23, 1966.

Ross, C.A., and Ross, J.P., Late Paleozoic sea levels and depositional sequences, in *Timing and Depositional History of Eustatic Sequences: Constraints on Seismic Stratigraphy*, edited by C.A. Ross, and D. Haman, Cushman Foundation for Foraminiferal Research, Special Publication No. 24, 137-149, 1987.

Sandberg, C.A., Gutschick, R.C., Johnson, J.G., Poole, F.G., and Sando, W.J., Middle Devonian to Late Mississippian geologic history of the Overthrust Belt region, western United States, *Geologic Studies of the Cordilleran Thrust Belt*, Rocky Mountain Association of Geologists, 2, 691-719, 1983.

Sando, W.J., Mississippian history of the northern Rocky Mountains Region, *Journal of Research*, United States Geological Survey, 4, 317-338, 1976.

Sando, W.J., Revised Mississippian time scale, western interior region, conterminous United States, *Bulletin of the United States Geological Survey*, 1605-A, A15-A26, 1985.

Sando, W.J., Madison Limestone (Mississippian) paleokarst: A geologic synthesis, in *Paleokarst*, edited by N.P. James and P.W. Choquette, Springer-Verlag, New York, 256-277, 1988.

Scotese, C.R., Bambach, R.K., Barton, C., Van der Voo, R., and Ziegler, A.M., Paleozoic Base Maps, *Journal of Geology*, 87, 217-277, 1979.

Scotese, C.R., and McKerrow, W.S., 1990. Revised World maps and introduction, in *Palaeozoic Palaeogeography and Biogeography*, edited by W.S. McKerrow and C.R. Scotese, Geological Society Memoir 12, 1-21.

Sloss, L.L., Tectonic evolution of the craton in Phanerozoic time, in *Sedimentary Cover, North American Craton; U.S.*, edited by L.L. Sloss, Geological Society of America, The Geology of North America, D-2, 25-51, 1988.

Smith, D.L., and Gilmour, E.H., The Mississippian and Pennsylvanian (Carboniferous) Systems in the United States -- Montana, *U.S. Geological Survey Prof. Paper*, 1110-X, 32 pp, 1979.

Smith, T.M., Diagenesis of shallow marine carbonate rocks: Isotopic and trace element constraints from the Mississippian Mission Canyon Formation, central and southwestern Montana: [unpublished Ph.D. dissertation], Texas A&M University, College Station, 126 p., 1991.

Van der Zwan, C.J., Boulter, M.C., and Hubbard, R.N.L.B., Climatic change during the lower Carboniferous in Euramerica, based on multivariate statistical analysis of palynological data, *Paleogeography, Paleoclimatology, Paleoecology*, 52, 1-20, 1985.

Vanstone, S.D., Early Carboniferous (Mississippian) paleosols from southwest Britain: Influence of climatic change on soil development, *Journal of Sedimentary Petrology*, 61, 445-457 1991.

Veevers, J.J., and Powell, C.Mc., Late Paleozoic glacial episodes in Gondwana, in *Gondwana Stratigraphy*, International Union of Geological Sciences, UNESCO, 551-588, 1987.

Witzke, B.J., Palaeoclimatic constraints for Palaeozoic palaeolatitudes of Laurentia and Euramerica, in *Palaeozoic Palaeogeography and Biogeography*, edited by W.S. McKerrow and C.R. Scotese, Geological Society Memoir 12, 57-73, 1990.

Wright, V.P., Paleokarsts and paleosols as indicators of paleoclimate and porosity evolution, in *Paleokarst*, edited by N.P. James and P.W. Choquette, Springer-Verlag, New York, 329-341, 1988.

Wright, V.P., Equatorial aridity and climatic oscillations during the Carboniferous, southern Britain, *Journal of the Geological Society*, London, 147, 359-363, 1990.

Wright, V.P., Vanstone, S.D., and Robinson, D.R., Ferrolysis in Arundian alluvial paleosols: evidence of a shift in the early Carboniferous monsoonal system, *Journal of the Geological Society*, London, 148, 9-12, 1991.

T.M. Smith and S.L. Dorobek, Department of Geology, Texas A&M University, College Station, TX 77083.

Principles and Applications of the
Noble Gas Paleothermometer

MARTIN STUTE

*Lamont-Doherty Geological Observatory, Columbia University
Palisades, New York*

PETER SCHLOSSER

*Lamont-Doherty Geological Observatory and Department of Geological Sciences,
Columbia University, Palisades, New York*

Deriving paleoclimate information from noble gas (Ne, Ar, Kr, Xe) concentrations in groundwater systems depends critically on two factors: (1) determining the relationship between air temperature and the initial noble gas concentration at the groundwater table, and (2) assessing how much these initial signals disperse during transit. Both aspects are treated quantitatively and the theoretical results are compared to published noble gas paleotemperatures. For typical conditions in a recharge area, the noble gas concentrations of recent groundwater closely reflect the mean annual temperature of the ground at the groundwater table. Simulations of the dispersion of high–frequency signals in aquifers with typical dispersion/advection coefficients were used to estimate the quality of groundwater flow systems as paleoclimate archives. Aquifers with dispersion/advection ratios D/v_x^2 between 100 and 1,000 years are well suited to archive the main low–frequency features of the paleotemperature record. A review of the available noble gas temperature records shows that the temperature differences between Holocene and last glacial maximum (LGM) range from about 5 °C (Texas) to 9 °C (Hungary), whereas those between Holocene and the interstadial preceding the LGM vary between 2.5 °C (Texas) and 6 °C (UK).

INTRODUCTION

Since the basic work of Mazor [1972] noble gas measurements in groundwater have been used in several studies to estimate paleotemperatures during the last glacial episode. The principle of the noble gas thermometer is that the solubility of noble gases in water is temperature dependent. The temperature at which a water parcel was equilibrated with the atmosphere can be calculated from its noble gas concentrations. The sensitivity of solubility to temperature increases with atomic mass. Xe has the highest sensitivity whereas Ne solubility shows only a very small temperature effect (Figure 1, [Weiss, 1970, 1971; Clever, 1979]). Noble gas concentrations in recent groundwater reflect the temperature of the ground at the water table. If ground temperature can be related quantitatively to mean annual air temperature, then noble gas measurements in dated groundwater provide a valuable paleoclimate record. To use this method to its full potential requires understanding of two critical factors: (1) processes controlling the noble gas concentrations of recent groundwater, and (2) preservation of this signal during subsurface flow of water in the aquifer.

A systematic treatment of the factors determining the initial noble gas concentration of recent groundwater is still missing. These factors include ground temperature, ground air composition, excess air in the groundwater, seasonal variations in the recharge, and depth of the groundwater table.

Once water leaves the groundwater table and gas exchange with ground air is suppressed, noble gas concentrations are modified mainly by dispersion and mixing. The quality of groundwater as paleoclimate archive depends critically on the extent to which the initial noble gas concentrations are changed by these processes.

The focus of this contribution is quantitative treatment of processes determining the initial noble gas concentrations at the groundwater table and their subsequent changes during transit of groundwater from the recharge to the discharge area. Additionally, the accuracy of groundwater dating is briefly discussed and noble gas records for six groundwater systems are reviewed, reconstructing Holocene and late Pleistocene paleotemperatures.

Climate Change in Continental Isotopic Records
Geophysical Monograph 78

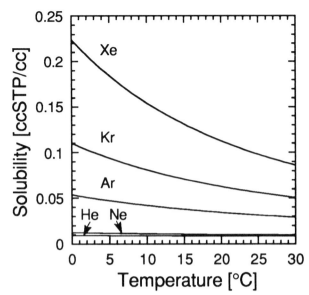

Fig. 1: Temperature dependence of the noble gas solubilities in distilled water expressed as Bunsen-coefficient [Weiss, 1970, 1971, Clever, 1979].

COMPOSITION OF GROUND AIR

Water vapor content

The initial condition for noble gases is set near the water table at the interface between saturated and unsaturated zone, where the air is characterized by a relative humidity of 100%.

Gravitational separation

Partial pressures of constituents of ground air may be influenced by gravitational separation of gases in a manner similar to that observed in ice cores [Craig et al., 1988, Schwander et al., 1988]. Although there are no suitable experimental studies of this effect in ground air, the expected effect can be estimated using the formula given by Craig et al. [1988]:

$$\frac{P_z}{P_0} = \exp\frac{mgz}{RT_G} \approx 1 + \frac{mgz}{RT_G} \qquad (1)$$

P_z ... partial pressure at depth z below surface
P_0 ... partial pressure at the surface
m ... mass of the gas [kg mole^{-1}]
g ... gravitational constant [9.81 m s^{-2}]
z ... depth below surface [m]
R ... gas constant [8.314 J mole^{-1} K^{-1}]
T_G ... ground temperature [K]

Calculation of the gravitational separation of Ne, Ar, Kr, and Xe as a function of depth (Figure 2) indicates that, considering the 1σ error of ± 1 to 2 % for the measurements of these gases, the effect of gravitational separation is significant only if the water table is more than 50 m below the surface. The effect of gravitational separation on the noble

gas temperature as a function of the depth z_w to the water table can be estimated using the following relationship:

$$\Delta T_{NG} = 0.013\frac{°C}{m} \cdot z_w \qquad (2)$$

Groundwater tables in the recharge areas of the aquifers discussed in this paper are less than 50 m below the surface and gravitational separation may be neglected.

Biological activity

Biological activity in the ground (root respiration, decomposition of organic material) generally consumes O_2 and produces CO_2. Under aerobic conditions oxygen consumption is compensated by equimolar production of CO_2 and will not change the partial pressures of the other gases in the ground. However, CO_2 concentrations may be lower than expected from the oxygen depletion, because of the higher solubility of CO_2 in water and the occurrence of chemical reactions between dissolved CO_2 and minerals (especially carbonates) in the ground. These reactions result in preferential removal of CO_2 from the ground gas [Schachtschabel et al., 1982]. However, except in tropical climates, CO_2 concentrations in the ground rarely exceed 2% during the growing season [Brook et al., 1983]. In most cases, biological activity is limited to the growing season, so that with the exception of tropical regions, average partial pressures of noble gases in ground air can be assumed to resemble closely (within a few %) the partial pressures in atmospheric air at a relative humidity of 100% at the elevation of the water table.

History of the noble gas composition of the atmosphere

Alteration of the atmospheric noble gas partial pressures on long timescales would complicate the application of the noble gas paleothermometer. Phillips [1981] discusses the various potential mechanisms and comes to the conclusion

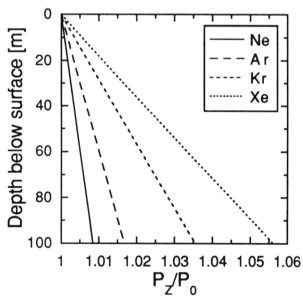

Fig. 2: Change of the partial pressure P of the noble gases in ground air as a function of depth due to gravitational separation.

that the atmospheric noble gas abundances and isotope ratios have remained essentially constant for at least the last 3 million years.

Processes determining noble gas concentrations in groundwater

Noble gas concentrations that can be measured in recent groundwater are influenced by several processes: solution of air according to Henry's law, dissolution of small air bubbles, and addition of non–atmospheric noble gases. For determination of the noble gas temperature, these components have to be separated.

Solubility equilibrium

During recharge, water percolates through the unsaturated zone and gases are exchanged continuously between water and ground air. In the saturated zone, exchange is limited to diffusion across the water table. The concentration of a dissolved gas depends on its solubility, which is a function of temperature [Weiss, 1970, 1971, Clever, 1979] and salt content of the water [Smith and Kennedy, 1983], and on its partial pressure in the gas phase. The salt content of recently-formed groundwater is usually fairly low (of the order of $2\,mmole\,l^{-1}$). Such low salt contents have practically no influence on the solubility of gases in water. Even a NaCl content of $25\,mmole\,l^{-1}$ would reduce the noble gas solubilities by less than 1% [Smith and Kennedy, 1983], which would not affect the accuracy of the noble gas paleothermometer.

Excess air

The second process influencing the concentrations of all dissolved gases is excess air formation due to dissolution of small air bubbles caused by a very rapid infiltration in karstic environments or fluctuations of the groundwater table [Herzberg and Mazor, 1979; Heaton and Vogel, 1981]. Because noble gas solubility and its temperature dependence increase with atomic mass, He and Ne concentrations are most sensitive to the addition of excess air, whereas Xe concentrations are mainly controlled by temperature.

Noble gases of non–atmospheric origin

During its underground passage, groundwater accumulates non–atmospheric noble gases that are produced in the ground matrix by nuclear processes or that have migrated upward from deeper layers of the crust or mantle. Ne, Ar, Kr, and Xe excesses due to this process are rarely detected. Helium may show concentrations several orders of magnitude above the solubility equilibrium level and the amount of atmospheric He cannot be separated from the total He concentration. However, due to the small temperature dependence of its solubility, He is not used for the determination of paleotemperatures.

Separation of the individual components

Non–atmospheric contributions to Ne, Ar, Kr and Xe concentrations of groundwater samples can be estimated by isotope ratios. The remaining atmospheric Ne, Ar, Kr, and Xe concentrations can then be divided into two components, the solubility equilibrium component and the excess air component (under the assumption that the salt content of the infiltrating water is negligible and that the composition of ground air is the same as that of humid (100% relative humidity) atmospheric air at the same elevation). The dissolved noble gas concentration ratios of water in solubility equilibrium with the atmosphere deviate significantly from those in the atmosphere as a result of the different solubilities of different gases. If it is assumed that the gases of the excess air component are not fractionated relative to the atmosphere, the solubility equilibrium component can be separated from the excess air component by a set of equations with the ground temperature and the amount of excess air dissolved in the groundwater sample as unknowns. Actual calculation of the noble gas temperature is an iterative process [Rudolph, 1981]. In a first step, the noble gas temperatures for Ne, Ar, Kr, and Xe are calculated from the measured concentrations. If there is a significant excess air component in the water sample, there will be a systematic trend in the noble gas temperatures, with Ne giving the lowest and Xe giving the highest temperature. In the next step, the concentrations of the individual noble gases are decreased according to their atmospheric ratios by a desired amount and the noble gas temperatures are recalculated. This step is repeated until all four noble gas temperatures match within analytical error.

Partial degassing of the water, for example during sample collection, can lead to significant changes in the noble gas ratios. The same holds true for fractionation during (partial) dissolution of air bubbles. If the noble gases dissolved in a water sample were significantly fractionated, the above described iterative procedure would yield inconsistent noble gas temperatures for the individual noble gases. In special cases, fractionation of excess air can produce noble gas ratios similar or equal to those of equilibrated water. In these cases, the excess air cannot be subtracted properly from the measured noble gas concentration. However, comparison of noble gas temperatures of young groundwater with measurements of the ground temperatures in the recharge areas (as discussed below) indicates that this process is of minor importance and that the algorithm described above is adequate for calculations of noble gas temperatures. For example, excess air concentrations of 12 Holocene groundwater samples collected in the Great Hungarian Plain vary between $0.2 \cdot 10^{-3}$ and $5.7 \cdot 10^{-3}\,ccSTPg^{-1}$ (mean value $(3.1 \pm 1.8) \cdot 10^{-3}\,ccSTPg^{-1}$). These excess air concentrations are not correlated with noble gas temperatures [Stute and Sonntag, 1992].

Impact of ground temperature and recharge variations on noble gas temperatures

Water percolating through the unsaturated zone equilibrates continuously with ground air until it reaches the capillary fringe where the saturated zone begins. For most recharge areas the thickness of the capillary fringe is negligible and the noble gases dissolved in groundwater may be assumed to reflect the ground temperature at the water table [Herzberg and Mazor, 1979; Phillips, 1981]. Temperature and precipitation in the recharge area vary with time. Below, we estimate the impact of these variations on noble gas temperatures.

Ground temperature variations

Daily variations in atmospheric temperature penetrate only to a depth of about one meter into the ground and can be neglected, because the water table is usually located deeper than that. The seasonal fluctuation of the ground temperature as a function of depth can be described by a sinusoidal oscillation around the mean ground temperature $\bar{T}_G$ with an amplitude T_a decreasing exponentially with depth [Kappelmeyer, 1968]:

$$T_G(z,t) = \bar{T}_{G(0)} + T_a \exp(-\frac{z}{\bar{z}})\sin(\omega t - \frac{z}{\bar{z}} + \phi) + \frac{\Delta T}{\Delta z}z \quad (3)$$

$\bar{T}_{G(0)}$ ...   average ground temperature at the surface ($z = 0$)

T_a ...   amplitude of the annual temperature variation at $z = 0$

ω ... 2π year^{-1}

$\bar{z}$ ...   average penetration depth of the seasonal temperature variation

ϕ ...   phase angle

$\Delta T/\Delta z$ ...   geothermal gradient (typical 3 °C/100 m)

Assuming values of 10 °C for $\bar{T}_G$ and T_a, and 2.5 m for $\bar{z}$ (typical numbers for central Europe, [Stute and Sonntag, 1992]), the amplitude of the temperature oscillation at a depth of 5 m is about 1.4 °C (Figure 3). Below 10 m there is almost no seasonal variation ($T_a < 0.2$ °C) and, the noble gas temperature $\bar{T}_{NG}$ reflects the mean annual temperature at the water table:

$$\bar{T}_{NG} = \bar{T}_{G(0)} + \frac{\Delta T}{\Delta z}z_w = \bar{T}_G(z) \quad (4)$$

These calculations show that for typical aquifers with water tables more than 10 m below the surface, noble gases record a temperature close to the mean annual ground temperature at the water table.

Infiltration pattern

The infiltration rate (I) is also subject to seasonal variations. In order to estimate the influence of variations in I on the noble gas temperature, two limiting cases will be discussed: (1) the total annual infiltration occurs in one event, and (2) the infiltration rate fluctuates sinusoidally as follows:

$$I(t) = \bar{I} + I_a\sin(\omega t + \psi) \quad (5)$$

$\bar{I}$ ...   mean annual infiltration rate

I_a ...   amplitude of the seasonal variation of the infiltration

ψ ...   phase angle

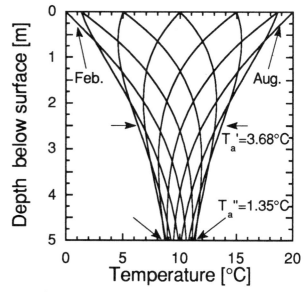

Fig. 3: Simulated annual variation of the monthly averaged ground temperature profile as a function of depth below surface for a typical soil. The following parameters were used: $\bar{T}_G = T_a = 10$ °C, $\bar{z} = 2.5$ °C. At 2.5 and 5 m depth, the amplitude of the temperature oscillation is reduced to 3.68 and 1.35 °C, respectively.

In a first order approximation, the average noble gas temperature $\bar{T}_{NG}$ can be estimated to be the ground temperature at the water table weighted for seasonality of infiltration. In the first case (instantaneous infiltration), $\bar{T}_{NG}$ is simply the temperature at the water table at the time of infiltration: $\bar{T}_{NG} = T(z_w, t)$. In the second, more realistic, case (sinusoidal variation of I); $\bar{T}_{NG}$ can be expressed as:

$$\bar{T}_{NG} = \frac{\int_0^{1\,year} T_G(z_w,t)I(t)dt}{\int_0^{1\,year} I(t)dt} \quad (6)$$

$$= \bar{T}_G + \frac{T_aI_a\exp(-\frac{z_w}{\bar{z}})}{2\bar{I}}\cos(\frac{z_w}{\bar{z}} + \psi - \phi) \quad (7)$$

The above estimate of $\bar{T}_{NG}$ does not account for diffusive (dispersive) gas exchange between groundwater and ground air across the water table that occurs after percolating water has reached the saturated zone. This process may be described by the following one–dimensional diffusion/advection equation for T_{NG} with time dependent vertical flow velocity and the seasonal variation in ground temperature as boundary condition:

$$D\frac{\partial^2 T_{NG}}{\partial z^2} - v_z\frac{\partial T_{NG}}{\partial z} = \frac{\partial T_{NG}}{\partial t} \quad (8)$$

where D is the vertical dispersion coefficient and v_z the vertical flow velocity. In the following, we discuss an example of the solution of this equation: the recharge temperature is calculated by assuming that infiltration occurs with a phase shift of π with respect to the temperature variation at the water table (maximum recharge during coldest periods).

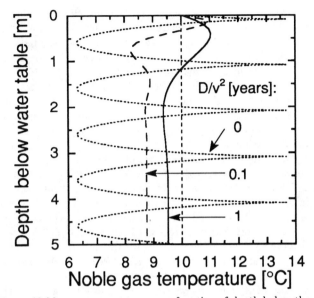

Fig. 4: Noble gas temperatures as a function of depth below the water table calculated for different values of $D/\bar{v}_z^2$. Infiltration and temperature at the water table is assumed to vary sinusoidal with the same frequency but a phaseshift of π. The difference between mean annual ground temperature ($10\,°C$) and average noble gas temperature is reduced with increasing influence of diffusion (dispersion) from $1.84\,°C$ (mean value for $D = 0$) over $1.24\,°C$ ($D/\bar{v}_z^2 = 0.1$ years) to $0.50\,°C$ ($D/\bar{v}_z^2 = 1$ year) at a depth of about 3 to 5 m below the water table (for explanation, see text).

To estimate a maximum value of the difference between the noble gas temperature and the mean annual ground temperature, the depth of the groundwater table is chosen as $2.5\,m$. The average vertical flow velocity $\bar{v}_z = \bar{I}/n$ (v_z equals the infiltration I devided by the porosity n, see equation 5) as well as its amplitude (I_a/n) is set to $1\,m\,year^{-1}$. Further, we assume that $\bar{z} = 2.5\,m$ and $\bar{T}_G = T_a = 10\,°C$ (at ground surface; this amplitude is reduced to $3.68\,°C$ at $2.5\,m$ depth below surface (equation (3)). The ratio $D/\bar{v}_z^2$ is a measure of the relative importance of diffusion (dispersion) and advection and was set in three different solutions to 0, 0.1 and 1 year. $D/\bar{v}_z^2 = 1\,year$, for example, is a typical value for recharge in humid climates where annual infiltration rates are of the order of $100\,mm\,year^{-1}$ [Schlosser et al., 1989].

The case $D/\bar{v}_z^2 = 0$ means that diffusion is neglected. For this case, the amplitude of the noble gas temperature is $3.68\,°C$ (Figure 4). Averaging of the temperature difference over depth in the groundwater yields a mean temperature difference between $\bar{T}_{NG}$ and $\bar{T}_G$ of $1.8\,°C$. This difference between recorded recharge temperature $\bar{T}_{NG}$ and mean annual ground temperature $\bar{T}_G$ is reduced to $1.2\,°C$ for $D/\bar{v}_z^2 = 0.1\,year$ and $0.5\,°C$ for 1 year (Figure 4). $\bar{T}_{NG} - \bar{T}_G$ is a complex function of all the parameters listed above, especially of the phases ϕ and ψ. In order to simplify the calculation, an upper limit for $|\bar{T}_{NG} - \bar{T}_G|$ was determined by varying the difference between ϕ and ψ. $|\bar{T}_{NG} - \bar{T}_G|$ can be expressed as a function of the depth of the groundwater

table z_w, the characteristic penetration depth of the temperature signal in ground air $\bar{z}$, and a factor F which accounts for the smoothing effect of the temperature difference in the groundwater:

$$|\bar{T}_{NG} - \bar{T}_G| = \frac{T_a I_a}{\bar{I}} \exp(-\frac{z_w}{\bar{z}}) F(\frac{D}{\bar{v}_z^2}) \qquad (9)$$

for sinusoidal variation of the infiltration, and

$$|\bar{T}_{NG} - \bar{T}_G| = T_a \exp(-\frac{z_w}{\bar{z}}) F(\frac{D}{\bar{v}_z^2}) \qquad (10)$$

for instantaneous infiltration.

The dependence of the factor F on $D/\bar{v}_z^2$ was calculated numerically for the two cases mentioned above, instantaneous and sinusoidal infiltration (Figure 5). Even if the infiltration occurs during a single day, gas exchange across the groundwater table considerably reduces the difference between $\bar{T}_{NG}$ and $\bar{T}_G$. If the influence of the phase difference between ground temperature and infiltration were taken into account, $|\bar{T}_{NG} - \bar{T}_G|$ would be even further reduced.

For practical purposes, it may be concluded from these theoretical calculations that the noble gas temperature closely resembles the mean annual ground temperature at the water table, except in cases where the water table is very close to the surface (1 to 2 m), the infiltration rates are very high (exceeding several hundred $mm\,year^{-1}$), or the recharge does not have sufficient time for equilibration with the ground air (e.g. in karstic areas).

The annual mean ground temperature is then converted into an annual mean air temperature by using local rela-

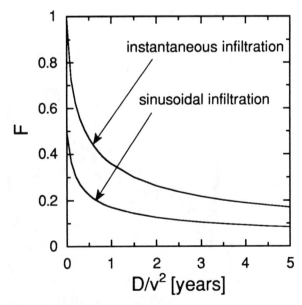

Fig. 5: Factor F as a function of D/v_z^2 for instantaneous infiltration (annual infiltration occurs in a single day) and sinusoidal variation of infiltration. F is a measure of the extent to which the difference mean annual ground and noble gas temperature is reduced (high F means large differences between $\bar{T}_G$ and $\bar{T}_{NG}$) by diffusive (dispersive) gas exchange across the water table.

tionships [e.g. Woeikof, 1904; Smith et al., 1964; Toy et al., 1978]. The paleoclimatic interpretation of noble gas concentrations in groundwater is more complicated if the depth of the water table is greater than about 40 m so that the geothermal gradient has to be taken into account. In that case a shift in noble gas temperature may not indicate a change in air temperature but a fluctuation of the depth of the water table caused by a considerably drier or more humid climate.

PRESERVATION OF CLIMATE SIGNALS IN GROUNDWATER SYSTEMS

After leaving the zone near the water table, groundwater is often isolated from the atmosphere and migrates to the discharge area, carrying information on past climate that was imprinted in the recharge area. Mixing processes in the aquifer, (macrodispersion) and leakage between aquifers (megadispersion) smooth the recorded climate fluctuations. Before the noble gas record in groundwater can be evaluated, the degree to which it is influenced by dispersion has to be estimated. In the following discussion, we address this problem concentrating on confined aquifers in which mixing processes are limited.

Macrodispersion

In order to study the effect of macrodispersion on the resolution of the climate record, smoothing of a high–frequency climate signal in a typical confined aquifer was simulated. Transport of dissolved gases in a confined aquifer can be described by the one–dimensional dispersion/advection equation [equation (8); Freeze and Cherry, 1979, Bear and Verruijt, 1987]. Groundwater is assumed to flow with a constant

velocity v_x to the discharge area carrying climate information recorded in the recharge area. As a proxy for (high frequency) ground temperature variations that occured over the last 150,000 years, a $\delta^{18}O$ record of benthic foraminifera obtained from deep sea cores [Shackleton et al., 1983] was used (this record mainly reflects fluctuations of the ice volume during the Pleistocene). The smoothing of the signal is again a function of the relative importance of dispersion and advection, $D/v_x^2 = \alpha_l/v_x$, with α_l being the longitudinal dispersivity.

For a large scale (of the order of 50 km), confined aquifer characterized by a typical flow velocity of $v_x = 1$ m year^{-1}, α_l is roughly 100 m [Domenico and Schwartz, 1990] resulting in $D/v_x^2 = 100$ years. The extent to which a climate signal is changed by dispersion can be derived from the comparison of the original climate record with the modeled record (Figure 6). The calculations show that smoothing of the climate signal increases with age. High–frequency fluctuations occurring on a time scale of less than about 1000 years have already completely disappeared at the very beginning of the record, while the main features of the glacial/interglacial transitions are still preserved, especially for the more realistic choice of $D/v_x^2 = 100$ years. Based on these results, it can be expected that for the last 30,000 years, the time scale accessible by ^{14}C dating, features related to the last glacial maximum and the preceding interstadial may be resolved, even if $D/v_x^2 = 1000$ years (Figure 6). The above model certainly simplifies the problem but it provides a first–order approximation.

Another way to express the function of a groundwater flow system as a low–pass filter is to calculate the characteristic time constant τ after which the amplitude of an original

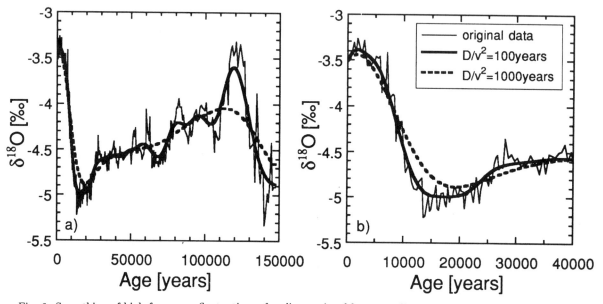

Fig. 6: Smoothing of high frequency fluctuations of a climate signal by macrodispersion in a regional groundwater flow system as a function of $D/v_x^2 = \alpha_l/v_x$ for time scales of 150,000 (a) and 40,000 years (b). $\delta^{18}O$ data for benthic foraminifera from Shackleton et al. [1983].

sinusoidal climate oscillation (frequency ω) is reduced to $1/e$:

$$\tau = \frac{v_x^2}{D\omega^2} = \frac{v_x}{\alpha_l \omega^2} \qquad (11)$$

The amplitude of the shortest Milankovich cycle, characterized by a period of 23,000 years, would be reduced to $1/e$ within 134,000 years, if $D/v_z^2 = 100$ years. This cycle will therefore be preserved on the ^{14}C timescale of 30,000 years, which is consistent with the results of the model discussed above.

Davidson and Airey [1982], while neglecting longitudinal dispersion, discuss the smoothing of climate signals caused by diffusive transversal exchange between flowing groundwater in an aquifer and its confining beds. The attenuation of climate fluctuations is a function of the transversal diffusion coefficient, the thickness and porosity of the aquifer and the aquiclude as well as the frequency of the climate fluctuation. For a diffusion coefficient of $0.01 \text{m}^2\text{year}^{-1}$, identical thickness, but different effective porosity of aquifer (0.2) and aquiclude (0.05), the amplitude of the 23,000 years cycle would be reduced to $1/e$ after 200,000 years. This is the same order of magnitude as the attenuation obtained by consideration of the effect of longitudinal dispersion.

Megadispersion

In addition to macrodispersion, mixing between aquifers frequently occurs. This can affect climate signals in the groundwater archive. Another source of problems are leaky well casings or wells with large open intervals which can result in mixing of groundwater of different age and origin. A careful study of chemical parameters and isotopic indicators (e.g. ^3H) of the groundwater flow system provides the basis for evaluating whether the climate signal obtained from a confined aquifer is disturbed by such processes [Mazor, 1991].

GROUNDWATER DATING

Measured noble gas temperatures can be converted into a paleoclimate record only if a reasonable chronology can be established for the groundwater flow system. The ^{14}C method is the only well established dating tool for groundwater on time scales extending to the late Pleistocene. Dating of groundwater with ^{14}C is complicated by the complex hydrochemistry of carbon that can alter the initial ^{14}C content. In order to derive reliable ^{14}C ages, the geochemical evolution of groundwater has to be taken into account and correction models have to be applied [e.g. Pearson and White, 1967, Tamers, 1967, Fontes and Garnier, 1979, Plummer et al., 1991]. Studies comparing different correction models [e.g. Phillips et al., 1989] imply that an error of at least ± 2000 years has to be assigned to ^{14}C ages of groundwater. For low ^{14}C concentrations corresponding to ages on the order of 30,000 years and higher, contamination of the sample and the limited accuracy of measurement gain importance, resulting in an increase of the errors to several thousand years. The limited resolution of the time axis, together with the effect of dispersion, prevent the resolution

of high–frequency climate signals in aquifers. However, the major emphasis of our noble gas temperature studies is to determine reliable temperature differences between the last glacial and the present interglacial which can be used to calibrate other paleothermomenters with higher time resolution. New dating methods (^{36}Cl [Bentley et al., 1986], ^{81}Kr [Lehmann et al., 1985]) might extend the time scale accessible by ^{14}C.

COMPARISON OF HOLOCENE NOBLE GAS TEMPERATURES
WITH GROUND TEMPERATURES

To check if the noble gas concentrations do indeed reflect mean annual ground temperature as suggested by model calculations discussed above, the noble gas temperatures of Holocene groundwaters from eight studies were compared with recent instrumental temperature records. For this comparison, we focus on studies of confined, ^{14}C-dated aquifers yielding climate records covering the Holocene and late Pleistocene period with a reasonable number of data points. The individual studies apply different models in order to correct for the complex hydrochemistry affecting the ^{14}C age. No attempt has been undertaken in this study to reassess the derived ^{14}C ages. Some studies incorporate only the accuracy of measurement, others the model error in the calculation of the age error bars. The use of error bars would therefore be misleading, and they were omitted from the following plots.

The individual noble gas records are discussed to provide background for comparison of the measured Holocene recharge temperatures and the ground temperatures estimated for the recharge areas (see Figure 7 and Table 1). As an example of the procedures used to compare ground and noble gas temperatures, the Great Hungarian Plain study is discussed in detail whereas the other aquifers are only discussed briefly.

1. Great Hungarian Plain (Hungary) [Stute, 1989, Stute and Deák, 1989, Deák et al., 1987, Stute and Sonntag, 1992]: The groundwater flow system of the Great Hungarian Plain consists of two aquifers: an intermediate flow system in Quaternary and a regional flow system in Pliocene layers. Noble gas temperatures were derived from Ne, Ar, Kr and Xe concentrations and hydrochemical corrections were performed on the ^{14}C ages.

Twelve samples were taken in the main infiltration area of the Quaternary flow system (Nyírség) representing groundwater that infiltrated during the last 10,000 years (^{14}C age) under climatic conditions similar to today's. Noble gas temperatures of those 12 samples may be compared to mean annual ground temperatures in the recharge area. The mean annual air temperature in the recharge area, measured by 6 meteorological stations is $9.7 \pm 0.4\,°C$ [Dési, 1967]. Using the average difference between mean annual ground and air temperatures of $1.7 \pm 0.5\,°C$ (ground temperature measured at 2 m depth at 12 stations distributed throughout Hungary [Dési, 1967]), the ground temperature for the recharge area may be estimated to be $11.4\,°C$. The average noble

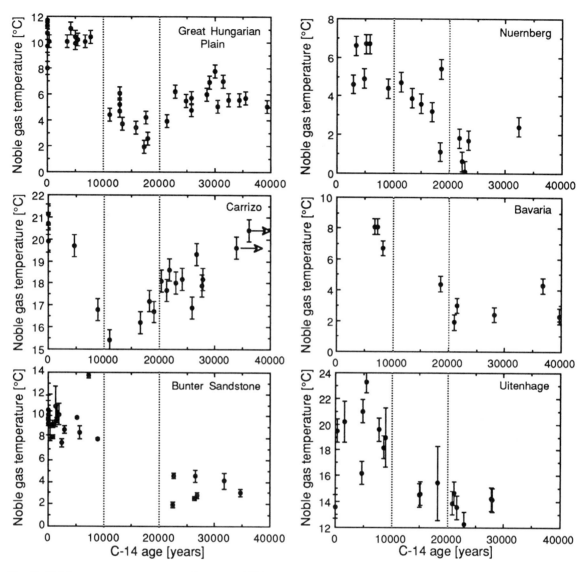

Fig. 7: Noble gas temperatures as a function of ^{14}C age for selected field studies. The dotted lines indicate the separation into 3 stages in analogy to the isotopic stages in ocean cores.

gas temperature of the 12 samples, (10.6 ± 0.7) °C, is very close to the present mean annual ground temperature, and its standard deviation, ±0.7 °C, nearly equals the measurement precision of ±0.5 °C. The difference between ground temperature and noble gas temperature of 0.8 °C may reflect the influence of vegetation on ground temperatures. The sites at which ground temperatures were measured have little vegetation. During most of the Holocene, the recharge area of the Quaternary flow system was probably covered with forests where ground temperatures would be more than 1 °C lower than ground temperatures at barely covered sites [Kappelmeyer, 1968]. In addition, it is possible that the mean annual air and ground temperature during the Holocene was lower than today's.

2. Carrizo aquifer (USA) [Stute et al., 1992, Pearson and White, 1967]: The Carrizo sandstone is part of the Gulf coastal plain in southern Texas and is confined by the slightly permeable Recklaw formation. Recharge temperatures were calculated using Ne, Kr and Xe data. ^{14}C-data for the samples are not yet available. The chronology was established on the basis of the ^{14}C ages determined by Pearson and White [1967].

3. Bunter Sandstone (UK) [Andrews and Lee, 1979, Bath et al., 1978]: The Bunter Sandstone aquifer of Triassic age is confined by thick overlying marls. Ar and Kr concentration were used, without performing a correction for excess air, to derive noble gas temperatures. However, the data indicate that the influence of this correction on the

TABLE 1. Noble Gas Temperature and Ground Temperature Data

Aquifer	location	1σ prec. [°C]	T_G[°C]	T_{NG}[°C] stage 1	T_{NG}[°C] stage 2	T_{NG}[°C] stage 3	ΔT_{NG}[°C] stage 1-stage 2	ΔT_{NG}[°C] stage 1-stage 3
Great Hungarian Plain Stute [1989]	21°E 47°45'W	±0.5°C	11.4 ±1	10.6±0.7 $N=12$	1.9±0.5 $N=1$	5.8±1 $N=14$	8.6±0.9	4.7±1.2
Carrizo Stute et al. [1992]	98°45'W 29°8'N	±0.5°C	21.8 ±1	20.6±0.6 $N=3$	15.4±0.5 $N=1$	18.1±0.6 $N=9$	5.2±0.8	2.5±0.8
Bunter Sandstone Andrews and Lee [1979]	0°30'E 53°25'N	±0.8°C	9.6 ±1	9.6±1.4 $N=18$	–	3.4±1.0 $N=7$	–	6.2±1.7
Nürnberg Rudolph et al. [1984]	11°E 49°30'N	±0.5°C	7.5 ±1	–	(0.1) $N=1$	(2.4) $N=1$	(7.4)	(5.1)
Bavaria Rudolph et al. [1984]	12°E 48°N	±0.5°C	7.5 ±1	8.1±0.5 $N=2$		3.0±1.1 $N=6$	5.1±1.2	
Uitenhage Heaton et al. [1986]	25°30'E 33°45'S	±0.9°C	19.0 ±1	19.5±2 $N=8$		14±1 $N=9$	5.5±2.2	
Kassel Rudolph et al. [1984]	9°30'W 51°20'N	±0.5°C	8 ±1	8.0±0.6 $N=25$				
Bocholt Stute and Sonntag [1992]	7°25'E 51°50'N	±0.5°C	10.7 ±0.6	10.7±0.9 $N=6$				

results would have been minor. ^{14}C ages were determined by Bath et al. [1978].

4. Nürnberg (Germany) [Rudolph et al., 1984, Rath, 1984, Geyh et al., 1984]: The hydrogeology of the Triassic sandstone near Nürnberg is fairly complex. δ^{18}O values, recharge temperatures derived from Ne, Ar, Kr and Xe, and conventional ^{14}C ages have to be interpreted as being the result of two–component mixing of Holocene and Pleistocene groundwaters, each with its own characteristics [Geyh et al., 1984]. For the Holocene and Pleistocene endmembers, noble gas temperatures of 7.5 and 0.1 °C were determined, suggesting a cooling of at least 7.4 °C between the two periods [Rath, 1984]. Paleotemperature estimates at this site depend on assumptions concerning the mixing ratios, and this introduces additional uncertainties.

5. Bavaria (Germany) [Rudolph et al., 1984, Eichinger et al., 1984]: The upper Tertiary aquifer system investigated in this study is located in southern Bavaria. Ne, Ar, Kr and Xe were used to determine the noble gas temperature. ^{14}C ages are based on an initial ^{14}C content of 85 pmc [Vogel, 1967].

6. Uitenhage (South Africa) [Heaton et al., 1986, Talma et al., 1984]: Fractured Ordovician quarzites crop out north of Uitenhage, and are overlain and confined to the east by impermeable Jurassic/Cretaceous shales and sandstones. Recharge temperatures were calculated from concentrations of dissolved atmospheric N_2 and Ar. Denitrification and ^{40}K–decay contributions to the nitrogen and argon concentrations seem to be insignificant. ^{14}C ages were corrected for closed system carbonate dissolution [Talma et al., 1984].

7. Kassel aquifer (Germany) [Rudolph et al., 1984, Rath, 1984]: The groundwater in the sandstone aquifer in the Kassel area is mainly of Holocene origin, as suggested by ^{14}C data. Only two data points seem to plot in the transition zone between Holocene and Pleistocene. The average of the remaining 25 noble gas temperatures (based on Ne, Ar, Kr and Xe) is (8.0±0.6) °C, which is identical to the present mean annual ground temperature in the area of (8±1) °C.

8. Bocholt aquifer (Germany) [Stute and Sonntag, 1992]: Samples for the Bocholt study were obtained from a multi-level well tapping a phreatic sand aquifer from the water table to a depth of 50 m. Tritium/^{3}He data indicate that the water is of recent origin [Schlosser et al., 1989]. The deeper groundwater was formed in an area covered with fields and meadows and yields a noble gas temperature of (10.7±0.9) °C ($N=6$). This value is practically the same as the ground temperature measured at a nearby site covered with the same type of vegetation (10.7±0.6) °C. However, the shallow groundwater that infiltrated beneath a forest is characterized by a noble gas temperature lower by 2.2 °C, which exceeds typical temperature differences between forests and fields [1.1 °C; Kappelmeyer, 1968] by a factor of about two. This indicates that not all the details of the impact of vegetation in the recharge area on the noble gas temperatures are fully understood.

Ground temperature measurements are available for the Great Hungarian Plain, and the Bocholt, Bunter sandstone and Uitenhage aquifers. In the other cases, the mean annual air temperatures were converted into mean annual ground temperatures using the relationships given by Woeikof [1904] for Germany and Smith et al. [1964] for the United States. The concordance of noble gas temperature and mean annual ground temperature is very high (Figure 8). This demonstrates that noble gas concentrations in groundwater are quantitative paleothermometers if well selected aquifers with favorable geohydraulic parameters are used as archives.

LATE PLEISTOCENE RECHARGE TEMPERATURES

In this section, published noble gas records are used to estimate mean annual ground temperatures during the last ice age. The most detailed record has been established for the Great Hungarian Plain. In this region, noble gas temperatures measured for the end of the Pleistocene period (10,000 to 40,000 years ago) are significantly lower than those measured for the Holocene. Noble gas temperatures at the time

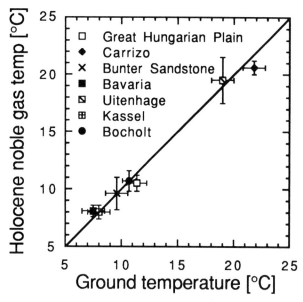

Fig. 8: Average Holocene noble gas temperatures for the field studies in comparison to recent ground temperatures in the corresponding recharge areas.

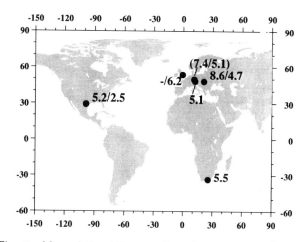

Fig. 9: Map of the differences in paleotemperature between Holocene (stage 1) and Pleistocene (stages 2, 3). The first number corresponds to the temperature difference between Holocene and last glacial maximum (stage 2), the second number to the difference between Holocene and stage 3. If only one number is given, no differentiation between stages 2 and 3 is possible.

of maximum glaciation, around 18,000 years ago, suggest a mean annual temperature at least 8.6 °C lower than the present temperature. Temperature during the preceding interstadial (28,000 to 35,000 years ago) was 4.7±1 °C lower than today (Figure 7).

Similar results were obtained in other extensive dissolved gas studies. These are presented in Table 1 and Figure 7. Based on the marine carbonate oxygen isotope chronology [Emiliani, 1966], the ^{14}C time scale in Figure 7 has been divided into the intervals 0 – 10,000 years (stage 1), 10,000 – 20,000 years (stage 2), and 20,000 – 40,000 years (stage 3).

In all the records, the temperature was distinctly lower during stage 2 and/or stage 3 than during the Holocene. In the Bunter Sandstone and possibly in the Bavaria record, data are available only for stages 1 and 3. This is probably because the ice sheets covering the U.K. and southern Germany during the last glacial maximum (stage 2) minimized or prevented infiltration of groundwater. In the other studies, noble gas data exist for stage 2 indicating that recharge was not interrupted for extended periods. However, only in the Great Hungarian Plain and the Carrizo studies is a temperature minimum during stage 2 observed. In the other records, either the difference between stages 2 and 3 was smoothed out, or the temperature in the recharge areas did not differ significantly between stages 2 and 3. Averaged paleotemperatures (and 1 σ errors) for the three stages as derived from the individual studies are summarized in Table 1, together with temperature changes between the stages for each aquifer. Values for the temperature difference between Holocene and last glacial maximum range from about 5 °C (Texas) to 9 °C (Hungary) and between the Holocene and the preceding interstadial from 2.5 °C (Texas) to 6 °C

(UK) (Figure 9 and Table 1). These data permit an independent reconstruction of paleoclimatic conditions during the late Pleistocene.

CONCLUSION

We have shown that the parameters influencing noble gas concentrations in groundwater can be understood quantitatively, justifying the use of noble gas measurements in paleotemperature studies. An advantage of this approach is that the concentrations of dissolved noble gases represent a direct physical indicator of climate and that they reflect true mean annual ground temperatures of high accuracy (1σ error ≈ ±0.5°C). These ground temperatures can be converted into mean annual air temperatures. The disadvantage of the noble gas paleothermometer is the fairly limited time resolution of the record due to mixing processes in the groundwater body and limited resolution of current groundwater dating methods. Paleoclimate studies using groundwaters as archives therefore require detailed understanding of the hydrogeology of the aquifer.

In combination with classical paleoclimatological methods (pollen analyses, diverse isotopic indicators of climate in lacustrine sediments, tree rings, speleothems, etc.) that are usually characterized by lower accuracy but better time resolution, the noble gas paleothermometer has a great potential to improve our knowledge of climate during the last ice age.

Acknowledgements. We are grateful to W.S. Broecker, C. Sonntag, and T. Stocker for helpful discussions and to S.M. Savin and two anonymous reviewers for their valuable comments. This study was supported by the Alexander von Humboldt Foundation (M.S.), the National Science Foundation (contract no. ATM 91–05538) and the Department of Energy's NIGEC program (contract no. HarvardCu01238101). L-DGO contribution no. 4991.

REFERENCES

Andrews, J.N. and D.J. Lee, Inert gases in groundwater from the Bunter Sandstone of England as indicators of age and paleoclimatic trends, *J. Hydrol. 41*, 233–252, 1979.

Bath, A.H., W.M. Edmunds, and J.N. Andrews, The hydrochemistry of a triassic sandstone aquifer, United Kingdom, in: *Int. Symp. on Isotope Hydrology*, IAEA, Vienna, 545–566, 1978.

Bear, J. and A. Verruijt, Modeling groundwater flow and pollution. D. Reidel Publishing Company, Dordrecht, 1987.

Bentley, H.W., F.M. Phillips, and S.N. Davis, Chlorine–36 in the terrestrial environment, in *Handbook of Environmental Isotope Geochemistry, Vol. 2B*, edited by P. Fritz and J.Ch. Fontes, Elsevier, 427–480, 1986.

Brook, G.A., M.E. Folkoff, and E.O. Box, A world model of soil carbon dioxide, *Earth Surface Processes and Landforms, 8*, 79–88, 1983.

Clever, H.L., (Ed.), Krypton, Xenon and Radon – Gas Solubilities, *Solubility data series, Vol. 2*, International Union of pure and Applied Chemistry, Pergamon Press, Oxford, 1979.

Craig, H., D. Burtner, and R.C. Wiens, Gravitational separation of gases and isotopes in Polar ice caps. *Science, 242*, 1675–1678, 1988.

Davidson, M.R. and P.L. Airey, The effect of dispersion on the establishment of a paleoclimatic record from groundwater. *J. Hydrol., 58*, 131–147, 1982.

Deák, J., M. Stute, J. Rudolph, and C. Sonntag, Determination of the flow regime of Quaternary and Pliocene layers in the Great Hungarian Plain (Hungary) by D, ^{18}C, ^{14}C and noble gas measurements, in *Isotope techniques in water resources development*, IAEA, Vienna, 335–350, 1987.

Dési, F.E.T., Magyarország Éghalati Atlasza, II. Kötet, A Dattár, Budapest, 1967.

Domenico, P.A. and F.W. Schwartz, Physical and Chemical Hydrogeology, John Wiley and Sons, 1990.

Eichinger, L., W. Rauert, W. Stichler, B. Bertleff, and R. Egger, Comparative study of different aquifer types using environmental isotopes, in *Isotope Hydrology*, IAEA, Vienna, 271–289, 1984.

Emiliani, C., Paleotemperature analysis of Carribean cores P6304-8 and P6304-9 and a generalized temperature curve for the past 425,000 years, *J. Geol., 74*, 109–126, 1966.

Fontes, J.C. and J.M. Garnier, Determination of the initial ^{14}C activity of the total dissolved carbon-A review of the existing models and a new approach, *Water Resour. Res., 15*, 399–413, 1979.

Freeze, R.A. and J.A. Cherry, Groundwater. Prentice–Hall, Englewood Cliffs, 1979.

Geyh, M.A., G. Backhaus, G. Andres, J. Rudolph, and H.K. Rath, Isotope study on the Keuper sandstone aquifer with a leaky cover layer, in *Isotope Hydrology*, IAEA, Vienna, 499–513, 1984.

Heaton, T.H.E. and J.C. Vogel, 'Excess air' in groundwater, *J. Hydrol., 50*, 201–216, 1981.

Heaton, T.H.E., A.S. Talma, and J.C. Vogel, Dissolved gas paleotemperatures and ^{18}O variations derived from groundwater near Uitenhage, South Africa, *Quaternary Res., 25*, 79–88, 1986.

Herzberg, O. and E. Mazor, Hydrological applications of noble gases and temperatures measurements in underground water systems: Examples from Israel, *J. Hydrol., 41*, 217–231, 1979.

Kappelmeyer, O., Beiträge zur Erschliessung von Thermalwässern und natürlichen Dampfvorkommen, *Geol. Jahrbuch 85*, 708–808, 1968.

Lehmann, B.E., H. Oeschger, H.H. Loosli, G.S. Hurst, S.L. Allman, C.H. Chen, S.D. Kramer, M.G. Payne, R.C. Phillips, R.D. Willis, and N. Thonnard, Counting ^{81}Kr atoms for analysis of groundwater, *J. of Geophysical Res., 90*, B13, 11547–11551, 1985.

Mazor, E., Paleotemperatures and other hydrological parameters deduced from noble gases dissolved in groundwaters, Jordan Rift Valley Israel, *Geochim. Cosmochim. Acta 36*, 1321–1336, 1972.

Mazor, E., Applied chemical and isotopic groundwater hydrology, John Wiley and Sons, 274pp, 1991.

Pearson, F.J., Jr. and D.E. White, Carbon-14 ages and flow rates of water in the Carrizo Sand, Atascosa County, Texas, *Water Resour. Res., 3*, 251–261, 1967.

Phillips, F.M., Noble gases in ground water as paleoclimatic indicators, Thesis, University of Arizona, 1981.

Phillips, F.M., M.K. Tansey, and L.A. Peeters, An isotopic investigation of groundwater in the central San Juan Basin, New Mexico: Carbon 14 dating as a basis for numerical flow modeling, *Water Resour. Res., 25*, 2259–2273, 1989.

Plummer, L.N., E.C. Prestemon, and D.L. Parkhurst, An interactive code (NETPATH) for modeling net geochemical reactions along a flow path, *Water-Resources Investigation Report 91-4078*, US Geological Survey, 1991.

Rath, H.K., Paläotemperaturbestimmungen aus Edelgasgehalten ^{14}C-datierter Grundwässer in Deutschland. Thesis, University of Heidelberg, 1984.

Rudolph, J., Edelgastemperaturen und Heliumalter ^{14}C--datierter Paläowässer. Thesis, University of Heidelberg, 1981.

Rudolph, J., H.K. Rath, and C. Sonntag, Noble gases and stable isotopes in ^{14}C-dated paleowaters from central Europe and the Sahara, in *Isotope Hydrology*, IAEA, Vienna, 467–477, 1984.

Schachtschabel, P., H.P. Blume, K.H. Hartge, and U. Schwertmann, Lehrbuch der Bodenkunde, Enke, Stuttgart, 1982.

Schlosser, P., M. Stute, C. Sonntag, and K.O. Münnich, Tritiogenic ^{3}He, in shallow groundwater, *Earth Planet. Sci. Lett., 94*, 245–256, 1989.

Schwander, J., B. Stauffer, and A. Sigg, Air mixing in firn and the age of the air at pore close-off, *Ann. Glaciol., 10*, 141–145, 1988.

Shackleton, N.J., J. Imbrie, and M.A. Hall, Oxygen and carbon isotope record of East Pacific core V19-30: Implications for the formation of deep water in the late Pleistocene North Atlantic, *Earth and Planet. Sci. Lett., 65*, 233–244, 1983.

Smith, G.D., F. Newhall, L.H. Robinson, and D. Swanson, Soil temperature regimes - their characteristics and predictability, *Rep. SCS-TP-144*, US Depart. Agricul., Soil Conservation Service, 1964.

Smith, S.P. and B.M. Kennedy, The solubility of noble gases in water and in NaCl brine, *Geochim. Cosmochim. Acta, 47*, 503–515, 1983.

Stute, M., Edelgase im Grundwasser – Bestimmung von Paläotemperaturen und Untersuchung der Dynamik von Grundwasserfließsystemen, PhD Thesis, University of Heidelberg, 1989.

Stute, M. and J. Deák,, Environmental isotope study (^{14}C, ^{13}C, ^{18}O, D, noble gases) on deep groundwater circulation systems in Hungary with reference to paleoclimate, *Radiocarbon, 31*, 902–918, 1989.

Stute, M. and. C. Sonntag, Paleotemperatures derived from noble gases dissolved in groundwater and relation to soil temperature, in *Isotopes of noble gases as tracers in environmental studies*, IAEA, Vienna, 111–122, 1992.

Stute, M, P. Schlosser, J.F. Clark, and W.S. Broecker, Paleotemperatures in the Southwestern United States derived from noble gas measurements in groundwater, *Science, 256*, 1000–1003, 1992.

Talma, A.S., J.C. Vogel, and T.H.E. Heaton, The geochemistry of the Uitenhage artesian aquifer: Carbonate solution in a closed system, in *Isotope Hydrology*, IAEA, Vienna, 481–497, 1984.

Tamers, M.A., Radiocarbon ages of groundwater in an arid zone unconfined aquifer, in: *Isotope Techniques in the Hydrologic Cycle*, edited by G.E. Stout, Geophys. Monogr. Ser., 11, 143–152, AGU, Washington, D.C., 1967.

Toy, T.J., A.J. Kuhaida, Jr., and B.E. Munson, The prediction of mean monthly soil temperature from mean monthly air temperature, *Soil Science, 126*, 181–189, 1978.

Vogel, J.C., Investigation of groundwater flow with radiocarbon, in *Isotope Hydrology*, IAEA, Vienna, 235–237, 1967.

Weiss, R.F., The solubility of nitrogen, oxygen and argon in water and seawater, *Deep-Sea Res., 17*, 721–735, 1970.

Weiss, R.F., Solubility of helium and neon in water and seawater, *J. Chem. Eng. Data, 16*, 235–241, 1971.

Woeikof, A., Probleme der Bodentemperatur. Typen ihrer vertikalen Verbreitung. Verhältnis zur Lufttemperatur, *Meteorologische Zeitschrift 21*, 50–62, 1904.

M. Stute and P. Schlosser, Lamont-Doherty Geological Observatory of Columbia University, Palisades, NY 10964.

Oxygen-Isotope Record in Recent Carbonate Sediments from Lake Greifen, Switzerland (1750 - 1986): Application of Continental Isotopic Indicator for Evaluation of Changes in Climate and Atmospheric Circulation Patterns

JUDITH A. MCKENZIE

Geological Institute, ETH-Zentrum, Switzerland

DAVID J. HOLLANDER

Department of Geological Sciences, Northwestern University, Evanston, IL 60208

Sedimentary sequences deposited in carbonate precipitating lakes can contain oxygen-isotope records of continental climate change. The isotope record in a short core from peri-Alpine Lake Greifen in northeastern Switzerland shows a decrease of about 2.5‰ in the $\delta^{18}O$ value of the lacustrine chalk from the end of the Little Ice Age at about 1865 to the present. We relate variations in this record primarially to changes in the isotopic composition of the run-off in the limited catchment area feeding the lake and, thus, to changes in the isotopic composition of local atmospheric precipitation. We attribute the observed longer term isotopic trend to a change in the dominant atmospheric circulation pattern in central Europe with a shift from prevailingly northwesterly circulation during the cooler climate of the Little Ice Age to a warmer climate with moisture being transported by more vigorous westerly to southwesterly winds.

INTRODUCTION

Over the past approximately 25 years, a new field of research - paleoceanography - evolved with the production of numerous stable isotope records from marine sediments in conjunction with systematic studies of the natural processes controlling stable isotopic fractionation in modern oceans. The emergence of this new field was officially recognized at the First International Conference on Paleoceanography in Zürich in 1983 and with the publication of a new AGU journal entitled "Paleoceanography" beginning in 1986. Cenozoic paleoclimates are now routinely interpreted from high-resolution isotope stratigraphies derived from sedimented biogenic carbonate.

During this same period, many stable isotope laboratories produced continental isotopic records of paleoclimate, but without the same co-ordination of effort seen in the field of paleoceanography. Perhaps, this is because continental records are contained in very diverse materials and a unifying system, comparable to the globally interconnected system of the world's oceans, has not been recognized for the continental environment. A major conclusion of the Chapman Conference on Continental Isotopic Indicators of Climate held in Jackson Hole Wyoming in June 1991 was that the study of lake sediments would undoubtedly yield the most comprehensive and easily obtainable and interpretable isotopic records of past continental climates. In fact, the subdiscipline of stable isotope stratigraphy within the field of paleolimnology could furnish continental paleoclimatic records in a fashion similar to those derived from paleoceanographic studies.

With this paper, we propose to follow-up on this conclusion and discuss the possible influence of climate change on the isotopic composition of precipitation and, as a consequence, lake water. Further, we propose that oxygen isotopic investigations of sedimentary sequences from a network of globally distributed lakes could serve as a continental equivalent to the oceans with the lakes being an interconnected system linked together through atmospheric circulation. To illustrate this point, we present below a case study from Lake Greifen, Switzerland, that shows the possible influence of changing climate and atmospheric circulation patterns over central Europe on the oxygen isotope record stored in the lacustrine sequence from the end of the Little Ice Age to the present.

ISOTOPIC PATTERNS IN ATMOSPHERIC PRECIPITATION

The evolution of the isotopic composition of atmospheric precipitation from the oceanic reservoir to the ice sheets at the poles can be depicted as a Rayleigh distillation process at liquid-vapor equilibrium [Epstein and Mayeda, 1953;

Climate Change in Continental Isotopic Records
Geophysical Monograph 78
Copyright 1993 by the American Geophysical Union.

Sigenthaler, 1979]. With increasing distance from the coast and at higher altitudes and latitudes, precipitation becomes progressively depleted in oxygen-18 [Dansgaard, 1964; Gat, 1980; Siegenthaler and Oeschger, 1980]. Other effects, such as precipitation temperature and amount of precipitation, help determine the isotopic composition. Warmer temperatures produce rainfall enriched in oxygen-18, whereas cooler precipitation is more depleted in oxygen-18. Also, the larger the amount of precipitation in an event, the more depleted the water will be. As a result of a combination of these effects, the isotopic composition of rainfall tends to vary systematically on a global scale.

Since 1961, the distribution of the isotopic composition of

rain water in a global network of meteorological stations has been monitored by the International Atomic Energy Agency (IAEA) in cooperation with the World Meteorological Organization (WMO). Contoured maps of the mean $\delta^{18}O$ values of precipitation collected from this global network show the distribution of modern isotopic patterns [Yurtsever and Gat, 1981; Rozanski et al., this volume]. These isotopic patterns can be related to modern patterns of atmospheric circulation. Figure 1 contours the long-term arithmetic mean $\delta^{18}O$ values for precipitation collected at the IAEA/WMO stations in central Europe [Rozanski et al., this volume]. In general, this figure shows that, with increasing distance from

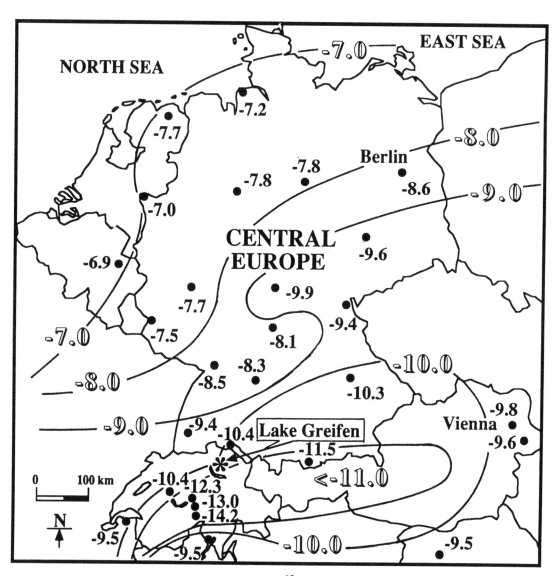

Fig. 1. Map showing the distribution of long-term arithmetic mean $\delta^{18}O$ values for precipitation collected at IAEA/WMO stations in central Europe. Data are from Rozanski et al. (this volume) and station locations are indicated by dots. The contours of $\delta^{18}O$ values graphically depict the continental and altitude effects, which result in isotopically lighter precipitation as the moisture moves inland away for the coasts and crosses mountain ranges. Note the location of Lake Greifen in the foreland region north of the Swiss Alps.

the coast to more inland stations, the $\delta^{18}O$ values become increasingly more negative, illustrating the continental effect. In addition, the $\delta^{18}O$ distribution mimics the regional topography with markedly more negative values associated with the mountain chains. This altitude effect is particularly noticeable along the east-west trending Alps of Switzerland and Austria (Figure 1).

The stable isotopic composition of local or regional precipitation regulates the composition of run-off, river and groundwater feeding a lacustrine system. The $\delta^{18}O$ value of a water body is further modulated by evaporation, relative humidity and its residence time. Autochthonous carbonates precipitated in isotopic equilibrium will reflect the isotopic composition of the lake water. Thus, oxygen-isotope stratigraphy in lacustrine sequences can provide a record of isotopic variations in the composition of the aqueous reservoir.

If evaporation is minimal and residence times short, the isotopic record can be related to changes in the isotopic composition of the inflowing water. In specific cases, this chemostratigraphic approach can provide a means of identifying changes in rainfall patterns that may be related to climatically induced changes in atmospheric circulation patterns, as well as placing such changes in a more or less calibrated time frame. In this case study, we present results from such a system, Lake Greifen, Switzerland, which is located in a small drainage basin north of the Swiss Alps (Figure 1). The long-term arithmetic mean $\delta^{18}O$ values for precipitation collected at two, relatively nearby IAEA/WMO stations located north of the Swiss Alps at Bern and Konstanz are both -10.4 ‰ [Rozanski et al., this volume]. A water profile from Lake Greifen sampled in Spring, 1978, after the winter overturn had thoroughly mixed the water column, gave $\delta^{18}O$ values between -9.8 and -10.0 ‰ [Weber, 1981]. The approximate equivalency of the isotopic composition of the precipitation and the lake water indicates that Lake Greifen, geographically situated between the two stations, contains water with a homogenized $\delta^{18}O$ value of the regional rainfall.

ISOTOPIC RECORD IN LACUSTRINE CARBONATES

In many lakes, the "planktonic" specimens used for oxygen and carbon isotope studies are authigenic carbonates that precipitate due to bicarbonate-carbonate disequilibrium resulting from photosynthetic activity, sometimes referred to as "bio-induced" authigenic carbonate precipitation [Kelts and Hsü, 1978]. As recorders of the carbon and oxygen isotopic compositions of surface waters, the precipitates can be used to monitor changes in surface-water conditions, such as changing temperature or salinity, and changes in the factors ultimately controlling the actual isotopic values of the water, such as changes in productivity or regional hydrology [McKenzie, 1985; Talbot, 1990]. In addition, their carbonate mineralogy can be related to environmentally controlled changes in water chemistry [Eugster and Hardie, 1978]. Carbonate secreting, bottom-dwelling organisms store the benthic isotopic signals and, together with the precipitate data, provide information on surface-to-bottom water isotope gradients.

The carbon and oxygen isotope compositions of carbonate precipitates from many lakes show a strong covariance with time. The occurrence of this isotopic covariance in lacustrine carbonates probably reflects the isotopic evolution of the lake water versus the recharge rate, whereby the strongest covariance is observed in closed systems or water bodies with extremely long residence times (Talbot, 1990; Talbot and Kelts, 1990). In contrast, weak to no covariance represents opens systems with short residence times, as has been observed for Lake Greifen with a residence time of about 1.5 years.

Assuming that authigenic and biogenic carbonates precipitate in isotopic equilibrium with the environment, their $\delta^{18}O$ values are a function of both the precipitation temperature and $\delta^{18}O$ value of the water. Frequently, in deeper lakes, bottom temperatures remain fairly constant permitting estimates of the oxygen-isotope composition of the water from the $\delta^{18}O$ values of shells produced by bottom-dwelling ostracods and mollusks [Lister, 1988]. In contrast, the temperature of the surface waters can vary considerably over an annual cycle depending upon the lake's geographic location, and, thus, the precipitation temperature for authigenic carbonate cannot be considered a constant. However, if it can be estimated that the bulk of the carbonate precipitates at approximately the same time during the annual cycle when the range of the surface-water temperatures is about the same each year, variations in the $\delta^{18}O$ values of the sedimented precipitate may primarily represent past changes in the isotopic composition of the water. Thus, changes in the $\delta^{18}O$ value of the water due to changes in the rate of evaporation or source of the water are often larger than can be accommodated by changes in the precipitation temperature. In general, lakes respond quickly to environmental pressures and, because of the smaller size of the reservoirs, the isotopic signals of such perturbations are amplified [McKenzie, 1985]. Based on these general assumptions for isotopic fractionation in lacustrine systems, this paper presents a case study from Lake Greifen, Switzerland, that demonstrates the application of oxygen isotope stratigraphy to document recent changes in the continental climate of central Europe from the end of the Little Ice Age to the present.

LAKE GREIFEN STUDY

Description

Lake Greifen is a small, shallow, peri-Alpine, "hard-water" lake located at altitude of 435.3 m in a densely populated area approximately 8 km east of Zürich, Switzerland (Figure 2). The lake trends northwest-southeast and is approximately 6.5 km long and 0.8 to 1.85 km wide. It has a volume of 151 x 10^6 m^3, surface area of 8.5 x 10^6 m^2, and maximum depth of 32 m [Jung, 1969; Weber, 1981]. Water discharge from Lake Greifen is controlled by the Glatt River which continues flowing northwestwards to join the Rhine. The residence time of water in the lake is 505 days. The regional climate is classified as temperate with an average annual mean temperature for the period 1864-1988 being 8.2°C [Chen, 1990]. The mean summer season temperature for the same period was 16.5°C. Rainfall data collected from 1901-1988 record an average annual precipitation of 1124 mm/yr, with slightly less than half (515 mm/yr) falling during the winter half-year [Chen, 1990].

LAKE GREIFEN
SWITZERLAND

Fig. 2. Location map of Lake Greifen in northeast Switzerland. The coring site, marked by a square, is situated at the deepest point of the lake at 32 m, where primarily pelagic sedimentation occurs.

The basin catchment area is a glacially eroded valley comprising carbonate-bearing Quaternary lacustrine, moraine gravel, and middle Miocene fresh-water molasse deposits. The two major rivers entering the lake, the Uster-Aa and Mönchaltorfer-Aa, represent the drainage of over 80 % of the total catchment area of about 150 km^2 and account respectively for 51 % and 28% of the inflowing water [Pleisch, 1970]. The maximum elevation within the drainage region is about 1200 m. The influence of snow melt or winter precipitation to the run-off can be seen in the seasonal changes in the $\delta^{18}O$ values of the river waters. The Uster-Aa has a $\delta^{18}O$ value of -10.8 ‰ in the winter and -9.7 ‰ in the summer months, whereas Mönchaltorfer-Aa has a $\delta^{18}O$ value -11.4 ‰ in the winter and between -10.0 and -10.3 ‰ in the summer months [Weber, 1981]. During the summer months, the lake becomes themally stratified and the epilimnion $\delta^{18}O$ value is decoupled from that of the hypolimnion, which retains the winter turnover value of between -9.8 and -10.0 ‰. Surface water data collected between 1976 and 1978 indicate a $\delta^{18}O$ deviation from a winter value of -9.8 ‰ to a maximum summer value in September of -9.1 ‰ [Weber, 1981].

During the past 100 years, Lake Greifen has undergone major changes in the chemistry of the water column and sediments in response to an augmented nutrient influx resulting from increased agriculture and industrialization in the catchment area surrounding the lake. Nutrient loading has promoted increasing productivity and progressive eutrophication of the lake, which reached a maximum in 1974. Since that date, the Swiss government has imposed strict regulations on the nutrient influx into the lake leading to diminished eutrophic conditions but with continued severe annual anoxia in the hypolimnion.

Sedimentation History

Short gravity cores taken from the center of the lake at a depth of 32 m (Figure 2) contain a pelagic record of its changing redox state during the past 250 years, which is observed in the changing sedimentation pattern [Weber, 1981; McKenzie, 1982; Hollander, 1989]. The lowermost sediments are weakly laminated marls followed by a transition interval to seasonally varved sediments. The transition from marl to varve sedimentation is accompanied by an increase in the percent carbonate (Figure 3), as well as by large increases in the content of organic carbon from an average of about 1.5 wt. % in the marls to values as high as 5.0 wt. % in the varves deposited during the years when the lake was most eutrophic with highest rates of productivity (1970-77). The core can be divided into three distinct zones: (1) varved zone (0-25.5 cm), (2) transition zone (25.5-35 cm) and (3) marl zone (35-70 cm).

The varved zone contains laminae couplets which consist of grey to pale yellow inorganically precipitated calcareous layers rich in organic matter, rhombohedral calcite crystals, diatoms and pyrite alternating with black gelatinous layers which are relatively richer in clays and poorer in organic matter and carbonate. The laminae in the varved and uppermost transition zones were separated for geochemical analyses into light ("summer") and dark ("winter") lamina. The assignment of "winter" and "summer" laminae is based on water-column flux studies [Weber, 1981]. A comparison of the percent carbonate content for the discrete lamina shows the tendency for larger amounts of authigenic carbonate precipitation during the spring and summer months [Hollander, 1989]. Because the calcite precipitates as a result of biologically induced disequilibrium during photosynthetic blooms, the percent carbonate content reflects the intensity of the annual productivity cycle.

The transition zone comprises yellow-grey calcareous laminae with progressively diminishing black laminae down section. The marl zone occurs below 35 cm and is clearly distinguished from the other zones by the lack of black, metastable sulfide striations and the degree of compaction. These sediments are yellow-grey marl which are weakly laminated to the base of the sequence but rarely contain pyrite. Although the sediments in the marl zone show no characteristic sediment structures indicating macro or microbioturbation, the fine mm scaled varving documented in the overlying zones was not observed. At 57 cm, there is a distinctive 1 cm-thick reddish-brown turbidite horizon that can be use to correlate cores from various locations within the lake.

The dating of the studied core is based on varve counting in the varved zone (1960-86) and correlation with previously dated lithologic changes in Lake Greifen cores, which relied on radiometric methods (^{14}C, ^{210}Pb and ^{137}Cs) for age determinations [Gäggeler et al., 1976; Emerson and Widmer, 1978; Tschoop, 1979; Wan et al., 1987]. The varve ages, as well as the radiometric ages, are shown in Figure 3. In the studied core, the uppermost summer varve was deposited in 1986 and the base of the varved zone was radiometrically dated at 1930. The varved zone represents the onset of intense eutrophication with extreme bottom-water anoxia and sedimentation of seasonal laminae, which has continued to the present. The transition zone appears to have been deposited under increasingly anoxic conditions with progressively increasing eutrophication between the base of the zone, dated at 1887, and its top. The marl zone with sediments deposited under oxygenated conditions continues to the base of the core. This latter zone shows an upward increase in percent carbonate

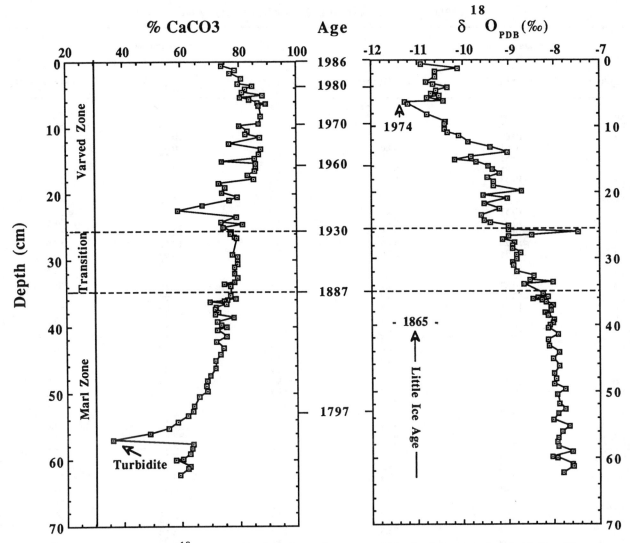

Fig. 3. Percent carbonate and δ¹⁸O stratigraphies for a short sediment core taken at the deepest point in Lake Greifen. The ages of the lithologic zones and specific events were obtained by varve counting and correlation to lithologic changes that were previously dated using radiometric techniques.

content (Figure 3) indicating that production of authigenic calcite may have been increasing while the bottom waters remained basically oxygenated during the entire year. A sample from 4 cm above the turbidite layer gave a ^{210}Pb radiometric age of 1797. Using an average sedimentation rate of 0.2 cm/yr calculated for the interval deposited between 1797 and 1887, an extrapolated age of 1750 is obtained for the base of the core.

Carbonate and Oxygen Isotope Stratigraphy (1750 - 1986)

The high rates of productivity in the modern Lake Greifen water column provide ideal conditions for studying the influence of productivity on the carbon isotope cycle within an eutrophic lake [McKenzie, 1982, 1985; Hollander, 1989; Hollander et al., 1992]. From our studies of Lake Greifen, we developed lake models for interpreting carbon-isotope

variations in the sedimented authigenic calcite and autochthonous organic matter which are applicable to both lacustrine and marine environments. Our research efforts concentrated on the record of carbon isotope changes, ignoring the corresponding oxygen-isotope record in the recent sediments. This paper discusses the equally interesting oxygen-isotope record and its implications for continental climate change during the approximately last 250 years.

In the varved zone of the Lake Greifen core, it was possible to separate the light and dark laminae and measure their discrete carbonate and isotopic compositions (Table 1). The percent carbonate and oxygen isotope data determined on bulk samples from the transition and marl zones are given in Table 2. In general, the sediment can be designated a lacustrine marl to chalk with the percent carbonate increasing from 60 to 75 % upsection within the marl zone and reaching about 80 % in the transition zone. "Summer" varves can have carbonate values

TABLE 1. Data for Lake Greifen Light and Dark Varve Couplets

Light Laminae			Dark Laminae		
Depth (cm)	$\delta^{18}O_{PDB}$ (‰)	$CaCO_3$ (wgt%)	Depth (cm)	$\delta^{18}O_{PDB}$ (‰)	$CaCO_3$ (wgt%
0.6	-10.09	73.9	0.8	-9.95	72.4
1.3	-10.11	78.7	1.6	-9.89	51.3
1.7	-10.60	76.9	2.4	-10.21	49.3
2.5	-10.61	80.6	3.1	-10.08	69.9
3.3	-10.81	79.8	3.5	-9.90	44.3
3.7	-10.65	84.9	3.9	-9.78	64.4
4.1	-10.33	82.3	4.2	-9.68	53.1
4.6	-10.58	81.3	4.8	-10.01	52.3
5.1	-10.69	88.4	5.3	-10.34	65.4
5.4	-10.51	80.4	5.5	-10.31	33.9
5.7	-10.78	83.7	5.9	-10.39	61.9
6.1	-10.43	86.5	6.3	-10.17	67.1
6.4	-11.27	89.5	6.6	-10.51	60.6
6.7	-11.21	87.0	7.1	-10.64	66.3
			7.9	-10.20	61.1
8.2	-10.77	87.6	8.9	-10.16	42.6
9.3	-10.37	86.9	9.6	-9.85	51.0
9.7	-10.41	80.0	10.1	-9.69	45.4
10.4	-10.41	82.8	10.5	-9.86	56.9
10.9	-10.34	82.4	11.2	-9.57	74.7
11.4	-10.09	87.1	11.7	-9.23	65.1
12.4	-9.88	76.5	12.6	-9.19	84.4
13.1	-9.40	87.4	13.4	-8.77	70.1
13.9	-9.02	86.9	14.4	-9.13	63.1
14.6	-9.82	85.2	14.8	-9.43	67.1
15.0	-10.18	74.2	15.2	-9.44	54.7
15.4	-9.72	85.7	15.8	-8.92	78.2
16.0	-9.44	86.0	16.3	-8.77	67.2
16.4	-9.36	85.5	16.6	-8.96	73.8
17.1	-9.19	82.9	17.3	-9.00	77.0
17.7	-9.47	85.2	17.9	-8.87	61.4
18.3	-9.34	72.9	18.5	-8.73	66.2
19.0	-9.34	75.3	19.2	-8.62	65.3
19.8	-8.72	74.2	19.9	-8.74	65.0
20.5	-9.56	79.4	20.1	-8.99	73.7
20.9	-9.03	76.6	20.7	-8.68	69.2
21.7	-9.54	67.3	21.1	-8.71	62.8
22.5	-9.21	58.8	21.9	-8.91	72.8
23.4	-9.60	79.1	22.7	-9.04	58.8
24.3	-9.53	73.8	23.5	-8.84	77.7
24.6	-9.39	81.1	24.4	-9.05	68.7
25.1	-9.00	74.4	24.8	-8.88	64.5

approaching 90 %, whereas the "winter" varves are poorer in carbonate (Table 1). The carbonate mineral is predominantly calcite. The proportion of the calcite that may be of detrital origin can be roughly estimated from the amount of detrital dolomite present, as determined by X-ray diffraction analysis. The ratio of calcite to dolomite in the average detrital flux to the lake was determined to be 2:1 (McKenzie, unpublished data). The low dolomite contents determined from the X-ray diffraction analysis indicate that the bulk carbonate is primarily composed of an authigenic calcite with detrital carbonate being never more than 15 % of the sample and most frequently less than 5 %.

Isotopic measurements were made on CO_2 gas derived from the bulk carbonate. The $\delta^{18}O$ values of bulk carbonate in the entire measured profile range from -7.6 ‰ near the base to -11.3 ‰ near the top. Within the marl zone, the $\delta^{18}O$ values

steadily decrease from -7.58 ‰ at 62 cm to -8.45 ‰ at 35 cm. The decrease continues through the transition zone reaching -9.01 ‰ at 25 cm. Within the varved zone, the decrease in the $\delta^{18}O$ value appears to accelerate with the carbonate in the "summer" varves being more negative than in the "winter" varves. More specifically, the most negative $\delta^{18}O$ values of -11.27 ‰ and -10.51 ‰ were obtained in the 1974 "summer" and "winter" varves, respectively. Since 1974, the $\delta^{18}O$ value has become progressively less negative, increasing to -10.09 ‰ in the "summer" laminae of 1986.

The average $\delta^{18}O$ values for the detrital calcite and dolomite flux to the lake are -8.2 and -3.5 ‰, respectively. Considering these values and the low percent of detrital carbonate in the sediments, the observed trend towards increasingly more negative $\delta^{18}O$ values in the section cannot be explained as a

TABLE 2. Data for Lake Greifen Marls

Depth (cm)	$\delta^{18}O_{PDB}$ (‰)	$CaCO_3$ (wgt%)
25.6	-9.00	77.2
26.0	-7.47	76.8
26.4	-8.48	78.2
26.6	-8.99	78.9
27.1	-9.14	------
27.5	-8.87	------
28.0	-8.90	------
28.5	-8.91	------
29.2	-8.74	77.5
29.4	-8.82	79.4
30.1	-8.82	79.3
30.6	-8.91	79.3
31.1	-8.88	78.4
32.0	-8.82	78.2
32.7	-8.45	79.5
33.2	-8.54	78.3
33.6	-8.03	74.8
33.9	-8.66	77.0
35.3	-8.24	77.0
35.8	-8.14	78.5
35.9	-8.36	75.9
36.1	-8.46	74.8
36.3	-8.26	69.8
36.6	-8.17	75.4
37.1	-8.03	72.1
37.4	-8.06	71.6
37.9	-8.07	72.7
38.2	-8.21	71.7
38.6	-8.13	78.0
39.2	-8.00	72.4
39.7	-8.02	73.7
40.0	-8.09	75.4
40.5	-8.14	72.2
41.5	-7.91	75.4
42.3	-8.14	72.1
43.3	-8.11	74.3
44.2	-7.89	73.3
45.1	-8.03	71.4
46.2	-7.88	71.5
47.3	-8.01	69.7
48.1	-7.95	68.6
49.0	-8.01	68.2
49.8	-7.76	68.6
50.5	-7.94	65.8
51.9	-7.89	64.2
52.8	-7.75	63.8
53.4	-7.92	62.1
54.3	-8.02	58.5
55.2	-7.66	55.0
56.1	-7.83	48.7
57.0	-7.91	35.8
57.7	-7.94	63.7
58.3	-7.90	63.2
59.0	-7.60	62.5
59.8	-8.04	60.2
60.1	-7.93	57.8
60.9	-7.61	62.5
61.3	-7.58	62.0
62.2	-7.80	59.2

function of varying detrital flux but probably represents a change in environmental conditions at the time of calcite precipitation. In addition, there is no statistically significant correlation between the percent carbonate and $\delta^{18}O$ values of the samples. Thus, the observed isotope trend towards increasingly more negative values reflects either a change in

the precipitation temperature or the isotopic composition of the water.

The offset between the two data sets for "summer" and "winter" varves (Table 1) is certainly a temperature phenomena because, based on sediment-trap studies, most of the authigenic calcite is precipitated during seasonal blooms in the warmer summer months (June-July), when near surface-water temperatures are between 18 and 25°C (ave. 21.5°C) [Weber, 1981]. The calcite of the "winter" varves probably reflects cooler precipitation temperatures between 16 and 20°C (ave. 18°C) during September-October when fall blooms occur. The mean average difference between the $\delta^{18}O$ values for the "summer" and "winter" laminae listed in Table 1 is 0.52±0.25 ‰. Assuming a change in $\delta^{18}O$ of 0.2 ‰ represents a temperature change of approximately 1°C [Epstein et al., 1953], the measured difference between the varves represents a difference in the precipitation temperature of 2.6±1.2°C, a value approximately equal to the average measured difference of about 3.5°C in near surface-water temperatures at the time of the two seasonal blooms.

Precipitation Temperature vs. Isotopic Composition of Water

During photosynthetic blooms, authigenic calcite precipitates from the surface waters of Lake Greifen as a result of biologically induced disequilibrium among the dissolved inorganic carbon species. Under conditions of isotopic equilibrium, the precipitating calcite incorporates an oxygen isotope ratio which is dependent on the temperature and isotope ratio of the water. In order to interpret isotopic records from the geologic past in terms of changing temperature or $\delta^{18}O$ of the water, it is necessary to separate these two effects. This is not always an easy process, unless one or the other effect can be considered constant. In the case of the Lake Greifen oxygen isotope record, both effects are probably important and must be considered.

The Lake Greifen sequence (Figure 3) shows a decrease in mean average $\delta^{18}O$ value from -7.99±0.20 ‰ in the marl zone (1750-1887) to -8.67±0.40 ‰ in the transition zone (1887-1930). Within the varved zone, the period from 1930 to 1964 has a mean average $\delta^{18}O$ value of -9.40±0.34 ‰, which afterwards trends towards increasingly more negative obtaining a minimum of -11.3 ‰ in 1974. Since then there was a shift toward less negative values, which would correspond to the observed trend measured in regional precipitation in recent years [Rozanski et al., 1992]. The mean average $\delta^{18}O$ value in the more recent varves from 1964 to 1986 is -10.55±0.32 ‰. Overall, the average $\delta^{18}O$ value in the short core has decreased by about 2.5 ‰ since the end of the last century. If the isotopic composition of the water remained constant and this decrease were only due to temperature, this trend would represent a steady increase of up to 12.5°C in the surface-water temperatures at the time of maximum calcite precipitation.

Such a large temperature increase seems very unlikely even considering the increased nutrient flux and increased productivity in the lake with larger amounts of calcite precipitating in the warmer summer months. Although the seasonal primary productivity cycle can begin in late April when the surface-water temperature is about 12°C, the precipitation of large amounts of calcite occurs primarily in June and July when the surface waters are the most saturated with respect to calcite and temperatures range from 18 to 25°C [Weber, 1981]. Smaller amounts of calcite can continue to precipitate throughout August into the latter half of September or early October, a period with a similar but decreasing temperature range. Thus, considering the extreme end situations, i.e. if all of the calcite had formerly precipitated at the cooler end of the range (12°C) and now precipitates at the warmer end (25°C), the observed 2.5 ‰ decrease in the $\delta^{18}O$ value of the carbonate could theoretically be fully accommodated by a temperature change at the time of precipitation.

A temperature scenario alone, however, seems unlikely because the data for the varves indicate that changes in the $\delta^{18}O$ value of the water have occurred and could account for at least part of the observed long-term decrease. For example, the consistently parallel trends observed in the $\delta^{18}O$ values for both the "summer" and "winter" varves (Table 1) indicate that the recorded change cannot simply be interpreted as a temperature phenomena. The temperature differences at the time of precipitation can be observed between the varve couplets. As mentioned above, the measured difference between the varve couplets represents an average difference in the precipitation temperature of 2.6±1.2°C, a value approximately equal to the average measured difference of about 3.5°C in water temperature at the time of the two seasonal blooms. The laminae are recording an averaged temperature over the bloom seasons and not the extremes. Thus, the offset between the two data sets for "summer" and "winter" varves is certainly a temperature phenomena, but the overall $\delta^{18}O$ trends in the couplets overprint this seasonal difference. We, therefore, propose that there has been a change in the $\delta^{18}O$ value of the water during the period of varve sedimentation and, most likely, the longer term trend towards more negative values begin at the end of the Little Ice Age around 1865 (Figure 3).

This interpretation is contrary to an earlier interpretation of a more limited data set collected only for summer laminae deposited between 1955 and 1978 [McKenzie, 1985]. A strong negative correlation between the $\delta^{18}O$ values and the corresponding $\delta^{13}C$ values, as well as % TOC, for laminae deposited during this period was previously interpreted to mean that the calcite precipitation continued to occur during the warmer months with the increasing availability of nutrients to supply longer productivity periods. This interpretation was supported by the fact that temperature data collected during this period showed no change in the annual warming-cooling cycle of the surface water and there was no indication of a dramatic warming during the time of calcite precipitation (Ambühl, unpublished EAWAG data). With the more extensive data set presented in this paper, we reinterpret the significance of the oxygen-isotope record in Lake Greifen. While conceding that the anthropogenic eutrophication can extend the period of calcite precipitation during the warmer summer months, it cannot account for the total isotope change, and we propose that the long-term Lake Greifen isotope record in the short core is primarily monitoring the

evolution of the isotopic composition of the lake's water body over the past 250 years.

CLIMATE CHANGE AND ISOTOPIC COMPOSITION OF LAKE GREIFEN

Isotopic Composition of Rainfall vs. Temperature

As mentioned above, the isotopic composition of atmospheric moisture is dependent on a number of factors, including temperature, amount, latitude and altitude of precipitation. These factors combine to produce distinct isotopic precipitation patterns that can be traced on a global or regional scale, such as central Europe (Figure 1). In order to change the isotopic composition of the water infilling Lake Greifen by 2.5 ‰ since the end of the last century, it would be necessary to modify the combination of factors which determine the isotopic composition of the local precipitation. Two possible ways to accomplish this would be to change the mean annual surface air temperature and/or timing of precipitation.

Surface air temperature is a very important controlling factor as shown by the pronounced increase of about 1.5 to 2.0 ‰ in the $\delta^{18}O$ values of precipitation, which has been measured in recent years at the IAEA/WMO stations in central Europe and is attributed to an apparent warming of about 1.5°C.[Rozanski et al., 1992]. The 30-year IAEA/WMO record further indicates that long-term changes in $\delta^{18}O$ values of precipitation follow long-term changes in surface air temperature with an average $\delta^{18}O$-temperature coefficient of about 0.6 ‰ / °C [Rozanski et al., 1992].

If the postulated 2.5 ‰ decrease in the $\delta^{18}O$ value of precipitation feeding Lake Greifen over the considered period (1887-1986) were the result of temperature change, a 4.2°C decrease in annual air surface temperature would be required, as calculated using the above coefficient. In fact, the temperature records from 1850-1980 indicate that there has been a long-term increase of about 0.8°C in the mean annual temperature of Switzerland [Chen, 1990]. On the other hand, precipitation records for the period 1860-1980 show an increase in the amount of annual precipitation by 180 mm, of which 110 mm falls during the winter half-year [Chen, 1990]. As cooler precipitation is isotopically lighter, a contribution of more winter rain to Lake Greifen could account for some of the postulated decrease in the isotopic composition of the water. It is questionable, however, whether the total 2.5 ‰ decrease could be accommodated by this explanation considering that the difference between winter and summer $\delta^{18}O$ values for the two main rivers entering Lake Greifen is only about 1.1 ‰ [Weber, 1981]. A larger difference in the isotopic composition of source waters is required to explain the observed 2.5 ‰ decrease.

Changing Atmospheric Circulation Patterns

Another potential mechanism associated with climate change that could significantly alter the isotopic composition of the water infilling Lake Greifen would be to modify the predominant atmospheric circulation pattern that transports moisture into the region. A climatic cooling, known as the Little Ice Age, is well documented in Europe during the historical period from the middle of the 13th century until about 1850. Within Switzerland, historical records of glacial advances and retreats indicate that climate has changed during the 250-year period represented by the Lake Greifen short core. For example, the advance and retreat of the lower Grindelwald glacier in central Switzerland, which has been monitored since around 1600, may be one of the strongest and most sensitive indicators of climate change. The lower Grindelwald glacier underwent major, long lasting extensions during the Little Ice Age. Compilation of historical records indicates that these advances coincide with an uninterrupted series of years having cold, wet summers and overall lower seasonal mean temperatures [Messerli et al., 1978]. Since about 1860, the lower Grindelwald glacier has been receding, which is consistent with the increase of more than 0.8°C in the mean annual temperature of Switzerland. Although the historical temperature records going back to 1755 indicate that overall summer temperatures have remained fairly stationary, spring and fall temperatures have tended to be warmer since 1860 [Messerli et al., 1978; Chen, 1990]. Also, occasional hot and dry summers seem to promote glacier regressions. In the last century, the total ice-covered area in the Alps has diminished from 4368 ± 54 km^2 in the 1870s to 2909 km^2 in the 1970s [Chen, 1990].

We propose that the advance and retreat of the lower Grindelwald glacier in central Switzerland follows a pattern that could be compared with the long-term Lake Greifen oxygen isotope record with a particular emphasis on the significant change in isotopic composition in the marl zone beginning at about 1865 (Figure 3). In addition, we propose that changes in the amount and seasonality of precipitation as evidenced by glacier advances and retreats can be correlated with changes in the isotopic composition of concurrent rainfall. Hypothetically, more rainfall during colder summers or winters should result in isotopically more negative precipitation than during warmer, perhaps drier periods. If correct, this would indicate that the Lake Greifen trend towards more negative values during the modern warmer period is not a function of temperature but may be a result of changes in the general atmospheric circulation pattern, i.e. changes in the predominant wind direction carrying the moisture and, hence, the path along which it has evolved isotopically. Evidence for such an atmospheric circulation change can be reconstructed from historical records of barometric pressure measurements.

Using available barometric pressure at sea level data, monthly mean maps for January and July during specific time intervals have been reconstructed to define the character of atmospheric circulation prevailing during extreme cold phases of the Little Ice Age to compare with extreme warm periods within the 20th century [Lamb, 1977]. These reconstructions show that mild winters in 20th century Europe are associated with strong circulation giving rise to vigorous flow of westerly to southwesterly winds with markedly anticyclonic patterns over western and central Europe during warm summers. In contrast, the prevailing west wind flow over Europe was weakened during cold winters and the transport of oceanic air from the west was less frequent. This led to winds of variable direction, sometimes becoming easterlies with lower average winter temperatures. During the cooler summers, the windstream over central Europe took on a more northwesterly prevalence with less anticyclonic influence [Lamb, 1977].

In light of these reconstructions, we conclude that major changes in atmospheric circulation patterns over central Europe from the end of the Little Ice Age into modern times strongly influenced the isotopic composition of the rainfall and was the predominant cause of the long-term 2.5 ‰ change in the isotopic composition of meteoric water flowing into Lake Greifen. In order to produce the increasingly more negative values in the younger sediments, we would suggest that the oxygen-18 depleted moisture carried by the westerly to southwesterly winds into the Swiss forelands (modern $\delta^{18}O = $ ~-10 ‰) has contributed increasingly more to the annual precipitation north of the Alps (Figure 4). This current meteorologic situation is in contrast to that proposed for the Little Ice Age when the predominantly northwesterly winds (modern $\delta^{18}O = $ ~-8 ‰) could have carried isotopically heavier moisture emanating from the lowlands to the northwest of Switzerland (Figure 4). Although the absolute isotopic composition of the Little Ice Age precipitation in central Europe may not have been the same as the modern values depicted in Figure 4, the relative isotopic difference (~2 ‰) between the two sources and distribution pattern may have been similar. Thus, predominantly northwesterly winds would have brought isotopically heavier rain to the peri-Alpine region of northern Switzerland than under the present conditions of westerly to southwesterly dominance. A change from the predominance of one metrological system to the other could account for a large protion of the observed 2.5 ‰ decrease in the long-term isotopic record from Lake Greifen. The record (Figure 3) indicates that the proposed isotopic transition from a northwesterly to westerly-southwesterly wind dominance may have begun rather gradually at the end of the Little Ice Age. After about 1865, the rate of change increases with more extreme variability in the signal. More detailed isotopic records from central European lakes are needed to test these preliminary conclusions concerning the influence of changing atmospheric circulation on the isotopic composition of local meteoric water.

CONCLUSIONS

Deciphering the global significance of isotope records from modern and ancient lakes located in various climatic regions remains a challenge requiring coordinated studies and an imaginative approach to understanding the multiple factors controlling isotopic fractionation in each new lacustrine system. The intriguing oxygen isotope record from Lake Greifen presented in this case study demonstrates the potential for applying isotope stratigraphy to lacustrine carbonates in order to obtain continental records of possible modifications of atmospheric circulation associated with climate change.

As with marine sediments, stable isotope stratigraphies developed from lacustrine sequences can contain critical records of past climate changes that will become increasingly more important as we strive to understand and predict future climates. A stratigraphic isotopic approach, equivalent to that used for the study of marine sediments, can indeed be applied to lacustrine sequences for the evaluation of changes in continental paleoclimates, as well as atmospheric paleocirculation patterns. Analogous to paleoceanographic studies, it is proposed that isotope investigations in numerous lakes widely distributed over the continents should be

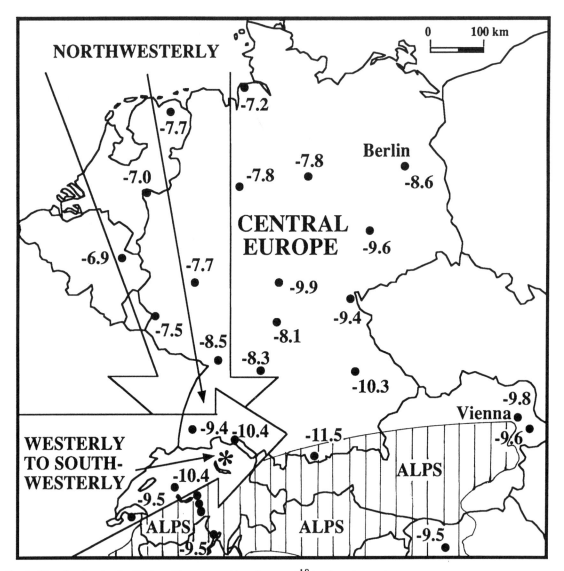

Fig. 4. Map showing the distribution of long-term arithmetic mean $\delta^{18}O$ values for precipitation collected at IAEA/WMO stations in central Europe. Data are from Rozanski et al. (this volume) and station locations are indicated by dots. The large arrows indicate dominant wind directions towards (northwesterly) and along (westerly to southwesterly) the Alpine chain. They enclose the possible isotopic composition of the accompanying moisture delivered into the catchment area of Lake Greifen (starred location) from the two different wind directions.

coordinated to generate records of paleo-isotopic patterns in precipitation. These records could potentially be used to reconstruct atmospheric paleocirculation patterns for specific time slices. To test the proposed isotope approach to continental records stored in lacustrine sequences and achieve maximum results with an initial coordinated project to reconstruct atmospheric paleocirculation patterns, a time slice with available historical data, such as the Little Ice Age to the present, should be selected.

Acknowledgements. This study on Lake Greifen oxygen-isotope stratigraphy could never have been accomplished without the continued excellent co-operation between members of the Lake Research Group at the Geological Institute of the ETH and the Swiss Federal Institute for Water Resources and Water Pollution Control (EAWAG). In addition, it represents the cumulative work of numerous individuals over the past 15 to 20 years, all of whom are gratefully acknowledged. Extremely constructive reviews by D. Hodell and W. Showers helped to significantly improve the manuscript. This work was funded in part by Swiss Nationalfond Grant No. 2.843-0.88 and ETH Research Grant No. 0.330.082.37/1.

REFERENCES

Dansgaard, W., Stable isotopes in precipitation, *Tellus, 16,* 436-468, 1964.
Chen, J., Changes of Alpine climate and glacier water resources, Ph.D. thesis, Swiss Federal Institute of Technology (ETH), Zürich, 1990.
Emerson, S., and G. Widmer, Early diagenesis in anerobic lake sediments--II. Thermodynamic and kinetic factors controlling the

formation of iron phosphate, *Geochem. Cosmochem. Acta, 40*, 925-934, 1978.

Epstein, S., and T. K. Mayeda, Variations of O^{18} content of waters from natural sources, *Geochim. Cosmochim. Acta, 4*, 213-224, 1953.

Epstein, S., R. Buchsbaum, H. A. Lowenstam, and H. C. Urey, Revised carbonate water isotopic temperature scale, *Geol. Soc. Am. Bull., 64*, 1315-1326, 1953.

Eugster, H. P., and L.A. Hardie, Saline lakes, in *Lakes: Chemistry, Geology, Physics*, edited by A. Lerman, pp. 237-293, Springer-Verlag, New York, 1978.

Gat, J. R., The isotope of hydrogen and oxygen in precipitation, in *Handbook of Environmental Isotope Geochemistry*, vol. 1, pp. 221-47, Elsevier, Amsterdam, 1980.

Gäggeler, H., H. R. von Gunten, and U. Nyffeler, Determination of ^{210}Pb in lake sediments and in air samples by direct gamma-ray measurements, *Earth and Planetary Science Letters, 33*, 119-121, 1976.

Hollander, D. J., Carbon and nitrogen isotopic cycling and organic geochemistry of eutrophic Lake Greifen: Implications for preservation and accumulation of ancient organic carbon-rich sediments, Ph.D. thesis, Swiss Federal Institute of Technology (ETH), Zürich, 1989

Hollander, D. J., J. A. McKenzie, and H. Lo ten Haven, A 200-year sedimentary record of eutrophication in Lake Greifen (Switzerland): Implications for the origin of organic-carbon-rich sediments, *Geology, 20*, 825-828, 1992.

Jung, G.P., Beiträge zur Morphogenese des Zürcher-Oberlandes im Spät- und Post glazial, *Vjschr. natf. Ges. Zürich, 114*, 293-403, 1969.

Kelts, K., and K. J. Hsü, Freshwater carbonate sedimentation, in *Lakes: Chemistry, Geology, Physics*, edited by A. Lerman, pp. 295-323, Springer-Verlag, New York, 1978.

Lamb, H. H., *Climatic History and the Future*, vol. 2, 835 pp., Princeton University Press, Princeton, New Jersey, 1977.

Lister, G. S., Stable isotopes from lacustrine ostracoda as tracers for continental palaeoenvironments, in *Ostracoda in the Earth Sciences*, edited by P. DeDeckker, J.P. Colin and J.P. Peypouquet, pp. 201-218, Elsevier, Amsterdam, 1988.

McKenzie, J. A., Carbon-13 cycle in Lake Greifen: A model for restricted ocean basins, in *Nature and Origin of Cretaceous Carbon-rich Facies*, edited by S. O. Schlanger and M. Cita, pp. 197-208, Academic Press, London, 1982.

McKenzie, J. A., Carbon isotopes and productivity in lacustrine and marine environments, in *Chemical Processes in Lakes*, edited by W. Stumm, Wiley-Interscience, New York, pp.99-118, 1985.

Messerli, B., P. Messerli, C. Pfister, and H. J. Zumbuhl, Fluctuations of climate and glaciers in the Bernese Oberland, Switzerland, and their geoecological significance, 1600-1975, *Arctic and Alpine Research, 10*, 247-260, 1978.

Pleisch, P., Die Herkunft eutrophierender Stoffe beim Pfäffiker- und Greifensee, *Vjschr. natf. Ges. Zürich, 115/2*, 127-229, 1970.

Rozanski, K., L. Araguás-Araguás, and R. Gonfiantini, Relation between long-term trends of oxygen-18 isotope composition of precipitation and climate, *Science, 258*, 981-985 1992.

Siegenthaler, U., Stable hydrogen and oxygen-isotopes in the water cycle, in *Lectures in Isotope Geology*, edited by E. Jäger and J. C. Hunziker, pp. 264-273, Springer-Verlag, Berlin, 1979.

Siegenthaler, U., and H. Oeschger, Correlation of O-18 in precipitation with temperature and altitude, *Nature, 285*, 314-317, 1980.

Talbot, M. R., A review of the palaeohydrological interpretation of carbon and oxygen isotopic ratios in primary lacustrine carbonates, *Chemical Geology (Isotope Geoscience Section), 80*, 261-279, 1990.

Talbot, M. R. and K. Kelts, Paleolimnological signatures from carbon and oxygen isotopic ratios in carbonates from organic carbon-rich lacustrine sediments, in *Lacustrine Basin Exploration - Case Studies and Modern Analogs*, edited by B. J. Katz, *AAPG Memoir, 50*, pp. 99-112, 1990.

Tschopp, J., Die Verunreinigung der Seen mit Schwermetallen: Modelle für die Regulierung der Metalkonzentrationen, Ph.D. thesis, Swiss Federal Institute of Technology (ETH), Zürich, 1979.

Wan, G. J., P. H. Santischi, M. Sturm, K. Farrenkothen, A. Lueck, E. Werth, and C. Schuler, Natural (^{210}Pb, ^7Be) and fallout (^{137}Cs, $^{239,240}Pu$, ^{90}Sr) radionuclides as geochemical tracers of sedimentation in Greifensee, Switzerland, *Chemical Geology, 60*, 181-196, 1987.

Weber, H., Sedimentologische und Geochemische Untersuchungen im Greifensee (Kanton Zürich, Schweiz), Ph.D. thesis, Swiss Federal Institute of Technology (ETH), Zürich, 1981.

Yurtsever, Y., and J. R. Gat, Atmospheric waters, in *Stable Isotope Hydrology: Deuterium and Oxygen-18 in the Water Cycle*, edited by J.R. Gat and R. Gonfiantini, IAEA Technical Reports Series No. 210, chapt. 6, pp. 103-142, Vienna, Austria, 1981.

J.A. McKenzie, Geological Institute, ETH-Zentrum, Switzerland.

D.J. Hollander, Department of Geological Sciences, Northwestern University, Evanston, IL 60208.

Coupled Stable-Isotope and Trace-Element Measurements of Lacustrine Carbonates as Paleoclimatic Indicators

ALLAN R. CHIVAS[1], PATRICK DE DECKKER[2], JOSEPH A. CALI[1], AUDREY CHAPMAN[1], ELMER KISS[1] AND J. MICHAEL G. SHELLEY[1]

We describe a technique for performing sequential stable-isotope and trace-element analyses on the same aliquot of carbonate using as little as 30 μg $CaCO_3$. The technique producing combined Sr/Ca, Mg/Ca ratios and $\delta^{18}O$ and $\delta^{13}C$ values is applied to a suite of individual nektic ostracodes from the past 10,000 years from Lake Keilambete, a crater lake in western Victoria, Australia. Emphasis is placed on those periods of the lake's history where salinity and evaporation do not covary. We show that a combination of measurements, particularly $\delta^{18}O$ values and Sr/Ca ratios, can elucidate the salinity, evaporation and chemical history of this simple lake system but that each technique in isolation fails to adequately describe those changes. The combined technique may be applied to more complex lacustrine basins in order to produce more realistic histories from which paleoclimatic inferences could be drawn.

INTRODUCTION

The application of both trace-element and stable-isotope measurements to lacustrine authigenic or biogenic carbonates lags behind equivalent practice in the marine environment. In part, this results from the added complexity of lacustrine systems wherein greater variability may occur in the amount, and, where applicable, the chemical and isotopic compositions of river inflow and outflow, groundwater inflow or loss, rainfall, evaporation, temperature and biological productivity. One approach in seeking paleoenvironmental records from lakes is to investigate the simplest situations first, such as a perched crater lake, with little catchment beyond its own lake area, and isolated from regional groundwater. In this case the principal variables governing lake levels (and thus salinity) are precipitation (rainfall) and evaporation, and a reliable chemical or physical proxy for past lake levels would provide a recorder for changes in the evaporation/precipitation ratio.

Trace-element and oxygen-isotope variations can provide useful information that bears on this problem although, in isolation, each technique produces results that are a function of several variables. One of the purposes of this paper is to demonstrate that a combination of both techniques greatly enhances the precision of their interpretation. In a simple lacustrine setting, such combined analyses may uniquely resolve several situations — (1) changes in lacustrine chemistry, independent of changes in salinity; (2) changes that increase salinity independently of evaporation rate as may be caused by dissolution of a salt crust, or in a more complex lake system by an increased solute load entering a lake; and (3) changes that decrease salinity largely independently of evaporation rate, such as by increased groundwater flow to a lake. The other aim of this paper is to describe an analytical technique that permits both stable-isotope and trace-element analyses to be performed on the *same* carbonate sample, in this case individual ostracode valves with shell weights as small as 30 μg $CaCO_3$.

GENERAL PRINCIPLES

A fairly complete matrix of lacustrine carbonate media for geochemical study might include authigenic precipitates and the remains of both nektic and benthic organisms. Use of such material potentially permits investigation of the entire water column and the use of species (e.g. ostracodes) with known seasonal life histories might allow seasonal variations to be addressed. To date,

[1]Allan R. Chivas, Joseph A. Cali, Audrey Chapman, Elmer Kiss and J. Michael G. Shelley, Research School of Earth Sciences, The Australian National University, Canberra ACT 0200, Australia.
[2]Patrick De Deckker, Department of Geology, The Australian National University, Canberra ACT 0200, Australia.

Climate Change in Continental Isotopic Records
Geophysical Monograph 78

such a degree of sophistication has not been achieved, and most studies involve one or two elements of this matrix. The present work is concerned with the chemistry of nektic (i.e. free swimming) ostracodes, whereas most Northern Hemisphere studies of lacustrine ostracodes have concentrated on benthic forms. However, the rules of operation outlined below are generally applicable to a variety of lacustrine carbonates.

Trace Elements

The partitioning of Mg and Sr between a dilute bicarbonate solution (i.e. lake water) and precipitating inorganic or biogenic $CaCO_3$ is commonly governed by Henry's Law

$(M/Ca)_{CaCO_3} = K_D[M]_T (M/Ca)_{H_2O}$;   where
M = Mg, Sr, Ba (or other compatible trace element)

wherein the M/Ca ratio of the precipitating carbonate phase is proportional to the same ratio in the host water. For inorganic calcite and calcitic biogenic carbonates (e.g. ostracodes), the distribution coefficient, K_D, for Mg/Ca is dependent on temperature (T), and possibly the rate of precipitation, whereas the K_D for Sr/Ca uptake is essentially independent of temperature. This is because calcite and $MgCO_3$ are both trigonal. By contrast, aragonite and $SrCO_3$ are both orthorhombic, and thus the Sr/Ca ratio of lacustrine and marine aragonitic organisms may have use as a paleothermometer (e.g. for corals; Smith et al., 1979).

For ostracodes, the Mg/Ca ratio carries both a temperature signal and a record of the Mg/Ca ratio of the host water. Care must be taken to use fully calcified ostracode shells because the Mg content is very high and unrepresentative of the environment during the initial stages of calcification [Chivas et al., 1983, 1986a]. The Sr/Ca ratio of ostracodes contains a record of the past Sr/Ca ratio of the host water, which may be quantified if the appropriate $K_D[Sr]$ value is known. Chivas et al., [1983, 1985, 1986a, b] made both field measurements and laboratory cultures to establish that ostracode $K_D [Sr]$ values are highly species dependent, but have an empirical relationship with ostracode phylogeny.

A knowledge of the variation of the Sr/Ca ratio of past water bodies provides information on several parameters. The Sr/Ca ratio may change by virtue of a change in the chemistry of the lake (i.e. due to precipitation, or cessation of precipitation, of a particular phase such as carbonates or sulfates), and is thus dependent on the bulk chemistry and 'chemical pathway' [Eugster and Hardie, 1978; Eugster and Jones, 1979] of the lake. Such causes may result from evaporitic concentration, or in the case of a lake with surface inflow, from changes in the chemistry of catchment waters or groundwater. This is the generalized case; a more specific case is provided by Lake Keilambete, Victoria [Chivas et al., 1985], a perched crater lake

containing a 10,000-year record of authigenic carbonate sediments, no gypsum, and water of the Na-Mg-Cl-HCO_3-type. Here, the reconstructed Sr/Ca ratio of the water provides a strong indication of past salinity because of the constancy and simplicity of authigenic precipitation and water chemistry with time, wherein the Ca^{2+} activity was continually high (close to $CaCO_3$ saturation). In this case, and refering to the equation for Henry's Law; the Ca^{2+} activity of the water is high and approximately constant, and the Ca content of an ostracode is constant. Thus the Sr content of the ostracodes are proportional to the Sr content of the water body at past times, and Sr is a conservative component of this system.

It is worth counselling caution in the application of the Sr/Ca and Mg/Ca ratios of ostracodes as direct paleosalinity indicators for lakes with more complex chemistry. Nevertheless, the technique has provided useful interpretations in a variety of additional environments including a marine/lacustrine sequence from northern Australia [De Deckker et al., 1988] and lacustrine materials from North Africa [Gasse et al., 1987], Spain [Anadón and Julià, 1990], North Dakota, [Engstrom and Nelson, 1991] and Kashmir [Holmes et al., 1992].

Stable Isotopes

The oxygen-isotope composition of any oxygen-bearing mineral depends on its temperature of precipitation and the oxygen-isotope composition of its host solution. For biogenic carbonates, there are so called "vital effects" (i.e. biologically mediated microenvironments of $CaCO_3$ precipitation) that give rise to offsets in the oxygen-isotopic fractionation between $CaCO_3$ and H_2O at the same temperature for a variety of organisms. There are thus differences in $\delta^{18}O_{CaCO_3-H_2O}$ for molluscs (Epstein et al., 1953; Craig, 1965), barnacles [Killingley and Newman, 1982] and within coral species [Weber and Woodhead, 1972]. We have performed experiments to determine such fractionations for several species of ostracodes and these results will be reported elsewhere. However, most studies of ^{18}O in lacustrine ostracodes have simply assumed that the usual temperature sensitivity for carbonates (i.e. -0.23‰ per 1°C) applies and that most of the larger $\delta^{18}O$ variations that are recorded from stratigraphic sequences are the result of changes in the $\delta^{18}O$ value of the lake water, caused by changes in evaporation, rainfall, or surface- or groundwater conditions [e.g. Fritz et al., 1975; Lister 1988; Benson et al., 1991; Hodell et al., 1991; Eyles and Schwarcz, 1991].

The factors that govern the probable $\delta^{18}O$ variations in lacustrine ostracodes are complex [e.g. Lister, 1991 p. 150-153], and even the progressively increasing $\delta^{18}O$ values of lake water caused by evaporation alone are dependent on variations in temperature, relative humidity and wind stress [Gilath and Gonfiantini, 1983]. There is a basic difficulty in ascribing small time-dependent changes in the $\delta^{18}O$ values of ostracodes from cored sediments to either temperature changes in the water column and/or changes in the $\delta^{18}O$ value of the water (i.e. precipitation/evaporation ratio in closed basin lakes).

Interpretation Using Combined Trace-Element and Stable-Isotope Data

Some of the basic uncertainties in interpreting changes in lacustrine temperature and salinity (or solute composition) may be alleviated by the previously suggested combination of techniques [Chivas, et al., 1986a]. For the simplest closed lake situation

(1) The Sr/Ca of an ostracode is a function of solute chemistry (or salinity in a lake of constant and simple solute composition);

(2) The Mg/Ca of an ostracode is a function of temperature *and* solute chemistry (or salinity); and

(3) The $\delta^{18}O$ value of an ostracode is a function primarily of evaporation/precipitation (i.e. salinity) and, to a lesser degree, of temperature; although in lake margins or in shallow lakes, large temperature fluctuations may contribute significantly to variations in the recorded $\delta^{18}O$ values of ostracodes.

Items (2) and (3) may appear to be similar, but the direction of response is different, e.g. an increase in temperature increases Mg/Ca, but decreases $\delta^{18}O$; and an increase in salinity increases Mg/Ca, Sr/Ca and $\delta^{18}O$. Application of these two relationships, plus the independent indicator of salinity change (Sr/Ca) should in principle resolve, or greatly assist interpretation of simple lacustrine systems where several parameters vary independently. After description of our analytical techniques for combined $\delta^{18}O$, $\delta^{13}C$, Mg/Ca, Sr/Ca analysis of carbonates, the above rules will be applied to new data from Lake Keilambete.

ANALYTICAL TECHNIQUES

Isotopic Analysis

The basic carbonate-reaction apparatus for our system is the "acid on individual carbonate" (or Kiel) preparation device manufactured by Finnigan MAT of Bremen, Germany. In this system the solid carbonate samples are placed in glass reaction vessels or "thimbles" and phosphoric acid dosed sequentially into each thimble. The evolved CO_2 and water vapour are continuously trapped by liquid nitrogen during the acid reaction, later separated and the CO_2 is admitted from a 0.25-ml microvolume cold finger on the carbonate device via a separate capillary directly to the ion source thus by-passing the standard large-volume inlet. In our system, the mass spectrometer is a Finnigan MAT 251 interfaced to the microprocessor control system of the carbonate

reaction device so that up to 44 carbonate samples can be reacted and analysed for their $\delta^{13}C$ and $\delta^{18}O$ values in one unattended operation lasting 24 hours.

There are nine such systems in the world currently (Kiel, Sunbury-on-Thames, Bremerhaven, Bremen, Ann Arbor, Gif-sur-Yvette, Tokyo, Tohoku). The instrument in Canberra has been modified to allow subsequent chemical analysis of the acid-reaction residue for, in this case, Ca, Mg and Sr. This is achieved by replacement of the borosilicate glass reservoir (for H_3PO_4) and the glass reaction thimbles with materials made of pure silica, because our previous experiments showed that borosilicate glass is a source of Mg, Sr and Ca contamination. The silica thimbles are longer (now 66 mm) than the original design to allow sufficient dilution volume for later trace-element analysis. The stroke of the pneumatic piston that presents each thimble to the acid-dosing valve has been adjusted accordingly. The acid-dosing valves are constructed of stainless steel which could be replaced by or lined with polytetrafluorethylene if subsequent Mn or Fe analyses were desired.

The carbonate samples for isotopic analysis typically weigh between 30 and 200 µg, and rarely up to 250 µg. Thus individual microfossils such as ostracodes and foraminifers and even individual chambers of foraminifers can be analysed. Most samples are reacted using 3 drops of 107% H_3PO_4 for 10 minutes at 70°C. The isotopic calcite standards, NBS-18 and NBS-19 are analysed as samples (100 µg) at the beginning and end of each batch of 20 to 40 samples and are assumed to have the recommended values of $\delta^{18}O_{PDB} = -23.00‰$, $\delta^{13}C_{PDB} = -5.00‰$ (NBS-18) and $\delta^{18}O_{PDB} = -2.20‰$, $\delta^{13}C_{PDB} = +1.95‰$ (NBS-19) [Coplen et al., 1983; Gonfiantini et al., 1990]. The acid reaction time for the equant grains of such standards is adjusted to 20 minutes, as a longer reaction time is needed than for the thin-walled microfossils. The different reaction times are analytically correct as no water equilibration takes place between the evolved CO_2 and H_2O. We have found that some locally produced isotopic standards of carbonates, including our own, are isotopically heterogeneous at the microgram-scale. Fortunately, NBS-18 and NBS-19 appear to be isotopically homogeneous. The internal precision of measurements for both $\delta^{18}O$ and $\delta^{13}C$ is 0.03 - 0.09‰ (mean 0.05‰) and the external precision (between runs) is ≤ 0.1 ‰.

Preparation of Phosphoric Acid

The key to subsequent successful trace-element analysis of the acid-reaction residue is the cleanliness of the apparatus in contact with the acid and its reaction products and the purity of the orthophosphoric acid. Our phosphoric acid is prepared in a laminar-flow filtered air (< 0.3µm) workstation within an ultraclean positive-pressure laboratory wherein incoming air is filtered (< 0.3µm), and may be scrubbed through deionized water to remove soluble gaseous contamination. The acid is dehydrated in a silica dish to a density of 1.95 g cm^{-3} (~107% H_3PO_4) using the combined heating of a ceramic-coated hot plate and a silica overhead infrared evaporator. Density measurements are made with a Paar® DMA 35 electromagnetic densitometer. The first batch of acid (220 ml) permitted the analysis of 2100 samples and had a combined reagent and reaction blank of 14 ng/g (ppb) Ca, 2 ng/g Mg, and 0.04 ng/g Sr and is pure enough to allow reliable analysis for these three elements in carbonate samples and for subsequent analysis of the $^{87}Sr/^{86}Sr$ ratio. The contribution of the total blank to the analysis represents, for a typical 20 μg $CaCO_3$ sample, about 0.3% of the measured Ca, 4% of the Mg and 0.05% of the Sr. The blank level of the acid is monitored by analysing acid dosed to an empty silica thimble in the normal course of every third sample run.

Trace-Element Analysis

The silica thimbles containing the few drops of acid-reaction products are also used directly as the sampling vessels for trace-element analysis. This minimizes the possibility of contamination and saves time as there is no transfer of material between vessels. The acid residue is diluted in three stages with 18 MΩ water, accompanied by externally-vibrated stirring, to a volume of 2.5 ml using an autopipette. The thimbles are placed directly in the 100-position sample holder of an inductively-coupled argon plasma (ICP) atomic emission spectrometer for unattended Ca, Mg, Sr and occasionally Ba, analysis. The limits of detection of Ca, Mg and Sr are respectively 36, 24 and 32 pg/g (parts per trillion) in solution. The mean precision of the analysis is about 3%. There is no effective lower-limit to the size of samples that can be analysed — as little as 0.5 μg $CaCO_3$ can be analysed for Ca, Mg, Sr, although this is below the effective potentially lower sample size of ~5 to 10 μg $CaCO_3$ required for $\delta^{18}O$ and $\delta^{13}C$ analysis. Because the acid-reaction residue has been diluted to a known volume, not only can the ratios Mg/Ca and Sr/Ca be determined, but also the absolute quantities of Ca, Mg, and Sr derived (usually expressed in μg), and thence the calculated weight of $CaCO_3$. This latter quantity is termed the nominal shell weight [Chivas et al., 1986a], because microfossils, such as ostracodes, are typically 95% $CaCO_3$ with the remainder being Mg, Sr and organic matter, commonly a chitinous sheath.

Maintenance of Analytical Reliability

The most probable sources of error in the $\delta^{18}O$ analysis are due to incomplete acid reaction, and incomplete transfer of CO_2 to the mass spectrometer; the latter can occur if the automated liquid nitrogen trap or its heating cycle operates incorrectly. Water vapour can also be a source of error if it were inadvertently also directed, with CO_2, to the mass spectrometer. These possible errors can be detected by automated monitoring of pressures at the appropriate stages and by comparing the yields and thus reliability of both the stable-isotope and trace-element analyses, and by checking for loss of material from the initial loading of samples into the silica thimbles through to the ICP analysis.

The sample weight can be determined at three stages of the complete analytical procedure. Firstly, by weighing each sample, using a microbalance (± 0.1 μg) before loading into silica thimbles; secondly, by measuring the pressure of CO_2 entering the mass spectrometer (i.e. checking for a CO_2 yield appropriate to the weighed amount of $CaCO_3$); and thirdly, at the ICP as described in the section above, which checks for complete dissolution, retention of acid residue, and retention of material between microbalance and loading the thimble. In the second step, a measure of the volume of CO_2 liberated (pressure in μbar) from the transfer cold trap, immediately prior to freezing in the mass spectrometer cold finger is recorded automatically during every batch of analyses. This is a more reliable indicator of CO_2 yield, after calibration, than the major-ion (m/e = 44) beam during the actual analysis.

The reaction yields from a suite of analysed ostracodes from Lake Keilambete (discussed below) are presented in Figure 1 which shows, for a series of individual ostracode valves, their microbalance weights and their ICP-calculated weights versus manometer pressure during transfer of CO_2 to the mass spectrometer. Data that plot well away from either curve would need scrutiny for reliability; one or more of the analytical or handling

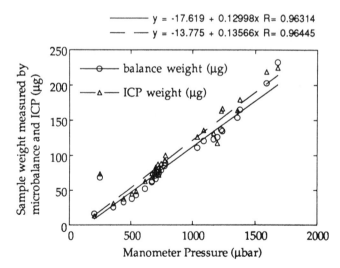

Fig. 1. Comparison of the measured weights (by microbalance, ± 0.1 μg) and calculated nominal shell weights (Ca content by ICP) for a series of individual ostracode valves and fragments of valves (from core KN, Lake Keilambete) versus the pressure (i.e. a function of volume) of CO_2 delivered to the mass spectrometer after reaction in a modified Finnigan individual-carbonate device.

procedures may be at fault, and the analytical results may need to be discarded.

The combined stable-isotope and trace-element procedure described above builds upon the technique of Coleman et al. [1989] which withdraws an aliquot of reacted acid from a conventional side-arm reaction vessel for later trace-element analysis by ICP. The sample size as described for this technique is typically 0.1 to 10 mg $CaCO_3$.

LAKE KEILAMBETE

Lake Keilambete is a near-circular maar lake (area 2.7 km^2) in western Victoria (190 km west of Melbourne) that is perched above the regional groundwater table, although there are minor springs that deliver small volumes of shallow groundwater generated outside the surface catchment area (i.e. within the crater rim) of 4.2 km^2 [Bowler, 1970]. The present water depth in the center of the lake is about 10.5 m, and has declined almost continuously throughout this century. The lake sediments are composed of organic-rich carbonate muds that have been extensively cored to depths of about 4 meters that span ~10,000 yr of sedimentation. These cores have been subject to paleoclimatic interpretation using sedimentology [Bowler, 1970, 1981; Bowler and Hamada, 1971], palynology [Dodson, 1974], paleomagnetism [Barton and Polach, 1980; Barton and McElhinny, 1981] and microfauna [De Deckker, 1982]. Stratigraphic correlation amongst various cores has been achieved using the distinctive sequence of white aragonite and calcite microlaminae (e.g. Figure 2, right hand column), with chronology provided by 18 radiocarbon dates.

Paleosalinity

A paleosalinity record for Lake Keilambete based on sediment grain-size analysis [Bowler, 1981] calibrated using modern sediment grain-size as a function of modern water depths is shown in Figure 2a. This transfer function is based upon the demonstrated positive relationship between water depth and textural variations (expressed as the proportion of the sediment with a grain size >44 μm) in Lake Keilambete. The adjacent Figure 2b shows a similar curve based on the Sr/Ca ratio of individual ostracode valves. The data in this latter figure include values from Chivas et al., [1985], derived using HCl dissolution; and new data using the combined trace-element/stable-isotope analytical method using phosphoric acid for dissolution. There is clearly strong agreement between the results of the two chemical analytical procedures. The similarity between the paleosalinity determinations derived from independent physical (grain size) and chemical (Sr/Ca) methods is reassuring, and in considering the $\delta^{18}O$ results we will assume, for the moment, that the Sr/Ca and grain size results are an adequate and reliable proxy for salinity and thus lake level. The very large inferred variations in past salinities

from ~12‰ at 6000 yr BP to >100‰ at 2700 yr and 9000 yr BP are useful extremes against which other techniques can be tested. The implications for paleoclimatic reconstructions based on the detailed lake-level variations, as can be derived from paleosalinity determinations, is presented in Bowler [1981] and will not be repeated here.

New Results

Figure 3 shows results from the combined Sr/Ca, Mg/Ca, $\delta^{18}O$, $\delta^{13}C$ analytical method (i.e. H_3PO_4) of individual ostracodes plotted as a function of age. These data form a reconnaissance study of the analytical technique applied to a relatively simple lake system, and a fuller exposition with more data for Lake Keilambete is intended. Because of incomplete analytical data at many stratigraphic levels (i.e. analyses of multiple samples) we have refrained from comment on possible short-term (intrastratum) variability in lake conditions, although the production of such data is clearly the strength of the described analytical technique.

Oxygen Isotopes

There is no simple correlation between the $\delta^{18}O$ curve and the paleosalinity curve derived from the Sr/Ca ratio of ostracodes. The $\delta^{18}O$ curve cannot be interpreted as a function of simple evaporation from a closed system that would in turn be related to lacustrine salinity. Other controls including temperature variations and possible additional sources of water may be important in determining the contemporaneous $\delta^{18}O$ values.

A significant difference between the trends of the Sr/Ca and $\delta^{18}O$ curves occur in the interval from 9800 to 8200 yr BP. At this time, from a variety of evidence, including the ostracode fauna [De Deckker, 1982] and palynology [Dodson, 1974], the lake was initially saline (»100‰) and with a very low volume that increased, possibly to overflowing (with a salinity of ≤12‰) by ~7000 yr BP. Interpretation of the $\delta^{18}O$ record of this interval, simply as an evaporation (more positive $\delta^{18}O$ values)/dilution (less positive $\delta^{18}O$) trend would produce almost the opposite interpretation. However, there are a variety of good reasons why the salinity (e.g. Sr/Ca) and $\delta^{18}O$ records are decoupled. Firstly, the lake was completely dry from 18,000 yr BP until sometime after 14,000 yr BP [Bowler, 1981] and the resumption of lacustrine conditions at ~10,000 yr BP produced a shallow saline swamp [Bowler, 1981; De Deckker, 1982]. The high salinity may have been established by dissolution of salt (or brine) retained in the floor of the lake from the previous drying cycle. However, the $\delta^{18}O$ value of the lake water at this time was clearly low, because little, if any, of the highly evaporated water remained from the previous drying event. The principal source of water, rainfall, is unevaporated (currently $\delta^{18}O_{SMOW}$ = ~-6‰), and the springs (also unevaporated rainwater recharge) which today provide a negligible

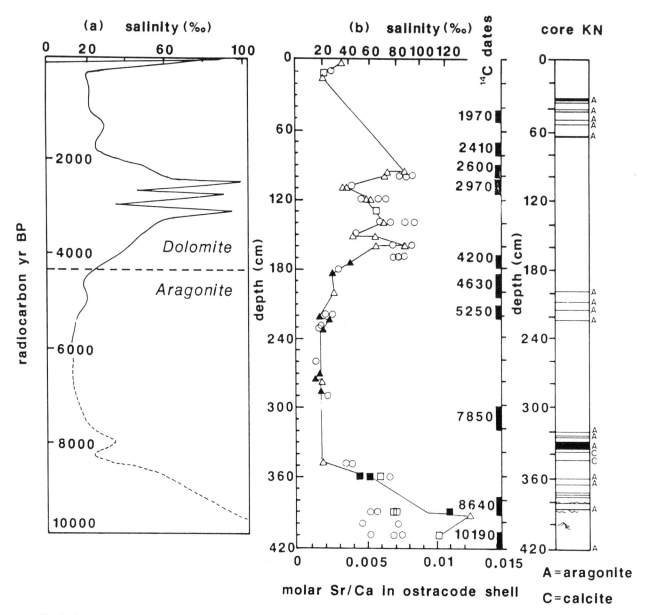

Fig. 2. Paleosalinity curves for the past 10,000 yr (~4 m of sedimentation) based on (a) carbonate sediment grain-size analysis [from Bowler, 1981, redrawn by Chivas et al., 1985]; and (b) Sr/Ca ratios of the nektic ostracodes *Australocypris robusta* and *Mytilocypris praenuncia* which have the same K_D [Sr] [Chivas et al., 1986b]. The Sr/Ca data for Fig. 2b are derived from both earlier results using HCl [Chivas et al., 1985] and new analyses using H_3PO_4 as part of the combined Ca-Mg-Sr-δ^{18}O-δ^{13}C scheme. The line connecting data is drawn through only the new analyses. The right hand column shows the positions of aragonite and calcite laminations that are commonly used to correlate among cores from the lake.

Key to symbols on Figures 2 and 3:
(A)   analyses using HCl
    o   *Australocypris robusta* and *Mytilocypris praenuncia*

(B)   analyses using H_3PO_4
    △   *Mytilocypris praenuncia* (whole valve)
    ▲   *Mytilocypris praenuncia* (fragment or fragments of valve)
    ■   *Australocypris robusta* (whole valve)
    □   *Australocypris robusta* (fragment or fragments of valve)

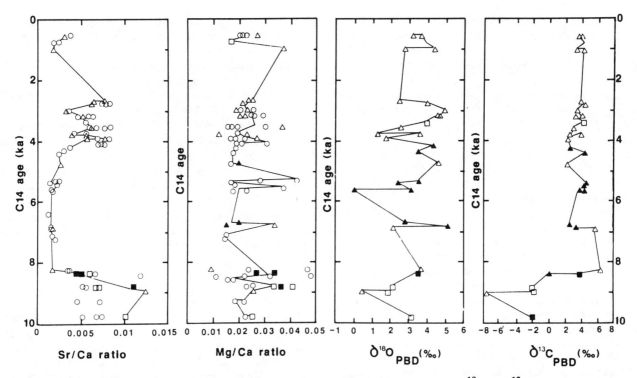

Fig. 3. Results of new analyses (H_3PO_4 combined technique) of Sr/Ca and Mg/Ca ratios, and $\delta^{18}O$ and $\delta^{13}C$ values from Lake Keilambete (core KN) plotted as a function of age. The Sr/Ca and Mg/Ca curves also contain older data [Chivas, et al., 1985; 1986b] derived using HCl. There are data from the interval 1000 to 2600 years [Chivas et al., 1985; Chivas et al., 1986b] which, for simplification, are not shown at this stage because they derive from ostracode species with a variety of K_D [Sr] values.

source of water to the lake, were they then in existence, may have produced a proportionately greater (but not necessarily absolutely greater) contribution of unevaporated water to the lake. Evaporation from the swamps would have been impeded by the vegetation cover, with loss of water from transpiration returning largely isotopically unfractionated water vapor to the atmosphere.

At ~3800 yr BP, there is a consistent trend of increasing $\delta^{18}O$ values that corresponds to an increase in salinity interpreted from both Sr/Ca values and grain-size analysis. In this case the conventional interpretation of $\delta^{18}O$ values in terms of evaporation and relative lake levels is well supported. The $\delta^{18}O$ and Sr/Ca records for Lake Keilambete are also broadly indicative of coeval salinity and evaporation variations for the period 2400 yr BP to the top of our combined record at 500 yr BP.

Of more interest is the more abrupt change in Sr/Ca ratios at 4200 yr BP that corresponds to the change in authigenic mineralogy from aragonite to calcite. This might be interpreted also as a fairly rapid increase in salinity, but the $\delta^{18}O$ record, and the grain-size data do not support this. The latter parameter indicates a smoother change, and the $\delta^{18}O$ values actually decrease at this level. So, in isolation, the two chemical indicators appear

contradictory. However, the resolution may lie in a shift in the chemistry of the lake at this time. The distribution coefficient for Sr between carbonate mineral and dissolved bicarbonate, at 25°C, is very different for aragonite (K_D = 1.13; Kinsman and Holland, 1969), calcite (K_D = 0.14; Kinsman, 1969) and dolomite. A change to authigenic dolomite precipitation would consume less Sr from the water column and lead to higher levels of Sr within ostracodes and other calcitic organisms. If this explanation is correct, the Sr/Ca record of salinity even in this simple system is not fully robust; with a step function introduced at ~4200 yr BP and a period thereafter to reestablish equilibrium.

Carbon Isotopes

Some portions of the curve showing the $\delta^{13}C$ values of ostracodes (Figure 3d) are negatively correlated with the Sr/Ca curve; particularly for example between 9800 and 7000 yr BP during a marked decrease in salinity. However, periods of high salinity between 4000 and 2600 yr are not discernible in the $\delta^{13}C$ record.

Stratigraphic sequences from several other lakes show total variations of typically 2 to 5‰ in $\delta^{13}C$ values of authigenic or biogenic carbonates (e.g. Great Salt Lake, Spencer et al., 1984; Hassi el Mejna, Algeria, Gasse et al.,

1987; Lake Zürich, Lister 1988; Walker Lake, NV, Benson et al., 1991; Lake Turkana, Johnson et al., 1991). The δ^{13}C values of Lake Keilambete nektic ostracodes show a 15‰ shift from -8‰ at 9200 yr (shallow, saline water) to +6.5‰ at 8200 yr (deep, fresh water). While it is tempting to interpret this shift as being due to a progressive increase in productivity in a near-closed system [cf. McKenzie, 1982, 1985], there may be additional factors at work. For example, as previously mentioned, Lake Keilambete completely dried during the Last Glacial Maximum, leaving a hard floor (that is currently impenetrable to piston coring) covered by an organic-rich soil with sedge-like vegetation [Bowler, 1981]. It is probable that the early shallow saline water contained isotopically light bicarbonate generated in part by oxidation of organic matter within the lake sediments and from contemporaneous swamp vegetation. The bulk of the large shift in δ^{13}C may be due to recovery from this early history and progression to a regime where dissolved bicarbonate became closely equilibrated isotopically with atmospheric carbon dioxide. Superimposed on this shift and thereafter, are small (2 to 3‰) variations in the δ^{13}C values of ostracodes that may be related to minor changes in lacustrine productivity. These minor variations in δ^{13}C values are generally correlated with known salinity variations, such that at periods of higher salinity the δ^{13}C values of ostracodes are lower, i.e. productivity is lower [cf. McKenzie, 1985]. Even though Lake Keilambete is a nearly hydrologically closed lake system there is no necessary correlation between the δ^{18}O and δ^{13}C values of ostracodes [cf. Talbot, 1990].

CONCLUSIONS

The well defined salinity and lake-level history of Lake Keilambete for the past 10,000 yr determined by sediment grain-size analysis and palynology has been examined by the application of combined trace-element and stable-isotope analysis of ostracodes. In isolation each techniques provides some information; together they commonly reinforce e.g. simultaneous increases in δ^{18}O and Sr/Ca that are interpreted as an increase in salinity and a decrease in lake-level, and thus a decrease in precipitation/evaporation ratio. In isolation, the trace-elements provide a better indicator of past salinities than does the δ^{18}O record.

Of more interest are those situations where the techniques do not reinforce the interpretation of a simple situation, but appear to provide reliable information on several common independently variable parameters within lacustrine systems. These situations could not be adequately deduced by the application of one of the chemical techniques alone.

Salinity and evaporation in the lacustrine environment do not positively covary where

(i)   dissolution of saline materials occurs, either within

the lake itself (e.g. a salt crust or groundwater brine pool), or within its surface or groundwater catchment,

(ii)   there is a modification of the chemistry of the lake water due to changes in authigenic precipitation. Such changes may also effect the Sr/Ca or Mg/Ca ratios of the lake water and will be recorded in the trace-element ratios of biogenic and authigenic carbonates. The δ^{18}O records of such precipitates may not reveal these changes.

These are all common changes that occur during the history and evolution of lacustrine basins. Application of combined chemical, isotopic and traditional sedimentological and faunal analyses will be required to elucidate these variations before inferences on paleoclimatic causes can be successfully drawn.

APPENDIX

Preparation of Ostracodes from Lake Keilambete

Sediment samples with volumes of 3 to 5 ml each were treated in 3% H_2O_2 principally to disaggregate the clays and remove any possible organic matter adhering to the ostracode valves. This process lasted on average 2 to 4 weeks. Samples were then washed through a 100 µm brass sieve with a gentle flow of distilled water. The fraction coarser than 100 µm was then dried in an oven at 60°C, and stored in glass vials.

Ostracodes were picked from this fraction with a fine (triple zero) brush under a binocular microscope and kept in standard plastic micropaleontological slides. Ostracodes which had been selected for chemical analysis were then further cleaned in a few drops of 10% H_2O_2 for a few seconds to remove any adhering clays, and occasionally fine tungsten needles were used to remove unsuitable (e.g. dirty) portions of the ostracode valves. The valves were rinsed twice in distilled water in black glass petri dishes and then immersed in ethanol. This renders the ostracode valve translucent/transparent, which allows one to see if any part of the ostracode valve is still dirty. Finally, the valve is left to dry before being weighed and placed in the silica reaction vessels.

Acknowledgments. We thank W.J. Showers and A. Cohen for constructive reviews of the manuscript.

REFERENCES

Anadón, P., and R. Julià, Hydrochemistry from Sr and Mg contents of ostracodes in Pleistocene lacustrine deposits, Baza Basin (SE Spain), *Hydrobiologia, 197*, 291-303, 1990.
Barton, C.E., and M.W. McElhinny, A 10000 yr geomagnetic secular variation record from three Australian maars, *Geophys J. Roy. Astronom. Soc., 67*, 465-485, 1981.
Barton, C.E., and H.A. Polach, ^{14}C ages and magnetic stratigraphy in three Australian maars, *Radiocarbon, 22*, 728-739, 1980.
Benson, L.V., P.A. Meyers, and R.J. Spencer, Change in the size of

Walker Lake during the past 5000 years, *Palaeogeogr. Palaeoclimatol. Palaeoecol.*, 81, 189-214, 1991.

Bowler, J.M., Late Quaternary environments: A study of lakes and associated sediments in southeastern Australia, PhD thesis, The Australian National University, Canberra, 1970.

Bowler, J.M., Australian salt lakes. A palaeohydrological approach, *Hydrobiologia*, 82, 431-444, 1981.

Bowler, J.M., and T. Hamada, Late Quaternary stratigraphy and radiocarbon chronology of water level fluctuations in Lake Keilambete, Victoria, *Nature*, 232, 330-332, 1971.

Chivas, A.R., P. De Deckker, and J.M.G. Shelley, Magnesium, strontium, and barium partitioning in nonmarine ostracode shells and their use in paleoenvironmental reconstructions — a preliminary study, in *Applications of Ostracoda*, edited by R.F. Maddocks, pp. 238-249, Univ. Houston Geosci., 1983.

Chivas, A.R., P. De Deckker, and J.M.G. Shelley, Strontium content of ostracods indicates lacustrine palaeosalinity, *Nature*, 316, 251-253, 1985.

Chivas, A.R., P. De Deckker, and J.M.G. Shelley, Magnesium content of non-marine ostracod shells: A new palaeosalinometer and palaeothermometer, *Palaeogeogr. Palaeoclimatol. Palaeoecol.*, 54, 43-61, 1986a.

Chivas, A.R., P. De Deckker, and J.M.G. Shelley, Magnesium and strontium in non-marine ostracod shells as indicators of palaeosalinity and palaeotemperature, *Hydrobiologia*, 143, 135-142, 1986b.

Coleman, M.L., J.N. Walsh, and R.A. Benmore, Determination of both chemical and stable isotope composition in milligramme - size carbonate samples, *Sedimentary Geology*, 65, 233-238, 1989.

Coplen, T.B., C. Kendall, and J. Hopple, Comparison of stable isotope reference samples, *Nature*, 302, 236-238, 1983.

Craig, H., Measurement of oxygen isotope paleotemperatures, in *Stable Isotopes in Oceanographic Studies and Paleotemperatures*, edited by E. Tongiori, pp. 161-182, Spoleto, CNR, Lab. Geol. Nucl., Pisa, 1965.

De Deckker, P., Holocene ostracods, other invertebrates and fish remains from cores of four main maar lakes in southeastern Australia, *Proc. Roy. Soc. Victoria*, 94, 183-220, 1982.

De Deckker, P., A.R. Chivas, J.M.G. Shelley, and T. Torgersen, Ostracod shell chemistry: A new palaeoenvironmental indicator applied to a regressive/transgressive record from the Gulf of Carpentaria, Australia, *Palaeogeogr. Palaeoclimatol. Palaeoecol.*, 66, 231-241, 1988.

Dodson, J.R., Vegetation and climatic history near Lake Keilambete, western Victoria, *Aust. J. Botany*, 22, 709-717, 1974.

Engstrom, D.R., and S.R. Nelson, Paleosalinity from trace metals in fossil ostracodes compared with observational records at Devils Lake, North Dakota, USA, *Palaeogeogr. Palaeoclimatol. Palaeoecol.*, 83, 295-312, 1991.

Epstein, S., R. Buchsbaum, H.A. Lowenstam, and H.C. Urey, Revised carbonate-water isotopic temperature scale, *Bull. Geol. Soc. Am.*, 64, 1315-1326, 1953.

Eugster, H.P., and L.A. Hardie, Saline lakes, in *Lakes — Chemistry, Geology, Physics*, edited by A. Lerman, pp. 237-293, Springer-Verlag, New York, 1978.

Eugster, H.P., and B.F. Jones, Behavior of major solutes during closed-basin brine evolution, *Am. J. Sci.*, 279, 609-631, 1979.

Eyles, N., and H.P. Schwarcz, Stable isotope record of the last glacial cycle from lacustrine ostracodes, *Geology*, 19, 257-260, 1991.

Fritz, P., T.W. Anderson, and C.F.M. Lewis, Late-Quaternary climatic trends and history of Lake Erie from stable isotope studies, *Science*, 190, 267-269, 1975.

Gasse, F., J.C. Fontes, J.C. Plaziat, P. Carbonel, I. Kaczmarska, P. De Deckker, I. Soulié-Marsche, Y. Callot, and P.A. Dupeuble, Biological remains, geochemistry and stable isotopes for the reconstruction of environmental and hydrological changes in the Holocene Lakes from North Sahara, *Palaeogeogr. Palaeoclimatol. Palaeoecol.*, 60, 1-46, 1987.

Gilath, C., and R. Gonfiantini, Lake Dynamics, in *Guidebook on Nuclear Techniques in Hydrology*, Int. Atomic Energy Agency, Vienna, 1983.

Gonfiantini, R., K. Rozanski, and W. Stichler, Intercalibration of environmental isotope measurements: The program of the International Atomic Energy Agency, *Radiocarbon*, 32, 369-374, 1990.

Hodell, D.A., J.H. Curtis, G.A. Jones, A. Higuera-Gundy, M. Brenner, M.W. Binford, and K.T. Dorsey, Reconstruction of Caribbean climate change over the past 10,500 years, *Nature*, 352, 790-793, 1991.

Holmes, J.A., P.E. Hales, and F.A. Street-Perrott, Trace-element chemistry of non-marine ostracods as a means of palaeolimnological reconstruction: An example from the Quaternary of Kashmir, northern India, *Chem. Geol.*, 95, 177-186, 1992.

Johnson, T.C., J.D. Halfman, and W.J. Showers, Paleoclimate of the past 4000 years at Lake Turkana, Kenya, based on the isotopic composition of authigenic calcite, *Palaeogeogr. Palaeoclimatol. Palaeoecol.*, 85, 189-198, 1991.

Killingley, J.S., and W.A. Newman, ^{18}O fractionation in barnacle calcite: a barnacle paleotemperature equation, *J. Marine. Res.*, 40, 893-902, 1982.

Kinsman, D.J.J., Interpretation of Sr^{+2} concentrations in carbonate minerals and rocks, *J. Sediment. Petrol.*, 39, 486-508, 1969.

Kinsman, D.J.J., and H.D. Holland, The co-precipitation of cations with $CaCO_3$—IV. The co-precipitation of Sr^{2+} with aragonite between 16° and 96°C, *Geochim. Cosmochim. Acta*, 33, 1-17, 1969.

Lister, G.S., Stable isotopes from lacustrine Ostracoda as tracers for continental palaeoenvironments, in *Ostracoda in the Earth Sciences*, edited by P. De Deckker, J.-P. Colin and J.-P. Peypouquet, pp. 201-218, Elsevier, Amsterdam, 1988.

Lister, G.S., K. Kelts, Chen Ke Zao, J.-Q. Yu, and F. Niessen, Lake Qinghai, China: closed-basin lake levels and the oxygen isotope record for ostracoda since the latest Pleistocene, *Palaeogeogr. Palaeoclimatol. Palaeoecol.*, 84, 141-162, 1991.

McKenzie, J.A., Carbon-13 cycle in Lake Greifen: A model for restricted ocean basins, in *Nature and Origin of Cretaceous Carbon-rich Facies*, edited by S.O. Schlanger and M.B. Cita, pp. 197-207, Academic Press, 1982.

McKenzie, J.A., Carbon isotopes and productivity in the lacustrine and marine environment, in *Chemical Processes in Lakes*, edited by J. Stumm, pp. 99-118, John Wiley & Sons, 1985.

Smith, S.V., R.W. Buddemeier, R.C. Redalje, and J.E. Houck, Strontium-calcium thermometry in coral skeletons, *Science*, 204, 404-407, 1979.

Spencer, R.J., M.J. Baedecker, H.P. Eugster, R.M. Forester, M.B. Goldhaber, B.F. Jones, K. Kelts, J. McKenzie, D.B. Madsen, S.L. Rettig, M. Rubin, and C.J. Bowser, Great Salt Lake, and precursors, Utah: the last 30,000 years, *Contrib. Mineral. Petrol.*, 86, 321-334, 1984.

Talbot, M.R., A review of the palaeohydrological interpretation of carbon and oxygen isotopic ratios in primary lacustrine carbonates, *Chem. Geol. (Isotope Geosci. Sect.)*, 80, 261-279, 1990.

Weber, J.N., and P.M.J. Woodhead, Temperature dependence of oxygen-18 concentration in reef coral carbonates, *J. Geophys. Res.*, 77, 463-473, 1972.

A.R. Chivas, J.A. Cali, A.Chapman, E. Kiss and J. M.G. Shelley, Research School of Earth Sciences, The Australian National University, Canberra ACT 0200, Australia.

P. De Deckker, Department of Geology, The Australian National University, Canberra ACT 0200, Australia.

Carbon and Oxygen Isotopes in African Lacustrine Stromatolites : Palaeohydrological Interpretation

JOEL CASANOVA[1,2] AND CLAUDE HILLAIRE-MARCEL[2]

[1] BRGM, Departement Géochimie, B.P. 6009, 45060 Orléans Cedex 2, France

[2] GEOTOP-Université du Québec à Montréal, C.P. 8888 Suc A, H3C 3P8 Montréal, Canada

Microbial reefs formed considerable primary carbonate deposits in the littoral to sublittoral environments of Pleistocene-Holocene African lakes. Oxygen isotopic ratio changes in these stromatolitic carbonates respond to the residence time of paleolake waters in relation to evaporation/water supply budget. The inferred oxygen isotopic composition of some paleolake waters, when compared to the values for modern waters and carbonates, may be used as a proxy indicator of the composition of regional rainfall. In the case of the stromatolites associated with lacustrine maxima in Africa, the low species diversity of the benthic microbial communities and the ecological impact of the rainfall pattern, permit to link the microbial growth rhythm to a seasonal cycle. It is possible to estimate quantitatively the seasonal contrast of each cycle from the ratio of mean lamina thickness corresponding to growth during the rainy and dry season respectively. This ratio provides an index of the seasonal distribution of paleorainfall. Oxygen isotopic profiles in the lamina couplets display large amplitude shifts which reflect the alternation between periods of steady conditions and phases of increased runoff. These records provide a high resolution record of hydrological changes in paleolakes that may improve the interpretation of continental paleoclimates.

INTRODUCTION

Fossil lacustrine stromatolites define continuous shorelines that can be used to reconstruct contemporaneous lake levels in times past [Johnson, 1974; Abell et al., 1982; Casanova, 1986a]. These paleolakes were near saturation with respect to calcite [Casanova and Hillaire-Marcel, 1992b] and their levels remained stable during the growth of the stromatolites. Continuous ecological zonations show that the potential bathymetric extension of the stromatolites commonly ranged from 0 to -30 m (exceptionally down to -50 m in Lake Tanganyika). All hard substrates are colonized by Benthic Microbial Communities (BMC) and mineralization results from in situ carbonate precipitation. Bioherms are the morphologies most commonly developed in lacustrine environments, and reach their maximum development in surface waters, where conditions are favourable in terms of light intensity, oxygenation and nutrient supply [Casanova, 1986b; Casanova, in press]. Most of the

stromatolites are made of low-Mg calcite, with a few aragonitic occurrences, and do not contain dolomite or evaporites common to saline lakes elsewhere [see Buchbinder, 1981 and references therein]. Fabrics and micro-organism associations of lacustrine stromatolites are distinctive indicators of freshwater environments [Casanova, 1986a]. Stromatolites may be absent during arid periods, characterized by alkaline rivers and lakes, because of the Ca deficiency of siliciclastic basins and/or biochemical inhibition. African lacustrine stromatolites (Figure 1) are characterized by : a large diversity of forms and assemblages; well-defined gross morphologies recording clear relationships with ecological parameters; and exceptional preservation of micro-organisms and fabrics, due to the absence of diagenetic recrystallization [Casanova, 1991].

The chemical composition of the lakes is closely related to the hydrologic budget and bedrock geology of each basin. The climatic variability in tropical Africa in its simplest terms can be distinguished in two extreme situations: "humid" and "arid" periods. Humid periods are time intervals characterized by high lake levels and elevated input of terrigenous organic matter. Because of forest expansion and increased terrestrial organic productivity, much of the organic matter accumulating in lakes is

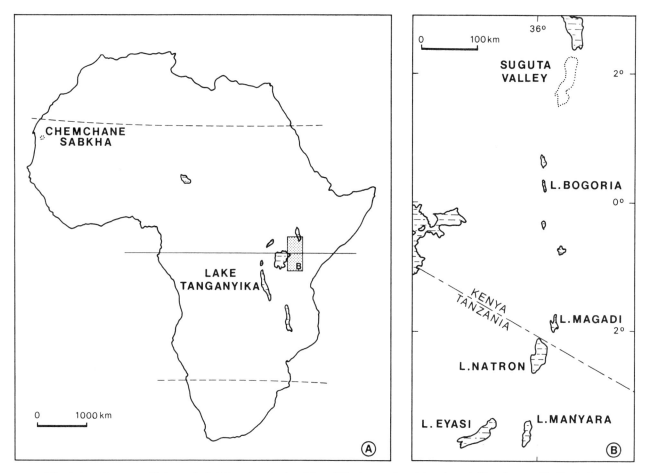

Fig. 1. Location map of Intertropical Africa showing localities of Quaternary lacustrine stromatolites discussed in the text.

of continental origin [Tiercelin et al., 1987]. The presence of a dense vegetation cover reduces fine particle supply to the basins. During arid periods, evaporites precipitate in shallow hypersaline lakes deprived of stratification due to winds being able to mix the entire water column. Related to both the forest regression and the nutrient concentration in waters, the organic matter that accumulates in these lakes is mostly of aquatic origin [Yuretich, 1982; Casanova, 1986b; Tiercelin et al., 1987].

Several paleontological [Kendall, 1969; Gasse and Street, 1978; Haberyan and Hecky, 1987; Tiercelin et al., 1987; Lézine and Casanova, 1990] and geochemical [Eugster, 1980; Talbot and Kelts, 1986; Halfman and Johnson, 1988; Fontes and Gasse, 1991; Aucour, 1992] techniques have been applied in the examination of paleolake deposits to reconstruct the Late Quaternary hydrological changes in Africa. Among primary lacustrine carbonates, stromatolites prove to be an especially promising material for paleoclimatic reconstruction. Stromatolites are biologically precipitated from ambient waters over long periods of time (of the order of 10^3-10^4 years), in conditions close to thermodynamic equilibrium [Chafetz et al., 1991; Merz, 1992], in an environment where seasonal variations of isotopic composition of water and dissolved carbon dioxide are considerable. This paper reviews the behaviour of

carbon and oxygen isotopes in african lacustrine stromatolites (Figure 1) and reconsiders some of the original data sets in the light of recent advances in the interpretation of stable isotopic data from lacustrine carbonates [Talbot, 1990; Talbot and Kelts, in press].

THE SEASONAL RECORD OF LACUSTRINE STROMATOLITES

The commonest microstructure observed in lacustrine stromatolites [Casanova, in press] results from the superposition of doublets composed of light-coloured sparitic laminae and dark micritic laminae (Figure 2). Each light-coloured lamina is 20-1500 μm thick and contains numerous erect filaments distributed in grasslike mats. The dominant species of the BMC is represented by elongate filaments, 1-1.6 μm in diameter, forming interlaced tufts. Filaments are typically found together in a common sheath and resemble modern *Schizothrix*. In most cases, they are also accompanied by isolated filaments, 1.8-2.5 μm and 3-8 μm in diameter. These oscillatoriacean forms likely represent different species of the modern *Phormidium*. The dark laminae range from 5 to 900 μm in thickness and are composed of one or several micritic films rich in organic matter. Under the scanning electron microscope, these films are composed of transport of clastic grains in suspension as well as increased dissolved nutrients, promoting microbial growth. Waters were frequently renewed by meteoric precipitation depleted in heavy isotopes. At the other extreme, dark laminae that are at least as thick as the light ones commonly rest unconformably on the light laminae. This suggests that microbial growth responsible for the light lamina was interrupted. A period of erosion followed, before formation of the overlying dark lamina. In these cases, the doublets are characterized by small percentages of detrital grains (ca. 3.5%) and organic matter (ca. 0.6%), while the carbonate is relatively enriched in ^{13}C ($4.0 < \delta^{13}C < 5.4‰$ PDB) and ^{18}O ($3.5 < \delta^{18}O < 5.4‰$ PDB). Consequently, each dark lamina is interpreted to reflect a break in microbial productivity in quite unrenewed water. The high ^{18}O content indicates a water residence time long enough to enrich the surrounding waters in heavy isotopes through evaporation [Hillaire-Marcel and Casanova 1987]. The high ^{13}C content indicates an increase in photosynthetic activity, which preferentially consumes $^{12}CO_2$ and $H^{12}CO_3^-$ [Rau, 1978; Herczeg and Fairbanks, 1987] and leads to an enrichment of heavy carbon in interstitial waters and in the precipitated carbonates [Hillaire-Marcel et al, 1986; Merz, 1992].

The basic doublet represents an ecological cycle of the microbial mat, which is interpreted in terms of seasonal contrast : the light lamina corresponds to the rainy season and the dark lamina to the dry season. Such a seasonal record in BMC has previously been observed in living stromatolites from southern France [Casanova and Lafont, 1985]. Similar isotopic trends of seasonal laminae comprising modern stromatolites have been reported from the Everglades [Monty and Hardie, 1976] and in the Arbuckle Mountains of Oklahoma [Chafetz et al., 1991]. It is possible to estimate quantitatively the seasonal contrast of each cycle from the ratio of mean lamina thickness corresponding to the growth during the rainy and the dry season respectively. Variations in this ratio are related to changes in water balance and thus provide information on the annual distribution of precipitation in paleorainfall regimes [Casanova, 1986b].

LAKE BOGORIA (KENYA)

A Holocene (5300-4700 yr B.P.) high lake episode is recorded by a double belt of stromatolites that occurs all around the lake and is particularly well developed along the western shore. The earlier stromatolites occur at up to 999 m, the later ones at up to 995 m (i.e. 9 m and 5 m above the present lake level, respectively). Lacustrine expansion resulted from conditions wetter than at present. Significant runoff is suggested by the presence of conglomerates consisting of well-rounded pebbles which have been individually encrusted by BMC and then cemented together. Palynological assemblages in stromatolites and correlative lacustrine sediments indicate a transition from humid forest to dry coniferous forest with *Podocarpus*, *Juniperus* and *Olea* [Vincens et al., 1985]. The seasonality ratio (SR) recorded by stromatolites is less than one, suggesting a long dry season interrupted by thundestorms and heavy rains distributed over a short rainy season [Tiercelin et al, 1987]. The period around 4500 yr B.P. is one of major climatic change over

Fig. 2. Lacustrine stromatolite microfacies from Lake Natron. Basic doublets are composed of a light-coloured sparitic laminae, formed during the rainy season, and of a dark micritic laminae, formed during the dry season.

clusters of bacteria, mucilaginous gels and rare cyanobacterial filaments. The sparitic microstructure of the light laminae is interpreted to have been generated by the periodic growth of a single population dominated by a *Schizothrix/Phormidium* association, similar to that observed in living stromatolites from freshwater lakes. Light laminae, therefore, represent growth phases favourable to Cyanobacteria, whereas the dark laminae correspond to a slowing down or a cessation of cyanobacterial growth, but relatively extensive bacteria blooms.

Most stromatolites display a microstructural gradation between two end members depending on the thickness of the dark laminae, as is the case in the example from Natron-Magadi (Figure 3). Where the light laminae are thicker than the dark ones, cyanobacterial filaments pass though the micritic films, and the doublets may contain high percentages of detrital particles (>20%) and organic matter (>1.5%). The carbonate is relatively depleted in ^{13}C ($3.1 < \delta^{13}C < 3.8‰$ PDB) and ^{18}O ($2.4 < \delta^{18}O < 3.7‰$ PDB). Accordingly, the light laminae formed during periods of extensive runoff, which favoured both

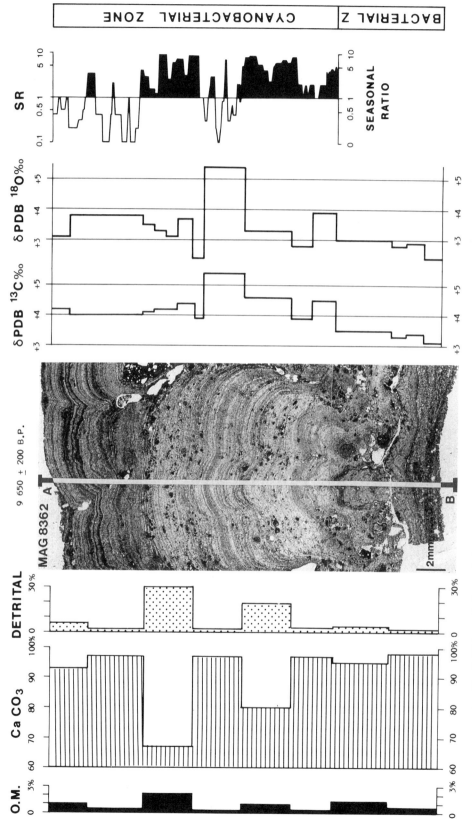

Fig. 3. High resolution analysis of a lacustrine stromatolite from Natron-Magadi basin : sedimentology, microfacies, isotopic contents and seasonal contrast variation recorded by the growth of the benthic microbial communities (SR=rainy season/dry season). See text for discussion.

LAKE BOGORIA STROMATOLITES

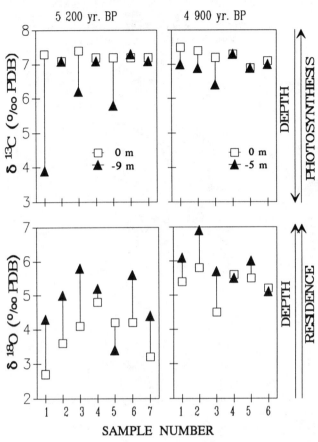

Fig. 4. Stable carbon and oxygen isotope behaviour in stromatolites from the Lake Bogoria high stands, with respect to their bathymetric distribution. Paleolake Bogoria stromatolites are distributed in two generations that are easily distinguishable on the basis of successive encrustings and through their specific morphologies.

East Africa, and marks the onset of aridification in this part of the African continent [Kendall, 1969; Richardson, 1972; Gasse et al., 1974; Street and Grove, 1976; Gasse and Street, 1978].

Figure 4 explores the stable carbon and oxygen isotope behaviour in stromatolites with respect to their bathymetric distribution. Samples come from various locations on the two paleolake shorelines. Because the BMC photosynthetic activity is closely related to light intensity, the ^{13}C enrichment of total dissolved inorganic carbon (TDIC) in the surface waters is uniform, while it is site-specific and usually smaller in deeper areas of the paleolake. As a consequence, contemporaneous lacustrine stromatolites display a variable ^{13}C depletion with increasing water depth of the growth site. Oxygen isotopic composition of surface waters in a closed basin varies both through space and time, according to local hydrological conditions. The ^{18}O content of littoral stromatolites (Figure 4) reflects this variability. Moreover, the isotopic trend of ^{18}O

enrichment with depth, observed in stromatolites, reflects increasing residence time of the paleolake waters with depth.

LAKE NATRON-MAGADI (KENYA-TANZANIA)

Today Lakes Natron and Magadi (southern Gregory Rift Valley, Figure 1) are shallow hypersaline water bodies with extensive trona crusts [Eugster, 1980; Vincens and Casanova, 1987]. The depression has been filled by a larger freshwater body on several occasions during Late Pliocene and Pleistocene times [Hillaire-Marcel et al, 1986]. During recent humid episodes, stromatolites were built along paleolake margins some 60m above modern water level. Three generations of stromatolites are observed, many of the more recent ones coating pebbles and boulders eroded from the older stromatolites. The oldest exceeds the limits of the Th/U dating method (>200 kyr); the intermediate one is dated at about 130 kyr and the most recent between 13000 and 9000 yr B.P. [Hillaire-Marcel et al, 1986].

The relatively high $\delta^{18}O$ values (-0.6 < $\delta^{18}O$ < +5.4 ‰ PDB) and $\delta^{13}C$ (+1.1 < $\delta^{13}C$ < +6.4‰ PDB) (Figure 5) reflect a long residence time for the paleolake water, which would have been favorable to the establishment of an isotopic equilibrium between atmospheric CO_2 and the dissolved inorganic carbon, despite hydrothermal carbon inputs related to carbonatitic volcanism [Hay, 1983]. A study of the ^{18}O, ^{2}H and ^{13}C contents in hydrothermal springs, perennial rivers and lacustrine brines

NATRON-MAGADI STROMATOLITES

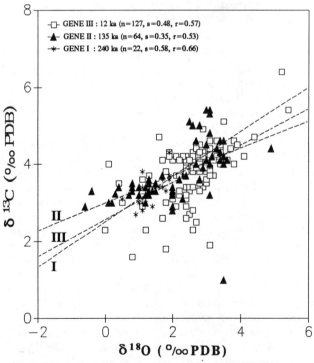

Fig. 5. Isotopic analyses of stromatolite carbonate from Pleistocene-Holocene high stands of the Natron-Magadi Lake (data replotted from Casanova, 1986a; Hillaire-Marcel and Casanova, 1987).

[Hillaire-Marcel and Casanova, 1987] indicates that, during the last period of significant renewal of the deep groundwater (i.e. contemporaneous with the third stromatolite generation), precipitation was depleted in heavy isotopes when compared to modern precipitation. This difference corresponds to a drop of about 600-1000 m in altitudinal hydroclimatic zones [Hillaire-Marcel, 1987]. The seasonal ratio recorded in the laminae of the third stromatolite generation averages 10, and can reach 16. The corresponding hydrological episode (9000-13000 yr B.P.) was characterized by fine drizzle, regularly distributed throughout a long rainy season, and by a short dry season [Hillaire-Marcel and Casanova, 1987]. Paleoprecipitation of ca. 750 mm/yr at lake level has been estimated [Casanova, 1986b]; this represents ca. 175% of the modern value (427 mm/yr).

The $\delta^{18}O$ and $\delta^{13}C$ ranges defined by the three generations cannot really be distinguished from each other (Figure 5). The data show considerable scatter and the correlations are only moderate ($0.53 < r < 0.66$). A comparison between the modern and paleohydrological data leads us to believe that the stabilization of the paleolake shorelines at the same altitude can be attributed to control by the water table level in the aquifer of the basaltic plateaux east of the basin. The fact that the various generations of stromatolites show comparable stable carbon and oxygen contents, as well as being located at the same paleolake stabilization levels, indicates that specific hydrological conditions are necessary for the development of the BMC responsible for stromatolite formation. A comparison with similar stromatolitic units from the nearby Lake Manyara basin [Casanova and Hillaire-Marcel, 1992a] shows that such conditions occurred during only a few of the Late Quaternary humid episodes known in Eastern Africa.

LAKE SUGUTA (KENYA)

The Suguta Rift is the most arid area of Kenya (Figure 1) and is characterized by a strongly negative hydrological budget. Several major lacustrine phases related to Late Quaternary climatic changes are, however, recorded by thick fluvio-lacustrine deposits [Casanova et al., 1988].

High lake levels are indicated by the presence of stromatolites, up to 1 m thick and exclusively formed of calcite of bacterial origin. This is unusual in Africa where stromatolites are commonly of algal origin [Abell et al., 1982; Casanova, 1991; in press]. The encrustations indiscriminately cover hard substrates, including bedrock surfaces and some vegetal remains. They show a variety of growth forms including large, flattened oncolites, tabular crusts and hemispheric bioherms. At least two stromatolite generations are present. The older one, in the southeastern part of the basin, is covered by a thick sequence of hyaloclastites and yielded infinite ^{14}C ages. Around the margin of the depression, a second generation yielded ^{14}C ages between 27000 and 30000 yr B.P. [Casanova et al., 1988]. It defines a paleoshoreline ranging from 437 to 460 m, deformed as a consequence of recent tectonic movements. At the time of growth, the paleolake Suguta reached an area of 1390 km² and had a maximum depth of ca. 200 m.

The Suguta Valley stromatolites (Figure 6) define two isotopic

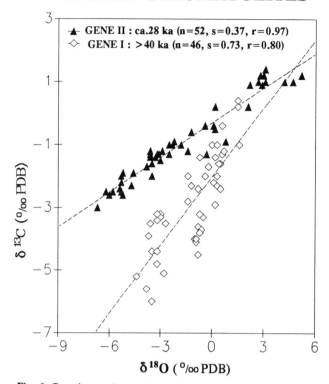

SUGUTA STROMATOLITES

Fig. 6. Covariant carbon and oxygen isotopic trends in lacustrine stromatolites from the Suguta Valley high stands (data replotted from Casanova, 1986a).

covariant trends (rI = 0.80; rII = 0.97), corresponding to the two generations. The distinct origins of these trends, as well as the slopes of the two regression lines (sI = 0.73; sII = 0.37), clearly indicate that the two generations of stromatolites formed in different waterbodies [Talbot, 1990; Talbot and Kelts, in press]. Despite being characterized by very similar surface area/depth ratios, the two paleolakes correspond to different hydroclimatic situations. The more recent stromatolites (Figure 6) have a larger $\delta^{18}O$ range ($-6.7 < \delta^{18}O < +5.4$ ‰ PDB) than the older ones ($-4.4 < \delta^{18}O < +1.6$‰ PDB), indicating more ^{18}O depleted meteoric precipitations alternating with more arid conditions. The pollen content of these limestones is characterized by an abundance of montane conifer forest elements (*Podocarpus, Juniperus*...) and a paucity of arid steppic taxa, compared to modern vegetation [Casanova et al., 1988]. Such pollen spectra suggests a vegetation different from the modern one. Savanna, bushland and montane forest very probably occupied most of the area, due to a drop in altitudinal vegetation zones. With reference to the modern climate, more humid conditions and a strong seasonnal contrast are inferred. Moreover, the $\delta^{13}C$ value of the origin on the isotopic trends is more negative for the first generation ($\delta^{13}C = -6‰$) than for the second generation ($\delta^{13}C = -3‰$), suggesting either a higher proportion of C3 vs C4 plants or more organic matter in soils from groundwater recharge zones, or both [Hillaire-Marcel et al., 1989; Talbot, 1990; Aucour, 1992].

LAKE TANGANYIKA (BURUNDI)

Lake Tanganyika, situated in the Western Branch of the East African Rift System (Figure 1), has a small outflow through the Lukuga River, dependent mainly on the Ruzizi River inflow from Lake Kivu in the northern part of the basin [Hecky and Degens, 1973; Hecky, 1978]. Conditions more arid than now closed the Kivu Basin from 3800 to 1300 yr B.P. During this period, stromatolites formed as thick incrustations on hard substrates and are found at depths between 6 and 60 m below modern lake level [Cohen and Thouin, 1987; Casanova and Thouin, 1990]. A major low stand at 10m below present level is identified from 3400 to 1900 yr B.P., bounded by a rapid recession and gradual rise to modern levels [Casanova and Hillaire-Marcel, 1992b].

The $\delta^{18}O$ values in stromatolites range between ca. 1.6 and 6‰ (mean $\delta^{18}O$ value = 3.2 $\pm$1.3‰ PDB; n=47; Casanova and Hillaire-Marcel, 1992b, Figure 7). In terms of isotopic composition of paleolake waters, these values would correspond to ca. 3.8 and 8.2‰ (vs SMOW) respectively, assuming present-day surface water temperature (27°C) and carbonate precipitation in isotopic equilibrium with the water. This is higher than today ($\delta^{18}O$ = 3.5 $\pm$0.2‰ vs SMOW, n=19; Casanova and Hillaire-Marcel, 1992b, Figure 2), therefore suggesting a longer water residence time.

The $\delta^{13}C$ values of stromatolites range between ca. 1.3 and 5.8‰ PDB (mean $\delta^{13}C$ value = 2.8 $\pm$1.1‰; n=47). Carbon 13 contents in stromatolites generally fluctuate in parallel with ^{18}O, although they are not exactly covariant [Casanova and Hillaire-Marcel, 1992b, Figure 7]. This is due to the greater variability of ^{13}C contents in TDIC, which is strongly influenced by BMC photosynthetic activity [Hillaire-Marcel and Casanova, 1987], compared to the more stable oxygen reservoir (i.e., the paleolake water). As a basis for comparison, calcite in isotopic equilibrium with present epilimnion water would have a $\delta^{18}O$ value of ca. 1.3‰ PDB, and a $\delta^{13}C$ value of 2.4‰ PDB [Bottinga, 1968; Gonfiantini et al., 1968; Mook et al., 1974].

Figure 7 depicts the $\delta^{18}O$ changes of the paleolake waters inferred from the $\delta^{18}O$ variations in the stromatolites through time. Stromatolite growth started ca. 3400 yr B.P., when paleolake Tanganyika stabilized at ca. 10 m below its present shoreline. A $CaCO_3$ deposition peak in the Southern Basin dates the beginning of the aridity crisis (due to a regional P:E ratio decrease) at ca. 3800 yr B.P. [Greene and Jones, 1970]; the lake level drop was certainly abrupt, reflecting the cutting off of inflow from the Ruzizi river [Hecky and Degens, 1973; Hecky, 1978]. The calculated value for the $\delta^{18}O_{WATER}$ when the paleolake stabilized at ca. 10 m below its present level (at ca. 3400 yr B.P.) is ca. 5.7‰ vs SMOW. With time, the ^{18}O content continued to increase (up to 8.2‰ SMOW) until ca. 2500 yr B.P., indicating a reduced rate of water renewal. From 2500 yr B.P., the calculated isotopic composition for the lake waters decreased more or less regularly to its present value. Until at least 1900 yr B.P., the paleolake remained stable at ca. -

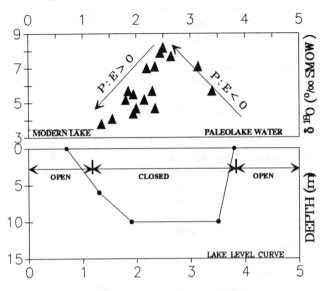

LAKE TANGANYIKA

Fig. 7. Lake Tanganyika level fluctuations during Late Holocene (below) and ^{18}O content of paleolake waters (above) inferred from shoreline stromatolites (data replotted from Casanova and Hillaire-Marcel, 1992b).

10 m level below its present level, although $\delta^{18}O$ data suggest changes in the P:E budget. A possible explanation for this decoupling mechanism could be found in the link between the lake and the large aquifers of the Rift escarpments (2000 m above lake level): the -10 m level may likely have represented an hydraulic equilibrium between regional groundwaters and lake waters.

During the period of stable lake level (i.e., from 3400 to 1900 yr B.P.), the P:E ratio fluctuations inferred from $\delta^{18}O$ data indicate changes in the rainfall regime, very likely in the amount of precipitation. At ca. 2500 yr B.P., a maximum enrichment of paleolake water ($\delta^{18}O$ ca. 8.2‰ vs SMOW) is observed, which correlates with a maximum in $CaCO_3$ deposition in the North and South Basins [Hecky and Degens, 1973; Haberyan and Hecky, 1987] and probably marks the height of aridity in the area. Stromatolites exhibit a clear tendency to be isotopically heavier with increasing depth, suggesting incomplete vertical homogenization of the paleolake waters. This trend is confirmed by serial analyses from two contemporaneous stromatolites sampled respectively at -20 m and -10 m [Casanova and Hillaire-Marcel, 1992b, Figure 4]. The former is enriched in ^{18}O by 1 to 3‰ compared to the latter.

As early as 2400 yr B.P. (Figure 7), a trend towards depletion in ^{18}O of paleowaters is observed (down to ca. 3.8‰ vs SMOW), indicating increasing precipitation (and decreasing residence time of the paleolake water), but it was only after 1900 yr B.P., that the lake started to rise to the -6 m level. This time-lag between the isotopic signal and the paleolake response may again be due to the time needed to recharge the regional aquifers controlling the lake level. The stromatolite $\delta^{18}O$ values show a very homogeneous isotopic composition ($\delta^{18}O = 2.0 \pm 0.3‰$ PDB, n=16). This indicates the restoration of an efficient vertical mixing in the paleolake surface waters ($\delta^{18}O$ ca. 4.2‰ vs SMOW). Around 1400 yr B.P., the positive regional P:E ratio finally resulted in the overflow of Lake Kivu into Lake Tanganyika. When these overflowing waters mixed with the surface waters of Lake Tanganyika, they put a definite stop to stromatolite growth [Casanova and Thouin, 1990].

SABKHA CHEMCHANE (MAURITANIA)

The Chemchane sabkha, a dry depression in SW Mauritania (Figure 1) provides an example of isotopic analyses from carbonate bearing sediments and contemporaneous stromatolites. A 7.5 m core from the western part of the depression was sampled at 10 cm intervals for sedimentological, isotopic and palynological analyses [Lezine et al., 1989]. Stable isotopes were measured on authigenic, inorganic calcite identified under SEM. Core sediments, between depth of 5.6 and 3.1 m, consist of lacustrine marls including a few centimeter-thick shelly layers. This unit is interpreted as evidence of sedimentation in a closed lake with relatively dilute water, as shown by the presence of *Cyperaceae* and *Typha* pollen [Lezine et al., 1989]. The basin was a sabkha before and after this lacustrine episode.

Maximum lake expansion between about 8300 and 6500 yr B.P. is recorded in a girdle of stromatolite carbonates. They range in elevation from 262 to 271 m a.s.l., the highest occurrences defining a paleoshoreline at about 20m above the paleolake floor. Stromatolites occur as bioherms and planar encrustations a few centimetres thick, comprising microbially precipitated low-Mg calcite and are thought to reflect growth in oligohaline waters [Casanova, 1986a].

During the lacustrine phase, paleohydrological oscillations are recorded by changes in *Typha* pollen percentages [Lezine et al., 1989] and ^{18}O and ^{13}C contents of primary carbonates in the core (Figure 8). Swamp vegetation spreads when lake levels are low and is reflected in decreasing $\delta^{13}C$ values, indicating the input of light CO_2 produced by the decay of organic matter. Rising $\delta^{18}O$ values indicate stronger evaporation of the paleolake water, which apparently induces more $CaCO_3$ precipitation, as shown by the carbonate curve (Figure 8). Opposite trends are observed when lake levels are high. The stromatolites display the succession of four growth stages characterizing oscillations of the lake level. Layers associated with a positive hydrological budget are depleted in ^{18}O, and layers formed during episodes of strong evaporation show $\delta^{18}O$ shifts toward higher values. The stromatolite isotopic profiles shown in Figure 8 come from the highest occurence containing the complete set of growth stages. The lowest ^{18}O values may, therefore, be indicative of the isotopic composition of feeding waters during the successive high stands. From base to top of the profile, these values regularly increase suggesting either a change in the isotopic

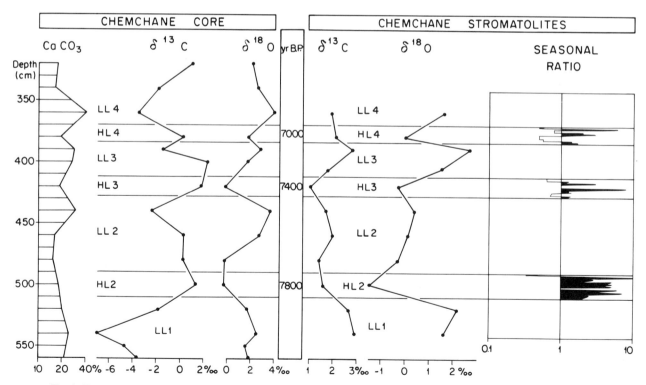

Fig. 8. The percentage carbonate and stable-isotope stratigraphy of the Sabkha Chemchane core, correlated with stable isotope data and seasonal ratio of the contemporary stromatolites (data replotted from Lézine et al., 1990). Paleohydrological oscillations are recorded by a succession of high levels (HL) and lowering levels (LL). See text for discussion.

CHEMCHANE SABKHA

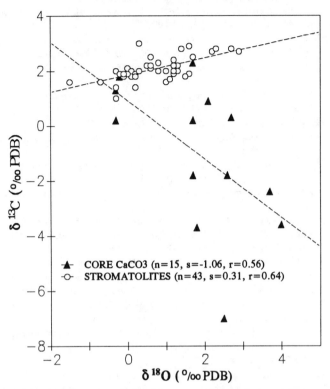

Fig. 9. Isotopic trends of shoreline stromatolites and core primary carbonates from Sabkha Chemchane (data replotted from Lézine et al., 1990).

enriched released by organic matter oxidation [Lee et al., 1987; Fontes and Gasse, 1991, Figure 2].

CONCLUSIONS

(1) In comparison to most other lake indicators, stromatolites contain a great deal of information. Paleoecological and isotopic analysis of these structures can provide precise data on the occurence and amplitude of paleoclimatic fluctuations in Africa. When present, stromatolites are precise indicators of high lake level limits; their morphology and structure, in relation to their paleobathymetric distribution, provide an insight into the paleoenvironment of the time; their isotopic composition ($\delta^{13}C$, $\delta^{18}O$) allows the assessment of paleohydrological conditions, with reference to modern hydrology.

(2) The biosedimentological analysis of lacustrine stromatolites provides information on the physicochemical state of the paleolake which may in part be directly related to climate. Growth rhythms in microbial mats, may preserve evidence of seasonality, allowing the reconstruction of paleorainfall regimes.

(3) The proportion of ^{13}C in lacustrine stromatolites provides information on sources of carbon to the TDIC in the surrounding waters, on which an isotopic enrichment effect, linked to the micro-organisms photosynthetic activity, is superimposed. Since intertropical lake waters typically show only small amplitude changes in temperature, the $\delta^{18}O$ composition of stromatolites depends primarily on the isotopic composition of feeding waters and on the residence time of paleolake waters. Covariance between $\delta^{13}C$ and $\delta^{18}O$ records during stromatolite growth is interpreted as variable development of microbial mats in relation to hydrological conditions in paleolake.

Acknowledgments. We wish to thank Mike Talbott and Tom Johnson for comments which greatly improved the manuscript. The isotopic analyses were financed by NSERC-Canada and conducted in the GEOTOP Research Center of the Université du Québec à Montréal. This work was supported by a BRGM research project. Contribution 92031 of the BRGM.

REFERENCES

Abell, P., S. M. Awramik, R. H. Osborne, and S. Tomellini, Plio-Pleistocene lacustrine stromatolites from Lake Turkana, Kenya: morphology, stratigraphy and stable isotopes, *Sedim. Geol.*, 32, 1-26, 1982.

Aucour, A. M., Composition isotopique de la matière organique de sédiments de tourbières du Burundi (0-40.000 ans BP). Relation avec les changements climatiques. Doc. Sc. Thesis, UQAM, 139 p., 1992.

Bottinga, Y., Calculation of the fractionation factor of carbon and oxygen exchange in the system calcite - carbon dioxyde - water, *J. Phys. Chem.*, 72, 800-808, 1968.

Buchbinder, B., Morphology, microfabric and origin of stromatolites of the Pleistocene precursor of the Dead Sea, Israel, in *Phanerozoic Stromatolites: Case histories*, edited by C. L. V. Monty, pp. 180-196, Springer Verlag, Berlin, 1981.

Casanova, J., Les stromatolites continentaux : paléoécologie, paléohydrologie, paléoclimatologie Application au rift Gregory. Doc. Sc. Thesis, Aix-Marseille II, 256pp., 1986a.

composition of precipitation, or a trend towards overall increasing residence time of paleolake waters. Concurrently, the seasonal ratio recorded by laminae formed during high lake levels indicates a progressive shortening of the rainy season. The early Holocene humid episode in western north Africa is known to be related to the maximum intensity of Atlantic monsoon influx [Kutzbach and Street-Perrott, 1985; Lézine and Casanova, 1989]. The front of moist air progressed from South to North and its influence decreased abruptly after 7500 yr B.P. The Chemchane stromatolite record defines more accurately the appearance of the marked dry season which characterizes the middle Holocene rainfall pattern in intertropical Africa.

The isotopic trends defined by inorganic carbonates and stromatolites do not correlate (Figure 9). The slope of the stromatolite trend (s=0.31) is normal for such a sabkha with large surface area/depth ratio [Talbot, 1990], while the negative slope of the core carbonate trend (s=-1.06) is quite unusual. This may be explained by the different water layers in which the two sets of carbonates formed [Eugster and Kelts, 1983; Lee et al., 1987, Figure 1]. Stromatolites grow in the uppermost waters in which TDIC was mainly controlled by isotopic exchanges with atmospheric CO2 and by photosynthesis. On the other hand, inorganic carbonates most likely precipitated from underlying waters in which TDIC was fully influenced by the carbon-12

Casanova, J., East African Rift Stromatolites, in *Sedimentation in the African Rifts*, edited by L. E. Frostick, et al., Geological Society Spec. Publ., 23, pp. 195-204, 1986b.

Casanova, J., Biosedimentary processes in Intertropical Quaternary stromatolites, *Journal of African Earth Sciences*, 12, 409 -415, 1991.

Casanova, J., Stromatolites from the East African Rift : a synopsis, in *Phanerozoic Stromatolites II*, edited by C., Monty, and J. Sarfati, Springer Verlag Berlin , in press.

Casanova, J., and C., Hillaire-Marcel, Chronology and Paleohydrology of Late Quaternary High Lake Levels in the Manyara Basin (Tanzania) from Isotopic Data (^{13}C, ^{18}O, ^{14}C, U/Th) on Fossil Stromatolites, *Quaternary Research*, 38, 1-22, 1992a.

Casanova, J., and C., Hillaire-Marcel, Late Holocene hydrological history of Lake Tanganika, East Africa, *Paleogeogr., Paleoclimat., Paleoecol.*, 91, 35-48, 1992b.

Casanova, J., C. Hillaire-Marcel, N. Page, M. Taieb, and A. Vincens, Stratigraphie et paléohydrologie des épisodes lacustres du Quaternaire récent du Rift Suguta, *C. R. Acad. Sci. Paris*, 307, 1251-1258, 1988.

Casanova J., and R. Lafont, Les cyanophycées encroûtantes du Var (France), *Verh. Internat. Verein . Limnol.*, 22, 2805-2810, 1985.

Casanova, J., and C. Thouin, Biosédimentologie des stromatolites holocènes du Lac Tanganyika (Burundi). Implications hydrologiques, *Bull., Soc. Géol. Fr.*, 4, 647-656, 1990.

Chafetz, H. S., N. M. Utech, and S. P. Fitzmaurice, Differences in the δ^{18}O and δ^{13}C signatures of seasonal laminae comprising travertine stromatolites, *J. Sed. Pet.*, 61, 1015-1028, 1991.

Cohen, A. and C. Thouin, Nearshore carbonate deposits in Lake Tanganyika, *Geology*, 15, 414-418, 1987.

Eugster, H. P., Lake Magadi, Kenya and its precursor, in *Hyper-saline brines and evaporitic environments*, edited by A. Nissenbaum, Developments in Sedimentology, 28, pp. 195- 230, Elsevier, Amsterdam, 1980.

Eugster, H. P., and K. Kelts, Lacustrine chemical sediments, in *Chemical sediments and Geomorphology*, edited by A. Goudie, and K. Pye, Academic Press, pp. 321-368, 1983.

Fontes, J. C., and F. Gasse, PALHYDAF (Palaeohydrology in Africa) program : objectives, methods, major results, *Palaeogeogr., Palaeoclimatol., Palaeoecol.*, 84,191-215, 1991.

Gasse, F., J. C. Fontes, and P. Rognon, Variations hydroclimatiques et extension des lacs holocènes du désert Danakil, *Palaeogeogr., Palaeoclimatol., Palaeoecol.*, 15, 109-148, 1974.

Gasse, F., and F. A. Street, Late Quaternary lake-level fluctuations and environments of the northern Rift Valley and Afar Region (Ethiopia and Djibouti), *Palaeogeogr., Palaeoclimatol., Palaeoecol.*, 24, 279-325, 1978.

Gonfiantini, R., C. Panichi, and E. Tongiorgi, Isotopic disequilibrium in travertine deposition, *Earth Planet Sci. Lett.*, 5, 55-58, 1968.

Greene, C. and E. N. Jones, Physical and chemical properties of lake Tanganyika. Tech. Memorandum 2213 - 331 - 70. National Underwater Systems Center, New London, 1970.

Haberyan, K. A. and R. E. Hecky, The late Pleistocene and Holocene stratigraphy and paleolimnology of lakes Kivu and Tanganyika, *Palaeogeogr., Palaeoclimatol., Palaeoecol.*, 61, 169-197, 1987.

Halfman, J. D., and T. C. Johnson, High-resolution record of cyclic climatic change during the past 4 ka from Lake Turkana, Kenya, *Geology*, 16, 496-500, 1988.

Hay, R. L., Natrocarbonatite tephra of Kerimasi volcano, Tanzania, *Geology*, 11, 599-602, 1983.

Hecky, R. E., The Kivu-Tanganyika basin: the last 14 000 years, *Pol. Arch. Hydrobiol.*, 25, 159-165, 1978.

Hecky, R. E., and E. T. Degens, Late Pleistocene-Holocene chemical stratigraphy and paleolimnology of the Rift Valley lakes of Central Africa. Technical Report WHOI-73-28, Woods Hole Oceanographic Institution, Woods Hole, 93 p., 1973.

Herczeg, A. L. and R. G. Fairbanks, Anomalous carbon isotope fractionation between atmospheric CO_2 and dissolved inorganic carbon induced by intense photosynthesis, *Geochim. et Cosmochim. Acta*, 51, 895-899, 1987.

Hillaire-Marcel, C., Hydrologie isotopique des lacs Magadi (Kenya) et Natron (Tanzanie), *Sci. Géol.*, 40, 111-120, 1987.

Hillaire-Marcel, C., O. Carro, and J. Casanova, ^{14}C and Th/U dating of Pleistocene and Holocene stromatolites from East African palaeolakes, *Quaternary Research*, 25, p. 312-329, 1986.

Hillaire-Marcel, C., and J. Casanova, Isotopic hydrology and palaeohydrology of the Magadi (Kenya) - Natron (Tanzania) basin during the Late Quaternary, *Palaeogeogr., Palaeoclimatol., Palaeoecol.*, 58, 155-181, 1987.

Hillaire-Marcel, C., A.M. Aucour, R. Bonnefille, G. Riollet, A. Vincens, and D. Willamson, ^{13}C / palynological evidence of differential residence times of organic carbon prior to its sedimentation in East african rift lakes and peat bogs, *Quat. Sci. Reviews*, 8, 207-210, 1989.

Johnson, G. D., Cenozoic lacustrine stromatolites from hominid-bearing sediments East of lake Rudolf, Kenya, *Nature*, 247: 520-523, 1974.

Kendall, R. L., An ecological history of the Lake Victoria basin, *Ecol. Monogr.*, 39, 121-176, 1969.

Kutzbach, J. E., and F. A. Street-Perrot, Milankovitch forcing of fluctuations in the level of tropical lakes from 18 to 0 kyr B.P., *Nature* 317, 2130-2134, 1985.

Lee, C., J. A. McKenzie, and M. Sturm, Carbon isotope fractionation and changes in the flux and composition of particulate matter resulting from biological activity during a sediment trap experiment in Lake Greifen, Switzerland, *Limnol. Oceanogr.*, 32, 83-96, 1987.

Lézine, A. M., and J. Casanova, Pollen and hydrological evidences for the interpretation of past climates in Tropical West Africa during the Holocene, *Quat. Sci. Reviews*, 8, 45-56, 1990.

Lézine, A. M., J. Casanova, and C. Hillaire-Marcel, Across an early Holocene humid phase in Western Sahara : pollen and isotope stratigraphy, *Geology*, 18, 264-267,1990.

Merz, M. U. E., The biology of carbonate precipitation by Cyanobacteria, *Facies*, 26, 81-102, 1992.

Monty, C. L. V., and L. A. Hardie, The geological signifiance of the freshwater blue-green algal calcareous marsh, in *Stromatolites*, edited by M. R. Walter, Development in Sedimentology, 20, Elsevier, Amsterdam, pp. 447-477, 1976.

Mook, W. G., J. C. Bommerson, and W. H. Staverman, Carbon isotope fractionation between dissolved bicarbonate and gaseous carbon dioxide. *Earth Planet. Sci. Lett.*, 22, 169-176, 1974.

Rau, G., Carbon-13 depletion in a subalpine lake : Carbon flow implications, *Science*, 201, 901-902.

Richardson, J. L., Palaeolimnological records from rift lakes in central Kenya, *Palaeoecol. Afr.*, 6, 131-136, 1972.

Street, F. A., and A. T. Grove, Environmental and climatic implications of Late Quaternary lake-level fluctuations in Africa, *Nature*, 261, 385-390, 1976.

Talbot, M. R., A review of the palaeohydrological interpretation of carbon and oxygen isotopic ratios in primary lacustrine carbonates, *Chem. Geol. (Isot. Geosci. Sect.)*, 80, 261-279, 1990.

Talbot, M. R., and K. Kelts, Primary and diagenetic carbonates in the anoxic sediments of Lake Bosumtwi, Ghana, *Geology*, 14, 912-916, 1986.

Talbot, M. R., and K. Kelts, Palaeolimnological signatures from carbon and oxygen isotopic ratios in carbonates from organic carbon-rich lacustrine sediments, in *Lacustrine exploration: case studies and modern analogues*, edited by B. J. Katz and B. R. Rosendahl, AAPG Studies in Geology, in press.

Tiercelin, J. J., A. Vincens, C. E. Barton, P. Carbonel, J. Casanova, G. Delibrias, F. Gasse, E. Grosdidier, J. P. Herbin, A. Y. Huc, S. Jardiné, J. Le Fournier, F. Mélières, R. B. Owen, P. Pagé, C. Palacios, H. Paquet, G. Péniguel, J. P. Peypouquet, J. F. Raynaud, R. W. Renaut, P. Renéville, J. P. Richert, R. Riff, P. Robert, C. Seyve, M. Vandenbroucke, and G. Vidal, Le demi-graben de Baringo-Bogoria, Rift Gregory, Kenya: 30 000 ans d'histoire hydrologique et sédimentaire. *Bull Centres Rech. Explor.-Prod Elf-Aquitaine*, 11, 249-540, 1987.

Vincens, A., J. Casanova, and J. J. Tiercelin, Palaeoclimatology of lake Bogoria (Kenya) during the 4500 B.P. high lacustrine phase, in *Sedimentation in the African Rifts*, edited by L. E. Frostick, et al., Geological Society Spec. Publ., 23, 315-322, 1986.

Vincens, A., and J. Casanova, Modern background of Natron-Magadi Basin (Tanzania-Kenya) : physiography, climate, hydrology and vegetation, *Sci. Geol.*, 40, 9-21, 1987.

Yuretich, R. F., Possibles influences upon lakes development in the East African Rift Valley, *J. Geol*, 90, 329-337, 1982.

J. Casanova, BRGM, Departement Géochimie, B.P. 6009, 45060 Orléans Cedex 2, France.

C. Hillaire-Marcel, GEOTOP-Université du Québec à Montréal, C.P. 8888 Suc A, H3C 3P8 Montréal, Canada.

An Isotopic and Trace Element Study of Ostracods from Lake Miragoane, Haiti: A 10,500 year Record of Paleosalinity and Paleotemperature Changes in the Caribbean

JASON H. CURTIS
DAVID A. HODELL

Department of Geology, University of Florida, Gainesville, FL

We report a high-resolution climate reconstruction for the Caribbean based on isotopic and trace element analysis of freshwater ostracod shells from Lake Miragoane, Haiti. By combining oxygen isotopes with Sr/Ca and Mg/Ca ratios, we are able to determine qualitative changes in temperature and salinity of this small, deep lake from the very late Pleistocene (10,500 years BP) to the present. During the latter part of the Younger Dryas Chronozone from ~10,500 to 10,000 years BP, isotopic, trace element, and pollen results suggest that climate was arid and temperature was cooler than today in the Caribbean region. Similar interpretations of lake level lowering and increased aridity have been made for African lakes during this period. During the last deglaciation (Termination 1b), from ~10,000 to 7,000 years BP, temperature increased and salinity decreased as lake levels rose during the early Holocene. Minimum salinity conditions are recorded between 7,000 and 4,000 years BP, which coincides with the early Holocene moist period when lake levels were consistently high in the tropics. From ~4,000 to 2,500 years BP, lake level declined and salinity increased with the onset of a dry climate that generally prevailed throughout the late Holocene. The interval from ~2,500 to 1,500 years BP was marked by an exceptionally dry period when all parameters indicate high salinity. This severe dry period persisted until ~1000 years BP when wetter conditions briefly returned. The last millennia has been marked by a general trend toward increased salinity and inferred drier conditions.

INTRODUCTION

Sediment cores from low-latitude, low-elevation lakes provide one of the best opportunities to investigate details of tropical paleoclimate in continental and island settings. Lake levels and solute concentrations in tropical closed-basin lakes fluctuate in response to the ratio of evaporation to precipitation (E/P). Shifts in the E/P ratio are driven, in turn, by climatic changes on time scales ranging from years to millennia. We are fortunate in that changes in a lake's hydrologic budget are faithfully recorded in the isotopic and trace element compositions of ostracods preserved in lake sediments [Lister, 1988; Chivas et al., 1985]. Here we present isotopic and trace element results of ostracod calcite from Lake Miragoane, Haiti, to reconstruct the history of changing salinity and temperature of lake water during the past 10,500 years.

SITE DESCRIPTION

Lake Miragoane is a small, deep-water lake located about 5 km from the coast on the north side of Haiti's southern peninsula (Fig. 1). It is one of the largest natural freshwater lakes in the Caribbean with an area of 7.06 km^2 and a maximum depth of 42 m. Residence time for water in the lake has been

estimated to be ~3 years [Brenner and Binford, 1988]. Lake Miragoane is believed to be a warm, monomictic system that mixes during the winter and stratifies during the summer [Brenner and Binford, 1988]. Hydrologic input to the lake includes direct precipitation, slope runoff, and probably groundwater from springs. Water is lost from the lake by evaporation and through a small, shallow outflow (1 cubic meter/sec) to the north and possibly by groundwater seepage. Because the lake surface is only 20 m above mean sea level, lake level is sensitive to the height of sea level, which controls the position of the phreatic aquifer. There is no evidence, however, for salt water intrusion into Lake Miragoane during its history as there is for other lakes in the Caribbean (e.g. Lake Enriquillo, Dominion Republic) [Taylor et al., 1985; Mann et al., 1984].

Lake Miragoane is not a closed basin lake in the strict sense because it possesses an outflow and may leak via groundwater seepage. However, it is likely that the lake has been "effectively" closed throughout a large part of its history. Although Lake Miragoane possesses a shallow outflow today, this outflow would not have existed during times of even slightly lowered lake level as the sill depth is very shallow, averaging ~0.23 m. Furthermore, Lake Miragoane's sediments are very clay-rich which may act to seal the lake from groundwater seepage. We suggest, therefore, that the dominant mechanism of water loss from Lake Miragoane throughout most of its history has been via evaporation.

Mean annual temperature averages 26-27°C at meteorological sites in lowland coastal areas near Lake

Climate Change in Continental Isotopic Records
Geophysical Monograph 78

LAKE MIRAGOANE, HAITI

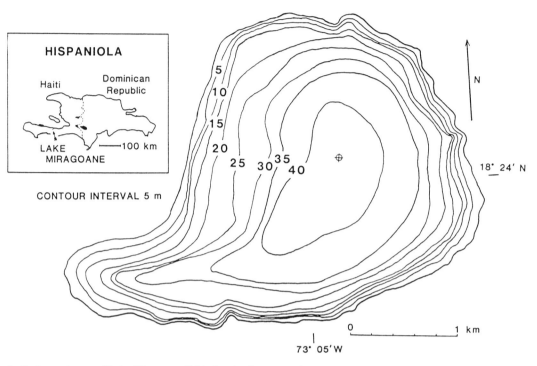

Fig. 1. Bathymetric map of Lake Miragoane, Haiti, showing 5m contour interval. Cross hair symbol in the central deep area of the lake denotes the location of the coring site. Inset map shows the island of Hispaniola and the location of Lake Miragoane on the northern side of Haiti's southern peninsula. After Brenner and Binford, 1988.

Miragoane and decrease with altitude. Although much of Haiti is in the rain shadow of mountains to the East in the Dominican Republic and is thus arid, the area surrounding Lake Miragoane is relatively wet, receiving between 1000 and 2000 mm of annual rainfall [Woodring et al., 1924]. The rainy season in the Caribbean region occurs during the northern hemisphere summer half-year when the Intertropical Convergence Zone (ITCZ) moves north. In Haiti, two rainy seasons occur: the first lasts from April through June with peak rainfall in May and the second occurs in the autumn with peak rainfall in October [Woodring et al., 1924]. The dry season, conversely, occurs during northern hemisphere winter and lasts from December through March [Woodring et al., 1924].

METHODS

The cores for this study were taken from the center of Lake Miragoane in 43 m of water during the summer of 1985 by M. Brenner and M. Binford. They consist of a long core (6.6 m), a short core (2.8 m), and a mud/water interface core (72 cm). The mud/water interface core was taken using a 4cm diameter piston corer while the short and long cores were taken using a square-rod piston corer.

The lithology of nearly all the core is dominated by $CaCO_3$ with lesser amounts of SiO_2, organic matter, $MgCO_3$, and Fe_2O_3 [Brenner et al., in press]. X-ray diffraction patterns of bulk sediment reveal that calcite is the dominant carbonate phase

with little detectable aragonite in the sediment. Ostracods are extremely abundant throughout the core and we selected a single taxon, *Candona* sp., for analysis. This taxon was identified by R. Forester and represents a new species that has not been formally described in the literature. Because the genus *Candona* is known to be a burrowing form that spends its life on the bottom, it will record bottom-water conditions.

Stable Isotopes

High-resolution samples for isotopic analysis were taken every centimeter throughout the core equivalent to an average of 1 sample every 20 years. Bulk samples were disaggregated in weak hydrogen peroxide (2-3%) and washed through a 106 μ sieve. After drying, samples of *Candona* sp. were picked from the >212 μ fraction under a binocular microscope. Ostracod valves were cleaned sonically in distilled water to remove adhering debris, soaked in hydrogen peroxide to remove organics, and rinsed in methanol before final drying. Sample weights for isotopic analysis were about 400 μg and consisted of ~40 ostracods valves. Ostracod shells were loaded into stainless steel carrying boats and gently crushed with a glass rod dipped in methanol. Samples were reacted in a common acid bath of 100% ortho-phosphoric acid (specific gravity =1.92) at 70°C using a VG Isogas autocarbonate preparation system. Isotopic ratios of purified CO_2 gas were measured on-line by a triple-collector VG Isogas PRISM mass spectrometer. All isotopic results are reported in standard delta notation and

Table 1 Radiocarbon dates of organic and inorganic carbon from the core taken in Lake Miragoane, Haiti.

Depth (m)	Sample Type	Target Number	Accession Number	Measured Age (yr)	Corrected Age (yr)
0.08	^{210}Pb			129±40	129
0.22	AMS ^{14}C/ostracod	WHG908	AA6703	1085±60	85
2.16	AMS ^{14}C/ostracod	WHG775	AA5814	2780±55	1780
2.33	AMS ^{14}C/ostracod	WHG909	AA6704	2680±60	1680
2.33	AMS ^{14}C/wood	WHG917	AA6705	1655±60	1655
3.21	AMS ^{14}C/ostracod	WHG777	AA5815	4110±60	3110
4.18	AMS ^{14}C/ostracod	WHG776	AA5816	4780±60	3780
5.20	AMS ^{14}C/ostracod	WHG782	AA5817	6945±65	5945
6.22	AMS ^{14}C/ostracod	WHG784	AA5818	9005±75	8005
6.71	AMS ^{14}C/ostracod	WHG720	AA5369	9700±90	8700
7.18	AMS ^{14}C/ostracod	WHG785	AA5952	10300±85	9300
7.53	Conventional ^{14}C/bulk organic carbon		GX13055	10230±160	10230

corrected to PDB (Appendix 1 and 2). Analytical precision, based on routine analysis of powdered carbonate standards (Carrara Marble, NBS-19, and NBS-20), was ±0.14 for δ^{18}O and ±0.06 for δ^{13}C (1-sigma; n=107).

Trace Elements

Samples for trace element analysis were taken every ten centimeters along the length of the core corresponding to an average sampling frequency of 1 sample per ~175 years. Sample processing was the same as that described for isotopic work except that samples for trace element analysis consisted of 2 to 6 valves of *Candona* sp. weighing between 30 and 50 μg. Samples were dissolved in 5ml of 0.5N distilled nitric acid in preparation for atomic absorption analysis. All preparatory work was done under a class 100 laminar flow hood to avoid contamination. All containers were thoroughly cleaned by soaking in 1.0N nitric acid for 24 hours and then twice rinsed in triple distilled water. Standards were mixed using similarly cleaned glass volumetric flasks and pipets. Standards and distilled nitric acid for sample dissolution were stored in pre-cleaned teflon bottles. Samples were dissolved and stored (for less than 3 days before measurement) in similarly cleaned polyethylene screw-top centrifuge tubes.

Sr and Mg concentrations were determined using standard graphite furnace techniques on a Perkin Elmer 3100 EDS atomic absorption spectrophotometer coupled to an HGA-600 graphite furnace and an AS-60 autosampler. Ca concentrations were measured with the same instrument using standard flame techniques. The Sr blank for the entire analytical procedure was close to the detection limit of the instrument (average=0.09 μg/l). The average blank for Mg was 0.89 μg/l. The average blank for Ca was 0.09 mg/l. Trace element results are reported in Appendix 1 and 2.

Chronology

The chronology of the core is based on 10 dated horizons (Table 1). These dates include eight accelerator mass spectrometer (AMS) ^{14}C measurements of ostracod calcite, one AMS ^{14}C of terrestrial wood, and one conventional ^{14}C

measurement on bulk organic carbon. In addition, a ^{210}Pb profile was measured for the top of the core [Brenner and Binford, 1988]. Because Lake Miragoane is a hard water lake, radiocarbon dates of shells and lacustrine organic matter are susceptible to error resulting from the dilution of ^{14}C by "old" carbon (i.e., devoid of ^{14}C) that is derived from the dissolution of limestone bedrock. To estimate the magnitude of this effect, we measured a pair of ostracod and wood samples from the same horizon (2.33m). The age difference was 1,025 years with the ostracod sample yielding the older date. This estimate of approximately 1000 years for the hard-water lake error is

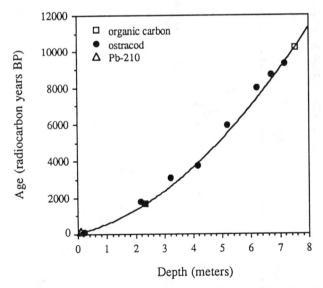

Fig. 2. Age-depth relationship used to establish chronology for Lake Miragoane core. The data are well described by the equation: Age (years BP) = 47.2 + 394.8 d + 127.2 d^2, where d is depth in meters. The equation is based on four datum levels as discussed in the text, and ostracod dates (once corrected for hard-water lake error) fall close to the age-depth curve.

supported by other results. For example, an extrapolation of the ^{210}Pb date to the level of the youngest ostracod date (0.22 m) yields an age of 355 years BP compared to 1,085 years BP for ostracods at this level, a difference of 730 years. Based upon the paired wood-ostracod age difference, we assume the hard-water lake error was constant throughout the core and subtract 1,000 years from all ostracod dates (Table 1).

The age-depth relationship in the core is well described by a second-order polynomial equation of the form: Age (years BP) = 47.2 + 394.8 d + 127.2 d², where d is depth in meters (Fig. 2). This equation is based on four datum levels: zero age at the top of the core, an estimated age of 129 years BP at 8 cm from the ^{210}Pb profile, an AMS ^{14}C date of 1,655 years BP on wood at 2.33 m, and a conventional ^{14}C date of 10,230 years BP on bulk organic carbon at 7.53 m. The hard-water corrected dates of ostracods are also in good agreement with the calculated age-depth curve (Fig. 2). Ages for each horizon in the core were calculated using this polynomial age-depth equation. The

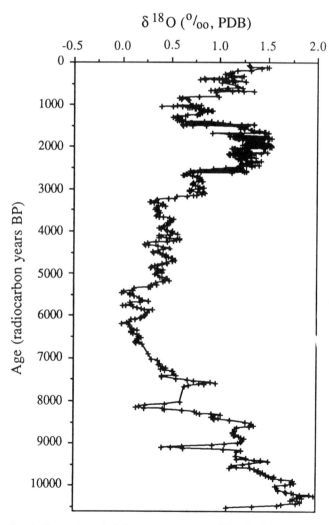

Fig. 3. Oxygen isotopic (relative to PDB) composition of calcitic shells of *Candona* sp. from sediments of Lake Miragoane, Haiti. Oxygen isotopic data were filtered using a 5-point running mean to dampen variability on time scales less than a century. After Hodell et al., 1991.

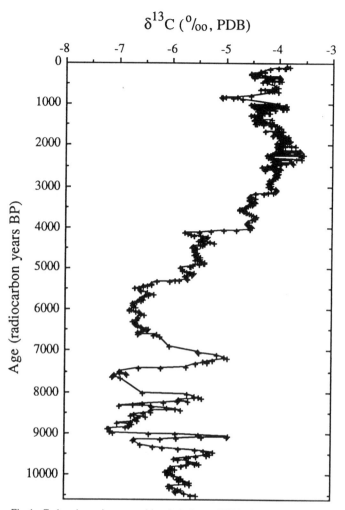

Fig 4. Carbon isotopic composition (relative to PDB) of calcitic shells of *Candona* sp. from sediments of Lake Miragoane, Haiti. Carbon isotopic data were filtered using a 5-point running mean to dampen variability on time scales less than a century.

second-order term in the equation probably reflects a combination of increased sedimentation rates and less compaction toward the top of the profile.

RESULTS

Stable Isotopic Record

The stable isotopic record consists of 624 data points that were smoothed using a 5-point running mean to dampen variability on time scales of less than a century (Fig. 3). The highest δ^{18}O values, averaging ~1.8°/$_{oo}$, are found near the base of the core between ~10,500 and 10,000 years BP. From 10,000 to ~7,000 years BP, there is a general trend toward decreasing δ^{18}O values, but this transition is not smooth. Superimposed upon the general decrease are at least two brief δ^{18}O minima at ~9,100 and ~8,100 years BP. The lowest δ^{18}O values are recorded between ~7,000 and 5,300 years BP and average ~0.2°/$_{oo}$. At 5,300 years BP, there is a slight increase in mean δ^{18}O values such that ratios average 0.4°/$_{oo}$ between 5,300 and 3,200 years BP (Fig. 3). This interval is followed by a two-step increase in δ^{18}O at ~ 3,200 and 2,600 years BP.

The sharp increase in $\delta^{18}O$ at ~2,600 years BP marks the beginning of an episode of consistently high values (average 1.3°/$_{oo}$) that lasts until ~1,500 years BP. Following this period, $\delta^{18}O$ values decrease again to approximately 0.6°/$_{oo}$, and remain relatively low until about 800 years BP when they begin to increase slowly toward the present. Recent $\delta^{18}O$ values from the top of the core average ~1.4°/$_{oo}$.

The carbon isotopic record was also smoothed by a 5-point running mean to reveal general trends (Fig. 4). The major trend in the $\delta^{13}C$ record is a long-term increase of 2.0°/$_{oo}$ that occurred between ~5,500 and 3,000 years BP. Before this shift, $\delta^{13}C$ values (between ~10,500 and 5,500 years BP) were quite variable and averaged about -6.0°/$_{oo}$. During this period, short term fluctuations were common. After the major $\delta^{13}C$ decrease from 5,500 to 3,000 years BP, the record becomes much less variable and values average -4.2°/$_{oo}$ from ~3,000 years BP to the present.

Sr/Ca Record

Samples for Sr/Ca analysis are not as closely spaced as those for stable isotopes and consist of 61 data points with an average sampling frequency of 1 sample per 175 years (Fig. 5). From 10,500 to ~9,000 years BP, Sr/Ca ratios are moderately high and average 1.9×10^{-3}. Sr/Ca values begin to decrease at 9,000 years BP and reach minimum values of 1.25×10^{-3} at 6500 years BP. Sr/Ca values remain low from ~6,500 to

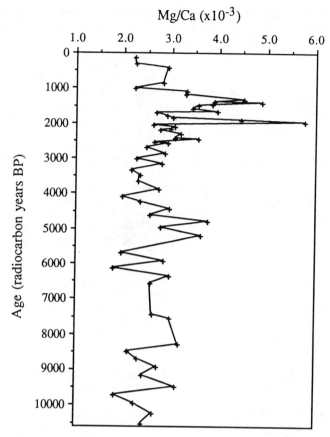

Fig. 6. Mole ratio Mg/Ca ($\times 10^{-3}$) in calcitic shells of *Candona* sp. from sediments of Lake Miragoane, Haiti.

~4,000 years BP with values averaging 1.5×10^{-3}. Beginning at ~4000 years BP, Sr/Ca ratios begin to increase again until ~2,500 years BP when the maximum values (averaging ~2.0 x 10^{-3}) are recorded between 2500 and 1000 years BP. Sr/Ca values then decline from ~1,500 until 1,000 years BP. Although only 3 data points exist for the last 1000 years, they appear to show a sharp increase in Sr/Ca towards the top of the profile.

Mg/Ca Record

The Mg/Ca record shows the least variability of all three records and averages ~2.5 x 10^{-3} throughout most of the record (Fig. 6). The only major excursion in Mg/Ca ratios occurs between ~2,000 and ~1,000 years BP when Mg/Ca values suddenly increased to between ~3 and 6 x 10^{-3}. This interval of high Mg/Ca coincides with similar increases in $\delta^{18}O$ and Sr/Ca.

RATIONALE FOR THE METHOD

Interpretation of our results requires an understanding of the factors controlling changes in the $\delta^{18}O$, Sr/Ca, and Mg/Ca of ostracod calcite in Lake Miragoane.

Stable Isotopes

The oxygen isotopic composition of authigenic calcite or aragonite is dependent upon the temperature and oxygen

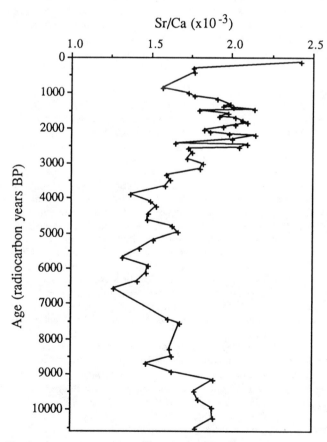

Fig. 5. Mole ratio of Sr/Ca ($\times 10^{-3}$) in calcitic shells of *Candona* sp. from sediments of Lake Miragoane, Haiti.

isotopic ratio of water from which the carbonate was precipitated. In tropical lakes under a seasonally dry climate where a substantial fraction of the water budget is lost via evaporation, the $^{18}O/^{16}O$ of lake water is mainly controlled by the ratio of evaporation to precipitation (E/P) (Fontes and Gonfiantini, 1967; Gasse et al., 1990). Because the lighter isotope, $H_2^{16}O$, has a higher vapor pressure than $H_2^{18}O$, ^{16}O is preferentially incorporated into the gaseous phase during evaporation leaving the remaining water enriched in ^{18}O. The magnitude of this fractionation effect is controlled by the temperature and relative humidity of the atmosphere at which evaporation occurs (i.e., the fractionation factor increases with decreasing temperature and humidity). Because Lake Miragoane loses a substantial fraction of its water by evaporation, long-term changes in $\delta^{18}O$ of shell calcite should mainly reflect changing E/P. Although temperature can also influence the $^{18}O/^{16}O$ of precipitated calcite, long-term temperature variations in the tropics are likely to be small relative to the effects of evaporation and precipitation.

The oxygen isotopic record from Lake Miragoane has been interpreted previously by Hodell et al. [1991] in conjunction with pollen analysis [Higuera-Gundy, 1991]. They assumed that extended periods of high evaporation and/or low precipitation would be reflected by an increase in $^{18}O/^{16}O$ ratios, whereas times of increased rainfall and/or reduced evaporation would be marked by a decrease in $^{18}O/^{16}O$ ratios. Recent advances in the understanding of the factors that control trace element ratios (i.e., Sr/Ca, Mg/Ca) in ostracod shells provide a new tool that can be used to test the inferences based upon $\delta^{18}O$ measurements. Here we compare the $\delta^{18}O$ results with variations in Sr/Ca and Mg/Ca ratios of ostracod calcite to qualitatively separate the effects of changing salinity and temperature of lake water during the past 10,500 years.

Trace Elements

Sr/Ca ratios of ostracod calcite cultured under controlled conditions and in natural lakes have been shown to be a function of the Sr/Ca content of the host water [Chivas et al. 1983, 1985, 1986b; Engstrom and Nelson, 1991]. The Sr/Ca of calcite is related to Sr/Ca of water by a partition coefficient (K_d) that is independent of temperature. The partition coefficient is taxon specific and ostracods belonging to the same genus appear to have the same K_d [Chivas, 1986a]. Engstrom and Nelson [1991] determined K_d [Sr] for *Candona rawsoni* to be 0.406 and we assume that this partition coefficient also applies to *Candona* sp. used in this study. In lakes where the concentration of Ca is held constant by inorganic precipitation of calcite, the Sr/Ca ratio measured in ostracod calcite can be used to monitor changes in paleosalinity. This technique works because the partition coefficient for inorganic calcite precipitation (K_d [Sr] 0.055 - 0.14; Engstrom and Nelson, 1991) is significantly less than that of biogenic calcite precipitation (K_d [Sr] ostracods 0.176 - 0.475; Engstrom and Nelson, 1991). As evaporation proceeds, therefore, the Sr content of the water increases relative to the Ca concentration which is held constant by calcite precipitation. Lake Miragoane's water is dominated by calcium and bicarbonate ions today [Brenner and Binford, 1988] and surface waters collected in 1985 were slightly oversaturated with respect to calcite (Langelier Index of +0.04; Langelier Index = pH_a - pH_s, where pH_a is the actual pH of the water tested and $pH_s = pK_{a,2} - pK_{so} + p[Ca^{2+}] + p[HCO^{3-}] - \log \gamma_{Ca^{2+}} - \log \gamma_{HCO^{3-}}$

; Snoeyink and Jenkins, 1980). If lake level volume was reduced slightly, calcite saturation would be greatly exceeded and $CaCO_3$ would precipitate. Indeed, the high calcite content of the sediments (long core $CaCO_3$ average=73.0%) of Lake Miragoane suggests the occurrence of authigenic precipitation of $CaCO_3$.

Unlike Sr, the incorporation of Mg into ostracod calcite has been shown to be a function of both the Mg/Ca ratio of host water and temperature [Chivas et al., 1983, 1986a, 1986b; Engstrom and Nelson, 1991]. The Mg/Ca of ostracod calcite increases as a function of increased temperature and/or increased salinity (assuming Ca remains constant due to precipitation of calcite). Engstrom and Nelson [1991] determined K_d [Mg] for *Candona rawsoni* to be $9.68 \times 10^{-5} \times T$ (°C).

As first proposed by Chivas et al. [1985], the combined use of $\delta^{18}O$ and trace element data (Sr/Ca, and Mg/Ca) should allow separation of temperature and salinity (E/P) effects in these proxy variables. Because the magnitude and direction of responses in these parameters are not the same, various combinations of temperature and salinity change will produce different patterns in the $\delta^{18}O$, Sr/Ca, and Mg/Ca signals. In Fig. 7, we follow the lead of Chivas et al. [1985] and present a hypothetical model describing the expected responses of these three variables to changing salinity and temperature. The signals show the expected responses of $\delta^{18}O$, Sr/Ca, and Mg/Ca to nine combinations (cases) of changing salinity and/or temperature. Next, we use these cases to interpret the observed relationships among $\delta^{18}O$, Sr/Ca, and Mg/Ca to reconstruct the temperature and salinity history of Lake Miragoane during the last 10,500 years.

INTERPRETATION

Our starting point is the base of the record from ~10,500 to 10,000 years BP (Fig. 8). This interval contains the greatest $\delta^{18}O$ values of the entire record and Sr/Ca values are high, but are not the highest of the record. Mg/Ca ratios are equal to the long-term average of ~2.5 x 10^{-3}. These data suggest that lake water was saline, but this was not the time of maximum salinity. The maximum in the $\delta^{18}O$ values reflect a number of constructive effects. Decreased temperature and increased E/P ratios during the latest Pleistocene would both act to increase the $\delta^{18}O$ of precipitated calcite. In addition, decreased temperature and lower relative humidity would increase the evaporative fractionation factor, leading to heavier residual water in the lake. Lastly, the $\delta^{18}O$ of rain may also have been higher during the latest Pleistocene because the ocean, the source of precipitation for Lake Miragoane, was ~1.3°/$_{oo}$ heavier in $\delta^{18}O$ at ~10,000 years BP than today because of increased global ice volume. In the case of Mg/Ca for this period, the expected increase due to elevated salinity was offset by a decrease in Mg/Ca due to lowered temperature, producing a damped signal. Pollen data from this portion of the record also indicate that climatic conditions were cool and arid. Arboreal pollen was scarce and plant communities were dominated by montane shrubs and xeric palms [Higuera-Gundy, 1991].

From ~10,000 to ~7000 years BP, $\delta^{18}O$ values decrease as do Sr/Ca ratios, but Mg/Ca ratios remain relatively constant. This pattern indicates a period of decreasing salinity and increasing temperature (Case G). Part of the $\delta^{18}O$ decrease during this period may have also been related to a decrease in the $\delta^{18}O$ of

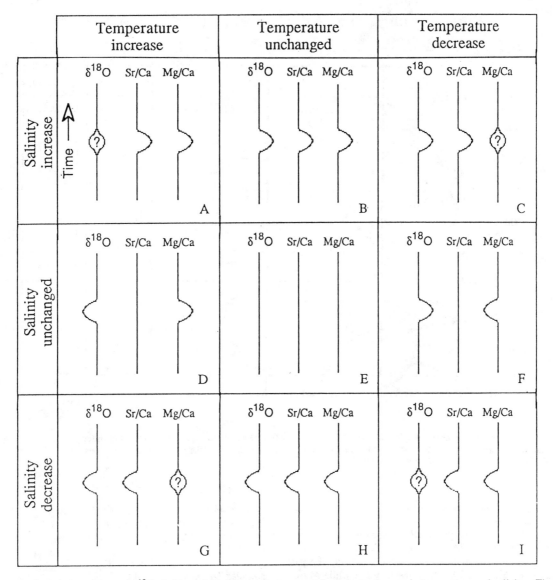

Fig. 7. Expected responses of $\delta^{18}O$, Sr/Ca, and Mg/Ca signals in ostracod calcite to changes in temperature and salinity. Time increases in an upward direction and the curves represents the predicted response of each indicator from initial conditions, to a transient change, and back to initial conditions. Figure modified after Chivas et al., 1986b.

precipitation as oceanic $\delta^{18}O$ decreased in response to deglaciation. The pollen data for this time period are generally consistent with our interpretation of decreasing E/P. Dry conditions are inferred until 8,200 years BP when a sharp increase in Cheno.-Ams. marks a rise in lake level and the development of an extensive littoral zone [Higuera-Gundy, 1991].

From ~7,000 to ~4,000 years BP, temperature and salinity conditions were fairly constant as reflected by unchanging $\delta^{18}O$, Sr/Ca and Mg/Ca values (Case E). The salinity of the lake was lowest during this period as evidenced by minimum Sr/Ca and $\delta^{18}O$ values (Fig. 8). During this interval, the pollen assemblages became increasingly dominated by arboreal taxa as mesic forests expanded in Miragoane's watershed [Higuera-Gundy, 1991].

From ~4,000 to 2,500 years BP, increases in all three parameters ($\delta^{18}O$, Sr/Ca, and Mg/Ca) indicate increasing salinity and inferred dry conditions (Case B). This return to dry conditions is supported by pollen evidence that shows a gradual replacement of mesic forest taxa by dry forest taxa and weeds between 2,800 and 2,100 years BP [Higuera-Gundy, 1991].

From ~2,500 to 1,500 years BP, unchanging $\delta^{18}O$, Sr/Ca, and Mg/Ca suggest constant conditions (Case E). The highest Sr/Ca and Mg/Ca ratios are recorded during this period indicating maximum salinity and inferred arid conditions (Fig. 8). In addition, $\delta^{18}O$ values are high, further suggesting high E/P ratios. Pollen evidence from the early part of this period

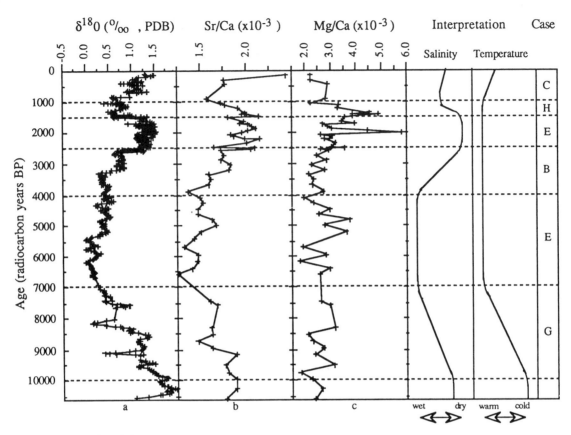

Fig. 8. Summary of stable isotopic and trace element results from Lake Miragoane, Haiti, including interpretation of changing salinity and temperature for the last 10,500 years. Interpretations are based upon the observed signal responses in $\delta^{18}O$, Sr/Ca, and Mg/Ca according to the cases outlined in Fig. 7.

also points to arid conditions as evidenced by increases in dry forest taxa (especially Celtis).

From ~1,500 to ~1,000 years BP, decreases in all three proxies ($\delta^{18}O$, Sr/Ca, and Mg/Ca) suggest decreasing salinity and relatively constant temperatures (Case H). During the past 1,000 years, both Sr/Ca and $\delta^{18}O$ values increase while Mg/Ca ratios remain constant. This pattern indicates increasing salinity and decreasing temperature during the past millennia (Case C). No climatic inferences are made from the pollen record for the last millennia because of the increasing impact of human settlement and agricultural practices on vegetation.

Interpretation of the carbon isotopic record (Fig. 4) is not as straightforward as the oxygen isotopic or trace element results. The carbon isotopic composition of lake water is controlled by the fluxes and isotopic composition of carbon to and from the lake as well as gas exchange with the atmosphere. The two major sources of carbon into the lake are oxidized organic carbon and dissolved bedrock limestone that are supplied to the lake by rivers and groundwater. Similarly, output fluxes also include organic carbon fixed by organisms during photosynthesis and carbonate carbon formed by either biogenic or authigenic precipitation. We do not have a good explanation for the increase observed in $\delta^{13}C$ ratios between ~5,500 and ~3,000 years BP (Fig. 6), but it was most probably related to a fundamental change in the lake's carbon budget such

as a decrease in the $\delta^{13}C$ of water supplied to the lake or an increase in removal of organic carbon to the sediments, perhaps resulting from an increase in lake productivity.

DISCUSSION

Younger Dryas Chronozone

From 10,500 to 10,000 years BP, the isotopic, trace element, and pollen results suggest that climate was arid and temperature was cooler than today in the Caribbean region (Fig. 8). If the chronology at the base of the core is correct, this period coincides with the latter part of the of the Younger Dryas chronozone dated between 11,000 and 10,000 years BP [Broecker et al., 1988]. The Younger Dryas marked a brief return to cooler temperatures and interrupted the general warming trend associated with the last deglaciation. The Younger Dryas was first documented in terrestrial records from Europe [Mercer, 1969] and later in marine records from the North Atlantic [Ruddiman and McIntyre, 1981]. Recent studies point to a more hemispheric, if not global, extent of the Younger Dryas including evidence from the Gulf of Mexico [Flower and Kennett, 1990; Spero and Williams, 1990], the Gulf of California [Keigwin and Jones, 1990], the equatorial Atlantic [Molfino and McIntyre, 1990], the Sulu Sea [Linsey and Thunell, 1990] and China [Lister et al., 1991].

Evidence for cool and/or dry conditions during the Younger

Dryas has also been reported for other tropical lakes. Studies of North African lakes demonstrate increased aridity and lowered lake levels at this time. For example, Sebkha Mellala in the Northern Sahara showed maximum aridity from 10,600 to 9,300 years BP while lake levels in the Sahel (at Bougdouma, Niger) were lowest from 10,300 to 10,000 years BP [Gasse et al., 1990]. Other African lakes also suggest arid conditions during the Younger Dryas including Lake Bosumtwi (Ghana), Lake Chad, and Lake Ziway-Shala (Ethiopia) [Roberts, 1990; Street-Perrott and Roberts, 1983; Talbot and Delibrias, 1980].

Why should the tropics of Africa and the Americas be marked by lowered lake levels and drier conditions during the Younger Dryas Chronozone? Street-Perrott and Perrott [1990] suggested that increased aridity in the African tropics may be linked to injection of fresh water and reduced thermohaline convection in the North Atlantic. Carbon isotopic measurements and Cd/Ca ratios of benthic foraminifer suggest that North Atlantic Deep Water (NADW) production was reduced during the Younger Dryas Chronozone [Boyle and Keigwin, 1987]. The reduction of NADW production may have been caused by glacial meltwater input into the North Atlantic [Broecker et al., 1988], but this mechanism has been challenged by Fairbanks [1989]. Evidence suggests that the reduction in NADW may have been accompanied by a decrease in the cross-equatorial heat flux in the equatorial Atlantic. During the Younger Dryas Chronozone, Mix et al. [1986] found an asymmetric pattern of sea-surface temperatures (SST) with cooler water north of the equator and warmer water south of the equator. This type of thermal anomaly and inferred southerly position of the Intertropical Convergence Zone (ITCZ) has been shown to be associated with lowered rainfall and increased aridity in the northern tropics [Parker et al., 1988; Hastenrath, 1976, 1984] and may explain the cool, arid conditions observed in our record between 10,500 and 10,000 years BP.

Deglaciation

From 10,000 to 7,000 years BP, temperature increased and salinity decreased in Lake Miragoane as climate became warmer and wetter (Fig. 8). This interval coincided with the second step of the last deglaciation (i.e., Termination 1b) in marine oxygen isotope records, which lasted from 10,000 to 7,000 years BP [Duplessy et al., 1986]. Two intervals of rapid rise in sea level have also been inferred from the Barbados sea level record [Fairbanks, 1989] with the younger rise (melt-water pulse 1b) centered at 9,500 years BP. Similar to the two-step deglaciation observed in marine records, the transition from arid to humid conditions in North Africa also occurred in two main steps, suggesting that common factors in the ocean-atmosphere system may have controlled marine and terrestrial changes during deglaciation [Gasse et al., 1990].

The decreased salinity and rise in Lake Miragoane's water level from 10,000 to 7,000 years BP is attributed to both rising sea level and increased moisture availability. Rising sea level would have affected the water level of Lake Miragoane because it is located only 5 km from the coast in karstic terrain. As sea level rose during deglaciation, so would have the level of the phreatic aquifer. The most rapid rate of sea level rise in the neotropics occurred between 11,000 and 8,000 and the rate of rise lessened by 7,300 when sea level reached an equilibrium [Covich and Stuiver, 1974 and references therein]. These dates

agree fairly well with the inferred rise in Lake Miragoane's level between 10,000 and 7,000 years BP (Fig. 8).

The decrease in $\delta^{18}O$ and Sr/Ca beginning at 10,000 years BP marks a switch from arid to humid conditions in the Caribbean region (Fig. 8). Today, the rainy season in the Caribbean occurs during the northern summer half year when the ITCZ moves north and displaces the North Atlantic subtropical high and weakens the easterly trade winds [Hastenrath 1976, 1984]. We suggest that the decrease in salinity and rise in Lake Miragoane's level during the early Holocene (10,000 to 7,000 years BP) was related, in part, to a similar process. Evidence suggest that the ITCZ was located further south (over South America) during the last Ice Age and it shifted north upon deglaciation as the trade winds weakened [Prell and Hays, 1976; Mix et al., 1986; Peterson et al., 1991; among others]. For example, a recent study of sediment cores from the Cariaco Basin (southern Caribbean) showed that upwelling intensity decreased markedly beginning at 10,000 years BP, because of a weakening of trade winds as the ITCZ shifted north during the earliest Holocene [Peterson et al., 1991]. These meteorological conditions have been shown to be associated with increased precipitation in the Caribbean region [Hastenrath, 1984] and are consistent with the results from Lake Miragoane.

The switch from arid-to-humid conditions during deglaciation (i.e., Pleistocene/Holocene boundary) has been documented in many lakes throughout the tropics including Africa [Gasse et al., 1990; Street-Perrott and Perrott, 1990; Bonnefille et al., 1990; Gasse et al., 1989; Talbot and Delibrias, 1980; Street and Grove, 1976], the Americas [Leyden et al., this volume; Piperno et al., 1990; Markgraf, 1989; Leyden, 1985; Leyden, 1984; Deevey et al., 1983; Bradbury et al., 1981; Watts, 1975], Europe [Duplessy et al., 1981], and China [Lister et al., 1991]. This arid-to-humid transition signals a fundamental change in the hydrometeorological cycle that may have affected transport of water vapor through the atmosphere and, possibly, thermohaline circulation of the oceans [Stocker and Wright, 1991]. In fact, one of the major differences in oceanic circulation between glacial and interglacial time was the flux of NADW [Broecker and Denton, 1989, 1990]. The rate of NADW production increased rapidly from ~9000 to 8000 years BP (Termination 1b), following a period of diminished rates between 11,500 and 9,000 years BP [Boyle and Keigwin, 1987]. The increase in NADW production coincides with the development of warm temperatures in the western Atlantic between 2 and 8° N and with an inferred increase in the cross-equatorial heat flux [Mix et al., 1986]. The ITCZ assumed a more northerly position at this time and rainfall increased in the Northern Hemisphere tropics. This trend is recorded by a decrease in salinity and rise in Lake Miragoane's level during the early Holocene from 10,000 to 7,000 years BP (Fig. 8).

Early-Mid Holocene Moist Period

Relatively low $\delta^{18}O$ and Sr/Ca ratios from ~7,000 to ~4,000 years BP suggest low salinity and moist conditions during this interval (Fig. 8). This wet period coincides with the early- to mid-Holocene moist period that has been documented throughout the Neotropics [Piperno et al., 1990; Leyden, 1985; Leyden, 1984; Deevey et al., 1983; Bradbury et al., 1981; Covich and Stuiver, 1974] and Africa [Street-Perrott and

Perrott, 1990; Kutzbach and Street-Perrott, 1985; Street-Perrott and Harrison, 1985; Street and Grove, 1976, 1979].

Hodell et al. [1991] suggested that increased precipitation in the tropics of the Northern Hemisphere during the early Holocene was related to an increase in the intensity of the annual cycle. The intensity of the annual cycle in the northern tropics has progressively increased during the early Holocene because of changes in seasonal insolation forced by the Earth's precessional cycle. A measure of the intensity of the annual cycle is reflected by the seasonal insolation difference at the top of the atmosphere at 10°N between summer (August) and winter (February) (Fig. 9). Meteorological studies of interannual variability of rainfall in the tropical Atlantic sector during the twentieth century have found a strong correlation between precipitation anomalies and intensity of the annual cycle [Hastenrath, 1984]. Abundant rainy seasons are coincident with an enhancement of the annual cycle when the ITCZ moves farther north during the summer (rainy season) and farther south during the winter (dry season). In contrast, low rainfall is associated with a reduction in the annual cycle.

Hodell et al. [1991] showed that the long-term (i.e., millennial) changes in circum-Caribbean climate were related to the relative intensity of the annual cycle. The general pattern of arid conditions of the late Pleistocene to moist conditions of the early Holocene to dry conditions of the late Holocene follows the expected changes in the intensity of the annual cycle (Fig. 9). We attribute, therefore, the early Holocene moist period to a time of relatively high intensity of the annual cycle caused by the Earth's orbital geometry. Similarly, high lake levels in Africa during the early Holocene have also been attributed to a precessional amplification of the annual cycle, which strengthens the monsoons and increases rainfall over parts of the African continent [Kutzbach and Guetter, 1984; Kutzbach and Street-Perrott, 1985].

Late Holocene Drying

Following the early Holocene moist phase, climate in the Caribbean began to dry beginning as early as ~4,000 years BP and reached maximum aridity between ~2,500 and 1,500 years BP (Fig. 8). Further indication of a return to dry conditions is a considerable decline in mesic forest taxa (such as Moraceae, *Cecropia*, and *Trema*) and the establishment of open, dry forest communities associated with abundant weeds [Higuera-Gundy, 1991]. A return to drier conditions and lowered lake level during the late Holocene has also been observed in many parts of Africa, Australia, India, and South America [Street and Grove, 1979], but the exact timing appears to vary from region to region. In general, the interval from 4500 years BP to present is considered to be a period of lowered lake levels and dry climate [Street and Grove, 1979].

Although the long-term drying trend of the late Holocene in Lake Miragoane can be explained by a reduction in the intensity of the annual cycle (Fig. 9), the abrupt nature of the increased salinity between ~2,500 and 1,500 years B.P. must have its origin in a mechanism other than slow orbital forcing. Evidence from other lakes in the circum-Caribbean tend to support a regional drying within the past 3,000 years. For example, oxygen isotopic evidence from Lake Chichancanab (Yucatan, Mexico) supports an increase in E/P during the past ~2,500 years [Covich and Stuiver, 1974]. In Lake Valencia (Venezuela), proxy indicators indicate a trend toward higher salinity conditions from ~2,800 years BP to the present [Bradbury et al., 1981; Binford, 1982; Leyden, 1985]. Further study of lakes in the circum-Caribbean is needed to better constrain the nature and timing of this drying trend during the late Holocene.

Recent Desiccation

The oxygen isotope and Sr/Ca records from Lake Miragoane reveal a trend towards higher salinity conditions during the last millennium (Fig. 8). Proxy indicators in modern sediments of Lake Miragoane suggest conditions as dry as any since the early Holocene. A change in pollen and carbonate content of the sediment has also been noted during the last 500 years and attributed to human disturbance [Brenner and Binford, 1988; Higuera-Gundy, 1989]. Prior to European influence, sediments were organic-rich and contained pollen evidence of intact, dry and mesic forest [Brenner and Binford, 1988; Higuera-Gundy, 1989]. During the post-Columbian period, pollen abundance decreased markedly, weeds increased at the expense of arboreal taxa, and organic content declined while carbonate content increased. These changes have been interpreted as resulting from both deforestation and higher sedimentation rates from soil erosion [Brenner and Binford, 1988; Higuera-Gundy, 1989]. It is unclear whether increased desiccation observed

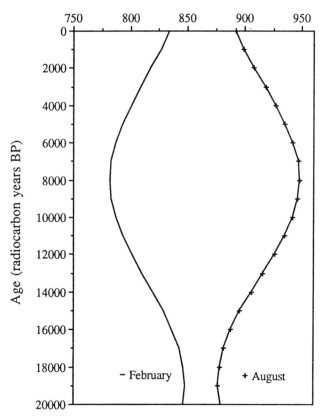

Insolation at 10°N (Langleys)

Fig. 9. Insolation at the top of the atmosphere at 10°N for the months of August and February [Berger, 1978]. The distance between the curves is a measure of the intensity of the annual cycle.

toward the top of our profile was a result of human influence or whether it was attributable to natural climatic causes. Today, Lake Miragoane's watershed is completely deforested and the soils are rapidly eroding, leading to desertification.

Records from other lakes in the circum-Caribbean region also contain evidence for recent desiccation. For example, Lake Valencia (Venezuela) has been desiccating steadily and increasing in salinity for the past two hundred years. Bradbury et al. [1981] suggest this may be due to an extrinsic shift toward drier climate during the last 2,800 years that has been magnified over the last 250 years by human activity. Additional study of recent sediments in Caribbean lakes is needed to determine the extent, if any, man is responsible for recent desiccation.

CONCLUSION

Using isotopic, trace element, and pollen data, we have documented relative changes in temperature and salinity of Lake Miragoane, Haiti, for the last 10,500 years. The results suggest that climatic conditions were cool and dry during the latest Pleistocene from ~10,500 to 10,000 years BP, coinciding with the latter half of the Younger Dryas Chronozone. During the last deglaciation from ~10,000 to 7,000 years BP (Termination 1b), lake level and temperature increased as salinity decreased. Warm, wet conditions culminated during the early- to middle-Holocene moist phase (~7,000 to 4,000 years BP). The late Holocene marked a return to dry conditions that began as early as 4,000 years BP and culminated in the driest phase between ~2,500 and 1,500 years BP. Moisture availability increased slightly at 1000 years BP and was followed by a drying trend until the present.

The long-term (i.e., millennial) changes in Caribbean climate are well explained by orbitally-induced (Milankovitch) variations in seasonal insolation that modified the intensity of the annual cycle. Climate was dry in the Caribbean when the intensity of the annual cycle was reduced and the ITCZ occupied a southerly position on average, such as during the last Ice Age, Younger Dryas Chronozone, and late Holocene. Wet conditions were associated with times of an enhanced annual cycle and northerly position of the ITCZ, such as during the early Holocene moist phase. Superimposed upon the long-term changes are more abrupt climatic events that cannot be explained by orbital forcing.

The millennial changes of climate in the Caribbean are generally similar to the pattern reported for Africa and from other lakes in Central and South America, suggesting that climate of these regions may be linked. Future research should focus on identifying the common mechanisms that may be driving these changes in the ocean-climate system.

APPENDIX 1. Stable isotopic and trace element results of *Candona* sp. from Lake Miragoane short core

Depth m	Age years	$\delta^{18}O$ o/oo, pdb	$\delta^{13}C$	Mg ppm	Sr	Mg/Ca mole ratio/1000	Sr/Ca
0.010	51	1.45	-4.33				
0.035	61	1.54	-4.09				
0.075	78	1.05	-3.38				
0.130	101	1.16	-3.86				

APPENDIX 1. (continued)

Depth m	Age years	$\delta^{18}O$ o/oo, pdb	$\delta^{13}C$	Mg ppm	Sr	Mg/Ca mole ratio/1000	Sr/Ca
0.170	118	1.30	-3.71				
0.190	127	1.50	-4.05				
0.210	136	1.56	-4.49				
0.230	145	1.98	-3.47	466	1830	2.228	2.428
0.360	206	1.04	-4.46				
0.380	216	1.06	-4.43				
0.470	261	0.95	-4.57				
0.540	297	0.88	-4.92				
0.560	308	1.59	-4.29				
0.580	319	0.97	-4.51	548	1560	2.231	1.762
0.590	324	1.01	-4.24				
0.600	330	1.13	-4.19				
0.620	341	0.97	-4.17				
0.660	363	1.48	-3.84				
0.670	369	1.61	-3.59				
0.680	374	0.91	-4.15				
0.700	386	0.76	-4.15				
0.710	392	0.56	-4.26				
0.720	397	1.26	-3.83				
0.730	403	0.70	-4.46				
0.740	409	0.67	-4.49				
0.750	415	1.03	-4.18				
0.760	421	1.00	-4.32				
0.770	427	1.53	-3.96				
0.780	433	1.01	-4.25				
0.790	438	1.08	-4.03	716	1568	2.912	1.770
0.800	444	1.00	-4.20				
0.810	450	1.11	-3.57				
0.820	456	1.50	-4.06				
0.830	463	1.60	-3.98				
1.360	819	0.56	-4.75				
1.370	827	0.67	-4.89				
1.380	834	0.57	-5.07				
1.390	842	0.50	-5.29				
1.400	849	0.62	-5.06				
1.410	857	0.70	-5.03				
1.420	864	0.69	-4.99				
1.430	872	0.47	-4.97				
1.440	879	0.54	-5.30	761	1533	2.815	1.573
1.610	1012	0.90	-4.14				
1.620	1021	0.78	-4.10				
1.630	1029	0.88	-4.22	639	1806	2.209	1.732
1.640	1037	0.65	-4.68				
1.650	1045	0.81	-4.27				
1.660	1053	0.74	-4.49				
1.670	1061	0.67	-4.29				
1.680	1069	0.56	-4.77				
1.690	1078	0.63	-4.57				
1.700	1086	0.95	-4.22				
1.720	1102	0.64	-4.33				
1.730	1110	0.98	-4.19	1075	2060	3.323	1.767
1.740	1119	0.73	-4.57				
1.750	1128	0.71	-4.59				
1.760	1136	0.93	-4.10				
1.780	1153	0.81	-3.94				
1.790	1161	0.88	-4.17				

APPENDIX 1. (continued)

APPENDIX 1. (continued)

Depth m	Age years	$\delta^{18}O$	$\delta^{13}C$	Mg	Sr	Mg/Ca	Sr/Ca
		o/oo, pdb		ppm		mole ratio/1000	
1.800	1170	0.79	-4.22				
1.810	1178	0.94	-4.50				
1.820	1187	0.67	-4.61				
1.840	1204	0.74	-4.45	810	1690	3.297	1.910
1.850	1213	0.76	-4.36				
1.860	1221	0.94	-3.98				
1.870	1230	0.80	-4.13				
1.880	1239	0.60	-4.50				
1.890	1248	0.85	-4.10				
1.910	1265	0.64	-4.23				
1.920	1274	0.41	-4.62				
1.930	1283	0.69	-4.42				
1.940	1292	0.64	-4.57				
1.950	1301	0.42	-4.53				
1.960	1310	0.39	-4.59				
1.970	1318	0.72	-4.32				
1.990	1336	0.65	-4.41	1423	2262	4.512	1.990
2.000	1345	0.58	-4.21				
2.010	1355	0.55	-4.35				
2.020	1364	0.50	-4.48				
2.030	1373	0.78	-4.26				
2.040	1382	0.50	-4.43				
2.050	1391	0.45	-4.64				
2.060	1400	0.64	-4.47				
2.070	1409	0.72	-4.39				
2.080	1419	0.69	-4.50	1370	2037	4.880	2.013
2.090	1428	0.70	-4.46				
2.100	1437	0.76	-4.41				
2.110	1446	0.65	-4.47				
2.120	1456	0.67	-4.13				
2.130	1465	0.69	-4.45				
2.140	1474	0.64	-4.25				
2.150	1484	0.61	-4.44				
2.160	1493	0.64	-4.60				
2.170	1503	0.47	-4.58	923	1683	3.557	1.799
2.180	1512	0.74	-4.37				
2.360	1687	1.12	-4.16				
2.370	1697	1.30	-3.81				
2.380	1707	0.99	-4.38	750	2043	2.677	2.023
2.390	1717	1.31	-3.76				
2.400	1727	1.25	-3.79				
2.410	1737	1.23	-4.10				
2.420	1747	0.98	-4.19				
2.430	1757	1.16	-4.15				
2.440	1768	1.08	-4.31				
2.450	1778	1.23	-3.80				
2.460	1788	1.07	-4.17				
2.470	1798	1.28	-3.76				
2.480	1808	1.14	-3.89	729	1865	2.909	2.064
2.490	1819	1.21	-4.06				
2.500	1829	1.27	-3.91				
2.510	1839	1.30	-3.89				
2.520	1850	1.26	-4.07				
2.530	1860	1.27	-4.14				
2.540	1870	1.41	-3.83				
2.550	1881	1.42	-3.91				
2.560	1891	1.33	-3.73				
2.570	1902	1.06	-4.02				
2.580	1912	1.35	-3.99				
2.590	1923	1.40	-3.78	1125	1840	4.452	2.020
2.600	1933	1.42	-3.64				
2.610	1944	1.14	-3.57				
2.620	1954	1.22	-4.12				
2.630	1965	1.23	-4.13				
2.640	1976	1.26	-4.05				
2.650	1986	1.32	-3.85				
2.660	1997	1.35	-4.11				
2.670	2008	1.29	-4.04				
2.680	2019	1.18	-4.01				
2.690	2029	1.21	-4.16				
2.700	2040	1.25	-3.86				
2.710	2051	0.82	-4.10				
2.720	2062	1.16	-4.13	658	1658	2.617	1.830
2.730	2073	1.32	-3.95				
2.740	2084	1.32	-3.85				
2.750	2095	1.45	-4.12				
2.760	2106	1.28	-4.04				
2.770	2116	1.10	-3.95				
2.780	2127	1.31	-3.93	738	1619	3.067	1.866
2.790	2139	1.12	-3.72				
2.800	2150	1.18	-4.24				
2.810	2161	0.97	-3.95				
2.820	2172	1.07	-3.96				
2.830	2183	1.14	-4.12				
2.840	2194	1.13	-4.05				
2.850	2205	1.18	-4.07				
2.860	2216	1.15	-4.15	655	1845	2.748	2.149
2.870	2228	1.20	-4.05				
2.880	2239	1.26	-4.29				
2.890	2250	1.25	-4.11				
2.900	2262	1.18	-4.15				
2.910	2273	1.27	-4.22				
2.920	2284	0.96	-4.31				
2.930	2296	1.28	-4.17				
2.940	2307	1.25	-3.93				
2.950	2318	1.37	-4.37				
2.960	2330	1.21	-4.02	774	1750	3.199	2.007
2.970	2341	1.26	-4.11				
2.980	2353	1.26	-4.07				
2.990	2364	1.48	-3.95				
3.000	2376	1.23	-4.11				
3.010	2388	1.29	-4.09				
3.020	2399	1.41	-4.19				
3.030	2411	1.16	-3.70				
3.040	2423	1.02	-4.42				
3.050	2434	1.31	-4.24				
3.060	2446	1.27	-3.45				
3.070	2458	1.21	-4.09				
3.080	2469	1.39	-3.91				
3.090	2481	1.19	-4.21	970	2060	3.554	2.094
3.100	2493	0.99	-4.20				
3.110	2505	1.28	-3.85				
3.120	2517	1.15	-4.28				
3.130	2529	1.32	-4.23				
3.140	2541	1.12	-4.17				
3.150	2553	1.28	-4.09	597	1681	2.626	2.050

APPENDIX 1. (continued)

Depth m	Age years	δ^{18}O o/oo, pdb	δ^{13}C	Mg ppm	Sr	Mg/Ca mole ratio/1000	Sr/Ca
3.160	2565	1.07	-4.34				
3.170	2577	1.37	-3.81				
3.180	2589	1.30	-3.90				
3.190	2601	1.37	-3.84				
3.200	2613	1.27	-3.95				

APPENDIX 2. Stable isotopic and trace element results of *Candona* sp. from Lake Miragoane long core

Depth m	Age years	δ^{18}O o/oo, pdb	δ^{13}C	Mg ppm	Sr	Mg/Ca mole ratio/1000	Sr/Ca
0.440	246	0.96	-4.69				
0.525	290	0.71	-4.95				
0.685	377	1.30	-4.04				
0.845	472	1.00	-4.08				
1.005	572	0.98	-4.63				
1.060	609	1.08	-3.91				
1.070	615	1.55	-3.71				
1.080	622	1.52	-3.84				
1.110	642	0.26	-4.68				
1.130	656	0.93	-4.18				
1.140	663	0.69	-4.37				
1.150	669	1.31	-4.34				
1.160	676	1.30	-4.22				
1.170	683	1.71	-3.91				
1.180	690	1.22	-3.87				
1.210	711	1.21	-4.73				
1.610	1012	0.20	-3.81				
1.630	1029	0.43	-3.84				
1.640	1037	0.49	-3.75				
1.650	1045	0.44	-4.08				
1.660	1053	0.43	-3.84				
1.690	1078	0.76	-3.87				
1.710	1094	0.92	-3.95				
1.730	1111	0.77	-3.70				
1.740	1119	0.82	-3.94				
1.750	1128	0.84	-3.92				
1.760	1136	1.06	-4.13				
1.770	1144	0.84	-3.69				
2.060	1400	1.07	-4.02				
2.070	1409	0.75	-4.64	1140	2060	3.889	1.950
2.080	1419	0.84	-4.30				
2.090	1428	0.69	-4.73				
2.130	1465	1.54	-4.11				
2.140	1474	1.31	-4.04	1043	2087	3.860	2.142
2.150	1484	1.12	-4.15				
2.160	1493	1.45	-3.88				
2.170	1503	1.30	-4.32				
2.200	1531	1.22	-4.29				
2.200	1531	1.33	-4.07				
2.210	1541	0.96	-4.15				
2.220	1550	1.17	-3.99				
2.250	1579	1.40	-3.92				
2.260	1589	1.37	-4.07				
2.270	1599	1.14	-3.93				

APPENDIX 2. (continued)

Depth m	Age years	δ^{18}O o/oo, pdb	δ^{13}C	Mg ppm	Sr	Mg/Ca mole ratio/1000	Sr/Ca
2.280	1608	1.18	-3.96	958	1986	3.447	1.982
2.290	1618	1.48	-3.98				
2.300	1628	1.48	-4.18				
2.320	1648	1.46	-4.19				
2.340	1667	1.49	-4.04				
2.350	1677	1.49	-3.84				
2.360	1687	1.44	-3.99	1036	1821	3.948	1.926
2.420	1747	1.52	-3.99				
2.430	1757	1.56	-3.79				
2.440	1768	1.31	-4.02				
2.450	1778	1.54	-4.03				
2.460	1788	1.39	-4.06				
2.470	1798	1.63	-3.90				
2.480	1808	1.47	-3.82				
2.490	1819	1.52	-3.90				
2.500	1829	1.49	-4.04				
2.530	1860	1.52	-3.45				
2.540	1870	1.20	-4.06	774	1929	3.033	2.097
2.550	1881	1.31	-3.88				
2.560	1891	1.59	-3.91				
2.570	1902	1.47	-3.91				
2.580	1912	1.46	-4.14				
2.590	1923	1.42	-3.81				
2.600	1933	1.49	-3.91				
2.620	1954	1.66	-3.83				
2.630	1965	1.44	-3.95				
2.640	1976	1.46	-3.92				
2.650	1986	1.53	-4.10	1531	1865	5.770	1.950
2.660	1997	1.56	-4.09				
2.670	2008	1.58	-3.87				
2.680	2019	1.51	-3.87				
2.690	2029	1.45	-3.97				
2.700	2040	1.45	-3.83				
2.710	2051	1.35	-3.90				
2.800	2150	1.52	-3.31				
2.810	2161	1.46	-3.46				
2.820	2172	1.65	-3.61				
2.840	2194	1.15	-3.97	712	1712	2.976	1.986
2.850	2205	1.31	-3.60				
2.880	2239	1.17	-3.52				
2.890	2250	1.23	-3.69				
2.900	2262	1.36	-3.43				
2.910	2273	1.59	-3.50				
2.980	2353	1.48	-3.54				
2.990	2364	1.20	-3.77				
3.000	2376	1.49	-3.63				
3.060	2446	1.16	-3.88	826	1609	3.056	1.651
3.070	2458	1.11	-4.00				
3.080	2469	1.84	-3.44				
3.100	2493	1.41	-3.73				
3.110	2505	1.09	-4.30				
3.120	2517	0.61	-4.44				
3.140	2541	0.33	-4.61				
3.160	2565	1.20	-4.03				
3.170	2577	0.74	-4.20				
3.180	2589	0.59	-4.15	802	1708	2.930	1.731
3.190	2601	0.62	-4.33				

APPENDIX 2. (continued)

Depth m	Age years	$\delta^{18}O$ o/oo, pdb	$\delta^{13}C$ o/oo, pdb	Mg ppm	Sr ppm	Mg/Ca mole ratio/1000	Sr/Ca mole ratio/1000
3.210	2625	0.86	-3.80				
3.220	2637	0.82	-3.95				
3.230	2649	0.68	-4.22				
3.240	2661	0.51	-3.95				
3.250	2673	0.46	-4.16				
3.260	2686	0.79	-3.99				
3.270	2698	0.67	-3.87				
3.280	2710	0.73	-4.07				
3.290	2722	0.92	-4.24	595	1524	2.473	1.757
3.300	2735	0.75	-3.97				
3.320	2760	0.92	-3.98				
3.330	2772	0.51	-4.15				
3.340	2784	0.87	-4.01				
3.350	2797	0.95	-4.00				
3.370	2822	0.77	-3.97				
3.380	2834	0.81	-4.21				
3.390	2847	0.72	-4.31				
3.400	2859	0.82	-4.13				
3.410	2872	0.53	-4.04	645	1408	2.851	1.727
3.430	2897	0.78	-4.19				
3.450	2923	0.74	-4.14				
3.460	2936	0.74	-4.31				
3.470	2948	0.95	-4.10				
3.490	2974	0.64	-4.27				
3.500	2987	0.92	-4.12				
3.510	3000	0.93	-4.05				
3.520	3012	0.67	-4.18				
3.530	3025	0.74	-4.36				
3.540	3038	0.71	-4.15	618	1803	2.257	1.825
3.550	3051	0.48	-4.09				
3.560	3064	0.81	-4.07				
3.570	3077	0.70	-4.07				
3.580	3090	0.80	-3.98				
3.590	3103	1.05	-3.98				
3.600	3116	0.82	-4.09				
3.610	3130	0.70	-4.14				
3.630	3156	0.85	-4.10				
3.640	3169	0.64	-4.02				
3.650	3182	0.81	-4.19	773	1795	2.803	1.807
3.660	3196	0.65	-4.60				
3.690	3235	0.39	-4.48				
3.700	3249	0.29	-4.93				
3.710	3262	0.55	-4.42				
3.720	3276	0.46	-4.30				
3.730	3289	0.11	-4.59				
3.740	3302	0.32	-4.53				
3.750	3316	0.29	-4.37				
3.760	3329	0.29	-4.74				
3.770	3343	0.38	-4.53	645	1711	2.164	1.593
3.780	3356	0.35	-4.49				
3.790	3370	0.54	-4.43				
3.800	3384	0.49	-4.62				
3.820	3411	0.35	-4.50				
3.830	3425	0.33	-4.30				
3.840	3438	0.47	-4.36				
3.850	3452	0.21	-4.61				
3.860	3466	0.36	-4.67				
3.880	3493	0.39	-4.55				
3.890	3507	0.31	-4.53	696	1726	2.352	1.618
3.900	3521	0.38	-4.56				
3.920	3549	0.42	-4.82				
3.930	3563	0.30	-4.76				
3.940	3577	0.34	-4.85				
3.950	3591	0.52	-4.67				
3.970	3619	0.36	-4.64				
3.980	3633	0.33	-4.45				
3.990	3647	0.22	-4.69				
4.000	3662	0.31	-4.63				
4.010	3675	0.51	-4.55	582	1446	2.301	1.587
4.030	3703	0.53	-4.44				
4.040	3718	0.55	-4.48				
4.060	3746	0.53	-4.48				
4.070	3760	0.49	-4.46				
4.090	3789	0.39	-4.28				
4.100	3803	0.43	-4.51				
4.110	3818	0.49	-4.44				
4.140	3861	0.67	-4.63				
4.150	3876	0.40	-4.63				
4.170	3905	0.32	-4.58	544	985	2.731	1.372
4.180	3919	0.25	-4.61				
4.190	3934	0.44	-4.51				
4.200	3948	0.48	-4.70				
4.220	3978	0.49	-4.55				
4.230	3992	0.32	-4.52				
4.240	4007	0.51	-4.35				
4.260	4037	0.53	-4.49				
4.280	4066	0.73	-4.62				
4.290	4081	0.42	-4.72				
4.300	4096	0.64	-4.50				
4.310	4111	0.14	-5.64	380	1043	1.967	1.497
4.320	4126	0.52	-5.69				
4.330	4141	0.17	-5.81				
4.340	4156	0.53	-5.84				
4.350	4171	0.53	-5.84				
4.360	4186	0.51	-5.39				
4.370	4201	0.67	-5.66				
4.380	4216	0.73	-5.56				
4.390	4231	0.52	-5.46				
4.400	4246	0.37	-5.32				
4.410	4261	0.35	-5.20	458	1083	2.340	1.535
4.430	4292	0.34	-5.16				
4.440	4307	0.14	-5.60				
4.450	4322	0.11	-5.83				
4.460	4337	0.17	-5.52				
4.480	4368	0.44	-4.99				
4.490	4383	0.47	-5.63				
4.500	4399	0.62	-5.38				
4.510	4414	0.38	-4.93				
4.530	4445	0.54	-5.14				
4.540	4460	0.38	-5.85	489	880	2.968	1.482
4.550	4476	0.20	-5.84				
4.560	4492	0.33	-5.55				
4.570	4507	0.39	-5.70				
4.580	4523	0.26	-5.17				
4.590	4538	0.44	-5.70				
4.600	4554	0.53	-5.54				

APPENDIX 2. (continued)

Depth m	Age years	δ¹⁸O o/oo	δ¹³C pdb	Mg ppm	Sr ppm	Mg/Ca mole ratio/1000	Sr/Ca
4.610	4570	0.45	-5.63				
4.620	4585	0.36	-5.76				
4.630	4601	0.49	-5.32				
4.640	4617	0.38	-5.74	315	657	2.548	1.476
4.650	4632	0.59	-5.38				
4.660	4648	0.52	-5.71				
4.680	4680	0.55	-5.52				
4.690	4696	0.51	-5.35				
4.700	4711	0.58	-5.79				
4.710	4727	0.41	-5.56				
4.730	4759	0.53	-5.43				
4.740	4775	0.22	-5.24				
4.750	4791	0.59	-5.61				
4.760	4808	0.31	-5.74	1000	1565	3.761	1.633
4.770	4824	0.41	-5.33				
4.780	4840	0.14	-5.71				
4.790	4856	0.39	-5.44				
4.800	4872	0.24	-5.11				
4.810	4888	0.48	-5.39				
4.820	4904	0.22	-5.63				
4.830	4921	0.42	-5.49				
4.840	4937	0.44	-5.37				
4.860	4969	0.51	-5.55				
4.870	4986	0.24	-6.05	821	1772	2.787	1.669
4.880	5002	0.42	-5.74				
4.890	5018	0.44	-5.64				
4.900	5035	0.14	-6.20				
4.930	5084	0.42	-5.44				
4.940	5101	0.40	-6.02				
4.950	5117	0.55	-5.70				
4.960	5134	0.52	-5.30				
4.980	5167	0.41	-5.50				
4.990	5183	0.38	-5.81				
5.000	5200	0.53	-5.69	940	1413	3.626	1.512
5.010	5217	0.35	-5.89				
5.040	5267	0.40	-5.72				
5.050	5284	0.08	-5.70				
5.060	5301	0.44	-5.68				
5.070	5317	0.25	-5.56				
5.080	5334	0.30	-6.68				
5.090	5351	0.25	-6.14				
5.120	5402	0.21	-6.19				
5.130	5419	-0.42	-6.79				
5.140	5436	0.20	-6.08		1737		1.426
5.150	5453	-0.15	-6.71				
5.160	5470	0.22	-6.54				
5.170	5487	0.10	-6.54				
5.180	5504	0.10	-6.86				
5.200	5538	0.10	-6.48				
5.220	5573	-0.16	-7.09				
5.230	5590	0.44	-6.06				
5.250	5625	0.38	-6.40				
5.270	5659	0.08	-6.29				
5.280	5677	0.28	-6.47				
5.290	5694	0.16	-6.49	542	1323	1.948	1.320
5.310	5729	0.01	-6.57				
5.320	5746	-0.16	-6.73				

APPENDIX 2. (continued)

Depth m	Age years	δ¹⁸O o/oo	δ¹³C pdb	Mg ppm	Sr ppm	Mg/Ca mole ratio/1000	Sr/Ca
5.330	5764	0.19	-6.57				
5.340	5781	-0.03	-6.59				
5.360	5816	0.01	-6.38				
5.370	5834	0.51	-6.43				
5.380	5852	0.10	-6.92				
5.390	5869	0.36	-6.97				
5.410	5905	0.17	-6.78				
5.430	5940	0.42	-6.57	845	1595	2.832	1.483
5.440	5958	0.22	-6.45				
5.450	5976	0.13	-6.80				
5.470	6011	0.24	-6.82				
5.480	6029	0.16	-6.91				
5.490	6047	0.37	-6.83				
5.500	6065	0.09	-6.64				
5.510	6083	0.12	-6.80				
5.520	6101	0.16	-6.16				
5.540	6137	-0.09	-6.69				
5.550	6155	0.15	-6.73	431	1278	1.783	1.468
5.570	6191	0.02	-6.46				
5.590	6228	-0.01	-6.56				
5.610	6264	-0.10	-6.78				
5.620	6282	0.25	-6.70				
5.640	6319	0.22	-6.83				
5.650	6337	0.12	-6.46				
5.660	6355	0.02	-6.89				
5.670	6374	0.14	-6.51				
5.680	6392	0.12	-6.76	812	1396	2.968	1.415
5.690	6410	0.06	-6.64				
5.710	6447	0.12	-6.55				
5.720	6466	0.10	-6.68				
5.730	6484	0.35	-6.21				
5.740	6503	0.21	-6.67				
5.750	6521	0.11	-6.10				
5.760	6540	0.02	-6.78				
5.770	6559	0.29	-6.54				
5.780	6577	0.15	-6.64	635	1125	2.570	1.264
5.790	6596	0.28	-6.79				
5.800	6615	0.01	-6.43				
5.810	6633	-0.02	-6.67				
5.820	6652	0.23	-6.69				
5.830	6671	0.38	-4.86				
5.960	6917	0.11	-7.09				
6.020	7032	0.33	-5.77				
6.040	7071	0.28	-5.77				
6.070	7129	0.38	-4.09				
6.090	7167	0.75	-4.77				
6.110	7206	0.17	-5.42				
6.130	7245	0.31	-5.00				
6.140	7265	0.49	-5.51				
6.150	7285	0.51	-5.42				
6.160	7304	0.52	-5.32				
6.170	7324	0.35	-5.81				
6.210	7403	0.72	-4.61				
6.220	7422	0.54	-6.61				
6.230	7442	0.64	-6.28	698	1552	2.609	1.610
6.260	7502	-0.22	-7.74				
6.270	7521	0.37	-7.83				

APPENDIX 2. (continued)

Depth m	Age years	δ^{18}O o/oo, pdb	δ^{13}C	Mg ppm	Sr	Mg/Ca	Sr/Ca mole ratio/1000
6.280	7541	1.48	-6.36				
6.290	7561	1.08	-6.63	729	1479	2.985	1.680
6.300	7581	0.99	-6.35				
6.310	7601	0.63	-7.14				
6.320	7621	0.62	-7.65				
6.330	7641	0.63	-7.63				
6.340	7661	1.35	-6.66				
6.520	8027	0.21	-6.43				
6.540	8068	0.43	-6.42				
6.560	8109	0.39	-5.55				
6.570	8130	-0.18	-3.47				
6.580	8150	0.36	-5.96				
6.590	8171	-0.12	-6.47				
6.600	8192	0.25	-5.98				
6.610	8212	0.83	-5.23				
6.620	8233	0.80	-5.67				
6.630	8254	0.42	-5.96				
6.640	8275	0.80	-5.62				
6.650	8296	0.88	-6.93	938	1725	3.171	1.619
6.660	8317	0.97	-6.56				
6.670	8337	0.95	-7.58				
6.680	8358	1.03	-6.85				
6.690	8379	1.23	-6.90				
6.700	8400	0.52	-4.02				
6.710	8421	0.89	-6.44				
6.720	8442	1.03	-6.20				
6.730	8463	1.29	-6.07				
6.750	8506	1.03	-6.39	618	1737	2.094	1.632
6.760	8527	1.34	-6.82				
6.780	8569	1.67	-6.44				
6.790	8590	1.35	-6.88				
6.800	8611	1.26	-6.97				
6.810	8633	1.17	-6.64				
6.820	8654	1.11	-5.49				
6.850	8718	1.06	-7.15	565	1304	2.296	1.470
6.860	8739	1.23	-7.11				
6.870	8761	1.29	-6.81				
6.880	8782	1.06	-7.18				
6.890	8804	1.11	-6.76				
6.900	8825	1.18	-5.99				
6.910	8847	1.16	-6.92				
6.920	8868	1.34	-6.99				
6.930	8890	1.28	-7.32				
6.940	8911	1.18	-7.08				
6.960	8955	1.33	-7.57	693	1500	2.718	1.632
6.970	8976	1.10	-7.00				
6.980	8998	1.32	-6.68				
6.990	9020	1.20	-7.39				
7.000	9041	1.02	-6.81				
7.010	9063	0.66	-4.26				
7.020	9085	0.45	-4.48				
7.030	9107	-0.20	-4.55				
7.030	9107	0.10	-4.59				
7.040	9129	1.01	-6.93				
7.050	9150	1.16	-6.63				
7.060	9172	1.09	-6.78	641	1815	2.412	1.894

APPENDIX 2. (continued)

Depth m	Age years	δ^{18}O o/oo, pdb	δ^{13}C	Mg ppm	Sr	Mg/Ca	Sr/Ca mole ratio/1000
7.070	9194	1.91	-6.09				
7.120	9304	0.99	-7.02				
7.140	9348	0.77	-7.06				
7.150	9370	1.11	-5.98				
7.160	9393	1.63	-5.51				
7.170	9415	1.46	-5.28				
7.180	9437	1.86	-5.45				
7.200	9481	1.13	-5.52				
7.210	9504	1.44	-5.16	737	1513	3.112	1.773
7.220	9526	0.69	-4.61				
7.230	9548	1.13	-5.95				
7.240	9571	1.28	-5.12				
7.250	9593	1.39	-5.84				
7.260	9615	1.07	-6.12				
7.270	9638	1.76	-5.27				
7.280	9660	1.35	-6.23				
7.290	9683	1.47	-6.30				
7.300	9705	1.35	-5.43				
7.310	9728	1.34	-5.12	402	1424	1.830	1.798
7.320	9750	1.68	-5.27				
7.330	9773	1.62	-5.47				
7.340	9796	1.46	-6.03				
7.350	9818	1.49	-5.87				
7.360	9841	1.59	-5.91				
7.370	9863	1.66	-5.80				
7.380	9886	1.61	-6.41				
7.390	9909	1.88	-5.92				
7.400	9932	2.06	-5.63				
7.410	9954	1.54	-6.29	602	1830	2.241	1.888
7.430	10000	1.77	-5.98				
7.440	10023	1.45	-6.70				
7.450	10046	1.62	-5.27				
7.470	10092	1.54	-6.11				
7.480	10115	1.66	-5.67				
7.490	10138	1.83	-5.31				
7.500	10161	1.44	-6.81				
7.500	10161	1.96	-5.48				
7.510	10184	1.61	-5.94				
7.520	10207	1.95	-5.28				
7.530	10230	1.94	-5.72				
7.540	10253	1.73	-5.81	714	1845	2.638	1.891
7.550	10276	2.46	-5.91				
7.550	10276	1.87	-5.57				
7.560	10299	1.07	-7.14				
7.570	10322	1.93	-5.77				
7.580	10346	1.98	-5.43				
7.590	10369	1.99	-5.89				
7.600	10392	1.74	-5.92				
7.610	10415	1.55	-6.33				
7.620	10439	1.96	-5.45				
7.630	10462	1.77	-6.07				
7.650	10509	1.22	-5.72				
7.660	10532	1.51	-5.19				
7.670	10556	0.29	-5.75	556	1486	2.398	1.780
7.670	10556	0.63	-4.96				

Acknowledgments. We thank G. Jones for AMS ^{14}C analyses, R. Forester for ostracod identification, A. Higuera-Gundy for pollen analysis, and M. Brenner and M. Binford for core recovery and discussion. D. Engtrom and G. Lister reviewed the manuscript and provided constructive criticism that substantially improved the paper. This work was supported by a Presidential Young Investigator award (NSF grant #OCE 88-58012) to DAH. JHC thanks AGU for student travel support to attend the Chapman Conference on Continental Isotopic Indicators of Climate (CIIC).

REFERENCES

Berger, A., Numerical values of caloric insolation from 1,000,000 YBP to 100,000 YAP (astronomical solution of Berger, 1978), *Contribution 37*, Institute d'Astron. et de Geophy., University Catholique, Louvain-la-Neuve, Belgium, 1978.

Binford, M.W., Ecological history of Lake Valencia, Venezuela: Interpretation of animal microfossils and some chemical, physical, and geological features, *Ecological Monographs*, 52, 307-333, 1982.

Bonnefille, R., J. C. Roeland, and J. Guiot, Temperature and rainfall estimates for the past 40,000 years in equatorial Africa, *Nature*, 346, 347-349, 1990.

Boyle, E. A., and L. Keigwin, North Atlantic thermohaline circulation during the past 20,000 years linked to high-latitude surface temperature, *Nature*, 350, 35-40, 1987.

Bradbury, J. P., B. W. Leyden, M. Salgado-Labouriau, W. M. Lewis Jr., C. Schubert, M. W. Benford, D. G. Frey, D. R. Whitehead, and F. H. Weibezahn, Late Quaternary environmental history of Lake Valencia, Venezuela, *Science*, 214, 1299-1305, 1981.

Brenner, M, J.H. Curtis, A. Higuera-Gundy, D.A. Hodell, G.A. Jones, M.W. Binford, and K.T. Dorsey, Lake Miragoane, Haiti (Caribbean), in *Global Geological Record of Lake Basins*, edited by K. Kelts and E. Gierlowski-Kordesch, Cambridge Univ. Press, Cambridge, in press.

Brenner, M., and M. W. Binford, A sedimentary record of human disturbance from Lake Miragoane, Haiti, *Journal of Paleolimnology*, 1, 85-97, 1988.

Broecker, W. S., M. Andree, W. Wolfli, H. Oeschger, G. Bonani, J. P. Kennett, and D. Peteet, The chronology of the last deglaciation: Implications to the cause of the Younger Dryas Event, *Paleoceanography*, 3, 1-19, 1988.

Broecker, W. S., and G. H. Denton, The role of ocean-atmosphere re-organizations in glacial cycles, *Geochimica et Cosmochimica Acta*, 53, 2465-2501, 1989.

Broecker, W. S., and G. H. Denton, What drives glacial cycles?, *Scientific American*, 49-56, 1990.

Chivas, A. R., P. De Deckker, and J. M. G. Shelley, Magnesium, strontium, and barium partitioning in nonmarine ostracode shells and their use in paleoenvironmental reconstructions-a preliminary study, in *Applications of Ostracoda*, edited by R. F. Maddocks, pp. 238-249, University of Houston Geoscience, Houston, 1983.

Chivas, A. R., P. De Deckker, and J. M. G. Shelley, Strontium content of ostracods indicates lacustrine palaeosalinity, *Nature*, 316, 251-253, 1985.

Chivas, A. R., P. De Deckker, and J. M. G. Shelley, Magnesium and strontium in non-marine ostracod shells as indicators of palaeosalinity and palaeotemperature, *Hydrobiologia*, 143, 135-142, 1986a.

Chivas, A. R., P. De Deckker, and J. M. G. Shelley, Magnesium content of non-marine ostracod shells: A new palaeosalinometer and palaeothermometer, *Palaeogeography, Palaeoclimatology, Palaeoecology*, 54, 43-61, 1986b.

Covich, A., and M. Stuiver, Changes in oxygen 18 as a measure of long-term fluctuations in tropical lake levels and molluscan populations, *Limnology and Oceanography*, 19, 682-691, 1974.

Deevey, E. S., M. Brenner, and M. W. Binford, Paleolimnology of the Peten lake district, Guatemala, *Hydrobiologia*, 103, 211-216, 1983.

Duplessy, J. C., M. Arnold, P. Maurice, E. Bard, J. Duprat, and J. Moyes, Direct dating of the oxygen-isotope record of the last deglaciation by ^{14}C accelerator mass spectrometry, *Nature*, 320, 350-352, 1986.

Duplessy, J. C., G. Delibrias, T. J.L., C. Pujol, and J. Duprat, Deglacial warming of the Northeastern Atlantic Ocean: Correlation with paleoclimatic evolution of the European continent, *Palaeogeography, Palaeoclimatology, Palaeoecology*, 35, 121-144, 1981.

Engstrom, D. R., and S. R. Nelson, Paleosalinity from trace metals in fossil ostracodes compared with observational records at Devil's Lake, North Dakota, USA, *Palaeogeography, Palaeoclimatology, Palaeoecology*, 83, 295-312, 1991.

Fairbanks, R. G., A 17,000 year glacio-eustatic sea level record: Influences of glacial melting rates in the Younger Dryas event and deep-ocean circulation, *Nature*, 342, 637-642, 1989.

Flower, B. P., and J. P. Kennett, The Younger Dryas cool episode in the Gulf of Mexico, *Paleoceanography*, 5, 949-961, 1990.

Fontes, J. C., and R. Gonfiantini, Comportement isotopique au cours de l'evaporation de deux bassins sahariens, *Earth and Planetary Science Letters*, 3, 258-266, 1967.

Gasse, F., V. Ledee, M. Massault, and J. Fontes, Water-level fluctuations of Lake Tanganyika in phase with oceanic changes during the last glaciation and deglaciation, *Nature*, 342, 57-59, 1989.

Gasse, F., R. Tehet, A. Durand, E. Gibert, and J. Fontes, The arid-humid transition in the Sahara and the Sahel during the last deglaciation, *Nature*, 346, 141-146, 1990.

Hastenrath, S., Variations in low-latitude circulation and extreme climatic events in the tropical Americas, *Journal of the Atmospheric Sciences*, 33, 202-215, 1976.

Hastenrath, S., Interannual variability and annual cycle: Mechanisms of circulation and climate in the tropical Atlantic sector, *Monthly Weather Review*, 112, 1097-1107, 1984.

Higuera-Gundy, A., Antillean Vegetational History and Paleoclim te Reconstructed from the Paleolimnological Record of Lake Miragoane, Haiti, PhD, University of Florida, Gainesville, 1991.

Higuera-Gundy, A., Recent vegetation changes in southern Haiti, in *Biogeography of the West Indies: past, present, and future*, edited by C.A. Woods, pp. 191-200, Sandhill Crane Press Inc., Gainesville, FL, 1989.

Hodell, D. A., J. H. Curtis, G. A. Jones, A. Higuera-Gundy, M. Brenner, M. W. Binford, and K. T. Dorsey, Reconstruction of Caribbean climate change over the past 10,500 years, *Nature*, 352, 790-793, 1991.

Keigwin, L. D., and G. A. Jones, Deglacial climatic oscillations in the Gulf of California, *Paleoceanography*, 5, 1009-1023, 1990.

Kutzbach, J. E., and P. J. Guetter, The sensitivity of monsoon climates to orbital parameter changes for 9,000 years BP: Experiments with the NCAR general circulation model, in *Milankovitch and Climate, Part 2*, edited by A. L. Berger et. al., pp. 801-820, Riedel, Dordrecht, 1984.

Kutzbach, J.E., and A. Street-Perrott, Milankovitch forcing of fluctuations in the level of tropical lakes from 18 to 0 kyr bp, *Nature*, 317, 130-134, 1985.

Leyden, B. W., Guatemalan forest synthesis after Pleistocene aridity, *Proceedings of the National Academy of Science, USA*, 81, 4856-4859, 1984.

Leyden, B. W., Late Quaternary aridity and Holocene moisture fluctuations in the Lake Valencia Basin, Venezuela, *Ecology*, 66, 1279-1295, 1985.

Leyden, B. W., M. Brenner, D. A. Hodell, and J. H. Curtis, Late Pleistocene climate in the Central American lowlands, (this volume).

Linsey, B. K., and R. C. Thunell, The record of deglaciation in the Sulu Sea: Evidence for the Younger Dryas event in the tropical western Pacific, *Paleoceanography*, 5, 1025-1039, 1990.

Lister, G. S., K. Kelts, C. K. Zao, J. Yu, and F. Niessen, Lake Qinghai, China: Closed-basin lake levels and the oxygen isotope record for ostracoda since the latest Pleistocene, *Palaeogeography, Palaeoclimatology, Palaeoecology*, 84, 141-162, 1991.

Lister, G.S., Stable isotopes from lacustrine ostracoda as tracers for continental palaeoenvironments, in *Ostracoda in the Earth Sciences*, edited by P. De Deckker et al., pp. 201-218, Elsevier, Amsterdam, 1988.

Mann, P., F. W. Taylor, K. Burke, and R. Kulstad, Subaerially exposed Holocene coral reef, Enriquillo Valley, Dominican Republic, *Geological Society of America Bulletin*, 95, 1084-1092, 1984.

Markgraf, V., Palaeoclimates in Central and South America since 18,000 BP based on pollen and lake-level records, *Quaternary Science Reviews*, 8, 1-24, 1989.

Mercer, J. H., The Allerod oscillation: A European climatic anomaly?, *Arctic and Alpine Research*, 1, 227-234, 1969.

Mix, A. C., W. F. Ruddiman, and A. McIntyre, Late Quaternary paleoceanography of the tropical Atlantic, 1: Spacial variability of annual mean sea-surface temperatures, 0-20,000 years B.P., *Paleoceanography*, 1, 43-66, 1986.

Molfino, B., and A. McIntyre, Nutricline variation in the equatorial Atlantic coincident with the Younger Dryas, *Paleoceanography*, 5, 1990.

Parker, D. E., C. K. Folland, and M. N. Ward, Sea-surface temperature anomaly patterns and prediction of seasonal rainfall in the Sahel region of Africa, in *Recent Climatic Change*, edited by S. Gregory, pp. 166-178, Belhaven, London, 1988.

Peterson, L. C., J. T. Overpeck, N. G. Kipp, and J. Imbrie, A high-resolution late Quaternary upwelling record from the anoxic Cariaco basin, Venezuela, *Paleoceanography*, 6, 99-119, 1991.

Piperno, D. R., M. B. Bush, and P. A. Colinvaux, Paleoenvironments and human occupation in Late-Glacial Panama, *Quaternary Research*, 33, 108-116, 1990.

Prell, W. L., and J. D. Hays, Late Pleistocene faunal and temperature patterns of the Colombia Basin, Caribbean Sea, in *Investigation of Late Quaternary Paleoceanography and Paleoclimatology*, edited by R. M. Cline and J. D. Hays, pp. 201-220, Memoirs of Geological Society of America, Boulder, 1976.

Roberts, N., Ups and downs of African lakes, *Nature*, 346, 107, 1990.

Ruddiman, W. F., and A. McIntyre, The North Atlantic Ocean during the last deglaciation, *Palaeogeography, Palaeoclimatology, Palaeoecology*, 35, 145-214, 1981.

Snoeyink, V. L., and D. Jenkins, *Water Chemistry*, 463 pp., John Wiley and Sons, New York, 1980.

Spero, H. J., and D. F. Williams, Evidence for seasonal low-salinity surface waters in the Gulf of Mexico over the last 16,000 years, *Paleoceanography*, 5, 963-975, 1990.

Stocker, T. F., and D. G. Wright, Rapid transitions of the ocean's deep circulation induced by changes in surface water fluxes, *Nature*, 351, 729-732, 1991.

Street, F. A., and A. T. Grove, Environmental and climatic implications of late Quaternary lake-level fluctuations in Africa, *Nature*, 261, 385-390, 1976.

Street, F. A., and A. T. Grove, Global maps of lake-level fluctuations since 30,000 yr B.P., *Quaternary Research*, 12, 1979.

Street-Perrott, F.A., and N. Roberts, Fluctuation in closed-basin lakes as an indicator of past atmospheric circulation patterns, in *Variations in the Global Water Budget*, pp. 331-345, Reidel, Dordrecht, 1983.

Street-Perrott, F. A., and S. P. Harrison, Lake levels and climate reconstruction, in *Paleoclimate Data and Modeling*, edited by A. D. Hecht, pp. 291-340, Wiley, New York, 1985.

Street-Perrott, F. A., and R. A. Perrott, Abrupt climate fluctuations in the tropics: The influence of Atlantic Ocean circulation, *Nature*, 343, 607-612, 1990.

Talbot, M. R., and G. Delibrias, A new late Pleistocene-Holocene water-level curve for Lake Bosumtwi, Ghana, *Earth and Planetary Science Letters*, 47, 336-344, 1980.

Taylor, F. W., P. Mann, S. Valastro, and K. Burke, Stratigraphy and radiocarbon chronology of a subaerially exposed Holocene coral reef, Dominican Republic, *Journal of Geology*, 93, 311-332, 1985.

Watts, W. A., A late Quaternary record of vegetation from Lake Annie, south-central Florida, *Geology*, 3, 344-346, 1975.

Woodring, W. P., J. S. Brown, and W. S. Burbank, *Geology of the Republic of Haiti*, 631 pp., Republic of Haiti Department of Public Works. Geological Survey of the Republic of Haiti, Port-Au-Prince, 1924.

J. Curtis and D. Hodell, Department of Geology, 1112 Turlington Hall, University of Florida, Gainesville, FL 32611

Seasonal Change In Paleogene Surface Water $\delta^{18}O$:
Fresh-Water Bivalves Of Western North America

DAVID L. DETTMAN AND KYGER C LOHMANN

Department of Geological Sciences, University of Michigan

Monthly to fortnightly variations in environmental conditions of the Paleogene Powder River Basin, Montana and Wyoming, were investigated through microanalysis of growth banding of fresh-water bivalves. Annual amplitudes of up to 9.8 ‰ PDB in carbonate $\delta^{18}O$ precipitated in equilibrium with surface waters reveal large seasonal changes in fresh-water $\delta^{18}O$. Both Late Paleocene and Early Eocene waters are highly variable in $\delta^{18}O$ intra-annually. Numerous instances of waters that are highly depleted in ^{18}O throughout the year occur in the Paleocene; this does not occur in the Eocene. Surface water $\delta^{18}O$ for all Late Paleocene samples ranges from -8 to -23 ‰ SMOW (±2). Early Eocene water $\delta^{18}O$ ranges from -8.5 to -16 ‰ SMOW (±2). Such highly negative surface waters suggest a setting of seasonal or year-round runoff of high altitude snow-melt into the basin on which seasonal temperature variation is superimposed.

INTRODUCTION

General agreement exists on the pattern of climate change during the Paleogene. A trend of warm Early Paleocene, cooler Late Paleocene, warming to a temperature optimum in the Early to Middle Eocene is seen in many paleoclimate disciplines: paleobotany [Wing, 1987; Wing et al. 1991; Wolfe and Poore, 1982; Wolfe, 1987], marine stable isotope studies [Savin, 1977; Miller et al. 1987], mammalian species range and diversity studies [Rose, 1980; Kraus and Mass, 1990], and climate modeling [Barron, 1987, 1985]. The details and chronology of this record are still debated, and as more regional studies are undertaken [e.g. Fox, 1990] this apparent concord may prove to be temporary. One area of considerable disagreement is the nature of seasonality in the interiors of continents. Climate modelers argue for large (>40°C) seasonal temperature amplitudes, with a significant sub-zero component [Sloan and Barron, 1992; Crowley et al. 1986], while paleobotanical studies argue for no sub-zero temperatures and a maximum thermal seasonality of 20 to 25°C [Hickey, 1980]. This study brings a new methodology to the study of continental interior seasonality by focusing on stable isotope measurement of accretionary structures in fresh-water biogenic carbonate.

As well as being a time of change in mean annual temperature, the Paleocene-Eocene transition of the western interior of North America is a time of great faunal turnover [Gingerich, 1989]. It has been suggested that this extinction/immigration was triggered by a change in the nature of thermal seasonality and the stability of seasonal extremes [Gingerich, 1989; Rose 1980]. Paleobotanical studies infer a reduction in seasonality from 25°C to 20°C across the Paleocene-Eocene boundary in the western interior [Wing, 1987]. Guided by these indications of changing climate, this study focuses on the Paleocene - Eocene transition in the Powder River Basin, comparing records of seasonality in fresh-water bivalves for the Late Paleocene with those of the Early Eocene.

Paleoclimatology has long made use of the stable isotope thermometry of marine pelagic carbonates as its primary record of Late Mesozoic and Cenozoic climate change. However, even in areas of high sedimentation rates, records of sub-annual temperature change are lost due to both bioturbation and the seasonal growth patterns of planktonic organisms. In contrast, inter- and intra-annual records of climate variation are preserved in skeletal growth structures of numerous taxa. Oxygen isotope analysis of annual growth bands has been used to extract paleoclimate records from corals [Pätzold, 1984; Aharon, 1991], molluscs [Emiliani et al., 1964; Krantz et al., 1987; Jones et al., 1983; Romanek and Grossman, 1989], otoliths [Kalish, 1991; Patterson et al., this volume] and mastodon tusks [Koch et al., 1989]. Sub-sampling a series of growth bands in one individual organism allows study of climate records on an annual and sub-annual time scale in contrast to the numerous time-averaged geological records currently utilized in paleoclimate studies. Bivalves are ideal for the study of inter- and intra-annual climate variations because of their simple growth geometry, clear annual growth increments (annuli) and high rates of shell accretion which allow resolution of daily to monthly time increments. In

Climate Change in Continental Isotopic Records
Geophysical Monograph 78

this study, annual growth bands are used as independent time markers against which isotopic variations will be evaluated.

This study examines fresh-water bivalves (unionids) to document intra-annual variation in $\delta^{18}O$ and $\delta^{13}C$, stable isotope records of weekly to monthly changes in aragonite precipitated in the fluvial or lacustrine environment. While fresh-water bivalves do not grow rapidly, growth rates of 0.5 to 2 mm per year during early ontogeny yield enough carbonate for a series of at least 12 sub-annual $\delta^{18}O$ analyses. Use of a new micro-drilling method to extract very fine resolution samples from the shells of these unionid bivalves enables collections of approximately 40 samples per millimeter of shell growth. Isotopic analysis of these samples provides a record of sequential changes in surface water temperature/composition throughout most, if not all, of the Late Paleocene and Early Eocene calendar year.

MATERIALS AND METHODS

Modern Unionids

Living unionids were collected from both riffle and ponded waters of the Huron River of Michigan in late 1989. Eight different species were represented in the collections. One of these unionids has been sectioned and analyzed for $\delta^{18}O$ and $\delta^{13}C$ variation using the methods described below. Beginning in 1987, water samples from the Huron River have been collected and water temperatures recorded on a biweekly to monthly

schedule when the river was not frozen. A selection of these water samples was analyzed for $\delta^{18}O$ using conventional CO_2-H_2O equilibration techniques.

To evaluate settings where seasonality is characterized by large $\delta^{18}O$ variation in surface waters combined with minimal seasonal temperature fluctuation, a specimen of *Hyria corrugata* from the Amazon Basin was analyzed. This specimen, provided by the University of Michigan Museum of Natural History, was collected sometime before 1927 from the Purus River, Brazil. While environmental parameters are not well known for this specimen, the climatic setting associated with equatorial rain forests provides a contrast to temperate settings.

Paleogene

Paleogene unionid bivalves (*Plesielliptio* spp.) were collected from fluvial and lacustrine deposits near the geographic center of the Powder River Basin. These deposits occur within the Fort Union Formation (Tongue River Member) and Wasatch Formation, and comprise interbedded sandstones, shales, and coals. Coals are very abundant, and the stratigraphic order in which these samples are placed is based on the coal stratigraphy of numerous workers in the basin [Bryson and Bass, 1973; Olive, 1957]. Molluscan remains are found in very fine grain sediments and as lag deposits of shells which typically show minor abrasion due to transport. Additional samples for bulk analysis were removed from the collections of John Hanley (U.S.G.S.).

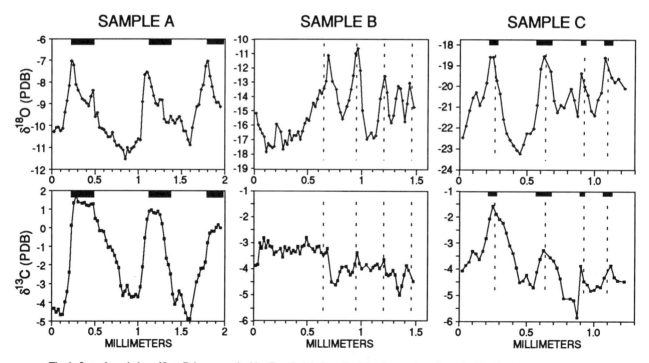

Fig. 1. Isotopic variation of Late Paleocene unionids. Sample A is from fluvial sediments 1 m above the Machin-Walker Coal in the lower half of the Tongue River Member of the Fort Union Formation (sec. 20, T. 3 S., R. 49 E., Powder River Co., Montana). Sample B is from fluvial sediments 120 m below the Elk Coal in the Tongue River Member (sec. 5, T. 4 S., R. 46 E., Powder River Co., Montana). Sample C was collected 8 m above the Roland Coal from a lacustrine molluscan coquina marking the boundary of the Fort Union and Wasatch Formations (sec. 10, T. 56 N., R. 75 W., Campbell Co., Wyoming). Where visible, cloudy shell areas are shown as shaded boxes and thickened conchiolin layers are represented by vertical dotted lines.

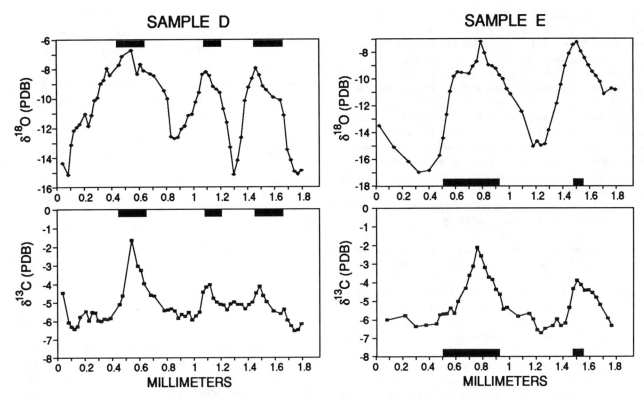

Fig. 2. Isotopic variation of Early Eocene unionids. Samples D and E were collected 5 m above the Ulm #1 coal high in the Wasatch Formation (sec. 31, T. 56 N., R. 80 W., Sheridan Co. Wyoming). Cloudy shell areas shown as shaded boxes.

Only two species of unionids survived into the Early Cenozoic, *Plesielliptio priscus* and *Plesielliptio silberlingi* [Hanley and Flores, 1987]. These two species are very closely related (having virtually identical dentition and beak sculpture) and were the only two living during the time interval under consideration. Five well preserved unionids were examined in this study utilizing microsampling techniques for stable isotope analysis. Sample A (Fig. 1) was collected from fluvial sediments 1 meter above the Machin-Walker Coal in the lower half of the Tongue River Member of the Fort Union Formation. Sample B (Fig. 1) was collected from fluvial sediments 120 meters below the Elk Coal in the Tongue River Member. Sample C (Fig. 1) was collected from lacustrine sediments 8 meters above the Roland Coal; this molluscan coquina marks the boundary of the Fort Union and Wasatch Formations. Samples D and E (Fig. 2) are two Early Eocene bivalves collected 5 meters above the Ulm #1 Coal [Culbertson and Klett, 1975], high in the Wasatch Formation. Both bivalves were taken from the same bed where molluscs were concentrated at the base of a fluvial channel deposit.

Preservation of Shell Mineralogy and Structure

Samples are excellently preserved. Nacreous layers of all shells are unaltered, although outer prismatic layers were either dissolved or calcitized. X-ray diffraction confirmed complete preservation of primary aragonite shell material and absence of diagenetic calcite. Additionally, specimens were examined under high power cathodoluminescence microscopy (CL) to evaluate possible diagenetic alteration as fine-scale intergrowth of secondary calcite cement. Under CL, specimens exhibited a pale green-gold luminescence which is characteristic of modern aragonitic fresh-water molluscan carbonate [Sommer, 1972]. Because these shells retain their primary aragonite mineralogy, we conclude that the original $\delta^{18}O$ and $\delta^{13}C$ composition of the shell is preserved.

Methods for High Resolution Sampling and Analysis

An essential element of our approach is the ability to mechanically sample fine-scale growth lamellae, to accurately analyze small amounts of carbonate (10 to 20 μgrams) and thus to document original compositional heterogeneities present within the shells. Fine scale samples were obtained by milling sequential increments as defined by internal growth structures between annual bands. Isotopic variation is plotted relative to these visible annuli by tracking sampling paths as they are drilled from the shell. Because the sampling increment averages 25 μm in thickness and annual growth of the unionids used in this study ranges between 0.2 and 1.0 mm per year, micro-sampling provides an optimal resolution of less than two weeks of shell growth. While the rate of shell growth is expected to vary with the seasons, we are unable to identify periodic sub-annual growth features that would allow us to estimate true temporal resolution. Sample resolution, therefore, ranges from 9 to 36 samples per year based solely on annual bands.

Annual growth bands in modern unionids are clearly visible as a thickening of the outer prismatic shell layer and an alternation of cloudy and clear areas of shell [Day, 1984]. In Paleogene unionids from the Powder River Basin the prismatic layer is usually missing and less conclusive markers, such as cloudy/clear shell and thickened organic layers (conchiolin), must be used to identify annual increments. In temperate environments, cloudy zones are precipitated during the cooler months before and after hibernation [Neves and Moyer, 1988].

The method for physical separation of samples is as follows. The position and curvature of accretionary banding in the first few years of growth were precisely digitized from a photomicrograph of a 100 μm thick section of the shell. Additional curves were interpolated between digitized bands to provide a computer representation of the sequential sampling traverses. Typical increments between traverses ranged from 10 to 40 μm. The coordinates of the digitized sampling paths were translated to congruent paths on the surface of the thin section using standard Cartesian transformations. Under computer control, samples averaging 25 μm wide, 50 μm deep and 4 mm long were sequentially milled from the section. The mixing of material from different times of growth was minimized by positioning planes of growth perpendicular to the thick section surface. These microsamples of carbonate, 10-20 μg in size, were analyzed for $\delta^{18}O$ and $\delta^{13}C$ on a Finnigan MAT 251 gas ratio mass spectrometer coupled to a CarboKiel, an automated carbonate reaction device. Precision was monitored through daily analysis of standard carbonates of equivalent size. Precision for $\delta^{18}O$ and $\delta^{13}C$ is ±0.1 ‰ and ±0.05 ‰, respectively.

$\delta^{18}O$ AND $\delta^{13}C$ VARIATION IN PALEOGENE UNIONIDS

Fine scale data for Paleogene unionids exhibit a pronounced annual cyclicity in both $\delta^{18}O$ and $\delta^{13}C$ composition (Figs. 1 and 2; samples are in stratigraphic order, lowest to highest from A to E). Samples A, B, and C (Fig. 1) are Late Paleocene in age; samples D and E (Fig. 2) are Early Eocene in age. The x-axis of these plots indicates the accumulated thickness of shell nacre drilled with each sample, and while related to time, has not been corrected for possible differences in growth rate which may occur during a year. The plots, therefore, illustrate $\delta^{18}O$ and $\delta^{13}C$ variation relative to the thickness of shell growth. Because unionids precipitate new shell across the entire inner surface of both valves as they grow, earliest growth is on the outer surface of the shell near the umbo and later growth lies toward the interior of the shell and away from the umbo.

Possible morphological indicators of annual growth intervals are indicated in the figures when they are visible in thick-section. Dotted vertical lines indicate thickened organic layers or conchiolin. Many of the shells were highly fractured which resulted in multiple partings parallel to growth structures,

TABLE 1. $\delta^{18}O$ of microsampled unionid specimens

Unionid	$\delta^{18}O$ amplitudes.	Volumetric mean $\delta^{18}O$	Age
A.	3.3 * 4.2 3.6	-9.5	Paleocene
B.	6.7 5.0 5.4	-15.2	Paleocene
C.	3.9 * 4.6 2.3	-20.8	Paleocene
D.	9.8 7.8	-11.7	Eocene
E.	8.4 5.3 7.0	-10.7	Eocene

obscuring the presence or absence of a thickened annual conchiolin layer. Shaded boxes indicate approximate areas of cloudy shell when viewed in 100 μm thick-section. Cloudy growth zones are identified by qualitative visual inspection which compares sections of ancient shell with sections of modern temperate climate unionids.

Sub-annual $\delta^{18}O$ and $\delta^{13}C$ resolution is highest where annual growth intervals are thickest. The first full year of growth of sample A, measured between $\delta^{18}O$ maxima, is subdivided into 34 samples. The same resolution was accomplished in the first partial year of growth in sample B. For at least part of these years of shell growth, $\delta^{18}O$ and $\delta^{13}C$ measurements record environmental conditions with a weekly to biweekly time resolution. Excellent spatial resolution of shell isotopic composition is also maintained; note in sample A (Fig. 1) the 6 ‰ shift in $\delta^{13}C$ over a distance of 120 μm (between 140 μm and 260 μm), indicating that zones of differing isotopic composition in the shell have been clearly separated.

Table 1 summarizes the amplitudes and mean $\delta^{18}O$ value for each sample. Amplitudes are the difference between each $\delta^{18}O$ minimum and the average $\delta^{18}O$ of the adjacent maxima. In the case of the first incomplete cycle the amplitude is simply the difference between first $\delta^{18}O$ maximum and minimum. When the shell composition has clearly not achieved a minimum value for the first incomplete cycle, the amplitude may be underestimated. These underestimated cycles are flagged with an asterisk in Table 1. Amplitudes are presented for the first three years of growth. $\delta^{18}O$ mean values are volumetric averages for the area of shell sampled and should approximate the average $\delta^{18}O$ value for the entire bivalve. Although D and E (Fig. 2) have greater amplitudes than the Late Paleocene unionids, the amplitude of sample B is nearly as large, and much larger than A or C. Because samples D and E were collected three years of growth, an average $\delta^{18}O$ for the shell comparable to mean $\delta^{18}O$ values listed in Table 1. Samples used for fine scale analysis are plotted as the letters A - E. Samples at the same level were taken from the same bed of shells. Note that the Late Paleocene samples exhibit greater variability in $\delta^{18}O$ than the Early Eocene samples.

CALIBRATION OF MODERN UNIONIDS

Before interpreting annual $\delta^{18}O$ and $\delta^{13}C$ cycles, the question of whether unionids precipitate shell carbonate in isotopic equilibrium must be addressed. Is there a biological control on the $\delta^{18}O$ of the shell (vital effect) or are the shell $\delta^{18}O$ values controlled solely by ambient water $\delta^{18}O$ and temperature? Preliminary data strongly suggest that, if present, vital effects are small. Analysis of $\delta^{18}O$ variation of modern unionids in the Huron River of Michigan suggests that shell carbonate precipitates in isotopic equilibrium. This is illustrated by the close relationship of 1978-1979 shell aragonite $\delta^{18}O$ with that predicted by environmental measurements for 1987 (Fig. 4). A study of Huron River unionids is currently underway to more precisely define the aragonite-water $\delta^{18}O$-temperature relationship. The overlap of measured $\delta^{18}O$ and predicted $\delta^{18}O$ and the agreement in the shape of the annual cycle validates the temperature-aragonite $\delta^{18}O$ relationship. The relationship used here to predict aragonite $\delta^{18}O$ is modified from that of Grossman and Ku [1986]:

$$10^3 \ln\alpha_{arag-water} = -0.2160T°C + 34.959 \qquad (1)$$

This modification uses fractionation factors for all calculations to eliminate problems introduced by mixing the SMOW and PDB scales.

Bulk carbonate samples from modern unionids of four different sub-families (following the classification of Vokes [1980]) from the same environment lie within 0.23 ‰ (-7.26 for a *Strophitus undulatus*; -7.30 for a *Cyclonaias tuberculata*, -7.30 for a *Villosa iris*, and -7.49 for a *Lampsilis siliquoidea*), indicating no vital effect offset specific to a single sub-family. Thus, $\delta^{18}O$ measurements of Paleogene taxa are interpreted as equilibrium isotopic values.

MODELS FOR INTERPRETATION

In ancient samples absolute temperatures cannot be independently assigned to any shell $\delta^{18}O$ value unless water $\delta^{18}O$ is known. However, environmental change within and among years can be quantified based on the patterns of annual $\delta^{18}O$ variation present in fresh-water bodies. Modeling of seasonal temperature change, its effect on shell growth, and related changes in surface water $\delta^{18}O$ constrains patterns of environmental variation that shape the annual cycle of shell $\delta^{18}O$. Simply conceived, systems are temperature-dominated or water $\delta^{18}O$-dominated with respect to environmental controls on shell $\delta^{18}O$ with each factor acting independently. The isotopic composition of surface water is, in effect, a baseline upon which temperature effects are superimposed. A decrease in ambient

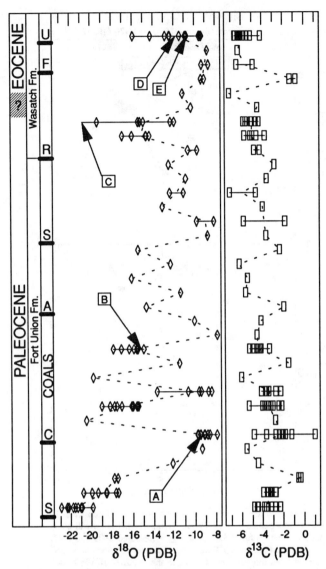

Fig. 3. Bulk shell $\delta^{18}O$ for unionids from the Powder River Basin. Dotted line connects average values for each bed. Coals (from early to late) are: S=Sawyer, C=Cashe, A=Anderson, S=Smith, R=Roland, F=Felix, U=Ulm.

from the same deposit of shell material these would be expected to show very similar patterns and amplitudes. Mean $\delta^{18}O$ values do not exhibit a simple trend stratigraphically.

Bulk unionid shell $\delta^{18}O$ values from the Powder River Basin are highly variable (Fig. 3), and are represented by 108 specimens placed in stratigraphic order relative to the major coals in the basin. The exact stratigraphic order of these samples cannot be determined and some samples may be out of ordered sequence by one or two positions, but major coal horizons are well defined. Vertical position in the figure has no relation to thickness of the sediment column. Each $\delta^{18}O$ and $\delta^{13}C$ analysis is a single sample drilled through multiple annual bands near the umbo with a 0.5 mm diameter drill bit and represents at least

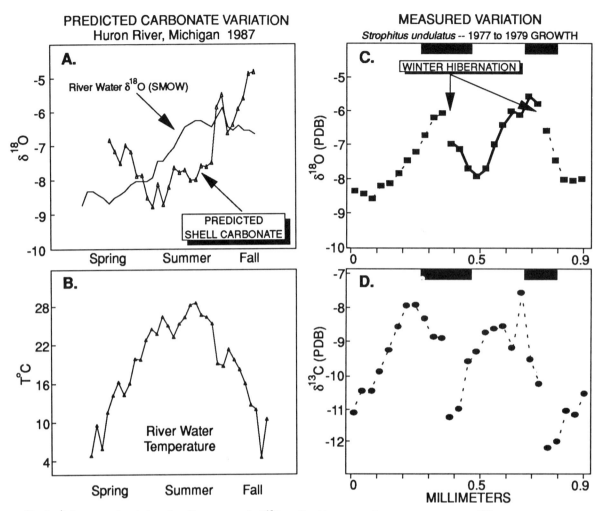

Fig. 4. **A**. Intra-annual variation of molluscan aragonite $\delta^{18}O$ predicted by measured temperatures and water $\delta^{18}O$ of the Huron River in 1987. The points plotted are weekly predictions using the measured $\delta^{18}O$ and temperature of the Huron River. Predicted growth stops when water temperatures drop below 12°C. **B**. Huron River shallow water temperatures for 1987. **C**. and **D**. Measured $\delta^{18}O$ and $\delta^{13}C$ for the first ~ 2 1/2 years of growth of a 12 year old *Strophitus undulatus* collected live in late 1989 from the Huron River. Shell was aged using the thin-section method of Neves and Moyer [1988]. Growth occurred from the summer of 1978 to the summer of 1980. All samples were taken from the nacreous layer of the shell. Later years have not been analyzed because growth bands are much thinner (less than 200 μm). Cloudy (slow growth) portions of the shell are indicated by shaded boxes. Winter hibernation events are located by projection of sub-annual growth bands from the end of the late fall thickening of the prismatic layer into nacreous layer [Day, 1984].

temperature by 4.7°C will enrich aragonite by +1.0 ‰ PDB. This temperature effect is the basic construct of our models.

Temperature Dominated Environments with Constant $\delta^{18}O$ Water

The simplest temperature-dominated system is one in which water $\delta^{18}O$ remains constant. Under these conditions, primary variations in shell $\delta^{18}O$ directly record changes in surface temperature. An example of this is the annual record derived from a near shore bivalve growing in a large lake or marine setting. This also applies to other large bodies of water even though $\delta^{18}O$ of precipitation may vary seasonally. For example, in large fluvial and lacustrine systems, or in rivers dominated by

groundwater input, seasonal variation in the composition of precipitation is dampened resulting in small changes in $\delta^{18}O$ of surface water. In such cases, temperature seasonality is the dominant factor controlling variation in shell $\delta^{18}O$.

Figure 5 illustrates hypothetical $\delta^{18}O$ patterns characteristic of bivalves growing over a range of latitudes under constant $\delta^{18}O$ water conditions. An increase in water temperature is recorded by lower shell $\delta^{18}O$ values, which reach a minimum during the period of summer maximum temperature (Fig. 5B). With cooling, $\delta^{18}O$ values increase. In effect, the measured amplitude of $\delta^{18}O$ variation directly records the range in seasonal temperature fluctuation. Seasonality of temperature is thus calculated by applying a factor of 4.7°C for each 1 ‰

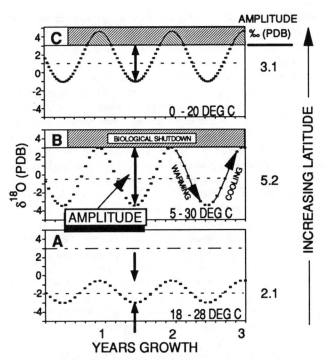

Fig. 5. Schematic patterns of δ18O variation in the growth of bivalves at different latitudes. δ18O of the water is constant and temperature variation is based on an hypothetical transect from low latitude **A** to high latitude **C** sites. δ18O recorded in shell carbonate (filled squares); cross-hatched areas define cessation of growth due to cold weather shutdown.

amplitude. However, measured δ18O seasonality should be considered only a minimum estimate (Fig. 5C) due to the effects of winter hibernation or shutdown in shell accretion [Day, 1984]. Similar growth cessations can also occur during summer warm months, due to reproductive activities or environmental disturbance [Harrington, 1989]. These can, however, be discerned from the location of the shell growth hiatus relative to the seasonal isotopic cycle. To minimize the impact of reproductive stress shutdowns, this study has sampled only early years of growth.

Amplitudes of variation are greatest and record the full range of seasonal temperature variation at sites where summer thermal maxima occur (Fig. 5B) and winter minima do not drop below the biological shutdown temperature. At sites where winter minimum temperatures drop below the shutdown threshold, a hiatus in shell growth occurs (Fig. 5C), mean annual temperature decreases, and the amplitudes of seasonality decrease. In this case, amplitudes serve as proxies for summer maximum temperatures, recorded as the difference between summer thermal maxima (low δ18O) and the biological shutdown temperature (high δ18O).

Temperature Dominated Environments with Variable δ18O Water

What happens when water isotopic composition changes during the year? In most settings temperature and water δ18O

variations are interrelated. For example, warm temperature causes rainfall to be enriched in 18O and can result in evaporative enrichment of 18O in surface water while having an opposite effect on shell δ18O. This opposing effect is illustrated in figure 6, as deviations from a mean carbonate δ18O value. In figure 6 both A and B have an identical seasonal temperature range of 28°C (5.9 ‰ PDB). The seasonal surface water δ18O cycle is larger in B (2 ‰ SMOW) than A (0.4 ‰ SMOW). Thus, seasonal variation in water composition typically dampens the effect of temperature, such that measured intra-annual δ18O variation is a minimum estimate of seasonal temperature range.

Environments Dominated by Seasonally Variable δ18O Water

In some settings, seasonal variations in water isotopic composition overwhelm temperature effects. As outlined above, warm temperature precipitation is enriched in 18O, while cold weather precipitation is depleted. The antagonistic relationship between temperature and water composition effects on shell δ18O should therefore be preserved in settings where changes in water composition are dominant. In contrast to temperature dominated environments, where seasonal change in water δ18O dampens the effects of temperature, the roles of temperature and water isotopic composition are reversed. A portion of the cycle in shell δ18O established by seasonal water δ18O variation is canceled out by superimposed temperature change, although we cannot assume that this antagonistic relationship is completely symmetrical.

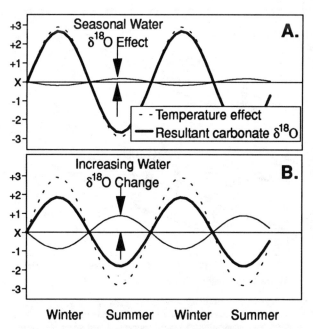

Fig. 6. Simple model of seasonal change in shell δ18O expressed as a deviation from a mean value (X). The effects of temperature and water δ18O are combined to produce carbonate δ18O. Temperature effects are identical in **A** and **B**. Change in water δ18O is larger in **B** than in **A**, a contrast that could be caused by the size of the water body, increased summer evaporation at **B** or greater seasonal precipitation at **B**.

Environments dominated by water $\delta^{18}O$ variation are only obvious in settings where the seasonal water $\delta^{18}O$ change exceeds the maximum possible temperature effect. The maximum range of temperature for shell growth should be approximately 30°C, based on organismal thermal tolerances. This would produce a maximum amplitude of 6.4 ‰ in shell $\delta^{18}O$. Thus, if the amplitude of shell $\delta^{18}O$ exceeds 6.4 ‰, seasonal changes in water composition must be invoked. Only in such cases is it possible to clearly distinguish between temperature-dominated and water-isotope-composition-dominated shell $\delta^{18}O$ amplitudes (Fig. 7).

Most of today's large amplitude seasonal cycling in precipitation $\delta^{18}O$ occurs in areas with significant snowfall. Temperate climate precipitation can easily create seasonal cycles in surface waters in excess of 6 ‰ (e.g. the 20 ‰ range in Chicago precipitation, 1974-1978 [I. A. E. A. / W. M. O. Database, 1991]). Snow melt additions to surface waters occur as surface temperatures warm above freezing and much of this water exits the drainage before the river system warms above the hibernation threshold of unionids. It is therefore unlikely that shell $\delta^{18}O$ will record local, short-lived snow melt events. However, in basins surrounded by highlands, snow melt influx into lower elevation basins can continue well after surface waters have warmed. The presence of snow melt waters which have warmed above the hibernation threshold temperature would be manifested as a large $\delta^{18}O$ offset between shell formed in the late fall, prior to hibernation, and in the spring when shell

growth resumed. Because the threshold temperature is the same at fall shutdown and spring startup, differences in shell $\delta^{18}O$ directly reflect differences in water $\delta^{18}O$ associated with snow melt inputs.

Another setting where large annual changes in water $\delta^{18}O$ change occur is exemplified by the interior of the Amazon Basin. While monthly mean temperatures in this basin vary by only 4°C, from 24°C to 28°C, water $\delta^{18}O$ changes from summer to winter by up to 10 ‰ [Salati and Marques, 1984]. As the central basin is less than 200 meters in altitude, altitude effects are negligible. Seasonal variation in precipitation $\delta^{18}O$ is primarily caused by the difference in the amount of water removed from air masses moving over the Amazon Basin. Dry season precipitation at Benjamin Constant (4°S lat., 70 W long.) is approximately -1 ‰ SMOW; wet season monthly average for May 1973 was -12.6 ‰ [Salati et al., 1979]. Modeling of air mass water content and $\delta^{18}O$ using simple Rayleigh distillation processes suggests no vapor loss in the air masses during the dry season and wet season vapor loss in excess of 80% [Grootes et al., 1989]. Model results suggest a 10 ‰ seasonal cycle in the basin center and a 17 ‰ cycle at the western edge of the basin, although no measurements have been made to validate the largest cycles.

The $\delta^{18}O$ of unionids living in areas characterized by invariant temperatures and wet-season/dry-season precipitation cycles is likely controlled by variation in water composition. Moreover, bivalves living in small drainages should exhibit pronounced seasonal cycles because these drainages are more quickly flushed by individual precipitation events. Even the eastern Amazon River (measured at ~51°W long.), integrating precipitation over the entire basin, changes by ~3.5 ‰ seasonally [Reis, 1977].

INTERPRETATION OF SHELL CARBONATE $\delta^{18}O$ VARIATION

The modern unionid from the Huron River examined in this study (Fig. 4) lived in a temperature-dominated environment. Water temperature of the Huron River varies by 28°C (0 to 28°C) while water $\delta^{18}O$ changes by 2.8 ‰ (-6.0 to -8.8 ‰ SMOW). In this case, environmental parameters measured for 1987 are taken as representative for the annual pattern of change in the Huron River and are compared to 1978 -1979 shell $\delta^{18}O$. Temperature introduces a potential 6 ‰ change to the shell $\delta^{18}O$ while water $\delta^{18}O$ variation attenuates much of that change with a 2.8 ‰ shift from spring to fall. However, during the summer season, water $\delta^{18}O$ change is not symmetrical relative to the temperature cycle. During the part of the year when water temperatures are above the hibernation threshold (12°C for Huron River unionids), river water $\delta^{18}O$ rises from -8.8 ‰ to -6.0 ‰ and then decreases to -6.8 ‰ before the onset of hibernation. This results in an offset in shell $\delta^{18}O$ carbonate formed in late fall and early spring. The amplitudes of the shell $\delta^{18}O$ cycles are approximately 2.5 ‰ PDB for both the first partial year and the first full annual cycle (heavy line in Fig. 4), corresponding to a minimum seasonal temperature cycle of 12-14°C. Overall, the patterns of $\delta^{18}O$ change clearly record the combined effects of spring snow melt runoff and seasonal temperature variations typical of a temperate climate region.

To evaluate a setting dominated by changes in $\delta^{18}O$ of waters

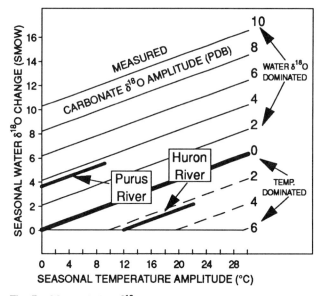

Fig. 7. Measured shell $\delta^{18}O$ amplitudes (diagonal lines) compared to seasonal changes in temperature and water $\delta^{18}O$. Amplitudes are plotted for symmetrical and opposing interaction of temperature and water $\delta^{18}O$ effects on shell $\delta^{18}O$. Minimum temperature change represented by shell $\delta^{18}O$ amplitude is calculated by following the amplitude line to the x-axis (i.e., point of no water $\delta^{18}O$ change). Minimum water $\delta^{18}O$ change is calculated by following diagonals in the water-dominated field to the y-axis. Modern samples from the Huron River (Michigan) and the Purus River (Amazonia) are shown for reference.

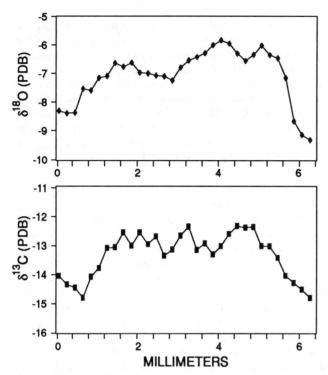

Fig. 8. Coarse sampling transect across one year of growth for a *Hyria corrugata* from the central Amazon Basin.

rather than temperature, a specimen of *Hyria corrugata* from the Purus River, Amazon Basin, was coarsely sampled in a transect for one year of growth (Fig. 8). The true complexity of the isotopic cycles may have been blurred by the use of a 0.5 mm burr for sample drilling. This unionid lived in a water-dominated environment with an annual temperature range of 4°C [Salati and Marques, 1984]. Although no measurements have been made for the Purus River, the $\delta^{18}O$ of water of the Solimões River has been measured approximately 200 km downstream from the confluence of the Purus and the Solimões. The amplitude of annual change of Solimões waters is 2.75 ‰ (varying from -5.3 to -8.05 ‰ SMOW) during the interval from Sept. 1971 to Dec. 1973 [Matsui et al. 1976]. If the $\delta^{18}O$ variation of the Solimões is representative of the Purus then a 2.75 ‰ amplitude in shell $\delta^{18}O$ would be expected. Figure 8 shows that a 3.5 ‰ amplitude in $\delta^{18}O$ is present in the shell, which is here interpreted as a minimum change in river water $\delta^{18}O$. Many rivers in the Amazon Basin have water isotope cycles with amplitudes larger than the Solimões: the Rio Napo, 4.3 ‰ SMOW; Rio Negro, 4.8 ‰; Rio Madeira, 4.7 ‰; Rio Tapajós, 4.3 ‰; Rio Amazonas, 3.5 ‰ (~51°W long.) [Salati et al., 1980]. Overall, the seasonal variation of shell $\delta^{18}O$ is quite compatible with conditions characteristic of the Amazon Basin.

Because it is not possible to independently ascertain whether a paleo-environment was temperature-dominated or water-dominated, interpretation of isotopic variation present in Paleogene unionids requires evaluation of the patterns of variation, the amplitudes of annual isotopic change, and absolute

$\delta^{18}O$ compositions. For example, the shell $\delta^{18}O$ cycles of samples A and C could reflect changes in either temperature or water composition. If temperature were the controlling factor, minimum thermal seasonality could be calculated for each complete cycle of shell $\delta^{18}O$. This results in amplitudes of 19.7° and 16.9°C for sample A and 21.6° and 10.8°C for sample C. In samples B, D, and E, amplitudes greater than 6 ‰ would correspond to thermal seasonalities up to 46°C, temperatures clearly exceeding the tolerance of unionids, which in the Huron River today grow between 12° and 28°C. In these samples, variation in $\delta^{18}O$ of shell carbonate must represent a combination of temperature and water effects.

Both the amplitude and mean values of some Paleogene unionids require waters highly depleted in ^{18}O and likely reflect the presence of snow. Snow $\delta^{18}O$ is often in excess of 10 ‰ more negative than mean annual rainfall for a region (for example, I. A. E. A. precipitation data for Chicago, IL and Gimli, Manitoba, [I. A. E. A. / W. M. O. Database, 1991]) and could create the amplitudes seen in samples B, D, and E. The very negative values of bivalve C are also a very strong indication of snow runoff; however, the consistently negative shell carbonate $\delta^{18}O$, where all values are less than -18.4 ‰ PDB, requires a special setting. If the range of possible water isotope compositions are considered over a temperature range of 0°C and 30°C, ambient water $\delta^{18}O$ must have remained quite negative throughout the year. The temperature - aragonite $\delta^{18}O$ relationship places shell-$\delta^{18}O_{PDB}$ equal to water-$\delta^{18}O_{SMOW}$ at 20.9°C and for every 4.7°C deviation from 20.9°C, a 1 ‰ difference is present between shell-$\delta^{18}O$ and water-$\delta^{18}O$. Therefore for any temperature between 0 and 30°C, the water-$\delta^{18}O_{SMOW}$ must lie within a range from 4.4 ‰ more negative than shell-$\delta^{18}O_{PDB}$ to 1.9 ‰ more positive than shell-$\delta^{18}O_{PDB}$.

Given the range of carbonate $\delta^{18}O$ of sample C (Fig. 1), the water in which this bivalve lived must have been less than -16.8 ‰ $\delta^{18}O_{SMOW}$ throughout the year. These negative $\delta^{18}O$ waters could not have had their source in local snowfall because snow melt waters derived within the basin would not have warmed above the hibernation threshold. If waters depleted in ^{18}O originated from snowfall, then these waters must have been sourced in adjacent highlands (i.e. alpine glaciers), whose melt waters warmed as they flowed across the basin. This is supported by paleoflow directions of the Paleocene rivers of the Powder River Basin, which indicate that rivers crossed the basin from the Big Horn mountains in the west to the eastern central area of the basin before turning and draining to the north [Flores, 1986]. Such a mechanism could provide isotopically negative waters, derived from high altitude precipitation, to fluvial systems in the eastern and central parts of the basin. Note that warm weather rains within the basin must not have been sufficient to mask the snow melt signal present in the shell carbonate of sample C.

This scenario is consistent with other samples which show large cycles in shell $\delta^{18}O$. Seasonal snowfall and seasonal variation in the mixing of high altitude precipitation with local precipitation within the basin could set up strong cycles in the $\delta^{18}O$ of surface waters. Annual shell $\delta^{18}O$ cycles with most positive values in the -6 to -11 range are compatible with warm season waters of -4 to -9 ‰ SMOW, values not unexpected for summer precipitation in continental interiors. The model based on Amazon Basin precipitation patterns, however, could also apply to these samples. Precipitation at Benjamin Constant varies by more than 10 ‰ $\delta^{18}O_{SMOW}$ annually, with little temperature change. While such environmental conditions could contribute to the annual cycles in shell $\delta^{18}O$, the $\delta^{18}O$ values of surface waters required for sample C are much more negative than the most negative waters recorded in the Amazon Basin or any other warm temperature precipitation. During seasons where shell growth records the most negative $\delta^{18}O$ values, waters must be less than -21 ‰ SMOW (applying the +1.9 ‰ temperature offset at 30°C discussed above). High altitude and/or cold temperature precipitation is the only source of precipitation less than -20 ‰ in the I. A. E. A. database [I. A. E. A. / W. M. O. Database, 1991].

Although it is not possible to uniquely extract intra-annual paleotemperatures from $\delta^{18}O$ variation in these Late Paleocene and Early Eocene bivalves, the data require that large changes in surface water $\delta^{18}O$ must have occurred. Paleoclimate systems proposed for the Paleogene interior of North America must accommodate very negative $\delta^{18}O$ values for surface waters, large seasonal changes in the isotopic composition of these waters, and therefore the presence of snow in highland areas.

Long term variations in surface water $\delta^{18}O$ are suggested by the highly negative bulk carbonate values obtained for many of the Late Paleocene bivalves (Fig. 3). In contrast to the fine-scaled sampling employed to discern intra-annual variation, bulk carbonate $\delta^{18}O$ measurements can be used to constrain average $\delta^{18}O$ of surface waters. The most conservative estimate of water $\delta^{18}O$ assumes a range of temperature for shell growth from 0 to 30°C. Under these conditions, the water could range from 4.4 ‰ more negative to 1.9 ‰ SMOW more positive than the shell carbonate $\delta^{18}O_{PDB}$. However, these estimates can be further refined by considering the bulk carbonate precipitated during each year as an integration of shell growth across a spectrum of seasonal temperatures. Bivalves tend to grow more rapidly in warmer water than in colder water up to a point where an optimal growth temperature is reached [Bayne and Newell, 1983]. The weighted average growth temperature based on the quantity of shell produced at each temperature should be within the range of 12 to 30°C, given that Huron River unionids become dormant below temperatures of 12°C and that unionids show significant reduction in growth rate above 30°C. Because shell-$\delta^{18}O_{PDB}$ = water-$\delta^{18}O_{SMOW}$ at 20.9° C, and a 4.7°C change is reflected as a 1 ‰ change in shell $\delta^{18}O$, average shell $\delta^{18}O$ values precipitated over a range of 12 to 30°C must equal water $\delta^{18}O$ values within ±2 ‰. Therefore, bulk shell-$\delta^{18}O_{PDB}$ values are proxies of surface water-$\delta^{18}O_{SMOW}$ within ± 2 ‰. This can be seen in bulk shell $\delta^{18}O$ values of the Huron River bivalves which average -7.3 ‰ PDB while water $\delta^{18}O$ varies from -8.6 to -6 ‰ SMOW. Bulk shell $\delta^{18}O$ measurements indicate that Early Eocene surface waters varied from -16 to -8.5 ‰ SMOW (±2) and Late Paleocene surface waters ranged from -23 to -8 ‰ SMOW (±2).

Bulk shell $\delta^{18}O$ and $\delta^{13}C$ vary significantly through the stratigraphic interval examined for this study (Fig. 3). It is clear that several bivalves throughout the Paleogene section possess highly negative shell $\delta^{18}O$ values which suggest that snow melt waters were present at numerous times during the evolution of the basin. While more data are necessary to refine the pattern of change across the Paleocene-Eocene transition, there is a significant reduction in the variability of shell $\delta^{18}O$ and, by implication, in $\delta^{18}O$ of surface waters present during the deposition of the Early Eocene Wasatch Formation.

CONCLUSIONS

Measurement of fine scale $\delta^{18}O$ and $\delta^{13}C$ variation in Paleogene fresh-water bivalves provides a record of monthly to fortnightly change in carbonate $\delta^{18}O$ composition reflecting changes in surface water temperature and $\delta^{18}O$ composition. If seasonal variability can be classified as either temperature dominated or water $\delta^{18}O$ dominated with respect to environmental effects on shell $\delta^{18}O$, then minimum seasonal change in temperature or in water $\delta^{18}O$ can be directly calculated from the amplitudes of $\delta^{18}O$ change recorded in the shell. Where the amplitude of shell $\delta^{18}O$ varies by less than 6 ‰ PDB, the factors controlling seasonal change in shell $\delta^{18}O$ can not be uniquely determined. However, in samples which have amplitudes greater than 6 ‰, large seasonal changes in water $\delta^{18}O$ must have occurred.

Bulk shell $\delta^{18}O$ measurements in the Late Paleocene Powder River Basin reflect a number of events in which very negative (<-20 ‰ SMOW) surface waters were present in the basin. Unionids of Early Eocene age indicate that some surface waters had isotopic compositions less than -14 ‰, although no evidence exists for surface waters <-20 ‰, as is seen in several unionids of Late Paleocene age.

Acknowledgments. We would like to thank William A. Cobban and Bob O'Donnell for access to the collections and notes of the late John H. Hanley. Bruce H. Wilkinson made numerous suggestions that improved this manuscript. This project was supported by the National Science Foundation [Grant No. EAR91-05903 (KCL) and a NSF Graduate Fellowship (DLD)] and the University of Michigan Scott Turner Fund.

REFERENCES

Aharon, P., Recorders of reef environment histories: stable isotopes in corals, giant clams, and calcareous algae, *Coral Reefs*, *10*, pp. 71-90, 1991.

Barron, E. J., Eocene Equator-to-pole surface ocean temperatures: a significant climate problem?, *Paleoceanography*, *2*, pp. 729-739, 1987.

Barron, E. J., Explanations of the Tertiary global cooling trend, *Palaeogeography, Palaeoclimatology, Palaeoecology*, *50*, pp. 45-61, 1985.

Bayne, B. L., and R. C. Newell, Physiological Energetics of Marine Molluscs, in *The Mollusca, Vol. 4:Physiology, Part 1*, edited by A. S. M. Saleuddin and K. M. Wilbur, pp. 407-515, Academic Press, New York, 1983.

Bryson, R. P., and N. W. Bass, *Geology of Moorhead coal field, Powder River, Big Horn, and Rosebud counties, Montana*, U.S.G.S. Bulletin 1338, Washington, DC, 1973.

Crowley, T. J., D. A. Short, J. G. Mengel, and G. R. North, Role of seasonality in the evolution of climate during the last 100 million years, *Science*, *231*, pp. 579-584, 1986.

Culbertson, W. C., M. C. Klett, *Preliminary geologic map and coal sections of the SR Springs quadrangle, Sheridan County, Wyoming*, 1:24,000 Miscellaneous Field Studies Map, MF-727, U. S. G. S., 1975.

Day, M. E., The shell as a recording device: growth record and shell ultrastructure of *Lampsilis radiata radiata* (Pelecypoda: Unionidae), *Canadian Journal of Zoology*, *62*, pp. 2495-2504, 1984.

Emiliani, C., L. Cardini, T. Mayeda, C. McBurney, and E. Tongiorgi, Paleotemperature analysis of fossil shells of marine mollusks (food refuse) from the Arene Candide Cave, Italy and the Jaua Fteah Cave, Cyrenaica, in *Isotopic and Cosmic Chemistry*, edited by H. Craig, S. Miller, and G. Wasserburg, pp. 133-156, North-Holland Publishing Co., 1964.

Flores, R. M., Styles of coal deposition in Tertiary alluvial deposits, Powder River Basin, Montana and Wyoming, in *Paleoenvironmental and tectonic controls in coal-forming basins in the United States*, edited by P. C. Lyons, and C. L. Rice, pp. 79-104, Special Paper 210, Geological Society of America, 1986.

Fox, R. C., The succession of Paleocene mammals in western Canada, in *Dawn of the age of mammals in the northern part of the Rocky Mountain Interior, North America*, edited by T. M. Bown, and K. D. Rose, pp. 51-70, Special Paper 243, Geological Society of America, 1990.

Gingerich, P. D., *New earliest Wasatch mammalian fauna from the Eocene of northwestern Wyoming: composition and diversity in a rarely sampled high-floodplain assemblage*, 97pp., Papers on Paleontology No. 28, University of Michigan, 1989.

Grootes, P. M., M. Stuiver, L. G. Thompson, and E. Mosley- Thompson, Oxygen isotope changes in tropical ice, Quelccaya, Peru, *Journal of Geophysical Research*, *94*, pp. 1187- 1194, 1989.

Grossman, E. L., and T.-L. Ku, Oxygen and carbon isotope fractionation in biogenic aragonite: temperature effects, *Chemical Geology (Isotope Geosciences Section)*, *59*, pp. 59-74, 1986.

Hanley, J. H., and R. M. Flores, Taphonomy and paleoecology of nonmarine mollusca: indicators of alluvial plain lacustrine sedimentation, upper part of the Tongue River Member, Fort Union Formation (Paleocene), northern Powder River Basin, Wyoming and Montana, *Palaios*, *2*, pp. 479-496, 1987.

Harrington, R. J., Aspects of growth deceleration in bivalves: clues to understanding the seasonal $\delta^{18}O$ and $\delta^{13}C$ record - a comment on Krantz et al. (1987), *Palaeogeography, Palaeoclimatology, Palaeoecology*, *70*, pp. 399-407, 1989.

Hickey, L. J., Paleocene stratigraphy and flora of the Clark's Fork Basin, in *Early Cenozoic paleontology and stratigraphy of the Bighorn Basin, Wyoming*, edited by P. D. Gingerich, pp. 33-50, Papers on Paleontology No. 24, University of Michigan, 1980.

I.A.E.A./W.M.O., *Isotopes in Precipitation*, Database, on Diskette in spreadsheet format, 1991.

Jones, D. S., D. F. Williams, and M. A. Arthur, Growth history and ecology of the Atlantic surf clam, *Spisula solidissima* (Dillwyn), as revealed by stable isotopes and annual shell increments, *Journal of Experimental Marine Biology and Ecology*, *73*, pp. 225-242, 1983.

Kalish, J. M., Oxygen and carbon stable isotopes in the otoliths of wild and laboratory-reared Australian Salmon (*Arripis trutta*), *Marine Biology*, *110*, pp. 37-47, 1991.

Koch, P. L., D. C. Fisher, and D. L. Dettman, Oxygen isotope variation in the tusks of extinct proboscideans: a measure of season of death and seasonality, *Geology*, *17*, pp. 515-519, 1989.

Krantz, D. E., D. F. Williams, and D. S. Jones, Ecological and paleoenvironmental information using stable isotope profiles from living and fossil molluscs, *Palaeogeography, Palaeoclimatology, Palaeoecology*, *58*, pp. 249-266, 1987.

Krause, D. W., and M. C. Maas, The biogeographic origins of late Paleocene - early Eocene mammalian immigrants to the Western Interior of North America, in *Dawn of the age of mammals in the northern part of the Rocky Mountain Interior, North America*, edited by T. M. Bown, and K. D. Rose, pp. 71-105, Special Paper 243, Geological Society of America, 1990.

Matsui, E., F. Salati, I. Friedman, and W. L. F. Brinkman, Isotopic hydrology in the Amazonia: 2. relative discharges of the Negro and Solimões rivers through $\delta^{18}O$ concentrations, *Water Resources Research*, *12*, pp. 781-785, 1976.

Miller, K. G., R. G. Fairbanks, and G. S. Mountain, Tertiary oxygen isotope synthesis, sea-level history, and continental margin erosion, *Paleoceanography*, *2*, pp. 1-19, 1987.

Neves, R. J., and S. N. Moyer, Evaluation of techniques for age determination of freshwater mussels (Unionidae), *American Malacological Bulletin*, *6*, pp. 179-188, 1988.

Olive, W. W., *The spotted horse coalfield, Sheridan and Campbell counties, Wyoming*, U.S.G.S. Bulletin 1050, Washington, DC, 1957.

Pätzold, J., Growth rhythms recorded in stable isotopes and density bands in the reef coral *Porites lobata* (Cebu, Philippines), *Coral Reefs*, *3*, pp. 87-90, 1984.

Reis, C. M., A. C. F. N. S. Tancredi, E. Matsui, and E. Salati, Caracterização das águas da região de Marajó através de concentrações de O-18 e D, *Acta Amazonica*, *7*, pp. 209-222, 1977.

Romanek, C. S., and E. L. Grossman, Stable isotope profiles of *Tridacna maxima* as environmental indicators, *Palaios*, *4*, pp. 402-413, 1989.

Rose, K. D., Composition and species diversity in Paleocene and Eocene mammal assemblages: an empirical study, *Journal of Vertebrate Paleontology*, *1*, pp. 367-388, 1981.

Salati, E., E. Matsui, A. Cervellini, and I. Friedman, Isotope Hydrology in Amazonia Part 1: $\delta^{18}O$ and δD in river waters, *Energ. Nucl. Agric.*, *2*, pp. 101-132, 1980.

Salati, E., A. Dall'Olio, E. Matsui, and J. L. Gat, Recycling of water in the Amazon Basin: an isotopic study, *Water Resources Research*, *15*, pp. 1250-1258, 1979.

Salati, E., and J. Marques, Climatology of the Amazon region, in *The Amazon, limnology and landscape ecology of a mighty tropical river and its basin*, edited by H. Sioli, pp. 85- 126, Dr. W. Junk Publishers, Boston, 1984.

Savin, S. M., The history of the earth's surface temperature during the past 100 million years, in *Annual Review of Earth and Planetary Sciences*, pp. 319-355, Vol. 5, 1977.

Sloan, L. C., and E. J. Barron, A comparison of Eocene climate model results to quantified paleoclimatic interpretations, *Palaeogeography, Palaeoclimatology, Palaeoecology*, *93*, pp. 183-202, 1992.

Sommer, S. E., Cathodoluminescence of carbonates, 2. Geological Applications, *Chemical Geology*, *9*, pp. 275-284, 1972.

Vokes, H. E., *Genera of the bivalva: a systematic and bibliographic catalogue (revised and updated)*, Paleontological Research Institution, Ithaca, New York, 1980.

Wing, S. L., Eocene and Oligocene floras and vegetation of the Rocky Mountains, *Annals of the Missouri Botanical Gardens*, *74*, pp. 748-784, 1987.

Wing, S. L., T. M. Bown, and J. D. Obradovich, Early Eocene biotic and climatic change in interior western North America, *Geology*, *19*, pp. 1189-1192, 1991.

Wolfe, J. A., Late Cretaceous-Cenozoic history of deciduousness and the terminal Cretaceous event, *Paleobiology*, *13*, pp. 215-226, 1987.

Wolfe, J. A., and R. Z. Poore, Tertiary marine and nonmarine climatic trends, in *Climate in earth history*, edited by Geophysics Study Committee, pp. 154-158, National Academy Press, Washington, DC, 1982.

D. L. Dettman and K. C Lohmann, University of Michigan, Department of Geological Sciences, 1006 C. C. Little Bldg., Ann Arbor, Michigan, 48019-1063.

Late Pleistocene Climate in the Central American Lowlands

Department of Geology, University of South Florida, Tampa, FL

MARK BRENNER

Department of Fisheries and Aquatic Sciences, University of Florida, Gainesville, FL

DAVID A. HODELL
JASON H. CURTIS

Department of Geology, University of Florida, Gainesville, FL

Pollen, sedimentologic, and isotopic data from the basal 12 m of a 19.6-m core taken in Lake Quexil, Guatemala, were used to infer late Pleistocene-early Holocene climatic conditions in the lowlands of Central America. The Pleistocene record is divided into three stages that coincide with changes in continental ice volume as reflected in the marine oxygen isotopic signal. Interstadial Stage 3 (~36 to 24 kyr BP) was characterized by cooler temperatures (4.7-6.5 °C) and somewhat drier climate than present. During Stage 2 (~24 to 12 kyr BP), all proxy indicators of climate suggest further temperature decline (6.5-8.0 °C < present) and extremely arid conditions. Lowest lake levels are inferred during this period. With the onset of deglaciation (~12.5 kyr BP), dry conditions began to ameliorate and lake level began to rise. The Pleistocene-Holocene transition (after 10.5 kyr BP) was marked by rapid filling of Lake Quexil under lingering cool conditions (3.0-4.7 °C < present). By ~9 kyr BP, warm and moist conditions were established and persisted throughout the early Holocene.

INTRODUCTION

Proxy climatic indicators from low-latitude lakes are rich sources of information about continental, tropical climate change since the late Pleistocene. Numerous studies on sediment cores from African lakes have provided detailed paleoclimatic reconstructions spanning the period from the last Glacial to present. Unfortunately, lacustrine records of Pleistocene age are scarce from the Neotropics and especially rare for the Circum-Caribbean region. A few late Pleistocene sediment sequences have been recovered from Florida [Watts, 1975; Watts and Stuiver, 1980; Watts and Hansen, 1989], the Guyanas [van der Hammen, 1974], Venezuela [Bradbury et al., 1981; Binford, 1982; Leyden, 1985], Haiti [Hodell et al., 1991], Panama [Bartlett and Barghoorn, 1973; Bush and Colinvaux, 1990; Piperno et al., 1990], and Costa Rica [Martin, 1964]. Here we present results from the basal 12 m of a 19.6-m core taken in Lake Quexil, Peten, Guatemala. This core is unique in that it was recovered from a low-elevation lake

(110 m above sea level), it possesses a near-continuous record of sedimentation spanning the last ~36 kyr, and it was sampled in detail for pollen, sedimentologic, and stable isotopic analyses. We focus on the late Pleistocene from 27 to 10 kyr BP and use palynological and geochemical data to interpret the climatic history of the Neotropical lowlands for this period.

Global climate models require input of proxy data from both continental and marine records. In some cases, these two sources of information yield disparate climatic inferences. For example, estimates of SST [CLIMAP, 1976, 1981] used to reconstruct climate during the last Ice Age fail to explain the magnitude of temperature depression and increased aridity inferred from continental records [Rind and Peteet, 1985]. In this paper, we use proxy data from Lake Quexil to infer relative estimates of temperature change and moisture availability during the last climatic cycle. We conclude that temperature depression during the last Ice Age was 6.5 to 8.0°C in the lowlands of Central America and that climate was extremely arid. We discuss the ramifications of our results with respect to the magnitude of temperature change in the tropics during the last Ice Age. In addition, our findings of increased aridity may have implications for changes in atmospheric vapor transport between the Atlantic and Pacific during the last glacial.

Climate Change in Continental Isotopic Records
Geophysical Monograph 78

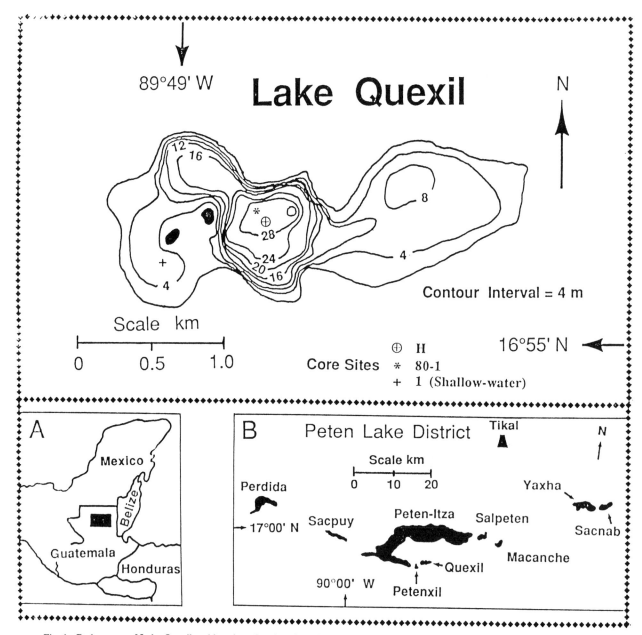

Fig. 1. Bathymetry of Lake Quexil and location of coring sites. Data from shallow-water Core 1 showed that areas beyond the deep central sink first flooded in the Holocene [Deevey et al. 1979]. Deepwater Core H provided a near-complete Holocene sediment record [Brenner 1983, Vaughan et al. 1985]. The terminal Pleistocene-early Holocene record from Core 80-1 has been reported [Leyden 1984]. Inset A) Location of the Peten Lake District in northern Guatemala. Inset B) Site of Lake Quexil..

STUDY AREA

The Peten lowlands of Guatemala are characterized by deeply dissected karst, developed on Cretaceous and Tertiary limestones that lie between 100 and 300 m above msl [West, 1964; Vinson, 1962]. Mollisols are the predominant soil type in the area, but clay-rich savanna soils occupy about 10% of the landscape [Simmons et al., 1959]. Lowlands extend northward across the entire Yucatan Peninsula, but the isolated Maya Mountains rise more than 600 m above msl about 60 km to the east. Extensive highlands to the south and west are more than 100 km away. The central Peten lake district contains a series of deep, closed basins along *en echelon* faults [Deevey et al., 1979]. Lake Quexil lies in the center of the lake district at 110 m above msl. The lake (Fig. 1) has an area of 2.1 km^2 and maximum depth of 32 m [Deevey et al., 1980]. Modern waters of Lake Quexil are dilute (103.9 mg L^{-1} TDS) and ionic composition is dominated by calcium and bicarbonate [Deevey et al., 1980].

Annual precipitation in the Peten averages 1600 mm, and the

mean annual temperature for the region is 25°C [Deevey, 1978]. In Central America, lapse rates below 1500 m average 5.9°C km⁻¹ [Schwerdtfeger, 1976]. Peten's climate is seasonal with a dry period lasting from January to May, and the regional landscape supports tropical semi-evergreen forest rather than true rainforest [Lundell, 1937]. *Brosimum alicastrum* ("ramon") is a dominant tree in most associations, and contributes about 40-60% of the modern pollen rain in the lake district [Vaughan et al., 1985; Leyden, 1987]. Grasslands with *Curatella* and *Byrsonima* occur south of the central lake district, and a small savanna with *Quercus oleoides* (oak) lies adjacent to Lake Quexil. The tropical *Pinus caribaea* (pine) is calciphobic [Perry, 1991], and was probably never a significant component of Peten vegetation. Lowland pine-oak associations with Melastomataceae, *Miconia*, and Ericaceae occur in Belize on less calcareous soils [Romney, 1959]. Mesic temperate forests with *Alfaroa* occur at >1200 m in Huehuetenango, Guatemala, more than 200 km to the south [Standley and Steyermark, 1958]. Subhumid temperate pine-oak forests are found above 1300 m in Chiapas, Mexico, about 250 km away [Breedlove, 1973]. Temperate thorn scrub with *Juniperus* and the non-calciphobic *Pinus cembroides* occurs above 1500 m at greater distance in Veracruz, Mexico [Gomez-Pompa, 1973].

Previous paleoecological investigations in the Maya Lowlands focused on impacts of prehistoric, human land use in a tropical karst environment [Cowgill et al., 1966; Deevey, 1978; Deevey and Rice, 1980; Deevey et al., 1979; Brenner, 1983; Brenner et al., 1990; Binford, 1983; Rice et al., 1983, 1985; Vaughan et al., 1985; Leyden, 1987; Bradbury et al., 1990; Hansen, 1990]. Near-complete Holocene sediment sequences were recovered from several Peten lakes, but only two basins (Lakes Quexil and Salpeten) yielded lacustrine deposits of Pleistocene age [Deevey et al., 1983]. These ancient deposits provided some of the first insights into climatic and ecological changes at the terminal Pleistocene/early Holocene boundary [Leyden, 1984, 1987]. In this study we report data from Lake Quexil, and extend the temporal scope of the Peten paleoclimatic record beyond 27 kyr BP.

METHODS

Field Work

Near-complete Holocene sections were raised from Lake Quexil in the 1970s, using a Livingstone piston corer or modified Kullenberg gravity corer [Brenner, 1983; Vaughan et al., 1985]. In order to extend the record into the late Pleistocene, we collected a 12-m core (Quexil 80-1) in 29 m of water (Fig. 1) that contains sediment from the interval between 7.6 m and 19.6 m below the lake bottom. We were assisted in the field by a Guatemalan well drilling company, Daho Pozos, S.A. Samples were collected in sections using a 51-cm-long, split-spoon corer. The apparatus possesses no piston, but has a 5-cm diameter, hollow steel barrel that was lined with a 51-cm long, 1/8"-wall acrylic tube. On retrieval of each section, the plastic liner was removed and the tube ends sealed for storage and shipment. Iron casing pipe was set into the lake bottom to guarantee return of the drill stem into the same bore hole on successive drives.

Twenty-one soil pits ("pozos") were excavated at selected locations in the Quexil watershed (Fig. 2). Pits were dug to maximum depths ranging from 30 to 100 cm. With the exception of hydromorphic and clay-rich savanna soils from sites 1, 5, and 8-11, pits typically bottomed on carbonate bedrock or carbonate-rich "sascab." Samples were removed from the exposed profiles at 10-

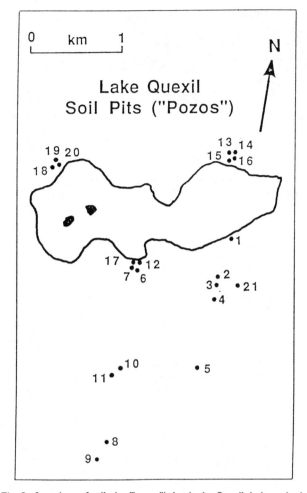

Fig. 2. Locations of soil pits ("pozos") dug in the Quexil drainage basin. δ¹³C analyses were run on soil carbonates in samples taken from the base of the soil profiles. Clayey samples from pits 1, 5, and 8-11 were not analyzed because of their low carbonate content.

cm intervals, each sample representing a composite of soil collected over the full 10 cm. Samples for isotopic analysis were chosen from the carbonate-rich bottoms of the profiles.

Radiocarbon Dating

Four radiocarbon dates were obtained from the Quexil 80-1 core (Table 1). Three samples were dated between 8.94-9.01 m, near the Pleistocene-Holocene boundary. They include a conventional ¹⁴C date on wood, and paired AMS measurements on terrestrial wood and mixed gastropod shells (*Biomphalaria* sp., *Pyrgophorus* sp., *Stenophysa* sp.). The fourth date is an AMS measurement on *Biomphalaria* sp. shells from 15.96 m depth.

Pollen and Charcoal Analysis

One-cm³ subsamples were processed by chemical extraction [Whitehead, 1981], and mounted in silicone oil. Entire slides were scanned at 400x for minimum counts of 200 grains. Pollen identifications were based on slide references and published accounts. Pollen slides were also scanned to estimate the density

Table 1. Radiocarbon dates from Quexil Core 80-1.

Depth (m)	Material	Accession No.	Method	^{14}C Age (years BP)
8.94-9.01	Wood	SI-5257	Conventional	10,750 ± 460
8.95	Wood	AA-3062	AMS	10,300 ± 110
8.95	Shells	AA-3063	AMS	10,630 ± 110
15.96	Shells	AA-3064	AMS	27,450 ± 500

of charred particles (cm^2 per cm^3 of wet sediment). Charcoal content was estimated using the point count method of Clark [1982].

Geochemistry and Mineralogy

Total carbon was measured with a LECO induction furnace. Inorganic carbon was assessed with a UIC/Coulometrics Model 5011 coulometer [Huffman, 1977] in conjunction with a System 140 preparation line that uses 2N HCl to evolve CO$_2$ from carbonates. Total sulfur was measured turbidimetrically [APHA, 1975] on a Model 14 Coleman spectrophotometer following digestion in a 2:1 nitric acid-perchloric acid mixture. Iron content in the digestate was assessed by atomic absorption using standard techniques. Calcium content in acid-digested soils was also measured by atomic absorption spectrophotometry.

X-ray diffraction analyses on whole sediment samples were done following air drying and grinding to a fine powder. Samples

were scanned using a Cu target tube at 45 kV, 30 mA, with steps of 0.02° and count time of 0.20 sec, from 3° to 60° using a Philips APD 3600. Samples were dispersed with calgon, shaken, and allowed to stand for 2 hr. Clay fractions (<4μ) were collected from 10 cm depth in settling cylinders and were air dried on glass slides. X-ray scans were made on untreated material and on glycol-saturated samples, after heating to 100°C, 300°C and 600°C. Scans were done at 45 kV, 30 mA, with steps of 0.02° at 0.20 sec, from 2° to 32° using the Philips APD 3600.

Stable Isotopes

Oxygen and carbon isotopic ratios were measured on bulk carbonate in Core 80-1 from Lake Quexil and in modern soil carbonates surrounding the lake. Samples were treated with 15% hydrogen peroxide to oxidize organic carbon and reacted in a common acid bath of orthophosphoric acid at 90°C using a VG Isogas autocarbonate preparation system. Isotopic ratios of purified CO$_2$ gas were measured on-line with a triple-collector VG Isogas PRISM Series II mass spectrometer. Isotopic results are expressed in standard delta notation relative to PDB (Tables 2 & 3). Analytical precision, based on routine analysis of an internal working standard (Carrara Marble), was ± 0.07‰ for δ^{18}O and ± 0.03‰ for δ^{13}C.

RESULTS

Chronology

Datable organic carbon is scarce in the Pleistocene section of the Quexil 80-1 core. Furthermore, both lacustrine organic matter and carbonates in bulk lake sediments from karst districts are

Table 2. Carbonate and stable isotopic results from Quexil Core 80-1

Depth	%CaCO$_3$	δ^{18}O	δ^{13}C	Depth	%CaCO$_3$	δ^{18}O	δ^{13}C
7.60	5.2			13.65	12.8	1.61	-5.71
7.75	0.0			13.80	10.6	1.37	-4.89
7.95	0.0			14.00	12.2	1.75	-5.20
8.20	0.1			14.20	12.9	1.93	-6.12
8.40	0.2			14.40	16.7	1.65	-4.13
8.65	23.2	1.86	-0.02	14.70	6.4		
8.85	73.7			15.00	35.9	2.13	2.69
9.05	25.7	0.69	2.58	15.20	16.4	2.74	1.44
9.20	27.6	0.80	2.25	15.40	0.0		
9.40	25.6	0.90	1.81	15.55	0.0		
9.65	25.0	1.33	0.69	15.80	33.5	2.59	2.29
9.80	25.2	1.30	-1.18	16.00	24.7	0.99	2.30
10.00	18.2	2.66	-7.34	16.20	25.8	1.25	1.01
10.20	20.3			16.40	23.3	1.81	0.21
10.40	17.1	3.60	-6.45	16.55	31.1	1.64	3.09
10.60	14.9	3.66	-12.22	16.80	14.8	1.42	0.57
10.80	11.8	2.14	-4.71	17.00	12.0	0.77	-0.55
11.00	4.8			17.25	15.5		
11.20	19.5	3.08	-9.18	17.40	22.4	1.66	3.45
11.40	2.2			17.55	21.9	1.45	1.38
11.65	10.9	2.00	-5.35	17.80	10.7	0.42	-1.68
11.80	2.5			18.00	16.2	1.48	-0.17
12.00	4.7			18.25	14.8	1.64	0.14
12.20	3.2			18.45	15.9	1.27	0.32
12.50	6.2			18.60	19.4	1.47	-0.26
12.80	7.1			18.80	12.6	1.34	-0.51
13.00	13.2	2.42	-4.69	19.00	33.5	1.44	-0.56
13.20	6.8			19.30	19.9	0.75	-0.25
13.40	14.3	2.25	-6.06	19.60	23.4	1.67	-1.95

Table 3. Stable isotope results of carbonates in soils surrounding Lake Quexil. Locations of soil pits are shown in Figure 2.

Sample #	Depth (cm)	$\delta^{18}O$ (per mil)	$\delta^{13}C$
Pozo 2	60-70	-4.83	-8.94
Pozo 3	40-50	-5.02	-8.21
Pozo 4	20-30	-4.74	-7.86
Pozo 6	60-70	-4.78	-10.42
Pozo 7	60-70	-4.71	-10.49
Pozo 12	50-60	-4.73	-9.27
Pozo 13	70-80	-5.13	-10.96
Pozo 14	60-70	-4.89	-10.40
Pozo 15	50-60	-4.69	-10.59
Pozo 16	90-100	-4.23	-10.22
Pozo 17	70-80	-5.10	-8.79
Pozo 18	90-100	-5.06	-8.17
Pozo 19	80-90	-4.58	-8.00
Pozo 20	70-80	-5.04	-8.23
Pozo 21	70-80	-4.94	-9.07
Mean		-4.83	-9.31
Standard deviation		±0.24	±1.10

susceptible to hard-water-lake error [Deevey and Stuiver, 1964]. Although we cannot rule out this effect on radiocarbon dates, paired carbonate (shell) and terrestrial wood samples from the Pleistocene/Holocene boundary yield nearly identical AMS ^{14}C ages, suggesting negligible hard-water-lake error (Table 1). Dating resolution is admittedly poor and based upon only two dated horizons. Although the chronology is preliminary, the two dated levels allow us to divide the record into three sections (>27 kyr BP, 27-10.5 kyr BP, and <10.5 kyr BP), which correlate roughly to the boundaries between isotopic Stages 3, 2, and 1. We interpret the results within the limitations of the chronology and confine discussion to gross differences in pollen, mineralogy, and stable isotopes between Interstadial Stage 3, Glacial Stage 2, and Interglacial Stage 1.

Pollen and Charcoal

The pollen profile is divided into four major zones with boundaries estimated at ~24, 14, and 10.5 kyr BP (Figs. 3 & 4). Pollen is abundant in sediments older than 24 kyr BP, whereas charred particles are sparse (Fig. 3). Pine, oak and *Alfaroa* predominate. *Juniperus*-type pollen is common and other temperate arboreal taxa are consistently present in small amounts (Fig. 4). Herbaceous taxa are moderately well represented. In contrast, diagnostic pollen of lowland rainforest taxa and modern associates of lowland pine-oak forest are absent. Aquatic indicator taxa are moderately represented (Figs. 3 & 4).

From about 24 to 14 kyr BP, pollen concentrations are very low (Fig. 3). Mesic temperate taxa disappear and the record becomes dominated by pollen of herbaceous plants (Fig. 4). The high percentages of pine pollen are attributed to long-distance transport and minimal local pollen production. Initially, Malvaceae and grasses increase in abundance and Cactaceae are present (Fig. 4). Phytoplankton disappear coincident with a greater representation of littoral and wetland indicators [e.g., Chenopodiaceae-Amaranthaceae (Cheno-Ams), ferns, and *Typha* (cattails); Figs. 3 & 4]. Subsequently, the pollen is almost entirely herbaceous and is associated with an increase in the abundance of charred particles (Figs. 3 & 4).

The pollen assemblage from about 14 to 10.5 kyr BP remains primarily herbaceous, but there are significant shifts in the dominant taxa. At ~14 kyr BP, oak pollen increases and pollen of temperate hardwoods, *Ulmus*, *Liquidambar*, and *Alfaroa* are present (Fig. 4). Percentages for grasses, ferns, *Typha*, and charcoal concentrations all decline (Figs. 3 & 4). The mesic temperate taxa are replaced between ~12 and 10 kyr BP by Cheno-Ams and Malvaceae (Fig. 4). Aquatic taxa are represented by the submergent *Potamogeton* (pondweed). Charred particle concentrations are sharply higher during this interval (Fig. 3).

After ~10.5 kyr BP, pollen concentrations increase dramatically (Fig. 3) and there is a shift to predominantly arboreal taxa (Fig. 4). Pine, oak, and temperate hardwoods are present initially, but are replaced by lowland rainforest taxa. This represents the first appearance of lowland mesic flora in the pollen record that extends back to ~36 kyr BP.

Geochemistry and Mineralogy

Pleistocene-age deposits contain little organic matter and are dominated by montmorillonite clays with some kaolinite or halloysite, the weathering products of local mollisols [Deevey et al., 1983]. The dominant non-clay components include calcite, gypsum, and quartz (Fig. 5). Based on geochemical and mineralogical analyses, the Quexil 80-1 core was divided into four zones (Fig. 6). Sediments older than 26 kyr BP possess high carbonate concentrations. Between about 26 and 12.5 kyr BP, the sediments are rich in gypsum. Calcite returns as the principal non-clay component between 12.5 and 10 kyr BP. At the Pleistocene/Holocene boundary (~10 kyr BP), the proportion of organic matter in the sediments increases dramatically.

Stable Isotopes

Oxygen isotopic values of bulk carbonate from the Lake Quexil core range from a minimum of 0.42‰ to a maximum of 3.66‰ (Fig. 7). In sediments older than ~27 kyr BP, $\delta^{18}O$ values of bulk carbonate are relatively low and average about 1.3‰. Near 27 kyr BP, mean $\delta^{18}O$ values increase by ~1‰ and remain high (mean = 2.4‰) until ~12.5 kyr BP. Maximum $\delta^{18}O$ values of 3.66‰ occur at ~14 kyr BP and are about 2.5‰ greater than mean values during Stage 3 or early Stage 1. This interval of high $\delta^{18}O$ correlates with the interval of gypsum precipitation (Fig. 7). From 14 to 11 kyr BP, oxygen isotopic values decrease abruptly by ~3‰ and remain low (mean = 1.2‰) during the early Holocene. This decrease in $\delta^{18}O$ values coincides with the resumption of calcite precipitation (Fig. 7).

Carbon isotopic values range from a low of -12.2‰ to a high of 3.45‰. In sediments older than 24 kyr BP, $\delta^{13}C$ values average about 0.7‰. At about 24 kyr BP, $\delta^{13}C$ values decrease abruptly by 7‰ and average -6.3‰ throughout the interval of gypsum precipitation (Fig. 7). At 12.5 kyr BP, $\delta^{13}C$ values increase abruptly by more than 8‰ and average +1‰ during the early Holocene.

In addition to the measurements on bulk carbonate from Quexil Core 80-1, we also measured stable isotopes in modern soil carbonates surrounding the lake to provide an estimate of the isotopic composition of detrital carbonate supplied to the lake. Calcium concentrations ranged from 19.4 to 46.0%. Oxygen isotopic values of soil carbonates ranged from -5.13 to -4.23‰ and averaged -4.83 ± 0.24‰ (Table 3). Carbon isotopic values of soil carbonates ranged from -10.96 to -8.00‰ and averaged -9.31 ± 1.10‰.

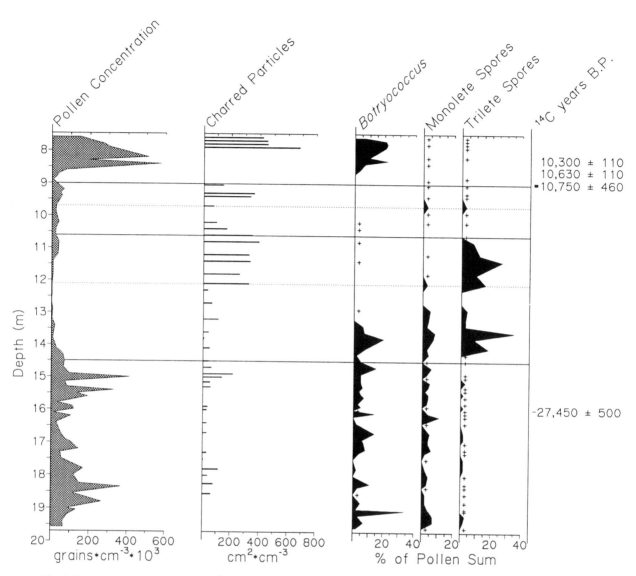

Fig. 3. Pollen concentration (grains * cm⁻³), charred particle abundance (cm² * cm⁻³), and percentages of algal (*Botryococcus*) and terrestrial fern spores for Lake Quexil Core 80-1. Algae and spores are expressed as a percent of the pollen sum. Formal pollen zones have not been assigned, but informal primary zones (solid lines) and secondary zones (dotted lines) are indicated.

DISCUSSION

General Climatic Context

We estimate that the sediments recovered in Quexil Core 80-1 span the period from ~36 kyr BP to the mid-Holocene. This interval includes the latter part of Interstadial Stage 3, all of Glacial Stage 2, and part of Interglacial Stage 1 as defined by the marine oxygen isotopic record. Interstadial Stage 3 was an intermediate period between full glacial and interglacial conditions when the climate was generally cool and ice volume was greater than today. The last glacial maximum occurred from ~22 to 14 kyr BP, with greatest ice volume recorded at ~18 kyr BP. Oxygen isotopic evidence suggests that the last deglaciation occurred in two steps, Termination IA and 1B [Duplessy et al., 1986]. Two discrete melting events are also indicated by sea level changes inferred from Barbados coral reef studies [Fairbanks, 1989]. The

two meltwater events and accompanying sea level rises were centered at 12.0 (melt-water pulse IA) and 9.5 kyr BP (melt-water pulse IB). The last deglaciation was interrupted by a brief cold period, the "Younger Dryas" event, that lasted from 11 to 10 kyr BP [Broecker et al., 1988].

Pollen

Prior to ~24 kyr BP during Stage 3, a humid to subhumid montane pine-oak forest existed in the Peten. *Alfaroa* grew in wetter areas, whereas *Juniperus*-type occupied more exposed and drier areas. This suggests a minimum depression of modern temperate forest zones by 800-1100 m, which yields a temperature decline of 4.7 to 6.5°C, assuming a lapse rate of 5.9°C km⁻¹ [Schwerdtfeger, 1976]. Oaks presently found near Lake Quexil

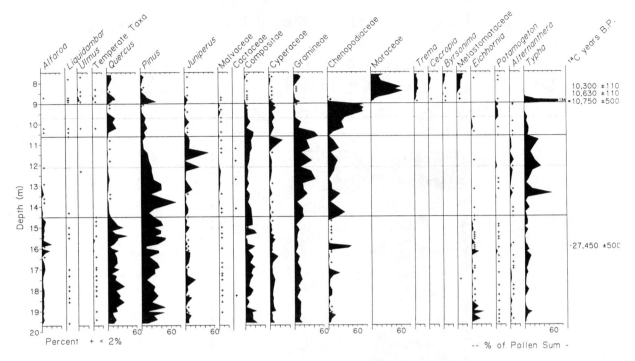

Fig. 4. Pollen percentage diagram of selected taxa for Quexil core 80-1. *Alfaroa, Liquidambar, Ulmus,* and Temperate Taxa are indicators of cool and moist conditions. The Temperate Taxa group includes *Betula, Cornus, Fraxinus, Hedyosmum, Ilex, Juglans,* Onagraceae, *Ostrya/Carpinus,* and *Rapanea*. Additional temperate and/or aridity indicators are presented next, followed by herbaceous taxa. Chenopodiaceae represent the Chenopodiaceae-Amaranthaceae pollen type. Warm tropical taxa, and aquatic taxa that have been excuded from the pollen sum complete the diagram. Formal pollen zone designations have not been assigned, but primary (solid lines) and secondary (dotted lines) zone boundaries are indicated.

may represent relict populations from this period [Gomez-Pompa, 1973].

Between about 24 and 14 kyr BP during Stage 2, sparse temperate thorn scrub replaced pine-oak forest (Fig. 4). A similar association now grows above 1500 m [Gomez-Pompa, 1973], suggesting that temperatures may have been 6.5 to 8°C colder than today during the last glacial. At the same time, aquatic indicators were replaced by marsh taxa, indicating that the lake may have been ephemeral. Greater concentrations of charred particles indicate that fires were more frequent during the latter part of this pollen zone (Fig. 3).

After ~14 kyr BP, temperate oak forest expanded, although the regional vegetation remained sparse and herbaceous. The central depression in the lake filled (Fig. 1), and was surrounded by a brackish community of Cheno-Ams as seen in other Neotropical lakes with extensive and fluctuating littoral zones [Leyden, 1985]. Prior to 10.5 kyr BP, there was a reversal of forest expansion, and an increased frequency of natural fires. By comparison, this period was more arid than the early Holocene, but not as dry as the last glacial maximum (LGM).

Lowland rainforest first appeared and became dominant in the early Holocene (Stage 1), although temperate taxa persisted while cooler temperatures prevailed [Leyden 1984, 1987]. The hypothesis that extensive, mesic refuges existed in this region during the LGM is untenable. Pollen from lowland rainforest taxa is below detectable levels for at least 17 kyr, and perhaps for more than 25 kyr (i.e. about 36 to 10.3 kyr BP), although small populations of mesic species must have persisted at isolated sites.

Geochemistry and Mineralogy

Calcite in Lake Quexil sediments has three possible origins: 1) erosion and transport of detrital carbonate from surrounding bedrock, 2) biogenic (shell) formation, and 3) authigenic precipitation. Lake Quexil lies in karsted lowlands, and some of the carbonate may be detrital. Nevertheless, detrital carbonate probably constituted a small fraction of the sediment during the Pleistocene. The carbonate content of glacial-age deposits is several times greater than concentrations measured for the late Holocene (3 to 0.4 kyr BP), when human disturbance caused rapid erosion [Binford et al., 1987]. A biogenic source for the calcite is improbable because shell material is scarce in the Pleistocene section of the core, with the exception of a 5-cm, shell-rich horizon near the Pleistocene-Holocene boundary. Finally, the presence of authigenic gypsum implies that calcite was also precipitated from the water column because gypsum solubility exceeds that of calcite. In Quexil's bicarbonate waters, calcite precipitation must have preceded gypsum precipitation. These arguments, combined with isotopic evidence presented below, suggest that Pleistocene carbonates are predominantly authigenic and were precipitated from solution under arid climatic conditions.

The stratigraphic succession of mineralogies in the core also supports evaporitic precipitation of calcite and gypsum from solution. The succession of non-clay mineralogies during the late Pleistocene represents an evaporitic sequence of calcite saturation during Stage 3, followed by gypsum saturation during maximum aridity associated with Stage 2, and back to calcite saturation with increased precipitation during deglaciation (Fig. 6). Across the

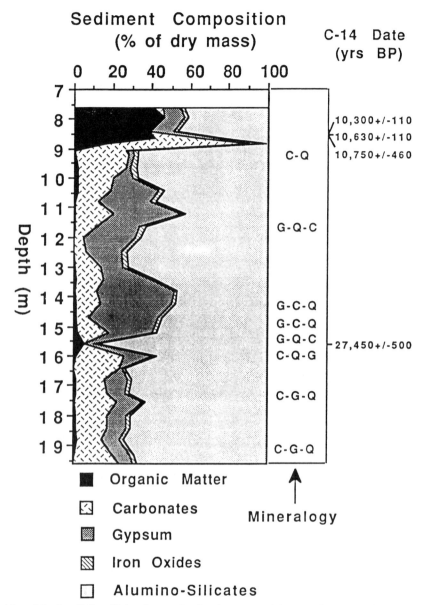

Fig. 5. Composition of the Quexil Core 80-1 sediments showing the percent of dry mass that each sediment constituent comprises. Organic carbon (OC) was calculated as total carbon (C_{tot}) minus inorganic carbon (IC). Organic matter was then figured as 2 x OC. Carbonate is the $CaCO_3$ equivalent of IC (i.e. 8.33 x IC). Gypsum is the $CaSO_4*2H_2O$ equivalent of total sulfur. Note that in the top meter of the section gypsum is overestimated, as most sulfur is probably bound in organic matter. Iron oxides represent the Fe_2O_3 equivalent of Fe content in the sediment. The sediment balance was figured by difference and is assumed to be primarily aluminosilicates (clays). Non-clay minerals C (calcite), G (gypsum), and Q (quartz) are shown in order of abundance.

Pleistocene/Holocene boundary, calcite abundance decreases markedly and reflects the cessation of evaporitic carbonate precipitation that accompanied the onset of wetter early Holocene conditions [Deevey et al., 1983; Leyden, 1984, 1987]. At this time, the organic content of the sediments rose dramatically (Figs. 5 & 6) as the lake filled and lacustrine productivity increased.

Stable Isotopes

Several lines of isotopic evidence suggest that stable isotopic measurements of bulk carbonate in Quexil Core 80-1 reflect authigenic calcite precipitation from lake water and that the influence of detrital carbonate was minimal. First, we measured the stable isotopic composition of carbonates from 15 soil pits

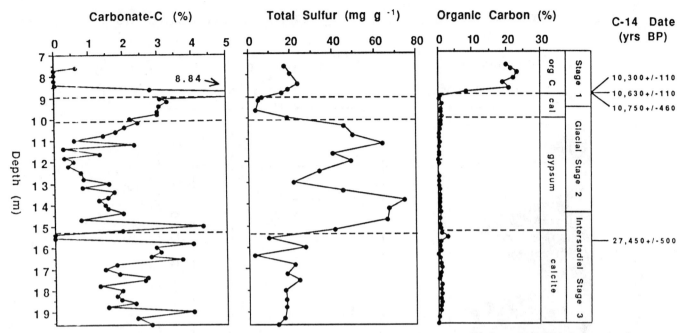

Fig. 6. Carbonate-carbon, total sulfur, and organic carbon in the Quexil 80-1 sediment core. The four principal lithologic zones of the core are labelled to the right of the organic carbon plot, and the zonal boundaries are indicated by dotted lines.

around the lake to derive an estimate of the expected isotopic values of detrital carbonate (Table 3). The average $\delta^{18}O$ from 15 soils around Lake Quexil was -4.83 ± 0.24‰ (1 standard deviation). Oxygen isotopic values of bulk carbonate from Quexil 80-1 range from 0.42 to 3.66‰ (Fig. 7), which is significantly greater than that expected if detrital carbonate constituted a substantial fraction of the sediment. Furthermore, the oxygen isotopic values measured from Quexil 80-1 are well within the range of values expected from calcite precipitated in equilibrium with lake water at a mean annual temperature of 25°C. Although we have no direct way to measure the $\delta^{18}O$ of water in Lake Quexil during the late Pleistocene, we estimate that it should be near +4 to +5‰ by means of comparison with Lake Chichancanab (Yucatan) to the north. Lake Chichancanab is near gypsum saturation today and may serve as a good analog for the late Pleistocene conditions of Lake Quexil. Calcite precipitated at 25°C in equilibrium with water having a $\delta^{18}O$ equal to +4‰ would have a $\delta^{18}O$ of about +2‰. Measured oxygen isotopic values of gastropods from Lake Chichancanab during the late Holocene range from 0.67 to 3.79‰, which is similar to the range observed during the late Pleistocene of Quexil Core 80-1 (Fig. 7). According to the geochemical and isotopic arguments outlined above, we assume that the stable isotopic measurements of bulk carbonate from Quexil 80-1 reflect authigenic calcite precipitation and that the influence of detrital carbonate has been minimal.

In closed basin lakes with seasonally dry climate, the $^{18}O/^{16}O$ ratio of lake water is controlled mainly by the ratio of evaporation to precipitation (E/P) [Gasse et al., 1990; Fontes and Gonfiantini, 1967], although changing temperature can also play a role. Periods of high evaporation and/or lower temperature should be marked by an increase in $^{18}O/^{16}O$ ratios of authigenic calcite. The 1‰ increase in $\delta^{18}O$ values at 27 kyr BP reflects a combination of increasing E/P and decreasing temperature (Fig. 7). The increase in $\delta^{18}O$ values at 27 kyr BP begins just before the onset of gypsum precipitation at ~24 kyr BP (Fig. 7) and the pollen change that indicates a shift from montane pine-oak forest to grassland (Fig. 3). Oxygen isotopic values remain high throughout Glacial Stage 2 with maximum values exceeding 3.5‰ (Fig. 7), which is about 2‰ greater than mean values during Stage 3 or the early part of Stage 1. These high $\delta^{18}O$ values reflect regionally arid conditions and cooler temperatures during the last Ice Age. Pollen evidence suggests that temperatures were about 1.5 to 1.8°C cooler during glacial Stage 2 than during interstadial Stage 3. The magnitude of this cooling could account for up to 0.5‰ of the $\delta^{18}O$ increase observed between Stages 3 and 2 with the remainder explained by increased evaporation under enhanced arid conditions. Upon deglaciation at ~12.5 kyr BP, $\delta^{18}O$ values decreased abruptly by up to 3‰ indicating climatic warming and a decrease in E/P (Fig. 7). At this time, gypsum deposition ceased and calcite precipitation resumed (Fig. 6), reflecting increased moisture availability.

The carbon isotopic composition of calcite precipitated from lake water reflects the $\delta^{13}C$ of dissolved inorganic carbon (DIC) at the time of precipitation. In turn, the carbon isotopic composition of lake DIC is controlled by the fluxes and isotopic composition of carbon inputs and outputs to and from the lake as well as gas exchange with the atmosphere. We consider two major sources of DIC to the lake. The first is carbon derived from the dissolution of marine limestone in the drainage basin, which should average about 1‰. The second major source of DIC to the lake is derived from the oxidation of organic carbon in plant biomass and soil humus. The $\delta^{13}C$ of CO_2 derived from the oxidation of organic carbon is highly depleted in ^{13}C, averaging -25‰ for C-3 plants and -12‰ for C-4 plants.

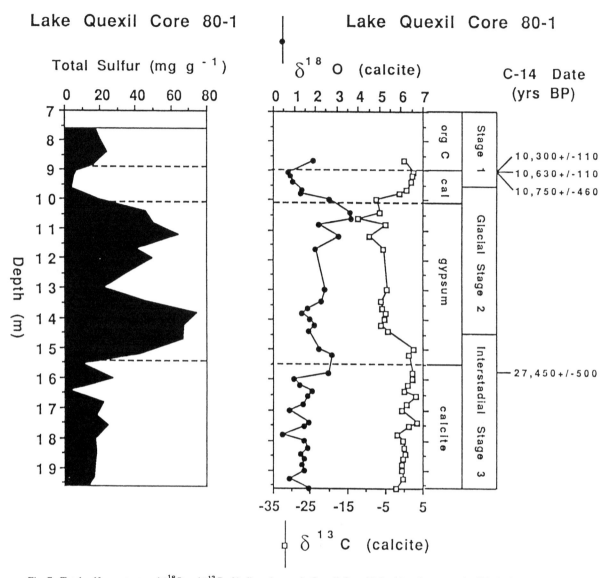

Fig. 7. Total sulfur content and $\delta^{18}O$ and $\delta^{13}C$ of bulk carbonate in Quexil Core 80-1. Also shown are the lithologic subdivisions of the core based upon the predominant component of the non-clay fraction of the sediment and approximate boundaries of marine oxygen isotopic stages. Note the general correspondence of high $\delta^{18}O$ values and low $\delta^{13}C$ values during the gypsum-rich interval deposited during Glacial Stage 2.

Within the lake, several processes can influence the carbon isotopic ratio of the DIC. Primary production in the epilimnion preferentially uptakes ^{12}C leaving the DIC enriched in ^{13}C. This biological pumping of ^{12}C from the epilimnion to the hypolimnion results in a carbon isotopic gradient between surface and deep waters when the water column is thermally stratified. Authigenic calcite precipitated from epilimnetic waters reflects the $\delta^{13}C$ of surface waters and may be used as a paleoproductivity indicator [McKenzie, 1982, 1985; Schelske and Hodell, 1991]. The average $\delta^{13}C$ of DIC in the whole lake is influenced by the proportion of carbon removed from the lake as carbonate carbon versus organic carbon. Increased organic carbon (orgC) burial over carbonate carbon will lead to a rise in whole-lake $\delta^{13}C$ values,

whereas a decrease in orgC/CaCO₃ burial will lead to a decrease in the $\delta^{13}C$ of the DIC. The final process that can affect the $\delta^{13}C$ of DIC is gas exchange with the atmosphere. Atmospheric CO_2 currently has a $\delta^{13}C$ value of -7‰. Upon dissolution in water, isotopic fractionation produces a $\delta^{13}C$ in bicarbonate (HCO_3^-) near zero.

The major change observed in the $\delta^{13}C$ record of bulk carbonate from Quexil 80-1 is a dramatic 7‰ decrease in $\delta^{13}C$ values between 24 and 12 kyr BP (Fig. 7). This interval of low $\delta^{13}C$ values coincides with Glacial Stage 2 and the interval of gypsum occurrence in the sediments (Fig. 7). The boundaries of the carbon isotopic shifts also coincide with major changes in both terrestrial and aquatic pollen assemblages (Fig. 4). The $\delta^{13}C$ decrease at 24 kyr BP coincides with a sharp decline in pollen concentrations (Fig. 3) and a switch from montane pine-oak forest during Stage 3 to grassland and thorn scrub during Stage 2 (Fig. 4). Also at this time, phytoplankton disappear and littoral and wetland indicators increase in abundance (Figs. 3 & 4). The $\delta^{13}C$ increase at 12 kyr BP coincides with the expansion of Cheno-Ams in Quexil's watershed together with aquatic taxa as lacustrine productivity resumed.

The dramatic decrease in carbon isotopic ratios during the last Ice Age reflects a greater contribution of oxidized organic carbon to the DIC pool of the lake. This may have resulted from a variety of processes that are considered below. The decrease in $\delta^{13}C$ may reflect a decrease in the $\delta^{13}C$ ratio of water supplied to the lake owing to increased organic carbon oxidation and/or decreased limestone dissolution in the drainage basin. Pollen data indicate a severe reduction in terrestrial biomass surrounding the lake during Stage 2 (Figs. 3 & 4). At this time, a C-3 forest ($\delta^{13}C$= -25‰) was replaced by C-4 grassland ($\delta^{13}C$=-12‰). The initial effect of deforestation and ecologic replacement of C-3 by C-4 biomass would have been to decrease $\delta^{13}C$ values of surface and groundwater (and hence lake DIC) because of the oxidation of the dead C-3 biomass and associated soil humus ($\delta^{13}C$=-25‰). However, this effect would have been transient and once the dead C-3 biomass was oxidized, the $\delta^{13}C$ of carbon input to the lake should have increased, having been derived almost exclusively from the dissolution of limestone. It is interesting to note that a similar decrease in $\delta^{13}C$ occurred in the oceanic carbon reservoir during the last glaciation, but the magnitude of the $\delta^{13}C$ shift was ~0.6‰, much less than that observed for Lake Quexil. Part of the decrease in the marine $\delta^{13}C$ record during the last glacial has been attributed to a decrease in tropical and temperate biomass and associated soil humus [Shackleton, 1977]. It is possible that we are observing the same effect during Stage 2 in Lake Quexil except that the carbon isotopic signal is amplified in the lake because of its small DIC pool compared to the enormous pool of DIC in the ocean.

A second explanation for the lowered $\delta^{13}C$ values during the last glacial may be related to lacustrine productivity. If the calcite in the sediments of Lake Quexil is the result of bio-induced authigenic calcite precipitation, then the $\delta^{13}C$ of the calcite will be enriched in ^{13}C relative to the mean $\delta^{13}C$ of lake water because of the preferential uptake of ^{12}C by primary producers. The greater the production, the more enriched the $\delta^{13}C$ of the DIC and the calcite precipitated from epilimnetic waters. During maximum stratification in late summer, when calcite saturation is at its highest, it is not unreasonable to find up to a 5‰ enrichment of $\delta^{13}C$ in epilimnetic waters compared to hypolimnetic waters [McKenzie, 1982, 1985]. A plausible explanation of the decrease in $\delta^{13}C$ at 24 kyr BP might be a severe reduction in lacustrine productivity in response to the development of hypersaline conditions during Stage 2. A reduction in limnetic productivity is supported by the near absence of *Botryococcus* during Stage 2 (Fig. 3). Furthermore, a decrease in primary productivity during Stage 2 would have also decreased the burial rate of organic carbon in the sediments. At this time, removal of carbon from the lake's DIC pool would have been mainly by authigenic calcite precipitation. The consequence of reduced burial of organic carbon

during Stage 2 would have been to decrease the $\delta^{13}C$ of DIC and precipitated calcite.

Although it is difficult to pinpoint the exact cause of the dramatic decrease in the $\delta^{13}C$ of Quexil sediments during Stage 2, the results clearly point to an increased contribution of oxidized organic carbon to the dissolved carbon pool of the lake. This may have resulted from a combination of factors including a decrease in the $\delta^{13}C$ of waters feeding the lake, a decrease in primary productivity under the hypersaline conditions of Stage 2, or a decrease in organic carbon burial in lake sediments. Whichever mechanism was the ultimate cause of the pronounced $\delta^{13}C$ shift, it is clear that the changes in aridity and temperature that accompanied the transitions from Interstadial Stage 3, to Glacial stage 2, to Interglacial stage 1 had a profound effect on the carbon budget of Lake Quexil.

SUMMARY OF PALEOCLIMATIC INTERPRETATIONS

Combined sedimentologic, isotopic, and palynological data from Lake Quexil Core 80-1 yield a consistent interpretation of climatic history for the Central American Lowlands during the late Pleistocene. The record is divided into three stages that coincide with changes in continental ice volume that are reflected in the marine oxygen isotopic record. During Interstadial Stage 3 (~36 to 24 kyr BP), lake level was lower than today and calcite was precipitated under moderately high E/P conditions. The vegetation was more temperate than today and marked by a montane pine-oak forest, suggesting cool-moist conditions, with temperatures 4.7 to 6.5 °C less than today.

Lake desiccation began after ~26 kyr BP, and extremely dry climate persisted in the region during the last Ice Age (Stage 2: 24 to 12.5 kyr BP). Enriched $\delta^{18}O$ values and gypsum precipitation indicate extremely arid conditions that dramatically reduced the lake volume. During Stage 2, forest disappearance and very low total pollen concentrations suggest greatly reduced terrestrial biomass. The pine-oak forest of Stage 3 was replaced by sparse thorn scrub, and the pollen assemblage indicates temperatures 6.5 to 8.0°C cooler than today.

With the onset of deglaciation (~12.5 kyr BP), dry climatic conditions began to ameliorate and E/P ratios decreased. Gypsum deposition was replaced by calcite precipitation, but evaporitic precipitation eventually ceased by 10.5 kyr BP, when autochthonous organic carbon became an important component of the sediments. Lake Quexil filled rapidly after 10.5 kyr BP, during lingering cooler temperatures (3.0-4.7°C < present). Lowland tropical forest was not fully developed in the watershed until after the onset of warmer, moister early Holocene conditions, ca. 9 kyr BP.

IMPLICATIONS FOR UNDERSTANDING CLIMATE CHANGE IN THE NEOTROPICS

Using changes in the pollen assemblage in Quexil 80-1 and an assumption of constant lapse rate, we estimate that temperature decline for the lowlands of Central America during the late Pleistocene was between 4.7 and 6.5°C for Interstadial Stage 3 and between 6.5 and 8°C for Glacial Stage 2. These temperature estimates are comparable to declines recorded at higher Neotropical elevations [Martin, 1964; Piperno et al., 1990; Bush and Colinvaux, 1990] and with revised estimates of SST decrease (-7.8°C) inferred from marine oxygen isotopic data in the Caribbean-equatorial Atlantic region [Emiliani and Ericson, 1991]. These estimates disagree with those of the CLIMAP Project [1976, 1981] that proposed little or no temperature change in the tropics. Our findings also preclude increased lapse rates as an explanation for temperature discrepancies between CLIMAP SST estimates and the much colder temperatures inferred from high-altitude sites in the tropics, because of Quexil's low elevation.

Our results confirm the well-known increase in late Pleistocene aridity that has been documented throughout much of the Neotropics [Markgraf, 1989 and references therein] and Africa [Gasse et al., 1990; Street-Perrott and Perrott, 1990; among others]. Gypsum precipitation in Lake Quexil was apparently confined to isotopic Stage 2 between ~26 and 12 kyr BP (Figs. 6 & 7). This inferred increase in E/P during the last Ice Age is consistent with the occurrence of high-salinity, low-productivity Sargasso Sea type water in the Colombia Basin (Caribbean Sea) during Stage 2 [Prell and Hays, 1976]. Prell and Hays [1976] suggested that these high salinity conditions resulted from higher E/P ratios because the ITCZ was located, on the average, further south over South America during the last glaciation. The southward shift of the ITCZ during Stage 2 has been attributed to an enhanced temperature gradient in the mid-latitudes of the Northern Hemisphere because of the presence of the Laurentide Ice Sheet [Newell, 1973].

Climate in the circum-Caribbean region is tightly linked to the position of the ITCZ and the intensity of the annual cycle because of the distinct seasonal pattern of rainfall in the northern hemisphere tropics [Hastenrath, 1976, 1984]. The rainy season occurs during the northern summer half-year when the ITCZ moves north, displacing the North Atlantic subtropical high, and weakening the easterly trade winds [Hastenrath, 1976, 1984]. In contrast, the dry season occurs during the winter when the ITCZ is located in the Southern Hemisphere over South America. Historically, anomalously dry years in the circum-Caribbean have been associated with a reduction in the intensity of the annual cycle when the ITCZ remains farther south on the average [Hastenrath, 1984]. Hodell et al. [1991] suggested that the long-term (millennial) changes in Caribbean climate could be explained partly by orbitally induced (Milankovitch) variations in seasonal insolation that modified the intensity of the annual cycle. According to this interpretation, climate was dry in the northern hemisphere tropics during the last Ice Age because the seasonal insolation difference at 10°N was reduced and the ITCZ occupied a southerly position on average [Curtis and Hodell, this volume].

Lake Quexil (Guatemala) is positioned in a critical region of vapor transport between the Atlantic and Pacific Oceans, because much of the excess evaporation in the Atlantic occurs between 10 and 25°N and this water vapor is exported to the Pacific by the trade winds across Central America [Warren, 1983; Weyl, 1968; Street-Perrott and Perrott, 1990]. Recent modelling studies predict that small changes in the flux of this fresh water transport through the atmosphere can cause major changes in oceanic thermohaline circulation [Broecker and Denton, 1989; Stocker and Wright, 1991]. In fact, one of the major differences in oceanic circulation between glacial and interglacial time was a significant reduction in the flux of North Atlantic Deep Water (NADW) during the last Ice Age [Broecker et al., 1985; Boyle and Keigwin, 1987]. Small reductions in the flux of vapor transport from the Atlantic to the Pacific may have been partly responsible for diminished production rates of NADW during glacial stages [Stocker and Wright, 1991]. Changes in relative humidity and sea-surface temperatures of the northern tropics are undoubtedly important variables for controlling evaporation rates and the atmospheric flux of water vapor between the Atlantic and Pacific. Our study, combined with other results from the Neotropics [Piperno et al., 1990; Markgraf, 1989; Leyden, 1984, 1985; Deevey et al., 1983; Bradbury et al., 1981], suggests that climate was cooler and drier in the circum-Caribbean region during the late Pleistocene. This implies that the atmosphere may have been less efficient at transporting water vapor from the Atlantic to the Pacific during the last Ice Age. This decreased inter-ocean vapor transport may explain, in part, the reversal that occurred in Atlantic thermohaline flow during the last glaciation [Stocker and Wright, 1991].

Upon deglaciation, an abrupt switch occurred from arid to humid conditions as has been documented in many localities throughout the tropics including the Americas [Piperno et al., 1990; Markgraf, 1989; Leyden, 1984, 1985; Deevey et al., 1983; Bradbury et al., 1981; Watts, 1975] and Africa [Gasse et al., 1990; Street-Perrott and Perrott, 1990; Bonnefille et al., 1990; Gasse et al., 1989; Talbot and Delibrias, 1980; Street and Grove, 1976]. In Quexil 80-1, we record the cessation of evaporitic sedimentation and a rapid rise in lake level beginning at ~10.5 kyr BP, but higher temperature and moisture availability were insufficient to allow the growth of lowland tropical forests until after ~9 kyr BP. These lingering cool conditions in the circum-Caribbean region during deglaciation may have been related initially to meltwater discharge into the Gulf of Mexico [Overpeck et al., 1989] and then later to discharge into the North Atlantic, which may have suppressed NADW, diminished the cross equatorial heat flux, and kept the ITCZ to the south [Street-Perrott and Perrott, 1990; Mix et al., 1986]. By ~9 kyr BP, SSTs warmed in the western equatorial Atlantic by 2 to 8°C, cross-equatorial heat flux increased, the trade winds weakened, and the ITCZ assumed a more northerly position [Prell and Hays, 1976; Mix et al., 1986; Peterson et al., 1991]. As a result, rainfall increased in the northern hemisphere tropics, Lake Quexil filled rapidly, evaporite deposition ceased and sediments became dominated by organic carbon deposition, and lowland tropical forest was re-established in Quexil's watershed.

CONCLUSIONS

Palynological, geochemical, and isotopic evidence from the Late Pleistocene of Lake Quexil (Guatemala) confirm previous results from the Neotropics indicating decreased temperature and increased aridity of the Central American lowlands during the last Ice Age. Based upon changes in pollen assemblages, we estimate a temperature decline of 6.5 to 8°C during glacial Stage 2 in agreement with estimates from higher Neotropical elevations [Martin, 1964; Piperno et al., 1980; Bush and Colinvaux, 1990] and with new interpretations of the marine oxygen isotopic record [Emiliani and Ericson, 1991]. This estimate disagrees, however, with the interpretation of little or no tropical SST change proposed by the CLIMAP Project [1976, 1981]. Decreased SST in the tropics during the last Ice Age helps explain the observed depression of mountain snowlines at low latitudes and the increase in tropical aridity caused by reduced evaporation rates from the ocean [Rind and Peteet, 1985]. The increased aridity of the Neotropics and Africa during the last Ice Age reflects a fundamental change in the Earth's hydrometeorologic cycle that may have influenced oceanic thermohaline circulation [Stocker and Wright, 1991]. Results from Lake Quexil suggest a reduction in water vapor transport from the Atlantic to the Pacific across Central America during Stage 2, which may partly explain suppression of NADW production during the last glaciation.

Acknowledgements. This work was supported by National Science Foundation grant EAR 79-26330 to the late E. S. Deevey, Jr., whose guidance is gratefully acknowledged, and by a Presidential Young Investigator Award to DAH (NSF Grant OCE-8858012). We thank Gail Murray and Frank N. Blanchard for mineralogical analyses. Atomic absorption determinations were done at the University of Florida IFAS Soil Testing Laboratory. Robert Stuckenrath kindly provided the SI radiocarbon date. We thank Daho Pozos, S.A. and Michael W. Binford for coring assistance.

REFERENCES

APHA (American Public Health Association), Standard Methods for the Examination of Water and Wastewater, 14th ed., American Public Health Association, Washington D.C., 1975.

Bartlett, A.S. and E.S. Barghoorn, Phytogeographic history of the Isthmus of Panama during the past 12,000 years (a history of vegetation, climate, and sea-level change), in *Vegetation and vegetational history of northern Latin America*, edited by A. Graham, pp. 203-299, Elsevier Scientific, New York, New York, 1973.

Binford, M.W, Ecological history of Lake Valencia, Venezuela: interpretation of animal microfossils and some chemical, physical and geological features, *Ecological Monographs*, 52, 307-333, 1982.

Binford, M.W, Paleolimnology of the Peten Lake District, Guatemala, I. Erosion and deposition of inorganic sediment as inferred from granulometry, *Hydrobiologia*, 103, 199-203, 1983.

Binford, M.W., M. Brenner, T.J. Whitmore, A. Higuera-Gundy, E.S. Deevey, and B.W. Leyden, Ecosystems, paleoecology, and human disturbance in subtropical and tropical America, *Quaternary Science Reviews*, 6, 115-128, 1987.

Bonnefille, R., J.C. Roeland and J. Guiot, Temperature and rainfall estimates for the past 40,000 years in equatorial Africa, *Nature*, 346, 347-349, 1990.

Boyle, E.A., and L. Keigwin, North Atlantic thermohaline circulation during the past 20,000 years linked to high-latitude surface temperature, *Nature*, 350, 35-40, 1987.

Bradbury, J.P., B. Leyden, M. Salgado-Labouriau, W.M. Lewis, Jr., C. Schubert, M.W. Binford, D.G. Frey, D.R. Whitehead, and F.H. Weibezahn, Late Quaternary environmental history of Lake Valencia, Venezuela, *Science*, 214, 1299-1305, 1981.

Bradbury, J.P., R.M. Forester, W.A. Bryant, and A.P. Covich, Paleolimnology of Laguna de Cocos, Albion Island, Río Hondo, Belize, in *Ancient Maya Wetland Agriculture: excavations on Albion Island, northern Belize*, edited by M.D. Pohl (ed.), pp. 119-154, Westview Press, Boulder, 1980.

Breedlove, D.E., The Phytogeography and vegetation of Chiapas (Mexico), in *Vegetation and vegetational history of northern Latin America*, edited by A. Graham, pp. 149-165, Elsevier Scientific, New York, New York, 1973.

Brenner, M., Paleolimnology of the Peten Lake District, Guatemala, II. Mayan population density and sediment and nutrient loading of Lake Quexil, *Hydrobiologia*, 103, 205-210, 1983.

Brenner, M., B. Leyden, and M.W. Binford, Recent sedimentary histories of shallow lakes in the Guatemalan savannas, *Journal of Paleolimnology*, 4, 239-251, 1990.

Broecker, W.S., and G.H. Denton, The role of ocean-atmosphere re-organizations in glacial cycles, *Geochimica et Cosmochimica Acta*, 53, 2465-2501, 1989.

Broecker, W.S., D.M. Peteet, and D. Rind, Does the ocean-atmosphere system have more than one stable mode of operation?, *Nature*, 315, 21-26, 1985.

Broecker, W.S., M. Andrec, W. Wolfli, H. Oeschger, G. Bonani, J.P. Kennett, and D. Peteet, The chronology of the last deglaciation: implications to the cause of the Younger Dryas Event, *Paleoceanography*, 3, 1-19, 1988.

Bush, M.B., and P.A. Colinvaux, A pollen record of a complete glacial cycle from lowland Panama, *Journal of Vegetation Science*, 1, 105-118, 1990.

Clark, R.L., Point count estimation of charcoal in pollen preparations and thin sections of sediments, *Pollen et Spores*, 24, 523-535, 1982.

CLIMAP Project Members, The surface of the ice-age earth, *Science*, 191, 1131-1137, 1976.

CLIMAP Project Members, Seasonal reconstructions of the earth's surface at the last glacial maximum, Geological Society of America Map and Chart Series, MC-36, 1981.

Cowgill, U.M., G.E. Hutchinson, A.A. Racek, C.E. Goulden, R. Patrick, and M. Tsukada, The history of Laguna de Petenxil, a small lake in northern Guatemala, *Memoirs of the Connecticut Academy of Arts and Sciences*, 17, 1-126, 1966.

Curtis, J.H., and D.A. Hodell, An isotopic and trace element study of ostracods from Lake Miragoane, Haiti: A 10.5 kyr record of paleosalinity and paleotemperature changes in the Caribbean. This volume.

Deevey, E.S, Holocene forests and Maya disturbance near Quexil Lake, Peten, Guatemala, *Polish Archives of Hydrobiology*, 25, 117-129, 1978.

Deevey, E.S. and M. Stuiver, Distribution of natural isotopes of carbon in Linsley Pond and other New England lakes, *Limnology and Oceanography*, 9, 1-11, 1964.

Deevey, E.S. and D.S. Rice, Coluviacion y retencion de nutrientes en el distrito lacustre del Peten central, Guatemala, *Biotica*, 5, 129-144, 1980.

Deevey, E.S., D.S. Rice, P.M. Rice, H.H. Vaughan, M. Brenner, and M.S. Flannery, Mayan urbanism: impact on a tropical karst environment, *Science*, 206, 298-306, 1979.

Deevey, E.S. , M. Brenner, M.S. Flannery, and G.H. Yezdani, Lakes Yaxha and Sacnab, Peten, Guatemala: limnology and hydrology, Archiv Fur Hydrobiologie, Supplement 57, 419-460, 1980.

Deevey, E.S., M. Brenner, and M.W. Binford, Paleolimnology of the Peten Lake District, Guatemala, III. Late Pleistocene and Gamblian environments of the Maya area, *Hydrobiologia*, 103, 211-216, 1983.

Duplessy, J.C., M. Arnold, P. Maurice, E. Bard, J. Duprat, and J. Moyes, Direct dating of the oxygen-isotope record of the last deglaciation by ^{14}C accelerator mass spectrometry, *Nature*, 320, 350-352, 1986.

Emiliani, C. and D.B. Ericson, The glacial/interglacial temperature range of the surface water of the oceans at low latitudes, in Stable Isotope Geochemistry: A Tribute to Samuel Epstein, edited by H.P. Taylor, Jr., R. O'Neil, and I.R. Kaplan, pp. 223-228, The Geochemical Society Special Publication No. 3, 1991.

Fairbanks, R.G, A 17,000 year glacio-eustatic sea level record: influences of glacial melting rates on the Younger Dryas event and deep-ocean circulation, *Nature*, 342, 637-642, 1989.

Fontes, J.C., and R. Gonfiantini, Comportement isotopique au cours de l'evaporation de deux bassins sahariens, *Earth and Planetary Science Letters*, 3, 258-266, 1967.

Gasse, F., V. Ledee, M. Massault, and J. Fontes, Water-level fluctuatrions of Lake Tanganyika in phase with oceanic changes during the last glaciation and deglaciation, *Nature*, 342, 57-59, 1989.

Gasse, F., R. Téhet, A. Durand, E. Gilbert, and J.C. Fontes, The arid-humid transition in the Sahara and Sahel during the last glaciation, *Nature*, 346, 141-146, 1990.

Gomez-Pompa, A., Ecology of the vegetation of Veracruz, in *Vegetation and Vegetational History of Northern Latin America*, edited by A. Graham, pp. 73-148, Elsevier Scientific, New York, New York, 1973.

Hansen, B.C.S., Pollen stratigraphy of Laguna de Cocos, in *Ancient Maya Wetland Agriculture: excavations on Albion Island, northern Belize*, edited by M.D. Pohl, pp. 155-186, Westview Press, Boulder, 1990.

Hastenrath, S., Variations in low-latitude circulation and extreme climatic events in the tropical Americas, *J. Atmos. Sci.*, 33: 202-215, 1976.

Hastenrath, S., Interannual variability and the annual cycle: mechanisms of circulation and climate in the tropical Atlantic sector, *Monthly Weather Review*, 112: 1097-1107, 1984.

Hodell, D.A., J.H. Curtis, G.A. Jones, A. Higuera-Gundy, M. Brenner, M.W. Binford, and K.T. Dorsey, Reconstruction of Caribbean climate change over the past 10,500 years, *Nature*, 352, 790-793, 1991.

Huffman, E.W.D., Jr., Performance of a new automatic carbon dioxide analyzer, *Microchemical Journal*, 22, 567-573, 1977.

Leyden, B.W., Guatemalan forest synthesis after Pleistocene aridity, *Proceedings of the National Academy of Sciences USA*, 81, 4856-4859, 1984.

Leyden, B.W., Late Quaternary aridity and Holocene moisture fluctuations in the Lake Valencia Basin, Venezuela, *Ecology*, 66, 1279-1295, 1985.

Leyden, B.W., Man and climate in the Maya lowlands, *Quaternary Research*, 28, 407-414, 1987.

Lundell, C.L., *The Vegetation of Peten*, Carnegie Institute, Washington, D.C., 244 pp, 1937.

Markgraf, V., Paleoclimates in Central and South America since 18,000 BP based on pollen and lake-level records, *Quaternary Science Reviews*, 8, 1-24, 1989.

Martin, P.A., Paleoclimatology and a tropical pollen profile, in *Report of the VIth International Congress on the Quaternary, Warsaw*, Vol. 2, 319-323, Paleoclimatology Section, Lodz, 1964.

McKenzie, J.A., Carbon-13 cycle in Lake Greifen: A model for restricted ocean basins, in *Nature and Origin of Cretaceous Carbon-rich Facies*, edited by S.O. Schlanger and M.B. Cita, pp. 198-207, Academic Press, London, 1982.

McKenzie, J.A., Carbon isotopes and productivity in the lacustrine and marine environment, in *Chemical Processes in Lakes*, edited by W. Stumm, pp. 99-118, Wiley, New York, NY, 1985.

Mix, A.C., W.F. Ruddiman, and A. McIntyre, Late Quaternary Paleoceanography of the tropical Atlantic, 1: Spacial variability of annual mean sea-surface temperatures, 0-20,000 years B.P., *Paleoceanography*, 1, 43-66, 1986.

Newell, R.E., Climate and the Galapagos Islands, *Nature, 245*: 91-92, 1973.

Overpeck, J.T., L.C. Peterson, N. Kipp, J. Imbrie, and D. Rind, Climate change in the circum-North Atlantic region during the last deglaciation, *Nature*, 338, 553-557, 1989.

Perry, J.P., Jr., *The Pines of Mexico and Central America*, pp. 231, Timber Press, Portland, Oregon, 1991.

Peterson, L.C., J.T. Overpeck, N.G. Kipp, and J. Imbrie, A high-resolution late Quaternary upwelling record from the anoxic Cariaco basin, Venezuela, *Paleoceanography*, 6, 99-119, 1991.

Piperno, D.R., M.B. Bush, and P.A. Colinvaux, Paleoenvironments and human settlements in late-glacial Panama, *Quaternary Research*, 33, 108-116, 1990.

Prell, W.L. and J.D. Hays, Late Pleistocene faunal and temperature patterns of the Colombia Basin, Caribbean Sea, in *Investigations of Late Quaternary Paleoceanography and Paleoclimatology*, edited by R.M Cline and J. D. Hays, pp. 201-220, Geol. Soc. Am. Memoir 145, Geol Soc. Am., Boulder, CO, 1976.

Rice, D.S., P.M. Rice, and E.S. Deevey, El impacto de los Mayas en el ambiente tropical de la cuenca de los lagos Yaxha y Sacnab, El Peten, Guatemala, American Indigena, 43, 261-297, 1983.

Rice, D.S., P.M. Rice, and E.S. Deevey, Paradise lost: Classic Maya impact on a lacustrine environment, in *Prehistoric lowland Maya environment and subsistence economy*, edited by M. Pohl, pp. 91-105, Peabody Museum Papers 77, Harvard Univ. Press, Cambridge, MA, 1985.

Rind, D. and D. Peteet, Terrestrial conditions of the last glacial maximum and CLIMAP sea-surface temperature estimates: are they consistent?, *Quaternary Research*, 24, 1-22, 1985.

Romney, D.H. (ed.), *Land in British Honduras*, 327 pp., Her Majesty's Stationery Office, Colonial Research Publications No. 24, London, 1959.

Schelske, C.L. and D.A. Hodell, Recent changes in productivity and climate of Lake Ontario detected by isotopic analysis of sediments, *Limnol. Oceanogr.*, 36, 961-975, 1991.

Schwerdtfeger, W. (ed.), *Climates of Central and South America*, 532 pp., World Survey of Climatology Volume 12, Elsevier Scientific Publishing Company, The Netherlands, 1976.

Shackelton, N.J., Carbon-13 in Uvigerina: tropical rainforest history and the Equatorial Pacific carbonate dissolution cycles in *The Fate of Fossil Fuel CO_2 in the Oceans*, edited by N.R. Anderson and A. Malahoff, pp.401-427, Plenum, New York, 1977.

Simmons, C.S., J.M. Tarano T., and J.H. Pinto, *Clasificacion de Reconocimiento de los Suelos de la Republica de Guatemala*, Ministerio de Agricultura, Guatemala City, 1000 pp., 1959.

Standley, P.C., and J.A. Steyermark, Flora of Guatemala, *Fieldiana*, 24, 353-354, 1958.

Stocker, T.F. and D.G. Wright, Rapid transitions of the ocean's deep circulation induced by changes in surface water fluxes, *Nature*, 351, 729-732, 1991.

Street, F.A., and A.T. Grove, Environmental and climatic implications of late Quaternary lake-level fluctuations in Africa, *Nature*, 261, 385-390, 1976.

Street-Perrott, F.A. and R.A. Perrott, Abrupt climate fluctuations in the tropics: the influence of Atlantic Ocean circulation, *Nature*, 343, 607-612, 1990.

Talbot, M.R., and G. Delibrias, A new late Pleistocene-Holocene water-level curve for Lake Bosumtwi, Ghana, *Earth and Planetary Science Letters*, 47, 336-344, 1980.

van der Hammen, T., The Pleistocene changes of vegetation and climate in tropical South America, *Journal of Biogeography*, 1, 3-26, 1974.

Vaughan, H.H., E.S. Deevey, and S.E. Garrett-Jones, Pollen stratigraphy of two cores from the Peten Lake District, in *Prehistoric Lowland Maya Environment and Subsistence Economy*, edited by M. Pohl, pp. 73-89, Peabody Museum Papers 77, Harvard Univ, Press, Cambridge, MA, 1985.

Vinson, G.L., Upper Cretaceous and Tertiary stratigraphy of Guatemala, *American Association of Petroleum Geologists*, Bulletin 46, 425-456, 1962.

Warren, B.A., Why is no deep water formed in the North Pacific?, *Journal of Marine Research*, 41, 327-347, 1983.

Watts, W.A., A late Quaternary record of vegetation from Lake Annie, south-central Florida, *Geology*, 3, 344-346, 1975.

Watts, W.A., and M. Stuiver, Late Wisconsin climate of northern Florida and origin of species-rich deciduous forest, *Science*, 210, 325-327, 1980.

Watts, W.A. and B.C.S. Hansen, Environments of Florida in the Late Wisconsin and Holocene, in *Wet Site Archaeology*, edited by B.A. Purdy, pp. 307-323, The Telford Press, Caldwell, New Jersey, 1988.

West. R.C, Surface configuration and associated geology of Middle America, in *Handbook of Middle American Indians Vol. 1: Natural Environment and Early Cultures*, edited by R.C. West, pp. 33-83, University of Texas Press, Austin, 1964.

Weyl, P.K., The role of the oceans in climatic change: a theory of the ice ages, *Meteorological Monographs*, 8, 37-62, 1968.

Whitehead, D.R., Late-Pleistocene vegetational changes in north-eastern North Carolina, *Ecological Monographs*, 51, 451-471, 1981.

B. Leyden, Department of Geology, University of South Florida, Tampa, FL 33620

M. Brenner, Department of Fisheries and Aquatic Sciences, University of Florida, 7922 NW 71st St., Gainesville, FL 32606

D. Hodell and J. Curtis, Department of Geology, University of Florida, Gainesville, FL 32611

Comparative Paleoclimatic Interpretations from Nonmarine Ostracodes Using Faunal Assemblages, Trace Elements Shell Chemistry and Stable Isotope Data

MANUEL R. PALACIOS-FEST, ANDREW S. COHEN, JOAQUIN RUIZ AND BRIAN BLANK

Department of Geosciences, University of Arizona, Tucson, AZ 85721

Lacustrine ostracodes (microcrustaceans) are powerful tools for reconstructing paleoclimates. Western North American Pleistocene pluvial lakes are ideal environments to study ostracode paleoecologic and geochemical attributes since many of them have proven to be responsive to climatic changes. The purpose of this paper is to illustrate how faunal assemblages, stable isotopes and trace elements present in ostracode carapaces, correlate in one Pleistocene lake basin in south-central Oregon, pluvial Lake Chewaucan, by comparing their response signals throughout the lake's history. The combined analysis of paleoecologic and geochemical profiles indicate that pluvial Lake Chewaucan underwent broad limnologic fluctuations during the late Pleistocene. Ostracode assemblages clearly demonstrate periods of increasing salinity and temperature alternating with cooler and less saline conditions. Covariant $\delta^{18}O$ and $\delta^{13}C$ values indicate that this lake remained a closed basin throughout its history. Trace elements (Mg/Ca and Sr/Ca molar ratios) indicate that salinity excursions were a major factor affecting pluvial Lake Chewaucan. Fluctuations in the $\delta^{18}O$ values during this interval suggest that temperature and humidity changed profoundly between the latest interglacial and the full glacial. Hydroclimate changes in Lake Chewaucan during the Pleistocene probably occurred in response to the greater influence of the jet stream south-branch as it shifted position during this period of time. As a result, greater and more frequent storms occurred in south-central Oregon.

INTRODUCTION

The reconstruction of ancient lake histories allows paleolimnologists to understand both local and regional processes of environmental and climatic change. Water temperature and the chemistry of lakes are sensitive to regional climate parameters (i.e. air-temperature, precipitation and evaporation). In the short-term (i.e. decade, yearly) other non-climatic factors or those factors which are indirectly related to climate, such as basin morphometry or input/output balance, influence seasonal limnologic processes [Forester, 1991a]. But in the long-term (i.e. hundreds to thousands of years) these effects are eclipsed by a climate signal which is likely to reflect regional processes. Some lacustrine organisms, like ostracodes, are highly responsive to environmental perturbations and may be used for paleohydrochemical and paleoclimatic interpretations.

The occurrence of lacustrine ostracodes appears to be primarily controlled by both water chemistry (both major dissolved ion content and total dissolved solids: TDS) and temperature. Ostracodes respond to environmental variations in several ways: biogeographic distribution [Forester, 1987; De Deckker & Forester, 1988], shell morphology [Delorme, 1969, and 1989] and shell chemistry [stable isotopes: Lister, 1988 and trace elements: Chivas et al., 1983]. As fossils these organisms provide a proxy record for interpreting paleoclimate [Forester, 1991a]. The purpose of this paper is to illustrate how these different signals correlate in one Pleistocene lake basin in south-central Oregon, pluvial Lake Chewaucan, by comparing three lines of evidence: faunal assemblages, shell chemistry and stable isotopes.

OSTRACODE/ENVIRONMENT RELATIONSHIP

Ostracode occurrence patterns and associations provide important tools for paleoenvironmental reconstructions. Delorme [1969 and 1989], Forester [1983, 1986, 1987, and 1991a,b] and Chivas et al. [1983, 1986] described a relationship between ostracodes and their environment and suggested that ostracode species distribution is defined primarily by thermal and chemical parameters. Forester [1991a] indicated that ostracode life-cycles are coupled to those parameters and that growth, maturation and reproduction can only be completed if specific conditions are satisfied for each species.

During growth ostracodes moult up to nine times to reach maturity. The calcification of a new carapace from ions in solution is thought to occur in thermal and chemical equilibrium with the host water [Chivas et al., 1983] . Water temperature, water chemistry and seasonal variability control the presence of many taxa [Forester, 1991b].

Limnocythere ceriotuberosa (the most common fossil species present in Lake Chewaucan sediments) today is a euryhaline (wide salinity tolerance range) organism, whereas *Limnocythere sappaensis* is a halobiontic (restricted salinity tolerance range) species. The co-occurrence of these two species brackets a specific environment; an evaporative basin. *Limnocythere bradburyi*, a species known only in the fossil record has been suggested to be a warm water indicator distributed from Lake Texcoco, central Mexico to Lake Estancia, New Mexico, and Silver (Mojave) Lake, California [Forester, 1987; Wells et al., 1989].

Candona caudata is also a common species in the Lake Chewaucan sediments. Today this species lives in freshwater to low salinity conditions and streams, within a wide temperature range (Forester, pers. comm. 1992). It appears to be associated with fairly deep waters (greater than 10m). In contrast, *Candona patzcuaro* a member of the *C. rawsoni* group lives in prairie lakes and pondsthat are permanent or ephemeral. Eventually these two species overlap within the Lake Chewaucan record. Another significant species, *Cypridopsis vidua*, is a eurythermic organism which prefers to live in warm wetlands (10-25°C: Palacios-Fest, unpublished data from Magdalena, Sonora and Agua Caliente Spring, Tucson, Arizona).

The replacement of any of these species by the cold/freshwater species *Cytherissa lacustris* (an occasional species present in the Lake Chewaucan sediments; with a present maximum temperature tolerance of 19°C and maximum salinity tolerance of 215 mg/l Na), indicates cooler conditions and a more positive water budget for the basin (increasing inflow and precipitation-evaporation) [Delorme, 1989]. Thus, species appearance/ disappearance trends are critical for determining the paleohydroenvironmental history of a lake.

Shell morphology also provides valuable information for interpreting lake paleoenvironments. Delorme [1989] indicates that changes in carapace growth may be controlled by environmental factors, such as salinity, solute composition, pH, and depth of water. Some eurytopic species may develop unusual ornamentations in response to water temperature, water chemistry and possibly seasonal variation. For example, *Cyprideis torosa*, a well-known brackish water species, generates a heavily ornamented carapace in freshwater but a smoother carapace as salinity increases [Teeter and Quick, 1990]. Ornamental changes within a species provide another useful criterion to understand paleohydroenvironmental fluctuations.

Ostracode shell chemistry is also sensitive to lake hydrochemistry. In 1971, Turpen and Angell demonstrated that calcium in solution is the exclusive source of Ca^{2+} for the calcification of ostracode valves. Later, Chivas et al. [1983] showed that some trace elements like magnesium and strontium are similarly extracted from ambient water during shell formation. They defined a relationship between water temperature, hydrochemistry and ostracode shell chemistry. Based upon these results Chivas and colleagues [1986a, b] discussed the applications of magnesium and strontium contents in ostracode carapaces as paleohydrochemical indicators. More recently, De Deccker and co-workers [1988a, b] have presented examples of the application of Mg and Sr trace element data for interpreting ostracodes from the Miocene of the Mediterranean and the late Pleistocene of the Gulf of Carpentaria (Australia). Ostracode trace-element geochemistry is becoming a major tool for paleohydrochemical reconstructions which in turn may be used for paleoclimatic interpretations.

Lister [1988] and Eyles and Schwarcz [1991] have pursued the utility of stable isotope (oxygen and carbon) geochemistry as another approach to paleoclimatic reconstructions from ostracode shells. In 1988, Lister reconstructed the latest Pleistocene-Holocene paleohydroenvironmental history of Lake Zurich, Switzerland (a deep, peri-alpine, open basin). The isotopic record from ostracode carapaces allowed him to establish the rate and timing of the latest Pleistocene Alpine deglaciation, Holocene climatic changes, and changing lacustrine productivity. Lake Turkana, Kenya (in a low latitude, closed basin) provides another important example of ostracode stable- isotope analysis; preliminary isotopic studies of specimens from the uppermost 12 m of annually laminated basinal sediments indicate a probable correlation with Holocene lake level changes [Lister, 1988]. Eyles and Schwarcz [1991] have recently analyzed the stable isotopic ($\delta^{18}O$ and $\delta^{13}C$) composition of two candonid ostracodes (*Candona subtriangulata* and *Candona caudata*). Using this method they determined the glaciolacustrine evolution of Lake Ontario during the last glacial cycle (Wisconsin).

HYDROLOGIC AND GEOLOGIC SETTING OF PLUVIAL LAKE CHEWAUCAN AND MODERN SUMMER LAKE

Allison [1982] has previously described the geomorphology and Pleistocene geology of Pluvial Lake Chewaucan. It is located in the northwest corner of the Great Basin. The basin contains two small relict lakes, Lake Abert and Summer Lake. Today Summer Lake is a topographically closed lake in the northwest corner of the Chewaucan Basin. Pluvial Lake Chewaucan consisted of a four-lobed body of water which reached a maximum area of about 1250 km² in the confluent structural basins of modern Summer Lake, Upper Chewaucan Marsh, Lower Chewaucan Marsh and Lake Abert in Lake County, south-central Oregon (Fig. 1). During the Pleistocene, shoreline terraces indicate a maximum recorded depth of 115 m, with lake

Pluvial Lake Chewaucan

Location Map

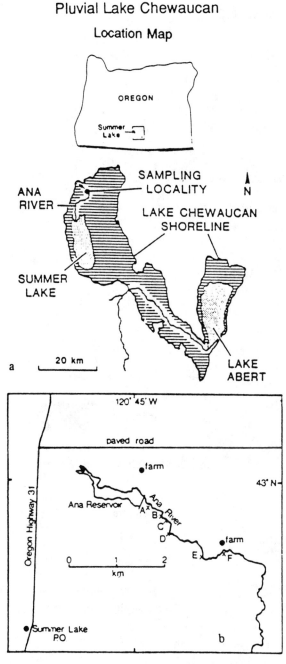

Fig. 1. Location map of the Pleistocene pluvial Lake Chewaucan Basin showing a) Holocene relicts: Summer Lake to the northwest and Lake Abert to the southeast, and b) R. Negrini's sampling localities along the Ana River canyon cut. Samples used for this study are from sections C and E. After Negrini et al. [1988].

levels reaching to about 1380 m above sea level (masl). Summer Lake is a playa lake with an area of ~180 km^2 and an elevation averaging 1264 masl. Its present average depth is less than 2 m.

Stream input has changed through time. At present,

Summer Lake receives most of its inflow from small streams from the surrounding highlands and from the Ana Spring, located in the northwest margin of the basin [Davis, 1985]. The Ana Spring drains southward via the Ana River to the lake. However, during the Pleistocene pluvial Lake Chewaucan was fed by the Chewaucan River which drained northward from the Lower Chewaucan Marsh. Davis [1985] have proposed that Summer Lake had a much higher lake level before the Holocene due to the continuous inflow from the Chewaucan River.

Van Denburgh [1975] presents the most complete study of present-day Summer Lake and Lake Abert (the remnants of pluvial Lake Chewaucan) water chemistry. Summer Lake is characterized by a large volume of solutes primarily derived from the groundwater supply through Ana Spring. The spring flow contains about 160 ppm of dissolved solids (mostly silica, sodium and bicarbonate). The influence of peripheral groundwater input is marked by brines with high salinity values (80,000-100,000 ppm). Measured salinities in the lake at 0.65 and 1.5 m depth are between 40,000 and 50,000 ppm. The lake brine is dominated by Na^+, Cl^- HCO_3^-, $CO_3^=$, K^+ and $SO_4^=$, and depleted in Ca^{2+}, Mg^{2+}, and Si^{4+}. The modern dissolved-solids content of Summer Lake varies significantly on both a seasonal and long-term basis.

Banfield et al. [1991a, b] have discussed the weathering and diagenetic processes affecting adjacent Lake Abert, which are probably similar to those of Summer Lake. These authors suggest that not only evapotranspiration but fractionation processes involving surrounding springs are significant in determining the composition of the lake water. Reactions between meteoric water and rocks in the vicinity of the lake also contribute to the lake water composition. Banfield et al. [1991a] indicate that the lithology of clay sediments (at Lake Abert) produced by weathering reflect their sources: pumice, ash, pyroclasts and felsic volcanics. The weathering alteration of olivine probably facilitated the redistribution of alkali and alkali-earth elements that enriched the lake water [Banfield et al., 1991a].

During the Holocene the lake level has dropped dramatically; the Ana River has cut a canyon, exposing a thick sedimentary sequence. A 20 m sequence of Pleistocene silty clays intercalated with tephra layers and fine ostracode-bearing sands are revealed along the canyon. Figure 2 presents a composite stratigraphic column of sections C, and E sampled by R.M. Negrini in 1987. Most of the sequence outcrops in section C but the youngest portion occurs at section E [Davis, 1985].

The upper 8 m of the sequence includes 21 tephra layers interspersed between marls. Fine laminae are rarely present, lacustrine marls are generally organized in beds a few centimeters thick. Ostracodes occur in high to very high concentrations throughout most of the lacustrine sediments. Several planar unconformities occur through the section, in part marked by degraded ostracode sands, tufa breccias and *in situ* carbonate cementation (R.M. Negrini, unpubl. manuscript). Others are marked by fine gravel lag or simple weathered horizons.

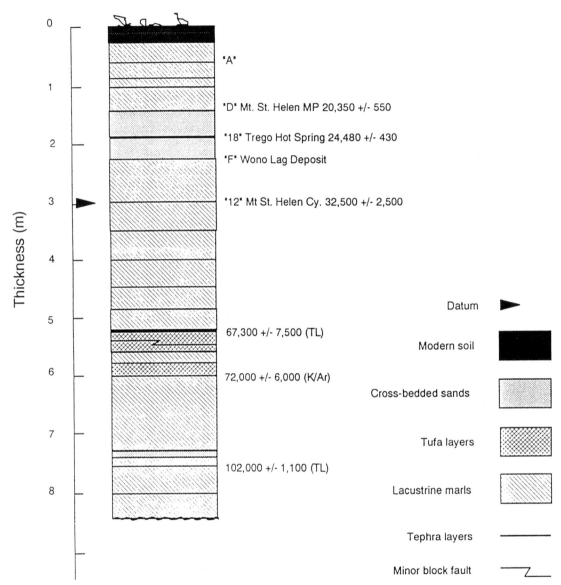

Fig. 2. Composite stratigraphic column of Pleistocene pluvial Lake Chewaucan deposits along the Ana River canyon cut. Sections C and E are integrated in this schematics. Section C includes the longest record and section E contains the youngest sediments. Note the unconformity at the base of the column which indicates a period of deflation of the Chewaucan Basin during isotope stage 5. No lake sediments were preserved around the Ana River during probably the last ~16,000 years (latest Pleistocene to Holocene).

GEOCHRONOLOGIC RECORD

A geochronologic framework has been defined for the Ana River section investigated in this study [Davis, 1985; Berger et al., 1990; Negrini and Davis, in press]. Tephra correlations and thermoluminescence age estimates were determined from tephra samples collected at sections C and E. Tephra 12 (Mt. St. Helens Cy), yielding an age of 32,500 +/- 2,500 yrs. B.P. provides the primary datum for this work. Other tephras like Mt. St. Helens Mp (20,350 +/- 550 yrs. B.P.), Wono and Trego Hot Spring (24,800 +/- 430 yrs. B.P.) provide additional constraints to establish the chronology of the upper pluvial Lake Chewaucan stratigraphic sequence [Negrini et al., 1988]. The correlation of the paleomagnetic record from Summer Lake with data from Lake Russell in east-central California, also provides excellent dating control for the interval between 30,000 and 16,000 years [Negrini and Davis, in press]. Four TL ages on the associated tephra also contribute to the age control of this interval; with ages of 24,300 +/- 2,700, 50,200 +/- 3,400, 67,300 +/- 7,500, and 102,000 +/- 1,100 yrs. B.P. [Berger et al., 1990]. Finally, a K/Ar age of 72,000 +/- 6,000 yrs. B.P. for tephra 1 [Negrini et al., 1988] constrains one of the oldest intervals considered for this study. Ages for all sampled intervals were estimated by linear interpolation from adjacent dated horizons.

METHODS

The samples used for this study were collected at 10 cm intervals from the upper 8 m of the high resolution record from the Ana River canyon exposed at sections C, and E (Fig. 1b). Paleomagnetic sampling cubes containing about 6 cm³ (2 cm X 2 cm X 1.5 cm) of lacustrine sediments collected by R.M. Negrini in 1987 (sampling procedure and sample description in Negrini et al., 1988), were processed to recover ostracodes.

Seventy seven samples were prepared according to a slightly modified version of Forester's [1991b] freeze and thaw procedure. Residuals were analyzed under a low power stereoscopic microscope. Routine paleontological study of all fossiliferous samples was performed to determine fossil content and faunal composition. Between 100 and 35,000 ostracode valves per gram of sample were recorded. Ostracode relative abundance and assemblages were defined by counting 300 specimens per sample. A qualitative "salinity index" was used to estimate salinity trends based on weighted average species proportions. This index is:

$$SI = 3(\% \text{ L. sappaensis}) + 2(\% \text{ L. bradburyi}) + \% \text{ L. ceriotuberosa} - (\% \text{ C. patzcuaro} + \quad (1)$$
$$2(\% \text{ C. caudata}) + 3(\% \text{ C. lacustris}))$$

The index positively weights species with incrementally higher salinity tolerances and negatively weights species with incrementally lower salinity tolerances.

Species identification, adult/juvenile and carapace/valve ratios were determined for all samples. Taphonomic features including fragmentation, encrustation, abrasion, oxidation and coating were also recorded to distinguish autochthonous from allochthonous individuals. These allowed us to select the ideal specimens for shell chemistry analysis. Eleven intervals did not contain any fossils; they consisted entirely of tephra. Of the remaining 66 samples 16 were selected for chemical analysis since they were associated with the best geochronologically controlled intervals. Individuals from the cleanest horizons were separated for shell chemistry analysis. Specimens from the

other 50 beds were either coated or encrusted with CaCO₃ or the samples had no significant number of adult individuals for analysis.

Fifteen pristine valves of Limnocythere ceriotuberosa, the most common species present at Pluvial Lake Chewaucan, were removed from each of the 16 stratigraphic horizons selected for spectrometric analysis. The specimens were thoroughly cleaned with ultra (4 times) distilled water and a fine (000) brush. Valves were weighed in a Cahn 29 electronic balance and then cut into two halves with a microsurgery scalpel. One half of each specimen was mounted into a micropaleon-tological slide per sample and sent to the Stable Isotope Research Laboratory at the University of Michigan for stable isotope ($\delta^{18}O$ and $\delta^{13}C$) analysis, using a Finnigan Mat 251 mass spectrometer. The other half of each specimen was kept at the Department of Geosciences of the University of Arizona for trace metal (Mg and Sr) study, using a VG inductively-coupled plasma mass spectrometer (ICP-MS). In an attempt to compare and correlate isotopic information, trace element analysis was performed only on the correlative shell fragments analyzed for stable isotopes. As a result of size and weight limitations a total of about 7 specimens per sample were jointly analyzed in this fashion. Calcium content in ostracode valves was determined stoichiometrically as a result of the instrument sensitivity to major elements like calcium. For purposes of consistency only those horizons from which both stable isotope and trace element data were secured were incorporated in the geochemistry portion of this analysis.

RESULTS
PALEOECOLOGIC DATA

Figure 3 summarizes the paleontologic and paleoecologic records for Pluvial Lake Chewaucan ostracodes. All fossiliferous samples are characterized by a relatively low diversity fauna comprising a variety of North American species. A total of eight species occur throughout the stratigraphic sequence. Figure 3 (a-i) synthesizes the total and relative abundance profiles for each species present. The ostracode abundance diagram (Fig. 3a) indicates the concentration of ostracodes of all species through the section. The relative abundance diagrams per species (Figs. 3b-i) are arranged according to decreasing tolerance to salinity, with the highest salinity tolerant species to the left and the more freshwater species to the right.

Based upon the relative abundance diagrams, six paleoecologic intervals are recognized. Limnocythere ceriotuberosa (the most common species) alternates with other species throughout the stratigraphic sequence. Interval 1, between 525 cm and 455 cm below datum (~110,000 to ~102,000 years B.P.), is characterized by the co-occurrence of L. ceriotuberosa, C. caudata, and C. patzcuaro. Interval 2, between 455-295 cm below datum (~102,000 to ~89,000 years B.P.), comprises almost entirely L. ceriotuberosa and L. sappaensis with minor occurrence

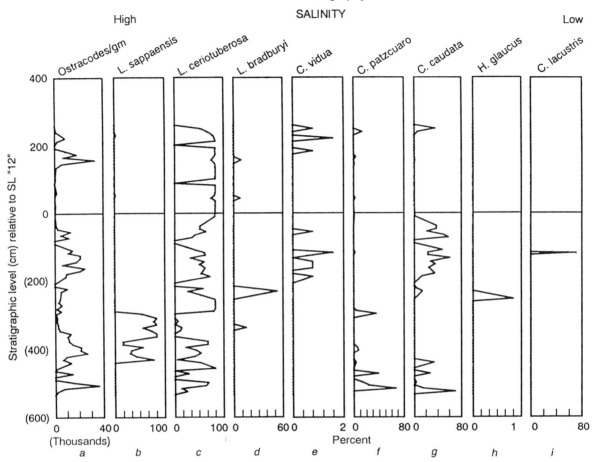

Pluvial Lake Chewaucan
Ostracode Stratigraphy

Fig. 3. Total (diagram a) and relative abundance (diagrams b-i) of ostracode species present at pluvial Lake Chewaucan during the Wisconsin. Note the strong dominance of *Limnocythere ceriotuberosa* and the ocassional occurrence of *Cytherissa lacustris* through the stratigraphic sequence. See text for explanation.

of *L. bradburyi* and *C. patzcuaro*. Interval 3, between 295 and 125 cm below datum (~89,000 to ~76,000 years B.P.), is dominated by *L. ceriotuberosa* and *C. caudata* with occasional occurrences of *Cypridopsis vidua*.

Interval 4, between 125-120 cm below datum and only ~2,000 years in duration (~76,000 to ~74,000 years B.P.) is defined by the presence of *Cytherissa lacustris*. Interval 5, between 120 cm below datum and 185 cm above datum (~74,000 to ~22,000 years B.P.) almost entirely consists of *L. ceriotuberosa* with minor occurrences of *L. bradburyi* and *C. patzcuaro*. Interval 6, between 185 to 261 cm above datum (~22,000 to 16,000 years B.P.), is characterized by a nearly monospecific assemblage of *L. ceriotuberosa* with erratic occurrences of *L. sappaensis, C. caudata* and *Cypridopsis vidua*.

Based upon the relative abundances shown in figure 3, a paleoenvironmental interpretation was generated (Fig. 4a).

Interval 1 assemblage suggests low to moderate but increasing salinity. Interval 2 assemblage dominated by *L. ceriotuberosa* and *L. sappaensis* (both eurythermic) suggests moderate to high salinity conditions. Interval 3 assemblage, characterized by *L. ceriotuberosa, C. caudata* and *Cypridopsis vidua*, strongly suggests low to moderate salinity conditions and increasing precipitation. The interval 4 assemblage, marked by the occurrence of *C. lacustris*, indicates the rapid development and disappearance of cold-, freshwater conditions. In interval 5 the occurrence of *L. ceriotuberosa* and the occasional appearance of *Cypridopsis vidua* and *L. bradburyi* suggests that salinity increased beyond *C. lacustris*'s maximum tolerance after ~74,000 years B. P. These conditions prevailed until ~22,000 B.P.; from this horizon (interval 6) to the end of the record an erratic pattern of ostracode species (*L. ceriotuberosa, C. vidua,* and occasional *C. caudata, C.*

Pluvial Lake Chewaucan
Paleoenvironmental Trends

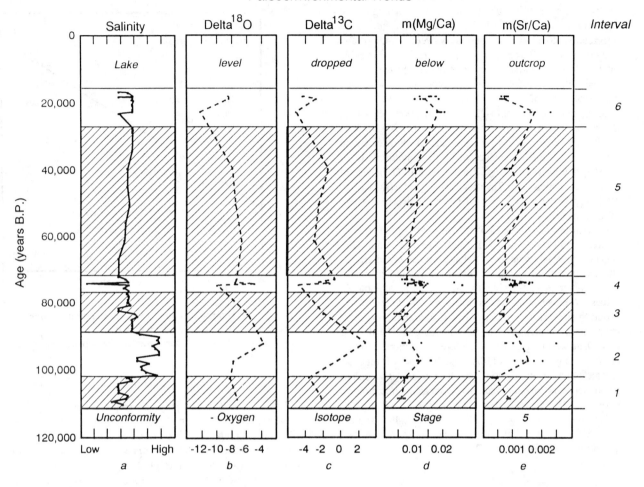

Fig. 4. Comparative trends from paleoecologic and geochemical profiles; a) salinity index generated from the appearance/disappearance pattern and the weigh average of the five most significant species; b) oxygen isotope trend; c) carbon isotope trend; d) Mg/Ca molar ratio trend; and e) Sr/Ca molar ratio trend of *Limnocythere ceriotuberosa* (the most common species present throughout the stratigraphic sequence). See text for explanation.

patzcuaro and *L. sappaensis*) suggests more unstable conditions which concluded with the lake level dropping below the Ana River section.

STABLE ISOTOPE DATA

Stable isotope values from *L. ceriotuberosa* are plotted in figures 4b and 4c. The six intervals recognized from the paleoecologic data are also evident from the isotope data. A covariant trend is evident through most of the sequence, except for a horizon at ~75,000 years B.P. and between ~22,000 and ~16,000 years B.P. where the trends are opposite. A secular isotopic trend for both the δ^{18}O and δ^{13}C values is discernible toward lower values (Fig. 4b).

OXYGEN

The δ^{18}O values during interval 1 remain stable within a one per mil range (-6°/$_{\infty}$ to -7°/$_{\infty}$) between ~110,000 and ~102,000 years B.P. (oxygen isotope stage 5d), suggesting stable temperatures and/or humidity through this term. Interval 2 between ~102,000 and 89,000 years B.P. (oxygen isotope stage 5c) is characterized by higher δ^{18}O values increasing from -7°/$_{\infty}$ to -4°/$_{\infty}$; this interval's pattern is consistent with the marine oxygen isotope stage 5 record [Martinson et al, 1987].

Interval 3 (~89,000 to ~76,000 years B.P.) marked by a decline towards more negative δ^{18}O values (-4°/$_{\infty}$ to -6°/$_{\infty}$) supports the conclusion of interglacial climate (oxygen isotope stage 5b and a). A minor reduction of δ^{18}O values

(-6°/$_{oo}$ to -7°/$_{oo}$) marks interval 4, the beginning of the last glacial (oxygen isotope stage 4). A strong drop in isotopic values (-6°/$_{oo}$ to -11°/$_{oo}$) which is consistent with the paleoecologic record, suggests the sharp freshening of the Chewaucan Basin. Interval 5 provides scarce but useful information for the time period between ~74,000 and ~22,000 years B.P. (oxygen isotope stage 3). During this interval little variation, if any, is recorded by the oxygen isotope; the isotopic data are consistent with the paleoecologic profile.

During interval 6, from ~22,000 to ~16,000 years B.P. (lower part of oxygen isotope stage 2), the δ^{18}O values reach their lowest values (-7°/$_{oo}$ to -11°/$_{oo}$) followed by an erratic trend of the δ^{18}O values which fluctuate from -11°/$_{oo}$ to -8°/$_{oo}$ and back to -10°/$_{oo}$. The range of δ^{18}O values are in good agreement with Taylor's [1974, in: Hoefs, 1980; p. 104] diagram typical of closed basins subject to intense evapotranspiration.

The long-term decrease in δ^{18}O values during the interval between 100,000 years B.P. and ~16,000 years B.P. suggests three alternatives: 1) changes in temperature, 2) changes in δ^{18}O of water, or 3) both. It is unlikely that a single factor is responsible for the variation of isotopic values. We assume that the decrease in δ^{18}O of lake waters is due to an increasing input of runoff and precipitation as temperature lowered and humidity increased. The decline in δ^{18}O shown in figure 4b is therefore a minimum estimate of the change in δ^{18}O of the lake because of regional climate variation during the late Pleistocene. The pollen record supports the interpretation of increasing precipitation prior to ~25,000 years B.P. [Mehringer, 1985, fig. 4; Beiswenger, 1991, fig. 6]. Isotopically lighter waters accumulated in the Lake Chewaucan Basin as the last glaciation progressed.

CARBON

The δ^{13}C values are covariant with the δ^{18}O (r=0.76) following the same interval pattern discussed above (Fig. 5). These δ^{13}C values are interpreted in terms of dissolved inorganic carbon (DIC) and alkalinity/salinity, rather than as paleoproductivity as is usually the case [Lister, 1988; Talbot and Kelts, 1990]. Values of δ^{13}C in closed basins are less influenced by primary productivity than are open basins [Stiller and Hutchinson, 1980]; rather variations in δ^{13}C in closed basins results from preferential outgassing of ^{12}C-rich CO_2 from the lake surface [Talbot and Kelts, 1990]. The characteristic covariant trend shown by δ^{18}O and δ^{13}C of pluvial Lake Chewaucan provides evidence for a long-term evolution of lake waters within a hydrologically closed basin environment.

The consistent covariant trends of both stable isotope curves is broken in two intervals (Fig. 4c). During interval 4 (~76,000 to ~74,000 years B.P.) the two curves are out of phase. A lack of covariance also occurs at the end of interval 6 (~22,000 to ~16,000 years B.P.). Rapid freshening and increased volume of Lake Chewaucan are probably responsible for these brief events.

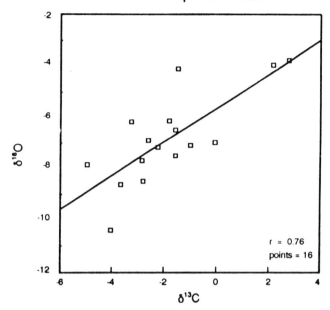

Fig. 5. Oxygen and carbon isotopes show a positive correlation coefficient which is consistent with a closed basin interpretation for pluvial Lake Chewaucan. See text for explanation.

TRACE ELEMENT DATA

Lake hydrochemistry plays an important role in pluvial Lake Chewaucan ostracode shell chemistry. Two variables control Mg uptake in ostracode shells, temperature and water chemistry, whereas Sr uptake is controlled only by water composition [Chivas et al., 1986a]. If the trends are covariant, as it is the case throughout most of the Lake Chewaucan record, trace element concentration in ostracode valves must be primarily driven by water chemistry. Alternating conditions are consistent with an evaporative basin subject to rapid fluctuations in salinity.

Trace metal concentrations from *L. ceriotuberosa* are plotted in figures 4d-e. Well-defined covariant trends for Mg/Ca and Sr/Ca ratios are evident throughout the section suggesting that salinity controlled the Mg and Sr content in ostracode valves. Tight clusters indicate small temperature and salinity variations more likely in a fairly deep and stable lake, whereas broad fluctuations of values imply variable conditions more probable under shallower lake conditions.

The Mg/Ca and Sr/Ca molar ratio trends generally show a zonation equivalent to that observed in assemblage and isotope data. Interval 1 from ~110,000 to ~102,000 years B.P. shows little variation in molar ratios indicating small salinity fluctuations within horizons meaning that Mg concentration rose as the lake level rose but Sr decreased probably in response to increasing precipitation.

Interval 2 (~102,000 to ~89,000 years B.P.) is

characterized by higher and variable values of the Mg/Ca and Sr/Ca ratios indicating rising salinity. This is consistent with paleoecologic and isotopic data. A decrease in both ratios and tighter clustering of values occurred during interval 3 suggesting a gradual decrease of salinity (~89,000 to ~76,000 years B.P.). This event was followed by a rapid drop in trace metal concentrations suggesting freshwater conditions (interval 4: ~76,000 to ~74,000 years B.P.). As with the stable isotope data we have little information for interval 5 (~74,000 to ~22,000 years B.P.) but available points suggest a similarity with the paleoecologic profile. Although at ~50,000 years B.P. lake conditions became more variable (as shown by the broad fluctuations of values) and probably more saline (as shown by higher Sr/Ca molar values).

During interval 6 (~22,000 to ~16,000 years B.P.) both Mg and Sr ratios increased erratically suggesting that salinity increased but kept fluctuating at this time. The apparent reversal between stable isotope and trace metal trends strongly suggests that dilute water entered a salt filled basin.

PALEOCLIMATIC HISTORY OF LAKE CHEWAUCAN

Pluvial Lake Chewaucan was a closed basin sensitive to regional climatic changes for the interval between ~110,000 and ~16,000 years B.P. Immediately after a period of aridity (prior to interval 1), before ~110,000, which produced deflation in the basin (unconformity at the base of the stratigraphic column associated with marine oxygen isotope stage 5), it began to fill as indicated by the ostracode species *L. ceriotuberosa*, *C. caudata* and *C. patzcuaro*.

A well-defined correspondence of the paleoecologic profile (increasing diversity and abundance) and the geochemical (stable isotopes and trace elements) patterns suggest temperature and salinity decreases (the latter probably associated with increasing humidity and higher precipitation and stream discharge) during interval 1.

The occurrence of *L. sappaensis* and *L. ceriotuberosa*, between 102,000 and 89,000 years B.P. (interval 2) associated with increasing δO^{18} isotopic values and variable but higher Mg/Ca and Sr/Ca ratios indicates that Lake Chewaucan was shrinking in response to climate change.

From 89,000 to 76,000 years B.P. (interval 3) climatic conditions began to shift towards cooler/wetter conditions as evidenced by a lower salinity faunal assemblages, more negative $\delta^{18}O$ and $\delta^{13}C$ values, and less variable Mg/Ca and Sr/Ca ratios.

Lake temperature and salinity declined briefly to very low values between ~76,000 and ~74,000 years B.P. as evidenced by all three profiles (interval 4). During this interval both the $\delta^{18}O$ values (from -6°/$_{oo}$ to -11°/$_{oo}$) and the low Mg and Sr content in ostracode valves are congruent with a cold-, freshwater environment. The brief lack of covariance between $\delta^{18}O$ and $\delta^{13}C$ at this time suggests that pluvial Lake Chewaucan underwent sudden

flooding as a result of rapidly increasing precipitation/evaporation ratios. It is possible that the lake overflowed into Alkali Basin at this time, although no evidence of surface connection between the Lake Chewaucan and Alkali Lake basins has been recognized, to date (D. Freidel, pers. comm., 1992).

Faunal data suggest that pluvial Lake Chewaucan remained a fairly stable basin during interval 5, from ~74,000 to ~22,000 years B.P., with smaller temperature fluctuations and slightly greater salinity variations than occurred in the prior ~50,000 years. However, a lack of geochemical information limits our confidence in interpretations of this interval. Geochemical and paleoecologic profiles suggest that rainfall and runoff were moderately high but "stable"; that is, water input versus output do not vary significantly for this time period.

By ~22,000 to ~16,000 years B.P. (interval 6) a sharp climatic change is recorded by paleoecologic and geochemical data which correlate with the cold-dry characteristics of the full glacial [Benson and Thompson, 1987; Davis, 1989; and Smith and Street-Perrot, 1983]. Drastic negative shifts in isotopic values associated with higher and wider spread trace metal values strongly suggest that pluvial Lake Chewaucan was subject to dilute water input in an evaporative basin probably during low temperature conditions. Towards the end of the record the paleoecologic and geochemical profiles are consistent. All data sets indicate that this lacustrine basin underwent fluctuating conditions, but remained at low temperatures. Another brief excursion towards fresher water conditions took place at this time although less intense than that at ~75,000 years ago (since salinity did not drop to a viable level for *Cytherissa lacustris*). The co-existence of *Candona caudata* and *Cypridopsis vidua* indicate that Lake Chewaucan received significant groundwater discharge or there is a wetland marginal to a stream, mechanisms which may be responsible for the relative dilution of the water chemistry and lake level rise to about 1330 masl [Allison, 1982].

DISCUSSION

The combined paleoecologic and geochemical (stable isotopes and trace elements) profiles provide a powerful tool for understanding Pleistocene climatic changes in western North America. For the first time paleoecologic records are combined with geochemical information generated from ostracodes to recognize the paleoclimatic signature of the Pleistocene on paleolakes. This research also allowed us to present one of the first approaches in using the same species and specimens to analyze trace elements and stable isotopes through mass spectrometry (the other is that of Chivas and co-workers in this volume). Although slightly different both approaches have proved to be highly reliable. The greates significance of this analysis is to demonstrate that it is possible to determine the paleohydrochemical history of a lake and to recognize the effects of climate on it. However, we acknowledge that more stratigraphic horizons at closer intervals would provide a more detailed record of Lake Chewaucan paleohydrochemistry.

The occurrence of environmentally sensitive species, like *L. sappaensis, L. ceriotuberosa, C. caudata, C. patzcuaro, Cypridopsis vidua* and *Cytherissa lacustris*, provide the basic criteria to determine paleohydrochemical changes in the Chewaucan Basin. Covariant trends of stable isotope data collected from *L. ceriotuberosa* strongly suggest that Lake Chewaucan remained a closed basin throughout its history with the possible exception of brief intervals at ~75,000 years B.P and between 22,000 and 16,000 years B.,P. Trace metal trends indicate that salinity played an important role in the hydrochemical evolution of Lake Chewaucan; these patterns are consistent with the paleoecologic profile and somehow to the stable isotope profiles. By combining these data it is possible to speculate about the climatic history of the Lake Chewaucan region.

Variations in local climate and Lake Chewaucan hydroclimate during the Pleistocene likely reflect regional changes in the western North America climate regime. Of primary importance in determining regional climate fluctuations, especially in temperate climates, is the position of the jet stream. Late Pleistocene western North America lake evolution suggests that the jet stream was driven mainly by the height and extent of the Laurentide ice-sheet [Benson and Thompson, 1987]. Accordingly, the ice-sheet produced a high pressure cell over the continent which in turn induced changes in the atmospheric circulation pattern forcing the jet stream southward [Kutzbach, 1987; COHMAP group, 1988]. The COHMAP group [1988] has proposed that the jet stream split into a north and a south branch which influenced climate during the full glacial (~18,000-20,000 years B.P.). If this model is correct, greater and stronger storms should be expected along western North America during episodes under the jet stream influence. The COHMAP model will be applied throughout the Lake Chewaucan record.

It is likely that during the last interglacial the jet stream had a position similar to the present. A gradual south shifting of the jet stream split south branch would introduce wetter and colder conditions to the area between ~110,000 and ~102,000 years B.P. As the jet stream moved south greater and stronger storms affected the Lake Chewaucan area. A year-round winter-like episode characterized the ~76,000 to ~74,000 years B. P. period.

If the COHMAP model is suitable, the jet stream probably retreated northward during the next ~50,000 years in response to the ice-cover reduction and then remained more or less stable until ~25,000 year. However, movement of the jet stream southward ~25,000 years B.P. is documented by woodrat midden data from further south. P.E. Wigand (pers. comm., 1992) reports that white bark pine pollen (an indicator of wetter conditions) occurs in woodrat middens near Reno, Nevada below 1400 masl prior to ~23,000 years

B.P. Today it is not present at that altitude. Further south in the Pahrangat Range of southeastern Nevada, whit fir drops to elevations below 1585 masl at the same time (P.E. Wigand, pers. comm. 1992). The paleoecologic profile from Lake Chewaucan suggests that minor climatic changes occurred in this region for that time interval. Clear, Grays and Tulare lakes provide strong evidence in support of our hypothesis since for the same time period both basins indicate transitional cold-wet to cold-dry to warm-dry (short interval) conditions [Sims et al. 1988; Atwater et al., 1986; Beiswenger, 1991]. An equivalent interpretation from Lake Chewaucan is presented in this paper. However, Lake Chewaucan suggests colder temperatures as a result of both latitudinal and topographic relief differences. The microclimatic effects shown by each basin discussed in this paper do not mask the greater influence of the jet stream throughout northwestern North America.

Our stable isotope values are in good agreement with isotopic data from distant basins [Lake Searles: Phillips, 1989 in: Benson et al., 1990] supporting the idea that a strong southerly jet stream incursion affected western North America during the Wisconsin glaciation. Similarly, our trace element data indicate alternating periods of water dilution and increasing total dissolved solids probably in response to changes in climate. In consequence, ostracode paleoecology and geochemistry contribute to make a high resolution reconstruction of Lake Chewaucan paleohydrologic and paleoclimatic evolution.

Acknowledgements. This project has been possible through NSF support (EAR-8803406). We thank Dr. R. Negrini for providing us with the sediment samples used for this study and helpful suggestions to the manuscript. Dr. K. C. Lohman and colleagues from the Stable Isotope Laboratory of the University of Michigan performed oxygen and carbon isotope analyses on Lake Chewaucan ostracode valves. Drs. A. Long and O.K. Davis from the Department of Geosciences of the University of Arizona, Dr. P. E. Wigand from the Desert Research Institute of the University of Nevada-Reno, Dr. Richard M. Forester from the U.S. Geological Survey at Denver and Dr. Charles G. Oviatt from the Department of Geology, Kansas State University, made helpful contributions to the improvement of this paper.

REFERENCES

Adam, D. P., Correlations of the Clear Lake, California, core CL-73-4 pollen sequence with other long climate records, *Geol. Soc. Amer., Spec. Paper 214*, pp. 81-95, 1988.

Allison, I. S., Geology of Pluvial Lake Chewaucan, Lake County, Oregon, *Oregon State University Press*, Corvallis, Oregon, 78 pp.,1982.

Atwater, B. F., D. P. Adam, J. P. Bradbury, R. M. Forester, R. K. Mark, W. R. Lettis, G. R. Fisher, K. W. Gobalet, and S. W. Robinson, A fan dam for Tulare Lake, California, and implications for the Wisconsin glacial history of the Sierra Nevada, *Geol. Soc. Amer. Bull., 97*, pp. 97-109, 1986.

Bachhuber, F. W., Quaternary history of the Estancia Valley, central New Mexico, *New Mexico Geol. Soc. Guidebook, 33rd Field Conference*, pp.343-346, Albuquerque County, 1982.

Banfield, J. F., B. F. Jones and D. R. Veblen, An AEM-TEM study of weathering and diagenesis, Abert Lake, Oregon: I. Weathering reactions in the volcanics, *Geochim. Cosmochim. Acta, 55,* pp. 2781-2793, 1991a.

Banfield, J. F., B. F. Jones and D. R. Veblen, An AEM-TEM study of weathering and diagenesis, Abert Lake, Oregon: II. Diagenetic modification of the sedimentary assemblage, *Geochim. Cosmochim. Acta, 55,* pp. 2795-2810, 1991b.

Benson, L. V., Fluctuations in the level of pluvial Lake Lahontan during the last 40,000 years, *Quat. Res., 9,* pp. 300-318, 1978.

Benson, L. V. and R. S. Thompson, Lake-level variation in the Lahontan Basin for the past 50,000 years, *Qtny. Res., 28,* pp. 69-85, 1987.

Benson, L. V., D. R. Currey, R. I. Dorn, K. R. Lajoie, C. G. Oviatt, S. W. Robinson, G. I. Smith and S. Stine, Chronology of expansion and contraction of four Great Basin lake systems during the past 35,000 years, *Paleogeogr., Paleoclimatol., Paleoecol., 78,* pp. 241-286, 1990.

Berger, G. W., J. O. Davis and R. M. Negrini, Thermoluminescence dating of tephra from Oregon and Nevada, *INQUA-ICCT Field Conference and Workshop on Tephrochronology abstracts,* p. 3, 1990.

Beiswenger, J.M., Late Quaternary vegetational history of Grays Lake, Idaho, *Ecol. Monogr., 61* (2), pp. 165-182, 1991.

Cameron, S. P. and R. F. Lundin, Environmental interpretation of the ostracode succession in late Quaternary sediments of pluvial Lake Cochise, southeastern Arizona, *Sixth Intnal. Ostracoda Symp.,* pp. 335-352, Saalfelden, 1977.

Chivas, A. R., P. De Deckker and J. M. G. Shelley, Magnesium, strontium, and barium partitioning in nonmarine ostracode shells and their use in paleoenvironmental reconstructions: A preliminary study, in: *Applications of Ostracoda,* edited by R. F. Maddocks, pp.238-249, Dept. Geosc., University of Houston, Houston, 1983.

Chivas, A. R., P. De Deckker and J. M. G.Shelley, Magnesium and strontium in non-marine ostracod shells as indicators of paleosalinity and paleotemperature, *Hydrobiologia, 143,* pp. 135-142, 1986a.

Chivas, A. R., P. De Deckker and J. M. G. Shelley, Magnesium content of non-marine ostracod shells: a new paleosalinometer and paleothermometer, *Paleogeogr., Paleoclimatol., Paleoecol., 54,* pp. 43-61, 1986b.

COHMAP, Climatic changes of the last 18,000 years: Observations and model simulations, *Science, 241,* pp. 1043-1052, 1988.

Currey, D. R., Quaternary paleolakes in the evolution of semidesert basins, with special emphasis on Lake Bonneville and the Great Basin, U. S. A., *Paleogeogr., Paleoclimatol., Paleoecol., 76,* pp. 189-214, 1990.

Davis, J. O., Correlation of late Quaternary Tephra Layers in a long pluvial sequence near Summer Lake, Oregon, *Qtny. Res., 23,* pp. 38-53, 1985.

Davis, O. K., The regionalization of climatic change in Western North America, in: *Paleoclimatology and paleometeorology: Modern and past patterns of global atmospheric transport,* edited by M. Leinen and M. Sarnthein, pp. 617-636, Kluwer Academic Publishers, 1989.

De Deckker, P. and R. M. Forester, The use of ostracodes to reconstruct paleoenvironmental records, in: *Ostracoda in the Earth Sciences,* edited by P. De Deckker, J. P. Colin and J. P. Peypouquet, pp. 175-200, Elsevier Scientific Publishers, Amsterdam, The Netherlands, 1988.

De Deckker, P., A. R. Chivas and J. M. G. Shelley, Paleoenvironment of the Messinian Mediterranean 'Lago Mare' from strontium and magnesium in ostracode shells, *Palaois, 3,* num. 3, pp. 352-358, 1988a.

De Deckker, P., A. R. Chivas, J. M. G. Shelley and T. Torgentsen, Ostracod shell chemistry: a new paleoenvironmental indicator applied to regressive/transgressive record from the Gulf of Carpentaria, Australia, *Paleogeogr., Paleocliamtol., Paleoecol., 66,* pp. 231-241, 1988b.

Delorme, L. D., Ostracodes as Quaternary paleoecological indicators, *Can. Jour. Earth Sci., 6,* pp. 1471-1476, 1969.

Delorme, L. D., Methods in Quaternary ecology #7. Freshwater ostracodes, *Geosci. Can., 16,* num. 2, pp. 85-90, 1989.

Eyles, N. and H. P. Schwarcz, Stable isotope record of the last glacial cycle from lacustrine ostracodes, *Geol., 19,* pp. 257-260, 1991.

Forester, R. M., Relationship of two lacustrine ostracode species to solute composition and salinity: implications for paleohydrochemistry, *Geol., 11,* pp. 435-438, 1983.

Forester, R. M., Determination of the dissolved anion composition of ancient lakes from fossil ostracodes, *Geol., 14,* pp. 796-799, 1986.

Forester, R. M., Late Quaternary paleoclimate records from lacustrine ostracodes, in: *North America and adjacent oceans during the last glaciation,* edited by W. F. Ruddiman and H. E. Wright Jr., *Geol. Soc. America,* The Geology of North America, *3-K,* pp. 261-276, 1987.

Forester, R. M., Pliocene-climate history of the western United States derived from lacustrine ostracodes, *Quat. Sci. Rev., 10,* pp. 133-146, 1991a.

Forester, R. M., Ostracode assemblages from springs in the western United States: Implications for paleohydrology. *Mem. ent. Soc. Can., 155,* pp. 181-201, 1991b.

Heusser, C. J., Palynology and phytogeographical significance of a Late-Pleistocene refugium near Kalaloch, Washington, *Quat. Res., 2,* pp. 189-201, 1972.

Hoefs, J., Stable isotope geochemistry, *Springer-Verlag,* Berlin, Germany, p. 104, 1980.

Kutzbach, J. E., Model simulations of the climatic patterns during the deglaciation of North America, in: *North America and adjacent oceans during the last deglaciation,* edited by W. F. Ruddiman and H. E. Wright, pp.425-446, *Geol. Soc. America,* The Geology of North America, v. K-3, Boulder, Colorado, 1987.

Lister, G. S., Stable isotopes from lacustrine Ostracoda as tracers for continental palaeoenvironments, in: *Ostracoda in the Earth Sciences,* edited by P. De Deckker, J. P. Colin and J. P. Peypouquet, pp. 201-218, Elsevier Science Publs., Amsterdam, The Netherlands, 1988.

Markgraf, V., J. P. Bradbury, R. M. Forester, W. McCoy, G. Singh and R. S. Sternberg, Paleoenvironmental reassessment of the 1.6 million-year-old record from San Agustin Basin, New Mexico, *New Mexico Geol. Soc. Guidebook, 34th Field Conference, Socorro Region II,* pp. 291-297, 1983.

Markgraf, V., J. P. Bradbury, R. M. Forester, G. Singh, and R. S. Sternberg, San Agustin Plains, New Mexico: Age and paleoenvironmental potential reassessed, *Quat. Res., 22,* 336-343, 1984.

Mehringer, P. J. Jr., Late-Quaternary pollen records from the interior Pacific northwest and northern Great Basin of the United States, in: *Pollen records of late-Quaternary North American sediments,* edited by V. M. Bryant and R. G. Holloway, *Amer., Assoc. Strat. Palynol. Found.,* pp.165-187, 1985.

Negrini, R. M., K. L. Verosub and J. O. Davis, The middle to late Pleistocene geomagnetic field recorded in fine-grained sediments from Summer Lake, Oregon, and Double Hot Springs, Nevada, U. S. A., *Earth Planet. Sci. Letters, 87*, pp. 173-192, 1988.

Negrini, R. M. and J. O. Davis, Correlation of two paleomagnetic secular variation records from western North America and its application as a high-resolution chronological tool for late Pleistocene, non-marine sedimentary sequences, *Quat. Res.,* in press.

Oviatt, C. G., Late Pleistocene and Holocene lake fluctuations in the Sevier Lake Basin, Utah, U. S. A., *Jour. Paleolimnol., 1,* pp. 9-21, 1988.

Oviatt, C. G., D. R. Currey, and D. M. Miller, Age and paleoclimatic significance of the Stansbury Shoreline of Lake Bonneville, Northeastern Great Basin, *Quat. Res., 33,* pp. 291-305, 1990.

Robinson, S. W., D. P. Adam, and J.D. Sims, Radiocarbon content, sedimentation rates, and a time scale for core CL-73-4 from Clear Lake, California, *Geol. Soc. Amer., Special Paper 214,* pp. 151-160, 1988.

Sims, J. D.(editor), Late Quaternary climate, tectonism, and sedimentation in Clear Lake, northern California Coast Ranges, *Geol. Soc. Amer., Special Paper, 214,* 221 pp., 1988.

Stiller, M., and G. E. Hutchinson, The waters of Merom: A study of Lake Huleh, part 1- Stable isotopic composition of carbonates of a 54 m core, paleoclimatic and paleotrophic implications, *Archiv fur Hydrobiologie, 89,* pp. 275-302, 1980.

Talbot, M. R. and K. Kelts, Paleolimnological signatures from carbon and oxygen isotopic ratios in carbonates from organic carbon-rich lacustrine sediments, in: *Lacustrine basin exploration: Cases studies and modern analogs,* edited by B. J. Katz, pp. 99-112, AAPG Mem. 50, 1990.

Teeter, J. W. and T. J. Quick, Magnesium-salinity relation in the saline lake ostracode *Cyprideis americana, Geol., 18,* pp. 220-222, 1990.

Thompson, R. S., L. J. Toolin, R. M. Forester, and R. J. Spencer, Accelerator-mass spectrometer (AMS) radiocarbon dating of Pleistocene lake sediments in the Great Basin, *Paleogeogr., Paleoclimatol., Paleoecol., 78,* pp. 301-313, 1990.

Turpen, J. B. and R. W. Angell, Aspects of moulting and calcification in the ostracod *Heterocypris, Biol. Bull., 140,* pp. 331-338, 1971.

Van Denburgh, A. S., Solute balance at Abert and Summer Lakes, south-central Oregon, *U. S. Geol. Survey Prof. Paper, 502-C,* 29 pp., 1975.

Wells, S. G., R. Y. Anderson, L. D. McFadden, W. J. Brown, Y. Enzel and J. L. Miossec, Late Quaternary Paleohydrology of the eastern Mojave River drainage, southern California: Quantitative assessment of the late Quaternary hydrologic cycle in large arid watersheds, *N. Mex. Water Resources Res. Inst., New Mexico State Univ., Rep. 242,* 253 pp., 1989.

B. Blank, A.S. Cohen, M.R. Palacios-Fest and J. Ruiz, Department of Geosciences, University of Arizona, Tucson, AZ 85721.

Continental Paleothermometry And Seasonality Using The Isotopic Composition Of Aragonitic Otoliths Of Freshwater Fishes

WILLIAM P. PATTERSON, GERALD R. SMITH, KYGER C LOHMANN

Department of Geological Sciences, University of Michigan

To investigate the applicability of oxygen isotope thermometry using fish aragonite, the $\delta^{18}O$ values of paired otolith and water samples were analyzed from six large modern temperate lakes. Otoliths are accretionary aragonitic structures which are precipitated within the sacculus of fish ears. Deep-water obligate benthic species from the hypolimnion of the Laurentian Great Lakes of North America and Lake Baikal, Siberia, provided cold-water end member values for aragonite $\delta^{18}O$. Warm-water values were obtained from naturally grown warm-water stenothermic species and from fish grown in aquaria under controlled conditions. These two groups, which represent growth over a temperature range of 3.2-30.3°C, were employed to determine the oxygen isotope temperature fractionation relationship for aragonite-water : $10^3 \ln\alpha = 18.56 \ (\pm 0.319) \cdot (10^3) \ T^{-1} \ K \ -33.49 \ (\pm 0.307)$.

Empirical calibration of a fish aragonite thermometry equation allows its direct application to studies of paleoclimate. For example, high-resolution sampling of shallow-water eurythermic species coupled with a knowledge of the isotopic composition of meteoric waters can be used to determine seasonal temperature variation. This approach was tested using a modern shallow-water eurythermic species from Sandusky Bay, Lake Erie. Temperatures calculated from carbonate composition agree with meteorological records from the Sandusky Bay weather station for the same time period.

INTRODUCTION

Studies of paleoclimate usually employ proxies which provide only time-average estimates of temperature; however, intra-annual variation in temperature, seasonality, is a significant causal variable when considering processes that control the distribution, evolution, and extinction of organisms [Fischer, 1960; Hibbard, 1960; Axelrod, 1967; Graham and Meade, 1987] and surficial geological processes in temperate regions. Organisms with rapid growth of accretionary carbonate structures offer potential for measurement of seasonality through detailed microsampling of oxygen isotopic variation preserved in growth bands [Killingley and Berger, 1979; Krantz et al., 1987; Aharon, 1991; Dettman and Lohmann, this volume]. Several studies have utilized such an approach for evaluating intra-annual records of seasonality for a variety of biotic carbonates including bivalves [Krantz et al., 1987; Erlenkeuser and Wefer, 1981; Jones et al., 1989], and corals [Pätzold, 1986]. In contrast, studies using fish otoliths have been limited and these rarely have been able to resolve other than average temperature during the life of the fish [Devereux, 1967; Degens et al., 1969;

Mulcahy et al., 1979; Radtke, 1984a, b; Wefer, 1985; Kalish, 1991a, b].

Teleost otoliths, accretionary aragonitic structures [Carlstrom, 1963] grown in fish ears, are ideal for reconstructing intra-annual temperature records. These occur in three pairs, the largest are usually the sagittae which are part of the hearing apparatus in the sacculae. Otolith accretion may exceed 1 mm/yr. in some species and develop a visible rhythmicity representing daily accretion [Pannella, 1971; Brothers et al., 1976; Campana, 1989]. These daily patterns and their extension to bands representing months to years allow reconstruction of the timing of events in the fish's life history [Bagenal, 1974]. In addition, differences in the ecology of fish allow multi-dimensional examination of temperature structure in ancient lacustrine settings. Stenothermic genera and species which live in environments of restricted temperature ranges, e.g., 3-5°C or 18-22°C, can provide information on changes in lake water chemistry. These are in contrast to eurythermic species which live and grow over a broad temperature range and can thus provide a record of seasonal variations in lake temperature. Because large lakes in temperate regions usually contain both deep- and shallow-water fauna including both stenothermic and eurythermic species, studies of such ancient lake systems have the potential to provide records of both deep and surface water conditions and their seasonal variations. Importantly, the

Climate Change in Continental Isotopic Records
Geophysical Monograph 78

otoliths of each species have diagnostic shapes enabling identification of fossil taxa in paleontology [Frizzell and Dante, 1965; Fitch, 1983; Nolf, 1985; Stringer, 1992], thus making paleo-ecological information available. Ecological tolerances and migrations of northern temperate species of fishes are sufficiently documented [Scott and Crossman, 1973] to enable evaluation of the expected depth-related variation relative to isotopic temperature estimates. Thus, the oxygen isotope variations of otolith growth bands in response to environmental conditions may provide a rich source of information about climate, limnology, ecology, and life history [Devereux, 1967; Degens et al., 1969; Mulcahy et al., 1979; Radtke, 1984a, b; Wefer, 1985; Kalish, 1991a, b].

BACKGROUND

Formation of Accretionary Banding

Understanding the factors that control the development of accretionary growth banding and their relation to environmental effects is necessary if sequential changes in otolith geochemistry are to be interpreted. Cessation of otolith growth can result from a number of causes and is expressed in the otolith as fibrous protein-rich discontinuities between zones of more continuous carbonate crystallites [Degens et al., 1969]. The resulting markers can be classified into two groups: 1) fine-scale discontinuities that are caused by daily cycles in temperature and nutrition; and 2) less frequent and more trenchant discontinuities that are caused by ecological and physiological stresses such as starvation, mating, or low winter temperatures [Boehlert and Yoklavich, 1983; Volk et al., 1984; Fowler, 1989; Smith et al., 1991; Zhang and Runham, 1992].

Daily increments record changes in the abundance of precipitated calcium carbonate relative to the protein matrix. Carbonate deposition dominates during the day and individual crystallites thin during the night producing banding which correspond to the darkened protein rich layers [Morales-Nin, 1987; Zhang and Runham, 1992]. Moreover, Wright (1991) observed that otolith accretion correlated primarily with metabolic rate rather than overall body growth rate and Umezawa and Tsukamoto (1991) noted that otolith growth ceases when food is withheld for several days. Although otolith carbonate may be resorbed under stress [Pannella, 1980], Mugiya and Uchimura (1989) showed that resorption occurs only rarely and is accompanied by carbonate deposition elsewhere on the otolith.

Otoliths as Potential Paleothermometers

The approach utilizing the isotopic chemistry of fish otolith carbonate for estimating aquatic temperatures is based on three assumptions. First, individual growth bands are assumed to form in isotopic equilibrium with the environmental waters at the temperature of precipitation. Because the enzyme-catalyzed exchange between carbonate and ear fluids is physiologically

rapid, on a scale of hours [Zhang and Runham, 1992], sequential bands within otoliths may yield information on a daily time scale. Second, isotopic exchange due to diagenesis is assumed to occur only during mineralogical transformation of aragonite to calcite; if the otolith mineralogy remains aragonite, then the primary chemical and isotopic record is assumed to remain unaltered. Finally, because of possible differences among species or genera, vital effects during isotope fractionation must be evaluated and, if necessary and possible, calibrated empirically for aragonitic fish otoliths [Horibe and Oba, 1972; Radtke, 1984a, b; Grossman and Ku, 1986; Kalish, 1991a, b] a principal objective of this paper.

Application of this empirical oxygen isotope geothermometer to ancient materials requires independent knowledge of the oxygen isotopic composition of the fresh, lake or river waters in which the fish lived. Unlike marine water, in most cases this information is not available for aquatic paleo-environments. However, a unique aspect of the approach to be developed in this study is the paired evaluation of stenothermic and eurythermic fish species. Stenothermic fish permit determination of the oxygen isotopic composition of the water; in turn, examination of eurythermic fish allows estimation of seasonal variations in surface water temperature. For example, cold-stenothermic bottom fish (e.g., deepwater sculpins, *Myoxocephalus* and the Baikal sculpin genera) are restricted to the hypolimnion of deep, temperate, lakes where water temperatures remain relatively constant at 4°C. If a biogenic fractionation relationship for fishes can be determined, the oxygen isotopic composition of the water can be directly calculated from the composition of otoliths derived from cold benthic stenothermic fishes. Once the isotopic composition of the water is known, variation in $\delta^{18}O$ among otolith growth bands of eurythermic organisms will record seasonal changes in surface water temperatures.

EXPERIMENTAL PLAN

To validate this approach, we have empirically determined oxygen isotope fractionations between water and aragonitic otoliths from freshwater fish grown in aquaria and collected from natural habitats with known temperature and water composition. We studied several large, deep, dimictic lakes that are isotopically homogenous. Bottom waters of these lakes are stable in temperature and range from 3.0°C in Lake Baikal to about 4-5°C in lakes Ontario, Superior, Michigan, and Huron (Table 1).

Fish collected from these lakes include cold, stenothermic obligate-benthic (i. e. deep-water sculpins), warm stenothermic (dace and minnows) and eurythermic (trout and drum) species. Deep-water sculpin from well constrained natural environments were utilized to determine fractionation at lower temperatures, while hatchery and laboratory raised fish provided fractionation relationships at higher temperatures. These fish (described

TABLE 1. Summary of All Data for Fish Otoliths Examined in this Study.

Genus	species	Location	Depth (m)	Temp C	$\delta^{18}O$	H_2O	$10^3\ln\alpha$
Obligate deep-water benthic specimens from natural lakes							
Comephorus	*baicalensis*	Baikal	980	3.2	17.81	-15.8[a]	33.58
Batrachocottus	*multiradiatus*	Baikal	700-720	3.3	17.58	-15.8[a]	33.35
Limnocottus	*griseus*	Baikal	700-720	3.3	17.71	-15.8[a]	33.49
Limnocottus	*bergianus*	Baikal	400-600	3.4	17.69	-15.8[a]	33.47
Limnocottus	*megalops*	Baikal	240-280	3.5	17.90	-15.8[a]	33.67
Limnocottus	*bergianus*	Baikal	150-270	3.5	17.65	-15.8[a]	33.42
Myoxocephalus	*thompsoni*	Superior	360	3.4	25.16	-9.2	34.07
Myoxocephalus	*thompsoni*	Huron	132	3.7	26.26	-7.1[b]	33.05
Myoxocephalus	*thompsoni*	Michigan	124	3.7	27.89	-6.4	33.93
Myoxocephalus	*thompsoni*	Superior	110	3.7	25.04	-9.2	33.96
Myoxocephalus	*thompsoni*	Ontario	110	3.7	26.82	-6.7[c]	33.19
Myoxocephalus	*thompsoni*	Michigan	80	3.8	27.42	-6.4	33.47
Myoxocephalus	*thompsoni*	Michigan	80	3.8	27.18	-6.4	33.24
Myoxocephalus	*thompsoni*	Michigan	80	3.8	27.33	-6.4	33.39
Myoxocephalus	*thompsoni*	Michigan	80	3.8	27.44	-6.4	33.49
Myoxocephalus	*thompsoni*	Michigan	80	3.8	27.63	-6.4	33.68
Myoxocephalus	*thompsoni*	Huron	75	3.8	26.30	-7.1[b]	33.09
Deep-water pelagic specimens from natural lakes							
Salvelinus	*namaycush*	Superior	100	3.8	24.74	-9.2	33.66
Salvelinus	*namaycush*	Superior	100	3.8	25.01	-9.2	33.92
Salvelinus	*namaycush*	Superior	100	3.8	24.88	-9.2	33.79
Stenothermic warm shallow-water species from natural lakes							
Leuciscus	*leuciscus*	Baikal	10	13.5	15.04	-15.8[a]	30.86
Couesius	*plumbeus*	Michigan	1	21.5	23.45	-6.4	29.60
Couesius	*plumbeus*	Michigan	1	21.5	23.25	-6.4	29.40
Stenthermic cold-water specimens (hatchery raised)							
Salvelinus	*namaycush*	Marquette		8	19.52	-13.3	32.76
Salvelinus	*namaycush*	Marquette		8	19.65	-13.3	32.89
Salvelinus	*namaycush*	Marquette		8	19.84	-13.3	33.07
Salvelinus	*namaycush*	Marquette		8	19.60	-13.3	32.84
Salvelinus	*namaycush*	Marquette		8	19.31	-13.3	32.56
Salvelinus	*namaycush*	Marquette		8	18.54	-13.3	31.80
Salvelinus	*namaycush*	Marquette		8	19.64	-13.3	32.88
Salvelinus	*namaycush*	Marquette		8	19.64	-13.3	32.88
Salvelinus	*namaycush*	USFWS		12	22.24	-9.1	31.13
Salvelinus	*namaycush*	USFWS		12	22.78	-9.1	31.65
Salvelinus	*namaycush*	USFWS		12	22.92	-9.1	31.79
Stenothermic warm-water specimens from laboratory tanks							
Poecilia	*reticulata*	U-M Museum		30.3	24.57	-3.2	27.45
Poecilia	*reticulata*	U-M Museum		30.3	24.73	-3.2	27.61
Poecilia	*reticulata*	U-M Museum		30.3	24.37	-3.2	27.25

Table 1. (continued)

Genus	species	Location	Depth (m)	Temp C	$\delta^{18}O$	H_2O	$10^3 \ln\alpha$
Poecilia	*reticulata*	U-M Museum		30.3	25.13	-3.2	28.00
Poecilia	*reticulata*	U-M Museum		30.3	24.56	-3.2	27.44
Poecilia	*reticulata*	U-M Museum		30.3	25.51	-3.2	28.37
Poecilia	*reticulata*	U-M Museum		30.3	24.88	-3.2	27.75
Poecilia	*reticulata*	U-M Museum		30.3	24.77	-3.2	27.65
Poecilia	*reticulata*	U-M Museum		30.3	24.81	-3.2	27.69
Poecilia	*reticulata*	U-M Museum		30.3	24.66	-3.2	27.54
Poecilia	*reticulata*	U-M Museum		30.3	24.89	-3.2	27.76
Poecilia	*reticulata*	U-M Museum		30.3	24.85	-3.2	27.73
Poecilia	*reticulata*	U-M Museum		30.3	24.71	-3.2	27.59
Poecilia	*reticulata*	U-M Museum		30.3	24.65	-3.2	27.53
Poecilia	*reticulata*	U-M Museum		30.3	24.97	-3.2	27.84

Marine specimens from other studies

Genus	species	Location	Depth (m)	Temp C	$\delta^{18}O$	H_2O	$10^3 \ln\alpha$
C.	*acrolepis* (23)[1]		991	4	34.38	-0.25	34.06
Ariomma[1]			200	15	31.73	-0.30	31.54
Nomeus[1]			200	15	30.81	-0.30	30.64
Cubiceps[1]			200	15	30.91	-0.30	30.74
Pampus[1]			50	15	31.12	-0.30	30.94
Peprilus[1]			50	15	29.78	-0.30	29.64
Stromateus[1]			50	15	31.84	-0.30	31.64
Psenopsis[1]			200	15	32.04	-0.30	31.84
Seriolella[1]			200	15	31.94	-0.30	31.74
Schedophilus[1]			200	15	30.81	-0.30	30.64
Centrolophus[1]			200	15	32.25	-0.30	32.04
Hyperglyphe[1]			200	10	30.70	-0.30	30.54
Stenotomus[1]			50	15	29.05	-0.03	28.67
Roccus[1]			50	25	26.37	-0.03	26.06
Centropristes[1]			50	25	31.01	-0.03	30.57
Prionotus[1]			50	15	30.70	-0.03	30.27
Merluccius[1]			50	12	32.35	-0.20	32.04
Melanogrammus[1]			150	12	32.46	-0.20	32.14
Gadus[1]			150	10	32.87	-0.20	32.54
Ceratoscopelus[1]			250	12	32.97	-0.20	32.64
Osmerus[1]			10	15	29.36	-0.03	28.97
Mugil	*cephalus* (2)[2]		0	23	29.21	-1.80	30.59
Thunnus	*thynnus* (6)[3]		50	25	29.55	-0.03	29.15
Pristomopoides	*filamentosus* (3)[4]		30	20	31.95	0.05	31.40
Arripis	*trutta* (25)[6]		3	16	31.17	0.05	30.64
Hoplostethus	*atlanticus* (10)[6]		1040	4	33.37	-0.25	33.08
Thyristes	*atun* (4)[6]		10	15	31.73	0.05	31.19
Nemadactylus	*macropterus* (4)[6]		10	15	31.30	0.05	30.77
Trachurus	*declivis* (2)[6]		10	15	31.99	0.05	31.44
Brama	*brama* (2)[6]		100	10	31.27	0.05	30.74
Macruronus	*novaezelandiae* (4)[6]		400	8	32.67	0.05	32.10
Psedophycis	*barbatus* (4)[6]		10	12	32.04	0.05	31.49
Thunnus	*maccoyi* (2)[6]		50	22	29.23	0.05	28.76
Arripis	*esper* (2)[6]		10	16	31.49	0.05	30.95
Notothenia	*squamifrons* (2)[6]		10	1	33.64	0.05	33.04

Table 1. (continued) 195

Genus	species	Location	Depth (m)	Temp C	$\delta^{18}O$	H_2O	$10^3\ln\alpha$
		*Shallow-water eurythermic species*					
				Calculated Temp			
Aplodinotus	*grunniens*	Erie	5	19.98	23.28	-6.8[d]	29.84
Aplodinotus	*grunniens*	Erie	5	14.66	24.48	-6.8[d]	31.01
Aplodinotus	*grunniens*	Erie	5	9.87	25.60	-6.8[d]	32.10
Aplodinotus	*grunniens*	Erie	5	12.91	24.89	-6.8[d]	31.40
Aplodinotus	*grunniens*	Erie	5	15.37	24.32	-6.8[d]	30.85
Aplodinotus	*grunniens*	Erie	5	15.14	24.37	-6.8[d]	30.90
Aplodinotus	*grunniens*	Erie	5	17.91	23.74	-6.8[d]	30.29
Aplodinotus	*grunniens*	Erie	5	20.86	23.09	-6.8[d]	29.65
Aplodinotus	*grunniens*	Erie	5	21.08	23.04	-6.8[d]	29.60
Aplodinotus	*grunniens*	Erie	5	19.47	23.40	-6.8[d]	29.95
Aplodinotus	*grunniens*	Erie	5	21.81	22.88	-6.8[d]	29.45
Aplodinotus	*grunniens*	Erie	5	22.67	22.69	-6.8[d]	29.26
Aplodinotus	*grunniens*	Erie	5	22.49	22.73	-6.8[d]	29.30
Aplodinotus	*grunniens*	Erie	5	22.26	22.78	-6.8[d]	29.35
Aplodinotus	*grunniens*	Erie	5	20.16	23.24	-6.8[d]	29.80
Aplodinotus	*grunniens*	Erie	5	10.27	25.51	-6.8[d]	32.01
Aplodinotus	*grunniens*	Erie	5	10.08	25.55	-6.8[d]	32.05
Aplodinotus	*grunniens*	Erie	5	13.35	24.78	-6.8[d]	31.31
Aplodinotus	*grunniens*	Erie	5	13.83	24.67	-6.8[d]	31.20
Aplodinotus	*grunniens*	Erie	5	16.01	24.17	-6.8[d]	30.71
Aplodinotus	*grunniens*	Erie	5	16.63	24.03	-6.8[d]	30.57
Aplodinotus	*grunniens*	Erie	5	16.97	23.96	-6.8[d]	30.50
Aplodinotus	*grunniens*	Erie	5	18.76	23.55	-6.8[d]	30.10
Aplodinotus	*grunniens*	Erie	5	18.74	23.56	-6.8[d]	30.11
Aplodinotus	*grunniens*	Erie	5	19.87	23.31	-6.8[d]	29.86
Aplodinotus	*grunniens*	Erie	5	17.40	23.86	-6.8[d]	30.40
Aplodinotus	*grunniens*	Erie	5	16.50	24.06	-6.8[d]	30.60
Aplodinotus	*grunniens*	Erie	5	12.69	24.94	-6.8[d]	31.45
		*Shallow-water sculpins from natural lakes*					
Cottus	*bairdi*	Superior	1	12.57	22.51	-9.2	31.48
Cottus	*ricei*	Michigan	10	8.75	26.28	-6.4	32.36
Cottus	*ricei*	Superior	50	7.27	23.77	-9.2	32.71
Cottus	*cognatus*	Superior	55	4.40	24.47	-9.2	33.39
Cottus	*ricei*	Superior	55	6.80	23.88	-9.2	32.82
Cottocomephorus	*inermis*	Baikal	30	8.41	16.65	-15.8[a]	32.44
Thymallus	*brevipinnus*	Baikal	10	8.97	16.52	-15.8[a]	32.31
Batrachocottus	*baicalensis*	Baikal	30	8.50	16.63	-15.8[a]	32.42

1 Mulcahy et al. (1979)
2 Degens (1969)
3 Radtke (1984a)
4 Radtke et al. (1987)
5 Radtke (1987)
6 Radtke (1983)
7 Kalish (1991b)
a Robert Seal pers. comm.
b David Rea pers. comm..
c Tom Edwards pers. comm.
d Martin Knyf pers. comm

All unfootnoted analyses are from this study

below) provide fractionation relationships over the temperature range from 3.2°C to 30.3°C.

SAMPLING PLAN: LOCATIONS AND DESCRIPTIONS

Stenothermic Fishes

Lake Baikal. Lake Baikal, at about 52° to 54° N in Siberia, is a 1620 m deep, dimictic lake showing positive stratification from March to November and inverse stratification through the winter. Mixing of the upper 300 m occurs in March and November at 4°C. Near 300 m depth, the lake is at 3.5°C for the entire year; temperatures range down to 3°C below that depth, elevating to 3.2°C near the bottom owing to geothermal heat and chemical processes [Bekman et al., 1979]. This relatively invariant thermal regime and known oxygen isotopic composition of the water ($\delta^{18}O$ = -15.8‰ ±0.2, [Robert Seal, pers. comm.]), provide an opportunity to measure the fractionation factors for benthic deep-water fishes.

Obligate benthic deepwater species, which are restricted to bottom waters year-round and experience a maximum temperature variation of about 1-1.5°C as adults, were chosen to represent the low end of the temperature scale. Whole otoliths of eight species of benthic sculpins, family Cottidae, were examined because their physiology limits them to deep habitats. Fish of this family lack hydrostatic organs and thus are unable to leave the bottom for more than a few minutes. Except for the shallow-water species, *Batrachocottus baicalensis*, the benthic sculpins have experienced temperatures between 3°C and 4°C for their entire adult lives and record average lake composition through several seasonal mixing events.

Two species of pelagic sculpins were also sampled. One of these (*Comephorus baicalensis*) cannot survive temperature variation of more than 5°C [Talyev, 1955] and is appropriate for estimating the fractionation factor α. The other pelagic species (*Cottocomephorus inermis*) experiences a wider range of temperatures [Talyev, 1955], so it and *Batrachocottus baicalensis* were utilized as examples of sculpins experiencing warmer and more variable temperatures. Grayling (*Thymallus brevipinnus*) and dace (*Leuciscus leuciscus*) are inshore, shallow-water species that enable assessment of the isotopic compositions of fish growing in Lake Baikal at 12-15°C.

The Laurentian Great Lakes.

The Great Lakes, lying between 41° and 47° N, are a stable lake system which has experienced little change in lake level or temperature over the last 50 years. Lake Superior water levels have fluctuated by only 0.6 m and Lakes Michigan and Huron by only 1.5 m since 1950. This suggests that long term variation in water isotopic composition has not occurred. Similarly, the lakes are relatively stable in temperature, with Lake Ontario showing a 2°C increase in average temperature between 1918 and 1958 [Beeton, 1961].

In the Great Lakes, adult deep-water sculpins, *Myoxocephalus thompsoni*, inhabit waters below about 70 m as adults. Lake Superior water temperature remains constant between 4-5.5°C below depths of 50 m [Beeton, 1959; Anderson and Rodgers, 1963]. Adult deep-water sculpins were sampled from Lake Superior from depths of 75-360 m. Adult sculpins were common below 70 m and were most abundant in waters of 4°C in May-June and 4.6°C in late August [Selgeby, 1988]. Slimy and spoonhead sculpins were common at depths of 50-100 m, being most abundant in waters of 4°C in May-June, 5.3°C in late August, and 5.5°C in late October.

In Lake Michigan, the depth of the isotherm varies with 4°C water present below 100 m in the south and below 70 m in the north. August surface waters averaged 21.5°C in the north; temperatures of the hypolimnion (below about 10 m) are less than 10°C in the south and 6°C in the north [Ayers et al., 1958]. The summer warming period ends in October (maximum T= 4.4°C below 70 m); by mid-December the offshore water temperature decreases from about 7°C at the surface to less than 5°C at depth; the lake is homothermous from January through March [Wells and McLain, 1973]. Deep-water sculpins range from 45-183 m, being most abundant at 73-91 m [Scott and Crossman, 1973].

Lake Huron is similar to Lake Michigan in temperature gradients and the population distribution of sculpins. Bottom waters of Lake Huron remain between 4.5-5°C below 45-60 m in Georgian Bay and the South Basin [Berst and Spangler, 1973]. October temperatures are highest, rising to 5.0°C at 40-70 m, and to 4.5°C below 70 m [Tom Todd, pers. comm., USFWS sculpin catch data].

The temperature of Lake Ontario below about 45 m is less than 5°C except for a brief time in November when waters up to 8°C extend downward to 70 meters depth at that time. From 70 to 110 m, the temperature gradually decreases from 7 to 5°C [Rodgers and Anderson, 1963; Allen, 1969]. Deep-water sculpins were collected by Dymond et al. (1929) at depths between 91-125 m in October 1927.

Specimens were also collected from other natural settings for which the habitat temperatures and timing of precipitation of otolith carbonate is known. Stenothermic lake chub, *Couesius plumbeus*, were also employed to calculate the temperature-fractionation relationship because they precipitate carbonate over a narrow temperature range [Smith, unpubl.]. These fish were captured in the Straits of Mackinac during August 1941 when the mean temperature was 21.5°C. Thirty-two Lake Michigan summer capture temperatures for *Couesius plumbeus* average 17.6°C (S.D.=2.4°C).

Eurythermic Fishes

Drum (*Aplodinotus grunniens*), a eurythermic fish species from Lake Erie, was examined to test our ability to reconstruct surface water temperature variations [Edsall, 1967]. Because Lake Erie is poorly stratified compared to the deeper Laurentian lakes, bottom water temperatures from stations in the western basin in 1958 ranged between about 15°C in May and October to about 22°C in July and August; surface temperatures were 0-3°C higher [Edsall, 1967]. In the summer surveys of 1959-60 surface temperatures ranged from 25-27°C while bottom temperatures dropped to 9°C in deeper basins where the hypolimnion was reduced to a thin layer of not more than 2 m [Beeton, 1963].

Laboratory Aquaria and Hatchery Cultures

To determine fractionation factors at higher temperature, a series of experimental aquaria were employed to raise fish under controlled temperatures and water compositions. Evaporation was prevented by sealing the aquaria, so that the isotopic composition of the water, which was determined at the beginning and end of the ten day experiment remained constant. Although hatchery cultures were not in sealed aquaria, these employed a constant flow-through water system utilizing local groundwater sources. Samples of the waters from each hatchery were obtained and analyzed.

Female guppies (*Poecilia reticulata*) were raised at a constant temperature of 30.3°C (S. D. = 0.29). Aquaria were maintained at these temperatures throughout the period of otolith formation and newborn fish sacrificed within hours of birth. Embryonic otoliths are accreted within the live-bearing female as the young fish develop. Two otoliths (1 mm diameter) were removed from each fish and analyzed together.

ANALYTICAL METHODS

Otoliths from the Laurentian Great Lakes, and Lake Baikal were available from fresh, dried, and ethanol-preserved fish. These samples represent a broad range of environmental conditions including differences in water isotopic chemistry and growth temperatures. Individual otoliths (usually the sagittae, 2 mm - 15 mm long) were removed from fish and 100 micron thick polished sections were prepared.

The accretionary growth structure was digitized from photographic enlargements of the otolith and multiple drilling paths were interpolated between visible growth bands. Using this image, a computerized drilling system comprising a custom bit and stepper motor controlled x-y-z micro positioning stage, sequentially removed age-specific samples paralleling growth banding. Samples of otolith carbonate typically represent a spatial resolution of 10-40 microns in width and a total mass of 10-30 micrograms.

Sample powders recovered from the polished otolith surface, were roasted at 200°C *in vacuo* to remove volatile organic contaminants and individually reacted at 73 °C with 4 drops of anhydrous phosphoric acid in a Finnigan "Kiel" carbonate preparation system coupled directly to the inlet of a Finnigan MAT 251 ratio mass spectrometer. Isotopic enrichments were

corrected for acid fractionation and ^{17}O contribution and reported in per mil relative to the PDB standard. Precision and calibration of data were monitored through daily analysis of NBS-18, and NBS-19 powdered carbonate standards. The $\delta^{18}O$ values of these bracketed those of the samples. Precision is better than ±0.1 ‰ for both carbon and oxygen isotopic compositions.

The oxygen isotopic compositions of lake, aquaria, and hatchery waters were determined on 2ml aliquots by CO_2-water equilibration and analyzed on a Finnigan Delta-S mass spectrometer. Reproducibility of replicate and standard analyses was better than ±0.1‰.

RESULTS

Calibration of Aragonite Thermometry in Fish Otoliths

One goal of this study is to develop an empirical relationship which describes the oxygen isotopic fractionation between fish otolith aragonite and water over a range of earth surface temperatures. This was accomplished by using a suite of obligate benthic species that occupy temperature constrained zones in six large, deep, temperate lakes whose waters differ in oxygen isotopic composition. Extension of the temperature range was accomplished by controlled laboratory aquaria and hatchery culture experiments. Limnological and ecological data from individual lakes indicate the temperature at which the aragonite was precipitated. Measured isotopic compositions of

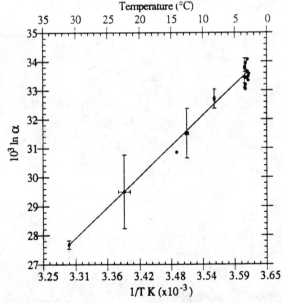

Fig.1. Mean values and standard errors of sample populations at the 95% probability level for calculated fractionation factors ($10^3 ln\alpha$) relative to temperature ($T^{-1}K$). Individual data are indicated by solid circles. A linear regression through the data yields the equation $10^3 ln\alpha = 18.56 \cdot (10^3) T^{-1}K - 33.49$ (R = 0.99).

waters were coupled with the temperature data and carbonate $\delta^{18}O$ compositions to determine the temperature dependence of the oxygen isotope fractionation factors for otolith aragonite. The fractionation factor (α) was determined for each specimen using the following relationship.

$$\alpha = \frac{\delta CaCO_3 + 1000}{\delta H_2O + 1000} \qquad (1)$$

A regression line through the data, plotted as $10^3 \ln\alpha$ vs. $T^{-1}K$ yields the relationship with standard errors at the 95% probability level:

$$10^3 \ln\alpha = 18.56\ (\pm 0.319) \cdot (10^3)\ T^{-1}K - 33.49\ (\pm 0.307) \qquad (2)$$

The slope of this relationship is statistically indistinguishable from that of Grossman and K (1986), who gave the relationship $10^3 \ln\alpha = 18.07\ T^{-1}\ K - 31.079$. However, the intercept is offset by approximately +2.5. Variation in the data about this line can be attributed to slight seasonal changes of bottom water temperatures and the isotopic composition of the water, analytical errors, and unexplained metabolic effects. Standard errors for the means (95% probability level) were calculated for each sample population (Fig. 1).

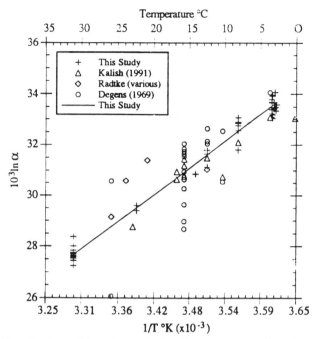

Fig. 2. Comparision of isotopic data from previous studies with empirically-derived fractionation relationship determined in this study.

DISCUSSION

Metabolic and Vital Effects

The possibility that fish otoliths are not formed in oxygen isotope equilibrium with ambient waters was suggested by Radtke (1984a, b, 1987). His experiments [Radtke, 1984a, b],

produced isotopic results that showed a 1.4-1.9.‰ depletion of ^{18}O in mullet and 3‰ enrichment of ^{18}O in cod raised at known temperatures in seawater. Offset of the data from predicted equilibrium fractionation values led the author to propose

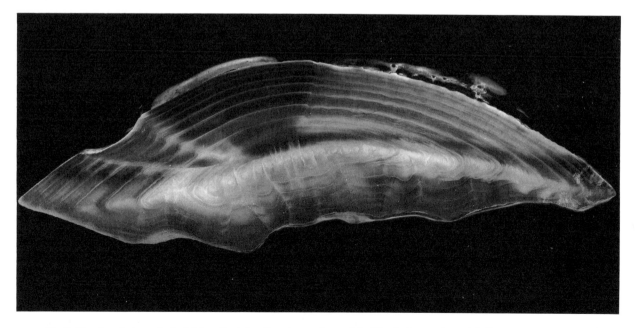

Fig. 3. Negative photograph of a thick section taken from the saggitae otolith of the freshwater drum *Aplodinotus grunniens* (Sandusky Bay, Lake Erie). Annual growth bands, representing 9 years of the specimen's life, are visible at this scale. The third and fourth yearly growth bands, accreted during 1950 and 1951, were sampled outward from the otolith center.

metabolic effects in otoliths of these poikilothermic fishes. However, the scatter in the data for marine fishes studied by Radtke (1984a, b) may reflect imprecise control of temperature and/or the isotopic composition of the water, neither of which was measured in his experiments. Radtke (1987) also invoked metabolic effects to explain depletion of heavy isotopes in tuna; however, because these fish maintain elevated body temperatures, tuna otoliths may be in isotopic equilibrium with body fluids while they are in disequilibrium with ambient water temperatures. In contrast to Radtke's findings, comparison of results of previous studies of marine fishes [Deveruex, 1967; Degens et al., 1969; Mulcahy et al., 1979; Kalish, 1991a, b] with the temperature fractionation relationship determined for freshwater fishes of this study (Fig. 2) suggest that this relationship is applicable to both marine and non-marine settings and that metabolic or generic/species effects are not significant.

Estimation of Seasonal Temperature Variation

To evaluate the hypothesis that the isotopic variation within otolith growth bands of shallow water species can be used to estimate seasonal temperature variation, we examined the otolith of a 9-year-old specimen (Fig. 3) of a eurythermic species of freshwater drum (*Aplodinotus grunniens*; 325 mm SL) collected from western Lake Erie in 1957. This fish species is particularly suited for this test because it remains in shallow

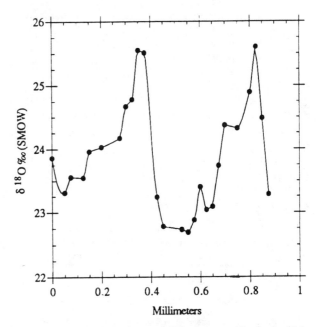

Fig. 5. Intra-annual variation in oxygen isotopic composition for the third and forth years of the freshwater drum (Fig. 3). Isotopic composition, $\delta^{18}O$ SMOW, is plotted relative to sample distance in millimeters beginning in the winter of 1949-50, from left to right.

water throughout the year and forms exceptionally large otoliths. Based on a detailed ecological study of this species in Lake Erie during 1957-58, Edsall (1967) determined that body growth was most rapid during the warmest part of the year.

Microscopic examination of daily growth banding indicates that aragonite accreted on otoliths during approximately 140 days of the year early in the life history of the fish. Erosional events during colder, more stressed time periods may be responsible for the removal of some daily bands (Fig. 4). SEM photomicrographs indicate mid-summer daily growth banding of approximately 4 microns width. Because sample paths are 25 microns in width, the record illustrated in Fig. 5 provides a resolution on an approximately weekly scale.

In order to calculate equilibrium oxygen isotope ratios of carbonates, seasonal variation in Lake Erie waters must be estimated. Surface and shallow bottom water temperatures of Sandusky Bay are correlated with monthly average air temperatures at Sandusky Ohio [Edsall, 1967]. Meteorological data from the Sandusky Bay weather station for 1950-51 are our most complete source of temperature data. From these, daily high and low temperatures were used to calculate weekly averages.

The $\delta^{18}O$ values (relative to SMOW) are plotted as a function of time in Figure 5. The record reveals a distinct seasonal signal. Temperatures calculated from the aragonite

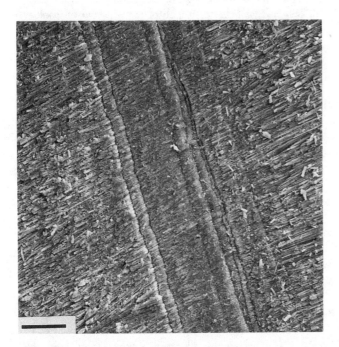

Fig. 4. SEM photomicrograph of winter banding observed in the otolith in Fig. 3. Erosional features are developed during periods of stress such as low winter temperatures or reproduction. This results in an incomplete isotopic record for these periods. Scale bar is 20 microns in length.

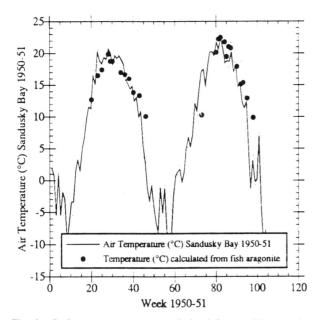

Fig. 6. Surface water temperature calculated from otolith aragonite precipitated during 1950-51. Also shown are weekly average air temperatures at Sandusky Bay. Isotopic data were correlated with the timing of aragonite precipitation by assuming the warmest calculated temperatures coincided with the warmest measured air temperatures. From this datum, adjacent values were adjusted to the time scale based on the width of daily growth banding measured for this otolith.

$\delta^{18}O$ data using equation (2) can be correlated with temperatures derived from Sandusky Bay weather data (Fig. 6). Since the accretion rate is related to temperature, the resolution is higher during warm intervals. Samples indicative of cool temperatures represent averages over longer times. Thus, the best climate record of these otoliths is that of warmest summer temperatures. Measurement of the variation in water temperature over the entire year is hampered because growth rate slows dramatically during winter months, making it difficult to retrieve adequate material for analysis.

CONCLUSIONS

Analyses of aragonitic fish otoliths from diverse species, over a broad range of naturally and experimentally controlled temperatures, yields the fractionation relationship: $10^3 \ln\alpha = 18.56 \cdot (10^3) \, T^{-1}K - 33.49$. Previous attempts to estimate the temperature fractionation relationship for fish have encountered apparent species-specific differences; these differences are possibly due to experimental error related to evaporation of water in which the fish were raised or lack of precise temperature control. Our results indicate that this fractionation relation can be applied to a diverse taxa of freshwater fish with calculated temperatures in close agreement with ecological data for the fish species studied.

The relationship between otolith aragonite isotopic composition and temperature described here indicates that otoliths are excellent proxies of paleoclimatic conditions in continental settings. Moreover, because deep temperate lake bottom waters are generally between 3.5-4.5°C year-round, the $\delta^{18}O$ of deep-water obligate benthic species can be used to determine the isotopic composition of the water in ancient lake systems. In settings containing both eurythermic and stenothermic fish populations, it is further possible to discern seasonal variations present in surface waters. Finally, the study of otoliths from large lake systems through time and across latitude should lead to greater understanding of continental thermal gradients and latitudinal variations in meteoric water compositions during periods of changing climatic regimes.

Acknowledgments. We would like to thank the Scott Turner Committee, Dept. Geol. Sci. for funding a portion of this work. Principal support for this research was provided by NSF-EAR-9105903 grant to K. Lohmann We are also grateful to Bruce Wilkinson for advice, encouragement, and partial financial support (Lakes grant NSF-EAR-901995). Jim Burdett provided laboratory assistance and helpful commentary. Aimee Dolan-Laughlin, Pedro Patterson, and Margaret Bickmore assisted with water analyses. We thank David Dettman for technical and conceptual help. Additional isotopic data were provided by: Bob Seal (Lake Baikal), Tom Edwards (Lake Ontario), Dave Rea (Lake Huron), and Martin Knyf (Lake Erie). Fish for this study were obtained from the collection of the Museum of Zoology, University of Michigan. This manuscript was greatly improved by reviews from Carl Drummond, Samuel Savin, Ethan Grossman and an anonymous reviewer.

REFERENCES

Aharon, P. Recorders of reef environment histories: stable isotopes in corals, giant clams, and calcareous algae, *Coral Reefs, 10*, pp. 71-90, 1991.

Allen, H. E. Chemical characteristics of Lake Ontario. pp. 1-18 in *Limnological survey of Lake Ontario, 1964*. Tech. Rept. 14, Great L. Fish. Comm., 1969.

Anderson, D. V., and Rogers, G. K. *A synoptic survey of Lake Superior.* Univ. Michigan Great Lakes Res. Div. 10: 79-90, 1963.

Axelrod, D. I. *Quaternary extinction's of large mammals.* Univ. California Pub. Geol. Sci. 74:1-42, 1968.

Ayers, J. C., Chandler D. C., Lauff, G. H., Powers, C. F., Henson, E. B.. Currents and water masses of Lake Michigan. *Great L. Res. Inst. Pub.* 3:1-169, 1958.

Bagenal, T. B. (ed.). *The ageing of fish.* Old Working, Surrey, England, 234 pp., 1974.

Beeton, A. M. Environmental changes in Lake Erie. *Trans. Amer. Fish. Soc. 90 (2)*:153-174, 1961.

Beeton, A. M. Limnological survey of Lake Erie 1959 and 1960. *Great L. Fish. comm. Tech. Rep.* 6:1-32, 1963.

Beeton, A. M., Johnson, J. H., and Smith, S. H. Lake Superior limnological data, 1951-1957. *U.S. Dep. Int. Spec. Sci. Rep. Fish.* 297:1-177, 1959.

Bekman, M. U., Votintsev, K. K., Verbolov, V. I., Galazy, G. I., Galkina, V. I., Ladeishcikov, N. P., Llut, B. F., Mazepova, G. F., Smirnov, V. V., Sidelev, G. N., Cherepanov, V. V. *Lake Baikal Guide-book.* Limnol. Inst. Siberian Branch Acad. Sci. USSR, Moscow, 1979.

Berst, A. H., and Spangler, G. R. Lake Huron: The ecology of the fish community and man's effects on it. *Great L. Fish. Comm. Tech Rept.* 21:1-41, 1973.

Boehlert, G. W., and Yoklavich, M. M. Larval and juvenile growth of sablefish, *Anoploma fibria*, as determined from otolith increments. *Fish. Bull.* 83: 475-481, 1983.

Brothers, E. B., Mathews, C. F., and Lasker, R. Daily growth increments in otoliths from larval and adult fishes. *Mar. Fish. Serv. Fish. Bull.* 74 (1): 1-8, 1976.

Campana, S. E. Calcium deposition and otolith check formation during periods of stress in coho salmon, *Oncorhynchus kisutch*. *Comp. Biochem. Physiol.* 75A : 215-220, 1989.

Carlstrom, D. A crystallographic study of vertebrate otoliths. *Biol. Bull.* 125:441-463, 1963.

Degens, E. T., Deuser, W. G., and Haedrich, R. L. Molecular structure and composition of fish otoliths. *Int. J. Life Oceans Coastal Waters 2* (2):105-113, 1969.

Devereux, I. Temperature measurements from oxygen isotope ratios of fish otoliths. *Science 155* : 1684-1685, 1967.

Dymond, J. R., Hart, J. L., Pritchard, A. L. The fishes of the Canadian waters of Lake Ontario. *Univ. Toronto Studies Biol. Ser. 33, Pub. Ontario Fish. Res. Lab.* 37:1-35, 1929.

Edsall, T. A. Biology of the freshwater drum in western Lake Erie. *Ohio J. Sci.* 67(6):321-340, 1967.

Erlenkeuser, H., and Wefer, G. Seasonal growth of bivalves from Bermuda recorded in their O-18 profiles. *Proceedings of the Fourth International Coral Reefs Symposium, Manila, Vol. 2*, 1981.

Fischer, A. G. Latitudinal variations in organic diversity. *Evolution 14 (1)*: 64-81, 1960.

Fitch, J. E., and Lavenberg, R. J. Teleost fish otoliths from Lee Creek Mine, Aurora, North Carolina (Yorktown Formation: Pliocene), pp. 509-529 in C. E. Ray (ed.), Geology and Paleontology of the Lee Creek Mine, North Carolina, I. Smithsonian Contr. *Paleobiol. 53*, 529 pp, 1983.

Fowler, A. J. Description, interpretation and use of the microstructure of otoliths from juvenile butterflyfishes (family Chaetodontidae). *Mar. Biol. 102* : 167-181, 1989.

Frizzell, D., and Dante, J. Otoliths of some early Cenozoic fishes of the Gulf Coast. *J. Paleontol. 39* : 687-718, 1965.

Graham, R. W., and Meade, J. I. Enviromental fluctuations and evolution of mammalian faunas during the last deglaciation in North America, in Ruddiman, W. F., and Wright, J. E., Jr. (eds.), North America and adjacent oceans during the last deglaciation. *Geol. Soc. Amer., The Geology of North America, vol. K-3*. Boulder, Colorado, 1987.

Grossman, E. L., and Ku, T. L. Oxygen and carbon isotopic fractionation in biogenic aragonite: Temperature effects. *Chemical Geology (Isotope Geoscience Section), 59*, 59-74, 1986.

Hibbard, C. W. Pliocene and Pleistocene climates in North America. *Ann. Rep. Mich. Acad. Sci. Arts, Letters. 62* : 5-30, 1960.

Horibe, Y., and Oba, T. Temperature scales of aragonite-water and calcite-water systems. *Fossils No. 23*: 71-80, 1972.

Jones, D. S., Arthur, M. A. and Allard, D. J. Sclerochronological records of temperature and growth from shells of *Mercenaria mercenaria* from Narragansett Bay, Rhode Island. *Marine Biology 102*, 225-234, 1989.

Kalish, J. M. Oxygen and carbon stable isotopes in the otoliths of wild and laboratory-reared Austrailian salmon (*Arripis trutta*). *Marine Biology 110*, 37-47, 1991

Kalish, J. M. ^{13}C and ^{18}O isotopic disequilibria in fish otoliths: metabolic and kinetic effects. *Marine Ecology Progress Series 75*: 191-203, 1991

Killingley J. S., and Berger, W. H. Stable isotopes in a mollusk shell: detection of upwelling events. *Science 205*: 186-188, 1979.

Krantz, D. E., Williams, D. F., and Jones, D. S. Ecological and paleoenvironmental information using stable isotope profiles from living and fossil molluscs. *Palaeogeography, Palaeoclimatology, Palaeoecology, 58*, pp. 249-266, 1987.

Morales-Nin, B. Ultrastructure of the organic and inorganic constituents of the otoliths of the sea bass, pp. 331-343, in C. S. Robert and G. E. Hall (eds.), *The age and growth of fish*. The Iowa State Univ. Press, Ames, 1987.

Mugiya, Y., and Uchimura, T. Otolith resorption induced by anaerobic stress in the goldfish, *Carassius auratus*. *J. Fish Biol.* 35 : 813-818, 1989.

Mulcahy, S. A., Killingley, J. S., Phleger, C. F., Berger, W. H. Isotopic composition of otoliths from a bentho-pelagic fish, *Coryphenoides acrolepis*. Macrouridae: Gadiformes. *Oceanol. acta 2*: 423-427, 1979.

Nolf, D. Otolithi piscium. *Handbook of Paleoichthyology, 10* : 1-145. Gustav Fischer Verlag, Stuttgart, 1985.

Pannella, G. Fish otoliths: Daily growth layers and periodical patterns. *Science 173* : 1124-1126, 1971.

Pannella, G. Growth patterns in fish sagittae, pp. 519-560 in D. C. Rhodes and R. A. Lutz (eds), *Skeletal growth of aquatic organisms*. Plenum, New York, 1980.

Pätzold, J. Temperature and CO_2 changes in tropical surface waters of the Philipines during the past 120 years: record in the stable isotopes of hermatypic corals, *Reports, Geoloisch-Palaeologisches Institute und Museum, No. 12*, 92 pp., Christian-Albrechts Univ. Press, Kiel, 1986.

Radtke, R. L., Williams, D. F., and Hurley, P. C. The stable isotopic composition of Bluefin Tuna (*Thunnus thynnus*) otoliths: Evidence for physiological regulation. *Comp. Biochem. Physiol. Vol. 87A, No. 3*, pp. 797-801, 1987.

Radtke, R. L., and Targett, T. E. Rhythmic structural and chemical patterns in otoliths of the Antarctic fish *Notothenia larseni*: Their application to age determination. *Polar Biol. 3*: 203-210, 1984

Radtke, R. L. Formation and structural composition of larval striped mullet otoliths. *Transactions of the American Fisheries Society, 113* : 186-191, 1984.

Radtke, R. L. Cod fish otoliths: Information storage structures. Pp. 273-298 in Dahl, D. S. Danielssen, E. Moksness, P. Solemdal (eds.) The propagation of cod *Gadus morhua L.* an international symposium, Arenda (Norway). *Floedevigen Rapp. 1, 1984.*

Rodgers, G. K., and Anderson, D. V. The thermal structure of Lake Ontario. *Proc. Sixth Conference on Great L. Res. Great L. Res. Div. 10:59-69, 1963.*

Selgeby, J. Comparitive biology of the sculpins of Lake Superior. *J. Great Lakes Res. 14 (1)* : 44-51, 1988.

Smith, K. M., and Kostlan, E. Estimates of age and growth of *Ehu Etelis carbunculus* in four regions of the Pacific from density of daily growth increments in otoliths. *Fishery Bulletin, U.S. 89*: 461-472, 1991.

Stringer, G. L. Late Pleistocene-Early Holocene teleostean otoliths from a Mississippi river mudlump. *J. Vert. Paleo. 12 (1)*: 33-41, 1992.

Scott, W. B., and Crossman, E. J. Freshwater fishes of Canada. *Bull. Fish. Res. Bd. Canada 184*: 1-966, 1973.

Talyev, D. N. *Sculpins of Baikal.* Acad. Sci. USSR, Moscow (in Russian), 1955.

Umezawa, A., and Tsukamoto, K. Factors influencing otolith increment formation in Japanese eel, *Anguilla japonica* T. & S., elvers. *J. Fish Biol. 39* : 211-223, 1991.

Volk, E. C., Wissmar, R. C., Simenstad, C. A., and Eggers, D. M. Relationship between otolith microstructure and the growth of juvenile chum salmon (*Oncorhynchus keta*) under different prey rations. *Can. J. Fish. Aquat. Sci. 41* : 126-133, 1984.

Wefer, G. Die Verteilung stabiliser Isotope in Kalkschalen mariner Organismen. *Geologisches Jarhbuch, Reihe A. Heft 82*, pp. 3-111, 1985.

Wells, L. R., and McLain, A. L. Lake Michigan: Man's effects on native fish stocks and other biota. *Great Lakes Fish. Comm. Tech. Rep. 20* : 1-55, 1973.

Wright, P. J. The influence of metabolic rate on otolith increment width in Atlantic salmon parr, *Salmo salar L. J. Fish. Biol. 38* : 929-993, 1991.

Zhang, Z., and Runham, N. W. Temporal deposition of incremental and discontinuous zones in the otoliths of *Oreochromis niloticus (L.). J. Fish Biol.* (in press), 1992.

W. P. Patterson, G. R. Smith, and K. C Lohmann, University of Michigan, Department of Geological Sciences, 1006 C. C. Little Bldg., Ann Arbor, Michigan, 48109-1063.

A Comparison of Stable Isotope Data with Pollen and Ostracod Faunal Data in Paleoclimate Reconstruction

KAREL L. ROGERS

Department of Biology and Environmental Science, Adams State College
Alamosa, Colorado

Sediments from 2300 m elevation in south-central Colorado and dating from 2.6–0.6 Ma have been recovered from surface outcrops and by core drilling. Three types of data (biogenic and inorganic stable isotope, pollen, and ostracod faunal) contributing to the two million year climatic record from this locality are compared against one another. Vadose zone carbonate $\delta^{13}C$ values are used to infer July minimum temperatures from percent of grasses using the C4 photosynthetic pathway. The inferred temperatures are consistent with the occurrence of *Limnocythere bradburyi,* an ostracod that lives today only south of the frostline in North America. Biogenic carbonate $\delta^{13}C$ data are interpreted as a measure of surface runoff, and $\delta^{18}O$ data as a measure of evaporation/precipitation. Stable isotope data are significantly different between temperature-sensitive species of ostracods, and less significant between salinity-sensitive species of ostracods. Biogenic isotope data are most negative when the abundance of *Picea, Pinus,* and *Artemisia* pollen indicates glaciation. The climate reconstruction derived from the combined interpretations of the Hansen Bluff data is in agreement with details of marine and terrestrial climatic records. Two major shifts in climate are apparent, one to more evaporative conditions at about 1.6 Ma and another toward cold, wet, glacial conditions starting at about 0.90 Ma.

INTRODUCTION

Isotopic studies of long, well-dated ocean cores have provided valuable insight into atmospheric, oceanic and orbital factors influencing global climate (Crowley, 1983). However, the history of terrestrial climate, especially in continental interiors, is poorly understood (Ruddiman and Raymo, 1988).

Several features of the marine climatic record have facilitated its study. First, sediment accumulation is often fairly uniform over extended time spans, making precise dating between stratigraphic datums possible. Second, similar kinds of proxy climate data have been obtained over long periods and broad geographic areas, which allows comparison between drilling sites and time intervals. In contrast to marine records, accumulation rates of terrestrial deposits are usually highly variable over short intervals, limiting the accuracy of dating techniques. Terrestrial deposits frequently contain a variety of types of proxy data that are difficult to compare with one another and that are present discontinuously both among and within localities.

In all but a few cases (e.g., the Chinese Loess, Kukla, 1987, Kukla et al., 1990; Lake Tekopa, Larson and Luiszer, 1991), terrestrial records spanning long time intervals have been studied in core section, so sediment quantity is too limited for recovery of most terrestrial fossils. Studies of diatoms, ostracods, pollen, magnetic susceptibility, sedimentology, and stable isotopes have proven useful for paleoclimatic reconstruction in subsurface studies.

These data provide insight into one or more aspects of terrestrial climate, but interpretation of each is subject to uncertainties. For example, pollen fluctuations can be caused by changes in inflow patterns (Jacobson and Bradshaw, 1981; Maher, 1977; McAndrews and Power, 1973), and ostracod faunal composition can vary with salinity changes (Delorme, 1969). Therefore, although all the above proxy data types seldom occur together at one locality, combinations of them can be used to develop a more reliable reconstruction of past climate than can any one alone.

The purpose of this paper is to compare pedogenic and biogenic $^{13}C/^{12}C$ and $^{18}O/^{16}O$ ratios with other data obtained in outcrop and in the subsurface at Hansen Bluff, Colorado. This comparison is useful for improving our understanding of the stable isotope data. If nuances of patterns can be understood properly, these data may become more useful as terrestrial paleoclimatic indicators.

GEOLOGIC AND STRATIGRAPHIC FRAMEWORK OF THE SECTION STUDIED

Data upon which this paper is based are taken from sediments of the Alamosa Formation, an outcrop of which, Hansen Bluff, lies at about 2300 m elevation in south-central Colorado. The section is a fossiliferous sequence of valley fill sediments dating from 2.67-0.67 Ma. Ages are based on magnetostratigraphy (Gauss through Brunhes Chrons) and tephrochronology (Huckleberry Ridge Ash, 2.02 Ma, and Bishop Ash, 0.74 Ma). See Figure 1. Ages were assigned to all sediments by interpolation and extrapolation, assuming constant sedimentation rates between magnetic control points. The fairly uniform sediment accumulation apparent in the Alamosa Formation appears to be due to tectonism associated with the Rio Grande Rift, to the paleodrainage pattern of the San Luis Valley, and to the location of the study site near the ancient Rio Grande.

The Alamosa Formation thickens from west to east in the San Luis Valley, suggesting that accommodation space for sediment was controlled by movement along the Sangre de Cristo fault, which borders the Sangre de Cristo Mountains at the east side of the valley. Substantive movement along other faults in the area had ceased by mid-Pliocene (Huntley, 1979). Movements of the Sangre de Cristo fault subsequent to the deposition of the Alamosa Formation occurred approximately every 10-50,000 years, with fault throws each time of 1.2 to 2.9 m (Kirkham and Rogers, 1981; McCalpin, 1983). There is no evidence of individual, large movements.

During deposition of the Alamosa Formation, accommodation space for sediment was provided periodically by small movements of the fault. Sedimentation does not appear to have been controlled by fault movement, however. Provenance of sediment in the Hansen Bluff core does not fit the stream migration patterns expected from tectonic control of sedimentation (Rogers et al., 1992).

The paleodrainage pattern and the location of the Rio Grande are further reasons for relatively uniform accumulation of sediment. The San Luis Valley was internally drained until between 0.67 and 0.3 Ma (Wells et al., 1987); the ancestral Rio Grande meandered north past the Hansen Bluff area, sometimes removing thin layers of sediment, other times depositing new material on top of young soil surfaces or floodplain ponds. Periodic thick, widespread clay layers, traceable between well logs (Burroughs, 1981; Huntley, 1979), attest to the flatness of the ancient valley floor. The volume of sediment reaching the valley floor varied primarily with the amount of precipitation in the mountains but was predictable in that all sediment leaving the adjacent mountain slopes eventually was deposited on the valley floor. Therefore, this unique combination of geological factors resulted in terrestrial deposits that can be accurately dated using interpolation and extrapolation.

Fossiliferous evidence in outcrop sediments (Rogers et al., 1985) and in continuous-sediment core (Rogers et al., 1992) has permitted reconstruction of climatic changes at Hansen Bluff. These reconstructions indicate that most of the Hansen Bluff sequence represents a variety of non-glacial climates:

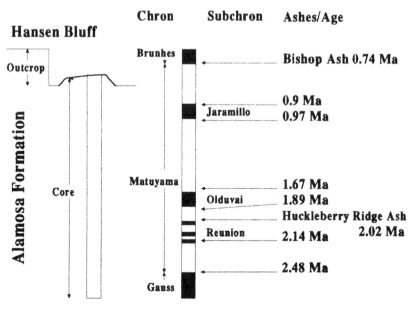

Fig. 1. Alamosa Formation sediments, studied in both the Hansen Bluff outcrop and in the Hansen Bluff core, have been dated by a combination of magnetostratigraphy and tephrochronology. Sediment between magnetic reversals has been dated by interpolation and extrapolation.

warm/wet, warm/dry, cool/wet, and cold/dry. Outcrop sediments contain evidence of major glaciation in south-central Colorado, starting ~ 0.80 Ma (Rogers et al., 1985). This glacial episode may be correlated with the classic Nebraskan Glaciation at its type locality as well as with Oxygen Isotope Stage 22 because all three are intervals of reverse magnetic polarity, all lie just below the Brunhes/Matuyama polarity boundary, and each was the period of coldest climate between the end of the Jaramillo and the beginning of the Brunhes (Boellstorff, 1978; Rogers et al., 1985; Ruddiman and Raymo, 1988). At Hansen Bluff, stable isotope, pollen, and ostracod data are of central importance to climatic reconstruction, especially in the subsurface section.

METHODS

The surface outcrop at Hansen Bluff was studied over a period of five years; fossils were retrieved by standard paleontological techniques. Subsequently in 1986, a continuous sediment core was recovered from the base of Hansen Bluff. There is about 6.4 m overlap between the lowest sediments studied in the outcrop and the top of the core. Drilling reached a total depth of 127 m, with average recovery of 72%. Core loss occurred primarily at the top of each run and in long intervals of sand.

Of the analyses done, all of which are reported in detail in Rogers et al. (1985) and Rogers et al. (1992), ostracods, molluscs, pollen, and stable isotope analyses of biogenic carbonate and inorganic carbonate are of importance to the present study.

Ostracods and molluscs were identified by means of comparative morphology; ostracod morphology changes with age because the animals shed their valves as they grow and metamorphose through a series of instar stages. These arthropods progress through a predictable yearly cycle, with adults in temperate climates alive only late in the season before the winter die-off. Ostracods are dispersed readily by traveling on birds; they are sensitive indicators of water chemistry and climate temperature (DeDeckker and Forester, 1988). Molluscs present in the section are mainly tiny aquatic clams and snails, with a few land snails present during the Stage 22 glaciation. Ostracods from outcrop samples were analyzed by R. Forester (US Geol. Survey, Denver). Molluscs from outcrop samples were studied by D. Taylor (Tiburon Center) and those from core samples by G. Mackie (U. Guelph).

Pollen samples were taken throughout the outcrop and core sections. Outcrop samples were processed and analyzed by S.A. Hall (U. Texas, Austin), and those of the subsurface by R.G. Baker (U. Iowa).

Isotopic analyses were made of inorganically precipitated calcium carbonate recovered from lacustrine, vadose, and groundwater zones in the core; no inorganic carbonates from the outcrop were sampled. Criteria for identification of carbonate influenced by soil-forming processes include absence of bedding, massive structure, presence of bioturbated layers with gradational contacts, root traces and root casts, and clay skin (T. Cerling, pers. comm., 1991). Petrographic

evidence indicates cementation by carbonate was not a diagenetic process; the extent and types of alteration throughout the core are uniform (P. Patterson, pers. comm., 1991). Samples were baked in vacuo at 450°C and then reacted with 100% phosphoric acid. The resulting CO_2 was cryogenically separated and analyzed on a mass spectrometer. The analysis was done in the lab of T. Cerling (U. of Utah).

Carbon and oxygen isotopic data were collected on ostracod and mollusc samples from all core layers containing concentrations of shell, and from samples derived from 15-cm washed segments (100-mesh screen). Isotopic compositions of biogenic carbonates from outcrops were also processed and analyzed. These samples consisted of single ostracod valves or of one fragment of mollusc shell. Stable isotope data from ostracods are affected by species and by developmental stage. In this study, ostracods were sorted by size but not by species before analysis. Most analyses were done on *Candona patzcuaro*, which was present in almost every sample and was the largest species available. Biogenic carbonate samples ranging from 100 to 300 μg were roasted at 380°C in a Finnigan "Kiel" extraction system coupled directly to the inlet of a MAT 251 ratio mass spectrometer. Isotopic compositions were corrected for acid fractionation and ^{13}C or ^{17}O contribution. All analyses were done in the laboratory of K.C. Lohmann (U. of Michigan). All isotopic data are reported in δ notation as per mil deviations from the PDB standard.

RESULTS

Ostracod faunal composition, pollen percentages of *Picea*, *Pinus*, and *Artemisia*, $\delta^{13}C$ and $\delta^{18}O$ values of inorganic carbonate, molluscs and ostracods are presented by depth and age in Appendix A. A plot of $\delta^{18}O$ versus $\delta^{13}C$ of inorganic carbonates is presented in Figure 2A; similar graphs for mollusc and ostracod samples are presented in Figures 3A and 3B, respectively.

INTERPRETATION

Faunal

Hansen Bluff ostracod faunal composition is indicative of two main climatic variables: temperature and water freshness (Rogers et al., 1992). In North America today, the ostracod, *Limnocythere bradburyi*, lives mainly south of the frostline and can be considered to be a warm indicator species (Forester, 1985). *Limnocythere ceriotuberosa* lives only north of the frostline (Forester, 1987), is present near the study area today, and was present during the Stage 22 glaciation. It, therefore, can be considered a cold indicator species. Morphological similarity and temporal distribution data indicate that *L. n. sp. A* is the ancestral species to *L. bradburyi* and thus probably had similar temperature requirements. Diatom and ostracod data from other fossil localities indicate *L. robusta* had temperature requirements intermediate between those of *L. bradburyi* and *L. ceriotuberosa* (Forester, pers. comm., 1990). *Limnocythere platyforma* is an indicator of water freshening (Forester, pers. comm., 1990) and

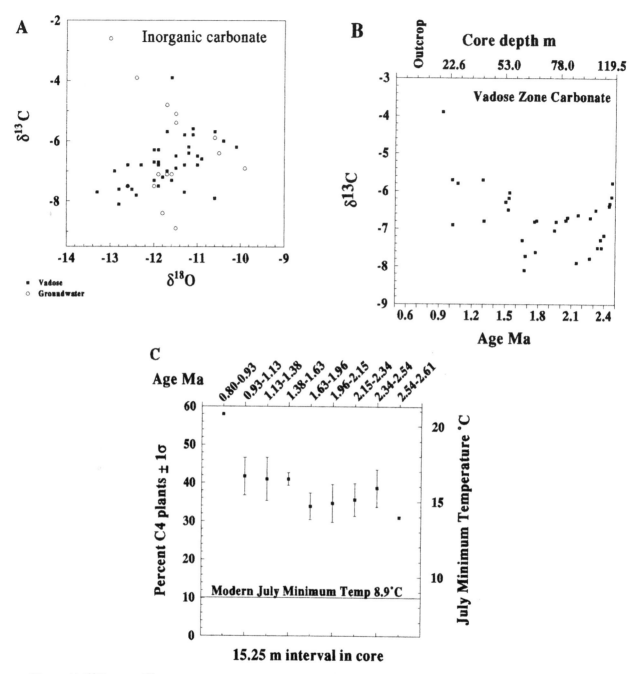

Fig. 2. A) δ^{13}C versus δ^{18}O of all inorganic carbonate occurrences in the Hansen Bluff core. B) Vadose zone δ^{13}C data plotted with respect to age and stratigraphic position. C) Mean $\pm 1\sigma$ of percent C4 plants derived from vadose δ^{13}C data within 15.25 m (50 foot) intervals of the Hansen Bluff core. July minimum temperature inferred from percent C4 plants is presented on the right Y axis.

may be an ecomorphotype of *L. ceriotuberosa. Candona patzcuaro* can tolerate seasonal drying (Delorme and Donald, 1969), but no other ostracod in the fauna can withstand desiccation even though all typically live in evaporative environments.

Floral

Pollen is present only in the top of the core and bottom of the outcrop in sediments deposited between ~0.91 and ~0.74 Ma. Its taxonomic composition reflects changing altitudes of biotic zones in the mountains surrounding

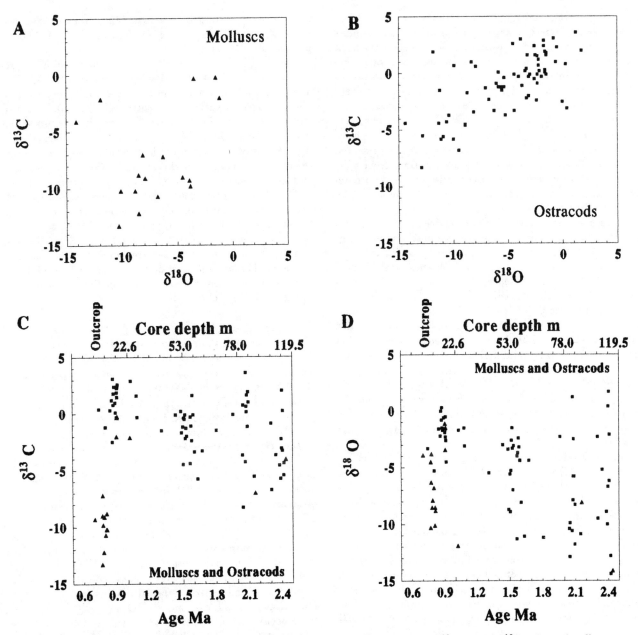

Fig. 3. A) Mollusc and B) Ostracod $\delta^{13}C$ values plotted against $\delta^{18}O$ values. C) $\delta^{13}C$ and D) $\delta^{18}O$ values of molluscs and ostracods plotted with respect to time in the Hansen Bluff combined core and outcrop section.

the study locality. In general, pine forests (*Pinus*) moved down to the valley floor during cool/wet periods, and sage (*Artemisia*) was dominant there during warm and/or dry periods. During the Stage 22 glaciation, spruce trees (*Picea*) lived on the valley floor along with the pines (Rogers et al., 1985; Rogers et al., 1992). The spruce/fir association occurs today near the peaks of mountains capped by tundra where the climate is characterized as having heavy precipitation and cold temperatures (H. Dixon, pers. comm., 1992). This interpretation of the pollen data is supported by a plethora of fossil vertebrate evidence (Rogers et al., 1985).

Stable isotopes

The $\delta^{13}C$ values of inorganic carbonate from the vadose zone reflect the contribution of CO_2 to the soil gas by C3 and C4 plants (Deines, 1980). Below 30 cm depth in soils, the effect of atmospheric $P(CO_2)$ on soil carbonate is negligible (Cerling, 1984; Quade et al., 1989). The model of Cerling (1984) was used to estimate the percentage of HCO_3^- contributed by C4 plants to inorganic carbonate formation. The percentage of C4 grasses in a flora correlates highly with July minimum temperature (Teeri and Stowe, 1976), and the percentage of C4 dicots correlates most highly with summer (May–October) pan evaporation (Stowe and Teeri, 1978). Because grasses make up 50% or more of the biomass in the study area today, estimates of climatic change at Hansen Bluff based on C4 grasses are more accurate than those based primarily on C4 dicots. However, an increase in summer temperature would also increase summer pan evaporation (Doesken, pers. comm., 1991), making the area more habitable for C4 dicots. Thus, the δ^{13}C values of inorganic carbonate from the vadose zone (Figure 2B) are the basis for reconstruction of July minimum temperatures presented in Figure 2C. Percentages of C4 grasses derived from δ^{13}C values of vadose carbonate were converted to July minimum temperature by means of a regression graph published by Teeri and Stowe (1976).

There is a significant difference between δ^{13}C ($p < .01$) and δ^{18}O ($p < .05$) of ostracod and mollusc samples (Figures 3A and 3B). However, the mollusc samples were deposited mainly during cold, glacial times and ostracods during warmer periods. When mollusc and ostracod samples are taken from the same sediments (10.67-11.28 m depth in the core), there is a difference of about 2 ^{o}/oo in δ^{13}C and about 1 ^{o}/oo in δ^{18}O; neither of which is significant. Control data from the Hansen Bluff ostracods indicate that variation between instar stages and different individuals of one species within one sedimentary unit accounts for a maximum variation of about 2.5 ^{o}/oo in δ^{13}C and about 2 ^{o}/oo in δ^{18}O. Interspecific variation from the same association tends to be no greater than intraspecific variation. Therefore, δ^{13}C and δ^{18}O values can fluctuate within about 2 ^{o}/oo from these variables.

The δ^{18}O values of inorganic carbonate vary by only 3.5 ^{o}/oo (Figure 2A) even though δ^{18}O values of biogenic carbonate extend over a 16 ^{o}/oo range (Figures 3A and 3B). The inorganic δ^{18}O data most likely reflect isotopic composition of meteoric waters. The water table in the San Luis Valley presently is at or near the surface; groundwater chemistry should be controlled by local infiltration, including surface and groundwater discharge from the mountains and valley precipitation (Bodrig, 1989). Thus, this difference between δ^{18}O of inorganic and biogenic carbonate indicates small changes in source-area oxygen isotopic composition while large changes are observed in biogenic carbonate δ^{18}O values.

Variations in δ^{18}O values of lacustrine biogenic carbonates in evaporative regimes are reasonably good recorders of first order changes in the δ^{18}O of lake water (G. Smith, pers. comm., 1991). Biogenic carbonate depleted in ^{18}O is formed from inflow of surface fresh-water during the growing season. Progressive evaporation during summer months results in ^{18}O enrichment of shallow lacustrine waters by as much as $+16$ ^{o}/oo (Detman, pers. comm., 1991; Fontes et al., 1970; Gonfiantini et al., 1973). However, summertime increases in the δ^{18}O values of precipitated carbonates are smaller than the summertime increases in the δ^{18}O values of the lakewater itself. This occurs because evaporation is typically higher in warmer months and because the oxygen

isotopic fractionation between calcium carbonate and water decreases with increasing temperature (K.C. Lohmann, pers. comm., 1991).

The ^{13}C/^{12}C ratios of biogenic carbonate primarily reflect the ^{13}C/^{12}C ratios of dissolved lacustrine HCO_3^-. The abundance of this anion is, in turn, controlled by local productivity (photosynthesis concentrates ^{12}C in plant tissues) and either influx of detrital organic material and its mobilization by decomposers or the influx of soil-gas bicarbonate from groundwater. Each of these would result in lowering the δ^{13}C values of dissolved carbonate available for ostracod and aquatic mollusc shells (G. Smith, pers. comm., 1991).

Therefore, Rogers et al. (1992) interpret the lower δ^{18}O values of biogenic carbonate to represent decreased evaporation/increased precipitation, and higher δ^{18}O values to represent the opposite. Correspondingly, lower δ^{13}C in this system would mean increased surface runoff, and higher δ^{13}C values would represent decreased surface runoff. See Figures 3C and 3D for variations with depth in the δ^{13}C and δ^{18}O values of biogenic carbonate.

Consistencies and inconsistencies of interpretations

In synthesizing the conclusions from the diverse types of data above, consistencies should be apparent if inferences are correct and if the magnitudes of changes inferred from the different types of data are comparable. Conclusions drawn from several data types should provide insights beyond interpretations gained from each type alone.

Different types of data were obtained from different parts of the stratigraphic section. As a result, pollen data can be compared with biogenic carbonate isotope ratios only during the time span between 0.91 and 0.74 Ma. The vadose zone carbonate δ^{13}C data and pollen data were not obtained on overlapping sediments. Sage (Artemisia), Pine (Pinus), and Spruce (Picea) pollen percentages are plotted with δ^{13}C and δ^{18}O of biogenic carbonate in Figures 4A, 4B and 4C, respectively. Correlations of the isotope data with the pollen data are not high, but δ^{18}O values tend to be most negative when pollen data indicate the climate to be coldest and wettest, especially between 0.82 and 0.74 Ma. High biogenic carbonate δ^{18}O and δ^{13}C values tend to co-occur with high percentages of sage pollen, and low δ^{18}O and δ^{13}C values with high percentages of pine and spruce pollen although high percentages of pine can occur with high δ^{18}O and δ^{13}C. Careful tracking of fluctuations of the stable isotope curves with those of the pollen curves demonstrates periods when the two signals vary together. For example, compare biogenic δ^{18}O values with Artemisia pollen percentages in Figure 4A.

Ostracod faunal differences caused by differences in water freshness should be reflected in the δ^{13}C and δ^{18}O values of the ostracod carbonate. There are no obvious correlations between abundances of any ostracod taxon and either δ^{18}O or δ^{13}C. To test the possibility of more subtle relationships, stable isotope data were grouped according to whether or not a species occurred in the unit from which the isotope

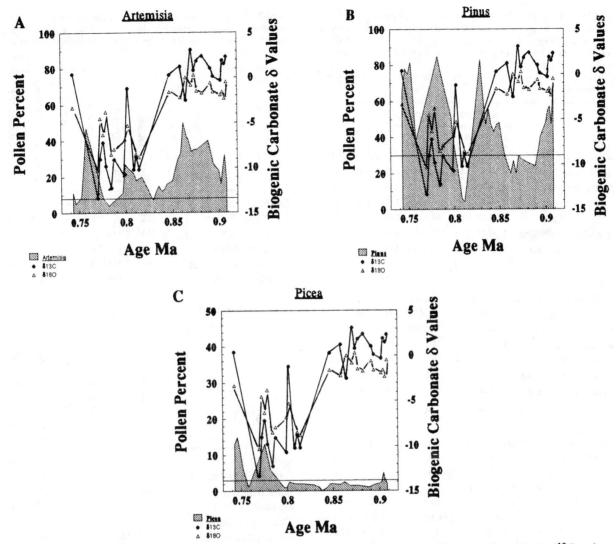

Fig. 4. A) Sage (*Artemisia*) B) Pine (*Pinus*) and C) Spruce (*Picea*) pollen percentages plotted with biogenic δ^{13}C and δ^{18}O data. Modern pollen levels of each genus at Hansen Bluff are indicated by horizontal lines.

sample was taken. Mean δ^{18}O and δ^{13}C values and standard deviations were calculated for the groups in which each species was present or absent, and a T-Test was performed to test for significance of the differences between the mean values. This approach was taken for each species except *Candona patzcuaro*; for *C. patzcuaro* the data were grouped according to whether the sample was monospecific or not. The results of this analysis are presented in Table 1. The greatest differences in δ^{18}O and/or δ^{13}C values between samples containing or not containing a species are often found for those species that are most temperature-sensitive. For example, there are significant differences between mean δ^{13}C and δ^{18}O values of samples containing *L. bradburyi* and samples from which the species is absent. Other temperature-sensitive species are associated with T-test values that lie just outside of the range of significance: *L. n. sp.* A comparison of δ^{13}C

is associated with $p \leq .07$ and *L. ceriotuberosa* δ^{18}O is associated with $p \leq .08$. These results may not contradict the interpretation of biogenic isotopes as evaporation/surface-runoff indicators nor of the ostracods as temperature indicators; it is more likely that the association of higher temperatures with increased evaporation caused the apparent relationship between the temperature-sensitive ostracods and the salinity-sensitive isotope data.

The results of T-tests between species-present data of all species are presented in Table 2. Statistically, the most significant differences again occur between species with the most divergent temperature requirements; eg., the comparison between *Limnocythere bradburyi* and *L. ceriotuberosa* is highly significant ($p \leq .01$). As expected, there is a significant difference between the δ^{18}O values of *C. patzcuaro* (monospecific), a desiccation indicator, and *L. platyforma*,

TABLE 1. Ostracod $\delta^{13}C$ and $\delta^{18}O$ data were grouped according to whether a given species was present or absent. Means and standard deviations were calculated, and a T-test was done between the species-present and species-absent data sets. In the case of *Candona patzcuaro*, isotope data were sorted into samples that were monospecific *C. patzcuaro* versus those that contained additional species.

| $\delta^{13}C$ | Species Occurrence | | | | | | T Test | |
| | Present | | | Absent | | | | |
	N	X̄	σ	N	X̄	σ	T	p≤
Candona patzcuaro	5	-0.78	1.47	65	-0.91	2.73	0.13	
Limnocythere bradburyi	10	0.60	1.61	60	-1.11	2.74	1.81	.05
Limnocythere n. sp. A	9	-2.34	2.40	61	-0.69	2.62	1.52	~.07
Limnocythere robusta	12	-0.12	2.45	58	-1.06	2.67	0.91	
Limnocythere ceriotuberosa	7	-1.57	2.90	63	-0.83	2.61	0.62	
Limnocythere platyforma	8	-1.25	2.81	62	-0.86	2.63	0.33	

| $\delta^{18}O$ | Present | | | Absent | | | T Test | |
	N	X̄	σ	N	X̄	σ	T	p≤
Candona patzcuaro	5	-3.92	1.55	65	-5.02	3.98	0.97	
Limnocythere bradburyi	10	-2.11	1.59	60	-5.40	3.93	3.22	.01
Limnocythere n. sp. A	9	-4.83	4.24	61	-4.95	3.81	0.08	
Limnocythere robusta	12	-4.65	4.87	58	-5.00	3.62	0.25	
Limnocythere ceriotuberosa	7	-7.03	5.12	63	-4.70	3.63	1.46	~.08
Limnocythere platyforma	8	-5.99	4.66	62	-4.80	3.73	0.79	

TABLE 2. Results of T-tests comparing mean $\delta^{13}C$ and $\delta^{18}O$ values of samples containing different ostracod species. Significance levels of the results of the T-tests are indicated by asterisks adjacent to T-test values.

$\delta^{13}C$	Candona patzcuaro	Limnocythere bradburyi	Limnocythere n. sp. A	Limnocythere robusta	Limnocythere ceriotuberosa	Limnocythere platyforma
Candona patzcuaro	—	*1.23	*1.24	0.55	0.58	0.35
Limnocythere bradburyi	*1.23	—	**2.50	0.63	*1.66	*1.46
Limnocythere n. sp. A	*1.24	**2.50	—	**1.77	0.53	0.79
Limnocythere robusta	0.55	0.63	**1.77	—	1.06	0.84
Limnocythere ceriotuberosa	0.58	*1.66	0.53	1.06	—	0.22
Limnocythere platyforma	0.35	*1.46	0.79	0.84	0.22	—

*** p≤.01
** p≤.05
* p≤.15

$\delta^{18}O$	Candona patzcuaro	Limnocythere bradburyi	Limnocythere n. sp. A	Limnocythere robusta	Limnocythere ceriotuberosa	Limnocythere platyforma
Candona patzcuaro	—	*1.58	0.61	0.49	**1.84	*1.30
Limnocythere bradburyi	*1.58	—	**1.91	**1.80	***3.04	**2.56
Limnocythere n. sp. A	0.61	**1.91	—	0.11	*1.16	0.64
Limnocythere robusta	0.49	**1.80	0.11	—	*1.26	0.74
Limnocythere ceriotuberosa	**1.84	***3.04	*1.16	*1.26	—	0.53
Limnocythere platyforma	*1.30	**2.56	0.64	0.74	0.53	—

*** p≤.01
** p≤.05
* p≤.15

an indicator of water freshening. *Limnocythere platyforma*, hypothesized to be an ecomorphotype of *L. ceriotuberosa*, exhibits $\delta^{13}C$ and $\delta^{18}O$ patterns similar to those of *L. ceriotuberosa*, consistent with the hypothesis that these two are the same species. Ostracod faunal changes observed in the section may be at least partially explained by comparisons of mean $\delta^{13}C$ values. *Limnocythere* n. sp. A does not co-occur temporally nor faunally with *L. bradburyi* or with *L. robusta*; its occurrence also decreases as *C. patzcuaro* becomes more abundant after about 2.10 Ma. These faunal changes are reflected in significant differences between mean $\delta^{13}C$ values of *L.* n. sp. A and the other three species. The data suggest that surface runoff was significantly greater while *L.* n. sp. A lived near Hansen Bluff than when the remaining species lived (or were abundant) there.

The $\delta^{13}C$ values of vadose zone inorganic carbonate and ostracod faunal data are difficult to compare with one another because ostracods live in lakes and ponds and carbonate influenced by pedogenic processes occurs in drained soil. Thus, the two types of data seldom occur in the same sediments. However, throughout the time interval represented by core sediments, reconstruction of July minimum temperature from $\delta^{13}C$ values in vadose carbonate indicates a substantially warmer climate than the modern one. Concurrently, *L.* n. sp. A and *L. bradburyi* are present whereas the range of *L. bradburyi* today is restricted to areas south of the frostline in the United States.

The most recent estimate of July minimum temperature based on $\delta^{13}C$ values of vadose zone carbonates is for approximately 0.93 Ma. That estimate, 20.5°C, is far higher than the modern value of 8.9°C. The percentage of sage (*Artemisia*) in modern pollen rain at the locality is between 5-10%. Because sage pollen is an indicator of warmth/dryness, the sage pollen percentages well above modern levels between 0.91-0.84 Ma are consistent with the estimate of July minimum temperature/summer pan evaporation from 0.93 Ma. Note, however, that the pollen analysis was done on sediment approximately 0.02 Ma younger than the sediment that was analyzed isotopically.

In summary, there is good general agreement between the diverse types of data discussed above. Taken together, these data form a coherent reconstruction of climatic events during the span from 2.6 to 0.6 Ma.

Summary of climate at Hansen Bluff

Throughout most of the core and outcrop sequence at Hansen Bluff, climate was substantially warmer than occurs in south-central Colorado today (Figure 2C). The time represented in the core above the Olduvai was warmer than pre-Olduvai times by 1–3°C. Such warming would cause mean annual temperature to rise about 3–5°C from present day, making temperatures at that time similar to those recorded just south of Albuquerque, New Mexico, today (Doesken, pers. comm., 1991). Cold periods within this warm time span are apparent from a combination of pollen and ostracod faunal data. Temperatures began to fluctuate between cold and warm toward the end of the Jaramillo (0.90 Ma) and degenerated to cold, glacial extremes during the time span from 0.80 to 0.74 Ma. Based on the occurrence of *Limnocythere ceriotuberosa* and other faunal data discussed in Rogers et al. (1992), other cold periods at Hansen Bluff occurred at about 2.35–2.43, 2.10, 1.5, and 0.89 Ma. In addition to temperature changes, shifts in moisture availability and precipitation are apparent in the climate sequence.

Prior to the end of the Olduvai (1.6 Ma), moisture was relatively abundant in the San Luis Valley. The time span from 2.6 to 1.6 Ma was punctuated by a short period during the middle Reunion event (2.06–2.09 Ma) when strongly evaporative times occurred. In that short interval, *Candona patzcuaro* was the only ostracod species represented in the sediments and a cluster of enriched ^{13}C values from biogenic carbonate are present (Appendix A). At about the end of the Olduvai (1.6 Ma), abundant summer precipitation became less prevalent (Figures 3C and 3D) and, after a transitional period of about 200,000 years, summers became strongly evaporative and surface runoff decreased. With the onset of the Stage 22 glaciation at ~0.80 Ma, the climate shifted toward distinctly wet conditions characterized by heavy surface runoff (Figures 3C and 3D). Other data, presented in Rogers et al. (1992), support both the temperature and moisture reconstructions outlined in this section.

Hansen Bluff climate compared to other climate records

The time span represented in the Hansen Bluff core and outcrop corresponds to oxygen-isotope stages 110 through 18 (Raymo et al., 1989; Rogers et al., 1985; Rogers et al., 1992; Ruddiman et al., 1986; Ruddiman et al., 1989). Most of the oceanic sequence in this time interval represents relatively warm, stable climates similar to those reconstructed at Hansen Bluff using vadose zone carbonate and ostracod faunal composition. In addition, the timing of cold periods is similar between the oceanic and Hansen Bluff sequences. The deep sea temperature reconstruction includes the onset of moderate-scale Northern Hemisphere glaciation at about 2.4 Ma (Ruddiman and Raymo, 1988). Additional pronounced cold periods in the oxygen isotope record are present at about 2.1 to 2.0, 1.9 Ma, and periodically between 1.5 and 1.8 Ma (Ruddiman and Raymo, 1988). Between 0.9 and 0.8 Ma, broad fluctuations are apparent in the marine record (Ruddiman et al., 1989), indicating that a long term reorganization of the climate system was occurring (Crowley and North, 1991). This reorganization resulted in the major Pleistocene glaciations that started with the Stage 22 glaciation about 0.80 Ma. Similarly, at Hansen Bluff, oscillations are apparent in both temperature and moisture between the end of the Jaramillo and the onset of the glaciation at 0.80 Ma.

Reconstruction of land and sea climates in the Arctic circle based on microtine rodents and deep sea cores (Repenning and Brouwers, in press) indicates a climate sequence in the Arctic circle compatible with the sequence found at Hansen Bluff. Notable in the sequence are the post-Olduvai precipitation changes observed at Hansen Bluff that are reflected as ice-cover oscillations in the Arctic.

Climate changes reconstructed from the Hansen Bluff core and outcrop can be correlated to terrestrial records from other localities. At Lake Tekopa, the climate was wet from 2.02 to 1.6 Ma and, between 1.6 and 1.0 Ma, precipitation was greatly reduced (Larson et al., unpublished manuscript). A similar shift in climate dated to the same time is recorded in the San Pedro Valley (Lindsay et al., 1990). Other correlations can be made but are beyond the scope of this paper. It is apparent, however, that the climatic history present in Alamosa Formation sediments and reconstructed with interdisciplinary methods reflects continental and global changes in the climate system. It is also apparent that the contributions of stable isotope data to the reconstruction are considerable.

CONCLUSIONS

Inorganic and biogenic stable isotope data are valuable for reconstructing terrestrial paleoclimate, but they frequently do not co-vary directly with the other types of paleoclimatic data in the samples studied. Consequently the isotopic data cannot be combined with other types of proxy data into a single sequence. However, there is general agreement between pollen floral reconstruction, ostracod faunal composition, and interpretations of the inorganic and biogenic stable isotopes. The latter are most reliable when used in conjunction with other types of proxy data.

The percentage of C4 plants can be reconstructed from vadose zone carbonate $\delta^{13}C$ values. Those, in turn, can be used to estimate July minimum temperature and/or summer pan evaporation. Such reconstructions are most valuable when combined with floral information that provide evidence of grass/dicot abundance and when combined with faunal information.

The relationship between ostracod faunal composition and stable isotopes of the shells should be studied in more detail. Correspondences exist but are difficult to understand, possibly because of the relationship between temperature and evaporation. Nevertheless, the potential benefits of understanding this system are great. Because ostracod valves are produced at predictable times of the year, seasonal climatic changes could be studied as well as comparisons between ages and localities.

Reconstruction of climate at the Hansen Bluff locality during the span from 2.6 to .6 Ma depends heavily on interpretations of inorganic and biogenic carbonate stable isotope data. The resulting climatic reconstruction agrees in general and in many details with the deep sea record as well as with terrestrial sequences from Arizona, Nevada, and the Arctic circle.

APPENDIX A. *Picea* (Pic), *Pinus* (Pin), and *Artemisia* (Art) pollen percentages by depth (m) and age (Ma). *Candona patzcuaro* (cp), *Limnocythere bradburyi* (lb), *L.* n. sp. A (sa), *L. robusta* (lr), *L. ceriotuberosa* (lc), and *L. platyforma* (lp) percentages in concentrations of ostracods. Inorganic carbonate (Typ: gw = groundwater or ps = vadose) stable isotope values (carbon = C13/T; oxygen = O18/T) are presented in $^0/oo$ notation relative to the PDB standard. Biogenic carbonate (Typ: M = mollusc; O = ostracod) stable isotope values (carbon = C13/K; oxygen = O18/K) are also presented in $^0/oo$ notation relative to the PDB standard.

Meters	M.y. AGE	Pic	Pin	Art	cp	lb	sa	lr	lc	lp	Typ C13/T	O18/T	Typ	C13/K	O18/K
-10.36	Recnt	0.7	15	5					X						
-8.53	0.70												M	-9.3	-3.9
-5.49	0.74	13	65	11									O	0.4	-3.3
-5.18	0.75	15	78	5											
-4.88	0.75	11	75	7											
-4.57	0.75	6	82	9											
-3.96	0.76	1	38	47											
-2.74	0.77												M	-13.3	-10.3
-2.44	0.77												M	-9	-4.5
-2.13	0.78	13	72	8									M	-7.2	-6.3
-1.83	0.78												M	-9.8	-3.8
-1.52	0.78	6	85	4											
-1.22	0.78												M	-12.2	-8.5
-0.91	0.79												M	-9.1	-7.9
0.00	0.80	0.7	51	11											
0.30	0.80	0.7	37	27									M	-10.7	-6.8
0.61	0.80	2.5	27	25									O	-1.2	-5.3
1.22	0.81	2	6	22									M	-10.2	-10.1
1.52	0.81	2	4	18									M	-8.8	-8.5
1.83	0.81												M	-10.2	-8.8
2.13	0.82	2	30	20											
3.35	0.83	1.7	83	7											
3.96	0.83	1	47	15											
4.27	0.84	0.1	56	12											
4.88	0.84	1	43	16											
5.18	0.85	2	47	17	87	13							O	0.3	-1.6
6.40	0.86	1.8	25	33									O	1.2	-2.2
6.71	0.86	2.5	20	50											
7.01	0.86	2	27	43	67	33							O	-2.5	-0.01
7.32	0.87	1.5	20	40											
7.62	0.87	1.5	30	34	74	26							O	3.1	-0.8
7.92	0.87				67		33						O	0.8	0.3
8.23	0.88				84	16							O	1.8	-1.5
8.84	0.88												O	2.4	-1.7
9.45	0.89	1	24	40											
9.75	0.89				24	51			25				O	1	-0.6
10.06	0.89				8	64	28						O	0.1	-1.6
10.36	0.90				90	10									
10.67	0.90	2.5	55	16	34				66				O	-0.3	-1.9
10.67	0.90												M	-2.01	-1.11
10.97	0.90	5	57	25	27				55	18			O	1.9	-1.6
10.97	0.90												M	-0.31	-3.44
11.28	0.91	3	47	32	23	57			17	2			O	1.5	-2.3
11.28	0.91												M	-0.23	-1.47
11.58	0.91	2	70	18	45				45	10			O	2.3	-0.5
11.58	0.91												O	3	-3.8

			Pollen		Ostracods						Inorg-Carbonate			Org-Carbonate		
Meters	M.y. AGE	Pic	Pin	Art	cp	lb	sa	lr	lc	lp	Typ	C13/T	O18/T	Typ	C13/K	O18/K
11.89	0.91				62			38						0	2.4	-2.6
12.19	0.91													0	2.6	-4.5
14.94	0.93										gw	-5.4	-11.5			
15.24	0.93										ps	-3.9	-11.6			
17.07	0.94										gw	-4.8	-11.7			
23.77	1.02										ps	-5.7	-10.6	M	-2.1	-11.9
24.08	1.02										ps	-6.9	-11.5			
24.38	1.03													0	2.9	-1.7
24.99	1.04				100											
25.91	1.05				73			27								
26.82	1.07										ps	-5.8	-11.3			
27.43	1.08													0	1.6	-1.5
27.43	1.08													0	0.7	-2.2
27.74	1.08				28	72					gw	-2.6	-13	0	-0.3	-3.1
28.04	1.09										gw	-3.9	-12.4			
40.84	1.30				10	70		20			ps	-5.7	-11.7			
41.15	1.30															
41.45	1.31				62					38	ps	-6.8	-12.3	0	-1.5	-5.5
45.42	1.37										gw	-5.1	-11.5			
49.07	1.43													0	-0.1	-3
52.12	1.48										gw	-8.4	-11.8	0	0.2	-3.4
52.43	1.49													0	-1.7	-8.7
52.73	1.49				92		8									
53.04	1.50				50	50								0	-1.2	-5.6
53.34	1.50										ps	-6.3	-12	0	-4.5	-8.9
53.64	1.51													0	-0.4	-5.3
53.64	1.51													0	-0.9	-6
53.95	1.51				97				3		gw	-7.1	-11.9	0	-0.5	-2.6
54.25	1.52													0	-0.1	-1.5
54.56	1.52													0	-2.2	-3.3
54.86	1.53										ps	-6.5	-11.5	0	-1.3	-7
55.17	1.53				100						ps	-6.2	-10.1	0	-2	-3.1
55.47	1.54										ps	-6	-10.4			
57.00	1.56													0	-4.4	-11.3
57.30	1.57				53			39	8					0	-0.3	-4
57.61	1.57													0	-1.1	-3.7
57.91	1.58				100						gw	-8.9	-11.5	0	-2.4	-2.4
58.22	1.58				48			41		11	gw	-7.6	-12.5	0	1.6	-3.2
58.83	1.59				86	14								0	-0.1	-4.4
59.13	1.60										gw	-6.9	-9.9			
59.44	1.60													0	-3.4	-8.1
61.26	1.63										gw	-6.4	-10.5	0	-5.8	-11.1
61.87	1.64					100										
62.48	1.65										ps	-7.3	-11.6			
63.40	1.68										ps	-8.1	-12.8	0	-3.3	-4.4
63.70	1.68										ps	-7.7	-13.3			
66.75	1.75				86		14									
67.67	1.77										ps	-6.8	-11.3			
67.97	1.77										ps	-7.6	-12.8			
68.58	1.79										ps	-6.8	-12.6			
69.49	1.81				93			7			gw	-7.1	-11.6	0	-1.5	-11.2
69.80	1.81				100											
70.71	1.83															
75.59	1.95										ps	-7	-11.7			
75.90	1.95															
76.20	1.96						100				ps	-6.8	-11	0	-0.1	-2.3
78.64	2.02				86	14										
81.08	2.04				86	14					gw	-5.9	-10.6	0	-3.7	-10.4
81.38	2.05													0	-8.3	-12.9
81.69	2.05										ps	-6.8	-11.9	0	0.7	-9.9
82.60	2.06										ps	-6.7	-12			
83.21	2.07													0	-4.3	-10.6
83.82	2.08													0	3.6	1.2

Meters	M.y. AGE	---Pollen-- Pic	Pin	Art	-----Ostracods--------- cp	lb	sa	lr	lc	lp	Inorg-Carbonate Typ	C13/T	O18/T	Org-Carbonate Typ	C13/K	O18/K
84.12	2.08													0	0.6	-7.9
84.43	2.08				100									0	0.1	-5.8
84.73	2.08				100									0	1.6	-2.5
85.04	2.09				100									0	-1.2	-5.8
85.34	2.09													0	1.9	-11.8
85.65	2.10				91			9						0	1	-8.3
90.83	2.14										ps	-7.9	-10.6			
91.14	2.15													0	-5.6	-10.9
92.05	2.16										ps	-6.6	-10.9	M	-7.03	-8.09
100.58	2.26										ps	-7.8	-12.4			
101.19	2.27										ps	-6.7	-11.9			
103.94	2.30													0	-0.9	-2.3
104.24	2.31													0	-6.8	-9.5
105.16	2.32										ps	-6.5	-11			
106.68	2.34										ps	-7.5	-12.6			
107.29	2.35				6		84		8					0	-3.7	-5.2
108.20	2.36										gw	-7.5	-12			
108.81	2.36										ps	-7.3	-12			
109.12	2.37										gw	-7.5	-12.6			
109.73	2.38										ps	-7.5	-11.9			
110.03	2.38										gw	-7.1	-11.7			
110.34	2.38				16		67	17						0	-4.6	-8.9
110.95	2.39				2		98							0	-5.8	-10
111.56	2.40				10		90				ps	-7.2	-11.8	0	-2.3	-6.7
111.86	2.40				6		94							0	-3.1	0.4
112.17	2.41				18		82							0	2	1.7
112.47	2.41													0	-3.3	-6.2
112.78	2.41				13		87							0	0.2	-2.1
113.39	2.42				3		64	17	16					0	-5.5	-12.8
113.69	2.42				9		45	14	31					0	-4.4	-14.4
114.00	2.43						100									
114.91	2.44													M	-4.07	-14.1
115.52	2.45										ps	-6.4	-11.2			
115.82	2.45										ps	-6.3	-11.9			
117.35	2.47										ps	-6.2	-11.2			
117.65	2.47										ps	-5.8	-11.1			
121.01	2.52										ps	-5.6	-11.1			
121.31	2.53										ps	-7	-12.9			
122.53	2.55										ps	-7.7	-11.3			

Acknowledgements. The manuscript was substantially improved through the suggestions of S. Savin, M. Margaritz, and an anonymous reviewer; their efforts are appreciated; I am especially indebted to G.R. Smith, T.E. Cerling, K.C. Lohmann, and Y. Wang who collaborated with me on stable isotope data. Others who have generously been involved in this interdisciplinary study are E.E. Larson, P. Patterson, G. Smith, D. Katzman, R.G. Baker, C.A. Repenning, R.M. Forester, G. Mackie, S.A. Hall, E. Anderson, and D. Taylor. I also thank students, colleagues, and staff at Adams State College who have helped in a variety of vital ways. Funding for the work was from the National Geographic Society (grant number 335-86) and the National Science Foundation (grant number ATM-8912517).

REFERENCES

Bodrig, P.D., An overview of a water budget for the San Luis Valley, Colorado. in E.J. Harmon, Ed. *Water in the valley; A 1989 perspective on water supplies, issues & solutions in the San Luis Valley, Colorado.* Colorado Ground-water Association, 1989, pp. 73–78.

Boellstorff, J., North American Pleistocene stages reconsidered in light of probable Pliocene–Pleistocene continental glaciation. *Science,* 202, 305–307, 1978.

Burroughs, R.L., A summary of the geology of the San Luis Basin, Colorado–New Mexico with emphasis on the geothermal potential for the Monte Vista Graben. *Colorado Geological Survey,* Special Pub. 17, pp. 1–30, 1981.

Cerling, T.E., The stable isotopic composition of modern soil carbonate and its relationship to climate, *Earth Planet. Sci. Lett.,* 71, 229–240, 1984.

Crowley, T.J., The geologic record of climatic change, *Review of Geophysics and Space Physics,* 21(4), 828–877, 1983.

Crowley, T.J. and North, G.R., *Paleoclimatology.* Oxford University Press, New York, 1991, pp. 110–131.

DeDeckker, P. and Forester, R.N., The use of ostracods to reconstruct continental palaeoenvironmental records. in P. DeDeckker, J.P. Collin and J.P. Peypouquet, Eds., *Ostra-*

coda in the earth sciences. Elsevier, Amsterdam, 1988, pp. 175–199.

Deines, P., in *Handbook of Environmental Isotope Geochemistry, 1. The Terrestrial Environment,* Elsevier, Amsterdam, 1980, pp. 329–406.

Delorme, L.D., Ostracodes as Quaternary palaeoecological indicators, *Canad. Journ. Earth Sci.,* 6, 1471–1475, 1969.

Delorme, L.D. and Donald, D., Torpidity of freshwater ostracodes, *Canad. Journ. Zool.,* 47(5), 997–999, 1969.

Fontes, J.C., Gonfiantini, R. and Roche, M.A. Deuterium et oxygene-18 dans les eaux du lac Tchad. in *Isotope Hydrology 1970.* IAEA, Vienna, pp. 387–404, 1970.

Forester, R.M., *Limnocythere bradburyi* n. sp.: A modern ostracode from central Mexico and a possible Quaternary paleoclimatic indicator, *Journ. Paleont.,* 59(1), 8–20, 1985.

Forester, R.M., Late Quaternary paleoclimate records from lacustrine ostracodes, in *The Geology of North America,* Vol. K-3, *North America and adjacent oceans during the last deglaciation,* Geol. Soc. Amer., pp. 261–276, 1987.

Gonfiantini, R., Bors, S., Ferrara, G. and Panichi, C., Isotopic composition of waters from the Danakil Depression. *Earth Plant. Sci. Lett.,* 18, 13–21, 1973.

Huntley, D., Cenozoic faulting and sedimentation in northern San Luis Valley, Colorado: summary. *Geol. Soc. Amer. Bull.,* Part I, 90, 8–10, 1979.

Jacobson, G.L. and Bradshaw, R.H.W., The selection of sites for paleovegetational studies, *Quat. Res.,* 16, 80–96, 1981.

Kirkham, R.M. and Rogers, W.P., Earthquake potential in Colorado, a preliminary evaluation. *Colorado Geol. Survey Bull,* 43, Open–File Report, 78-3, 1981.

Kukla, G., Loess stratigraphy in central China, *Quat. Sci. Rev.,* 6, 191–219, 1987.

Kukla, G., An, Z.S., Melice, J.L., Gavin, J., and Xiao, J.L., Magnetic susceptibility record of Chinese Loess, *Trans. Royal Soc. Edinburgh: Earth Sciences,* 81, 263–288, 1990.

Larson, E.E., and Luiszer, F., Paleoclimate record from 2.4 to 0.5 Ma, Tecopa Basin, SE Calinornia (Abstract), *Geol. Soc. Amer.,* 1991 Annual Meeting, San Diego, CA, 1991.

Lindsay, E.H., Smith, G.A., Vance Haynes, C., and Opdyke, N.D., Sediments, geomorphology, magnetostratigraphy, and vertebrate paleontology in the San Pedro Valley, Arizona. *Journ. Geol.,* 98, 605–619, 1990.

Maher, L.J. Jr., Palynological studies in the western arm of Lake Superior, *Quat. Res.,* 7, 14–44, 1977.

McAndrews, J.H. and Power, D.M., Palynology of the Great Lakes: the surface sediments of Lake Ontario, *Canad. Journ. Earth Sci.,* 10, 777–792, 1973.

McCalpin, J., Quaternary geology and neotectonics of the west flank of the northern Sangre de Cristo Mountains, south-central Colorado, *Colorado School of Mines Quarterly,* 77(3), 1–89, 1983.

Quade, J., Cerling, T.E., and Bowman, J.R., Systematic variations in the carbon and oxygen isotopic composition of pedogenic carbonate along elevation transects in the southern Great Basin, United States. *Geol. Soc. Amer. Bull.,* 101, 464–475, 1989.

Repenning C.A. and Brouwers, E.M., Late Pliocene–early Pleistocene ecologic changes in the Arctic Ocean borderland. *U.S. Geological Survey Bulletin,* 2036, in press, 1992.

Rogers, K.L., Repenning, C.A., Forester, R.M., Larson, E.E., Hall, S.A., Smith, G.R., Anderson, E., and Brown, T.J., Middle Pleistocene (Late Irvingtonian: Nebraskan) climatic changes in south–central Colorado, *National Geographic Research,* 1(4), 535–563, 1985.

Rogers, K.L., Larson, E., Smith, G., Katzman, D., Smith, G., Cerling T., Wang, Y., Baker, R., Lohmann, K., Repenning, C., Patterson, P., and Mackie, G., Pliocene and Pleistocene geologic and climatic evolution in the San Luis Valley of south–central Colorado, *Palaeogeography, Palaeoclimatology, Palaeoecology,* 94, 55–86, 1992.

Raymo, M.E., Ruddiman, W.F., Shackleton, N.J., and Oppo, D.W., Evolution of Atlantic–Pacific δ^{13}C gradients over the last 2.5 my, *Earth Planet. Sci. Lett.,* 97, 353–368, 1989.

Ruddiman, W.F. and Raymo, M.E., Northern Hemisphere climate regimes during the past 3 Ma: possible tectonic connections, *Phil. Trans. Royal Soc., London,* B318, 411–430, 1988.

Ruddiman, W.F., Raymo, M.E., and McIntyre, A., Matuyama 41,000 year cycles: north Atlantic Ocean and Northern Hemisphere ice sheets. *Earth Planet. Sci. Lett.,* 80, 117–129, 1986.

Ruddiman, W.F., Raymo, M.E., Martinson, D.G., Clement, B.M., and Backman, J., Pleistocene evolution: Northern Hemisphere ice sheets and North Atlantic Ocean, *Paleoceanography,* 4(4), 353–412, 1989.

Stowe, L.G. and Teeri, J.A., The geographic distribution of C4 species of the Dicotyledonae in relation to climate, *Amer. Nat.,* 112(985), 609–623, 1978.

Teeri, J.A. and Stowe, L.G., Climatic patterns and the distribution of C4 grasses in North America, *Oecologia,* 23, 1–12, 1976.

Wells, S.G., Kelson, K.I., and Menges, C.M., Quaternary evolution of fluvial systems in the northern Rio Grande Rift New Mexico and Colorado: implication for entrenchment and integration of drainage systems. in Menges, C., Ed., *Quaternary techtonics, landform evolution, soil chronologies and glacial deposits – northern Rio Grande Rift of New Mexico.* Field trip guidebook. Friends of the Pleistocene – Rocky Mountain Cell, 55–69, 1987.

K.L. Rogers, Department of Biology and Environmental Science, Adams State College.

Stable Carbon and Oxygen Isotopes in Soil Carbonates

THURE E. CERLING

Department of Geology and Geophysics, University of Utah, Salt Lake City, Utah 84112

JAY QUADE

Department of Geoscience, University of Arizona, Tucson, Arizona 85721

We explore the relationships of the stable carbon and oxygen isotopic composition of modern soil carbonates in relation to ecologic and climatic variables. The carbon system can be modeled using diffusion theory and the carbon isotopic composition of pedogenic carbonate is well correlated with the proportion of C_3 and C_4 photosynthesis in the local ecosystem. The oxygen isotopic composition of pedogenic carbonate is well correlated with the isotopic composition of local meteoric water, even if the latter is modified by differential filtration or evaporation. These properties make the isotopic composition of pedogenic carbonate a very powerful indicator of ecologic and climatic change on continents.

INTRODUCTION

Soil carbonates form in soils with a net water deficit, generally in soils where precipitation is less than about 100 cm per year [Jenny, 1980; Birkeland, 1984], which includes more than half of the world's land area. They are often preserved in continental sediments in fossil soils (paleosols) and therefore are a potentially important paleoenvironmental indicator of continental climatic and ecologic conditions [Retallack, 1990]. The purpose of this paper is to review the stable carbon and oxygen isotopic trends that have been observed in modern soils studied over the last decade [e.g., Cerling 1984; Cerling et al., 1989; Quade et al., 1989a, b].

The carbon isotopic composition of soil carbonates is controlled by CO_2 in soil gases, whose carbon isotopic composition is controlled by the proportion of C_3 and C_4 plants in the local ecosystem. The concentration and carbon isotopic composition of soil CO_2 is determined by soil productivity and by diffusion processes, and can be modeled by diffusion equations. Diffusion models for the isotopic composition of soil CO_2 make some specific predictions, many of which have been tested by field studies. The oxygen isotopic composition of soil carbonates is controlled by the isotopic composition of soil water which is related to local meteoric water.

Climate Change in Continental Isotopic Records
Geophysical Monograph 78

This paper examines the relationship between local ecology and climate and the isotopic composition of soil carbonates. Diffusion theory for CO_2 in soils is developed, and then is tested by field studies. The relationship between the oxygen isotopic composition of soil carbonate and meteoric water is presented by comparing them over a large range of climatic conditions. Finally, some applications of soil carbonate isotope geochemistry to local and global problems are presented.

ECOLOGY AND CARBON ISOTOPES

Plants can be grouped into three important groups which have different isotopic values: C_3 plants, C_4 plants, and CAM plants. The C_3 and C_4 plants use different photosynthetic pathways and have very different $\delta^{13}C$ values averaging about -26 and -12‰, respectively. CAM plants have intermediate values.

C_3 plants, which include trees, most shrubs and herbs, and cool season grasses have $\delta^{13}C$ values between about -25 and -32‰ [Deines, 1980; Figure 1]. The C_3 pathway is the older of the photosynthetic pathways and evolved under high atmospheric $P(CO_2)$ conditions [Ehleringer et al., 1991]. The $\delta^{13}C$ value of C_3 plants is variable for several reasons. Within the C_3 plants, there is a tendency for plants with high water-use efficiency to be enriched in $\delta^{13}C$. Therefore, C_3 plants growing in arid environments tend to have more positive $\delta^{13}C$ values compared to those growing under less water stressed conditions [Ehleringer, 1988]. The isotopic composition of C_3 plants

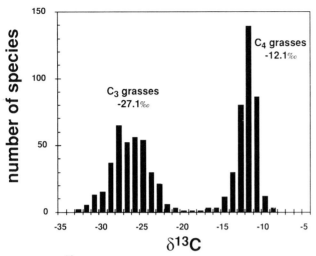

Figure 1. $\delta^{13}C$ values for C_3 and C_4 grasses. Compiled from Brown [1977], Winter et al., [1976], Smith and Brown [1973], and Vogel et al. [1976], and unpublished data.

also is controlled by the isotopic composition of the atmosphere, which is presently about -8‰ and depleted in ^{13}C relative to the pre-industrial (ca. 1850) value of -6.5‰ [Friedli et al., 1986; Marino and McElroy, 1991]. Under a closed forest canopy, the CO_2 content of the local atmosphere is higher than the global average and the $\delta^{13}C$ content of the atmosphere and of plants can be significantly depleted in ^{13}C relative to the average atmosphere, reaching values as negative as -14‰ and -37‰, respectively [Medina and Minchin, 1980; van der Merwe and Medina, 1989].

C_4 plants, which are adapted to high light and water stressed conditions, have $\delta^{13}C$ values between about -10 and -14‰. The C_4 pathway, which uses a CO_2 pump to enrich CO_2 internally within plants, is more efficient than C_3 photosynthesis when CO_2 levels drop below about 500 ppmV due to photorespiration. The C_4 photosynthetic pathway is thought to have evolved in the Tertiary in response to lower atmospheric CO_2 conditions [Ehleringer, et al., 1991]. Maize, sorghum, most prairie grasses of North America, and the savanna grasses of Africa and South America are well known examples of C_4 plants. Other important C_4 plants include members of the Euphorbiaceae and Chenopodiaceae families [Smith, 1982].

Crassalacean acid metabolism (CAM) plants have intermediate $\delta^{13}C$ values. Cacti and other succulents are examples of CAM plants [Smith, 1982]. CAM plants will not be specifically addressed

C_4 dominated ecosystems are favored by having the growing season in the warm part of the year, and are often under considerable water stress. The North American prairies are dominated by C_4 grasses from the Gulf of Mexico to about the USA - Canadian border, where C_3 grasses become important [Teeri and Stowe, 1976]. In East Africa, C_4 grasses dominate the low and intermediate altitude savannas, but the grasses

growing above 2700 meters are predominantly C_3 grasses [Tieszen et al., 1979]. Many desert or semi-desert adapted plants use the C_4 pathway, so that temperate and tropical desert ecosystems often have a substantial fraction of C_4 plants. The Sonoran Desert of the Southwestern USA, which receives significant summer rainfall, has a much higher proportion of C_4 plants than the nearby Mojave Desert, whose growing season is in the spring. Both regions, however, show strong altitudinal gradients with the proportion of C_3 plants increasing with increasing altitude. Closed canopy forests, including tropical forests, have very few C_4 plants; any understory grasses are usually C_3 grasses. The Mediterranean climate, in which the growing season is in the late winter to spring, generally has very few C_4 plants. The modern global distribution of C_3 and C_4 ecosystems is shown in Figure 2.

LOCALITIES AND CLIMATE SYSTEMS STUDIED

The soil carbonates discussed in this study come from all of the continents except Antarctica and were chosen to represent a wide range of climatic and ecologic conditions. Sites with Holocene parent material were selected so that the modern climate and ecologic conditions should be similar to those present during formation of the pedogenic carbonate. The oxygen and carbon isotopic composition of soil carbonate is essentially constant below 30 cm depth, and there is slight variation in the carbon isotopic composition of soil organic matter below about 10 cm depth (Figure 3). This means that only a single measurement of pedogenic carbonate or organic matter from depth in a soil or paleosol is needed to characterize the isotopic composition of the soil or paleosol. Localities, important climatic parameters, vegetation type, and the isotopic composition of soil organic matter, soil carbonate, and local meteoric water cited in this study are given in Table 1. The US Soil Survey classification system [Soil Survey Staff, 1975] was used, and precipitation and temperatures are from nearby local weather stations. The maximum and minimum soil temperatures were calculated using standard heat equations in soils [Carslaw and Jaeger, 1959; Baver et al., 1972; Cerling and Hay, 1986] and assuming that the mean soil temperature is one °C warmer than the mean surface temperature. Potential evapotranspiration was roughly estimated using the observation that potential evapotranspiration (in mm/day) is approximated by 0.2mm/°C/day*T(°C) for each month whose mean temperature is above 0°C [Oliver, 1987].

Several different climatic zones are used to illustrate different aspects of the soil carbonate system. Four different climatic regimes were chosen to establish the variability in the soil system for different ecosystems and climates.

Midwestern North American monsoon climate. The term monsoon was originally used to describe the changes in seasonal wind patterns associated with the change from the summer low pressure cell centered over Tibet to the winter high pressure cell also centered over Tibet. It has come to mean continental climate systems where precipitation falls predominantly in the warm season (summer) and the cool season (winter) is dry, and includes regions such as the

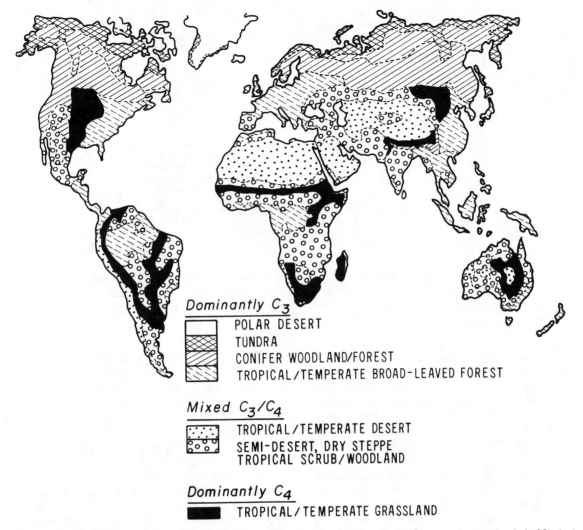

Figure 2. General distribution of C_3 and C_4 ecosystems. Note that the northern limit of temperate grasslands in North America and Asia include a significant proportion of C_3 grasses due to the cool growing season.

midwestern plains of North America and the southwestern Sonoran and Chihuahuan deserts of North America. The growing season in regions with monsoon climates is the summer, and therefore favor C_4 plant growth because of the adaptation to both high light and moisture stress. The North American midwestern prairie region has a monsoonal precipitation and temperature pattern (Figure 4). We examined soils over a transect from the C_4-dominated tall- and short-grass prairies of Kansas, Iowa, and Nebraska to the C_3-dominated boreal prairies of central Saskatchewan. This latitudinal transect also encompasses a wide range in the oxygen isotopic composition of meteoric water, from about -4‰ in Kansas to about -14‰ in Saskatchewan [Yurtsever and Gat, 1981]. The parent materials for these soils included latest Pleistocene glacial till and loess, and Holocene colluvium and lacustrine sediments.

Western USA: Mojave Desert and Great Basin Desert. The deserts of Nevada and Utah have precipitation distributed almost uniformly during the year (Figure 4), but the summers are so hot and dry that the growing season is predominantly in the spring. C_4 plants such as *Atriplex* sp.and several perennial grasses, and CAM plants (especially cactus) are found at lower elevations and decrease in abundance with increasing altitude. The oxygen isotopic composition of meteoric water decreases with increasing elevation [Smith et al., 1979]. To estimate the isotopic composition of meteoric water (Table 1) we used the empirical relationship of Smith and Friedman [in Quade et al., 1989b]:

$$\delta^{18}O_{SMOW} = -6.0 - 4.1 \times 10^{-3} Z_{msl} \qquad [1]$$

where Z_{msl} is the altitude in meters. The altitudinal transects were made to study the change in the isotopic composition of

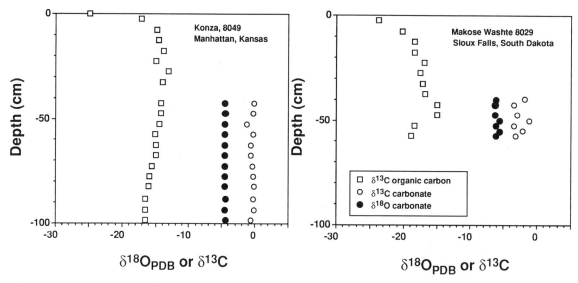

Figure 3. Isotopic composition of Holocene pedogenic carbonate from two soil profiles from Midwestern North America. The $\delta^{18}O$ and $\delta^{13}C$ values are essentially constant at depth for pedogenic carbonate. There is more variation in the isotopic composition of soil organic matter, especially near the soil surface, which may be due to differences in rooting depths of C_3 and C_4 plants.

soil carbonates with respect to changes in ecology and the isotopic composition of local meteoric water, both of which change as a function of elevation. Two different parent materials, limestone or volcanic alluvium, were studied to determine if there was any isotopic inheritance during dissolution/precipitation. All soils were from the innermost alluvial terrace in alluvial fans and are Holocene in age [Quade, 1986].

Equatorial climate. Several modern soils from equatorial regions were examined. Equatorial soils are distinguished by their small annual temperature range (Figure 4). This is particularly true for soils at depth, since below about 30 cm daily temperature fluctuations are insignificant. Soils from Bolivia and Kenya were from recent alluvial deposits estimated to be late Holocene in age, and the Tanzanian site is a calcrete developed on a 2200 year old carbonatite ash [Cerling and Hay, 1986].

Mediterranean climates. Soils from Mediterranean climates in southern Europe and southern Australia were studied as well. The Mediterranean climate (Figure 4) is distinguished by winter precipitation and dry summers, resulting is a spring growing season. Because the growing season is in the cool part of the year, very few C_4 plants are present in Mediterranean-type ecosystems. Table 1 shows that the soils from southern Europe have predominantly C_4 vegetation. All soils were from Holocene terraces.

PRECIPITATION OF PEDOGENIC CARBONATE

Pedogenic carbonate forms in soils when the solution becomes supersaturated with calcite. Equation [2]:

$$CaCO_3 + CO_2 + H_2O = Ca^{+2} + 2HCO_3^-   [2]$$

and $$K_{cal} = [a(Ca^{+2})] [a(HCO_3^-)]^2 / P(CO_2)   [3]$$

(where the terms are defined in Table 2) shows that a calcite-saturated solution can become supersaturated in several ways, including a decrease in $P(CO_2)$ in the soil, or an increase in the concentrations and activities of dissolved calcium and bicarbonate due to water loss by evapotranspiration or by direct evaporation. Furthermore, calcite could precipitate due to an increase in temperature of the solution because of its retrograde solubility, or due to soil solution composition changes during freezing of the soil. Microbial action may well play an important role in pedogenic carbonate formation. Unfortunately, soil solution composition, soil $P(CO_2)$ partial pressure, microbial activity, and temperature are variable throughout the year, so that it is difficult to determine which of the variables is responsible for carbonate precipitation. It is known that the rate of change in the $P(CO_2)$ and temperature of soils is slow [see Solomon and Cerling, 1987] which would favor equilibrium between the gaseous CO_2, dissolved HCO_3^-, and carbonate phase that precipitates from the soil solution. Solution composition changes occur due to evapotranspiration, but specific experiments monitoring soil solution chemistry and isotopic changes in the carbonate system have not established which mechanisms are largely responsible for carbonate precipitation in specific soils. However, measurements of soil CO_2 and carbonate formed in the soil in the late Holocene show that equilibrium isotopic fractionation between soil CO_2 and soil carbonate is attained or nearly attained (Figure 5).

INHERITANCE OF CARBON DURING DISSOLUTION/PRECIPITATION

It has been suggested that the dissolution of calcite in the parent material of soils and its subsequent precipitation as

Table 1. Characteristics of modern soils studied.

Locality	Soil name /number	Soil classification	Vegetation type	MAP (cm/yr) 1	PET (cm/yr) 2	MAT surface °C 3	min @50cm depth °C 4	max @50cm depth °C 5	$\delta^{13}C$ organic matter (PDB) 6	$\delta^{13}C$ carbonate (PDB) 7	$\delta^{18}O$ carbonate (PDB) 8	$\delta^{18}O$ meteoric water (SMOW) 9
Midwestern North America: monsoon climate												
Saskatchewan	Elstow	Mollisol	boreal prairie	35	39	1.6	-13.4	18.6	-24.2	-7.9	-14.0	-15.2
Saskatchewan	Sask.1	Mollisol	boreal prairie	35	39	1.6	-13.4	18.6	-24.1	-8.4	-14.2	-15.2
Saskatchewan	Weyburn	Mollisol	boreal prairie	40	41	2.1	-12.4	18.6	-22.1	-6.3	-9.6	-14.5
Minnesota	8018	Mollisol	prairie	50	54	4.9	-9.5	21.3	-17.8	-1.9	-7.5	-11.4
Minnesota	Hudderite	Mollisol	prairie	50	54	4.9	-9.5	21.3	-20.7	-5.4	-11.6	-11.4
South Dakota	Barnes	Mollisol	prairie	53	59	6.2	-7.3	21.7	-17.3	-2.0	-9.5	-10.1
South Dakota	#35	Mollisol	prairie	53	59	6.2	-7.3	21.7	-16.7	-1.9	-6.1	-10.1
South Dakota	#38	Mollisol	prairie	53	59	6.2	-7.3	21.7	-16.6	-1.2	-8.8	-10.1
South Dakota	8032	Mollisol	prairie	63	65	7.4	-5.6	22.5	-16.8	-3.6	-5.9	-9.0
South Dakota	8029	Mollisol	prairie	63	65	7.4	-5.6	22.5	-17.5	-2.7	-6.1	-9.0
Nebraska	8036	Mollisol	prairie	63	65	7.4	-5.6	22.5	-18.4	-1.7	-8.0	-9.0
Iowa	Clarion 1	Mollisol	prairie	73	69	8.7	-4.0	23.4	-18.5	-3.4	-5.0	-8.0
Nebraska	Burchard	Mollisol	prairie	68	84	11.4	-0.7	25.5	-17.3	-2.4	-4.6	-6.2
Kansas	Dwight	Mollisol	prairie	81	95	13.3	2.1	26.5	-15.7	-1.6	-4.2	-5.3
Kansas	Tully	Mollisol	prairie	81	95	13.3	2.1	26.5	-14.9	-0.4	-4.6	-5.3
Western USA: Mojave Desert - southern Great Basin												
S. Nevada	SM-6	Aridisol	conifer woodland	84	42	4.8	-3.0	14.6	-23.9	-8.5	-12.1	-17.2
S. Nevada	SM-5	Aridisol	conifer woodland	61	52	6.5	-1.6	16.5	-23.7	-8.5	-12.1	-16.2
S. California	PaM-4	Aridisol	conifer woodland	55	57	7.1	-1.0	17.2		-7.9	-12.1	-15.8
S. Nevada	SM-4	Aridisol	conifer woodland	46	62	8.0	-0.3	18.2	-23.4	-6.8	-10.5	-15.3
S. California	PaM-3	Aridisol	conifer woodland	40	67	8.7	0.3	19.0		-7.9	-12.5	-14.9
S. Utah	PiVM-2	Aridisol	conifer woodland	38	68	8.9	0.6	19.2		-7.8	-10.8	-14.7
S. Nevada	SM-3	Aridisol	conifer woodland	29	76	10.1	1.6	20.6		-5.4	-11.1	-14.0
S. Nevada	SM-3B	Aridisol	conifer woodland	27	78	10.4	1.9	21.0		-4.2	-9.9	-13.8
S. Nevada	GM-1	Aridisol	conifer woodland	24	81	10.9	2.3	21.5		-7.2	-11.0	-13.5
S. Utah	PiVM-1	Aridisol	conifer woodland	19	88	12.0	3.2	22.7		-6.1	-11.5	-12.9
S. Nevada	GM-2	Aridisol	conifer woodland	16	93	12.6	3.8	23.4		-5.9	-9.3	-12.5
S. Nevada	SM-2	Aridisol	desert shrubland	15	94	12.8	4.0	23.6		-4.0	-10.8	-12.4
S. Nevada	GM-3	Aridisol	desert shrubland	9	114	15.4	6.2	26.6		-1.6	-6.9	-10.8
S. Nevada	GM-4	Aridisol	desert shrubland	7	127	17.2	7.8	28.6		-3.3	-7.2	-9.7
S. Nevada	N-87	Aridisol	desert shrubland	6	129	17.5	8.1	29.0	-10.5	4.9	-2.5	-9.5
S. Nevada	SM-1	Aridisol	desert shrubland	6	130	17.6	8.1	29.0		-1.7	-8.2	-9.4
S. Nevada	GM-5	Aridisol	desert shrubland	6	142	19.1	9.5	30.8		1.1	-3.5	-8.5
S. Nevada	TC-1	Aridisol	desert shrubland	5	157	21.2	11.3	33.1		3.3	-3.7	-7.2
S. California	PaM-1	Aridisol	desert shrubland	5	157	21.2	11.3	33.1		-0.4	-3.4	-7.2
Western USA: northern Great Basin Desert												
N. Utah	Dimple Dell	Aridisol	desert shrubland	39	83	10.6	0.0	23.1	-24.5	-7.4	-11.2	-14.2
N. Utah	Parley Canyon	Aridisol	desert shrubland	39	83	10.6	0.0	23.1	-23.8	-8.8	-13.8	-14.2
N. Utah	Big Cottonwood	Aridisol	desert shrubland	39	83	10.6	0.0	23.1	-24.4	-7.5	-15.5	-14.2
Equatorial: South America and Africa												
Bolivia	Salla 1	Aridisol	desert shrubland	57	79	10.8	10.3	13.2	-22.8	-7.3	-7.4	-10.5
Bolivia	Salla 2	Aridisol	desert shrubland	57	79	10.8	10.3	13.2	-23.3	-8.5	-10.2	-10.5
Kenya	Simbi	Vertisol	wooded grassland	120	141	19.3	19.7	20.8	-16.0	-1.2	-2.9	-4.0
Tanzania	Olduvai	Mollisol	savanna grassland	57	166	22.8	23.0	24.6	-13.1	0.5	0.3	-1.0
Kenya	Baringo	Vertisol	wooded grassland	65	180	24.6	24.8	26.4	-14.0	-1.3	-2.0	-2.0
Kenya	Turkana.1	Aridisol	desert shrubland	18	213	29.2	29.5	30.8		-2.4	4.1	1.0
Mediterranean climate: Australia and Europe												
Australia	Gambier 1	Aridisol	desert shrubland	63	104	14.3	10.3	20.2	-20.7	-5.9	-4.2	-5.5
Australia	Gambier 2	Aridisol	desert shrubland	63	104	14.3	10.3	20.2	-20.1	-4.9	-3.8	-5.5
Greece	Axios	Mollisol	riparian forest	45	129	16.1	7.8	26.4	-23.7	-7.5	-5.6	-6.5
Greece	Samos	Mollisol	pine/scrub woodland	75	130	17.6	11.5	25.7	-25.7	-9.0	-5.0	-6.3
Turkey	Pasalar 1	Vertisol	woodland	70	132	18.1	10.9	27.2	-24.5	-10.0	-5.0	-7.0
Turkey	Pasalar 2	Vertisol	woodland	70	132	18.1	10.9	27.2	-24.5	-10.3	-6.0	-7.0
Northeastern USA												
New York	Howard	Alfisol	deciduous forest	92	62	7.7	-2.5	19.9	-25.6	-9.4	-8.6	-8.8

1. Mean annual precipitation.
2. Potential evapotranspiration (see text).
3. Mean annual temperature.
4. Minimum temperature at 50 cm (see text).
5. Maximum temperature at 50 cm (see text).
6. Average below 10 cm depth. Not measured for samples with significant organic matter inhomogeneity, such as from desert soils.
7. Average below 30 cm depth
8. Average below 30 cm depth
9. Estimated from regional IAEA records or using local data.

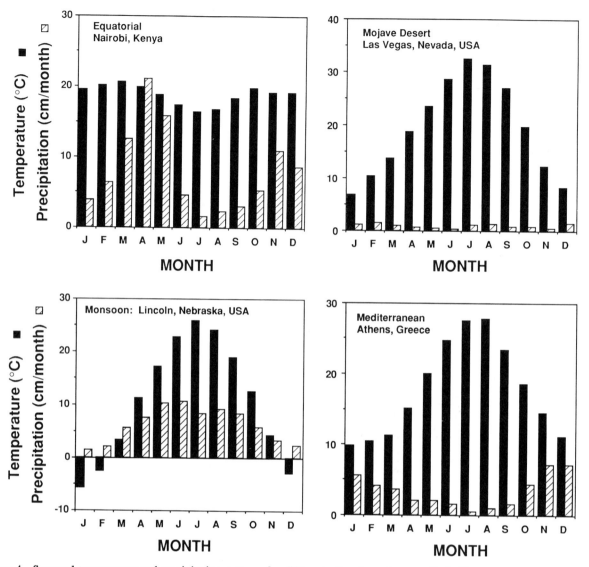

Figure 4. Seasonal temperature and precipitation patterns for different climatic regimes studied. Examples include: equatorial where seasonal temperature is essentially constant over the year, temperate desert where the growing season is in the spring due to winter and spring precipitation accompanied by moderate temperature, Mediterranean where the growing season is in the late winter and early spring due to winter and spring rains, and monsoonal where the precipitation maximum is in the summer season.

pedogenic carbonate results in a systematic enrichment of $\delta^{13}C$ of soil carbonate [Salomons et al., 1978; Salomons and Mook, 1989] according to equation [2] which shows that half of the dissolved carbon is from the original calcite. This statement is too simplistic, as discussed below. Mass balance considerations show that typical fluxes of CO_2 in soils [10^{-3} to 10^{-2} moles/cm²/year, Singh and Gupta (1977), Schlesinger (1977)], is several orders of magnitude higher than the rates of soil carbonate formation [10^{-6} to 10^{-5} moles/cm²/year, Arkley, 1963; Gile et al., 1981], which implies that the soil respired CO_2 will control the isotopic composition of dissolved inorganic carbon. Salomons et al. (1977) and Magaritz and

Amiel (1980) suggest that repeated dissolution and precipitation will result in precipitated soil carbonate being dominated by the vegetation isotopic signal. To test the hypothesis that part of the carbon isotopic signal of pedogenic carbonate can be inherited from the dissolution of pre-existing limestone in the terrain we selected two altitude transects in southern Nevada, one where limestone alluvium was the parent material and the other where igneous rocks were the parent material [Quade et al., 1989b]. The latter probably also had carbonate dust added to the surface of the soils. Figure 6 shows that isotopic gradients for the two transects have similar slopes indicating that inheritance of carbon isotopes by the

Table 2. Terms used in text.

soil CO_2	gas occupying pore space in a soil (ppmV)
soil-respired CO_2	flux of CO_2 through a soil (moles/cm²/s)
$a(Ca^{+2})$	activity of the calcium ion
$a(HCO_3^-)$	activity of the bicarbonate ion
C_a	concentration of CO_2 in the atmosphere (ppmV)
C_s^*	concentration of CO_2 in the soil (moles/cm³)
D_s^*	diffusion coefficient for CO_2 in soil (cm²/s):

$$D_s^* = D_{air}\ \varepsilon\ \rho$$

D_{air}	diffusion coefficient for CO_2 air (cm²/s)

$$D_{air} = D_{air}^o \left(\frac{P^o}{P}\frac{T}{T^o}\right)^{1.823}$$

(Bird, Stewart, and Lightfoot, 1960)

D_{air}^o	diffusion coefficient for CO_2 in air under standard conditions (STP) = (0.144 cm²/s)
D_s^{12}	diffusion coefficient for $^{12}CO_2$ (cm²/s)
D_s^{13}	diffusion coefficient for $^{13}CO_2$ (cm²/s)
L	lower no-flux boundary of soil at an impermeable barrier or at infinite depth
K_{cal}	equilibrium constant for calcite (equation [2])
$M(^{12}CO_2)$	atomic mass of $^{12}CO_2$
$M(^{13}CO_2)$	atomic mass of $^{13}CO_2$
$M(air)$	atomic mass of air
$P(CO_2)$	partial pressure of CO_2 (ppmV)
R_{PDB}	$^{13}C/^{12}C$ ratio in isotopic reference standard PDB
P	ambient pressure (bars)
P^o	standard pressure (1 bar)
S(z)	specific solution to equation [3]
T	temperature (K)
T^o	standard temperature (298.15 K)
z	depth in soil (cm)
$\bar{z}$	characteristic production depth of CO_2 in the soil (cm)
δ_s	permil value for soil CO_2
δ_a	permil value for atmospheric CO_2
δ_ϕ	permil value for soil-respired CO_2
ε	free air porosity in soil
$\phi_s^*(z)$	production rate of CO_2 (moles/cm³/s) as a function of depth
ρ	tortuosity (<1.0, Kirkham and Powers (1972))

subscript "s" refers to "soil"
superscript "*" refers to bulk CO_2 without isotopic distinction
Boundary conditions

1. CO_2 concentration at the soil-air interface being the same as the atmosphere:
    $$C_s^* = C_a^* \qquad \text{for } z = 0$$

2. the lower boundary is a no-flux boundary at an impermeable barrier (L) (a groundwater table approaches this condition) so that:
    $$\frac{\partial C_s^*}{\partial z} = 0 \qquad \text{for } z = L, \text{ or infinity}$$

dissolution/reprecipitation process is not significant. Amundson et al. [1988a, b] observed similar slopes for altitudinal gradients in the same region. Marion et al., [1990] studied soils in Alaska formed over the last 300 years which indicated that leaching of the upper part of the soil occurs in that time, and that the isotopic composition of carbonate precipitated in the B-horizon approached that expected for a pure C_3 vegetation source.

It is important to remember that the above discussion does not address the problem of carbonate detritus that is present in pedogenic carbonate. Samples with undissolved detrital carbonate mixed with pedogenic carbonate will fall on a mixing line between the two endmembers.

DIFFUSION

Diffusion of gases in soils is a well known phenomena and has been used to describe the mass transport of CO_2 and other gases in soils [Baver et al., 1972]. The concentration of soil CO_2 can be described by the diffusion-reaction equation:

$$\frac{\partial C_s^*}{\partial t} = D_s^* \frac{\partial^2 C_s^*}{\partial z^2} + \phi_s^*(z) \qquad [4]$$

where the terms are defined in Table 2. For simple production functions (e.g., constant from surface to depth L: $\phi_s^*(z)$ = constant to depth L; linear decrease from surface to depth L where $\phi_s^*(L) = 0$; or exponential decrease with depth $\phi_s^*(z) = \phi_s^*(0)\,e^{-(z/\bar{z})}$) the steady-state solution to this equation is straightforward (Cerling, 1984; Wood and Petraikis, 1984). For the boundary conditions listed in Table 2 the general solution to this equation is:

$$C_s^*(z) = S(z) + C_a^* \qquad [5]$$

where S(z) is the specific solution to equation [4] for a particular production function $\phi_s^*(z)$. For the case where $\phi_s^*(z) = \phi_s^*(0)\,e^{-(z/\bar{z})}$:

$$S(z) = \frac{\phi_s^*(0)\,\bar{z}^2}{D_s^*}\left(1 - e^{-(z/\bar{z})}\right) \qquad [6]$$

Figure 7 shows that the diffusion equation very closely models observed CO_2 concentrations in modern soils.

Similar equations can be derived for $^{12}CO_2$ and $^{13}CO_2$ using the diffusion coefficients D_s^{12} and D_s^{13} rather than the bulk

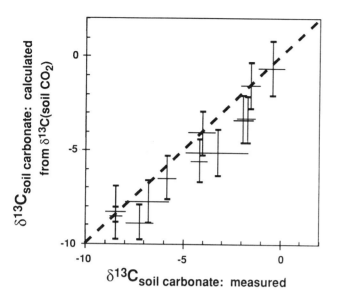

Figure 5. Comparison between the observed $\delta^{13}C$ for Holocene soil carbonate at 50 cm depth and the $\delta^{13}C$ value calculated from measurements of soil CO_2 at 50 cm depth from 12 soils in North America. Values were calculated using the minimum (above 0°C) and maximum soil temperatures at 50 cm depth in the soil using the isotopic fractionation factors of Romanek et al. [1992]. Multiple soil CO_2 measurements were made on the Midwestern soils 8018 (n = 10) and Tully (n = 5), whereas single soil CO_2 samples were collected in the spring during the growing season for soils SM-1 to -6, and GM-1 to -4.

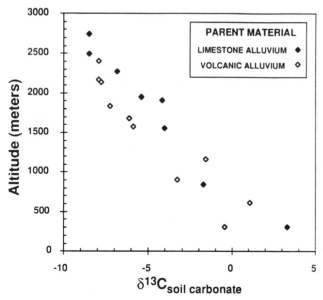

Figure 6. Average $\delta^{13}C$ values for soils along altitude transects in the southern Great Basin - Mojave Desert region [modified from Quade et al., 1989b]. Each point represents the average $\delta^{13}C$ value from 50 cm depth for 3 different soils at a single locality. All Holocene soils from alluvial terraces where limestone or igneous clasts made up the parent material.

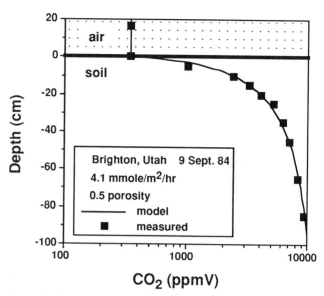

Figure 7. Measured CO_2 concentrations at different depths in a montane soil in the Wasatch Mountains, Utah [from Solomon and Cerling, 1987]. Modeled line is the steady state solution to the diffusion equation described in the text for the measured soil respiration rate on the same date as the soil CO_2 collection.

diffusion coefficient for CO_2. Using the standard δ notation, it is possible to derive the equation describing the variation of $\delta^{13}C(CO_2)$ in a soil with depth (see Cerling, 1984 for a specific solution). The general solution is:

$$\delta_s(z) = \left(\frac{1}{R_{PDB}} \left[\frac{S(z)\frac{D_s^*}{D_s^{13}}\widehat{\delta_\phi} + C_a^*\widehat{\delta_a}}{S(z)\left(1 - \frac{D_s^*}{D_s^{13}}\widehat{\delta_\phi}\right) + C_a^*\left(1 - \widehat{\delta_a}\right)} \right] - 1 \right) \times 1000$$

[7]

where:

$$\widehat{\delta_i} = \left[\frac{R_{PDB}\left(\frac{\delta_i}{1000} + 1\right)}{1 + R_{PDB}\left(\frac{\delta_i}{1000} + 1\right)} \right]$$

[8]

and the terms are as defined in Table 2.

It is important to note that the diffusion coefficients of $^{12}CO_2$ and $^{13}CO_2$ differ by [Jost, 1960]:

$$\frac{D_s^{12}}{D_s^{13}} = \left[\left(\frac{M(air) + M(^{12}CO_2)}{M(air) * M(^{12}CO_2)} \right) \left(\frac{M(air) * M(^{13}CO_2)}{M(air) + M(^{13}CO_2)} \right) \right]^{1/2}$$

$$= 1.0044$$

[9]

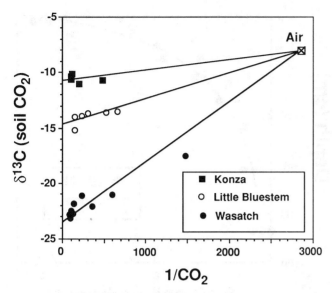

Figure 8. Soil CO_2 measurements from three soils in North America. The solid lines use the model described in the text for $\delta^{13}C$(soil-respired CO_2) values of -15.2‰, -19.1‰, and -27.7‰ for the Konza (Tully), Little Bluestem (8018), and Wasatch soils, respectively, which are 4.4‰ depleted in ^{12}C compared to the $1/CO_2$ intercept value. See Cerling et al. [1991] for further details.

This results in an enrichment in $^{13}CO_2$ in soil CO_2 compared to soil-respired CO_2: the isotopic composition of soil CO_2 is 4.4‰ <u>or more</u> enriched in $^{13}CO_2$ relative to soil-respired CO_2 at all levels in the soil (Cerling, 1984; Cerling et al., 1991b; also see Figure 8). Dörr and Münnich (1980) pointed out the soil CO_2 is enriched in ^{13}C compared to soil-respired CO_2 and attributed the difference to diffusion processes. The enrichment of soil CO_2 compared to soil-respired CO_2 can be studied by collecting soil CO_2, soil-respired CO_2, and soil organic matter. Figure 8 shows that the diffusion model closely approximates observed $\delta^{13}C$(soil CO_2) measurements in three different soils. Figure 9 shows that the endmember isotopic composition of soil CO_2 is 4.4‰ enriched compared to soil-respired CO_2 or to soil organic matter.

Equation [7] makes some specific predictions that can be tested by field experiments. Using conditions of STP (25°C and 1 bar pressure) and 300 ppmV CO_2 in the atmosphere with the pre-industrial carbon isotopic composition of -6.5‰ [Friedli and others, 1986] the predicted concentrations and isotopic composition of soil CO_2 display several important features (Figure 10). First, because there is no discontinuity in the concentrations of CO_2, $^{12}CO_2$, and $^{13}CO_2$ at the soil-atmosphere interface and because the concentrations all increase with depth, the $\delta^{13}C(CO_2)$ values for soil air vary continuously from the atmospheric value at the soil-atmosphere interface to more negative values at depth. Thus, the isotopic composition of soil CO_2 is not constant in a soil [Cerling, 1984; Quade, Cerling, and Bowman, 1989b]. The isotopic gradient is steepest in the upper few cm of soil (Figure

10) where it is extremely difficult to sample for soil gas samples, especially for sample sizes greater than 100 cm^3. Second, the gradient is flat below about 25 cm (Figure 10). The isotopic composition of soil CO_2 at greater depths therefore reflects the proportion of C_4 biomass unless productivity is extremely low. Because pedogenic carbonates are thought to precipitate under equilibrium conditions, the carbon isotopic composition of pedogenic carbonate may be a proxy indicator of soil CO_2 conditions. Isotopic fractionation between CO_2 and carbonate results in a temperature dependant enrichment of ^{13}C in the carbonate phase of about 12‰ at 0 °C to 9‰ at 25 °C [Romanek et al., 1992]. Including the minimum of 4.4‰ diffusional enrichment, the total enrichment between soil-respired CO_2 and soil carbonate should range from about 16.5 to 13.5‰ between 0 and 25°C.

The above observations led to several field experiments. The diffusional gradient was studied in two ways. First, we collected gas from snow-covered soils in the Wasatch Mountains, Utah during the winter when the soil respiration rate was extremely low [<0.2 mmol/m²/hr; Solomon and Cerling, 1987], and from several soils in the midwestern USA. The soils showed a diffusional enrichment of ^{13}C in the samples nearer the surface, which had lower CO_2 concentrations (Figure 8). The second way was to measure the $\delta^{13}C$ of soil

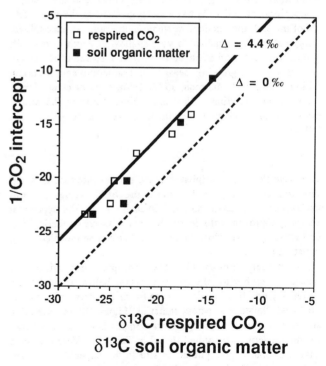

Figure 9. Isotopic composition of soil CO_2 and soil-respired CO_2 for soils from North America. The soil CO_2 $\delta^{13}C$ value is the endmember value based on the relationship shown in Figure 8. Soil-respired CO_2 and soil organic matter $\delta^{13}C$ values were measured directly. The solid line is the theoretical 4.4‰ enrichment predicted by the diffusion equation. See Cerling et al. [1991] for further details.

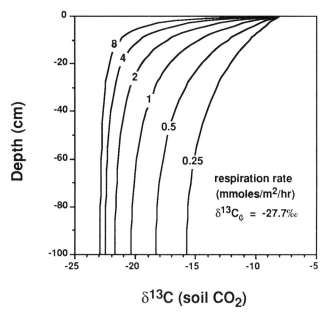

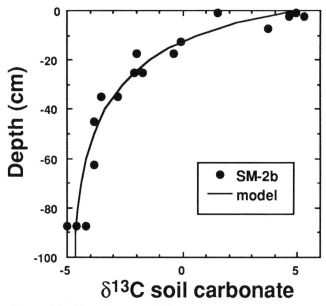

Figure 10. Isotopic composition of soil CO_2 using the model described in the text with a no-flux lower boundary at 1000 cm depth, one atmosphere total pressure, δ_a = -8‰ at 350 ppmV, δ_ϕ =-27.7‰, ρ = 0.5, ε = 0.6, T=15°C, using an exponential production term with $\bar{z}$ =25 cm. This shows that the isotopic composition of soil CO_2 has a steep gradient in the upper few cm of the soil, that the surface value is the same as atmospheric CO_2, and that the gradient is very shallow at depth in the soil. Values for soil respiration between 5 and 10 mmol/m²/hr are typical for the growing season in temperate and montane regions [Dörr and Münnich, 1987, 1990; Gaudry et al., 1990; Solomon and Cerling, 1987] and values between 0.5 and 2 mmol/m²/hr were measured in desert regions [Quade et al., 1989b].

Figure 11. Isotopic composition of soil carbonate in desert soil profile SM-2 near Las Vegas, Nevada, USA. Solid line is modeled for δ_a = -6.5‰ at 300 ppmV, δ_ϕ =-23.4‰, ρ = 0.5, ε = 0.6, T=13.6°C, using a production term with $\bar{z}$ = 50 cm. For details, see Quade et al. [1989b].

The third implication of the diffusion model is that there is a 13.5 to 16.5 permil enrichment in soil carbonate compared to soil-respired CO_2. Unfortunately, few measurements have been made on the isotopic composition of respired soil CO_2 for

carbonate that was precipitated to the soil-air interface in desert environments. Figure 11 shows the $\delta^{13}C$ values for soil carbonates with increasing depth in such a soil. The uppermost samples approach isotopic equilibrium with atmospheric CO_2, and all samples are close to a profile that can be modeled using equation [7].

Our model implies that the isotopic composition of pedogenic carbonate should be constant below about 25 cm depth (Figure 10). This is the case for midwestern soils formed on latest Pleistocene parent material (Figure 3); the variation shown is typical for pedogenic carbonates below 25 cm depth for the modern soils studied (listed in Table 1). Very few soils have carbonates precipitated *in situ* in the upper 25 cm; such soils are usually from desert climates or from groundwater discharge zones. Figure 11 also shows that the isotopic composition of pedogenic carbonate approaches a constant value with depth. These results suggest that if significant variation in the isotopic composition of pedogenic carbonate is apparant from a single soil, then a complex vegetation history is implied.

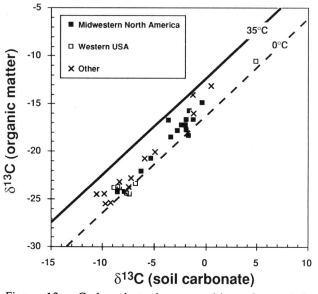

Figure 12. Carbon isotopic composition of co-existing pedogenic carbonate and organic matter from modern soils in Africa, Asia, Australia, Europe, North America, and South America. Soil carbonates are below 25 cm depth, and organic carbon is below 10 cm depth.

comparison to soil carbonate. However, the isotopic composition of soil-respired CO_2 is controlled by the isotopic composition of vegetation. We have used the isotopic composition of soil organic matter as a proxy for the isotopic composition of soil-respired CO_2. Figure 12 shows that there is a systematic enrichment in ^{13}C of soil carbonate compared to soil organic matter that is similar to that predicted by the diffusion considerations. Quade et al. [1989a, 1992], Cerling et al. [1991] and Wang et al., [1992] have found the same relationship in Miocene and Pliocene fossil soils from Pakistan, Kenya, and North America.

These field experiments show that the carbon isotopic composition of pedogenic carbonate is determined by the isotopic composition of vegetation and by the diffusional mass transport of gases within soils. With care, the isotopic composition of pedogenic carbonate in soils or paleosols can be used to determine vegetational histories of soils or of sedimentary sequences.

OXYGEN ISOTOPES

There have been no systematic studies of the oxygen isotopic composition of soil waters in relation to pedogenic carbonate formation. Any such studies must monitor the soil solution over several growing seasons to understand differential infiltration and attenuation of the isotopic signal with depth. The isotopic composition of meteoric water is strongly temperature dependant, and regions with strong seasonal temperature contrasts have a large range in the isotopic composition of meteoric water [Yurtsever and Gat, 1981; International Atomic Energy Agency, 1981].

We analyzed soil carbonates from a variety of environments

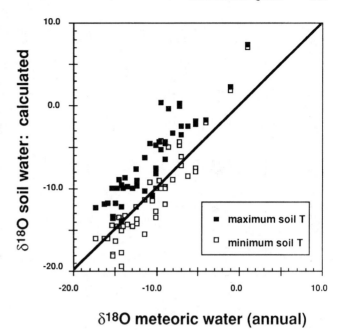

$\delta^{18}O$ meteoric water (annual)

Figure 14. Isotopic composition of soil water calculated from soil carbonate $\delta^{18}O$ values in Table 1 assuming the maximum and minimum soil temperatures, using the isotopic fractionation factors from Friedman and O'Neil [1977].

having a range in mean annual temperature of about 30°C. Figure 13 shows that there is a very good correlation between the isotopic composition of pedogenic carbonate and the mean isotopic composition of local meteoric water over a range from +1‰ to -15‰ (with respect to SMOW) for local meteoric water ($r^2 = 0.84$ for all samples). The $\delta^{18}O$ of pedogenic carbonate forms in equilibrium with soil water, whose isotopic composition is related to meteoric water. The isotopic composition of soil water can be significantly different than average local meteoric water for several reasons, including differential infiltration of waters of different isotopic composition, modification of meteoric water by evaporation, and modification of meteoric water by evapotranspiration. The last mentioned is thought not to cause isotopic fractionation, but Allison et al. [1984] have shown that evaporation from bare soil surfaces can result in isotopic fractionation in soil water. For the soils in Table 1, none of the soils from Midwestern North America had bare soil, but all the the soils from the Great Basin region and from Africa had some bare soil. Figure 14 shows the range of the isotopic composition of the soil water that could produced the pedogenic carbonates in this study. Evapotranspirational loss is highest in the growing season, which implies soil temperatures between the average annual soil temperature (for Mediterranean and equatorial climates) and the maximum soil temperature (for monsoon climates). Therefore, it is likely that most of the pedogenic carbonates formed from waters enriched relative to average annual meteoric water. This could be due to two reasons: direct evaporation of water in the upper levels of the soils resulting in enrichment of ^{18}O of the soil water, or differential

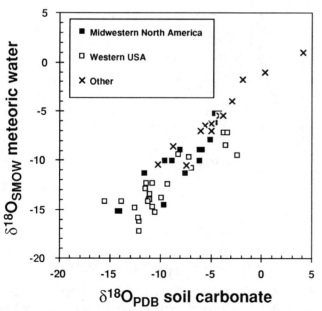

Figure 13. Isotopic composition of modern pedogenic carbonate compared to the isotopic composition of local average meteoric water.

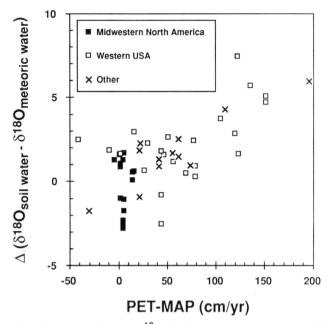

Figure 15. Enrichment of ^{18}O in soil water compared to local meteoric water for different soil moisture regimes. $\delta^{18}O$(soil water) was calculated assuming equilibrium isotopic fractionation between carbonate and soil water [Friedman and O'Neil, 1977]. The temperature for each site was assumed to be the average soil temperature at 50 cm depth during a four month growing season in the spring (Western USA and Mediterranean soils), in the summer (Midwestern USA) or the average annual soil temperature (Equatorial).

temperatures. Thus, in the midwestern plains of North America both ^{13}C and ^{18}O of soil carbonates are depleted at high latitudes, and North American altitude transects also show depletion in ^{13}C and ^{18}O with increasing elevation [Quade et al., 1989b], resulting in a strong correlation between ^{13}C and ^{18}O (Figure 16). However, the Mediterranean climate shows a different relationship because the growing season is in the cool part of the year which favors C_3 photosynthesis. Samples from sites with a Mediterranean climate in Australia, Greece, and Turkey (Table 1) do not show a correlation between $\delta^{13}C$ and $\delta^{18}O$ in pedogenic carbonate, although our data is sparse. Magaritz et al. [1981] and Salomons et al. [1976] also found that the $\delta^{13}C$ of recent pedogenic carbonate was fairly negative (-8 to -12‰) for pedogenic carbonates from Israel, Italy, and France.

APPLICATIONS

It is shown above that the carbon isotopic composition of pedogenic carbonate can be understood using diffusion theory and is directly related to the fraction of C_3 and C_4 biomass in the local ecosystem. Furthermore, the oxygen isotopic composition is well correlated with the isotopic composition of local meteoric water even though the latter may be slightly shifted from soils waters. These relationships can be used to interpret the paleoenvironments in which pedognic carbonate in fossil soils, or paleosols, formed [Cerling and Hay, 1986; Cerling et al., 1988, 1991; Quade et al., 1989a]. For example, Quade et al. [1989a] showed that the local ecosystem in Pakistan underwent a rapid transition from a C_3 ecosystem to a C_4 ecosystem between 7.3 and 6 million years ago which was

infiltration of ^{18}O enriched waters. The latter would imply that ^{18}O depleted waters, which generally fall in the cool season, have a higher rate of runoff than ^{18}O enriched waters. Some soils, especially those from desert and semi-desert soils with high potential evapotranspiration (Figure 15), probably lose water by direct evaporation from the soils. Quade et al. [1989] shows an example of a soil with a strong ^{18}O enrichment in pedogenic carbonate from the upper 20 cm of a desert soil.

It is preliminary to use the absolute $\delta^{18}O$ values of pedogenic or paleosol carbonate to estimate paleotemperatures. There are several variables that can modify the isotopic composition of soil water compared to meteoric water, as well as temperature related fractionation factors between carbonate and water. However, $\delta^{18}O$ values of pedogenic or paleosol carbonate can be used to mark environmental change in stratigraphic sequences or to compare different localities.

CORRELATION BETWEEN $\Delta^{13}C$ AND $\Delta^{18}O$ IN PEDOGENIC CARBONATE

In some regions there is a high correlation between the carbon and oxygen isotopic composition of soil carbonate. This is because as temperature decreases, the C_3 photosynthetic pathway is favored resulting in ^{13}C depletion in plants, and because meteoric waters are depleted in ^{18}O at cooler

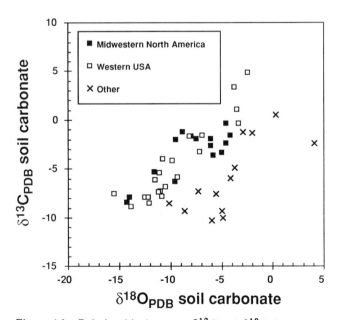

Figure 16. Relationship between $\delta^{13}C$ and $\delta^{18}O$ for modern soil carbonates. They are well correlated within some regions such as Midwestern North America or the Western USA, although other regions are not well correlated.

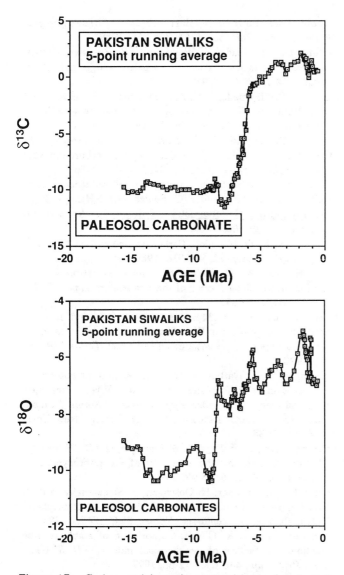

Figure 17. Carbon and isotopic compositions of pedogenic carbonates from the Siwaliks in Pakistan. Carbon isotopes show a shift from a C_3 dominated to a C_4 dominated ecosystem between 7.3 and 6 million years ago, whereas oxygen isotopes record an important shift at about 8 million years ago as well as a few smaller shifts at other times.

preceded by a significant shift in the $\delta^{18}O$ of pedogenic carbonate (Figure 17). The most significant faunal change in the Siwaliks was between 7 and 6 million years ago when browsers being replaced by grazers [Barry et al., 1982, 1985].

The local correlations between $\delta^{13}C$ and $\delta^{18}O$ for pedogenic carbonates also can be used to identify carbonates of unknown origin: Quade and Cerling [1990] showed that the isotopic composition of pedogenic carbonates in the western USA was very different than that of spring derived carbonates. They showed that carbonates of disputed origins found in fractures near the proposed high-level nuclear site at Yucca Mountain,

Nevada displayed $\delta^{13}C$ and $\delta^{18}O$ values characteristic of soil carbonates of the region.

A third interesting application is to study the history of CO_2 in the atmosphere. The diffusion equation that was developed above has as one of its boundary conditions the concentration of CO_2 in the atmosphere. Equation [5] shows that the atmospheric contribution to total soil CO_2 is low for virtually all modern soils, which typically have $P(CO_2)$ values between 5000 and 10,000 ppmV during the growing season. However, if the CO_2 content of the atmosphere were significantly higher, above 1000 ppmV, then there would be a significant shift in the carbon isotopic composition of soil CO_2 and therefore soil carbonate. Since C_4 plants are adapted to low $P(CO_2)$ conditions [Ehleringer et al., 1991] high atmospheric $P(CO_2)$ levels and C_4 photosynthesis are unlikely to co-exist. Therefore, Cerling [1991, 1992] suggested that the isotopic composition of pedogenic carbonate and co-existing organic matter could be an indicator past atmospheric CO_2 levels. The biggest problem in applying this technique so far is in establishing the temperature of carbonate formation and the amount of total CO_2 in the paleosols [Cerling 1992]. However, the modern range for $\Delta(\delta^{13}C_{soil\ carbonate} - \delta^{13}C_{organic\ matter})$ is between about 14 and 17‰ (Table 1). Ferguson et al. [1991] found that $\Delta(\delta^{13}C_{soil\ carbonate} - \delta^{13}C_{organic\ matter})$ changed from 19‰ to 13.3‰ across the Cretaceous-Tertiary boundary, indicating a significant drop in $P(CO_2)$ of the atmosphere near the K-T boundary.

CONCLUSIONS

The carbon isotopic composition of pedogenic carbonate can be modeled using diffusion equations. The diffusion equation for the isotopic composition of soil CO_2 makes some rather specific, testable predictions. The 4.4‰ enrichment in soil CO_2 compared to soil respired CO_2 predicted by theory is found in modern soils developed over the last few thousand years. The theory also predicts a gradient in the isotopic composition of CO_2 which is steepest in the upper 20 cm of soils. Such gradients are preserved in high porosity desert soils where carbonate is precipitated to the soil-air interface. The carbon isotopic composition of soil carbonate is essentially constant at depth in soils.

The oxygen isotopic composition of pedogenic carbonate is well correlated with the isotopic composition of meteoric water, although the isotopic composition of soil carbonate actually related the isotopic composition of soil water and the temperature of formation. Soil carbonates precipitated from regions with high potential evapotranspiration but low precipitation show are significantly enriched in ^{18}O compared to that expected from local meteoric water, indicating that evaporative isotopic enrichment may have taken place.

The $\delta^{13}C$ to $\delta^{18}O$ values of pedogenic carbonate are sometimes well correlated within regions. In some cases such relationships may be used to identify the origin of problematic carbonates. In buried paleosols, the $\delta^{13}C$ to $\delta^{18}O$ values of pedogenic carbonate can be used to indicate periods of significant climatic or ecologic change.

Acknowledgments. We are grateful to many colleagues for discussions and assistance as this study has evolved over the years, including J.R. Bowman, A.K. Behrensmeyer, F.H. Brown, J.R. Ehleringer, R.L. Hay, J.R. O'Neil, W.T. Parry, D.K. Solomon, and Y. Wang. We thank D. Dettman and J. R. Graney for thoughtful reviews. This work was supported by the National Science Foundation, the Research Corporation, Mifflin and Associates, and the University of Utah, and critical support was provided by BLS.

REFERENCES

Allison, G. B., C. J. Barnes, M. W. Hughes, and F. W. J. Leary, The effect of climate and vegetation on the oxygen-18 and deuterium profiles in soils. In: *Isotope Hydrology 1983*, International Atomic Energy Agency, Vienna, Austria, p. 105-123, 1984.

Amundson, R. G., O.A. Chadwick, J. M. Sowers, and H. E. Doner, Relationship between climate and vegetation and the stable carbon isotope chemistry of soils in the eastern Mojave, Nevada. *Quat. Res.* 29, 245-254, 1988a.

Amundson, R. G., O.A. Chadwick, J. M. Sowers, and H. E. Doner, The stable isotope chemistry of soils at Kyle Canyon, Nevada. *Soil Sci. Soc. Amer. Jour.* 53, 201-210, 1988b.

Arkley, R. J., Calculations of carbonate and water movement in soil from climatic data. *Soil Sci.* 96, 761-767, 1963.

Barry, J.C., E. H. Lindsay, and L. L. Jacobs, A biostratigraphic zonation of the Middle and Upper Siwaliks of the Potwar Plateau of Northern Pakistan. *Palaeogeogr., Palaeoclim., Palaeoecol.y* 37, 95-130, 1982.

Barry, J.C., N. M. Johnson, M. Raza, and L. L. Jacobs, Neogene mammalian faunal change in southern Asia: correlations with climatic, tectonic, and eustatic events. *Geology* 13, 637-640, 1985.

Baver, L.D., W. H. Gardner, and W. R. Gardner, *Soil Physics,* 4th Edition. Wiley, New York, 1972.

Bird, R.B, W. E. Stewart, and E. N. Lightfoot, *Transport phenomena* John Wiley, New York, 780 pp, 1960.

Birkeland, P.W., *Soils and Geomorphology.* Oxford University Press, New York, 1984.

Brown, W.V. The Kranz syndrome and its subtypes in grass systematics. *Mem.Torrey Bot. Club* 23, 1-97, 1977,.

Carslaw, H.S. and J. C. Jaeger, *The Conduction of Heat in Solids,* 2nd Edition. Oxford, New York, 1959.

Cerling, T.E., The stable isotopic composition of modern soil carbonate and its relationship to climate: *Earth Planet. Sci. Lett.,* 71, 229-240, 1984.

Cerling, T.E., Carbon dioxide in the atmosphere: evidence from Cenozoic and Mesozoic paleosols. *Amer. Jour. Sci.* 291, 377-400, 1991.

Cerling, T.E., Further comments on using carbon isotopes in paleosols to estimate the CO_2 content of the atmosphere. *Jour. Geol. Soc. London,* 149, 673-676, 1992.

Cerling, T.E. and R. L. Hay, An isotopic study of paleosol carbonates from Olduvai Gorge. *Quat. Res.* 25, 63-78, 1986.

Cerling, T.E., J. R. Bowman, and J. R. O'Neil, An isotopic study of a fluvial-lacustrine sequence: the Plio-Pleistocene Koobi Fora Formation, East Africa. *Palaeogeogr., Palaeoclimat., Palaeoecol.* 63, 335-356, 1988.

Cerling, T.E., J. Quade, Y. Wang, and J.R. Bowman, Carbon isotopes in soils and palaeosols as paleoecologic indicators: *Nature,* 341, 138-139, 1989.

Cerling, T.E., J. Quade, S. H. Ambrose, and N. E. Sikes, Fossil soils from Fort Ternan, Kenya: grassland or woodland? *Jour.Hum. Evol.n* 21: 295-306, 1991a.

Cerling, T.E., J. Quade, D. K. Solomon, and J. R. Bowman, On the carbon isotopic composition of soil carbon dioxide: *Geochim.Cosmochim. Acta* 55, 3403-3405, 1991b.

Deines, P., The isotopic composition of reduced organic carbon: in: P. Fritz and J. C. Fontes (eds.), Handbook of Environmental Geochemistry. Volume 1, Amsterdam, Elsevier, p. 329-406, 1980.

Dörr, H. and K. O. Münnich, Carbon-14 and carbon-13 in soil CO_2. *Radiocarbon,* 22, 909-918, 1980.

Dörr, H., and K. O. Münnich, Annual variation in soil respiration in selected areas of the temperature zone: *Tellus* 39B, 114-121, 1987.

Dörr, H., and K. O. Münnich, ^{222}Rn flux and soil air concentration profiles in West Germany. Soil ^{222}Rn as tracer for gas transport in the unsaturated soil zone: *Tellus* 42B, 20-28, 1990.

Ehleringer, J.R., Carbon isotope ratios and physiological processes in aridland plants: In Rundel, P.W., Ehleringer, J.R., and Nagy, K.A., eds. *Applications of Stable Isotopic Ratios to Ecological Research,* Springer-Verlag., New York, p. 41-54, 1988.

Ehleringer, J.R., R. F. Sage, L. B. Flanagan, and R. W. Pearcy, Climatic change and the evolution of C_4 photosynthesis, *Trends Ecol. Evol.,* 6, 95-99, 1991.

Friedli, H., H. Lotscher, H. Oeschger, U. Siegenthaler, and B. Stauffer, Ice core record of the $^{13}C/^{12}C$ ratio of atmospheric CO_2 in the past two centuries: *Nature,* 324, 237-238, 1986.

Friedman, I and J. R. O'Neil, Compilation of stable isotope fractionation factors of geochemical interest. *U. S. Geol. Surv. Prof. Pap.,* 440-KK. 12 pp, 1977.

Ferguson, K.M., T. H. Lehman, and R. T. Gregory, C- and O-isotopes of pedogenic nodules from two sections spanning the K-T transition in West Texas. *Geol. Soc. Ame. Abs. Prog.,* 23, 302, 1991.

Gaudry, A., G. Polian, B. Ardouin, and G. Lambert, Radon-calibrated emissions of CO_2 from South Africa: *Tellus,* 42B, 9-19, 1990.

Gile, L.H., J. W. Hawley, and R. B. Grossman, *Soils and geomorphology in the Basin and Range area of southern New Mexico - Guidebook to the Desert Project,* New Mexico Bureau of Mines and Mineral Resources Memoir 39, 1981.

International Atomic Energy Agency, *Statistical Treatment of Environmental Isotope Data in Precipitation,* International Atomic Energy Agency Technical Report Series 206. International Atomic Energy Agency, Vienna, Austria. 255 pp, 1981.

Jenny, H., *The Soil Resource,* Springer-Verlag., New York, 377 pp, 1980.

Jost, W., *Diffusion in Solids, Liquids, Gases,* Academic Press, New York. 558 pp, 1960.

Kirkham, D., and W. L. Powers, *Advanced Soil Physics,* Wiley-Interscience, New York, 534 pp, 1972.

Magaritz, M. and A. J. Amiel, Calcium carbonate in calcareous soil from the Jordon Valley, Israel: its origin revealed by the stable carbon isotope method, *Soil Sci. Soc. Amer. Jour.,* 44, 1059-1062, 1980.

Magaritz, M., A. Kaufman, and D. H. Yaalon, Calcium carbonate nodules in soils: $^{18}O/^{16}O$ and $^{13}C/^{12}C$ ratios and ^{14}C contents, *Geoderma,* 25, 157-172, 1981.

Marino, B.D., and M. B. McElroy, Isotopic composition of atmospheric CO_2 inferred from carbon in C_4 plant cellulose, *Nature,* 349, 127-131, 1991.

Marion, G.M., D.S. Introne, and K. Van Cleve, The stable isotope geochemistry of $CaCO_3$ on the Tanana River floodplain of interior Alaska, U.S.A.: composition and mechanisms of formation, *Chem. Geol. (Isotope Geosci. Sect.),* 86, 97-110, 1991.

Medina, E. and P. Minchin, Stratification of $\delta^{13}C$ values of leaves in Amazonian rain forests, *Oecologia,* 45, 377-378, 1980.

Oliver, J.E., Evapotranspiration, In J.E. Oliver and R.W. Fairbridge (Eds), *The Encyclopedia of Climatology,* Van Nostrand Reinhold, New York. p. 449-456, 1987.

Quade, J., Late Pleistocene environmental change in the upper Las Vegas Valley. *Quat. Res.,* 26, 340-357, 1986.

Quade, J. and T. E. Cerling, Stable isotopic evidence for a pedogenic origin of carbonates in Trench 14 near Yucca Mountain, *Science,* 250, 1549-1552, 1990.

Quade, J., T. E. Cerling, and J.R. Bowman, Development of Asian monsoon revealed by marked ecological shift during the latest Miocene in northern Pakistan, *Nature,* 342, 163-166, 1989a.

Quade, J., T. E. Cerling, and J. R. Bowman, Systematic variations in the stable carbon and oxygen isotopic composition of pedogenic carbonate along elevation transects in the southern Great Basin, USA, *Geol. Soc. Amer. Bull.,* 101, 464-475, 1989b.

Quade, J., T.E. Cerling, M.M. Morgan, D.R. Pilbeam, J. Barry, A.R. Chivas, J.A. Lee-Thorp, and N.J. van der Merwe, A 16 million year record of paleodiet using carbon and oxygen isotopes in fossil teeth from Pakistan, *Chem. Geol. (Isotope Geosci. Sect.),* 94, 183-192, 1992.

Retallack, G.J., *Soils of the Past,* Unwin Hyman, London, 520 pp, 1990.

Romanek, C. S., E. L. Grossman, and J. W. Morse, Carbon isotopic fractionation in synthetic aragonite and calcite: effects of temperature and precipitation rate, *Geochim. Cosmochim. Acta,* 56, 419-430, 1992.

Salomons, W., and W. G. Mook, Isotope geochemistry of carbonates in the weathering zone, In J.Ch. Fontes and P. Fritz (eds) *Handbook of Environmental Geochemistry Vol. 2,* Amsterdam, The Netherlands, p. 239-269, 1986.

Salomons, W., A. Goudie, and W. G. Mook, Isotopic composition of calcrete deposits from Europe, Africa, and India, *Earth Surface Proc.,* 3, 43-57, 1976.

Schlesinger, W.H., Carbon balance in terrestrial detritus, *Ann. Rev. Ecol. Syst.,* 8, 51-81, 1977.

Singh, J. S. and S. R. Gupta, Plant decomposition of soil respiration in terrestrial ecosystems, *Bot. Rev.,* 43, 449-528, 1977.

Smith, B.N., General characteristics of terrestrial plants (agronomic and forests) - C_3, C_4, and Crassulacean Acid Metabolism plants. In A. Mitsui and C.C. Black (Eds), *CRC Handbook of Biosolar Resources,* V. 1, Part 2, CRC Press, Boca Raton, Florida, 99-118, 1982.

Smith, B. N. and W. V. Brown, The Kranz syndrome in the Gramineae as indicated by carbon isotopic ratios. *Amer. Jour. Bot.,* 60, 505-513, 1973.

Smith, G.I., I. Friedman, H. Klieforth, and K. Hardcastle, Areal distribution of deuterium in eastern California precipitation, *Jour. App. Meteorol.,* 18, 172-188, 1979.

Soil Survey Staff, *Soil taxonomy. Handbook,* United States Department of Agriculture No. 436, 1975.

Solomon, D.K., and T. E. Cerling, The annual carbon dioxide cycle in a montane soil: observations, modeling, and implications for weathering: *Water Res. Res.,* 23, 2257-2265, 1987.

Teeri, J. and L. G. Stowe, Climatic patterns and the distribution of C_4 grasses in North America. *Oecologia,* 23, 1-12, 1976.

Tieszen, L.L., M. M. Senyimba, S. K. Imbamba, and J. H. Troughton, The distribution of C_3 and C_4 grasses along an altitudinal and moisture gradient in Kenya, *Oecologia,* 37, 337-350, 1979.

van der Merwe, N.J. and E. Medina, Photosynthesis and $^{13}C/^{12}C$ ratios in Amazonian rain forests, *Geochim. Cosmochim. Acta,* 53, 1091-1094, 1989.

Vogel, J.C., A. Fuls, and R.P. Ellis, The geographical distribution of Kranz grasses in South Africa, *South Afr. Jour. Sci.,* 74, 209-215, 1978.

Wang, Y., T.E. Cerling, G.A. Smith, J. Quade, E.H. Lindsay, and J.R. Bowman, Stable isotopes of paleosol carbonates and fossil teeth as Paleoecology and paleoclimate indicators: an example from the Upper Cenozoic non-marine sediments in the San Pedro Valley, Arizona. (P. Swart, J. A. McKenzie, and K. C. Lohmann, Eds.) *Continental Isotopic Indicators of Climate,* (this volume).

Winter, K, J. H. Troughton, and K. A. Card, $\delta^{13}C$ values of grass species collected in the northern Sahara, *Oecologia,* 25, 115-123, 1976.

Wood, W.W., and M. J. Petraikis, Origin and distribution of carbon dioxide in the unsaturated zone of the southern high plains of Texas, *Water Res. Res.,* 20, 1193-1208, 1984.

Yurtsever, Y., and J. R. Gat., Atmospheric waters. In J.R. Gat and R. Gonfiantini (Eds.), Stable Isotope Hydrology: *Deuterium and Oxygen-18 in the Water Cycle,* International Atomic Energy Agency Technical Report Series 210, International Atomic Energy Agency, Vienna, Austria. p. 103-142, 1981.

Thure E. Cerling, Department of Geology and Geophysics, University of Utah, Salt Lake City, Utah, 84112

Jay Quade, Department of Geosciences, University of Arizona, Tucson, Arizona, 85721

Stable Carbon Isotope Composition of Paleosols:
An Application to Holocene

Eugene F. Kelly and Caroline Yonker

*Department of Agronomy, Colorado State University,
Fort Collins, CO 80523*

Bruno Marino

*Department of Earth and Planetary Sciences,
Harvard University, Cambridge, MA 02138*

The carbon isotopic composition of soil organic matter, carbonate and opal phytoliths (biogenic silica) varies as a function of the relative proportion of C_3 and C_4 plants grow at that locality. This relationship provides important information regarding vegetative composition and prehistoric climate in terrestrial ecosystems. Isotopic equilibrium between soil organic matter and carbonate is evident in Holocene paleosols. Isotope signatures of various soil constituents indicate the dominance of C_4 vegetation in the latter part of the Holocene. These isotopic techniques, usually applied to more ancient environments, appear useful in evaluating Holocene paleoclimatic conditions.

INTRODUCTION

Soils are an integral part of the terrestrial ecosystem that evolve through time in response to the combined effects of climate, biota, geology and topography (Jenny, 1941). During this evolutionary process the composition and amount of terrestrial vegetation and animal life vary in response to change in temperature and precipitation and leave an isotopic record in the soil. The reliability and accuracy of paleoclimatic data and inferences made from soils has been questioned for a number of reasons. First, many climatically sensitive soil properties (organic carbon and nitrogen content, mineralogy) change once the soil is buried or is subjected to a different climatic cycle. Secondly, the study of past climates in terrestrial ecosystems is hindered by a lack of precise chronological control. Many studies of paleosols span differing timescales, and range in age from Pleistocene (Hay, 1976; Cerling and Hay, 1986) to Oligocene (Retallack, 1983). In many cases age quantification is difficult since datable materials are likely to be altered as a result of weathering

in the soil environment. Finally, spatial and temporal variations in modern soils indicate that extreme caution should be observed when interpreting these data for paleoclimatic reconstruction (Kelly et al, 1991b).

In spite of these potential pitfalls, a pivotal paper by Cerling (1984) demonstrated the potential of stable carbon isotope analysis of carbonate for paleoclimatic reconstruction. A coherent field of research has evolved, and quantitative methods for extracting paleoenvironmental information from soils are currently being developed and tested. Researchers have demonstrated the environmental significance of stable carbon isotope ratios of soil organic matter, and pedogenic and biogenic minerals (Amundson et al, 1989; Cerling et al, 1989; Cerling, 1990; Kelly et al, 1991a; Quade et al, 1989; Cerling, 1984). The basis for application of stable carbon isotope analyses to paleosols is related to the distribution of C_3 (Calvin-Benson Cycle during photosynthesis) and C_4 (Hatch-Slack photosynthetic pathway) plants. The relative proportions of C_3 and C_4 plants growing at a site reflect climatic conditions, with higher minimum summer temperature and lower soil moisture favoring higher proportions of C_4 plants (Terri and Stowe, 1976; Ehleringer, 1978; Tieszen et al., 1979). The $^{13}C/^{12}C$ ratios of C_3 and C_4 plants differ by a substantial amount, with the former being more ^{13}C

Climate Change in Continental Isotopic Records
Geophysical Monograph 78

depleted than the latter (Bender, 1968; Smith and Epstein, 1971). Studies of modern soil systems indicate that the $^{13}C/^{12}C$ of soil organic matter and pedogenic and biogenic minerals is closely related to the relative proportions of C_3 and C_4 plants at a specific site (Balesdent et al., 1987; Cerling et al., 1989; Kelly et al., 1991a; Kelly et al., 1991b).

Relative to the period of climatic instability that brought an end to the ice age, the Holocene is generally considered to be a period of climatic stability. When studied across modern climatic gradients, isotopic data from Holocene soils are generally used to interpret data from older paleosols (Cerling, 1990; Kelly, 1989). We will demonstrate the application of radiometric ^{14}C dating and stable carbon isotope ratios ($^{13}C/^{12}C$) to Holocene paleosols in the semiarid Great Plains of North America. To assure accurate paleoclimatic interpretation, we will demonstrate the need for concurrent analysis of several biotic and abiotic components in these geologically recent environments.

SITE LOCATION AND DESCRIPTION

Sampling sites are located at the Central Plains Experimental Range (CPER), a USDA-ARS research facility in the northeastern Colorado Piedmont section of the Great Plains Province (Figure 1). The Piedmont is a zone located between the Rocky Mountains to the west and the High Plains to the east (Thornbury, 1965). Mean

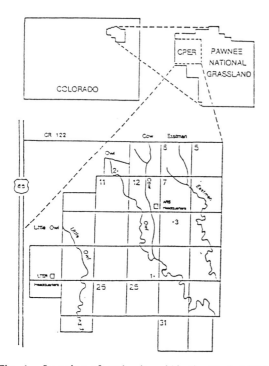

Fig. 1. Location of study site within the Central Plains Experimental Range, Colorado.

annual precipitation at the CPER is approximately 31 cm, 70% of which occurs between May and September. Mean annual temperature is 8°C. The average elevation is 1650 m. The natural vegetation of the area is characteristic of much of the shortgrass steppe. Dominant C_4 species include blue grama (Bouteloua gracilis) and buffalograss (Buchloe dactyloides), while the dominant C_3 species are fringed sagewort (Artemesia frigida) and western wheatgrass (Agropyron smithii).

The CPER is underlain by the Laramie Formation. Deposited in the late Cretaceous, the Laramie is approximately 975 m of interbedded sandstones and shales. The post-Pleistocene history of the area includes several soil-forming intervals, as evidenced by the abundance and variable age of paleosols (Yonker et al., 1988). A minimum of five types and ages of alluvium and two types and ages of eolian material have been suggested to occur in this area (Davidson, 1988). Superimposed on alluvial terraces are numerous deflation basins and vegetated dunes.

METHODS

Soil Sampling and Characterization

Detailed mapping and sampling of buried soils was conducted at the CPER as part of the Long Term Ecological Research (LTER) program. Soil pits were excavated at three sites to a depth of 3 meters; soils were then described morphologically and sampled by genetic horizon (Soil Survey Staff, 1991). Soil carbonate samples for isotopic study were collected within each genetic horizon. Nodules and powdery concretions were physically separated from disseminated carbonate in the field and stored in glass vials. Rock and gravel was returned to the laboratory where laminar coatings were physically separated, and stored in glass vials. Phytoliths were extracted from the sand and silt size fractions (2 to 200 microns) of surface and buried horizons using standard particle size sedimentation techniques (Jackson, 1969) and a zinc bromide heavy liquid separation procedure (Fredlund et al., 1985; Kelly, 1989). Upon recovery, all phytolith samples were washed with hot 6N HCl, rinsed with water, and freeze-dried. Soil organic matter samples used to determine the stable carbon isotope composition were passed through a 2 mm sieve, subjected to dispersion techniques and separated by density into heavy (associated with clay) and light (not associated with clay) fractions by sedimentation. Samples were then subjected to a series of acid-alkali washes to remove soluble organic acids (Ertel and Hedges, 1985).

Carbon-14 Dating

Conventional dating procedures and extended counting procedures were conducted by Beta Analytic Inc., Miami, Florida. Bulk soil organic carbon samples and charcoal fragments for radiometric dating were pre-treated as follows: Soil samples were sieved to remove fine rootlets, dispersed in hot acid to eliminate carbonate, repeatedly rinsed to neutrality, brought to dryness and subjected to multiple combustions under vacuum. Soil samples were then subjected to benzene syntheses and subjected to extended counting procedures. Charcoal samples were subjected to microscopy to remove fine roots and also treated to an acid-alkali-acid soaking series to remove carbonate and humic acids. The CO_2 collected was purified and reacted with hydrogen on a cobalt catalyst to produce graphite. Samples were then subjected to AMS (accelerator mass spectrometry) at ETH (Eidgenossische Technische Hochschule) University in Zurich, Switzerland.

Stable Isotope Analysis

Soil carbonate samples were pre-treated by baking samples at 425°C for 4 hours under vacuum. The gas samples for carbonate isotopic analysis of were obtained by reacting each sample with 100% H_3PO_4 for 24 hours at 25°C. Liberated CO_2 was purified cryogenically and carbonate content determined by measuring the quantity of CO_2 manometrically.

Prior to isotopic analysis soil phytolith samples were boiled in 18% H_2O_2 for a period of six hours, rinsed with water, freeze-dried and pulverized (Kelly, 1989). Soil organic carbon samples were acidified with 0.1N HCl to assure carbonate removal, rinsed with H_2O, and were oven-dried and then freeze-dried. The CO_2 samples for isotopic analysis of phytoliths and soil organic matter were obtained by combusting the samples at 875°C, purifying the resulting CO_2 cryogenically, and determining carbon quantities manometrically (Northfelt et al., 1981). The $^{13}C/^{12}C$ ratio of all CO_2 samples was determined by mass spectrometry utilizing a VG Prism (Series II) mass spectrometer and reported relative to Pee Dee Belemnite (PDB) carbonate. The precision of determination of carbon concentrations and $\delta^{13}C$ values are $\pm$ 0.01 wt.% and $\pm$ 0.10 o/oo, respectively.

ISOTOPIC COMPOSITION OF SOIL CARBON

Relationships among climatic parameters, vegetation and the isotopic composition of soil components in modern ecosystems provide the basis for interpreting paleosol data. Kelly et al. (1991a, 1991b) demonstrated a strong relationship between the $\delta^{13}C$ values of soil organic carbon, carbonate, plant phytoliths and the isotope signature of plant material growing at a specific site in the central Great Plains. Sites that have lower mean annual temperatures and more favorable soil moisture conditions favor C_3 vegetation. These climatic conditions result in soil components with $\delta^{13}C$ values more negative than sites dominated by C_4 vegetation.

The stable C isotope composition of paleosols at the CPER were compared to models established in modern soil environments (Figure 2). CPER data fall within the range of $\delta^{13}C$ values established for each of the soil components. Differences in $\delta^{13}C$ values between paleosols (P_1, P_2) and modern surfaces (S_1, S_2, S_3) are evident and vary depending

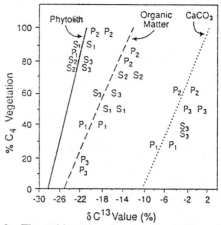

Fig. 2. The stable carbon isotope composition of modern soils (S) and paleosols (P) at CPER versus the percentage of C_4 vegetation. Data compared to models of isotope composition of modern soils and vegetation across the climatic gradients of the Great Plains (Models for phytoliths, organic matter and carbonates adopted from Kelly, 1989)

on the component analyzed. The environmental and paleoclimatic significance of these differences is discussed in the following sections.

Soil Organic Matter

The $\delta^{13}C$ values of soil organic matter as a function of soil depth for sites 1, 2 and 3 are shown in Figure 3. Assuming that microbial degradation could account for only a 1 to 2 o/oo enrichment in these values (O'Brien and Stout, 1978; Nadelhoffer and Fry, 1988), they reflect the long term inputs of C_3 and C_4 plants (Kelly et al, 1991a).

The $\delta^{13}C$ values of surface horizons (upper 10 cm) reflect input of organic materials from contemporary vegetation. Utilizing a simple mixing model and average isotope composition for C_3 ($\delta^{13}C$ value = -26.0 o/oo) and C_4 ($\delta^{13}C$ value -12.0 o/oo) plants, the $\delta^{13}C$ values for organic matter at the surface of sites 1 and 3 indicate a long term plant composition of approximately 61% and 63% C_4 vegetation, respectively. Based on the same calculation the plant composition at site 2 is 79% C_4 vegetation. Since organic matter in the upper 10 cm of the soil reflects relatively long-term accumulation, complete agreement between the $\delta^{13}C$ values and annual estimates of C_3 and C_4 plant abundance and biomass measurements is not necessarily expected. However, results from modern soils indicate the $\delta^{13}C$ values of soil organic matter are relatively good indicators of average vegetative composition (Kelly et al, 1991a). The apparent differences in the $\delta^{13}C$ values among surface horizons reflect the spatial and temporal variability of the modern vegetation in the shortgrass steppe. These differences can be attributed to factors such as topographic variation, soil texture and anthropogenic influences. For

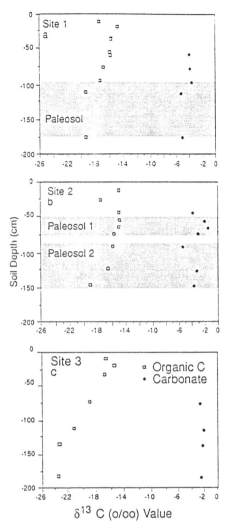

Fig. 3. The stable carbon isotope composition of soil organic matter and its relation to soil carbonate at CPER.

first buried horizon (100 cm) was approximately 3330 yr b.p. The $\delta^{13}C$ values of organic matter are -17.2 and -17.5 o/oo in the first buried horizon and surface horizon, respectively, indicating vegetative composition 3000 years ago was similar to present conditions. The lowest $\delta^{13}C$ value, -19.5 o/oo, for this profile is found at the base of the paleosol. This latter value indicates a lower proportion of C_4 vegetation than present at an undetermined time, presumably earlier than 3330 yr b.p.

Two paleosols are present in the site 2 profile. The radiocarbon date for charcoal extracted from the upper buried surface (50 cm) was approximately 4950 yr b.p. The more positive $\delta^{13}C$ values, approximately -14.6 o/oo, of soil organic matter in the upper paleosol reflect the largest proportion of C_4 vegetation in this study. The $\delta^{13}C$ value of the lower paleosol (75 cm) is slightly more negative, approximately -16.0 o/oo. The lowest $\delta^{13}C$ value for this site, approximately -19.0 o/oo, is at the base of this lower paleosol. The latter values for the lower, undated, paleosol are in agreement with those of sites 3 and the lower portion of site 1, indicating an increase in C_4 vegetation over the last several thousand years.

Soil Carbonate

The relationship between the $\delta^{13}C$ values of soil organic matter and soil carbonate for each site are shown in Figure 3. We calculated a weighted mean $\delta^{13}C$ value for each of the sites which reflects the average amount of organic matter present in all horizons. Based on previous reports and theoretical considerations, pedogenic soil carbonate in equilibrium with coexistent organic matter should be 14.0 to 16.0 o/oo higher (Cerling et al, 1989).

At site 1, carbonate in the upper part of the paleosol indicate the presence C_4 dominated flora. The 2.0 o/oo more negative $\delta^{13}C$ values in the lower portions of the paleosol is consistent with the $\delta^{13}C$ values of soil organic matter.

Carbonate in the upper paleosol of site 2 is in isotopic disequilibrium (i.e. not 14.0 to 16.0 higher than organic matter). This carbonate indicates that the vegetation had a lower C_4 component than predicted by the organic matter. Some of the noted discrepancy could be the result of climatic overprinting. For example, periods of C_4 rich flora could have masked the organic matter responsible for the isotope signature of the carbonate. This is likely since the site 2 buried layer is only 50 cm from the soil/atmosphere interface. Organic materials from plants at the surface may have become incorporated into these deposits. Also, this carbonate may be the result of atmospheric influx (Gile et al, 1967). Carbonate deeper in the profile would be less likely to undergo such alteration. The carbonate data for the lower paleosol correspond closely with the organic matter data.

The carbonate at site 3 may be considered in isotopic equilibrium (14.0 to 16.0 higher than organic matter) with the soil organic matter despite divergence of organic matter

example, in the prairies of eastern Colorado higher proportions of C_3 species can be found in landscape positions with higher moisture conditions such as depressions. This amount of spatial variability in the distribution of C_3 and C_4 plants is similar to micro-topographic variations noted by several investigators (Tieszen et al, 1979; Ehleringer, 1978).

The $\delta^{13}C$ values of soil organic matter from site 3 decrease with depth. This site does not contain a paleosol and may therefore be considered a continuous or uninterrupted record of the Holocene vegetative history at the site. If this is the case, the decrease in $\delta^{13}C$ value with depth can be attributed, in part, to a gradual increase in C_4 plants during the Holocene.

There is an irregular decrease in organic matter $\delta^{13}C$ values with depth at site 1 due to the presence of a paleosol. The radiocarbon date for organic matter from the values in the lower portions of the soil. The $\delta^{13}C$ values do not suggest a gradual increase in C_4 flora during the Holocene as did the organic matter.

Opal Phytoliths

The phytolith content of buried horizons from sites 1 and 2 averaged 4 mg and 7 mg per gram of soil, respectively. These quantities were similar to those found in modern surfaces at the CPER and across the Great Plains (Kelly et al, 1991a). It is important to note that these quantities are considerably higher than adjacent horizons within the paleosols. This supports the contention that these buried horizons were actually vegetated surfaces during the mid-Holocene.

The $\delta^{13}C$ values of phytoliths from each of the surfaces are very similar (Figure 2). This indicates that the contemporary vegetative composition has been similar. A significant departure in isotopic signal could indicate different plant composition (Kelly et al, 1991a). At site 1 the $\delta^{13}C$ values of occluded carbon within phytoliths from the surface and buried soil horizons are similar, -21.9 versus -22.4 o/oo, indicating similar vegetative conditions. However, the $\delta^{13}C$ values of phytoliths recovered from the site 2 modern surface and upper paleosol differ significantly, -21.4 versus -18.2 o/oo. The upper paleosol data are in aggreement with the organic matter data, which indicate a greater amount of C_4 vegetation during the mid-Holocene than at present.

REGIONAL PALEOCLIMATIC SIGNIFICANCE

C-14 dates for CPER paleosols fit a model of Holocene soil-dune formation proposed by Muhs (1985) which includes a period of eolian activity, regionally known as the Altithermal, at 8,000-5,000 yr b.p., a period of soil formation spanning 5,000-3,000 yr b.p., dune formation during 3,000-1,500 yr b.p., and soil formation from 1,500 yr b.p. to the present. Our earlier work indicates the early Holocene, 12,000-8,000 yr b.p., may be considered the first soil forming interval post-Pleistocene (Kelly et al., 1991c, Yonker et al., 1988). This Holocene model provides a chronological framework for the interpretation of stable carbon isotope analyses as applied to paleoclimatic reconstruction. Based on this chronology, the site 1 and 2 paleosols represent the mid-Holocene soil forming interval we have proposed.

Kelly et al (1991) suggests an increase in the quantity of C_4 vegetation through the Holocene, which may be attributed to increasing temperature and decreasing available soil moisture. Based on our organic matter data alone, it would be difficult to establish any paleoclimatic trend with confidence. Data from site 3 indicate a shift to more C_4 vegetation over an undetermined time interval. Dated buried horizons of site 1 indicate a similar C_3/C_4 composition to that of the surface, however, deeper buried horizons also indicate a shift to more C_4 vegetation with time. In contrast, data from the site 2 upper paleosol

indicate higher proportions of C_4 vegetation during the mid-Holocene than at present. The C isotope data derived from both phytoliths and organic matter are in aggreement at site 2.

What could account for the apparent descrepency within the same soil forming interval? Our concurrent anaylses of soil components of the paleosol from site 2 (4950 yr b.p.) indicates greater proportions of C_4 vegetation than present which would be indicative of warmer climatic conditions. It is possible that the the paleosol from site 2 reflects conditions in the earliest portion of the mid-Holocene soil forming period. Utilizing the same rational the paleosol from site 1 would represent conditions very similar to the present and the increase in $\delta^{13}C$ values of organic matter at site 1 below the dated paleosol reflect the climatic conditions of the early mid-Holocene period. If this is the case, these paleosols are providing a higher resolution of climatic reconstruction than expected.

When available data for the region are viewed collectively (Figure 4), there is an apparent trend toward C_4 dominance since the early Holocene. Based on the relationship between C_3/C_4 adaptability and climate, we conclude that the early Holocene soil forming interval was cooler than either the mid-Holocene or current soil forming intervals. Temperature does not appear to increase from the mid-Holocene to the present. Our data indicate considerable climatic variability within soil forming intervals.

We are currently developing a collection of C-14 dates and isotope data for paleosols in the region, that we might postulate with greater confidence the trends in climate through the entire Holocene, including those periods immediately preceding and following periods of eolian

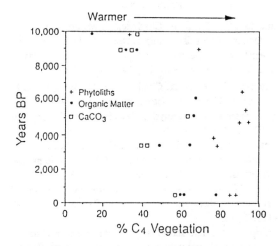

Fig. 4. The percentage of C_4 vegetation versus time as calculated from stable C isotope ratios and C-14 dates. Data represent the stable carbon isotope composition of soils across the Central Great Plains region. Data from this current study and adapted from Kelly, 1989; Kelly et al., 1991a.

activity. Chronologically these results appear to correspond to reported vegetation changes during the Holocene in other portions of the Great Plains although, to our knowledge, this is the first attempt to report mid-Holocene climatic conditions.

The following conclusions can be drawn regarding the application of stable carbon isotope characterization of paleosols to Holocene environments: 1. Carbonates may not be in isotopic equilibrium due to recent additions and masking by modern or more recent carbon inputs. Carbonates lower in the soil profile appear to alleviate this problem. 2. Definitive statements regarding Holocene climatic shifts cannot be made from the stable isotope composition of a single soil component. Concurrent analysis of several biotic and abiotic components is necessary. 3. Intepretation and regional extrapolation of climatic information derived from paleosols is dependent on knowledge of soil and landscape evolution within the region.

Acknowledgments. We thank Klaus Flach, Alan Price, Mike Petersen, Carolyn Olson, and Carol Wettstein for field assistance and Cheryl Plotner and Steve Blecker for laboratory assistance. Interactions and conversations with Thure Cerling at the Chapman Conference proved to be most helpful. The comments of K.C. Lohmann and Bruce Wilkinson were appreciated. This research was supported in part by Colorado State University Agricultural Experiment Station Funds and Faculty Research Grants, NSF Long Term Ecological Research Grant (BSR-9011659), USDA-SCS and USDA-ARS.

REFERENCES

Amundson, R. G., O. A. Chadwick, J. M. Sowers, and H. E. Doner, The relationship between modern climate and vegetation and the stable isotope chemistry of Mojave Desert soils. Quat. Res. 29, 245-254, 1988.

Amundson, R. G., O. A. Chadwick, J. M. Sowers, and H. E. Doner, The stable isotope chemistry of pedogenic carbonates at Kyle Canyon, Nevada. Soil Sci. Soc. Am. J. 53, 201-210, 1989.

Balesdent, J., A. Mariotti, and B. Guillet, Natural ^{13}C abundance as a tracer for studies of soil organic matter dynamics. Soil Biol. Biochem. 19,1, 25-30, 1987.

Bender, M. M., Mass spectrometric studies of carbon 13 variations in corn and other grasses. Radiocarbon 10,468-472, 1968.

Cerling, T. E., The stable isotopic composition of soil carbonate and its relationship to climate. Earth Planet. Sci. Lett. 71,229-240, 1984.

Cerling, T. E., and R. L. Hay, The stable isotopic composition of soil carbonate and its relationship to climate. Earth Planet.Sci. Lett. 71,229-240, 1986.

Cerling, T.E., J. Quade, Y. Wang, and J.R. Bowman, Carbon isotopes in soils and paleosols as ecology and

paleoecology indicators. Nature (London) 341, 138-139, 1989.

Davidson, J.M., Surficial geology and Quaternary history of the Central Plains Experimental Range, Colorado. M.S. Thesis, Colorado State University, Fort Collins, 1988.

Ertel, J.R. and J.I. Hughes, Source of sedimentary humic substances: vascular plant debris: Geochim. Cosmochim. Acta. 49, 2097-2107, 1984.

Ehleringer, J. R., Implications of quantum yield differences on the distribution of C_3 and C_4 grasses. Oecologia. 31,255-267, 1978.

Fredlund, G. G., W. C. Johnson and W. Dort, Jr., A preliminary analysis of opal phytoliths from the Eustis ash pit, Frontier County Nebraska: Institute for Tertiary-Quaternary studies, (TER-QUA, Symposium Series, The Nebraska Academy of Science) 1,147-162, 1985.

Gile, L. H., F. F. Peterson and R. B. Grossman, Morphological and genetic sequence of carbonate accumulation in desert soils. Soil Sci. 101,347-360, 1966.

Hay, R. L., Geology of the Olduvai Gorge. University of California Press, Berkeley, 1976.

Jackson, M. L., Soil Chemical Analysis - Advanced Course. University of Wisconsin Press, Madison, 1969.

Kelly, E.F., A Study of the Influence of Climate and Vegetation on the Stable Isotope Chemistry on Soils in Grassland Ecosystems of the Great Plains. Ph.D. dissertation. University of California, Berkeley, Calif., 1989.

Kelly, E.F., R.G. Amundson, B.D. Marino, and M.J. DeNiro, The Stable Isotope Ratios Carbon in Phytoliths as a Quantitative Method of Monitoring Vegetation and Climatic Change. Quat. Res. 35, 222-233, 1991a.

Kelly, E.F., R.G. Amundson, B.D. Marino, and M.J. DeNiro, Environmental and Geological Influences on the Stable Isotope Composition of Carbonate in Holocene Grassland Soils. Soil Sci. Soc Am. J. 55,1651-1658, 1991b.

Kelly, E.F., C.M. Yonker, M.L. Petersen, and B.D. Marino, Paleoindicators of Holocene Landscape Evolution, Shortgrass Steppe, Colorado. Am. Soc. Agr. Abst. p. 314, 1991c.

Muhs, D.R., Age and paleoclimatic significance of Holocene sand dunes in northeastern Colorado. Ann Amer. Assoc. Geog. 75,566-582, 1985.

Nadelhoffer, K.J. and B. Fry, Controls on natural nitrogen-15 and carbon-13 abundances in forest soil organic matter. Soil Sci. Soc. Am. J. 52,1633-1640, 1988.

Northfelt, D. W., M. J. DeNiro and S. Epstein, Hydrogen and carbon isotopic ratios of cellulose nitrate and saponifiable lipid fractions prepared from annual growth rings of California redwood. Geochimica et Cosmochimica Acta 42,1895-1898, 1981.

Obrien, B.J. and J.D. Stout, Movement and turnover of soil organic matter as indicated by carbon isotope measurements. Soil Biology and Biochemistry. 10,309-317, 1978.

Quade, J., T. E. Cerling, and J. R. Bowman, Systematic variations in carbon and oxygen isotopic composition of pedogenic carbonate along elevational transects in the southern Great Basin, United States. Geol. Soc. Am. Bull. 101,464-475, 1989.

Retallack, G. J., Late Eocene and Oligocene fossil soils from Badlands National Park, South Dakota, Geological Society of America Special Paper. 193, 82p., 1983.

Smith, B. N. and S. Epstein, Two categories of $^{13}C/^{12}C$ ratios for higher plants. Plant Physiology 47,380-384, 1971.

Soil Survey Staff., Soil Survey Manual. Issue 1, Directive 430-v-ssm, U. S. Govt. Printing Office, Washington, D.C., 1991.

Terri, J. A., and L. G. Stowe, Climatic patterns and the distribution of C_4 grasses in North America. Oecologia 23,1-12, 1976.

Thornbury, W.T., The regional geomorphology of United States. John Wiley and Sons, Inc., New York., 1965.

Tiezen, L. L., M. M. Senyimba, S. K. Imbamba, and J. H. Troughton, The distribution of C_3 and C_4 grass species along an altitudinal and moisture gradient in Kenya. Oecologia 37,337-350, 1979.

Yonker, C. M., D. S. Schimel, E. Paroussis and R. D. Heil, Patterns of organic carbon accumulation in a semi-arid shortgrass steppe, Colorado. Soil Sci. Soc. Am. J. 52,478-483, 1988.

E.F. Kelly and C. Yonker, Department of Agronomy, Colorado State University, Fort Collins, CO 80523.

B. Marino, Department of Earth and Planetary Sciences, Harvard University, Cambridge, MA 02138.

Stable Isotopes of Paleosols and Fossil Teeth as Paleoecology and Paleoclimate Indicators: An Example from the St. David Formation, Arizona

YANG WANG, T. E. CERLING, J. QUADE AND J. R. BOWMAN

Department of Geology and Geophysics, University of Utah, Salt Lake City, UT 84112

G. A. SMITH

Department of Geology, The University of New Mexico, Albuquerque, NM 87131

E. H. LINDSAY

Department of Geosciences, University of Arizona, Tucson AZ 85721

Stable isotopes of paleosol carbonates from the St. David Formation in Arizona indicate a major change in $\delta^{18}O$ and $\delta^{13}C$ from ~3.4 to 2.8 Ma ago, approximately the time of the onset of global cooling postulated from the deep-sea oxygen record. Another major shift in isotopic signature occurred at ~2.4 to 1.8 Ma ago. $\delta^{13}C$ values of paleosol carbonates also indicate that grasses were an important component of the local ecosystems at least since 3.4 Ma. $\delta^{13}C$ values of fossil herbivorous mammal teeth from the St. David Formation also indicate that grasses had been available to support these C_4 eaters (grazers) from 3.4 to 0.6 Ma, which is consistent with the paleosol carbonate results.

INTRODUCTION

The stable isotopic composition of paleosol carbonate and organic matter, if unaltered after burial, can be used to reconstruct certain aspects of the paleoenvironment of soil formation. Carbon isotopes yield paleoecological information because the carbon in the soil carbonate forms in isotopic equilibrium with local soil CO_2, whose isotopic composition is determined by local plant cover (Cerling, 1984, 1991; Cerling et al., 1986, 1988). Plants fall into two major carbon isotopic groups depending on their photosynthetic pathways, either C_3 or C_4. C_3 plants, which include all trees and almost all shrubs and cool-season grasses, have $\delta^{13}C$ values ranging from -20 to -35‰, with an average of -27‰ (Deines, 1980). Plants with C_4 photosynthetic pathway, which include warm season grasses and few shrubs (e.g. Euphorbia and Chenopodiacae), have $\delta^{13}C$ values between -9 to -17‰, with an average of about -13‰ (Deines, 1980).

The carbon isotopic composition of soil organic matter is determined by the proportion of C_3 and C_4 plants in the local biomass. The $\delta^{13}C$ values for soil CO_2 produced by a pure C_4 flora is about -8.6‰, and for a pure C_3 flora is about -22.6‰, due to diffusive isotopic effects (Cerling et al., 1984).

Soil carbonate formed in the presence of pure C_4 and C_3 biomass has $\delta^{13}C$ of about 2‰ and -12‰, respectively, due to isotope fractionation resulting from equilibrium between $CaCO_3$ and CO_2 and gaseous diffusion in the soil system (Cerling, 1984, 1991; Cerling et al., 1986, 1988). The enrichment factor in $\delta^{13}C$ of soil carbonate with respect to soil organic matter is about 15‰ at 25°C. Therefore carbon isotopes in pedogenic carbonates can provide estimate of the proportion of C_4/C_3 biomass: $C_3 = (2 - \delta^{13}C_{nodule})/14$.

Oxygen isotopes in soil carbonate are well correlated with the $\delta^{18}O$ of local rainfall (Cerling, 1984) and therefore can provide some information about the $\delta^{18}O$ of soil water that precipitated these carbonates. But it is not possible to uniquely determine the $\delta^{18}O$ of soil water without independent soil temperature estimates. The $\delta^{18}O$ of paleosol carbonate is principally controlled by temperature and the $\delta^{18}O$ of soil water which is related to the $\delta^{18}O$ of meteoric water, seasonal differential infiltration of water, and subsequent isotopic enrichment accompanying evaporation. If $\delta^{18}O$ values are to be utilized quantitatively, the effects of these parameters on the $\delta^{18}O$ of soil carbonate must be determined. Without detailed study of the processes controlling $\delta^{18}O$ of soil carbonate, it is premature to infer the exact reason for the $\delta^{18}O$ shift. We can say, however, that major changes in some aspects of the regional climate such as precipitation, temperature, or

Climate Change in Continental Isotopic Records
Geophysical Monograph 78

evapotraspiration are indicated by shifts in $\delta^{18}O$ of soil carbonate.

The San Pedro Valley, Arizona (Fig. 1), contains Pliocene-Pleistocene age valley fill sediments with an abundant fossil record that preserves more than 4 million years (Ma) of biotic changes (Lindsay, 1984). These sediments consist of more than 180 m of fluvial, lacustrine and swamp deposits (Gray, 1967). Pedogenic carbonates, which have previously been used for paleoecological and paleoclimatic studies (Cerling, 1984, 1991; Cerling et al., 1986, 1988; Quade et al., 1989b; Wang and Zheng, 1989), are common throughout the stratigraphic section.

GENERAL GEOLOGY

The St. David Formation, which is well exposed in badlands along the San Pedro River and the lower piedmonts adjacent to the Dragoon and Whetstone Mountains (Fig. 1), has been divided into lower, middle and upper members (Fig. 2) (Gray, 1967). The St. David Formation is capped by the 'granite wash', which is composed of reddish-orange fine to coarse gravels and coarse sands. These two sequences are separated by a significant erosional unconformity (Gray, 1967).

Most of the lower St. David Formation is characterized by a red clay and mudstone, and fine-grained sand lenses with variable gypsum content. The lower member was deposited in playas and on associated mudflats as shown by the fine-grained nature of these sediments, the generally tabular and thin-bedded nature of strata, locally abundant gypsum near the valley axis, and the abundance of clay drapes and associated mudcracks and

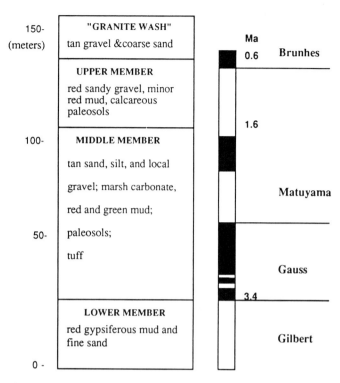

Fig. 2. Schematic stratigraphic diagram showing the St. David Formation in the upper San Pedro Valley (after Lindsay, Smith, and Haynes, 1990)

mudchip grains. During this time the basin was probably closed and aridity favored rapid evaporation of fluids and precipitation of evaporites (Lindsay, Smith and Haynes, 1990).

The middle St. David contains mostly fluvial sediments with local gravels, conglomerate, interlayered pedogenic carbonates, pond (marsh) carbonates, and minor tuff. Where exposed, the top of the middle member is marked by a laterally persistent, extensively bioturbated mudstone containing superimposed paleosols. The middle member was deposited following establishment of through-going drainage in the valley (Lindsay et al., 1990).

In its limited area of exposure on the east side of the valley, the upper member consists of scour-and-fill bedded, pebbly coarse-grained sands containing well developed calcic paleosols. The upper St. David records early to middle Pleistocene bajadas on the eastern piedmont. Middle and upper St David sediments are predominantly clastic and are of fluvial origin; however, the limestones, and green clays indicate a minor paludal component. Most ponds were spring fed along buried faults on the western piedmont. Some sediments, however, were deposited in ox-bow lakes and flood basins associated with the axial drainage (Lindsay, Smith and Haynes, 1990).

Paleosols are abundant in the middle and upper St. David Formation (Johnson et al., 1975) and typically contain red to orange Bt horizons that have been thoroughly bioturbated (i.e. original sedimentary texture has been destroyed and root traces/root casts are visible) and leached of carbonate, and have high concentrations of clay minerals. Carbonates associated

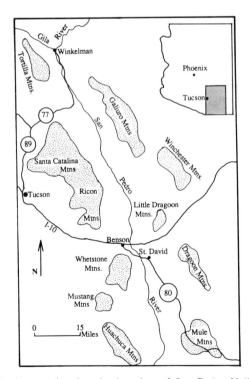

Fig. 1. A map showing the location of San Pedro Valley and St. David Formation near Benson, Arizona (after Lindsay, 1984).

TABLE 1 Stable isotope results of carbonates from St. David Formation

Sample	δ¹³C	δ¹⁸O	δ¹³C$_{O.M.}$	Estimated Age* (Ma)	Notes
Curtis Wash:					
SPV-1	-4.1	-6.2	-19.3	3.29	P.S.C**
SPV-11	-3.7	-7.8		2.81	P.S.C.
SPV-12	-3.8	-6.9		2.8	P.S.C.
SPV-10	-5.0	-7.3		2.77	P.S.C.
SPV-13	-2.3	-7.6		2.77	P.S.C.
SPV-16	-3.8	-8.2		2.64	P.S.C.
SPV-21	-3.8	-8.2		2.6	P.S.C.
SPV-17	-4.1	-8.7		2.59	P.S.C.(?)
SPV-18	-3.4	-7.1		2.56	P.S.C.
SPV-22	-2.8	-7.9		2.28	P.S.C.(?)
SPV-23	-3.3	-7.6		2.28	P.S.C.(?)
SPV-24	-2.7	-8.1		2.18	P.S.C.
SPV-25	-2.7	-7.5		2.08	P.S.C.
SPV-27	-1.0	-6.9	-16.7	1.38	P.S.C.
SPV-28	-0.8	-6.5		1.33	P.S.C.
SPV-30	-1.7	-6.8	-15.9	1.19	P.S.C.
SPV-31	-1.8	-7.1		1.09	P.S.C.
SPV-32	-0.7	-6.0	-16.4	0.94	P.S.C.
SPV-29	-1.1	-6.6	-16.9	0.83	P.S.C.
McRae Wash:					
SPV-33	-5.2	-5.9	-20.9	3.28	P.S.C.
SPV-34	-5.0	-5.3		3.22	P.S.C.
SPV-36	-6.9	-6.9		3.06	P.S.C.
SPV-38	-6.9	-7.3		2.92	P.S.C.(?)
SPV-40	-4.7	-7.0		2.87	P.S.C.
SPV-41	-4.9	-8.8		2.72	P.S.C.
SPV-42	-4.9	-9.0	-20.3	2.67	P.S.C.
SPV-43	-5.4	-9.3		2.65	P.S.C.

Sample	δ¹³C	δ¹⁸O	δ¹³C$_{O.M.}$	Estimated Age* (Ma)	Notes
SPV-44	-6.0	-8.6	-20.3	2.63	P.S.C.
SPV-46	-1.8	-7.9		2.57	P.S.C.
SPV-47	-4.0	-8.4	-20.1	2.52	P.S.C.
SPV-48	-4.7	-8.3		2.43	P.S.C.
SPV-49	-4.1	-8.7	-19.1	2.38	P.S.C.
Dragoon Wash:					
DW-7	-5.5	-6.4		3.37	P.S.C.
DW-12	-5.2	-6.2	-21.0	3.3	P.S.C.
DW-13	-6.3	-7.3		3.2	P.S.C.
DW-14	-6.0	-7.1	-22.2	3.14	P.S.C.
DW-15	-5.8	-7.6		3.03	P.S.C.
DW-17	-6.4	-7.6		2.9	P.S.C.
DW-18	-6.2	-8.8		2.88	P.S.C.
DW-20	-6.0	-7.3		2.82	P.S.C.
CAW-2	-3.2	-8.3	-16.3		Pond Cc.
CAW9001	-3.2	-7.0	-23.4		Pond Cc.
CAW-2-1	-1.9	-8.8	-20.8		Pond Cc.
SPV-2	-4.2	-7.3		3.0	Marl***
SPV-3	-4.7	-7.4	-22.5	3.0	Marl
SPV-4	-1.7	-6.9		2.9	Marl
SPV-6	-2.3	-7.5	-13.0	2.85	Marl
SPV-9	-3.7	-7.6	-20.6	2.85	Marl
SPV-14	-4.3	-8.5	-20.4	2.7	Marl
SPV-15	-4.3	-8.8		2.7	Marl
SPV-19	-3.7	-7.9		2.53	Marl
SPV-20	-4.0	-8.7		2.5	Marl

* Ages were estimated by linear interpolation.
** P.S.C.=paleosol carbonate.
***Marl has no genetic meaning here.
Cc = carbonate.

with paleosols are generally nodular and in zones located below a leached, bioturbated, reddish B horizon > 30cm thick. The depth to carbonate accumulation was probably greater than 50 cm assuming a minimum compaction of 40%. Studies of modern soils show that the downward diffusion of atmospheric CO_2 has only minimal influence below about 20 cm depth in modern soils (Cerling et al., 1989; Quade et al., 1989; Cerling, 1991; Cerling and Quade, 1992). We have also studied the isotopic variations in pond/marsh carbonates, occluded organic matter in carbonates, and fossil herbivorous mammal teeth from the St. David Formation. Pond carbonates were identified in the field by their occurrences as massive, sharp based limestones containing variably abundant ostracodes and less common gastropods, bivalves, and chacophytes.

ANALYTICAL METHODS

The isotope analyses of carbonate were performed on CO_2 prepared from the $CaCO_3$ by treatment with 100% phosphoric acid at 25°C. Prior to analysis the soil carbonate samples were roasted for 1 hour under vacuum at 400-450°C to carbonize any

organic matter present. Samples for the isotope analysis of organic matter occluded in carbonate were treated with 10% HCl to remove carbonate. Carbon isotope ratios of occluded organic matter in pedogenic carbonate were determined by analyzing the carbon dioxide resulting from the combustion of a sample with CuO and silver foil in an evacuated quartz tube at 900°C. Enamel and dentine were manually separated and ground into powder. The powder was allowed to react overnight in a weak solution of sodium hypochlorite (5%) to eliminate bacterial proteins and humates, after which it was filtered with micro-glass filter paper and thoroughly rinsed. This powder was then treated with 1 M acetic acid overnight under weak vacuum to remove diagenetic carbonates, washed and dried. The residue should thus consist mainly of larger crystals containing structural carbonate. CO_2 was produced by reacting hydroxyapatite with 100% phosphoric acid at 25°C, purified, and carbon and oxygen isotopes determined on mass spectrometer (Lee-Thorp and Van der Merwe, 1987; Koch, et al., 1992). The fractionation factor for oxygen in this reaction was assumed to be the same as that for caclite at 25°C. The carbon and oxygen isotopic ratios are reported in the standard notation relative to PDB. The analytical precision is better than 0.2 ‰.

RESULTS AND DISCUSSION

1. Isotopes of pedogenic carbonate:

The isotope data for carbonates and occluded organic matter from the St. David Formation are listed in Table 1. For 12 paleosols, the $\delta^{13}C$ of both the pedogenic carbonate ($CaCO_3$) and the occluded organic matter (O.M.) were measured (figure 3). The $\Delta^{13}C$ ($CaCO_3$-O.M.) values for these two phases varied between 14.2 to 16.1‰ and averaged 15.3±0.7‰, comparable to the range observed for modern soils (Cerling et al., 1989). These results suggest that samples have not suffered significant diagenetic alteration, because it would require identical alteration of two different phases, carbonate and organic matter, to maintain the observed $\Delta^{13}C$ ($CaCO_3$-O.M.) values, which is unlikely. Therefore, the $\delta^{13}C$ values of pedogenic carbonate from the St David Formation can be used to determine the fraction of C_4/C_3 biomass.

Figure 4 shows the carbon and oxygen isotopic compositions of paleosol carbonates from the St. David Formation and the 3-point-running average of the data. The age

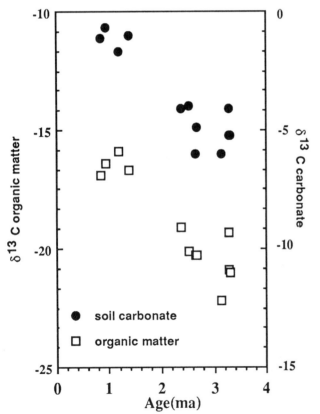

Fig. 3. The stable carbon isotope values of occluded organic matter and coexisting carbonate through time. The $\delta^{13}C$ of organic matter displays the similar shifts with time as observed in soil carbonate. The average difference between soil carbonate and occluded organic matter is between 14 to 16‰, which is the same as that observed in modern high respiration soils. This is a strong evidence against diagenetic alteration of either phase.

of each sample was determined by interpolation from polarity-reversal boundaries determined by magnetostratigraphy (Johnson et al., 1975; Geissman et al., in prep.) and the geologic time scale of Harland et al. (1990). It seems that the record from the St. David Formation contains five distinct periods:

(1) ~3.4 to 3.0-2.8 ma: $\delta^{13}C$ shifts from -3.8‰ to -6‰ (figure 4), which implies a decrease in C_4 biomass from ~60% to ~40% or a relative increase in C_3 biomass (trees, shrubs, or cool season grasses). $\delta^{18}O$ also decreases dramatically from -6 ‰ to -8.5 ‰ (figure 4). The observed $\delta^{18}O$ shift may indicate a decrease in the $\delta^{18}O$ of meteoric water (Cerling, 1984), which in turn might imply: (1) increased winter relative to summer precipitation, or (2) a decrease in local temperature, or (3) a change in global mean temperature (Dansgaard, 1964; Yurtsever, 1975), or a combination of the above. This carbon and oxygen shift occurred at about the same time as the onset of ocean cooling postulated from the deep sea oxygen record (Shackleton, et al , 1984).

(2) 3.0-2.8 to ~2.6 Ma: $\delta^{13}C$ of paleosol carbonate displays an increasing trend, which could result from an increase in the proportion of C_4 plants in the local biomass. Since rainfall in the warm season (late spring or summer) favors C_4 plants (VanDevender, 1990), the increase in C_4 biomass in local ecosystem may indicate more summer rain and/or higher temperature, which is also supported by the observed increase in $\delta^{18}O$.

(3) ~2.6 to 2.4 Ma: Both $\delta^{13}C$ and $\delta^{18}O$ shift to more negative values, suggesting increased winter rains relative to summer rains.

(4) 2.4 to ~1.8 Ma: $\delta^{13}C$ increases from -4‰ to -1 ‰, implying a dramatic increase in C_4 biomass from 55% to 85%. $\delta^{18}O$ values shift from about -8‰ to -6.5‰, which again might indicate an increase in $\delta^{18}O$ of meteoric water (Cerling, 1984). Both $\delta^{13}C$ and $\delta^{18}O$ indicate a shift toward higher summer rainfall. Cerling and Hay (1986) and Cerling et al. (1988) noted a significant change in climate and ecology in East Africa at about 1.7 to 1.8 Ma. The $\delta^{13}C$ and $\delta^{18}O$ minimum at ~2.4 Ma coincides with the first glacial maximum deduced from deep sea oxygen record (Shackleton, et al , 1984).

(5) ~1.8 to ~0.8 Ma: Almost all $\delta^{13}C$ values fall between -1.1 to -0.5 ‰, suggesting a C_4-dominant biomass (80% - 85% C_4). This indicates the presence of grasslands during this time interval because grasslands are the only known C_4 biomasses of large extent. The occurrence of C_4 grassland can be attributed to (1) higher summer rains/more monsoonal climate, which may be related to a change in the atmospheric flow pattern due to the uplift of the American west during Late Cenozoic period (Ruddiman, 1989); or (2) higher temperature. No pond or marsh carbonates have been found in this stratigraphic interval. Paleosol carbonates on average are more enriched in ^{18}O than previously observed, indicating increased summer rainfall relative to winter rainfall and/or higher temperature, which is consistent with the observations obtained from $\delta^{13}C$ values.

The carbon isotope data of pedogenic carbonates from the St.David Formation indicate that there was a mixture of C_3 and

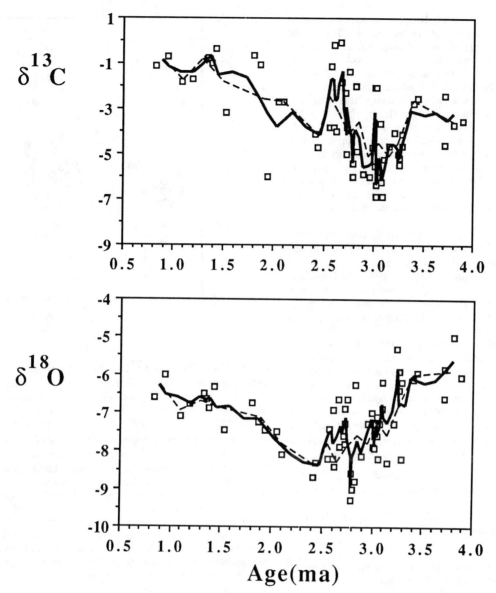

Fig.4. Carbon and oxygen isotopic compositions of pedogenic carbonate from the St. David Formation. The solid curves represent the three-point-running average of the data. The dashed lines are 0.1 Ma averages. It shows that there were major shifts in the $\delta^{18}O$ and $\delta^{13}C$ of paleosol carbonate from ~3.4 to ~3.0-2.8 Ma, which corresponds to the onset of global cooling indicated by deep sea oxygen isotope data. Another episode of major change in the isotopic compositions of soil carbonate occurred at ~2.4 to ~1.8 ma ago.

C_4 biomasses in local ecosystem before ~1.6 Ma ago with a gradual increase in C_4 biomass since ~2.9 Ma. The more positive carbon isotope values after ~1.8 Ma mark the expansion and dominance of C_4 plants in the local biomass. Because the most common C_4 plants are grasses, we may conclude that grasses have existed since the early deposition of the St. David Formation, increased in importance after ~2.9 Ma ago, and became dominant in the local ecosystem after ~1.8 Ma ago until the end of the St. David Formation deposition. The reason for the late Pliocene and early Pleistocene ecological

shift in San Pedro Valley may be related to a climate change that allowed the expansion of C_4 grasses, such as more summer rain (more monsoonal climate).

2. Stable isotopes of fossil mammal teeth:

Studies on the San Pedro Valley by Lindsay (1984) and Lindsay et al. (1990) show the occurrence of both 3-toed horse (*Nannippus*) and 1-toed horse (*Equus*) fossils in Blancan faunas, which span most of the Pliocene (about 4.5 to 2.0 Ma). Only 1-toed horse (*Equus*) fossils have been found, however, in Irvingtonian faunas, which span most of the Pleistocene (about

1.8 to 0.5 Ma). Shotwell (1961) believes these horses occupied two different habitats dependant on vegetation. Three-toed horses had woodland and/or forest habitats, whereas one-toed horses were adapted to open grassland. If Shotwell is correct, the disappearance of 3-toed horses and dominance of 1-toed horses in the Upper St. David Formation might suggest the expansion of open grassland, which is compatible with our isotope results. The dominance of grasslands after ~1.8 Ma indicated by our isotope results also coincides with the 1.6 Ma dispersal event (Lindsay, 1984) which marks the appearance of the elephant *Mammuthus* in North America. Intrigued by these observations, we analyzed carbon and oxygen isotopic compositions of fossil herbivorous mammal teeth (horses and proboscidean) from the St. David Formation to assess the paleodiet of these fossil mammals and the paleovegetation available to support them.

It is well known that the distinct isotopic signatures of different photosynthetic-type plants (C_3 vs. C_4) are passed along the food chain to animal tissues (teeth and bone etc.) with further isotopic fractionation (Parker, 1964; DeNiro and Epstein, 1978a, b; Quade et al., 1992). Living calcified tissues such as enamel, dentine and bone are all inorganic/organic composites. The inorganic (mineral) component accounts for about 69% to 97% by weight in these tissues. Most of this is in the form of hydroxyapatite $Ca_{10}(PO_4)_6(OH)_2$, which contains a small amount of "structural" carbonate, probably as carbonate-ion substitutions for hydroxyl or phosphate ions. Studies of carbon isotopes of modern and fossil teeth and bone indicate that the carbon incorporated into hydroxyapatite is consistently enriched by 13 to 14‰ relative to their diet (Lee-Thorp and Van der Merwe, 1987). Hydroxyapatite derived from pure C_3 diets has $\delta^{13}C$ of about -13.5‰, whereas that derived from pure C_4 diets has a $\delta^{13}C$ value around +1‰. Therefore, $\delta^{13}C$ of fossil bone and teeth, if preserved, can be used to determine the paleodiets of fossil mammals and the vegetation available to support them.

The isotope ratios of tooth enamel samples fall within the narrow range of modern grazers (Table 2) and are distinct

TABLE 2 Carbon and oxygen isotopic compositions of teeth from St. David Formation

Sample	$\delta^{13}C$	$\delta^{18}O$	Estimated age (ma)	Species, locality
Enamel:				
AZ-1-E	-1.6	-5.5	3.1	Cuvieronius, Post Ranch
AZ-3-E	-0.1	-4.5	3.0 - 2.9	Equus, Mendevil Ranch
AZ-8-E	0.8	-4.7	2.6	Nannippus, McRae Wash
AZ-7-E	0.0	-3.8	2.5 -2.6	Nannippus, Horsey Green
AZ-4-E	2.1	-4.5	2.4 -2.0	Cuvieronius, Lower Wash
AZ-2-E	1.2	-6.4	2.0- 1.86	Stegomastodan, Curtis Flats
AZ-5-E	0.7	-5.1	1.86	Equus, Curtis Ranch
AZ-6-E	1.1	-6.2	1.2	Equus, Noye's Prospect
AZ-Q-E	-0.9	-2.0	0.015-0.020	Equus, Coro Marl
Dentine:				
AZ-1-D	-4.4	-6.7	3.1	Cuvieronius, Post Ranch

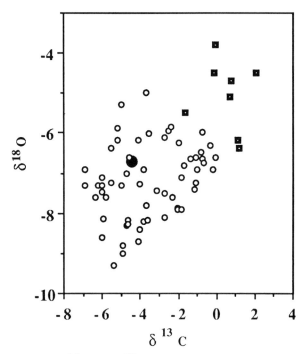

Fig. 5. The $\delta^{18}O$ and $\delta^{13}C$ values of carbonates and fossil teeth from the St. David Formation. Isotope ratios of tooth enamel (solid squares) are very different from those of carbonates (open circles=paleosol carbonates, triangles=pond carbonates), indicating little or no isotopic exchange with carbonates in the environment. $\delta^{13}C$ of tooth enamel ranges from -1.6 to +2.1‰, indicating a pure dietary intake of C_4 plants, which in turn suggests that grasses had been available to support these early proboscidean and horses during all of the St. David Formation time. However, the isotopic compositions of tooth dentine (solid dot) fall in the range of carbonate, suggesting a significant diagenetic modification of dentine, because any early or late diagenesis tends to change the isotopic compositions of tooth toward that of carbonate in the environment. In this case, tooth dentine failed to preserve the original C_4-dominated isotope signal.

from values for contemporaneous soil and pond carbonates (Fig. 5). Diagenesis does not appear to play a role in the ^{13}C content of these tooth enamel samples because the carbonate reservoir in the early and middle St. David Formations has $\delta^{13}C$ of -1.7 to -6.9‰ and exchange of this reservoir with the calcified mammal tissues would have depleted the hydroxyapatite in these mammal teeth in ^{13}C relative to modern ones. The consistency of the isotopic compositions of fossil tooth enamel throughout the St. David Formation and the fact that the prominent shifts in $\delta^{13}C$ of paleosol carbonates are not seen in that of teeth also suggest very limited, if any, diagenesis. The $\delta^{13}C$ of tooth enamel therefore appears to record the paleodiet of Cenozoic mammals. The isotope ratio of dentine ($\delta^{13}C$ = - 4.4, $\delta^{18}O$ =-6.7) is more negative than that of enamel and closer to that of carbonate of similar age (e.g. SPV-36 in Table 1 and Fig. 5), suggesting significant diagenetic modification of the dentine, because diagenesis tends to change the isotopic compositions of teeth toward that of carbonate in the environment. The difference in the preservation of the original isotope signal between enamel and dentine can be explained by the primary difference in the texture and composition between these two tissues (Williams et al., 1979). The dental enamel is almost entirely inorganic and mainly composed of hydroxyapatite (>96%) which appears to be entirely crystalline. The crystallites of enamel are considerably larger than those of dentine and are packed together forming a dense, crystalline mass. On the other hand, dentine contains a significant organic component (collagen: 17 -19%), besides inorganic hydroxyapatite (74 - 76%). The smaller apatite crystallites and the presence of organic component in dentine make it more vulnerable to diagenetic modification as seen in Figure 5.

The $\delta^{13}C$ values of fossil tooth enamel samples from the St. David Formation, which range from -1.6‰ to +2.1‰, are well within the $\delta^{13}C$ range of modern grazers from Africa (Lee-Thorp and Van Merwe, 1987) and suggest a pure dietary intake of C_4 plants of these mammals. Because the most common C_4 plants are grasses, the $\delta^{13}C$ of mammal tooth enamel indicate that grasses were available to support these mammals throughout the time interval of the St. David Formation deposition, which is consistent with the isotope results of paleosol carbonates. Our isotope data also suggest that even though the 3-toed horses occupied woodland habitat according to Shotwell (1961), they still had grasses as their diet -- they were grazers!

CONCLUSIONS

Stable isotope analyses are useful in reconstructing the paleoenvironments of paleosol formation and paleodiets of fossil mammals. The carbon and oxygen isotopic compositions of paleosol carbonate provide information on the proportion of C_4/C_3 biomass in the soil and the isotopic composition of soil or meteoric waters. The St. David Formation isotopic record indicates that there were major shifts in the isotopic compositions of soil water or meteoric water and in the proportion of C_4/C_3 biomass from ~3.4 to ~3.0-2.8 Ma. This corresponds to the onset of global cooling indicated by deep sea oxygen isotope data. Another episode of major change in isotopic compositions of soil and/or meteoric waters and vegetation occurred at ~2.4 to ~1.8 ma ago, close to the boundary between the middle and upper St. David Formation which is near to the Pliocene/Pleistocene boundary. $\delta^{13}C$ values of paleosol carbonate suggest that local biomass consisted of mixed C_4 and C_3 plants which could be mixed C_4/C_3 grassland, or grass woodland/shrubland (C_4 <50%) or wooded grassland (C_4 >50%) before ~1.8 Ma ago. Since then, grasslands with 80% - 85% C_4 grasses dominated the local ecosystem until the end of the St. David Formation deposition. The carbon isotope ratios of fossil herbivorous mammal teeth indicate that these late Cenozoic horses and proboscidean were grazers, and that C_4 grasses had been available to support them throughout the St. David Formation time. The changes in the ecological system in San Pedro Valley are probably related to changes in climate, such as shifts in the seasonal distribution of rainfall and/or change in temperature.

ACKNOWLEDGMENTS. This research was supported by National Science Foundation grant EAR8916355 to G. Smith. D. Katzman assisted in the stratigraphic studies and samples collection. We are indebted to J. W. Geissman and R. Livacarri for conducting the magnetostratigraphic studies that provided the geochronologic control for our data set.

REFERENCES

Cerling, T.E., The stable isotopic composition of modern soil carbonate and its relationship to climate, *Earth and Planetary Science Letters*, 71, 229-240, 1984..

Cerling, T.E., Hay R.L., An isotopic study of paleosol carbonates from Olduvai Gorge, Tanzania, *Quaternary Research*, 25, 63-78, 1986.

Cerling, T.E., Bowman, J.R. and O'Neil, J.R., An isotopic study of a fluvial-lacustrine sequence: the Plio-Pleistocene koobi fora sequence, east Africa, *Palaeogeography, Palaeoclimatology, Palaeoecology*, 63, 335-356, 1988.

Cerling, T. E., Quade, J., Wang, Y. and Bowman, J. R., Carbon isotopes in soils and paleosols as ecology and paleoecology indicators. *Nature*, 341, 138-139, 1989.

Cerling, Carbon dioxide in the atmosphere: evidence from Cenozoic and Mesozoic paleosols, *American Journal of Science*, 291, 377-400, 1991.

Cerling, T. E. and Quade, J., Stable carbon and oxygen isotopes in soil carbonates (this volume), 1992.

Dansgaard, W., Stable isotopes in precipitation, *Tellus*, 16, 436-468, 1964.

Deines P , The isotopic composition of reduced organic carbon. In: *Handbook of Environmental Isotope Geochemistry* (The Terrestrial Environment), 1, edited by Fritz P, Fontes JC, pp.329-406, Elsevier, Amsterdam, 1980.

DeNiro, M. J. and Epstein, S., Influence of diet on the distribution of carbon isotopes in animals, *Geochim Cosmochim. Acta*, 42, 495-506, 1978a.

DeNiro, M. J. and Epstein, S., Carbon isotopic evidence for different feeding patterns in hyrax species occupying the same habitat, *Science*, 201, 906-908, 1978b.

Friedman, I., and O'Neil, J. R., Compilation of Stable Isotope Fractionation Factors of Geochemical Interest. *United States Geological Survey Professional Paper* 440-KK, 1977.

Gray, R. S., Petrography of the upper Cenozoic non-marine sediments in the San Pedro Valley, Arizona, *Journal of Sedimentary Petrology*, 37, 774-789, 1967.

Harland, W. B., Armstrong, R. L., Cox, A. V., Craig, L. E., Smith, A. G., and Smith, D. G., A geologic time scale, Cambridge University Press, Cambridge, 263p, 1990.

Johnson, N. M., Opdyke, N. D. and Lindsay, E. H., Magnetic Polarity Stratigraphy of Pliocene-Pleistocene Terrestrial Deposits and Vertebrate Faunas, San Pedro Valley, Arizona. *Geological Society of America Bulletin*, 86, 5-12, 1990.

Koch, P. L., Zachos, J. C. and Gingerich, P. D., Correlation between isotope records in marine and continental carbon reservoirs near the Palaeocene/Eocene boundary, *Nature*, 358, 319-322, 1992.

Lee-Thorp, J. and van der Merwe, N. J., Carbon isotope analysis of fossil bone apatite. South African, *Journal of Science*, 83, 712- 715, 1987.

Lindsay, E. H., Windows to the Past: Fossils of the San Pedro Valley, *Fieldnotes from the Arizona Bureau of Geology and Mineral Technology*, 14, 1-9, 1984.

Lindsay, E. H., Blancan-Hemphillian Land Mammal Ages and Late Cenozoic Mammal Dispersal Events, *Annual Review of Earth and Planetary Science*, 12, 445-488, 1984.

Lindsay, E. H., Smith, G. A. and Haynes, C. V., Late Cenozoic depositional history and geoarcheology, San Pedro Valley, Arizona. In: Geologic Excursions Through the Sonoran Desert Region, Arizona and Sonora, *Arizona Geologic Survey Special Paper* 7. Edited by George E. Gehrels and Jon E. Spencer. pp 9-19, 1990.

Parker, P. L., The biogeochemistry of the stable isotopes of carbon in a marine bay, *Geochim. et Cosmochim. Acta*, 28, 1155-64, 1964.

Quade, J., Cerling, T. E. and Bowman, J. R., Systematic variations in the carbon and oxygen isotopic composition of pedogenic carbonate along elevation transects in the southern Great Basin, United States, *Geological Society of America Bulletin*, 101, 464-475, 1989a.

Quade, J., Cerling, T. E. and Bowman, J. R., Development of Asian monsoon revealed by marked ecological shift during the latest Miocene in northern Pakistan, *Nature*, 342, 163-166, 1989b.

Ruddiman, W. F., Forcing of Late Cenozoic Northern Hemisphere Climate by Plateau Uplift in Southern Asia and the American West, *Journal of Geophysical Research*, 94, 18409-18427, 1989.

Shackleton, N. J., Backman, J., Zimmerman, H., Kent, D. V., Hall, M. A., Roberts, D. G., Schnitker, D., Baldauf, J. G., Desprairies, A., Homrighausen, R., Huddlestun, P., Keene, J. B., Kaltenback, A.J., Krumseik, K. A. O., Morton, A. C., Murray, J. W. and Westberg-Smith, J., Oxygen isotopic calibration of the onset of icerafting and the history of glaciation in the North Atlantic region, *Nature*, 307, 620-623, 1984.

Shotwell, J. A., Late Tertiary Biogeography of Horses in the Northern Great Basin, *Journal of Paleontology*, 35, 203-217, 1961.

VanDevender, Thornas, R., Late Quaternary Vegetation and Climate of the Chihuahuan Desert, United States and Mexico, In: *Packrat Middens- The Last 40,000 Years of Biotic Change*, 1990.

Yurtsever, Y., Worldwide survey of stable isotopes in precipitation. In: *Stable isotope hydrology, deuterium and oxygen-18 in water cycle*, edited by J. R. Gat and R. Gonfiantini, p. 117. Vienna: International Atomic Energy Agency, 1975.

Wang, Y. and Zheng, S. H., Paleosol nodules as Pleistocene paleoclimatic indicators, Luochuan, P. R. China, *Palaeogeography, Palaeoclimatology, Palaeoecology*, 76, 39-44, 1989.

Wang, Y., T. E. Cerling, G. A. Smith, , J. W. Geissman, J. Quade, E. H. Lindsay & J. R. Bowman, Climatic and Ecologic Changes During the Pliocene and Early Pleistocene in Siutheastern Arizona: Stable Isotopic Records from the St. David Formation, *Geol. Soc. Amer. Abstracts* , 23, A300, 1991.

Williams R. A. D. & Elliott J. C., *Basic and applied dental biochemistry*, Edinburgh, Churchill Livingstone,. 1979.

Y. Wang, J.R. Bowman, T.E. Cerling and J. Quade, Department of Geology and Geophysics, University of Utah, Salt Lake City, UT 84112.

G.A. Smith, Department of Geology, The University of New Mexico, Albuquerque, NM 87131.

E.H. Lindsay, Department of Geosciences, University of Arizona, Tucson AZ 85721.

The Stable Isotopic Composition of Ancient Kaolinites of North America

JAMES R. LAWRENCE

Department of Geosciences, University of Houston, Houston, TX   77204-5503

JANET RASHKES MEAUX

Browning Ferris Industries, P.O. Box 3151, Houston, TX   77253

The D/H ratios of ancient North American kaolinites commonly do not reflect the D/H ratios of the meteoric waters in which they formed. Also, they often do not reflect isotopic reequilibration with present-day meteoric waters. The original D/H ratios of most kaolinites are partially preserved and are usually higher than those of kaolinites in isotopic equilibrium with present-day meteoric waters. This enrichment in D reflects the higher D/H ratios of meteoric waters at most locations in the past and is compatible with warmer climates in the past.

A detailed hydrogen isotope study of kaolinites from Lower Cretaceous formations in central Colorado indicates that the kaolinites underwent complete isotopic reequilibration during the early Tertiary and that the isotopic compositions acquired then have been preserved. This isotopic reequilibration occurred during burial and uplift during which time maximum temperatures reached about $100°$ C. The δD and $\delta^{18}O$ values of other ancient kaolinites from North America also suggest that δD values commonly do not reflect original conditions of formation of the kaolinites.

The original oxygen isotopic compositions of ancient North American kaolinites appear to be preserved. Therefore, the $\delta^{18}O$ values of ancient kaolinites from North America have been used to estimate the the $\delta^{18}O$ values of ancient meteoric waters. The range of $\delta^{18}O$ values of meteoric waters from southern Canada and the United States have increased from about -4 to -13 $°/_{oo}$ in the Cretaceous to -4 to -14 $°/_{oo}$ in the Tertiary to -4 to -18 $°/_{oo}$ at present. This is consistent with overall climatic cooling over North America since the Cretaceous.

The ranges of δD and $\delta^{18}O$ values of ancient kaolinites of North America are compatible with the hypothesis that little or no change has occurred in the isotopic compositions of the oceans since the Pennsylvanian. The oxygen isotopic compositions of Pennsylvanian kaolinites suggest no radical differences between the $^{18}O/^{16}O$ ratio of the Pennsylvanian ocean and the modern ocean.

INTRODUCTION

The stable hydrogen and oxygen isotopic compositions of kaolinites formed in weathering environments reflect the hydrogen and oxygen isotopic compositions of the waters in which they have formed [Savin and Epstein, 1970, Lawrence and Taylor, 1971,1972]. Hydrogen and oxygen isotopic compositions of meteoric waters vary systematically with climatic parameters such as temperature and amount of rainfall [Dansgaard,1964, Yurtsever and Gat, 1981, Lawrence and White, 1991]. The potential therefore exists to retrieve paleoclimatic information from the stable isotopic composition of ancient kaolinites.

Early isotopic studies defined the primary factors controlling the isotopic compositions of kaolinites [Savin, 1970, Savin and Epstein, 1970, Lawrence, 1970, Lawrence and Taylor, 1971, 1972]. Savin and Epstein [1970] postulated that kaolinites formed at Earth surface conditions would plot in a linear array on a graph of δD and $\delta^{18}O$. They called the line defined by the isotopic compositions the "kaolinite line". This line is approximately parallel to the "meteoric water line" [Craig, 1961], which defines the distribution of oxygen and hydrogen isotope ratios of meteoric waters on the Earth's surface. Isotopic studies of clays from Quaternary soils of North America by Lawrence and Taylor [1971,1972] essentially confirmed this hypothesis. Supergene clays from

Climate Change in Continental Isotopic Records
Geophysical Monograph 78

ore deposits also plot close to the kaolinite line (Sheppard et al., 1969).

Preservation of the isotopic ratios of ancient kaolinites subsequent to their formation is essential if they are to be used for paleoclimatic studies. Studies of ancient kaolinites by Lawrence [1970] clearly indicated that the isotopic compositions of ancient kaolinites were at least partially preserved. They did find, however, considerable scatter of isotopic compositions about the "kaolinite line", suggesting that incomplete isotopic exchange may have occurred after formation. In addition Lawrence [1970] showed that halloysite, a mineral in the kaolinite family with interlayer water, readily underwent underwent hydrogen isotopic exchange, at room temperature in the laboratory. Laboratory studies by O'Neil and Kharaka [1976] showed that kaolinite undergoes significant hydrogen isotopic exchange above 80°C, but that oxygen isotopic exchange is significant on a laboratory time scale only at temperatures in excess of 200° C. They concluded that hydrogen isotope exchange below 200° C occurs by proton transfer.

Bird and Chivas [1988, 1989] showed that a suite of ancient kaolinites from Australia had undergone hydrogen isotopic exchange on a geologic time scale at temperatures below 80° C. They concluded that the same kaolinites had undergone very little oxygen isotopic exchange. Bird and Chivas postulated that the increase in $\delta^{18}O$ values of Australian kaolinites since Paleozoic times reflected an increase in the $\delta^{18}O$ values of meteoric waters in Australia through time. This is the result of the equatorward migration of Australia through a variety of climatic zones.

Longstaffe and Ayalon [1990], in a study of diagenetic kaolinite in sandstones from western Canada, found evidence of hydrogen isotopic exchange at temperatures as low as 40° C. The kaolinite appears to have undergone hydrogen isotopic reequilibration with formation waters with no detectable oxygen isotopic exchange.

The hydrogen and oxygen isotopic compositions of kaolinites formed by weathering, and sampled from a large number of localities in North America, are presented in this study. Ages of samples range from Carboniferous through Quaternary. Most of samples were collected or obtained by the authors, although some of the isotopic data are from other studies including those of Savin and Epstein [1970], Hassanipak [1980] and Hassanipak and Eslinger [1985].

EXPERIMENTAL TECHNIQUES

Prior to isotopic analyses, kaolinite was concentrated by standard size settling techniques. The mineralogy of bulk samples and of particle size separates prepared by settling in a dilute solution of Calgon was examined using X-ray diffraction (XRD). Samples were packed into a standard geometry holder for XRD analysis. In many instances semiquantitative estimates of mineral abundances were obtained by comparison with diffractograms of standard mixtures. In most instances the only significant contaminant was quartz. In earlier studies, the

hydrogen yield obtained during hydrogen isotopic analysis of the sample provided a measure of the percent kaolinite [Lawrence, 1970]. In later studies, weight loss during dehydration provided a measure of the percentage of kaolinite (Rashkes,1988). More detailed descriptions of techniques can be found in Lawrence [1970] and Rashkes [1988].

Oxygen isotopic analyses were performed on dried and outgassed 10 to 40 mg samples using the technique of Clayton and Mayeda [1963]. In some instances the oxygen gas liberated was converted to CO_2, using an electrically heated carbon rod, before isotopic analysis. In other instances the O_2 liberated was analyzed directly. Repeated oxygen isotopic analyses of a standard kaolinite have a standard deviation of better than 0.2 per mil. Oxygen isotopic results are reported relative to the V-SMOW standard.

H_2 for isotopic analysis was obtained by dehydration of kaolinite at a temperature of about 1000° C [Savin, 1967, Lawrence, 1970] followed by reduction of the H_2O using either pure U [Friedman, 1953] or Zn metal [Coleman et al, 1982] . The 100-200 mg samples were outgassed at 150° C *in vacuo* before dehydration. Samples were dehydrated in the presence of CuO in a quartz tube. More detailed descriptions of the procedures and extraction systems can be found in Lawrence [1970] and Rashkes [1988]. The H_2O and, sometimes, trace amounts of CO_2 obtained were trapped using liquid nitrogen. Samples yielding measurable amounts of CO_2 were reanalyzed after treatment with hydrogen peroxide to remove organic matter. Repeated hydrogen isotopic analyses of a standard kaolinite had a standard deviation of better than 2 per mil. Hydrogen isotope ratios are reported relative to the V-SMOW standard.

RESULTS

Isotopic results are listed in Table 1. Two $\delta^{18}O$ values are listed for each sample. The first is the measured $\delta^{18}O$ value of the kaolinite concentrate. The second is the calculated $\delta^{18}O$ value of the pure kaolinite endmember. The calculation is based upon the estimated kaolinite content of the sample and an assumed oxygen isotope value for the non clay component, which in most cases is quartz. If the quartz in a sample was known to be residual from the weathering of an igneous rock then the $\delta^{18}O$ value of the contaminant was assumed to be 10 $^o/_{oo}$. If the source of the quartz was unknown the $\delta^{18}O$ value of the contaminant was assumed to be 14 $^o/_{oo}$. This is a median value between that of igneous quartz, 10 $^o/_{oo}$, and wind blown quartz silt , 18 $^o/_{oo}$ [Clayton et al, 1972]. We assume that quartz derived from most metamorphic and sedimentary rocks falls within this range. The $\delta^{18}O$ values calculated assuming these endmember contaminants are in parentheses.

DISCUSSION

Regional and Temporal Trends in Hydrogen Isotopic Composition

The δD values of all the ancient kaolinite samples are plotted on a map of North America in Figure 1. Some of the

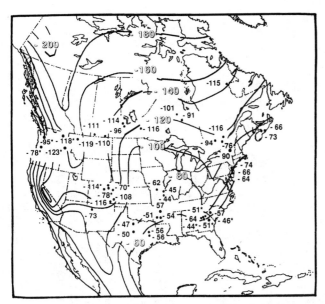

Fig. 1.   The δD values of ancient kaolinites from North America are plotted. Some numbers, denoted by an asterisk, are averages of a few to several samples from a single region. The contour lines give the ∂D values of kaolinites that would be in isotopic equilibrium with present-day meteoric waters. A close approximation of the hydrogen isotopic composition of present-day meteoric water can be obtained by adding 30 per mil to the contoured values. Note that almost all localities have hydrogen isotopic values that are enriched in comparison with kaolinites in isotopic equilibrium with present-day waters.

values plotted are averages of several samples from the same formation but from more than one outcrop. These are indicated by asterisks. The δD values of the contour lines are calculated values for kaolinites in isotopic equilibrium with present-day meteoric waters (calculated assuming a hydrogen isotopic fractionation factor between kaolinite and water of 0.97 [Savin, 1970, Lawrence and Taylor, 1971, 1972]). These contours can be drawn with confidence because the distribution of δD values of the meteoric waters of North America is well established [Friedman et al., 1964, Lawrence,1970, Kendall and Coplen, 1991].

The climate of North American during Pennsylvanian, Cretaceous and Tertiary times [Frakes, 1979] was much warmer than at present. In the Pennsylvanian (Carboniferous), North America was close to the equator and had a tropical climate. In the Early Cretaceous North America occupied a range of latitudes similar to today's, but was rotated clockwise, placing the east coast at somewhat lower latitudes [Irving, 1979]. Overall during the Cretaceous, elevations were lower, a large seaway crossed much of the continent [Cross and Pilger, 1974, Parrish et al., 1984, Kauffman, 1984], and the climate was distinctly warmer than at present [Frakes, 1979]. In the Tertiary the climate was, in general, intermediate between that of the Cretaceous and that existing today, reflecting the general cooling trend for the Earth since the Cretaceous [Savin, 1977].

The δD values of most of the kaolinites studied are higher

than those indicated on the contour lines of Figure 1. A few of the δD values, however, are very close to the equilibrium values. In addition there is considerable range in δD values of kaolinites at similar elevation, and of the same age collected within a few hundred kilometers of one another. This is contrary to expectation if the original δD values of the kaolinites are preserved.

As can be seen by the contour lines in Figure 1 the δD values of kaolinites in equilibrium with modern meteoric waters do not show large gradients except where large variations in elevation occur. Elevation ranges were very limited in the Tertiary and Cretaceous along the eastern coastal plain of North America and throughout the Canadian shield area of Canada. In Colorado, elevations were less than those of today but were still significant (Rashkes,1988). In the northwestern United States in the Tertiary, elevations were significant and varied (Dickinson, 1979).

The global climate in the Cretaceous and Tertiary was warmer than in the Quaternary. If it is assumed that the mean δD value of the oceans has not changed since the Cretaceous, it is reasonable to conclude that the δD values of ancient meteoric waters at any site were higher than those at the same site either today or during the glacial maxima of the Pleistocene. The δD values of ancient North American kaolinites in Figure 1 are compatible with formation during times of warmer climate, if the kaolinites have at least partially preserved their isotopic compositions.

Alternatively, the δD values of the ancient kaolinites would be higher if the δD value of the ancient ocean had been distinctly higher than that of today. The only well documented cause of change in the mean oceanic δD value on a 10^8 yr or shorter time scale is a change in the size of the polar ice caps. Shackleton [1967] calculated that the melting of the polar ice caps would decrease the $\delta^{18}O$ value of the ocean by 0.9 per mil, which corresponds to a decrease of 7 per mil for the δD value. Because we have no evidence for a δD value higher than today's for the pre-Pleistocene oceans we will assume that the high δD value of the Cretaceous and Tertiary kaolinites reflect warmer climates.

Five Pennsylvanian kaolinites, three from Missouri and two from north central Texas [see Table 1 and Figure 1], were formed very near the Pennsylvanian equator. Their δD values are 20 to 50 per mil higher than those of kaolinites in equilibrium with present day meteoric waters at the same sites. Such high values suggest preservation of original isotopic compositions but it is also possible that post-formational hydrogen isotopic exchange with meteoric water during the warmer climates of the Mesozoic and Cenozoic caused the higher δD values.

Post-Formation Isotopic Exchange

The measured δD values and the calculated $\delta^{18}O$ values of pure endmember kaolinites are shown in Figures 2, 3, 4, and 5. Kaolinites from Quaternary soils are shown in Figure 2.

TABLE 1. Hydrogen (δD) AND Oxygen (δ18O) Isotopic Composition of Ancient Kaolinites

State or Province#	Age	δD ‰$	Measured δ18O‰	% Kaolinite	Calculated δ18O‰*	Description@
Clearwater Idaho	Miocene	-125 ±1	13.0	81	12.8 (11.8-13.7)	1; u
Latah Idaho	Miocene	-122 ±2	12.8	91	12.7 (12.3-13.1)	1; u
Latah Idaho	Miocene	-122				1
Nez Perce Idaho	Miocene	-121				1
Clackamas Oregon	Miocene	-88 ±3	17.7	95	17.9 (17.7-18.1)	1; u
Marion Oregon	Miocene	-67 ±2	19.4	78	20.9 (19.8-22.1)	1; u
Bennington Vermont	Miocene	-90 ±1	19.5	72	21.6 (20.1-23.2)	1; u
Spokane Washington	Miocene	-117	13.4	81	13.3 (12.3-14.2)	1; u
Spokane Washington	Miocene	-110 ±1	14.3	98	14.3 (14.2-14.4)	1; u
Spokane Washington	Miocene	-128 ±1	13.9	96	14.1	1; g
Pierce Washington	Mid-Tert.	-98 ±1	13.9	86	14.5	1; a
Pierce Washington	Mid-Tert.	-92	11.8 ± 0.1	81	12.2	1; a
Saline Arkansas	Eocene	-57	17.7	77	20.0	1; g
Wilkinson Georgia	Eocene	-45	20.5			2; GK 11,12
Wilkinson Georgia	Eocene	-47	21.1			2; GK 14,15
Cherokee Texas	Eocene	-56 ±1	20.4	81	21.9 (21.0-22.8)	1; u
Cherokee Texas	Eocene	-56 ±1	19.6	86	20.5 (19.9-21.2)	1; u
Bibb Georgia	Paleocene	-49	21.5			2; GK 56
Wilkinson Georgia	Paleocene	-53	22.2			2; GK 57
Big Muddy V. Sakatch.	U. Cret.	-110 ±3	15.2 ± 0.2	58	19.0	3; g
Cactus Hills Sakatch.	U. Cret.	-111	15.9 ± 0.1	67	18.8	3; g
Pike Arkansas	U.Cret.	-51	22.4			4
Pike Arkansas	U.Cret.	-54 ±3	22.0	95	22.4 (22.2-22.6)	1; u
Aiken So. Carolina	U. Cret.	-57	21.7			4
Bibb Georgia	U. Cret.	-64	19.5			2; GK 51
Crawford Georgia	U.Cret.	-41	21.5			2; GK 1,2
Crawford Georgia	U. Cret.	-46	20.4			2; GK 5
Twiggs Georgia	U. Cret.	-52	22.3			2; GK 23,24
Twiggs Georgia	U. Cret.	-50	22.3			2; GK 41,42
Wilkinson Georgia	U. Cret.	-45	21.8			2; GK 62
Wilkinson Georgia	U. Cret.	-53	18.5			2; GK 63
Wilkinson Georgia	U. Cret.	-53	23.1			2; GK 65
James Bay Ontario	Cret.	-101	15.9 ± 0.1	71	18.3	5; g
St. Remi Quebec	Cret.	-76	17.2 ± 0.1	94	17.4 (17.1-17.7)	6; u
Brebeuf Quebec	Cret.	-116	15.6 ± 0.3	89	16.3	6; g
Point Comfort Quebec	Cret.	-94	17.7 ± 0.3	85	19.1	6; g
Schefferville Quebec	Cret.	-115				6
Swan River Manitoba	L. Cret.	-114 ±11	15.3 ± 0.5	82	15.6 (14.7-16.5)	3; u
Swan River Manitoba	L. Cret.	-96	14.8	85	14.9 (14.2-15.6)	3; u
Punk Island Manitoba	L. Cret.	-116	18.7 ± 0.5	97	18.8 (18.7-19.0)	3; u
Shubenacadie N.S.	L. Cret.	-73	16.2 ± 0.6	83	16.7 (15.8-17.5)	7, 8; u
Musquodoboit N. S.	L. Cret.	-66	17.4	86	18.0 (17.3-18.6)	7, 8; u

Locality	Age	δD	%		$\delta^{18}O$ (range)	Reference; Sample
Huerfano Colorado	L. Cret.	-116 ±2	62	13.5	13.2 (10.7-15.7)	1; u
Pueblo Colorado	L. Cret.	-108 ±2	91	16.3	16.5 (16.1-17.0)	1; u
Cascade Montana	L. Cret.	-119 ±1	52	15.5	16.9 (13.2-20.6)	1; u
Middlesex New Jersey	L. Cret.	-74	72	18.6	20.4 (18.8-21.9)	1; u
Middlesex New Jersey	L. Cret.	-66	77	18.4	19.7 (18.5-20.9)	1; u
Middlesex New Jersey	L.Cret.	-64 ±2	91	20.6	21.3 (20.9-21.6)	1; u
(Mesa Alta) New Mexico	L.Cret.	-73		18.7		4
Callaway Missouri	Penn.	-62 ±1	95	19.0	19.3 (19.0-19.5)	1; u
Franklin Missouri	Penn.	-45 ±2	100	20.2	20.2	1
Franklin Missouri	Penn.	-44 ±1	100	20.3	20.3	1
Eastland Texas	Penn.	-50	81	18.2	19.2 (18.2-20.1)	1; u
Young Texas	Penn.	-47 ±1	77	19.3	20.9 (19.7-22.1)	1; u
Adams Colorado (10)	L. Cret.	-64 ±5				9 ; DB-G,DB-H; Altered ash
Arapahoe Colorado (24)	L. Cret.	-74 ±2				9 ; DB-A,DB-B,DB-C; Clay bed
Boulder Colorado (6)	L. Cret.	-81 ±6				9 ; 67-70; Altered ash
Jefferson Colorado (8)	L. Cret.	-89				9 ; 59; Altered ash
Jefferson Colorado (8)	L. Cret.	-95				9 ; 57; Clay bed
Jefferson Colorado (14a)	L. Cret.	-76 ±2				9 ; 5-7,9; Altered ash
Jefferson Colorado (14a)	L. Cret.	-79 ±7				9 ; 2-4,8; Clay bed
Jefferson Colorado (14b)	L. Cret.	-80 ±7				9 ; 16-18; Altered ash
Jefferson Colorado (14b)	L. cret.	-77 ±1				9 ; 11-12; Clay bed
Jefferson Colorado (14b)	L. Cret.	-76 ±6				9 ; 13-15,19; Clay clast
Jefferson Colorado (15)	L. Cret.	-73 ±5				9 ; 21-25,27,29; Altered ash
Jefferson Colorado (15)	L. Cret.	-80 ±3				9 ; 20a-20d; Clay clast
Jefferson Colorado (17)	L. Cret.	-76 ±6				9 ; 31,34-36; Altered ash
Jeffereson Colorado (17)	L. Cret.	-89 ±8				9 ; 30,33; Clay bed
Jefferson Colorado (19)	L. Cret.	-74 ±4				9 ; 42-43,46-48,51-55; Altered ash
Jefferson Colorado (19)	L. Cret.	-84 ±6				9 ; 44-45; Clay bed
Jefferson Colorado (21)	L. Cret.	-77 ±2				9 ;1.86-4.86,7.86-8.86; Altered ash
Park Colorado (27)	L. Cret.	-114±7				9 ; 90,90a,91-92; Clay bed

Localities include county and state (USA) or town or region and province (Canada). Sample locality numbers given for Rashkes collection [Rashkes,1988].

* Determined by mass balance. A $\delta^{18}O$ value for the quartz contaminant of 10 °/oo or 14 °/oo was assumed. If the quartz was residual from granite (g) or andesite (a) a value of 10 °/oo was assumed. If the quartz was derived from an unknown (u) parent rock, a value of 14 °/oo was assumed. The range in parentheses represents the possible range for the kaolinite assuming a range for quartz of 10 °/oo to 18 °/oo

@ The first number in the sequence refers to a reference. 1= Lawrence [1970], 2= Hassanipak [1980], 3= Dean [1978], 4= Savin and Epstein [1970], 5= Telford and Verma [1978], 6= Brady and Dean [1966], 7= Dean [1975], 8= Bell et al. [1978], 9= Rashkes [1988]. The second number(s) in the sequence refer to the sample number(s). The parent rock identity is indicated as granite (g) or andesite (a) unless unknown (u). Sample types are given for the Rashkes' collection.

$ Mean and standard deviation of multiple analyses on a single sample except for collections of Hassanipak (1980) and Rashkes (1988) in which the errors represent the mean and standard deviation of multiple samples.

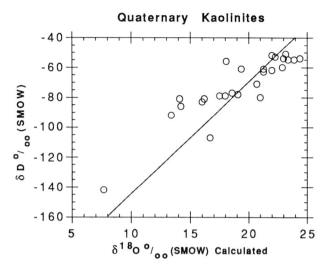

Fig. 2.   The δD values of Quaternary kaolinites plotted as a function of their δ¹⁸O values [Lawrence and Taylor, 1971]. The line drawn is the "kaolinite line" of Savin and Epstein [1970] for 20° C.

Kaolinites of Tertiary, Cretaceous and Pennsylvanian ages are shown in Figures 3, 4 and 5 , respectively. The δD values (see Figure 1) and the δ¹⁸O values of the pure end member kaolinites (see Figures 6,7 and 8) are also plotted on maps .

Pure kaolinites formed in isotopic equilibrium with meteoric

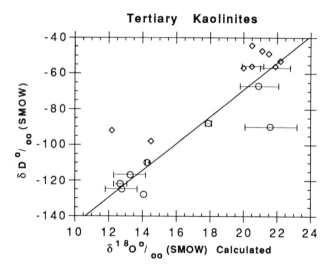

Fig. 3.   The δD values of Tertiary kaolinites plotted as a function of their δ¹⁸O values. The δ¹⁸O values represent the oxygen isotopic composition of pure kaolinite. Most samples contained quartz as a contaminant. Mass balance calculations were used to obtain the δ¹⁸O value of the pure endmember. The error bars given for some samples represent the uncertainty in the mass balance calculation due to the unknown δ¹⁸O value of the contaminant (see Table 1). The circles denote kaolinites of Miocene age and the diamonds denote kaolinites of Lower Tertiary age (see Table 1). The line drawn is the "kaolinite line" of Savin and Epstein [1970] for 20° C.

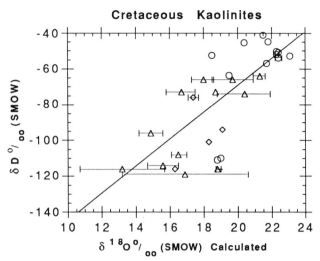

Fig. 4.   The δD values of Cretaceous kaolinites plotted as a function of their δ¹⁸O values. The δ¹⁸O values represent the oxygen isotopic composition of pure kaolinite. Most samples contained quartz as a contaminant. Mass balance calculations were used to obtain the δ¹⁸O value of the pure endmember. The error bars given for some samples represent the uncertainty in the mass balance calculation due to the unknown δ¹⁸O value of the contaminant (see Table 1). The circles denote kaolinites of Upper Cretaceous age, the diamonds kaolinites of Cretaceous age and the triangles kaolinites of Lower Cretaceous age (see Table 1). The line drawn is the "kaolinite line" of Savin and Epstein [1970] for 20° C.

waters at about 20° C would plot on the "kaolinite line" (see Figures 2, 3, 4 and 5) approximately parallel to the "meteoric water line" [Savin and Epstein,1970]. An increase of about 0.2 per mil in the δ¹⁸O value of kaolinite per degree centigrade cooling is expected as a consequence of the temperature sensitivity of the kaolinite-water oxygen isotope fractionation factor [Land and Dutton,1978 and Kulla and Anderson, 1978].

The oxygen and hydrogen isotopic compositions of kaolinites shown in Figures 2, 3, 4, and 5 scatter about the "kaolinite line". There are several possible causes for the observed scatter. First assume that the only cause of changes in the isotopic composition of the oceans through time is changes in the size of the polar ice caps. In that case the isotopic composition of meteoric waters would always plot along the "meteoric water line" and the "kaolinite line" would always reflect the isotopic compositions of kaolinites formed at about 20° C in isotopic equilibrium with meteoric waters.

Because the δ¹⁸O values of the kaolinites in Figures 2, 3, 4 and 5 were estimated using material balance calculations as described above, it is possible that the assumptions made in these calculations were in error. The δ¹⁸O values of the contaminants were assumed rather than measured. Error bars are shown for the data points for which the δ¹⁸O value of the contaminants were uncertain or unknown. Another source of uncertainty is in the estimation of the amount of contaminant.

Pennsylvanian Kaolinites

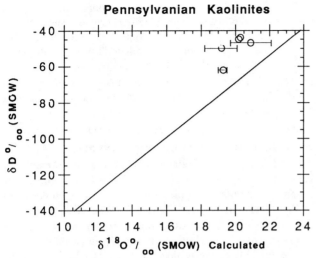

Fig. 5.   The δD values of Pennsylvanian kaolinites plotted as a function of their δ¹⁸O values. The δ¹⁸O values represent the oxygen isotopic composition of pure kaolinite. Most samples contained quartz as a contaminant. Mass balance calculations were used to obtain the δ¹⁸O value of the pure endmember. The error bars given for some samples represent the uncertainty in the mass balance calculation due to the unknown δ¹⁸O value of the contaminant (see Table 1). The line drawn is the "kaolinite line" of Savin and Epstein [1970] for 20° C.

In most instances estimates were made using the H₂ yield measured during the hydrogen isotopic analysis of the sample. If no hydrous minerals other than kaolinite were present in the sample (as was usually the case) this method should be

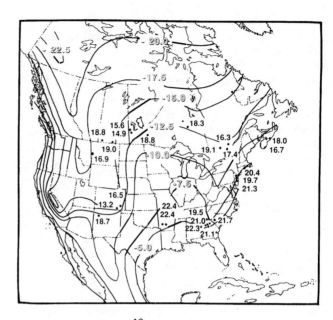

Fig. 7.   The calculated δ¹⁸O values of Cretaceous kaolinites from North America are plotted (see Table 1). Some numbers, denoted by an asterisk, are averages of a few to several samples from a single region. The contour lines give the δ¹⁸O values of present-day meteoric waters.

accurate. For a majority of the samples, X-ray diffraction analyses in fixed geometry holders employing standard mixtures were also conducted to determine the amount of contaminant. The δ¹⁸O values of samples with the highest percentages of kaolinite are the most reliable (see Table 1).

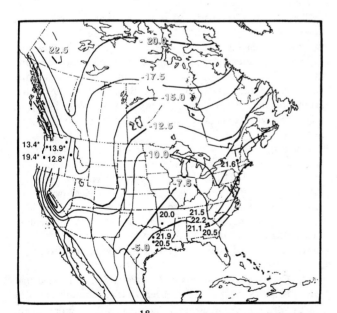

Fig. 6.   The calculated δ¹⁸O values of Tertiary kaolinites from North America are plotted (see Table 1). Some numbers, denoted by an asterisk, are averages of a few to several samples from a single region. The contour lines give the δ¹⁸O values of present-day meteoric waters.

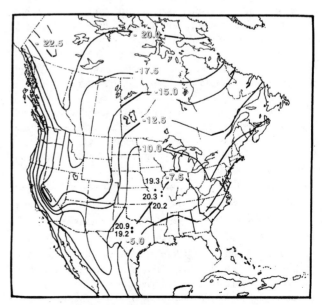

Fig. 8.   The calculated δ¹⁸O values of Pennsylvanian kaolinites from North America are plotted (see Table 1). The contour lines give the δ¹⁸O values of present-day meteoric waters.

Scatter in Figures 2, 3, 4 and 5 could also result from variation in the temperatures at which the kaolinites formed. Hassanipak and Eslinger [1985] concluded, in a study of kaolinites from Georgia of Tertiary and Cretaceous age (see Table 1), that those clays formed over a range of about 15° C. This translates into a range of $\delta^{18}O$ values of 3 per mil. Systematic changes in δD values of kaolinites over this temperature range would be relatively small, about 3 per mil [Sheppard et al., 1969, Lambert and Epstein, 1980, Liu and Epstein, 1984].

The $\delta^{18}O$ values of kaolinite from a sample can also vary with crystal size and crystallinity [Hassanipak and Eslinger 1985]. The $\delta^{18}O$ values of the Georgia kaolinites generally decrease with increasing crystallinity and with increasing crystallite size. Variations of 0.5 to 1.5 per mil were found among different size fractions of individual samples.

It is also possible that not all of the the kaolinites formed from meteoric waters. The isotopic compositions of formation waters deviate significantly from those of meteoric waters [Clayton et al., 1966]. Continental formation waters are commonly derived from meteoric waters but their $\delta^{18}O$ values may become more positive due to interaction with carbonate in the sediments [Clayton et al., 1966]. Formation waters commonly have very high salt contents, however, decreasing the likelihood that kaolinite would form in their presence. Geologic evidence shows that most of the kaolinites investigated in this study were formed by weathering at or near the Earth's surface. In some instances these kaolinites were buried and later exposed to formation waters.

Finally, as pointed out earlier partial hydrogen isotopic exchange without oxygen isotopic exchange in kaolinites has been documented in other studies. Examination of the hydrogen isotopic data in Figure 1 suggests partial hydrogen isotopic exchange in ancient kaolinites of North America has taken place. In that case, any point on Figures 3, 4, or 5 corresponding to a kaolinite that originally lay on the "kaolinite line" would move up or down from it's original position if it exchanged with a water having a different hydrogen isotopic value than the original water in which it was formed.

It is likely that some ancient kaolinite samples underwent post-formational exchange with D-depleted waters. These samples include one Lower Cretaceous kaolinite from Manitoba, two Upper Cretaceous kaolinites from Sakatchewan and a Miocene kaolinite from Vermont (see Table 1). The Canadian samples have $\delta^{18}O$ values between 18.8 and 19.0 per mil and δD values between -110 and -116 per mil (see Table 1 and Figures 1,4 and 7). The sample from Vermont has a $\delta^{18}O$ value of 21.6 per mil and a δD value of -90 per mil (see Table 1 and Figures 1,3 and 6) . All four of these samples clearly plot well below the "kaolinite line". For much of the last million years large glaciers have existed over much of the eastern half of Canada and the northeastern United States. The melt waters from these glaciers would have been highly depleted in D, and exchange with that water may have caused the downward shift in δD values.

The distribution of isotopic compositions of three Cretaceous kaolinites from Quebec, Canada suggest that differential hydrogen isotopic exchange has occurred. The samples were located near one another. They are nearly pure, containing between 85 and 94 % kaolinite. The $\delta^{18}O$ values are 16.3, 17.4 and 19.1 per mil (see Table 1 and Figure 7) and the corresponding δD values are -116, -76 and -94 per mil (see Table 1 and Figure 1). Two of the samples plot below the kaolinite line (see Figure 4) and the spread in δD values is greater than that expected from the observed spread in $\delta^{18}O$ values. Furthermore, the relationship between the δD and $\delta^{18}O$ values is not linear as would be expected for kaolinites in isotopic equilibrium with meteoric waters (see filled diamonds in Figure 4). Finally, the δD value of the most D-enriched sample is only slightly lower than those of five Cretaceous kaolinites from Nova Scotia and New Jersey (see Table 1 and Figure 1). This suggests that the other two kaolinites from Quebec have undergone hydrogen isotopic exchange with D-depleted meteoric waters. Other Cretaceous kaolinites from Canada also show a greater range of δD values than would be expected from their $\delta^{18}O$ values if post-formational exchange of hydrogen isotopes had not occurred (See Table 1 and Figures 1,4 and 7).

In summary, the displacement of some of the kaolinite data below the kaolinite line reinforces the conclusion of previous studies that at low temperatures little or no oxygen isotopic exchange may occur even though extensive hydrogen isotopic exchange has occurred.

Hydrogen Isotope Ratios of Cretaceous Kaolinites from Colorado

The Dakota Group and Mowry Shale of Colorado were deposited during the Early Cretaceous in fresh to brackish water environments [Waage, 1961]. These sediments represent a general transition from a nonmarine fluvial environment, that of the underlying Morrison formation, to a marine environment, that of the overlying Benton Group, Niobrara Formation and Pierre Shale [Leroy and Weimer, 1971]. A schematic geologic cross section of the field area is shown in Figure 9.

The kaolinites are of three types: altered ash beds in shales, clay beds in sandstones and clay clasts in sandstones [Rashkes, 1988]. Based on petrographic evidence [Mozley, 1985] the kaolinites from the Dakota Group and Mowry Shale formed over a range of time from late Jurassic to mid Tertiary. There is also considerable evidence for later diagenetic recrystallization in these formations [Mozley, 1985].

Rashkes [1988] made a detailed study of the hydrogen isotopic composition of a large number of kaolinites from Lower Cretaceous sediments of Colorado. She concluded that these kaolinites are not in hydrogen isotopic equilibrium with present day meteoric waters. Their δD values more likely reflect the D/H ratios of Tertiary ground waters than those of Early Cretaceous meteoric waters. Geologic evidence suggests that the kaolinites were buried during the Tertiary to depths great enough for the temperatures to exceed 80° C. Hydrogen

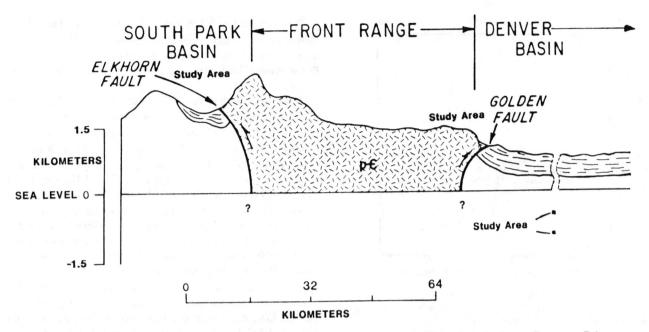

Fig. 9.   A schematic geologic cross section of the sample locality area for the kaolinites from the Lower Cretaceous Dakota Group and Mowry Shale of central Colorado [Rashkes, 1988]. The study areas include a sediment section adjacent to the Elkhorn Fault (4 samples), a small sediment wedge adjacent to the Golden Fault (58 samples) and sediments obtained from drill holes penetrating the Denver Basin (5 samples). The average hydrogen isotopic composition of the kaolinites from these three localities are -114 ±3, -78 ±6 and -70 ±6 per mil ,respectively (see Figure 1).

isotopic exchange must have occurred because this 80° C threshold applies to the time scale of laboratory studies [O'Neil and Kharaka, 1976]. On the time scale of the natural environment the temperature threshold probably is lower.

The δD values of kaolinites from the Dakota Group and Mowry Shale are too high to be in isotopic equilibrium with present day or Pleistocene meteoric waters. This can be seen in Figure 1 where the average δD values of the Cretaceous kaolinites from each of the three localities (see Figure 9) are shown. The mean δD values of the three groups are -70 ±6, -78 ±6 and -114 ±3 per mil. The δD values of kaolinites in isotopic equilibrium with present day meteoric waters would be much lower than -70 to -114 per mil (see contour lines in Figure 1). Hydrogen isotopic analyses of 21 local waters in Colorado from streams, lakes, springs and snows collected in July, 1985 and May, 1986 ranged from -104 to -184 per mil. Kaolinites in isotopic equilibrium with these water would range from -131 to -209 per mil [Rashkes, 1988].

In order to establish whether the D/H ratios of the kaolinites might at least in part reflect the isotopic composition of Cretaceous meteoric waters it is necessary to evaluate the paleoclimatic conditions likely to have existed during the deposition of the Dakota Group and Mowry Shale. The paleolatitudes during the time of deposition were a few degrees lower than those of today (see Figure 10). Elevations were very near sea level for the entire field area (Tweto, 1975). The climate was distinctly warmer than that existing today [Kauffman,1977,1984; Lloyd, 1982; Barron and Washington,

1982,1984]. Using this information and hydrogen isotopic analyses of modern precipitation from regions with climates similar to those existing in the mid-continent during the early Cretaceous, Rashkes [1988] estimated that the δD value of early Cretaceous meteoric waters was about -30 per mil. The δD value of the kaolinite formed in such a water would be about -

Paleolatitudes of 40°N, 105°W
From The Jurassic To The Holocene

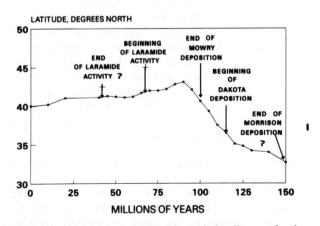

Fig. 10.   A paleolatitude history of sample locality area for the kaolinites from the Lower Cretaceous Dakota Group and Mowry Shale of central Colorado [Rashkes, 1988].

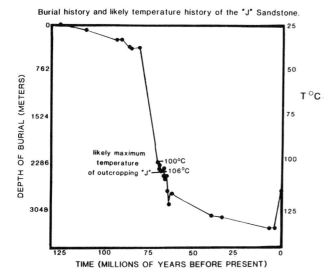

Burial history and likely temperature history of the "J" Sandstone.

Fig. 11.   Burial and temperature history of the outcropping and buried "J" Sandstone. Temperatures based on a 25° C surface temperature. Modified from Ottman [1981].

59 per mil. All of the kaolinites studied have δD values lower than -59 per mil.

Several lines of evidence suggest that the D/H ratios of the kaolinites reflect post Early Cretaceous meteoric waters. The kaolinite-bearing altered ash beds were originally deposited in sea water and were not exposed to meteoric waters until the late Cretaceous [de Rojas, 1980, Rashkes, 1988]. Alteration of the ash first to smectite, followed by later alteration of the smectite to kaolinite is suggested by the petrographic and clay mineralogical studies of de Rojas [1980]. The kaolinite-bearing clay beds in sandstones of the Dakota Group show evidence of recrystallization during and after the Laramide uplift [Mozley, 1985]. Finally, all the kaolinites were buried during the late Cretaceous to depths at which temperatures reached about 100° C (see Figure 11). Hydrogen isotopic exchange takes place readily at these temperatures.

Evidence that hydrogen isotopic exchange can take place without obvious petrographic changes comes from the δD values of clasts of kaolinite in the Dakota Group sandstones. These clasts were probably derived from the underlying Jurassic Morrison Formation [Rashkes, 1988]. The clasts of kaolinite have hydrogen isotopic values that overlap those of the altered ash beds and the kaolinite clay beds. It appears that all three types of kaolinite underwent hydrogen isotopic exchange with Tertiary meteoric waters.

Estimation of the Isotopic Composition of Ancient Meteoric Waters

Preservation or partial preservation of the isotopic compositions of ancient kaolinites allows estimations to be made of the isotopic composition of ancient meteoric waters. Several groups of kaolinites examined in this study lend themselves to such estimations.

The calculated $\delta^{18}O$ values of kaolinites from Table 1 are plotted on three maps along with contours of $\delta^{18}O$ values of modern meteoric waters. Tertiary, Cretaceous and Pennsylvanain kaolinites are separately displayed on Figures 6, 7 and 8.

If we assume that the $\delta^{18}O$ values of the Cretaceous kaolinites from Canada are largely preserved we can estimate for the $^{18}O/^{16}O$ ratio of Cretaceous meteoric waters over southern Canada. The $\delta^{18}O$ values range from 14.9 to 19.1 (see Figure 7) per mil. Using an oxygen isotopic fractionation factor for the kaolinite water system of 1.0265 [Lawrence and Taylor, 1971] we calculate a $\delta^{18}O$ value of meteoric waters between -7.2 and -11.1 per mil. These values are higher and their range narrower than observed over southern Canada today (see Figure 7). This probably reflects the much warmer and more uniform climate over Canada during the Cretaceous. In a similar fashion, using the entire range of $\delta^{18}O$ values of Cretaceous kaolinites of North America yields a range of $\delta^{18}O$ values of Cretaceous meteoric waters between -4.1 and -13.0 per mil. Only one Cretaceous sample from Colorado, yields a calculated $\delta^{18}O$ value of meteoric water lower than -11.1 per mil. The $^{18}O/^{16}O$ ratios of present-day meteoric waters for all of North America range from about -4 to -23 per mil (see Figure 7).

Unlike the Cretaceous kaolinites, the Tertiary kaolinites investigated in this study with two exceptions plot close to the "kaolinite line" (Figure 3). These kaolinites also have δD values that are from 0 to 40 per mil higher than those of hypothetical kaolinites in isotopic equilibrium with present day waters (Figure 1). The two exceptions are the kaolinite from Vermont mentioned earlier and a kaolinite from eastern Washington. The low $\delta^{18}O$ value of the kaolinite from eastern Washington may be the result of the high Fe_2O_3 content of the separate, 12% by weight. Amorphous iron was not removed from this kaolinite sample. The sample from Vermont appears to have undergone hydrogen isotopic exchange with D-depleted meteoric waters as described earlier.

If the $\delta^{18}O$ values of the Tertiary kaolinites are preserved, those values can be used to estimate the oxygen isotopic composition of Tertiary meteoric waters over some areas of the United States (see Figure 6). Using an oxygen isotopic fractionation factor of 1.0265 and the range of $\delta^{18}O$ values for Tertiary kaolinites of 12.2 to 22.2 per mil (see Table 1), the calculated range for the oxygen isotopic composition of Tertiary meteoric waters would be -4.0 to -14.1 per mil. This range is not as great as the range of $\delta^{18}O$ values observed over the United States at present (-4 to -18 per mil, see Figure 7) but is greater than the range observed during the Cretaceous. This is compatible with the general cooling of the climate over the Earth from the Cretaceous to the present.

Up to this point it has been assumed that the oxygen isotopic composition of the oceans has remained unchanged except for polar ice volume effects. Such an assumption is probably good for post Jurassic times as most investigators

[Muehlenbachs and Clayton, 1976, Savin, 1977, Gregory and Taylor, 1981] suggest only minor fluctuations in the oxygen isotopic composition of the oceans from that existing today. For earlier times there is disagreement. Veizer et al. [1986] suggest a $\delta^{18}O$ value of -3 per mil for the Permian and Brand and Veizer [1981] suggest a value as low as -5.5 per mil for the Silurian. Most other investigators suggest a value no lower than -2 per mil for the Paleozoic [Muehlenbachs and Clayton, 1976, Kahru and Epstein, 1986, Knauth and Beeunas, 1986, Muehlenbachs, 1986, Gregory, 1991].

It is difficult to assess the history of the hydrogen isotopic composition of the oceans in the past because of the occurrence of hydrogen isotopic exchange in clay minerals at temperatures as low as 80° C [O'Neil and Kharaka,1976]. An indication of what the δD values of ancient oceans might have been comes from this study. The generally higher δD values of ancient kaolinites than of kaolinites in isotopic equilibrium with modern meteoric waters suggests that the D/H ratios of ancient kaolinites from North America are at least partially preserved. This implies that the δD value of the ocean since the Cretaceous has been similar to or greater than that of today's ocean. An ancient ocean having mean $\delta^{18}O$ and δD values similar to today's could easily have been the source for meteoric waters in the Tertiary and Cretaceous. Such meteoric waters would yield the observed $\delta^{18}O$ and δD values found in ancient kaolinites if the climates during those times were warmer than the current climate.

The Pennsylvanian samples come from two locations: Missouri and Texas. The samples from Texas are in part composed of illite indicating burial and recrystallization in a later diagenetic environment while the samples from Missouri are pure kaolinite. The $\delta^{18}O$ values of the Missouri kaolinites are very close to 20 per mil (see Table 1 and Figure 8). These kaolinites are only 2 per mil lower than the most ^{18}O enriched kaolinites from younger sample localities (see Figures 6 and 7). This would suggest that if the $^{18}O/^{16}O$ ratios are pristine, the oxygen isotopic composition of the ocean in the Pennsylvanian was not radically different from that of today.

CONCLUSIONS

The D/H ratios of ancient kaolinites of North America are at least partially preserved and are higher than those of kaolinites in isotopic equilibrium with present-day meteoric waters. This enrichment in deuterium reflects higher D/H ratios of meteoric waters in the past. This enrichment is compatible with warmer climates in the past. Hydrogen isotopic exchange with younger meteoric waters, however, has occurred in many samples.

The relationship between the δD and $\delta^{18}O$ values of ancient kaolinites from North America and the kaolinite line indicates that some kaolinites have undergone significant hydrogen isotopic exchange with only minor amounts of oxygen isotopic exchange. Some kaolinites plot distinctly below the kaolinite line. This positioning suggests that these kaolinites

were formed in meteoric waters enriched in D relative to those existing today and have subsequently undergone isotopic exchange with meteoric waters of lower isotopic composition. The shift to a point below the kaolinite line suggests hydrogen isotopic exchange occurred without oxygen isotopic exchange.

A detailed study of kaolinites from the Lower Cretaceous Dakota Group and Mowry Shale formations of Colorado indicates that the δD values of the kaolinites do not reflect the isotopic composition of Cretaceous meteoric waters. Neither are these kaolinites in isotopic equilibrium with present-day or Quaternary meteoric waters. An evaluation of the petrologic relationships and geologic history of those kaolinites indicates that their δD values probably equilibrated last with meteoric waters which flushed through the formations during the Laramide orogeny in the early Tertiary.

Isotopic studies of ancient kaolinites including this study and those of Bird and Chivas (1988,1989) and laboratory experiments indicate that $^{18}O/^{16}O$ ratios of kaolinites are preserved in samples that have not been exposed to post-burial temperatures in excess of 200° C. If so the $\delta^{18}O$ values of ancient kaolinites can be to used to estimate the oxygen isotopic composition of ancient meteoric waters. Applying this approach to the suite of samples analyzed in this study reveals a progressive decrease in the $\delta^{18}O$ values of meteoric waters of southern Canada and the United States from the Cretaceous to the present. The range of $\delta^{18}O$ values for this area shifted from about -4 to -13 $^o/_{oo}$ in the Cretaceous to -4 to -14 $^o/_{oo}$ in the Tertiary to -4 to -18 $^o/_{oo}$ presently. This shift is compatible with the general cooling of the Earth since the Cretaceous (Savin, 1977).

The $\delta^{18}O$ values of kaolinites of Pennsylvanian age from Missouri are very close to 20 per mil. If these $\delta^{18}O$ values are pristine the oxygen isotopic composition of the ocean in the Pennsylvanian was not radically different from that of today.

Acknowledgements. This research was supported by two grants from the National Science Foundation; grant number GA-12945 awarded to Hugh P. Taylor of the California Institute of Technology which supported the research of the PhD Thesis of J.R. Lawrence and grant number EAR-81-08553 awarded to J.R. Lawrence when he was at Lamont Doherty Geological Observatory of Columbia University. Funding from the Houston Coastal Center and the Energy Lab of the University of Houston supported the research of the Masters Thesis of Janet Rashkes Meaux. She also received a grant from the Sigma Xi foundation which supported her field studies.

The authors thank Bruce Bohor, Robert Weimer, Ken Bell, Don Scafe, J.H. Hudson, Max Vos, Peter Telford, H.M. Verma, Henry McDougal and Jorgen Poschmann for assistance in obtaining samples and pertinent geologic information.

Critical reviews by S.M. Savin, F. Longstaffe and A.R. Chivas improved this report considerably.

REFERENCES

Barron, E.J. and Washington, W. M., Cretaceous climate: a comparison of atmospheric simulations with the geologic record, *Ecologae Geologicae Helvetiae*, 74, 443-470, 1982.

Barron, E.J. and Washington, W. M., The role of the geographic variables in explaining paleoclimates: Results from Cretaceous climate model sensitivity studies, *J. Geophys. Res., 89,* 1267-1279, 1984.

Bell, K.E., Brady, J.G. and Zemgals, L.K., Ceramic clays and shales of the Atlantic Provinces, *CANMET Rep., 78-21,* (Printing and Publishing Supply and Services Canada, Ottawa), 1978.

Bird, M.I. and Chivas, A.R., Stable-isotope evidence for low-temperature kaolinitic weathering and post-formational hydrogen-isotope exchange in Permian kaolinites, *Chem. Geol., 72,* 249-265, 1988.

Bird, M.I. and Chivas, A.R., Stable-isotope geochronology of the Australian regolith, *Geochim. Cosmochim. Acta, 53,* 3239-3256, 1989.

Brady, J. G. and Dean, R.S., The composition and properties of ceramic clays and shales of Quebec, *Mines Branch Res. Rep., R-187,* (Queens Printer, Ottawa), 1966.

Brand, U. and Veizer, J., Chemical diagenesis of a multicomponent system, 2. Stable isotopes, *J. Sediment. Petrol., 51,* 987-997, 1981.

Clayton, R.N., and Mayeda, T.K., The use of bromine pentafluoride in the extraction of oxygen from oxides and silicates for isotopic analysis, *Geochim. Cosmochim. Acta, 27,* 43-52, 1963.

Clayton, R.N., Friedman, I., Graf, D.L., Mayeda, T.K., Meents, W.F., Shimp, N.F., The origin of saline formation waters, 1, isotopic composition, *J. Geophys. Res.,71,* 3869-3882, 1966.

Clayton, R.N., Rex, R.W., Syers, J.K., and Jackson, M.L., Oxygen isotope abundance in quartz from Pacific pelagic sediments, *J. Geophys. Res.,77,* 3907-3915, 1972.

Coleman, M.L., Shepherd, T.J., Durham, J.J., Rouse, J.E. and Moore, G.R., Reduction of water with zinc for hydrogen isotope analysis, *Anal. Chem., 54,* 993-995, 1982.

Craig ,H., Isotopic variations in meteoric waters. Science, 133, 1702-1703, 1961.

Cross, T.A. and Pilger, R.H. Jr., Tectonic controls of Late Cretaceous sedimentation, western interior, USA, *Nature, 274,* 653-657, 1978.

Dansgaard, W. Stable isotopes in precipitation, *Tellus, 16,* 436-468, 1964.

de Rojas, I., Stratigraphy of the Mowry Shale (Cretaceous) western Denver Basin, Colorado, *MS Thesis, Colorado School Mines,* 1980.

Dean, R.S., Cursory mineralogical examination of ceramic clays and shales of the Prairie Provinces, *Mineral Sci. Lab Rep., MRPI MSL 78-169,* (CANMET, Ottawa), 1978.

Dean, R.S., Mineralogy of ceramic clays and shales of the Atlantic Provinces, *Scientific Bull.,Cm 75-8,* (Information, Canada, Ottawa), 1975.

Dickinson, W.R. Cenozoic plate tectonic setting of the Cordilleran region in the United States. In: Cenozoic paleogeography of the western United States, eds. J. M. Armentrout, M.R. Cole, H. Terbest Jr., *Pacif. Coast Paleogeo. Sym. , 3, SEPM ,* 1-13, 1979.

Frakes, L.A. Climates throughout geologic time (Elsevier, Amsterdam), 1979.

Friedman, I., Deuterium content of natural waters and other substances, *Geochim. Cosmochim. Acta, 4,* 89-103, 1953.

Friedman, I., Redfield, A.C., Schoen, B. and Harris, J., The variation of the deuterium content of natural waters in the hydrologic cycle, *Rev. Geophys., 2,* 177-224, 1964.

Gregory, R.T. and Taylor, H.P., An oxygen isotope profile in a section of Cretaceous oceanic crust, Samail Ophiolite, Oman; Evidence for $\delta^{18}O$ buffering of the ocean by deep (> 5km) seawater-hydrothermal circulation at mid-ocean ridges, *J. Geophys. Res., 86,* 2737-2755, 1981.

Gregory, R.T., Oxygen isotope history of seawater revisited: Timescales for boundary event changes in the oxygen isotopic composition of seawater, In: Stable Isotope Geochemistry: A Tribute to Samuel Epstein, The Geochem. Soc., Spec. Publ. No. 3, (eds. H.P. Taylor, J.R. O'Neil, I.R. Kaplan), 65-76, Lancaster Press, 1991.

Hassanipak, A.A. and Eslinger, E.V., Mineralogy, crystallinity, $^{18}O/^{16}O$ and D/H of Georgia kaolins, *Clays Clay Miner. , 33,* 99-106, 1985.

Hassanipak, A.A., Isotopic geochemical evidence concerning the origin of Georgia kaolinite deposits, *Ph.D. Thesis, Ga. Inst. Tech.,* 198 pp.,1980.

Irving, E., Paleopoles and paleolatitudes of North America and speculations about displaced terranes, *Can. J. Earth. Sci., 16,* 1979.

Karhu, J. and Epstein, S., The implication of the oxygen isotope records in coexisting cherts and phosphates, *Geochim. Cosmochim. Acta, 50,* 1745-1756, 1986.

Kauffman, E.G., Geological and biological overview: Western interior Cretaceous Basin, *The Mountain Geologist, 14,* 75-99, 1977.

Kauffman, E.G., Paleobiogeography and evolutionary response dynamic in the Cretaceous Western Interior Seaway of North America, In: Jurassic-Cretaceous biochronology and paleogeography of North America (ed. G.E.G. Westermann), *Geol. Assoc. Canada Spec. Pap., 27,* 273-306, 1984.

Kendall, C, and Coplen, T.B., Distribution of oxygen-18 and deuterium in river waters in the USA, *Chapman Conference, Continental Isotopic Indicators of Climate, Amer. Geophys. Un.,* 1991.

Knauth, L.P. and Beeunas, M.A., Isotopic geochemistry of fluid inclusions in Permian halite with implications for the isotopic history of ocean water and the origin of saline formation waters. *Geochim. Cosmochim. Acta, 50,* 419-433, 1986.

Kulla, J.B. and Anderson, T.F., Experimental oxygen isotope fractionation between kaolinite and water, Short pap., 4th Int. Conf. on Geochronology, Cosmochronology and Isotope Geology, Colorado, *U.S. Geol. Surv., Open-File Rep., 78-701,* 234-235, 1978.

Lambert, S.J. and Epstein, S., Stable isotope investigations of an active geothermal system in Valles Caldera, Jemez, New Mexico, *J. Volcanol. Geotherm. Res., 8,* 111-129, 1980.

Land, L.S. and Dutton, S.P., Cementation of a Pennsylvanian deltaic sandstone: isotopic data, *J. Sediment. Petrol., 48,* 1167-1176, 1978.

Lawrence, J. R. and Taylor, H.P., Hydrogen and oxygen isotope systematics in weathering profiles,*Geochim. Cosmochim. Acta, 36,* 1377-1393, 1972.

Lawrence, J.R. and Taylor, H.P., Deuterium and O^{18} correlation: clay minerals and hydroxides in Quaternary soils compared to meteoric waters, *Geochim. Cosmochim. Acta, 35,* 993-1004, 1971.

Lawrence, J.R. and White, J.W.C., The elusive climate signal in the isotopic composition of precipitation. In: Stable Isotope Geochemistry: A Tribute to Samuel Epstein, The Geochem. Soc., Spec. Publ. No. 3, (eds. H.P. Taylor, J.R. O'Neil, I.R. Kaplan),169-185, Lancaster Press,1991.

Lawrence, J.R., O^{18}/O^{16} and D/H ratios of soils, weathering zones and clay deposits, *PhD Thesis, Calif. Inst. Tech.,* 263 pp,1970.

Leroy, L. W. and Weimer, R.J. Geology of the Interstate 70 roadcut, Jefferson County, Colorado. *Prof. Contrib. Colorado Sch. Mines, No. 7,* 1971.

Liu, K. K. and Epstein, S., The hydrogen isotopic fractionation factor between kaolinite and water, *Isotope Geosci, 2,* 335-350, 1984.

Lloyd, C.R., The mid-Cretaceous Earth: Paleogeography; ocean circulation and temperature; atmospheric circulation, *J. Geol., 90,* 393-413, 1982.

Longstaffe, F. and Ayalon, A., Hydrogen-isotope geochemistry of diagenetic clay minerals from Cretaceous sandstones, Alberta, Canada: evidence for exchange, *Appl. Geochem., 5,* 657-668, 1990.

Mozley, P.S., Origin of kaolinite in the Dakota Group (Cretaceous Age), northern Front Range foothills, Colorado, *MS Thesis, Univ. of Colorado,* 51p, 1985.

Muehlenbachs, K. and Clayton, R.N., Oxygen isotope composition of the oceanic crust and its bearing on seawater, *J. Geophys. Res., 81,* 4365-4369, 1976.

Muehlenbachs, K., Alteration pf the oceanic crust and the ^{18}O history of seawater, In: Stable isotopes in high temperature geological processes, *Rev. Mineral., 16,* (Eds. J.W. Valley, H.P. Taylor, Jr., and J.R. O'Neil,), 425-442, 1986.

O'Neil, J.R. and Kharaka, Y.K., Hydrogen and oxygen isotope exchange reactions between clay minerals and water, *Geochim. Cosmochim. Acta, 40,* 241-246, 1976.

Ottman, J.D., Hydrodynamic entrapment of petroleum in the "J" Sandstone of the Denver Basin, northeastern Colorado, *MS Thesis, Univ. of Houston,* 99p, 1981.

Parrish, J.T., Gaynor, G.C., Swift, D.J.P., Circulation in the Cretaceous western interior seaway of North America, a review. In: The Mesozoic of Middle North America (eds. D.F. Stott and D. Glass), *Can. Soc. Petrol. Geol. Mem., 9,* , 221-231, 1984.

Rashkes, J. L., Hydrogen isotope ratios of kaolinites in the Lower Cretaceous Dakota Group and Mowry Shale, north-central Colorado : Evidence for a Tertiary equilibration event, *MS Thesis, Univ. Houston,* 141 pp,1988.

Savin, S.M., Oxygen and hydrogen isotope ratios in sedimentary rocks and minerals, *PhD Thesis, Calif. Inst. Tech.,* 1967.

Savin, S. M. and Epstein, S., The oxygen and hydrogen geochemistry of clay minerals, *Geochim. Cosmochim. Acta, 34,* 25-42, 1970.

Savin, S.M., The history of the earth's surface temperature during the past 100 million years, *Ann. Rev., 5,* 319-355, 1977.

Shackleton, N.J. Oxygen isotope analyses and Pleistocene temperatures reassessed, *Nature, 215,* 15-17, 1967.

Sheppard, S.M.F., Nielsen, R.L., and Taylor, H.P., Oxygen and hydrogen isotope ratios of clay minerals from Porphyry Copper Deposits, *Econ. Geol. 64,* 755-777, 1969.

Telford, P.G. and Verma, H.M., Cretaceous stratigraphy and lignite occurrences in the Smokey Falls area, James Bay Lowland, Preliminary lithological logs from the 1978 drilling program, *Ontario Geol. Surv.Open File Rep., 5255,* 1979.

Tweto, O. Laramide (Late Cretaceous-Early Tertiary) orogeny in the southern rocky Mountains, In: Cenozoic history of southern Rocky Mountains (ed. B. F. Curtis), *Geol. Soc. Amer. Memoir, 144,* 1-44, 1975.

Veizer, J., Fritz, P. and Jones, B., Geochemistry of brachiopods: Oxygen and carbon isotopic records of Paleozoic oceans, *Geochim. Cosmochim. Acta, 50,* 1679-1696, 1986.

Waage, K.M., Stratigraphy and refractory clayrocks of the Dakota Group along the northern Front Range, Colorado, *U.S. Geol. Surv. Bull., 1102,* 1961.

Yurtsever, Y. and Gat, J.R., Atmospheric Waters. In: Stable isotope hydrology. Deuterium and Oxygen-18 in the water cycle. (eds. J. R. Gat and R. Gonfiantini), *IAEA. Vienna Tech. Rept. 210,* 103-142, 1981.

J.R. Lawrence, Department of Geosciences, University of Houston, Houston TX 77204-5503.

J. Raskes Meaux, Browning Ferris Industries, P. o. Box 3151, Houston, TX 77253.

Application of Paleomagnetic and [10]Be Analyses to Chronostratigraphy of Alpine Glacio-fluvial Terraces, Sava River Valley, Slovenia

MILAN J. PAVICH

U.S. Geological Survey, Reston, VA

NATAŠA VIDIC

University of Ljubljana, Ljubljana, Slovenia

Correlation of glacial deposits is a key method in determining the distribution and synchroneity of terrestrial and marine responses to climate change. Many correlations of Quaternary deposits are, however, based on untested assumptions, or forcing of deposition ages into accepted chronologic frameworks. Improved dating methods for terrestrial deposits that pre-date precise ^{14}C dating are needed. Application of isotopic [10]Be dating and paleomagnetic analyses to soils developed on glacio-fluvial terraces has shown that the ages of terrestrial deposits can not be estimated solely by correlation to marine $\delta^{18}O$ record. The estimated ages of the four major glacio-fluvial terraces in the upper Sava River Valley in Slovenia do not correlate to the last four cold $\delta^{18}O$ stages. The oldest two terraces, mapped as Mindel and Günz, are older than Brunhes/Matuyama boundary based on our preliminary data. Although precise ages can not be obtained by these methods, the age-ranges of soils on these deposits can be constrained.

INTRODUCTION

The correlation of marine and terrestrial deposits is one of the fundamental problems in Quaternary geology. Predictions and age estimations for terrestrial deposits are often made based on Milanković's [1941] theory of orbital forcing of global climate. These predictions and estimations should be tested by independent dating methods. Even if the orbital forcing of marine $\delta^{18}O$ cycles is accepted unsceptically as a record of global climate cycles, and as a time scale for Quaternary climate oscillations, the problem of independent dating of terrestrial deposits remains unsolved. [14]C dating provides only a limited time scale, thus there is a crucial need for development of other geochronological techniques for the Pleistocene. Despite current uncertainties about time variation of cosmogenic [10]Be production and delivery, we have measured [10]Be inventories in soils with the goal of evaluating time stratigraphy. This study provides evidence that supports the application of cosmogenic [10]Be to dating terrace soils and paleomagnetic analyses to estimate the time-stratigraphy of soils developed on Quaternary deposits.

One of the areas of classic glacial geomorphology, the upper Sava River Valley (Ljubljana Basin, Figure 1) in Slovenia, was studied to illustrate why even imprecise ages can provide new

evidence that can be considered in discussion of the marine/terrestrial correlation problem. The construction of an Alpine glacial chronology is not simple. As noted by Šibrava [1986, p. 433]: "Many problems in the Alpine areas are derived from surviving quadriglacial concepts which contrast in general with generally accepted polyglacial schemes."

In the "worst case" scenario, each major Alpine valley could have a unique sequence of surviving moraines and glacio-fluvial terraces because of the variations of glacial sediment flux, interglacial-glacial weathering and erosion, fluvial response to varying uplift rates, or local controls on ice accumulation. Attractive features of the studied area were the distinct glacio-fluvial terraces mapped and described by Šifrer [1969], and the uniformity of the sandy gravel parent material for the terrace soils [Stritar, 1969]. The terrace gravels lack both volcanic ash and lava flows, and they cannot be easily dated by standard isotopic methods. To circumvent these problems we resorted to less definitive and relatively untested methods, [10]Be analyses and paleomagnetism of post-incisive chronosequence [Birkeland, 1984] of soils developed on the deposits.

The technique of accelerator mass spectrometry (AMS) measurement of [10]Be is straightforward, and the principles describing cosmogenic production, delivery and adsorption of the isotope onto soil exchange complexes are relatively well known [Lal and Peters, 1967; Monaghan and al., 1986; Pavich et al.,1986]. [10]Be accumulates over time in surface soils exposed to precipitation. [10]Be deposition fluxes are observed to vary linearly

Climate Change in Continental Isotopic Records
Geophysical Monograph 78

Fig. 1. Location of the Sava River in Europe. The field area is designated by the black square.

with annual - average precipitation fluxes [Monaghan et al., 1986; 1992]. Once infiltrated into the soil, [10]Be is adsorbed tightly onto the clay fraction. Its inventories (atoms/cm[2]) increase with the ages of well preserved morphostratigraphic units [Pavich et al., 1986]. Thus, [10]Be inventories accumulated since the initiation of pedogenesis can be compared with the age of the deposit. Uncertainties in calculating ages for long time scales are affected by variations in production and delivery rate ($\pm$ 20%) and loss of [10]Be by leaching or erosion from the soil.

Paleomagnetic analyses of soils are less straightforward. Soils generally acquire their magnetic remanence in a complex manner [Patterson and Larson, 1990]. In most cases the total remanence of surface soils is a resultant of at least two components, depositional (DRM) or post-depositional (PDRM) remanent magnetization or, and chemical remanent magnetization (CRM). DRM, which is carried by the detrital magnetic grains, is acquired shortly after deposition of the sediments. Sediments can get remagnetized afterwards, and acquire post-depositional remanent magnetization (PDRM) [Verosub, 1977].

CRM generally forms by weathering and pedogenesis. Fine grained iron oxides form secondarily by weathering of iron-bearing silicate minerals.   Clay-size iron oxide grains can be easily mobilized, transported and reprecipitated, but usually their size is below the critical grain size (that is, they are superparamagnetic and not ferrimagnetic), so they are unable to carry remanence. If their size continues to increase, the grains will eventually acquire stable CRM when their size exceeds the critical blocking volume [Tarling, 1983].

These components, isolated by proper magnetic cleaning techniques, can yield important information about the direction of Earth's magnetic field at the time of sediment deposition and during subsequent pedogenesis. The main purpose of present study was the attemt to improve the chronostratigraphy of the terrace deposits using [10]Be dating and paleomagnetic analyses of surface soils. A secondary goal was to test how the terrace ages correlate with other Alpine Stratigraphic sequences and the marine oxygen isotope record.

Previous Work

The terraces were mapped and described by Šifrer [1969] and the post-incisive chronosequence [Birkeland, 1984] of the soils developed on the deposits identified and described by Stritar

[1969]. Vidic et al. [1991] presented new data on the Quaternary geology and soils in the upper Sava Valley. Initial age assignments in that study were based on the assignments of previous authors [Penck and Brückner, 1909; Šifrer, 1969] and Milanković's astronomical theory [1941], supported by the correlations of European glaciations with the deep-sea record presented by Šibrava [1986]. All the terraces were considered to be younger than 700 ka, and were assigned the following ages: Würm terrace - 30 ka, Riss terrace - 140 ka, Mindel terrace - 425 ka, and Günz terrace - 575 ka. The names have been and are used as morphostratigraphic and not chronostratigraphic terms, as recommended by Kukla [1975]. Despite the recognition of polyglacial sequences elsewhere, for the purposes of this paper, we use the original morphostratigraphic terms to correspond with earlier studies of the area [Penck and Brückner, 1909; Šifrer, 1969].

Vidic et al. [1991] described the soils and their geochemistry. Differences in soil properties because of the progressive alteration of parent material were observed [Vidic et al., 1991]. Soil taxa range from Mollisols on the youngest Würm terrace, to Alfisols on the Riss, and Ultisols on the Mindel and Günz. Soil formation on the carbonate gravel deposits has been strongly influenced by solution and leaching of the limestone and dolomite pebbles (ca. 80% of deposit), and accumulation of insoluble residues. Dominant sources of silt and clay minerals of soils on such deposits are the weathering of the siliceous components of the deposits. No signs of eolian deposition were observed in the field. Major element and mineralogical analyses confirmed the parent material uniformity observed in field [Vidic et al., 1991].

The hypothesis of the in situ residual soil formation is fundamental to the concept of the terrace age being related to soil properties, and to the modeling of the relation of soil properties versus time. This working hypothesis, supported by field observations [Vidic et al., 1991] is an important basis for the present work.

METHODS

Field Methods

Sampling for geochemical and isotopic analyses. Two to three complete soil profiles were sampled on highest, least eroded parts of major terrace surfaces (Table 1). An average sample was collected from each described soil horizon.

Samples for ^{14}C analyses were obtained by hand auger by Anton Šercelj from a lacustrine deposit on Ljubljansko Barje, south of Ljubljana, where late Würm-Holocene palynology has been previously investigated [Šercelj, 1966]. The stratigraphically lowest ^{14}C sample was from a sandy clay 20cm above the youngest gravel. Palynology indicates that the local climate was cold in that stratigraphic interval. The gravel beneath this site is assumed to correlate with the youngest Würm terrace gravel.

Sampling for paleomagnetic analyses. Three sampling sites were chosen for the pilot study, two on the oldest, Günz terrace (G1, G2), and one on Mindel terrace (M2) (Figure 2). Two to three samples were collected from each soil horizon, 14 to 18 samples from each soil profile. Profile G1 was sampled to the depth of 1.5 m, profile G2 to the depth of 2 m, and profile M2 to the depth of 4.6 m. Plastic cubes, 1.9 cm on a side and 6.86 cm³ in volume, were pressed into moist soil and extracted with the aid of a knife.

TABLE 1. SOIL SAMPLING SITE LOCATIONS

Site	Terrace	Elevation	Location
M12	Würm III	485 m	Lesce, 0.7 km E from the river Sava
M13	Würm II	500 m	Lesce, 0.9 km E from Sava
M11	Würm I	495 m	Vrbnje, 1.2 km NE from Sava, near Würm moraine
M5	Würm I	425 m	Senožet, 1.5 km NE of Sava
M6	Würm I	428 m	Jurčkovo Polje, 1 km NE from river, near the N edge of 415-425 m terrace
M4	Riss	435 m	Senožet, 1.8 km NE from river
M9	Riss	500 m	Terrace of the tributary Tržiška Bistrica, 0.5 km W from Tržiška Bistrica, 4 km from Sava
M1	Mindel	360 m	Cerkovnica, 5 km NE from Sava
M2	Mindel	420 m	Pivka, 1.2 km NE from Sava
M10	Mindel	525 m	Terrace of the Tržiška Bistrica near Kovor, 1.5 km W from Tržiška Bistrica, 3.7 km NE from Sava
M14	Mindel	428 m	Gorenje Polje, 0.25 NE from Sava
M7	Günz	367 m	Žejski hrib, 0.5 km W from Sava
M8	Günz	570 m	Hudo, terrace of Tržiška Bistrica, 1.5 km from Tržiška Bistrica, 4.2 km NE from Sava.
G1	Günz	610 m	Terrace of Tržiška Bistrica, 2.3 km from Tržiška Bistrica, 5 km NE from Sava
G2	Günz	440 m	Strahinj, 2.5 km NE from Sava

After extraction, a plastic cover was fitted over the open end. Orientation of samples was determined using Brunton compass and hand level.

Laboratory Methods

Isotopic analyses. ^{14}C dates of samples from the core on Ljubljansko Barje were determined by accelerator mass spectrometry at the Isotrace Laboratory at University of Toronto.

Cosmogenic ^{10}Be was analyzed by accelerator mass spectrometry at the University of Pennsylvania Accelerator Mass Spectrometry Lab, and the Purdue University PRIME Lab. Beryllium was extracted from soil samples by the methods described by Brown et al. [1988].

Paleomagnetic analyses. Magnetic minerals were extracted from the soil dispersed in distilled water with a rare-earth ceramic magnet. They were made into a grain mount that was polished so that it could be observed by reflected light at magnification of 400x.

Curie temperature measurement was made on the same opaque grain separate by means of Curie balance. A small amount of separate was heated in a strong field, and the decay of saturation magnetization with increase of temperature was observed.

After measurement of natural remanent magnetization (NRM), the samples were subjected to various levels of magnetic cleaning. All paleomagnetic measurements were made at University of Colorado, using Schoensted SSM-1A spinner magnetometer, with

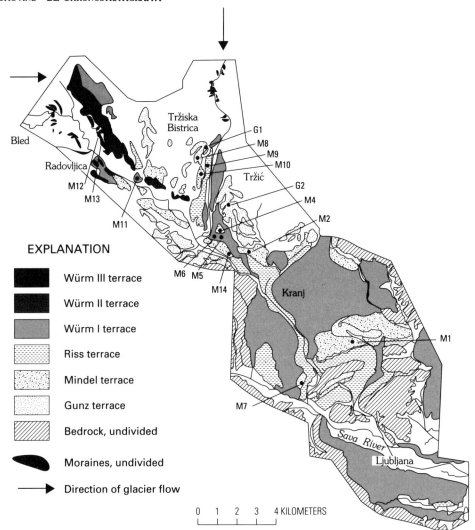

Fig. 2. Map of Quaternary deposits of the Sava River Valley showing sampling localities.

sensitivity level of 1x10⁻⁴ A/m and measurement error angles 95 for repeat measurements of 1-2 degrees. After the NRM was determined, all samples were subjected to demagnetization in progressively increased alternating magnetic field at peak fields of 5, 10 and 15 mT. Subsequently all samples were subjected to stepwise thermal cleaning.

In preparation for heating, the plastic cubes were opened, and placed in field-free space to allow the samples to dry out. After drying and shrinking the samples were easily removed from the plastic containers. Some of the samples were friable, and they required impregnation by water glass "Sauereisen" before the removal from the boxes. Thereafter, the samples were thermally demagnetized at temperatures of 150, 250, 350, 450, 550, 590, 610 and 640°C.

RESULTS

¹⁰Be analyses of soils

Table 2 presents ¹⁰Be concentrations of indivdual soil horizons and inventories of the soils. Inventories of ¹⁰Be are calculated

following the methods of Monaghan et al. [1983] and Pavich et al. (1986). To estimate minimum residence times of the soils we use the equation:

$$t = -1/\lambda \ \ln[1 - \frac{\lambda N}{q}] \qquad (1)$$

where t = age of soil, λ is the decay constant of ¹⁰Be, N is the ¹⁰Be inventory per cm², and q is the deposition flux of ¹⁰Be, we assume N = 0 at t = 0. The inventories for the best preserved sites increase progressively with relative age between the Würm and Mindel terraces, but decrease for the oldest terrace (Günz). Reduction of the isotope inventory by surface erosion was noted by Pavich et el. [1986] to be significant on a morphostratigraphic units older than 500 ka.

Comparison of ¹⁴C age of lacustrine sediments with ¹⁰Be age of the youngest Würm terrace.

Translating the ¹⁰Be inventories into assigned ages requires assumption of a constant deposition rate. The independently

TABLE 2 RESULTS OF [10]BE ANALYSES OF SOILS

WÜRM TERRACES

Site	Depth cm	Horizon	Atom/g x 10^6	ATOM/CM^2 X 10^{11}
M12	0-12	Ap	910	0.14
	12-24	AB	934	0.15
	24-33	(B)r	965	0.11
	33-45	C1	91	0.02
			Σ =	0.42
M13	0-12	Ap	1170	0.18
	12-25	AB	1049	0.18
	25-44	(B)r	706	0.17
	44-60	C	172	0.04
			Σ =	0.57
M6	0-16	A	369	.07
	16-38	Bt1	328	.09
	38-58	Bt2	323	.08
	58-80	Bt2	504	.14
	80-110	C1	134	.05
	110-135	C2	68	.02
			Σ =	0.47
M5	0-22	A	--	--
	22-41	Bt1	669	0.17
	41-122	Bt2	303	0.32
	122-172	Bt3(C1)	84	0.05
	172-200	Bt4(C2)	43	0.02
			Σ =	0.56
[a]M11	0-17	Ap	1300	0.27
	17-30	Bt1	1200	0.19
	30-43	Bt2	800	0.13
	43-55	BC1	1320	0.19
	55-63	C1	220	0.02
			Σ =	0.80

RISS TERRACES

Site	Depth cm	Horizon	Atom/g x 10^6	ATOM/CM^2 X 10^{11}
M9	0-61	Ap	946	0.37
	30-50	A/B	702	0.18
	50-90	Bt1	1101	0.57
	90-115	Bt2	1853	0.60
	115-185	Bt3	1500	1.37
	185-275	Bt4	1500	1.76
	275-365	C	1500	1.76
			Σ =	6.61
M4	0-61	A	--	--
	61-125	Bt1	706	0.59
	125-160	Bt2	2457	1.12
	160-192	Bt3	1107	0.46
	192-234	Bt4	2003	1.10
	234-309	Bt5	2115	2.06
			Σ =	5.33

MINDEL TERRACES

Site	Depth cm	Horizon	Atom/g x 10^6	ATOM/CM^2 X 10^{11}
M2	0-12	E	983	0.15
	12-19	Ab	--	--
	19-40	Bt1	1045	0.29
	40-59	Bt2	2277	0.56
	59-86	Bt3	2200	0.77
	86-164	Bt4	2118	2.15
	164-234	Bt5	2189	2.00
	234-329	BC1	2555	2.90
	329-429	BC2	1796	2.33
			Σ =	11.15
M10	0-36	A	867	.44
	36-70	Bt1	925	.31
	70-110	Bt2	1361	.76
	110-160	Bt3	1446	1.01
	160-190	Bt4	1407	.59
	190-224	Bt5	1687	.80
	224-340	Bt6	1606	2.61
	340-365	Bt7	1491	0.52
			Σ =	7.04
[a]M14	0-7	A	360	0.03
	7-30	B	870	0.26
	30-48	Bt1	807	0.19
	48-73	Bt2	880	0.28
	73-90	Bt3	903	0.21
	90-135	Bt4	1370	0.81
	135-196	Bt5	904	0.72
	196-233	Bt6	450	0.22
	233-269	Bt7	1713	0.81
			Σ =	3.53

GÜNZ TERRACE

Pedon	Depth cm	Horizon	Atom/g x 10^6	ATOM/CM^2 X 10^{11}
M8	0-30	Ap	--	-
	40-64	Bt1	1469	0.49
	64-87	Bt2g	1042	0.34
	87-118	Bt3g	1153	0.50
	118-174	Bt4g	1227	0.96
	174-226	Bt5g	1135	0.83
	226-320	Bt6g	884	1.16
	320-340	Bt7g	2393	0.67
			Σ =	4.95

[a] Data from Purdue PRIME Lab

determined ^{14}C age of the youngest terrace can be used to check previously calculated deposition rates. Even using this age to constrain the average [10]Be deposition rate, precision of the age assignments on Table 3 is probably ± 20% due to uncertainties of variance of [10]Be production rate, of [10]Be delivery by precipitation and erosional depletion. The youngest Würm terrace is assumed to be contemporaneous with the latest maximum cold period recorded in palynological analyses of a lake core [Šercelj,

TABLE 3 Age Assignments Based on [10]Be and Paleomagnetic Analyses

Morphostratigraphic unit	Terrace No. (Šifrer, 1969)	Age Assignment (method)	$\delta^{18}O$ stage [1,2]
Würm III	IV	$\geq$20 Ka (^{14}C), $\geq$32 Ka ([10]Be	2
Würm II	III	$\geq$44 Ka	?
Würm I	II	$\geq$44 Ka, $\geq$62 Ka ([10]Be)	4?
Riss	I	$\geq$450 Ka ([10]Be)	12
Mindel	IA	$\geq$780 Ka and $\leq$ 990 Ka(paleomag)	24?
		$\geq$1000 Ka ([10]Be)	26?
Günz	IB	$\geq$780 Ka (paleomag)	
		Jaramillo? (0.99-10.7 Ma; paleomag)	28?, 30?
		Olduvai? (1.77-1.95 Ma; paleomag)	64-70?

1) Shackleton and Opdyke, 1973
2) Shackleton et al, 1990

1966]. A disseminated organic sample from the same core in Ljubljansko Barje yield a ^{14}C date of 20,310 + 400 years B.P. [Vidic et al., 1991] for lacustrine sediments deposited prior to the transition from a cold-climate to warm-climate pollen assemblages. A Würm III correlative with deep-sea oxygen isotope stage 2 (32-13 ka), [Shackleton and Opdyke, 1973] would be consistent with this age.

The [10]Be inventory of soil profile M12 on Würm III terrace is 4.2×10^{10} atoms/cm². Using a nominal deposition rate of 1.2×10^6 (100 cm/year rainfall) reported by Monaghan et al. [1986], the minimum age of soil profile M12 calculated by equation 1 is 32 ka. This supports the conclusion that Würm III terrace is older than the youngest cool-climate episode recorded by pollen assemblage of lacustrine clay in Ljubljansko Barje, dated at 20,310 ± 400 years B.P.[Vidic et al. 1991]. Our calculated age of $\geq$ 32 ka for Würm III terrace is in reasonable agreement with the morphostratigraphic evidence that the site is early stage 2.

Alternatively, we could assume that this terrace (represented by profile M12) was deposited during youngest part of stage 2, divide the [10]Be inventory by ^{14}C age of 20 ka, and obtain a deposition flux of 2.1×10^6 atom/cm² year. This value is about 60% higher than the values calculated by Monaghan et al. [1986] and Brown et al. [1988] for 100 cm rainfall, but proportional to the present rainfall of 140 to 170 cm/year. We do not know if the present rainfall is representative of the past 20 ka. Palynological records of the area [Šercelj, 1966] for the last 20 ka do not indicate periods of aridity or major climatic variation.

We see no evidence that the deposition of [10]Be has been higher than the global average rate because of eolian deposition. Eolian dust contribution from the Sahara which has been observed elsewhere in the Mediterranean region [Pye, 1992], or another sediment source, would probably contain a relatively low concentration of the isotope, and decrease rather than increase the apparent deposition rate. Some of the [10]Be in the soil could be inherited from parent material. Low concentrations of the isotope in C-horizons of younger soils compared to other soil horizons is evidence against significant inheritance.

We do not have other independent evidence to assess the age of M12. It could be between 20ka to 32ka. In either case, the correlation with stage 2 is valid.

Paleomagnetic Studies of Günz and Mindel Terraces

The initial hypothesis that all of the terraces were younger than Brunhes/Matuyama boundary (780 ka), [Izett and Obradovich, 1991], based on the correlation of Penck and Brückner's model with Milanković's cycles [Eberl, 1930], was tested by paleomagnetic analyses of soils. Oriented samples were collected from the oldest two terraces.

Magnetic minerals. The detrital fraction of magnetic minerals was observed in grain mount from Bt horizon of the oldest soil. Magnetite grains with ilmenite lamellae along the octahedral planes were common. These grains indicated a volcanic source, in which the lamellae intergrowth formed because of high-temperature oxidation during initial cooling. The other major iron-oxide mineral in the separate was martite, a pseudomorphous replacement of magnetite by hematite. Martite was probably derived from Permian red beds in the source area. Fine-grained hematite was observed as replacement of labile iron-bearing silicate grains. Curie temperature of the same separate was 560°C, indicating that the most magnetic mineral in the sample was nearly pure magnetite.

Magnetic minerals in soils on the Mindel terrace have not been examined yet. A smaller decrease in intensity during the AF demagnetization, however is consistent with the idea that more of the remanence is carried by detrital hematite.

Paleomagnetism of soils on the Günz terrace. The direction of present geomagnetic field of the area is D=0°, I=60°, downward. The NRM directions of samples below 30 cm of depth from profile G1 (Figure 3) on this oldest terrace were grouped around a mean declination of 30° and a mean inclination of 55° downward, close to the expected axial dipole field for Brunhes epoch. AF cleaning to 15 mT resulted in little change in the directions; but the average intensity however, dropped by about 50%. Thermal cleaning was much more effective in producing the directional change, although most of the directions remained in normal polarity. Over some of the temperature intervals, several directions moved southward, but after heating at higher temperatures they returned to the north. Mean directions of demagnetization steps of this group of samples remain in normal polarity (Figure 3b).

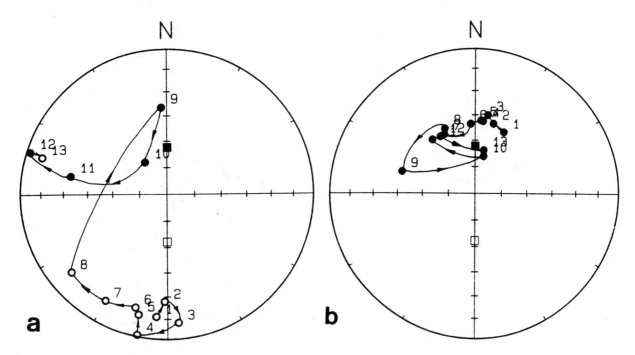

Fig. 3. Mean directions of two sets of samples from below layer 30 cm (30-150cm). Normal component seems to be stronger, probably because of the Brunhes normal overprint, keeping the inclinations of reversed samples shallow upward. Means of reversed group of samples migrate towards north during the demagnetization, suggesting that the sediment was deposited in a normal period, and secondarily overprinted with the reversed component. Natural remanent magnetization (NRM) intensities of samples were between 10^{-1} and 10^{-2} A/m orders of magnitude. Numbers denote progressive demagnetization steps: 1- NRM, 2-5 mT, 3-10 mT, 4-15 mT, 5-150°C, 6-250°C, 7-350°C, 8-450 o C, 9-550°C, 10-590°C, 11-610°C, 12-625°C, 13-640°C. Modern diaxial field direction is indicated by a square; solid-downward inclination, open-upward inclination.

Another group of samples collected form a surface layer 0-30 cm within the same soil profile (G1), in part displayed reversed component. Two samples out of five from this layer exhibited clearly reversed NRM directions and remained reversed up to the moderately high temperatures. Above 450°C, these two samples displayed normal directions (Figure 3a). Our interpretation of these results is that the normal polarity at high temperatures was carried by detrital hematite and acquired at the time of terrace deposition. Subsequently, during pedogenesis a secondary reversed component was acquired, followed by a secondary normal overprint of the Brunhes period.

Samples from another soil profile on this terrace, G2, displayed similar behavior (Figure 4). Most of the samples posses a clearly normal-polarity NRM direction. During stepwise thermal demagnetization, most directions move southward, and some return to a normal polarity during the final stages. Strongly reversely overprinted samples from the near surface are lacking at this locality, probably because of more intense surface mixing or erosion.

Mindel terrace. The NRM directions of samples from this terrace were of normal polarity; the mean declination was about -5° and inclination about 36°. During AF cleaning, some samples moved toward the south, but essentially all retained a downward

inclination. During intermediate stages of thermal cleaning several samples displayed intermediately reversed and reversed directions. At 640° C the mean direction was shallow reversed, at a mean declination of 193° and inclination of -5° (Figure 5).

Age Assignments Based on New Data

The summary of provisional age assignments based on ^{14}C, ^{10}Be inventories, and paleomagnetic analyses is presented in Table 3. The Würm terrace ages based on ^{14}C and ^{10}Be agree with previous age models [e.g. Milanković, 1941]. A large time gap exists between the oldest Würm terrace and the Riss terrace, which appears to be older than 450 ka based on ^{10}Be inventories of soils on this terrace. The minimum age for the Mindel terrace is probably the Brunhes/Matuyama boundary most recently dated at 780 ka, [Izett and Obradovich, 1991] based on the paleomagnetic history of soils, and greater than 1 Ma based on the ^{10}Be inventory of site M2. Use of M2 for minimum age assignment is based on the interpretation that the lower inventories at sites M10 and M14 are due to soil erosion, thus the maximum inventory observed at M2 defines the minimum age of the stratigraphic unit. The minimum age for the Günz terrace is about 1 Ma according to paleomagnetic analyses of soils, but it probably dates closer to the Pliocene/Pleistocene boundary, about 1.8 Ma, based on the

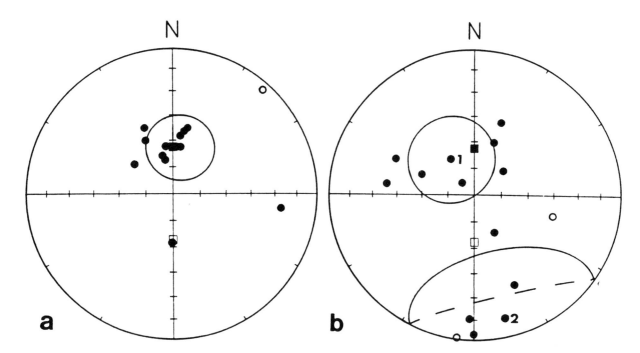

Fig. 4. Two steps in the progressive demagnetization of samples from site G2 showing the southward directional migration of some samples from this locality, which suggests that samples possess mixed components of remanence. a) NRM directions groped around normal mean direction (D=9, I=63 degrees, downward and α_{95} = 18.7 degrees). b) thermal demagnetization at 590°C, where the directions of samples fall into two groups: 1) normal group, around mean direction of D = 33, I = 65 degrees and α_{95} = 24.9 degrees. 2) shallow reversed group, around mean direction of D = 165, I = 14 degrees, downward, α_{95} = 42.5 degrees. Ellipses represent α_{95} cones of confidence. Normal component seems to be much stronger than reversed, keeping the downward or shallow upward inclinations of the samples. Strong normal component is reflected in means of the demagnetization steps, calculated from all samples from the locality, which remain in northern hemisphere throughout all demagnetization steps. The samples with intermediate directions migrating south through demagnetization provide evidence that this locality was exposed to the reversed magnetic field at some point of its history. NRM intensities of samples were in 10^{-2} A/m order of magnitude. Numbers denote progressive demagnetization steps: 1-NRM, 2-5 mT, 3-10 mT, 4-15 mT, 5-150°C, 7-350°C, 8-450°C, 9-550°C, 10-590°C, 11-610°C, 12-625°C, 13-640°C. Modern diaxial field direction is indicated by a square; solid-downward inclination, open-upward inclination.

interpretation of paleomagnetic data and >1 Ma age assignment for the Mindel terrace. Despite of lack of precision of these dates, the important result of these studies is that the time scale of deposition of studied deposits is greatly expanded compared to previous estimates.

DISCUSSION

The expansion of the time scale of deposition of the Sava Terraces is a significant revision of the accepted glacial time stratigraphy of this valley. As a check on the isotopic and paleomagnetic data, we can consider the relations of soil development to time. The studied soils are considered to be residual and no significant deposition of eolian sediment (e.g. loess derived from glacial deposits or Saharan dust) was observed [Vidic et al., 1991]. This is important, because the interpretation of the relations between soil characteristics (such as thickness, chemistry and mineralogy) and soil age is made simpler by choosing the best

preserved, least eroded residual sites on each terrace. These are not necessarily representative of the average site [Pavich et al., 1986], since the average site, particularly on older terraces, may be significantly thinned by erosion of A, E and upper Bt horizons. The justification for using the best preserved sites is illustrated by the age trend plots of various soil characteristics (Figure 6 and 7). These plots summarize several important results from this study, which require further discussion.

Pedogenesis As a Function of Time - Evidence for Residual Soil Formation

The graph of solum thickness versus age based on ^{10}Be inventories (Figure 6) and paleomagnetic analyses shows a log-linear relation for the best preserved, least eroded sites on the terraces. This systematic relationship could be due to chemical adsorption of ^{10}Be, mainly on Fe-oxides, Mn-oxides, and clay minerals but it also indicates that the processes that control

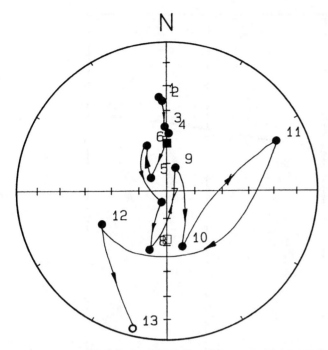

Fig. 5. Mean directions of samples from soil profile M2, Mindel terrace, displaying southward migration of the means. Reversed remanence component acquired shortly after the deposition was overprinted by normal of Brunhes period. NRM intensities were between 10^{-1} and 10^{-2} A/m orders of magnitude. Numbers denote progressive demagnetization steps: 1-NRM, 2-5 mT, 3-10 mT, 4-15 mT, 5-150°C, 6-250°C, 7-350^W°, 8-450°C, 9-550°C, 10-590°C, 11-610°C, 12-625°C, 13-640°C. Modern diaxial field direction is indicated by a square; solid-downward inclination, open-upward inclination.

thickness and ^{10}Be inventory may be consistent over a relatively long time period. The simple log-linear regression supports the argument that the soils develop by in situ processes. Carbonate dissolution and alteration of alumosilicate and iron silicate minerals of parent material are the processes of soil formation documented from field observations and laboratory data [Vidic et al., 1991]. A model for soil genesis will not be presented in this paper, but it should be stressed that a significant eolian or slope deposition of fine grained material would be expected to significantly alter the simple observed relationship. The in situ weathering source of silt and clay from carbonate-rich parent materials is supported by studies in Italy [Moresi and Mongelli, 1988] and on Crete [Pye, 1992]. Pye [1992] presents results from Crete that show that in situ bedrock weathering dominates over eolian inputs as the source of non-carbonate minerals for soil formation. Soil thicknesses and ^{10}Be inventories can be used to define the relative degree of preservation of sites on a chosen morphostratigraphic unit.

The graph of clay mass versus age (Figure 7) provides evidence for textural change related to mineralogical alteration of alumosilicate parent minerals such as feldspars and micas. Loss of $CaCO_3$ by dissolution leaves an alumosilicate plus quartz residuum that produces silty-clay soil textures. The simplicity of this

textural trend is consistent with the hypothesis of simple, in situ pedogenic processes as the mechanism of clay accumulation. Pye [1992] noted that recent mean annual dust deposition fluxes on Crete are well below the fluxes required to produce distinct loess layers. If eolian accumulation over 1 Ma were the primary source of these terrace soils, then a more linear trend of soil thickness and clay mass versus age might be expected. Eolian input of fine-silt and clay without erosion would not result in the observed trend of increasing clay mass with age.

Geomorphic evidence (e.g., increasingly defined karst topography on older terraces) also contradicts a significant high-magnitude, episodic eolian flux to these soils [Vidic et al., 1991]. Thick loess deposited during the most recent glacial period on the older landscape would be expected to cover or obscure the karst topography.

Age Assignments Based on Paleomagnetic Data

Paleomagnetic directions of the samples are correlated to certain chrons in the magnetic polarity timescale. The geomagnetic polarity time scale for Cenozoic is well defined [Berggren et al., 1985], but the ages of some Quaternary reversals have been recently revised based on laser-fusion 40Ar/39Ar dating [Baksi, 1991; Walter et al., 1991, Walter et al., 1992], and correlation to orbital variation cycles [Johnson, 1982; Shackleton et al., 1990; Hilgen, 1992]. Astronomically calibrated ages imply that K/Ar radiometric dating yields ages that are consistently too young by 5-7% [Hilgen, 1992]. We try to present both age ranges in this paper.

Günz terrace. On the basis of the paleomagnetic data, discussed in Results section, we concluded the following succession of events:
1) deposition of terrace sediment occurred during a normal polarity chron prior to the Brunhes,
2) the relatively strong reversed remanence component was probably acquired during Matuyama chron (0.73-2.47 Ma), [Berggren et al., 1985; 0.78-2.59/2.62 Ma; Hilgen, 1992], 3) a second overprinting occurred during the Brunhes normal chron. The duration of the reversed period between the normal polarity Jaramillo Subchron (0.91-0.98 Ma); [Berggren et al., 1985]; (0.99-1.07 Ma), [Shackleton et al., 1990] and the Brunhes/Matuyama boundary (0.73 Ma); [Berggren et al., 1985];(0.78), [Shackleton et al., 1990] may be too short for acquisition of the relatively strong reversed remanence component observed during demagnetization. It is possible that deposition of the gravel occurred during the earlier normal polarity Olduvai subchron (1.66-1.88, Ma), [Berggren et al.]; (1.78-1.98 Ma), [Shackleton et al., 1990, Walter et al., 1991].

Although several alternate interpretations of the data are possible, we conclude that Günz terrace is at least older than the Brunhes/Matuyama boundary (O.78-0.79 Ma); [Johnson, 1982; Shackleton et al., 1990]. More likely, it is older than 1 Ma, and possibly as old as 1.8 Ma. The 1.8 Ma age is supported by the best statistical fit of time dependent soil properties such as solum thickness and clay-mass (Figures 6 and 7).

Mindel terrace. The paleomagnetic data from soils on the Mindel terrace indicate that:

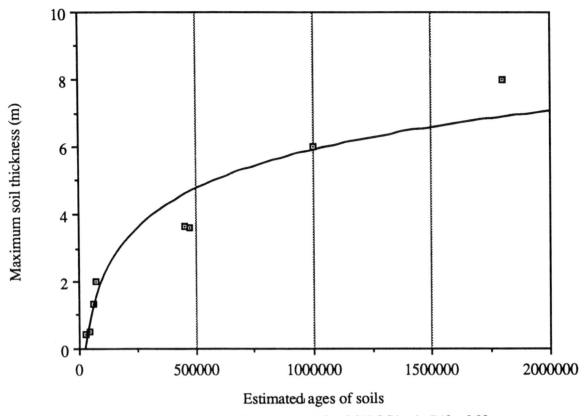

Fitted curve: Thickness= - 17.0 + 3.2*LOG(age); R^2 = 0.93

Fig. 6. Plot of solum thickness versus age is showing the systematic relationship between age and thickness. Improved age assignments used in this plot are listed in Table 3. The age estimate used for GÜnz is 1.8 Ma, based on 1 Ma estimate for Mindel.

1) the gravel was deposited during the Matuyama chron, probably after the Jaramillo subchron;
2) the section was overprinted during the normal polarity Brunhes chron.

The Mindel terrace was formed prior to the Brunhes/Matuyama boundary (>0.78 Ma) and probably after the Jaramillo subchron (ca. 0.99 Ma); [Shackleton et al., 1990]. The >1 Ma [10]Be date supports the paleomagnetic evidence that this terrace pre-dates the Brunhes/Matuyama boundary age.

Comparison of Paleomagnetic and [10]Be Data

The [10]Be inventories are consistent with increasing age of the morphostratigraphic order of terraces up to the Mindel terrace. The oldest, Günz terrace does not show the expected increase of [10]Be inventory. This lack of increase is most likely due to surficial erosion on the small, fluvially dissected remnants of the Günz. Simple chemical leaching is unlikely due to the very high distribution coefficient of adsorbed beryllium in soils of pH $\geq$ 5 [Pavich et al., 1986]. Movement of [10]Be complexed on manganese-rich colloids is a possible means of transport of [10]Be through the soil as discussed by [Pavich et al., 1989]. In either case the [10]Be inventories are partly dependent upon the preserved solum thickness.

Paleomagnetism, on the other hand, is not likely to be dependent upon solum thickness, or soil preservation, and it improves the age assignments for Mindel and Günz terraces, where it is difficult to find a completely preserved site required for [10]Be dating.

Comparison with Other European Time Stratigraphic Information

It is reasonable to consider the possible regional correlation of glacio-fluvial terraces. [Schluchter 1986, p.414] based on studies in Switzerland suggests the use of: "... the terms Günz, Mindel, Riss and Würm as informal, relative chronostratigraphic terms for the third and second before last glaciation, respectively."

Würm terraces. The youngest gravel deposit (Würm III) is probably older than the last cold-warm climate transition which according to pollen analyses [Šercelj, 1966] occurred after 20,310 $\pm$ 400 years B.P. based on ^{14}C analysis of plant fragments in the core of lacustrine sediments in Ljubljansko Barje [Vidic et al., 1991]. The dated zone is about 60 cm below the transition from a dominantly pine and grass cool-climate assemblage to a mixed pine and hardwood temperate-climate assemblage. Late glacial

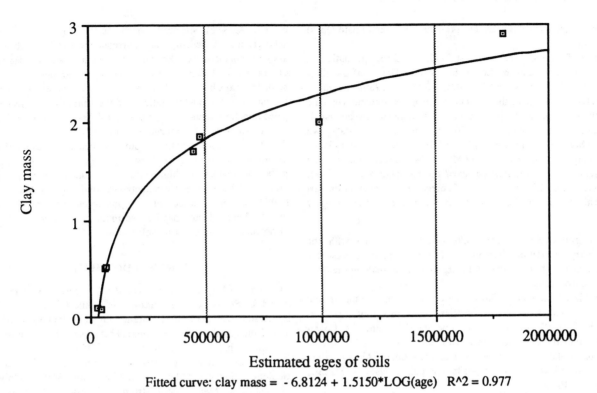

Fitted curve: clay mass = - 6.8124 + 1.5150*LOG(age)   R^2 = 0.977

Fig. 7. Graph of clay mass versus age shows progressive textural change with age, and provides evidence against significant eolian deposition.

deposition of gravel therefore predates 20 ka. No other ^{14}C or archeological dates are available to better constrain the age of the deposit.

A ≥32 ka age for the M12 site (Würm III) based on [10]Be inventory correlates with other regional ages for the onset of stage 2 [Shackleton and Opdyke, 1973] and supports the assumption that the gravel was deposited during, and not after, the major Alpine glacial advance period [c.f. Schluchter, 1986]. This age may correspond to the Würm 2 (meso-Würm) of the Jura range [Campy, 1986]. The 20 ka old Ljubljansko Barje lacustrine sediment is slightly younger than Würm 3 of Campy (22,400 ± 500 yrs B.P.); [Campy, 1986]. In the Jura, Würm 3 deposits are separated from Würm 2 (meso-Würm, period of maximum glaciation and sediment flux) by a period of pedogenesis. The base of Würm 2 is dated by ^{14}C as 29,500 ± 1400 yrs B.P. [Campy, 1986]. Würm III in the Sava Valley may thus be related to a regional period of fluvioglacial sedimentation at about 30 ka. The ≥44 ka age calculation for profile M13 on Würm II, and ≥62 ka for M11 on Würm I, may also correspond with the older Würm 1 identified by Campy [1986] in karst sediments of the Jura, which pre-date stage 2.

It is important to note that among the analyzed Würm terrace sites, the [10]Be inventories are not simply dependent upon solum thickness. Thus, carbonate dissolution and aluminosilicate accumulation may proceed at rates that are somewhat independent of [10]Be input over a time range of 30,000 years (i.e. 30ka to 60 ka). The terrace age assignments are based on the assumption that

the soil sampling sites represent the maximum profile development on that particular terrace. Further sampling and analyses may prove the age differences are erroneous.

The range of Würm ages calculated from the [10]Be data is reasonable for the region studied. Schluchter [1986] pointed out that the age of the last glacial maximum is still unclear. There is increasing evidence, however, that the maximum glacial advance was early (isotope stage 4, ≥50 ka) and that the stage 2 (~20 ka) was less extensive than stage 4. The map of areas for the Sava terraces (Figure 2) support the argument that older Würm I produced greater sediment flux than the latest, stage 2 (Würm III) cold period. The possible correlation of the Sava Würm II and I with Campy's [1986] Jura karst deposits may be fortuitous, but they do suggest the possibility of a broad regional correlation of the periodicity of Alpine Würm glacial advances.

The problem of large time gap between deposits mapped as Würm and Riss. Our data, summarized in Table 3, indicate that the terrace mapped as Riss, is at least 450 ka old. This is based on two independent soil profiles on this terrace, which give very similar results. Irrespective of the uncertainty about long-term deposition rate of [10]Be, the probable precision of ± 20% [Monaghan et al., 1986] makes it unlikely that this terrace correlates with stage 6 or 8 of deep sea sediments. The evidence that a pre-stage 8 glaciation was a major regional event can be found in other Alpine studies. Schluchter [1986] pointed out that pre-Riss interglacial deposits in Alpine valleys and terraces indicate that a more extensive and erosionally intensive "Big Event" pre-dated the penultimate (Riss) glaciation.

This pre-Riss event may correlate with the Elster glaciation of the Lower Rhine, and the Lower Interterrace Gravel of the Iller-Lech Plain [Brunnacker et al., 1982]. Šibrava [1986] correlated the Elster glaciation in the northern European lowland with the cold stage 12 in the deep-sea curve. The age of this deposit is estimated to be between 440-470 ka. Based on our data, we support Schluchters argument for the "Big Event" predating stage 8, possibly correlative with stage 12, and we suggest that the next younger major gravel flux occurred during stage 4 (64-75 ka), [Shackleton and Opdyke, 1973]. If correct, it appears that the last two major events in the Sava Valley were separated by roughly 400 ka.

Older glaciations. Our results can be compared with the Quaternary correlation chart for the Lower Rhine-Alps presented by Brunnacker et al., [1982]. Our expanded time scale in the Sava Valley shows that:
a) the Riss, Mindel and Günz do not correlate with the major cold cycles of Brunhes chron, b) the Riss terrace probably correlates with the Elster I cold cycle, (isotope stage 12) and the Lower Interterrace Gravel of the Iller-Lech Plain, and
c) the older glacial deposit remnants in this southern Alpine Valley may correlate with older deposits recognized elsewhere [e.g., Donau and Biber].

Our identification of significantly older Pleistocene gravel is important. Šibrava [1986] noted that, although inferred from Shotton's [1982] investigations, no mountain or continental ice deposits have been dated in the Matuyama epoch. Our data indicate that two glacio-fluvial deposits from Matuyama Chron, one of which could be from Olduvai subchron, occur in the Sava Valley.

Completeness of the Terrace Stratigraphic Record. It is possible, that the terraces identified by Šifrer [1969], and put in a time-stratigraphic framework through our work, represent incomplete preservation of all the Quaternary glacio-fluvial deposits that have occupied parts of the Sava River Valley. Older deposits could be obliterated by younger, more extensive deposits as suggested by Gibbons et al. [1984], who also showed that a three-fold terrace sequence is the most probable result of multiple glaciations in an Alpine valley undergoing uplift. A complete understanding of Alpine glacial stratigraphy will require independent dating of soils and deposits in several major river valleys. The resolution of the question of preservation must await correlation between exposed terrace deposits and the subsurface stratigraphy of buried glacio-fluvial deposits in the Ljubljana Basin.

CONCLUSIONS

Based on the results of ^{10}Be and paleomagnetic analyses of soils we conclude:
a) The glacio-fluvial terrace sequence of the upper Sava River in Slovenia (Ljubljana Basin) spans over a time interval much larger than 700 ka, possibly over the whole 1.8 Ma of the Quaternary. Günz and Mindel terraces, once considered to be younger than Brunhes/Matuyama boundary, may be older. However, their exact ages cannot be determined at this point.

b) The ages of specific morphostratigraphic units, including subdivisions of Würm, may correlate regionally with the age assignments made by other authors for major Alpine glaciations.
c) Simple correlation of these terraces ages with marine $\delta^{18}O$ not be made, perhaps because of preservation of an incomplete record of glaciations. For interpretations of the timing and correlation of paleoclimatic events, we cannot simply assume a correlation of such terraces with the marine $\delta^{18}O$ record.
d) ^{10}Be provides relatively simple means of estimating the minimum exposure age of residual soils on terraces. On older units, the results are highly dependent upon sampling the best preserved, least eroded remnants. Beyond about 1 Ma, it seems highly unlikely that completely preserved soil profiles can be found. These old sites may be discriminated, however, on the basis of detailed paleomagnetic analyses.

ACKNOWLEDGEMENTS

Present work was partially funded by Yugoslav-American Joint Board for Scientific Cooperation. We would like to thank Edwin E. Larson, University of Colorado, for considerable contribution to paleomagnetic interpretations. This research was also made possible with the considerable help of Milan Šifrer and Anton Šercelj, Slovenian Academy of Sciences; Ljubo Žlebnik, Geological Survey of Slovenia; and late Albin Stritar, University of Ljubljana. Many people from University of Ljubljana and U.S.G.S. Reston have contributed their time and effort to the present research, and we thank them all.

REFERENCES

Baksi, A.K, Hsu, V., McWilliams, M.O., Farrar, E., ^{40}Ar/^{39}Ar dating of the Brunhes/Matuyama geomagnetic field reversal; implications for the Milanković forcing frequency observed in deep-sea cores; *Geological Society of America Abstracts with Programs*, v. 23, p. A91-A92, 1991.

Berggren, W.A., Kent, D.V., Flynn, J.J., Van Couvering, J.A., Cenozoic geochronology, *Geol. Soc. of America Bulletin*,96/11, 1407-1418, 1985.

Birkeland, P.W., *Soils and Geomorphology*, Oxford Univ. Press, 372 p., 1984.

Brown, L., Pavich, M.J., Hickman, R.E., Klein, J. and Middleton, R., Erosion of the eastern United States observed with ^{10}Be, *Earth Surface Processes and Landforms*, 13, 441-457, 1988.

Brunnacker, K., Hosacher, M., Tillmans, W. and Urban, B., Correlation of the Quaternary terrace sequences in the lower Rhine Valley and Northern Alpine Foothills of Central Europe, *Quaternary Research* 18, 152-173, 1982.

Campy, M., Glaciations in Jura Range, In *Quaternary Glaciations in the Northern Hemisphere*, edited by V. Šibrava et al., Quaternary Science Reviews, 5, 403-406, Pergamon Press, Oxford, 1986.

Eberl, B., *Die eiszeitfloge in nordlichen Alpenforlande*. Dr. Beno Fliser, Augsburg, 1930.

Gibbons, A.B., Megaerth, J.D. and Pierce, K.L., Probability of moraine survival in a succession of glacial advances, *Geology*, 12, 327-330, 1984.

Hilgen, F.J., Astronomical calibration of Gauss to Matuyama sapropels in the Mediterranean and implication for the Geomagnetic Polarity Time Scale, *Earth and Planet. Sci. Lett.,* in press, 1992.

Izett, G. and Obradovich, J., Dating of the Matuyama-Brunhes boundary based on 40Ar-39Ar ages of the Bishop tuff and Cerro San Luis rhyolite, *GSA Annual Meeting Abstracts with Programs*, A106, 1991.

Johnson, R.G., Brunhes/Matuyama magnetic reversal dated at 790,000 years B.P. by marine-astronomical correlations; Quaternary Research, 17, 135-147, 1982.

Kukla, G.J., Loess stratigraphy of central Europe, in *After the Australopithecines*, edited by K.W. Butzer and G.L. Isaac, Monton Publishers, The Hague, 99-188, 1975.

Lal, D. and Peters, B., Cosmic ray produced radioactivity on the Earth, in *Handbuch der Physik*, vol. XLVI/2, pp. 551-612, 1967.

Milanković, M., Kanon der Erdbestrahlung und seine Andwenduref auf das Eiszeitenproblem, *Posebna izdanja Srpske Kraljevski Akademije* 133, Prirodnjaćki in matematićki spisi 33, 633 p., 1941.

Monaghan, M.C., Krishnaswami, S. and Turekian, K.K., The global average production of ^{10}Be, *Earth Planet. Sci. Letters*, 76, 279-287, 1986.

Monaghan, M.C., McKean, J., Dietrich, W., and Klein, J., ^{10}Be chronometry of bedrock-to-soil conversion rates, *Earth Planet Sci. Letters*, III, 483-492, 1992.

Moresi, M. and Monzelli, G., The relation between the Terra Rossa and the carbonate-free residue of the underlying limestones and Dolostones in Apulia, Italy, *Clay Minerals*, 23, 439-446.

Patterson, P.E. and Larson, E.E., Paleomagnetic studies and age assessment of a succession of buried soils in the type section of the Blackwater Draw Formation, northwestern Texas, *Geologic Framework and Regional Hydrogeology: Upper Cenozoic Blackwater Draw and Ogallala Formations, Great Plains*. edited by T.C. Gustavson, 233-244, Bureau of Economic Geology, The University of Texas at Austin, 1990.

Pavich, M.J., Brown, L., Harden, J., Klein, J., Middleton, R., ^{10}Be distribution in soils form Merced River Terraces, California, *Geochem. Cosmo. Act.*, 50, 1727-1735, 1986.

Pavich, M.J., Vidic, N. and Lobnik, F., Geochemistry of soils on glacio-fluvial terraces in the upper Sava River Valley, USGS/JF Project 763 Annual Reports, 1988, 1989.

Penck, A. and Brückner, E., *Die Alpen in Eiszeiten*, Tauchnitz, Leipzig, 1909.

Pye, K., Aeolian dust transport and deposition over Crete and adjacent parts of the Mediterranean Sea, *Earth Surface Processes and Landforms*, 17, 271-288, 1992.

Schluchter, C., The Quaternary Glaciations of Switzerland, with special reference to the Northern Alpine Foreland, in *Quaternary Glaciations in the Northern Hemisphere*, edited by V. Šibrava et al., Quaternary Science Reviews, 5, pp. 413-420, Pergamon Press, Oxford, 1986.

Sercelj, A., Pelodne analize pleistocenskih in holocenskih sedimentov Ljubljanskega Barja, *Razprave 9/9*, 4 raz., SAZU, Ljubljana, 1966.

Shackleton, N.J. and Opdyke, N.D., Oxygen isotope and paleomagnetic stratigraphy of equatorial Pacific core V28-238: oxygen isotope temperatures and ice volumes on a 10_5 and 10^6 year scale, *Quaternary Research* 3, 39-55, 1973.

Shackleton, N.J., Berger, A., Peltier, W.R., An alternative astronomical calibration of the lower Pleistocene timescale based on ODP site 677; *Transactions of the Royal Society of Edinburgh* Earth Sciences, v. 81, p. 251-261, 1990.

Shotten, F.W., A Lower Pleistocene Glaciation in England, IGCP Project 24, Report No. 7, pp. 203-213, Prague, 1982.

Šibrava, V., Correlation of European glaciations and their relation to deep sea record, in *Quaternary Glaciations in the Northern Hemisphere*, edited by V. Šibrava et al., 433-442, Quaternary Science Reviews, Pergamon Press, Oxford, 1986.

Sifrer, M., The Quaternary development of Dobrave in Upper Carniola (Gorenjska), *Acta Geographica IX, SAZU*, 101-120, Ljubljana, 1969.

Stritar, A., Neke sistematske jedinice tala na fluvioglacialnima šljuncima in konglomeratima Gorenjske, PhD thesis, 86 p., Zagreb, 1969.

Tarling, D.H., 1983, *Palaeomagnetism*. Chapman and Hall, 379 p.

Verosub, K.L., 1977, Depositional and post depositional processes in the magnetization of sediments. *Reviews of Geophysics and Space Physics*, 15/2, 129-143.

Vidic, N., Pavich, M.J., Lobnik, F., Statistical analyses of soil properties on a Quaternary terrace sequence in Upper Sava River Valley, Slovenia, Yugoslavia, *Geoderma*, 51, 189-212, 1991.

Walter, R.C., Manega, P.C., Hay, R.L., Drake, R.E., Curtis, G.H., Laser-fusion ^{40}Ar/^{39}Ar dating of Bed I, Olduvai Gorge, Tanzania, *Nature* 345, 1991.

Walter, R.C., Manega, P.C., Hay, R.L., Tephrochronology of Bed I, Olduvai Gorge: An application of laser-fusion ^{40}Ar/^{39}Ar dating to calibrating biological and climatic change, *Quaternary International* 13/14, 37-46, 1992.

Milan J. Pavich, U.S. Geological Survey, Reston, VA.
Nataša Vidic, University of Ljubljana, Ljubljana, Slovenia.

D/H Ratios of Supergene Alunite as an Indicator of Paleoclimate in Continental Settings

GREG B. AREHART[1] AND JAMES R. O'NEIL

Department of Geological Sciences, University of Michigan, Ann Arbor, MI 48109-1063

Supergene alunite is an important benchmark indicator of continental climates. Variations in $\delta^{18}O_{OH}$ and δD reflect the isotopic composition of regional meteoric waters, which in turn can be used to infer paleoclimatic information. Because alunite also contains K, it can be easily and accurately dated by K/Ar methods to determine an absolute age.

δD values of meteoric waters in Nevada as recorded by alunite increased from about -150 permil at 30 Ma to about -105 permil in mid-Miocene time. Since mid-Miocene, there has been a distinct decrease in δD values to present-day values near -130 permil. These changes are congruent with changes in other paleoclimatic indicators, both continental and oceanic. The changes in isotopic ratios for alunite suggest that climatic indicators from continental sites, and particularly continental interior sites, may be more sensitive to global changes than are oceanic indicators. However, caution must be taken to minimize or eliminate changes in isotopic compositions which may have been caused by local or regional climatic factors.

INTRODUCTION

Alunite $[KAl_3(SO_4)_2(OH)_6]$ and its isostructural equivalent, jarosite $[KFe_3(SO_4)_2(OH)_6]$ are two of the best common minerals available for isotopic studies because they host several elements of interest. Stable isotope analyses of S, O (both sulfate and hydroxyl), and H provide information on source materials and conditions of formation, and the presence of K allows age determinations to be made using the K/Ar or $^{40}Ar/^{39}Ar$ methods. Thus, supergene alunites formed near the surface of the earth may record the D/H and $^{18}O/^{16}O$ ratios of local meteoric water at a known time. These samples may therefore be proxies for the isotopic composition of meteoric water and, hence, paleoclimate in the region.

Although individual crystals of alunite may form geologically instantaneously and have been synthesized in laboratory times [Stoffregen and Alpers, 1990], the often massive, earthy nature of samples reported herein suggests formation over periods of time longer than a few years. This formation over longer time periods is critical to any paleoclimatological interpretation, as short-term irregularities in weather are effectively smoothed out yielding an average climatic signal for the time of formation of the mineral.

ENVIRONMENTS OF FORMATION OF ALUNITE

Alunite forms under highly acidic, oxidizing conditions, which may be present in a variety of environments, including primarily hydrothermal and supergene settings [for a detailed discussion of the dependence of stable isotope ratios on conditions of formation, see Rye et al., 1992]. The most common supergene occurrence is as a weathering product of highly pyritic rocks, such as those associated with mineral deposits or sulfidic, reduced sediments. Confirmation that a given specimen is of supergene origin is requisite for its utilization as a paleoclimatological proxy. The distinction between supergene and hypogene formation commonly can be made on the basis of the nature of the occurrence of alunite (e.g. field and textural relations), as well as the isotopic composition of the mineral and other associated minerals, such as kaolinite. The $^{34}S/^{32}S$ ratio of supergene alunite is in most cases nearly identical to that of precursor and/or coexisting sulfides, and often differs markedly from the $^{34}S/^{32}S$ ratio of hypogene sulfates [Steiner and Rafter, 1966; Field, 1966; Rye et al., 1992; Arehart et al., 1992](Figure 1). Although not a definitive test, supergene alunite generally is depleted in D and ^{18}O relative to alunite of other origins. The clearest stable isotope discriminant is unusually small values of $\Delta^{18}O(SO_4-OH)$ as described by Pickthorn and O'Neil [1985] and Rye et al. [1992]. An age for alunite that is

---

[1]Present address: Geosciences/CMT-205, Argonne National Laboratory, Argonne, IL 60439.

Climate Change in Continental Isotopic Records
Geophysical Monograph 78
Copyright 1993 by the American Geophysical Union.

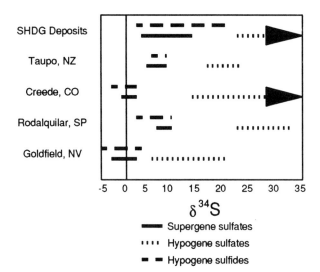

Fig. 1. Sulfur isotope signature of supergene alunite from a variety of locations. Data are from Steiner and Rafter, 1966 (Taupo); Rye et al., 1992 and Bethke et al., 1973 (Creede); Arribas et al., 1989 (Rodalquilar); Arehart et al., 1992 (sediment-hosted disseminated gold [SHDG] deposits); Jensen et al., 1971 (Goldfield).

younger than the age of primary mineralization can be diagnostic of a supergene origin.

Hydrogen isotope fractionation between ambient water and hydroxyl hydrogen in low-temperature alunite is minimal, with $\alpha \approx 1.004$ [Bird et al., 1989]. Bird et al. [1989] and Alpers et al. [1992] also suggested that iron-rich alunites and jarosites would have lower δD values than cogenetic iron-free alunites. On the other hand, alunites precipitated from highly saline evaporated waters may have δD values that are significantly higher than those of alunites deposited from less saline waters [Craig, 1963; Bird et al., 1989]. Of the samples utilized for this study, none was deposited from highly saline waters, and only one has a significant iron component. Therefore, δD values of alunite analyzed for this study should be within a few permil of ambient groundwater during mineral formation.

On the basis of extrapolated experimental data, Stoffregen [1990] suggested that some fine-grained alunites could undergo complete hydrogen isotope exchange with environmental waters at surficial temperatures in periods as brief as tens of thousands of years. Rye et al. [1992], however, suggested that nearly all alunites should retain their primary hydrogen isotope composition. The data presented in this study are in agreement with the latter suggestion, as δD values for many very old samples are considerably different from those of modern meteoric waters (see below). Stoffregen's suggestion is based on experiments in which the mineral is extremely fine grained and disaggregated, and in which the water/mineral ratio is very high. Such conditions are not representative of natural conditions prevalent in the Great Basin. In particular, many of the samples analyzed in this study were coherent minerals collected from (presently) unsaturated zones, where the lack of water would retard any isotopic exchange.

In the ideal situation, in which alunite forms only from water at near-surface temperatures, $\delta^{18}O_{OH}$ values are offset from the meteoric water line by a constant amount, resulting in an "alunite hydroxyl line" [Rye et al., 1992; see Figure 2] similar to the kaolinite line of Savin and Epstein [1970]. Shifts toward lower $\delta^{18}O$ values may occur as a result of elevated temperatures that arise because of exothermic oxidation reactions involving sulfides. However, because of the possibility of inclusion of atmospheric oxygen [$\delta^{18}O \approx 23.5$ permil; Kroopnick and Craig, 1972] in reactions which produce sulfate from sulfide, the $\delta^{18}O$ value of the sulfate radical may be shifted toward more positive values and do not form an array parallel to the meteoric water line [Rye et al., 1992]. Because of these uncertainties, paleoclimatological information can not be obtained from bulk oxygen isotope ratios of alunite. Analysis of hydroxyl oxygen provides information on the isotopic composition of the supergene fluid. However, the analytical uncertainties inherent in analyzing hydroxyl oxygen make such a measurement less accurate than a measurement of δD. On the other hand, combined hydrogen and oxygen isotope analyses of cogenetic kaolinite can provide useful paleoclimatological information [e.g. Lawrence and Taylor, 1971; 1972]. These analyses are being made at the present time for those cases where kaolinite samples are available.

ANALYTICAL TECHNIQUES

Alunites examined in this study were collected from a variety of locations in northern and central Nevada (Figure 3). All have been determined or judged to be supergene in origin [Arehart et al., 1992; E.H. McKee, personal communication; W. Pickthorn,

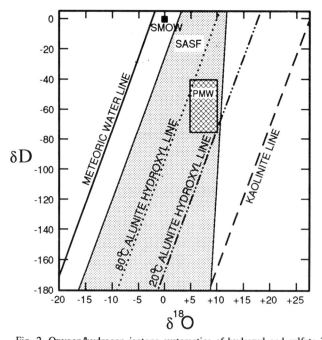

Fig. 2. Oxygen/hydrogen isotope systematics of hydroxyl and sulfate in supergene alunites (simplified from Rye et al., 1992).

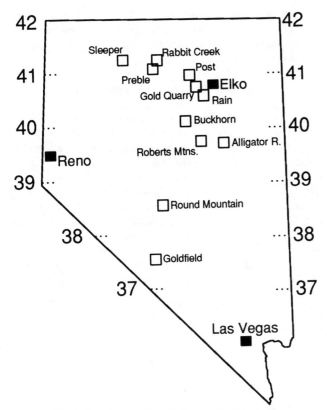

Fig. 3. Location map for alunites used in this study.

personal communication]. Each sample was cleaned in HF to eliminate clay minerals, and then X-rayed to verify its purity and determine its K/Na ratio. K/Ar ages and δD values are reported in Table 1. The age of alunite can be determined by either conventional K/Ar or $^{40}Ar/^{39}Ar$ fusion techniques. Ages of our alunite samples were determined at either the U.S. Geological Survey in Menlo Park, CA, or the Radiogenic Isotope Laboratory at The Ohio State University. Although it is well-established that hypogene alunite yields accurate K/Ar ages [Mehnert et al., 1973; Cunningham et al., 1984], it might be thought that supergene alunite could suffer from Ar loss because of the fine grain size of most samples. The available data, however, suggest that alunite retains Ar over tens of millions of years. An age of 10.4 Ma for supergene alunite from Goldfield, Nevada agrees with the K/Ar ages of pre- and post-supergene volcanic events determined by Ashley and Silberman [1976]. K/Ar ages (4.4 to 91.5 Ma) of supergene alunite from a variety of Australian settings determined by Bird et al. [1990] are geologically reasonable. Therefore, we presume that the alunites in this study have retained their radiogenic argon and yield correct ages of formation.

In addition to the hydroxyl group, hydrogen may also be present as H_3O^+ in the K site in alunite [e.g. synthetic alunites of Bayliss and Koch, 1955; Stoffregen and Alpers, 1990]. If significant quantities of H_3O^+ were present, and its isotopic composition differed from that of hydroxyl hydrogen, the measured D/H ratio could be misleading. Although a few natural samples that contain

measurable hydronium have been reported by Alpers et al. [1992], none of the samples analyzed for this study gave indication of more than trivial amounts of non-hydroxyl water, as indicated by comparison of measured and calculated water yields. Pure endmember alunite contains a calculated 13.0 weight percent H_2O, whereas endmember natroalunite contains 13.5% H_2O. Water contents are lower when Fe substitutes for Al (jarositic alunite); pure jarosite contains 10.8 % H_2O.

Extraction of hydrogen from alunite was performed in the manner described by Vennemann and O'Neil [1992], with some modifications. Application of standard techniques to very fine-grained supergene alunite may result in low yields and/or erratic δD values that are attributed to the formation of a visible low vapor pressure liquid during extraction [probably H_2SO_4; Wasserman et al., 1990]. For some samples, the generation of this liquid occurred no matter how slowly the temperature of the sample was increased. Inasmuch as we encountered significant acid production in over 30 percent of our samples, we presume it is a common occurrence and one that must be taken into account when making deuterium analyses of alunite. Failure to do so may result in errors of as much as 20 permil or more. Where acid generation is a problem, the resulting δD values are usually, but not always, too low.

Our initial attempt to eliminate this problem was to generate H_2 directly by reacting zinc metal with alunite in a closed quartz tube at temperatures ranging between 400 and 750°C for durations of 20 minutes to 20 hours. This technique was unsuccessful. Our effective modification consists of the addition of Zn metal on top of the alunite powder in the reaction vessel. Fairly rapid heating of the samples (from 150 to 650°C over 20 minutes) produces no

TABLE 1. Age, water yield, δD value, and corrected δD value for supergene alunites from Nevada.

Deposit	Sample	Age	H$_2$O Yield	δD[+]	δD$_{corr}$
Roberts Mtns	5-5-10	5.3	13.8	-112	-120
Sleeper	SL-1	5.4	14.1	-128	-128
Alligator R.	AR005A	8.3	12.5	-129	-136
Post	BP-79B	8.6	13.0	-137	-137
Round Mtn.	RM900-1	9.5	10.1	-94	-108
Post	BP-79P	9.5	13.7	-126	-126
Round Mtn.	RM32-1	9.8	12.1	-107	-119
Preble	PR007A	11.3	14.0	-104	-104
Goldfield	P3-15	11.6	13.0	-90	-118
Rabbit Creek	435-625	14.4	11.8	-106	-106
Buckhorn	GO5933	16.6	14.1	-102	-107
Rain[*]	9834A	18.8	13.5	-130	-130
Rain[*]	9835A	20.0	13.2	-141	-141
Rain	R-9	20.7	12.9	-125	-125
Preble	PR005A	23.0	13.1	-122	-122
Gold Quarry	GQ-5280	25.9	13.5	-132	-132
Gold Quarry	87GQ001	27.9	13.7	-149	-149
Gold Quarry	GQ89-2	28.9	12.6	-144	-144

[+]Uncertainty ± 2 permil. [*]Analyses of these samples were not reproducible and only the highest δD values are reported. See text for discussion.

visible liquid. When the temperature of the reaction chamber approaches 420°C, Zn metal begins to vaporize and deposit on the walls of the quartz tube where it exits the furnace. With continued heating of the sample and increase of temperature, all acid and most or all of the water evolved is immediately reduced by Zn to yield H_2 gas, which is then passed over heated (~700°C) CuO to reconvert the H_2 produced to water. This water is then transferred to a tube containing zinc metal and reacted in the normal manner to produce H_2 for stable isotope analysis. Using this method, we obtain yields near 100 percent and reproducible δD values for nearly all samples.

The isotopic composition of two alunites from Rain (9834A and 9835A; Table 1) were enigmatic. We have been unable to obtain satisfactory replicate analyses of either water content or δD for these samples. δD values of samples for which water yields were satisfactory range from -180 to -130 permil. A third sample of approximately the same age from Rain (R-9) yielded reproducible analyses and is taken to be more accurate. The highest δD values obtained for samples 9834A and 9835A are approximately

the same as those obtained from R-9. This suggests that the true values for samples 9834A and 9835A are similar to that of sample R-9. We have not been able to identify the nature of the problem with these two samples.

CHANGES IN δD OF ALUNITE WITH TIME

The samples range in age from 5.3 Ma to 29 Ma, and δD values span the range -90 permil to -150 permil. In addition to the measured values of δD, normalized values are reported in Table 1. These values are normalized to a latitude of 41 degrees, based on a latitudinal variation of 5 permil/degree latitude [Dansgaard, 1964], and agree with the variation in isotopic compositions of present-day meteoric water in Nevada [O'Neil and Silberman, 1974; Mariner et al., 1983; Arehart and O'Neil, unpublished data]. There is a significant correlation between the normalized δD values and age for samples of alunite from northern and central Nevada (Figure 4). The oldest samples have the lowest δD values, and δD values increase progressively

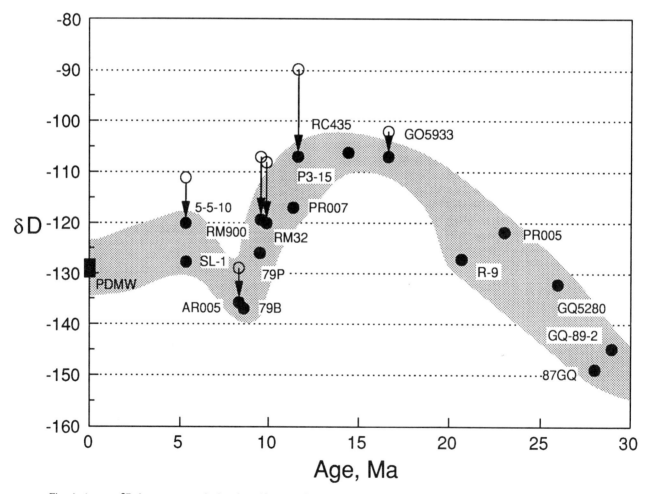

Fig. 4. Age vs. δD for supergene alunites from Nevada. Open circles are raw data and solid circles are latitude-corrected data. PDMW = present-day meteoric water compositions from several localities near 41° N latitude. Data from Table 1.

between 29 Ma and approximately 13 Ma. After 13 Ma, there is an abrupt decrease from about -110 permil to -135 permil over 4 m.y. At least one of the two samples having an age near 8.5 Ma (79B) has a measurable iron content, and therefore its isotopic composition may not fall on the equilibrium fractionation line of Bird et al. [1989]. The true meteoric water value for such a sample is undoubtedly higher than the value obtained, but the degree of that shift is unknown. Since 8.5 Ma, δD values of alunite and, by inference, meteoric water in the Great Basin apparently have not changed significantly.

POSSIBLE CAUSES OF CHANGES IN δD

From the available data, we infer that the isotopic composition of meteoric water in Nevada changed significantly over the past 30 m.y. Several processes may have caused the observed temporal variations in δD values of the alunite samples, including differences in paleolatitude, differences in paleoaltitude, or changes in paleoclimate. Changes in latitude required to affect the δD values are much larger than those determined for western North America over the time span represented by these samples [Denham and Scotese, 1988]. Stable isotope compositions of precipitation can vary significantly with altitude, but in most cases groundwaters are long-term averages such that local variations due to altitude, as well as seasonal variations, are eliminated [e.g. Thordsen et al., 1992]. Because the alunites analyzed are from deep weathering zones, we infer that changes in altitude and seasonal variations in the isotopic composition of the meteoric water from which the alunites formed are smoothed out. If local altitude effects were dominant, we would expect the data for this sample set to be more random. Time variations of δD values of samples from a single location (e.g. Gold Quarry, Table 1) or closely spaced locations (Figure 3, Table 1) are not likely to be ascribed to changes in altitude. This conclusion is supported by the fact that deuterium analyses of present-day springs across Nevada are very similar for any given latitude, regardless of the altitude of the spring [O'Neil and Silberman, 1974; Mariner et al., 1983; G. Arehart and J. O'Neil, unpublished data]. On the other hand, the average elevation of the Great Basin has been increasing since at least 17 Ma, and probably since the inception of Basin and Range extension at 40 Ma [Zoback and Thompson, 1978; Dickinson, 1981]. All else remaining constant, this type of change should result in a decrease in δD values of meteoric water through time. Exactly the opposite is seen in the data for 30-16 Ma, as δD increases through time. Whereas δD values of alunite decrease relatively rapidly between 13 and 10 Ma, the change in altitude has been relatively slow. Altitude changes caused by broad uplift are thus thought not to have played a major role in producing the changes in the isotopic composition of meteoric water that are recorded by alunite.

From the above arguments, we conclude that the systematic change in the isotopic composition of meteoric water from which these alunites precipitated is most likely due to major changes in paleoclimate. More specifically, it appears that local climates warmed significantly between 29 Ma and 13 Ma, and have cooled somewhat since that time.

CORRELATION WITH OTHER PALEOCLIMATOLOGICAL INDICATORS

Western North America

The data for alunite from Nevada are consistent with several other paleoclimatological indicators for western North America. Paleobotanical indicators have been interpreted to suggest that two major episodes of cooler temperatures occurred at ~36 Ma and ~15 Ma [Axelrod, 1968; 1990]. While the earlier episode is beyond the range covered by our data, it is consistent with the low δD values recorded at 29 Ma. The second is poorly constrained in time, but is broadly consistent with the climatic record as indicated by the alunite stable isotope data. Mid-Tertiary clay minerals from western North America have δD values that are approximately 20 permil greater than values of present meteoric water, suggesting that climates were warmer at that time [Lawrence, 1970].

O'Neil and Silberman [1974] showed that ore fluids in the epithermal deposits of the Great Basin of Nevada were dominated by meteoric water and suggested that information on paleoclimates could be obtained from stable isotope analyses of fluid inclusions in datable samples of adularia and cogenetic quartz in these deposits. On the basis of the data obtained, they suggested that climates in Nevada had not undergone significant changes through the Tertiary. In light of more recent advances made in understanding the systematics of fluid inclusion data, however, it now seems most likely that several of those samples analyzed contained significant proportions of secondary inclusions (given the abundant tectonic and hydrothermal activity in the Great Basin since the initial deposition of these minerals). The isotopic compositions of such inclusions would, of course, be similar to those of more modern fluids. In addition, it is also possible that ground waters that ultimately became the ore fluids were older than mid-Miocene (perhaps paleo-waters of early Miocene age?).

It remains that most of the O'Neil-Silberman data are in agreement with the alunite data presented here. Recent analyses of fluids extracted from native gold at the Tenmile deposit [Arehart and O'Neil, unpublished data] are in general agreement with the analyses of fluids from adularia reported by O'Neil and Silberman [1974]. These fluids have uniform and relatively high δD values of -85 to -96 permil, close to values measured for nearby alunite samples of the same age.

Tectonic events in western North America appear to have had minimal influence on the isotopic compositions of local meteoric water. Uplift and extension of the Basin and Range was initiated at approximately 40 Ma [Zoback et al., 1978]. Major extension began around 17 Ma [Zoback et al., 1978] and arc volcanism was at a maximum shortly thereafter [Dickinson, 1981]. Both of these events should have caused a decrease in δD values of meteoric waters in the Great Basin. From 30 Ma to 13 Ma, exactly the opposite effect is seen, as δD values of meteoric water increased. Although δD values of meteoric waters have decreased since 13 Ma, the decrease was relatively sharp (over a couple of million years) and δD values have remained essentially constant since 9 Ma.

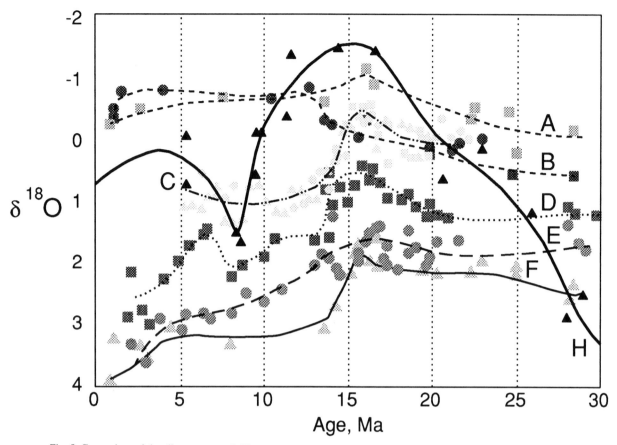

Fig. 5. Comparison of the climate as recorded by supergene alunite in Nevada with marine indicators of climate for the past 30 m.y. Marine data from Savin, 1977; Savin et al., 1985; Shackleton and Kennett, 1975. A, B = north Pacific planktic foraminifera; C = equatorial Pacific benthic foraminifera; D = north Pacific benthic foraminifera; E = subantarctic planktic foraminifera; F = subantarctic benthic foraminifera; H = western North America alunite. The alunite data have been converted to oxygen-equivalent values ($\delta^{18}O$ = 8δD) so that the scales are consistent. Note that the changes seen in alunite values record the less thermally buffered continental environment.

Worldwide climate change

The data presented herein are broadly congruent with other indicators of paleoclimate worldwide. Continental samples reflect changes in temperature that are likely to be exaggerated, much as present-day continental interiors have a much larger range of temperature variations compared to oceans. Paleotemperature data from New Zealand [Devereux, 1967] suggest cool temperatures in early Oligocene time followed by a warming trend which increased ocean temperatures by approximately 6°C. Similar trends are evident in the data for foraminifera from all three ocean basins [Savin, 1977; Savin et al., 1985; Kennett et al., 1985](Figure 5). Low temperatures in Oligocene times may have resulted from lowered sea levels [Ernst, 1981] or from changes in ocean circulation patterns. Three possible causes of changes in ocean circulation patterns have been suggested. Savin [1977] suggested as a possible cause the opening of the Drake Passage and concurrent changes in deep ocean circulation, whereas Kennett et al. [1985] suggest that the development of surface oceanic currents may have been important. More recently,

Woodruff and Savin [1989] speculated that the termination of the major influx of Tethyan warm saline water into the northern Indian ocean may have strongly affected global ocean circulation patterns.

A sharp drop in temperatures (~4°C) in mid-Miocene time recorded in the New Zealand data is followed by relatively constant temperatures until Plio-Pleistocene glaciation [Devereux, 1967]. Data from the subantarctic Pacific also are consistent with warmer ocean surface temperatures in mid-Miocene followed by a sharp decrease in temperature (~6°C) at ~13 Ma and relatively constant temperatures thereafter [Shackleton and Kennett, 1975]. This change is attributed to the development of permanent ice sheets on Antarctica. On the basis of $\delta^{13}C$ data from several locations in the ocean basins, Wright et al. [1991] suggested that oceanic circulation patterns began changing at ~10 Ma.

In addition to the marine record, other continental data also suggest that at least the sharp mid-Miocene change in climate was not restricted to western North America. Alpers and Brimhall [1988] present a variety of geological data from Chile which suggest that a rapid decrease in erosion rates occurred at approximately 14 Ma. They relate this change to cooling of the ancestral Humboldt current waters during the development of the Antarctic ice sheet. Cerling and Quade [1992] document important faunal changes in the Siwaliks of Pakistan at 13.5, 12, 9.5, and 7.5 Ma which are interpreted to have resulted from climate-induced changes in vegetation.

The alunite data are concordant with other data in suggesting that mid-Miocene climates were significantly warmer than present climates (Figure 5). The observed changes in δD values for alunite samples from western North America may be interpreted to suggest that climate changes have occurred somewhat after the sharp change in the marine record. At present, changes in the continental record cannot be as closely constrained because of the lack of resolution of our sample base. The marine record may reflect rapid changes almost immediately, whereas the continental record, as preserved in alunite, may be recording older water that resided in the ground for hundreds of thousands of years longer.

SUMMARY

δD values of meteoric waters in Nevada increased from -150 permil in late Oligocene to near -105 permil in mid-Miocene. This was followed by a sharp decrease in δD values between 13 and 9 Ma to near present-day values of -130 permil. These changes correspond to changes in other paleoclimatic indicators for western North America. The increase from Oligocene to Miocene probably represents a gradual warming of climates and transgression of sea level, and to the absence of significant topographic barriers to atmospheric circulation (i.e. the Sierra Nevada). The development of the Miocene island arc and uplift of the Sierra Nevada may have caused shifts in atmospheric circulation that locally enhanced extant global patterns of cooling.

Because of its chemistry, alunite has the potential to become a benchmark indicator of paleoclimate throughout the geological record. Although alunite is most common in the Tertiary, it is present in older records and could provide a very useful mineral with which to calibrate other paleoclimatological indicators. The combination of stable and radiogenic isotope measurements of a single mineral allows accurate calibration of other indicators which are more common and continuous through the record. The relatively long formation time obviates the problems of confusing meteorological and climatic records.

Acknowledgments. We thank E.H. McKee, W.H. Pickthorn, and B. Maher for providing 5 samples of dated alunites. K.A. Foland of Ohio State University kindly allowed access to his K/Ar laboratory for additional age determinations. T.W. Vennemann, J. Zachos, and L. Cirbus Sloan provided valuable discussion of various portions of the manuscript. The manuscript also benefitted from critical comments of R.O. Rye and S.M. Savin. This work was supported in part by NSF grant EAR-9005717.

REFERENCES

Alpers, C.N., and Brimhall, G.H., Middle Miocene climatic change in the Atacama Desert, northern Chile: Evidence from supergene mineralization at La Escondida, *Geological Society of America Bulletin, 100*, pp. 1640-1656, 1988.

Alpers, C.N., Rye, R.O., Nordstrom, D.K., White, D.L. and King, B.S., Chemical, crystallographic, and isotopic properties of alunite and jarosite from acid hypersaline Australian lakes, *Chemical Geology, 96*, pp. 203-226, 1992.

Arehart, G.B., Kesler, S.E., O'Neil, J.R. and Foland, K.A., Evidence for the supergene origin of alunite in sediment-hosted micron gold deposits, Nevada, *Economic Geology, 87*, pp. 263-270, 1992.

Arribas, A., Jr., Rytuba, J.J., Rye, R.O., Cunningham, C.G., Podwysocki, M.H., Kelly, W.C., Arribas, A., McKee, E.H. and Smith, J.G., preliminary study of the ore deposits and hydrothermal alteration in the Rodalquilar caldera complex, southeastern Spain, *U.S. Geological Survey Open-File Report 89-327*, 39 p., 1989.

Ashley, R.P. and Silberman, M.L., Direct dating of mineralization at Goldfield, Nevada, by potassium-argon and fission-track methods, *Economic Geology, 71*, pp. 904-921, 1976.

Axelrod, D.I., Age and origin of subalpine forest zone, *Paleobiology, 16*, pp. 360-369, 1990.

Axelrod, D.I., Tertiary floras and topographic history of the Snake River Basin, Idaho, *Geological Society of America Bulletin, 79*, pp. 713-734, 1968.

Bethke, P.M., Barton, P.B., Jr. and Rye, R.O., Hydrogen, oxygen and sulfur isotopic compositions of ore fluids in the Creede District, Mineral County, Colorado (abstract), *Economic Geology, 68*, p. 1205, 1973.

Bird, M.I., Andrew, A.S., Chivas, A.R. and Lock, D., An isotopic study of surficial alunite in Australia: I. Hydrogen and sulfur isotopes, *Geochimica et Cosmochimica Acta, 53*, pp. 3223-3237, 1989.

Bird, M.I., Chivas, A.R. and McDougall, I., An isotopic study of surficial alunite in Australia: 2. Potassium-argon geochronology, *Chemical Geology, 80*, pp. 133-145, 1990.

Brophy, G.P., Scott, E.S. and Snellgrove, R.A., Sulfate studies II: Solid solution between alunite and jarosite, *American Mineralogist, 78*, p. 73-90, 1962.

Craig, H., The isotopic geochemistry of water and carbon in geothermal areas, *Proceedings of Nuclear Geology on Geothermal Areas, Spoleto, Italy*, pp. 17-53, 1963.

Cunningham, C.G., Origins and exploration significance of replacement and vein-type alunite deposits in the Marysvale Volcanic Field, west central Utah, *Economic Geology, 79*, pp. 50-71, 1984.

Dansgaard, W., Stable isotopes in precipitation, *Tellus, 16*, pp. 436-468, 1964.

Denham, C.R. and Scotese, C.R., Terra Mobilis - a plate tectonics program for the macintosh: version 2.1, Earth in Motion Technologies, Houston, TX, 1988.

Devereux, I., Oxygen isotope paleotemperature measurements on New Zealand Tertiary fossils, *New Zealand Journal of Science, 10*, pp. 988-1011, 1967.

Dickinson, W.R., Plate tectonics and the continental margin of California, in *The Geotectonic Development of California (Rubey Volume I)* edited by W.G. Ernst, pp. 1-28, Prentice-Hall, 1981.

Ernst, W.G., Summary of the geotectonic development of California, in *The Geotectonic Development of California (Rubey Volume I)* edited by W.G. Ernst, pp. 601-613, Prentice-Hall, 1981.

Field, C.W., Sulfur isotopic method for discriminating between sulfates of hypogene and supergene origin, *Economic Geology, 71*, pp. 1428-1435, 1966.

Jensen, M.L., Ashley, R.P. and Albers, J.P., Primary and secondary sulfates at Goldfield, Nevada, *Economic Geology, 66*, pp. 618-626, 1971.

Kennett, J.P., Keller, G., and Srinivasan, M.S., Miocene planktonic foraminiferal biogeography and paleoceanographic development of the Indo-Pacific region, *Geological Society of America Memoir 163*, pp. 197-236, 1985.

Kroopnick, P., and Craig, H., Atmospheric oxygen: Isotopic composition and solubility fractionation: *Science, 175*, pp. 54-55, 1972.

Lawrence, J.R. and Taylor, H.P., Deuterium and oxygen-18 correlation: clay minerals and hydroxides in Quaternary soils compared to meteoric waters: *Geochimica et Cosmochimica Acta, 35*, pp. 993-1004, 1971.

Lawrence, J.R. and Taylor, H.P., Hydrogen and oxygen isotope systematics in weathering profiles, *Geochimica et Cosmochimica Acta, 36*, pp. 1377-1393, 1972.

Mariner, R.H., Presser, T.S. and Evans, W.C., Geochemistry of active geothermal systems in northern Basin and Range province, *Geothermal Resources Council Special Report #13*, pp. 95-119, 1983.

Mehnert, H.H., Lipman, P.W. and Steven, T.A., Age of mineralization at Summitville, Colorado as indicated by K-Ar dating of alunite, *Economic Geology, 68*, pp. 399-401, 1973.

O'Neil, J.R. and Silberman, M.L., Stable isotope relations in epithermal Au-Ag Deposits, *Economic Geology, 69*, pp. 902-909, 1974.

Parker, R.L., Isomorphous substitution in natural and synthetic alunite, *American Mineralogist, 47*, pp. 127-136, 1962.

Pickthorn, W.J. and O'Neil, J.R., ^{18}O relations in alunite mineral: potential single-mineral thermometer (abstract), *Geological Society of America Abstracts, 17*, pp. 686, 1985.

Rye, R.O., Bethke, P.M. and Wasserman, M.D., The stable isotope geochemistry of acid sulfate alteration, *Economic Geology, 87*, pp. 225-262, 1992.

Savin, S.M., Abel, L., Barrera, E., Hodell, D., Kennett, J.P., Murphy, M., Keller, G., Killingley, J. and Vincent, E., The evolution of Miocene surface and near-surface marine temperatures: Oxygen isotopic evidence, *Geological Society of America Memoir 163*, pp. 49-82, 1985.

Savin, S.M., The history of the earth's surface temperature during the past 100 million years, *Annual Reviews in Earth and Planetary Sciences, 5*, pp. 319-355, 1977.

Savin, S.M. and Epstein, S., The oxygen and hydrogen isotope geochemistry of clay minerals, *Geochimica et Cosmochimica Acta, 34*, pp. 43-64, 1970.

Shackleton, N.J. and Kennett, J.P., Paleotemperature history of the Cenozoic and the initiation of Antarctic glaciation: oxygen and carbon isotope analyses in DSDP sites 277, 279, and 281, in *Initial Reports of the Deep Sea Drilling Project, 29*, pp. 743-755, 1975.

Steiner, A. and Rafter, T.A., Sulfur isotopes in pyrite, pyrrhotite, alunite and anhydrite from steam wells in the Taupo volcanic zone, New Zealand, *Economic Geology, 61*, pp. 1115-1129, 1966.

Stoffregen, R.E. and Alpers, C.N., On the unit cell dimensions, water contents and hydrogen isotopes of natural and synthetic alunites (abstract), *GAC-MAC Annual Meeting Abstracts*, 1990.

Stoffregen, R.E., Stable isotope systematics of alunite, *Extended Abstracts, Third Symposium on Deep-Crust Fluids, Geological Survey of Japan*, pp. 27-30, 1990.

Vennemann, T.W. and O'Neil, J.R., A simple and inexpensive method of hydrogen isotope and water analyses of minerals and rocks based on zinc reagent, in press, *Chemical Geology*, 1992.

Wasserman, M.D., Rye, R.O., Bethke, P.M. and Arribas, A.A., Methods for separation of alunite from associated minerals and subsequent analysis of D, $^{18}O_{OH}$, $^{18}O_{SO4}$, and ^{34}S (abstract), *Geological Society of America Abstracts, 22*, p. A135, 1990.

Woodruff, F., and Savin, S.M., Miocene deepwater oceanography, *Paleoceanography, 4*, pp. 87-140, 1989.

Wright, J.D., Miller, K.G. and Fairbanks, R.G., Evolution of modern deepwater circulation: Evidence from the late Miocene southern ocean, *Paleoceanography, 6*, pp. 275-290, 1991.

Zoback, M.L. and Thompson, G.A., Basin and range rifting in northern Nevada: Clues from a mid-Miocene rift and its subsequent offsets, *Geology, 6*, pp. 111-116, 1978.

Greg B. Arehart and James R. O'Neil, Department of Geological Sciences, University of Michigan, 1006 C.C. Little Building, Ann Arbor, MI 48109-1063.

The Stable Isotope Geochemistry of Low Temperature Fe(III) and Al "Oxides" with Implications for Continental Paleoclimates

Crayton J. Yapp

Department of Geology, University of New Mexico, Albuquerque, New Mexico 87131, USA

Minerals which formed in highly evolved continental chemical weathering environments could be sources of stable isotope information on paleoclimates throughout much of the Phanerozoic. Systems such as lateritic paleosols, certain oolitic ironstones, bog ores, etc., can contain such phases as goethite, hematite, phosphate, boehmite, gibbsite and kaolinite. In some circumstances, single mineral paleothermometry using hydroxyl-bearing iron (III) and Al "oxides" may be practicable. A population of twenty-six widely distributed continental goethite samples exhibits a δD range from -220 to -113 per mil and a $\delta^{18}O$ range from about -15.5 to +1.5 per mil. Single mineral isotope temperatures for these diverse goethites range from about 7 °C to about 68 °C. A "surface domain" can be delineated for goethite on δD vs. $\delta^{18}O$ plots. Samples which plot within this domain could have formed in continental environments under conditions of thermal equilibrium with the Earth's ancient atmosphere.

Mineral-pair oxygen isotope geothermometry in these systems holds promise, but geothermometers need to be better calibrated by experiments. A preliminary use of the phosphate-goethite oxygen isotope pair from a Late Ordovician oolitic ironstone yields a temperature of about 23 °C and a water $\delta^{18}O$ value of -7.3 per mil. Seasonal (monsoonal?) ancient precipitation regimes are suggested for that time and place.

INTRODUCTION

Ancient climates can leave their imprint in the continental geologic record in many ways. From the standpoint of stable isotope geochemistry, ancient temperature and the isotopic composition of meteoric water are the climatically important variables most amenable to quantitative interpretation. Modern meteoric waters have spatial and temporal oxygen and hydrogen isotope distributions that are related to air mass history, air temperature, degree of rainout, post-condensation exchange and evaporation, etc. [e.g. Epstein and Mayeda, 1953; Dansgaard, 1964; Friedman, et al., 1964; Miyake, et al., 1968; Gat, 1980; Yapp, 1982; Joussaume, et al., 1984; Lawrence and White, 1991]. Thus, proxy indicators of the δD and $\delta^{18}O$ values of meteoric water can provide important information on continental paleoclimates.

Authigenic minerals formed in surficial continental environments include (but are not limited to) calcite, aragonite, kaolinite, boehmite, gibbsite, apatite, goethite, hematite and opaline silica. All of these minerals are potentially useful as isotopic indicators of ancient continental climates. The stable isotope geochemistry of calcite is by far the most studied and best understood for reasons which are related to both ease of analysis and widespread distribution in surficial environments. Calcite (and aragonite) $\delta^{18}O$ values depend on both the temperature and the $\delta^{18}O$ value of the water present at the time of mineral precipitation [Epstein, et al, 1953]. To deduce the value of one of these two paleoenvironmental unknowns from the measured $\delta^{18}O$ value of calcite (aragonite), the value of the second unknown must be assumed or independently estimated. However, if the oxygen isotope fractionation between calcite and another coprecipitated mineral were a known function of temperature, the $\delta^{18}O$ value of each mineral in this pair would permit the calculation of both the temperature of formation and the $\delta^{18}O$ value of the ambient water. Silica commonly coexists with calcite, but there are uncertainties in the silica-calcite fractionation curve. These uncertainties arise from differences among variously determined silica-water fractionation curves. These differences seem to be related to whether the silica is hydrous or anhydrous and whether it is biogenic or

Climate Change in Continental Isotopic Records
Geophysical Monograph 78

abiogenic [e.g. Matheney and Knauth, 1989]. The most widely utilized phosphate-water fractionation curve suggests that there is little or no temperature dependence for the calcite-apatite fractionation factor at low temperatures [e.g. Friedman and O'Neil, 1977]. Therefore, the problem of one equation and two unknowns remains a "fact of life" in many studies of calcite and aragonite oxygen isotope geochemistry. Fortunately, the climatic signals preserved in calcite $\delta^{18}O$ values are often constrained by other non-isotopic considerations, and carbonates will continue to be valuable sources of information in continental paleoclimate studies.

The importance of paleosol carbonates in the study of ancient continental climates by isotopic methods has been convincingly demonstrated [Cerling, 1984, 1991; Quade, et al., 1989a, 1989b]. However, as noted by Cerling [1984], weathering profiles which receive more than about one meter of rain per year are usually devoid of pedogenic carbonate because of the relatively high solubility of calcite. Thus, if paleosols which formed in very wet, tropical environments are to provide isotopic information on continental paleoclimates, it will be necessary to study the stable isotope geochemistry of the non-carbonate minerals which constitute these systems. Under modern conditions, lateritic soils tend to develop in tropical climates with seasonally varying rainfall totalling between about 1.3 and 1.7 meters per year [Tardy, et.al., 1990]. The authigenic minerals which can form in such intense chemical weathering environments include goethite, hematite, boehmite, gibbsite and kaolinite [e.g. Trolard and Tardy, 1987]. Temperature, activity of the water and the chemistry of the aqueous phase determine which of these minerals will coexist at equilibrium.

Stable isotope studies of the low temperature Fe(III) and Al "oxides" are recent and relatively few in number. This paper will discuss what is currently known about the isotopic systematics of these minerals and possible implications for studies of continental paleoclimates. Analytical methods will not be detailed here, because they are reported at length in the cited literature.

MINERAL ISOTOPIC SYSTEMATICS

Single-mineral Paleothermometers

The presence of both hydrogen and oxygen in the structures of goethite (α-FeOOH), boehmite (γ-AlOOH) and gibbsite (Al(OH)$_3$) suggests the possibility of using these phases as single-mineral, two-element isotope paleothermometers. This possibility was discussed for

clays by Savin and Epstein [1970] and for cherts by Knauth and Epstein [1976]. The success of the approach depends on a number of conditions: (1) the existence of mineral-water isotopic equilibrium at the time of mineral formation; (2) fractionation in the presence of ancient meteoric waters whose δD and $\delta^{18}O$ covaried in the same manner as those of modern meteoric waters; (3) high water/rock ratios; (4) knowledge of the mineral-water hydrogen and oxygen isotope fractionation factors as a function of temperature; (5) no isotopic exchange between the mineral and the environment subsequent to mineral formation.

If it is assumed that the modern "global" meteoric water line (MWL) of Craig [1961] is applicable to conditions in the Earth's past, the following relationship is expected between the δD and $\delta^{18}O$ values of a single hydroxyl-bearing mineral formed in the presence of such waters [Savin and Epstein, 1970]:

$$\delta D_m = 8 \frac{\alpha(D)}{\alpha(18)} \delta^{18}O_m + 1000 \left[8 \frac{\alpha(D)}{\alpha(18)} - 1\right]$$
$$- 6990 \ \alpha(D) \qquad (1)$$

where, δD_m = δD of the mineral

$\delta^{18}O_m$ = $\delta^{18}O$ of the mineral

$\alpha(D)$ = mineral-water hydrogen isotope fractionation factor

= $(D/H)_m/(D/H)_{water}$

$\alpha(18)$ = mineral-water oxygen isotope fractionation factor

= $(^{18}O/^{16}O)_m/(^{18}O/^{16}O)_{water}$

The values of $\alpha(D)$ and $\alpha(18)$ differ from unity by no more than a few percent in most mineral-water systems [e.g. O'Neil, 1986]. Therefore, the isothermal slope of a δD_m vs. $\delta^{18}O_m$ plot for a mineral system that satisfies the foregoing assumptions should be near eight. Because the intercept in eqn. 1 includes multiplicative factors on the order of 10^3, small differences in $\alpha(D)$ and/or $\alpha(18)$ due to differences in temperature will be most strongly manifested in the intercept of a δD_m vs. $\delta^{18}O_m$ plot rather than in the slope. Subparallel isotherms of δD_m vs. $\delta^{18}O_m$ can, in principle, be plotted and used to determine paleotemperatures from measured values of δD and $\delta^{18}O$ in hydroxyl-bearing minerals [e.g. Savin and Epstein, 1970; Knauth and Epstein, 1976]. Knowledge of $\alpha(D)$ and $\alpha(18)$ at different temperatures is necessary to the calculation of these isotherms from eqn 1.

Yapp and Pedley [1985] and Yapp [1987, 1990] have reported hydrogen and oxygen isotope fractionation

TABLE 1. Selected oxygen and hydrogen isotope mineral-water fractionation factors.

	T ≈ 25°C			
Mineral	α(D)	α(18)	*1000ln α(18)	Source
Goethite	0.905	1.0061	$1.63 \times 10^6/T^2 - 12.3$	(1)
Boehmite	0.975	1.0175	$2.11 \times 10^6/T^2 - 4.4$	(2,3)
Gibbsite	1.000	1.0160	-------	(4)
Phosphate	-----	1.0224	$2.32 \times 10^6/T^2 - 4.0$	(5)

* Probably valid in general for 273K ≤ T ≤ 333K. (1) Yapp [1990]; (2) Bird, et al. [1989]; (3) adapted from Savin and Lee [1988]; (4) Bird, et al. [1990]; (5) Calculated from carbonate-phosphate pairs reported in Karhu and Epstein [1986] for analysis of phosphate extracted as β - $BiPO_4$.

factors for the goethite-water system. They found little or no temperature dependence for the goethite-water hydrogen isotope fractionation factor over a range of temperatures from 25°C to 145°C. However, there is a strong temperature dependence for the goethite-water oxygen isotope fractionation factor. The goethite-water α(D) and α(18) values are listed in Table 1.

Hydrogen and oxygen isotope fractionation factors for boehmite vs. water were reported by Bird, et al. [1989]. The α(D) and α(18) values for the boehmite-water system were determined from natural samples at sedimentary temperatures. Combination of the data of Bird, et al. [1989] with those of Graham, et al. [1980] from an exchange experiment at 150°C suggests that the boehmite-water D/H fractionation factor has very little or no temperature dependence for temperatures less than about 150°C. Semi-empirical bond-type calculations of Savin and Lee [1988] suggest that the boehmite-water $^{18}O/^{16}O$ fractionation factor does vary with temperature. This suggestion has yet to be experimentally confirmed. α(D) and α(18) values for the boehmite-water system are in Table 1.

Chen, et al. [1988], Bird, et al. [1989] and Bernard [1978] analyzed δD and $δ^{18}O$ values of natural gibbsite samples. Gibbsite-water α(D) and α(18) values were calculated for the sedimentary temperatures represented by these samples. There are presently no published experimental data on the temperature dependence of gibbsite-water α(D) or α(18) values. The Al-OH bond-type calculations of Savin and Lee [1988] are the most uncertain of their results (Savin, personal communication). Thus, gibbsite-water α(D) and α(18) values are reported in Table 1 only at sedimentary temperatures. There is some disagreement about the correct value for the

gibbsite-water α(D) at sedimentary temperatures [Chen, et al., 1990; Bird, et al, 1990], but the difference between the proposed values is too small to affect the conclusions here.

The calculated relationships between $δD_m$ and $δ^{18}O_m$ for goethite, boehmite and gibbsite (eqn 1) are plotted as the 25°C isotherms in Fig. 1 using the α(D) and α(18) values of Table 1. The ~25°C "kaolinite line" of Savin and Epstein [1970] is shown for reference. The effect of mineral formation temperature on the relationship between $δD_m$ and $δ^{18}O_m$ is seen from comparison of the 25°C and 60°C isotherms for goethite. Goethite has been used to illustrate this in Fig. 1, because goethite-water fractionation factors at different temperatures have been determined by experiment [Yapp, 1987, 1990]. Note that the separation between the MWL and the mineral line of interest decreases as temperature increases. This is to be expected, since α values approach unity with increasing temperature in the absence of local maxima or minima [e.g. O'Neil, 1986].

Bird, et al. [1989] performed δD and $δ^{18}O$ analyses of purified natural gibbsites obtained from different locales. In addition they reported some gibbsite analyses performed by Bernard [1978]. These data are plotted in Fig. 2. Also plotted in Fig. 2 are δD and "uncorrected" $δ^{18}O$ data for some natural goethites [Yapp, 1987]. The more positive δD and $δ^{18}O$ values for the gibbsite samples

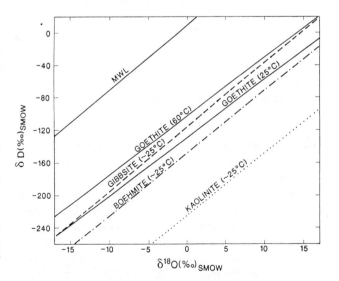

Fig. 1. Plots of the expected isothermal relationships between the δD and $δ^{18}O$ values in hydroxyl-bearing minerals which are in isotopic equilibrium with meteoric water. The ~25°C isotherms for gibbsite, boehmite and kaolinite are shown. Also, depicted are the 25°C and 60°C isotherms for goethite. Note the mutual proximity of the 25°C gibbsite, boehmite and goethite lines.

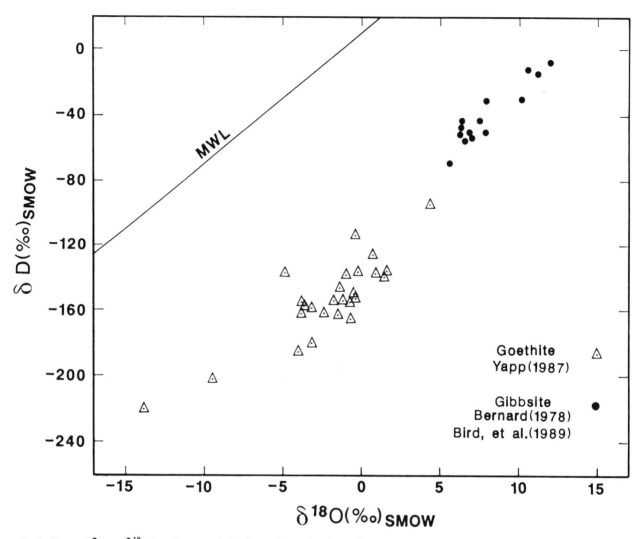

Fig. 2. Plots of δD vs. $\delta^{18}O$ data for natural gibbsites and goethites from widely separated continental locales. The apparent colinearity of the gibbsite and goethite data reflects the proximity of the gibbsite and goethite lines shown in Fig. 1. There are three new goethite data plotted here in addition to those reported by Yapp (1987) - e.g. the most D and ^{18}O - depleted data point at $\delta D = -220$.

in Fig. 2 generally reflect the fact that the $\alpha(D)$ and $\alpha(18)$ values of gibbsite are higher than those of goethite, rather than *necessarily* reflecting more D and ^{18}O-rich waters in the environments of gibbsite formation. The goethite samples of Fig. 2 represent a variety of sample locales and modes of goethite formation. The data set includes samples that are pseudomorphs after pyrite and siderite. With one exception, all of the goethite samples are from continental environments. The exception is a marine goethite from the Atlantis II Deep in the Red Sea [Yapp, 1987]. This marine sample is the goethite with the most positive δD and $\delta^{18}O$ values in Fig. 2. The continental origin of most of the goethites of Fig. 2 suggests that they could be a source of information on continental

paleoclimates. However, these natural goethite samples are not pure FeOOH and contain varying proportions of SiO_2.

The measured goethite $\delta^{18}O$ data of Yapp [1987] can be adjusted for SiO_2 content to obtain an estimate of the $\delta^{18}O$ values of endmember goethite. This adjustment is made by assuming that the SiO_2 and goethite had coprecipitated at isotopic equilibrium [see Yapp, 1987]. The measured goethite δD data can be adjusted for Mn content when Mn/(Mn+Fe) ratios are 0.01 or higher [Yapp and Pedley, 1985; Yapp, 1987]. The calculated "endmember" δD and $\delta^{18}O$ values of the continental goethites are plotted in Fig. 3. Because the amounts of SiO_2 in these samples are small, failure of the assumption

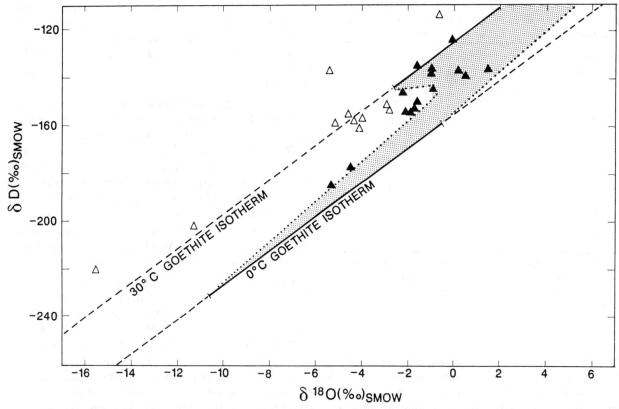

Fig. 3. Plot of δD vs. $\delta^{18}O$ for natural continental goethites of Fig. 2. However, in this instance the $\delta^{18}O$ values have been corrected to eliminate the contribution of silica impurities in the samples (see Yapp, 1987). The 0 °C and 30 °C goethite isotherms are shown for reference. The stippled area in the figure defines the "surface domain". This domain represents the goethite δD and $\delta^{18}O$ values for which a calculated goethite temperature of formation could plausibly be considered to be the same as that of the ancient surface air temperature (see text). Thus, the temperatures determined for goethites in the "surface domain" may directly reflect the paleoclimate. Filled triangles are those samples which, within analytical error, are considered to lie within the limits of the "surface domain". Open triangles represent samples whose temperatures of formation probably do not reflect paleoclimates (i.e., do not reflect surface air temperatures).

of equilibrium coprecipitation of SiO_2 and goethite would probably introduce an error of no more than about 0.5 per mil in the calculated endmember goethite $\delta^{18}O$ value.

Also plotted in Fig. 3 are the 0 °C and 30 °C isotherms for goethite calculated using eqn 1 and α values from Table 1. Among different locales with identical modern average annual surface air temperatures, there are differences in the average annual precipitation $\delta^{18}O$ (and δD) values. For example, at sites with an average annual surface temperature of 0 °C, the $\delta^{18}O$ values of modern annual precipitation range from about -20 per mil to -10 per mil among the sites [e.g. Yurtsever and Gat, 1981]. Thus, on the 0 °C goethite isotherm in Fig. 3, there is a 10 per mil $\delta^{18}O$ range within which samples would plot, if they had formed at 0 °C in equilibrium with average annual meteoric waters which had fallen as precipitation in 0 °C climates. This portion of the 0 °C isotherm is shown as a solid line in Fig. 3. Similarly, among a variety

of sites with the same average annual surface temperature of about 30 °C, modern $\delta^{18}O$ values of average annual precipitation range from about -8 to 0 per mil [e.g. Yurtsever and Gat, 1981]. Goethites which formed at 30 °C in equilibrium with meteoric waters that precipitated in 30 °C climates would plot only on the solid portion of the 30 °C goethite isotherm in Fig. 3 (including portions which would extend beyond the upper border of the diagram).

The same reasoning can be applied to the 5 °C, 10 °C, 15 °C, 25 °C, etc., goethite isotherms. Estimates of lower and upper limits for the $\delta^{18}O$ values of average annual precipitation at different sites with the same average annual surface air temperature, based on data in Yurtsever and Gat [1981], are listed in Table 2. These maximum and minimum $\delta^{18}O$ values of precipitation at each average annual surface temperature were used to determine the corresponding maximum and minimum

TABLE 2.   Estimates of the probable range of $\delta^{18}O$ values for modern average annual precipitation among different sites with a common average annual surface air temperature.

Ave. annual Surface air T(°C)	*Approximate Ave. annual $\delta^{18}O$ values	
	Minimum	Maximum
0	-20	-10
5	-17	-5
10	-10	-3
15	-8	-2
25	-8	0
30	-8	0

* Estimates based on data reported in Yurtsever and Gat [1981]. At temperatures > 15 °C, the range of average annual precipitation $\delta^{18}O$ values among different sites is probably controlled, at least in part, by the "amount effect" [Joussaume, et al., 1984].

goethite $\delta^{18}O$ values on the equivalent isotherm in Fig. 3. It was assumed that the loci of points for the upper and lower limits of these $\delta^{18}O$ each varied smoothly with temperature. These data and associated assumptions permitted the delineation of the stippled "surface domain" in Fig. 3. Goethite δD and $\delta^{18}O$ values would plot within this domain if the temperature of goethite formation were the same as the contemporaneous local average annual surface air temperature. The boundaries of this "surface domain" are only as accurate as the upper and lower limits on precipitation $\delta^{18}O$ listed in Table 2. Furthermore, the stippled "surface domain" of Fig. 3 assumes that all precipitation falls on Craig's [1961] meteoric water line (MWL), and that the goethite was formed in isotopic equilibrium with this precipitation.

Because of analytical uncertainties of about ±2 per mil for goethite δD and ±0.2 per mil for goethite $\delta^{18}O$ [Yapp, 1987] and unknown errors in the estimates of extremum $\delta^{18}O$ values for modern meteoric water in Table 2, the solid triangles of Fig. 3 are all presumed to represent samples within the "surface domain". As seen in Fig. 3, 15 of the 26 data points plot in this domain. These 15 goethite samples could have formed in thermal equilibrium with the atmosphere at the Earth's surface and thus could record paleoclimatic temperatures. Unfortunately, the age of formation of most of these "surface domain" goethites is not known, because most of the goethites in this sample population are secondary minerals in their respective host rocks [Yapp, 1987]. One

exception is a Late Ordovician oolitic ironstone (OPWis-6) with $\delta D = -139$ and $\delta^{18}O = -1.0$. This sample from a Late Ordovician weathering environment [Paull, 1977] plots within the surface domain of Fig. 3, and its calculated formation temperature of about 23 °C ± 4 °C is plausibly that of the ancient climate.

The eleven open triangles of Fig. 3 are all external to, and on the high temperature side of, the "surface domain". The $\delta^{18}O$ value calculated for the meteoric water in equilibrium with each of these eleven samples is more negative than the most negative $\delta^{18}O$ values observed for average annual precipitation at modern sites with corresponding *surface* air temperature. Furthermore, three of these open triangles indicate temperatures of formation in excess of 50 °C, which suggests environments of formation that were probably not in long-term thermal equilibrium with the Earth's surface atmosphere. Consequently, unless the $\delta^{18}O$ values of average annual precipitation under ancient climatic conditions exhibited a wider range among different sites with the same annual surface temperature, the eleven open triangles of Fig. 3 probably represent goethites which formed in warmer, subsurface environments. Alternatively, some of the eleven open triangles might represent goethites which crystallized during the warmer summer season in temperate climates. However, summer rain is usually enriched in ^{18}O relative to the average annual precipitation at temperate sites, and seasonal temperature variations tend to be considerably attenuated at depths greater than about two meters below the ground surface [Hillel, 1982]. Therefore, an explanation involving a seasonal temperature bias may be of limited applicability for these samples.

The concept of the "surface domain" illustrated in Fig. 3 could be applied to data sets from any hydroxyl-bearing mineral (e.g. kaolinite, gibbsite, boehmite, etc.). However, more work will be required to evaluate the validity of the ranges of modern meteoric water "isothermal" $\delta^{18}O$ values and the effect of changing climate on those ranges. Also, concerns about post-crystallization hydrogen isotope exchange of phyllosilicates [e.g. Bird and Chivas, 1988; Kyser and Kerrich, 1991] might be relevant to goethite. In the absence of dissolution and reprecipitation goethite does not show evidence of *oxygen* isotopic exchange [Yapp, 1991], but any hydrogen isotope exchange could negate the interpretation applied to affected samples in plots such as Fig. 3. For example, a positive shift in goethite δD values due to post-crystallization hydrogen isotope exchange with relatively deuterium-rich waters might have caused some of the open triangles of Fig. 3 to plot outside of the surface domain.

Mineral-pair Oxygen Isotope Paleothermometers

Oxygen isotopes in low temperature minerals such as kaolinite and goethite do not appear to be susceptible to post-crystallization exchange at sedimentary or diagenetic temperatures [Kyser and Kerrich, 1991; Yapp, 1991]. Co-precipitated oxygen-bearing minerals which form in isotopic equilibrium with the ambient water hold the greatest promise of serving as reliable paleothermometers in low temperature systems. Surficial geologic systems such as laterites which contain coexisting goethite, hematite, phosphate minerals, gibbsite, boehmite, and/or kaolinite, etc., are potentially rich sources of such paleotemperature information. The finely divided, intimately mixed nature of the constituent minerals in these systems poses problems for the determination of endmember mineral $\delta^{18}O$ values that are needed to calculate paleotemperatures. Selective dissolution methods have been applied with apparent success to this problem [Yapp, 1991; Bird, et al, 1992]. Therefore, the use of these mineral pairs to determine ancient continental climatic temperatures from laterites, etc., is limited largely by (1) the temperature dependence of the respective mineral-water fractionation factors, (2) establishment of criteria for ascertaining the presence of mineral-pair isotopic equilibrium, and (3) the possible influence of mutual cation substitution in the structures of the mineral pairs.

An example of the latter complication is shown by goethite which can exhibit substitution of up to 33 mole percent Al for Fe in the mineral lattice [e.g. Fitzpatrick and Schwertmann, 1982; Siehl and Thein, 1989]. This substitution corresponds to the solid solution of AlOOH in FeOOH. The effects of such solid solutions on the mineral-water isotopic fractionation factors for ideal solid solutions are shown in the following equations:

$$1000 \ln \alpha(D) = 1000 [\ln \alpha(D)_{Al} - \ln \alpha(D)_{Fe}] X_{Al}$$
$$+ 1000 \ln \alpha(D)_{Fe} \qquad (2)$$

$$1000 \ln \alpha(18) = 1000 [\ln \alpha(18)_{Al} - \ln \alpha(18)_{Fe}] X_{Al}$$
$$+ 1000 \ln \alpha(18)_{Fe} \qquad (3)$$

$\alpha(D)$ = mineral-water $\alpha(D)$ for the solid solution

$\alpha(D)_{Al}$ = mineral-water $\alpha(D)$ for pure AlOOH

$\alpha(D)_{Fe}$ = mineral-water $\alpha(D)$ for pure FeOOH

X_{Al} = mole fraction of the AlOOH component in the solid solution of AlOOH + FeOOH

Analogous definitions apply for $\alpha(18)$, $\alpha(18)_{Al}$ and $\alpha(18)_{Fe}$.

Diaspore is the AlOOH isomorph of goethite, but there are no low temperature isotope fractionation data for diaspore. To facilitate discussion I assume that the isotopic systematics of boehmite are the same as those of diaspore. The fractionation factors of goethite and boehmite at 25°C, from Table 1, were used to calculate values of $1000 \ln \alpha$ vs. X_{Al} for eqns 2 and 3. The results are plotted in Fig. 4. AlOOH is not completely miscible in FeOOH [Trolard and Tardy, 1987], so the lines in Fig. 4 are likely to be valid approximations only at low values of X_{Al}. These curves indicate that X_{Al} values in excess of about 0.04 have a measurable effect on goethite δD and $\delta^{18}O$ values. Since X_{Al} ranges up to 0.33 in some soil

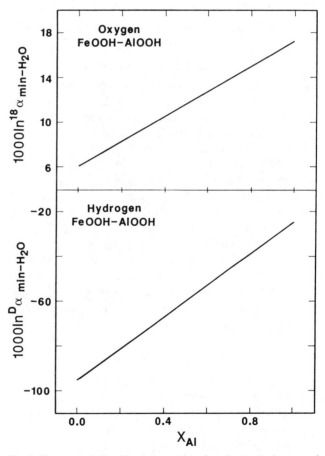

Fig. 4. Expected relationships between the mineral-water hydrogen and oxygen isotope fractionation factors and the mole fraction of AlOOH in the FeOOH + AlOOH system at <u>ca.</u> 25 °C. These relationships assume ideal solution behavior and probably do not have validity for $X_{Al} > 0.30$. However, they do show that Al substitution for Fe in goethite could be expected to have a measurable effect on goethite δD and $\delta^{18}O$ values for $X_{Al} > 0.04$.

goethites [e.g., Siehl and Thein, 1989], it is important to employ x-ray diffraction methods to determine X_{Al} values [Schulze, 1984] with which to adjust goethite δD and $\delta^{18}O$ results.

The expected temperature dependence of kaolinite-goethite, boehmite-goethite and phosphate-goethite oxygen isotope fractionation factors is illustrated by the following equations.

$$1000 \ln \alpha(18)_{K\text{-}G} = \frac{0.87 \times 10^6}{T^2} + 9.4 \qquad (4)$$

$$1000 \ln \alpha(18)_{B\text{-}G} = \frac{0.48 \times 10^6}{T^2} + 7.9 \qquad (5)$$

$$1000 \ln \alpha(18)_{P\text{-}G} = \frac{0.69 \times 10^6}{T^2} + 8.3 \qquad (6)$$

$\alpha(18)_{K\text{-}G} = \alpha(18)$ for the kaolinite-goethite pair $= R_K/R_G$
$\alpha(18)_{B\text{-}G} = \alpha(18)$ for the boehmite-goethite pair $= R_B/R_G$
$\alpha(18)_{P\text{-}G} = \alpha(18)$ for the phosphate-goethite pair $= R_P/R_G$
$R = {}^{18}O/{}^{16}O$ and T is in degrees Kelvin.

The kaolinite-goethite pair (eqn 4) is probably the best constrained by experiments [Yapp, 1987, 1990; Savin and Lee, 1988]. The equation for the boehmite-goethite pair (eqn 5) was calculated from the mineral-water equations in Table 1. The phosphate-water curve (Table 1) used to determine the phosphate-goethite equation was calculated from calcite-phosphate pairs analyzed by Karhu and Epstein [1986] and applies to the β-$BiPO_4$ compound prepared by the modified method of Tudge (1960). There is currently no systematic "correction" which can be applied to the earlier α-$BiPO_4$ analyses [e.g. Longinelli and Nutti, 1973] to make them consistent with β-$BiPO_4$ analyses. The mineral pairs of eqns 4, 5 and 6 can provide estimates of paleotemperatures for very wet, ancient continental systems where no other viable methods presently exist.

A PALEOCLIMATIC APPLICATION

An example of such a system is the Upper Ordovician Neda Fm. oolitic ironstone of the north central United States [Paull, 1977]. This goethite-dominated ironstone seems to have been subaerially weathered in the Late Ordovician [Paull, 1977]. The goethite contains no more than 4 mole % Al substituted for Fe. Details of the oxygen isotope geochemistry of the Neda Fm. deposit will be discussed elsewhere. The major point summarized here is the implication of $\delta^{18}O$ analyses obtained for coexisting goethite and phosphate in ooids in sample OPWis-6 [Yapp, 1991]. The goethite $\delta^{18}O$ value is about

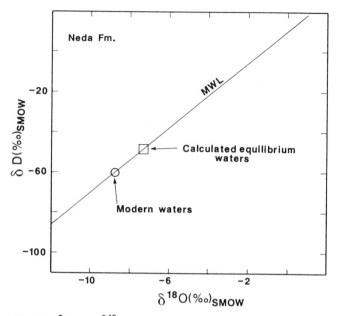

Fig. 5. δD and $\delta^{18}O$ values calculated for waters presumably in equilibrium with oolitic goethite from the Upper Ordovician Neda Fm. ironstone of the north central United States (see text). Note that these presumed Late Ordovician waters lie on the modern meteoric water line, but that they have different δD and $\delta^{18}O$ values than the modern waters in the region of the Neda Fm. The Neda Fm. meteoric waters of 440 million years ago might have originated as intense rainfall of a seasonal (monsoonal?) character.

-1.0 per mil, while the phosphate $\delta^{18}O$ value (as β-$BiPO_4$) is about +15.3 per mil. Substitution of these $\delta^{18}O$ values into eqn 6 yields a temperature of about 23°C ± 4°C. The calculated $\delta^{18}O$ value of the corresponding water is about -7.3 ± 0.5 per mil. If this water were to fall on the modern MWL, the δD value would be about -48 per mil. Interestingly, the measured δD value of the hydrogen in the oolitic goethite is about -139 per mil. Assuming that the goethite hydrogen did not experience post-crystallization isotope exchange, the calculated δD value of the original water would have been approximately -49 per mil. The agreement between the δD values calculated for the ancient water by these two methods is rather good. The calculated δD and $\delta^{18}O$ values of the Neda Fm. water are plotted in Fig. 5. Also shown in Fig. 5 are the values for modern water in the region as analyzed by the author. Given the number of uncertainties in the calculations it is best to exercise caution in the interpretation of the result.

However, if the calculated Neda Fm. waters accurately reflect the δD and $\delta^{18}O$ values of meteoric waters present at this site in the Late Ordovician, we might speculate that 440 million years ago, a low latitude, low altitude continental site was receiving precipitation, which (by comparison with δD and $\delta^{18}O$ values of modern tropical precipitation) originated as intense rainfall, perhaps with a highly seasonal (monsoonal?) character to it. Furthermore, the coincidence of this water with the modern MWL would suggest that the isotopic composition of the source ocean waters was similar to that of modern oceans (for both δD and $\delta^{18}O$). These results illustrate the potential for using the stable isotope geochemistry of low temperature iron and aluminum oxides to investigate continental paleoclimates not previously accessible to quantitative interpretation.

Acknowledgements. I thank H. Poths and M. Pedley for performing the δD analyses, and M. Sherman for typing the manuscript. The paper benfitted from the thoughtful reviews of R. Becker, S. Savin and Y. Shieh. This research was supported by NSF Grant EAR-9003108.

REFERENCES

Bernard, C., Composition isotopique des mineraux secondaires des bauxites. Problems de genese. Ph.D. dissertation, Univ. de Paris, Paris, 1978.

Bird, M.I., and A.R. Chivas, Stable isotope evidence for low temperature kaolinitic weathering and post-formational hydrogen isotope exchange in Permian kaolinites, *Chem. Geol. (Isotope Geosci. Sec.) 72*, 249-265, 1988.

Bird, M.I., A.R. Chivas, and A.S. Andrew, A stable isotope study of lateritic bauxites, *Geochim. Cosmochim. Acta 53*, 1411-1420, 1989.

Bird, M.I., A.R. Chivas, and A.S. Andrew, Reply to comment by C.H. Chen, K.K. Liu, and Y.N. Shieh on "A stable isotope study of lateritic bauxites", *Geochim. Cosmochim. Acta 54*, 1485-1486, 1990.

Bird, M.I., F.J. Longstaffe, W.S. Fyfe, and P. Bildgen, Oxygen isotope systematics in a multi-phase weathering system in Haiti, *Geochim. Cosmochim. Acta 56*, 2831-2838, 1992.

Cerling, T.E., The stable isotopic composition of modern soil carbonate and its relationship to climate, Earth Plan. Sci. Lett. 71, 229-240, 1984.

Cerling, T.E., Carbon dioxide in the atmosphere: evidence from Cenozoic and Mesozoic paleosols, Amer. J.Sci. 291, 377-400, 1991.

Chen, C.H., K.K. Liu, and Y.N. Shieh, Geochemical and isotopic studies of bauxitization in the Tatun volcanic area, northern Taiwan, Chem. Geol. 68, 41-56, 1988.

Craig, H., Isotopic variations in meteoric waters, Science 133, 1702-1703, 1961.

Dansgaard, W., Stable isotopes in precipitation, Tellus 16, 436-468, 1964.

Epstein, S., and T. Mayeda, Variation of ^{18}O content of waters from natural sources, Geochim. Cosmochim. Acta 4, 213-224, 1953.

Epstein, S., R. Buchsbaum, H.A. Lowenstam, and H.C. Urey, Revised carbonate-water isotopic temperature scale, Geol. Soc. Amer. Bull. 64, 1315-1326, 1953.

Fitzpatrick, R.W., and U. Schwertmann, Al-substituted goethite - an indicator of pedogenic and other weathering environments in South Africa, Geochim. 27, 335-347, 1982.

Friedman, I., and J.R. O'Neil, Compilation of stable isotope fractionation factors of geochemical interest, Data of Geochem., 6th Ed., Chap. KK, Geol. Surv. Prof. Pap. 440-KK, 1977.

Friedman, I., A.C. Redfield, B. Schoen, and J. Harris, The variation of the deuterium content of natural waters in the hydrologic cycle, Rev. Geophys. 2, 177-224, 1964.

Gat, J.R., The isotopes of hydrogen and oxygen in precipitation. In: P. Fritz and J. Ch. Fontes (Editors), Handbook of Environmental Isotope Geochemistry, Vol. 1, The Terrestrial Environment, Elsevier, Amsterdam, 21-47, 1980.

Graham, C.M., S.M.F. Sheppard, and T.H.E. Heaton, Experimental hydrogen isotope studies - I. Systematics of hydrogen isotope fractionation in the systems epidote-H₂O, zoisite-H₂O and AlOOH-H₂O, Geochim. Cosmochim. Acta 44, 353-364, 1980.

Hillel, D., Introduction to Soil Physics, Academic Press, New York, 364 pp., 1982.

Joussaume, S., R. Sadourny, and J. Jouzel, A general circulation model of water isotope cycles in the atmosphere, Nature 311, 24-29, 1984.

Karhu, J., and S. Epstein, The implication of the oxygen isotope records in coexisting cherts and phosphates, Geochim. Cosmochim. Acta 50, 1745-1756, 1986.

Knauth, L.P., and S. Epstein, Hydrogen and oxygen isotope ratios in modular and bedded cherts, Geochim. Cosmochim. Acta 40, 1095-1108, 1976.

Kyser, K.T., and R. Kerrich, Retrograde exchange of hydrogen isotopes between hydrous minerals and water at low temperatures, in: Stable Isotope Geochemistry: A Tribute to Samuel Epstein, Geochem. Soc., Spec. Pub. 3 (H.P. Taylor, Jr., J.R. O'Neil and I.R. Kaplan, eds.), 409-422, 1991.

Lawrence, J.R., and J.W.C. White, The elusive climate signal in the isotopic composition of precipitation, in: Stable Isotope Geochemistry: A Tribute to Samuel Epstein, Geochem. Soc., Spec. Pub. 3 (H.P. Taylor, Jr., J.R. O'Neil and I.R. Kaplan, eds.), 169-185, 1991.

Longinelli, A., and S. Nutti, Revised phosphate-water isotopic temperature scale, Earth Plan. Sci. Lett. 19, 373-376, 1973.

Matheney, R.K., and Knauth, L.P., Oxygen isotope fractionation between marine biogenic silica and seawater, Geochim. Cosmochim. Acta 53, 3207-3214, 1989.

Miyake, Y., O. Matsubaya, and C. Nishihara, An isotopic study of meteoric precipitation, Pap. Meteorol. Geophys. 19, 243-266, 1968.

O'Neil, J.R., Theoretical and experimental aspects of isotopic fractionation, in: Stable Isotopes in High Temperature Geological Processes, Rev. Mineral. 16, Mineral. Soc. Amer., 1-40, 1986.

Paull, R.A., The Upper Ordovician Neda Formation of eastern Wisconsin, in: Geology of Southeastern Wisconsin (Nelson, K.G., ed.), A Guidebook for the 41st Annual Tri-State Field Conference, C-1 to C-18, Wis. Geol. Nat. Hist. Surv., 1977.

Quade, J., T.E. Cerling, and J.R. Bowman, Systematic variations in the carbon and oxygen isotopic composition of pedogenic carbonate along elevation transects in the southern Great Basin, United States, Geol. Soc. Amer. Bull. 101, 464-475, 1989a.

Quade, J., T.E. Cerling, and J.R. Bowman, Development of Asian monsoon revealed by marked ecological shift during the latest Miocene in northern Pakistan, Nature 342, 163-166, 1989b.

Savin, S.M., and S. Epstein, The oxygen and hydrogen isotope geochemistry of clay minerals, Geochim. Cosmochim. Acta 34, 25-42, 1970.

Savin, S.M., and M. Lee, Isotopic studies of phyllosilicates, in: Hydrous Phyllosilicates (Exclusive of Micas); Rev. Mineral. 19, 189-223, Mineral. Soc. Amer., 1988.

Schulze, D.G., The influence of aluminum on iron oxides. VIII. Unit cell dimensions of Al-substituted goethites and estimation of Al from them, Clays Clay Minerals 32, 36-44, 1984.

Siehl, A., and J. Thein, Minette-type ironstones, in: Phanerozoic Ironstones, (Young, T.P. and W.E.G. Taylor, eds.) Geol. Soc. London, 175-193, 1989.

Tardy, Y., B. Kobilsek, C. Roquin, and H. Paquet, Influence of

Periatlantic climates and paleoclimates on the distribution and mineralogical composition of bauxites and ferricretes, Chem. Geol. 84, 179-182, 1990.

Trolard, F., and Y. Tardy, The stabilities of gibbsite, boehmite, aluminous goethites and aluminous hematites in bauxites, ferricretes and laterites as a function of water activity, temperature and particle size, Geochim. Cosmochim. Acta 51 945-957, 1987.

Tudge, A.P., A method of analysis of oxygen isotopes in orthophosphate - its use in the measurement of paleotemperatures, Geochim. Cosmochim. Acta 18, 81-93, 1960.

Yapp, C.J., A model for the relationships between precipitation D/H ratios and precipitation intensity, J. Geophys. Res. 87, C-12, 9614-9620, 1982.

Yapp, C.J., Oxygen and hydrogen isotope variations among goethites (α-FeOOH) and the determination of paleotemperatures, Geochim. Cosmochim. Acta 51, 355-364.

Yapp, C.J., Oxygen isotopes in iron (III) oxides. 1. Mineral-water fractionation factors, Chem. Geol. 85, 329-335, 1990.

Yapp, C.J., Oxygen isotopes in an oolitic ironstone and the determination of goethite $\delta^{18}O$ values by selective dissolution of impurities: the 5M NaOH method, Geochim. Cosmochim. Acta 55, 2627-2634, 1991.

Yapp, C.J., and M.D. Pedley, Stable hydrogen isotopes in iron oxides - II. D/H variations among natural goethites, Geochim. Cosmochim. Acta 49, 487-495, 1985.

Yurtsever, Y., and J.R. Gat, Atmospheric waters, in: Stable Isotope Hydrology, Deuterium and Oxygen-18 in the Water Cycle, Tech. Rep. Ser. 210, Internat. Atomic Ener. Agency, Vienna, 103-142, 1981.

C.J. Yapp, Department of Geology, University of New Mexico, Albuquerque, New Mexico 87131, USA.

An Oxygen-Isotope Study of Weathering in the Eastern Amazon Basin, Brazil

MICHAEL I. BIRD[1]*, FRED J. LONGSTAFFE[1], WILLIAM S. FYFE[1], BARBARA I. KRONBERG[2] AND AUGUSTO KISHIDA[3]

The oxygen-isotope composition of a suite of weathering minerals from regolith deposits in the Carajás and Paragominas regions of the eastern Amazon Basin, Brazil, has been determined. A range of physical and chemical (partial dissolution) techniques were employed to obtain $\delta^{18}O$ values for pure mineral phases (hematite/goethite, kaolinite, gibbsite, cryptomelane and anatase). The majority of regolith minerals from the Carajás region exhibit $\delta^{18}O$ values in equilibrium with waters similar in $\delta^{18}O$ to modern waters in the region, or up to ~2‰ lower. The lower than modern $\delta^{18}O$ values calculated for some of the waters can be explained without invoking any large-scale changes in climate in the region.

While some minerals from the Paragominas bauxite are also in isotopic equilibrium with waters similar in isotopic composition to modern waters in the area, gibbsite from the major bauxite horizon has an extremely low $\delta^{18}O$ value, in equilibrium with waters of approximately -9‰ (about 4‰ lower than modern waters). Such behaviour suggests that the major phase of bauxitization occurred when a strongly monsoonal climatic regime prevailed in the region.

A mineral-water oxygen-isotope fractionation factor of 1.0073 ± 0.0009 is proposed for cryptomelane at approximately 23°C.

INTRODUCTION

Much of the land surface of the Amazon Basin is mantled by a thick lateritic regolith that has probably been developing since at least the early Tertiary [Melfi et al., 1988; Vasconcelos, 1992]. During this time, the Amazon Basin has been in an equatorial position, ensuring that comparatively warm and wet conditions have prevailed for extended periods of time. The long-term tectonic stability of the region has facilitated preservation of the regolith. As a result of this long weathering history, the Amazon Basin contains some of the largest bauxite and lateritic iron-ore deposits in the world, as well as deposits of titanium, nickel, manganese, gold and kaolin.

Stable-isotope studies of regolith deposits can provide information about the continental palaeoclimates responsible for their formation [Bird and Chivas, 1988, 1989; Bird et al., 1989]. In the past, the regolith has received comparatively little attention, due mainly to three problems: (a) the paucity of well-defined mineral-water fractionation factors for minerals at surficial temperatures, (b) the difficulty in obtaining pure samples for analysis from complex mixtures of fine-grained minerals, and (c) the ambiguities that are commonly involved in drawing palaeoclimatic inferences from calculated $\delta^{18}O$ values of ancient meteoric waters. Lateritic deposits are amenable to oxygen-isotope investigations because they have comparatively simple mineralogy, making it possible to obtain relatively pure mineral separates for analysis.

Temperature estimates for the Amazon Basin during the last glacial maximum range from similar to the present to ~5°C cooler than modern temperatures [e.g. Colinvaux, 1989], while marine equatorial temperature estimates for the Cretaceous and Tertiary range from around 5°C below, to a few degrees above those presently observed [Barron, 1983; Savin et al., 1985; Shackleton and Boersma, 1981]. Such a range of temperatures could cause variations of ~1‰ for kaolinite and less for minerals such as gibbsite, anatase and hematite/goethite, which have less temperature dependent fractionation factors.

[1]*Michael. I. Bird, Fred J. Longstaffe and William S. Fyfe, Department of Geology, University of Western Ontario, London, Ontario, N6A 5B7, Canada.
[2]Barbara I. Kronberg, Department of Geology, Lakehead University, Thunder Bay, Ontario, P7B 5E1, Canada.
[3]Augusto Kishida, Rio Doce Geologia e Mineração S.A., Av. Pres. Wilson 231, 20030 Rio de Janeiro, RJ, Brazuk.

Climate Change in Continental Isotopic Records
Geophysical Monograph 78

In this paper, the $\delta^{18}O$ values for a variety of regolith minerals from deposits in the eastern Amazon Basin are used as the basis for estimating the isotopic composition of the meteoric waters from which they formed, and compared with analyses of modern waters in the region. From such comparisons information concerning the climate prevailing at the time at which the deposits formed can be deduced. Mineral-water fractionation factors at surficial temperatures are now known, with varying precision, for a range of minerals, including kaolinite [Land and Dutton, 1978], hematite/goethite [Yapp, 1990] and gibbsite [Bird et al., 1989; 1990; Chen et al., 1988]. This allows direct comparison of $\delta^{18}O$ results obtained from different minerals. Hence, it is possible to establish the degree to which isotopic equilibrium has been attained, both among different minerals and with the modern weathering environment. Lack of isotopic equilibrium among minerals indicates a complex polygenetic history.

GEOLOGICAL BACKGOUND

Carajás region

The Carajás region comprises a prominent series of plateaux, 500km south of Belém, the capital of Pará (Figure 1). The plateaux may be relics of the late Cretaceous - early Tertiary Sul-Americana Surface of King [1956], now at an elevation of around 600m, and rising ~100m above the surrounding countryside. Renewed uplift during the Late Tertiary 'Velhas cycle' led to the incision of the older plateau surface, and the development

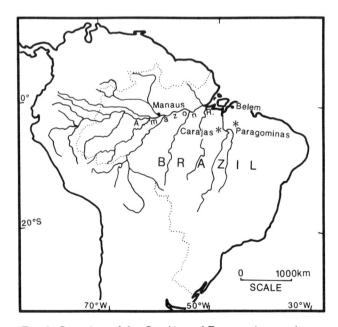

Fig. 1. Location of the Carajás and Paragominas regions.

of a lower surface in the region which is itself deeply weathered. The modern climate is comparatively mild, with a mean annual temperature of 23°C, and an annual rainfall of 1,500 to 2,000mm per annum, concentrated in the period December to March.

Within a radius of 50km, a wide variety of regolith types have been formed by intense weathering of Archaean and Proterozoic basement, and several of these were chosen for this study:

N4E iron-ore deposit. The N4E deposit is one of several major iron-ore deposits in the Carajás region, with total reserves of 18 billion tons of high-grade hematite ore [Melfi et al.,1988; Hirato, 1982]. The deposit developed by intense weathering of the Carajás Formation, an early Proterozoic banded iron formation, interbedded with mafic volcanic rocks. Silica and most other major cations in the original jaspilite protore have been leached out, enriching the deposit from an original 40% Fe to 66% Fe. The supergene ore averages 200m in thickness, and several types have been recognized. Capping the deposit is an indurated layer of 'canga', consisting of fragments of the ore cemented by secondary iron oxides with minor gibbsite and clay minerals. This unit grades down into brownish 'wet' hematite ore, consisting of martite crystals (hematite pseudomorphs after magnetite) in a friable goethitic matrix.

'Wet' hematite grades into 'soft' hematite, which is dark-grey in colour with conspicuous banding inherited from the original jaspilite. This material rapidly disaggregates into loose martite octahedra and primary hematite platelets. A small amount of goethite is sometimes present as a fine coating on the surface of martite crystals, and some martite crystals retain relic magnetite cores. A fourth ore-type, known as 'hard' hematite, is composed of massive metasomatic hematite bodies formed during metamorphism. They are commonly localized along fractures and shear zones which served as conduits for hydrothermal fluids.

Potassium-bearing manganese minerals in the deposit have recently been dated by $^{40}Ar/^{39}Ar$ and yielded ages of 40-50Ma [Vasconcelo, 1992].

N-5 bauxite deposit. Bauxite caps the plateau immediately east of the N-4 deposit, and is developed on volcanic units of the early Proterozoic Graó Pará Group, with a well defined vertical sequence. The surface consists of 3 to 5m of earthy, friable, orangy-brown bauxite that becomes reddish, and contains scattered ferruginous nodules toward the base. This is followed by 8 to 10m of hard, porous and sometimes pisolitic laterite, grading downward into mottled zone clays and saprolite at depth [da Silva Alves, 1985]. The age of the bauxite is not known.

Azul Manganese deposit. The Azul deposit formed by the lateritic weathering of the Azul member of the early Proterozoic Rio Fresco Formation - a manganiferous shale [Bernadelli, 1982; Beauvais et al., 1987]. Acidic, reducing weathering solutions caused the dissolution of primary rhodochrosite in the protore and subsequent reprecipitation of manganese as a variety of secondary minerals, including cryptomelane, nsutite and pyrolusite. The texture of the ore is variable, from pisolitic to manganiferous breccia ('canga'), to blocky or pelitic. The deposits of highest grade are in the blocky ore, which is an enrichment blanket formed above the protore and beneath the surface pisolitic material. Dating of potassium bearing manganese minerals have revealed a complex polygenetic weathering history extending back to at least 70Ma [Vasconcelo, 1992].

Bahia lateritic gold deposit. The Igarapé Bahia deposit is centred on an auriferous breccia pipe in sediments, pyroclastics and basic volcanic rocks of the early Proterozoic Igarapé Bahia Group. Gold has been concentrated by intense weathering of the chloritized, sericitized and carbonatized breccia, to produce a thick sequence of canga, pisolitic laterite and gossan, with comparatively little argillaceous material. Outside the breccia pipe, a thick lateritic profile comprises 5-12m of canga, underlain by around 90m of mottled zone material (predominantly kaolinite and iron oxides), grading downward into a thin (~10m) pallid zone and finally, saprolite.^{40}Ar/^{39}Ar ages of potassium bearing manganese minerals have yielded ages ranging from 0 to 40Ma [Vasconcelo, 1992].

Paragominas region

Paragominas lies approximately 200km SE of Belém, at an elevation of ~200m (Figure 1). The climate is hot and humid with a mean annual temperature of 25°C and an average annual rainfall of 2,500mm. The bauxite has been the subject of numerous studies that have resulted in conflicting conclusions regarding its genesis [Kotschoubey and Truckenbrodt, 1981; Kronberg et al., 1982; Hieronymus et al., 1990]. The bauxite is developed on a thick, deeply weathered sequence of interbedded fluvial sediments, originally thought to be the Barreiras Group, but now considered to be the Itapecuru Formation, of probable Cretaceous age [e.g. Hieronymus et al., 1990].

The weathering profile is typical of most bauxites in the Amazon Basin, with 1 to 15m of yellowish clay known as the Belterra Clay capping the deposit. This unit was originally thought to be of lacustrine origin [Sombroek, 1966], then an allochthonous deposit deposited by mudflows [Truckenbrodt and Kotschoubey, 1981], but is now viewed as an integral part of the bauxite profile [Kronberg et al., 1982; Hieronymous et al., 1990]. The Belterra Clay overlies a 10m-thick bauxitic horizon which consists of a pisolitic bauxite layer grading downward into ferruginous bauxite and then a basal massive bauxite layer. The bauxite is underlain by mottled and pallid kaolinitic clays.

The age of the bauxite is not known. The dissected surface on which the bauxite is formed was considered to be the equivalent of the Plio-Pleistocene Velhas surface by King [1956] and this coupled with a Plio-Pleistocene age for the Barreiras Group has led some workers to propose a Plio-Pleistocene age of formation [see Melfi et al., 1988 for discussion]. However, recognition that the underlying rocks may be much older, coupled with estimates of the time-scales involved in the formation of such an extensive deposit [e.g. Kronberg et al., 1982], suggest that it is probably considerably older.

EXPERIMENTAL TECHNIQUES

A range of pretreatments, both chemical and physical, was employed to concentrate, to the extent possible, individual minerals for isotopic analysis. The mineralogy of each separate was determined by X-ray diffraction (XRD). The morphology of iron-oxide phases in samples from the N4E deposit was investigated by scanning electron microscope and in polished thin section. The degree of aluminium substitution in the iron oxides was determined by XRD, using the d-spacings of the goethite-111 and hematite-110 diffractions according to the techniques of Schwertman et al. [1979] and Schulze [1984].

To correct for the presence of (usually minor) contaminants in some samples, a weighed aliquot of the contaminated sample was treated to destroy the major phase, concentrating the impurities in the insoluble residue. This residue was then washed, dried and re-weighed to determine the percentage of impurities in the original sample. The reproducibility of weight losses incurred during partial dissolution treatments is generally better than ±2% [Hodges and Zelazny, 1980; Bird et al., 1992]. The oxygen-isotope compositions of both the bulk sample and the residue were then determined and the isotopic composition of the pure mineral calculated by mass balance, based on the oxygen yield of bulk sample and residue. The details of the procedure employed for each mineral of interest are as follows:

A: Kaolinite pretreatment - Each sample was ultrasonically disaggregated and the <2μm fraction collected by standard sedimentation techniques. The sample was then treated overnight at room temperature with 5% sodium hypochlorite solution to destroy organic matter, washed, and boiled for one hour in 6N HCl, to remove gibbsite and iron oxide impurities. The sample was washed and dried, and an aliquot was boiled for one hour in 5N NaOH to destroy kaolinite [Kampf and Schwertman, 1982; Yapp, 1991]. The residue from this process normally contained one or more of the following: quartz, mica, anatase, rutile and boehmite.

B: Iron oxide pretreatment - Each sample from the N4E mine were crushed if necessary and passed through a 125μm sieve. The <125μm fraction was passed through a Franz magnetic separator to remove non-magnetic impurities (mostly quartz) and then ultrasonically disaggregated to separate fine-grained iron-oxide coatings from large euhedral martite crystals. The fine coatings were decanted, concentrated by centrifugation and dried. In the case of sample N4E-1, gibbsite impurities were destroyed by boiling for 2.5 minutes in 0.5N NaOH. Maghemite and magnetite concentrates were obtained by repeated grinding in water, followed by magnetic separation with a hand magnet after each grinding.

C: Bauxite pretreatment - Each sample was crushed and ultrasonically disaggregated. Large detrital grains (ilmenite, quartz, zircon etc.) were separated by settling and decanting the fine material. Each sample was then treated overnight at room temperature with 5% sodium hypochlorite solution to destroy organic matter. Iron oxide impurities were then removed by agitating the sample overnight at room temperature in 40mls of 0.2M ammonium citrate solution containing approximately 1g of sodium dithionite (S. Short pers. commun.). An aliquot of the purified bauxite sample was boiled for 2.5 minutes in 0.5N NaOH to remove gibbsite (several treatments were sometimes required to completely remove coarsely crystalline gibbsite). An aliquot of the gibbsite-free material was boiled in 5N NaOH to remove kaolinite, leaving a residue of anatase ± rutile ± quartz. In some cases iron oxides from a bulk bauxite sample were concentrated by destroying gibbsite and kaolinite in boiling 5N NaOH [Kampf and Schwertman, 1982; Yapp, 1991]. The iron oxides from an aliquot of this concentrate were then removed by boiling in 6N HCl, to determine the proportion and isotopic composition of the insoluble residue, and to allow calculation of the isotopic composition of the pure iron oxides.

D: Manganese mineral pretreatment - The sample was crushed to <125μm and an aliquot treated for 30 minutes at room temperature with 0.1M hydroxylamine hydrochloride to remove poorly crystalline, easily reducible Mn and Fe oxides [Chao, 1972]. An aliquot of the treated sample was then treated with the ammonium citrate -- dithionite solution as described above to dissolve crystalline Mn-minerals leaving an insoluble silicate residue.

The use of chemical pretreatments, particularly at elevated temperatures, leads to concern that the oxygen-isotope composition of the sample to be analysed will have been modified by isotopic exchange during the treatment. Yapp [1991] has demonstrated that iron oxides are resistant to isotopic exchange during repeated boiling in 5N NaOH to remove silicate impurities provided dissolution - reprecipitation doesn't occur. Yeh [1980] showed that acid treatment, followed by hydrogen-peroxide treatment, followed by dissolution of iron oxides using the sodium citrate-dithionite technique of Mehra and Jackson [1960] did not affect the hydrogen-isotope (and by implication oxygen-isotope) composition of chlorite, kaolinite, illite and smectite mixtures. The ammonium citrate technique employed in this study is even less likely to cause oxygen-isotope exchange because the treatment is conducted at room temperature rather than 80°C.

The use of sodium hypochlorite (both at room temperature and 50°C) to destroy organic matter, and of boiling HCl to destroy gibbsite and iron oxide impurities have been previously employed with no adverse effects reported [Ayalon and Longstaffe, 1988; Giral, 1989]. The effect of hydroxylamine hydrochloride on the oxygen-isotope composition of manganese minerals has not been investigated, although in general, it appears that most minerals are resistant to oxygen-isotope exchange, provided dissolution and reprecipitation reactions do not occur [O'Neil, 1987]. However, while it is likely that partial dissolution techniques do not cause direct isotopic exchange, few affect only a single mineral. Some dissolution of other phases can be expected, and this has the potential to affect the accuracy of results.

Oxygen for isotopic analysis was liberated quantitatively from dried 10-15mg samples by reaction with bromine pentafluoride at 550°C [Clayton and Mayeda, 1963]. Prior to reaction, samples were outgassed for two hours in vacuo at temperatures which varied from 110°C for gibbsite and iron oxides, to 200°C for kaolinite and titanium oxides. Oxygen was converted to carbon dioxide by reaction with an incandescent carbon rod, and the $\delta^{18}O$ value of the resultant gas measured using a Micromass 602D mass spectrometer. Results are reported in per mil (‰) relative to Vienna Standard Mean Ocean Water (V-SMOW). A mean value of +9.60 ± 0.08‰ (1s; n=7) was obtained for the NBS#28 quartz standard over the period during which samples were analysed for this study. The reproducibility of the sample analyses is lower than obtained for the standard and the average deviation from the mean is estimated to be ±0.25‰. The assignment of errors in the calculation of pure mineral $\delta^{18}O$ values by mass balance is discussed in the footnote to Table 2. The $\delta^{18}O$ values of water samples were determined by standard CO_2 equilibration techniques, with a reproducibility of ±0.1‰.

RESULTS

Modern waters

Oxygen-isotope results for modern ground- and surface-waters in both the Carajás and Paragominas regions are presented in Table 1. While the number of analyses is small, the data from each region show a very small range, and are consistent with published data for the eastern Amazon Basin. Salati et al. [1979] report annual weighted-mean $\delta^{18}O$ values for rainfall at stations in the eastern Amazon region ranging from -3.4 to -5.0‰. Two samples of waters from the Paragominas area yielded values of -4.6 and -4.7‰. Waters from the Carajás region have slightly lower values from -5.3 to -5.6‰. This is consistent with the Carajás Range being both at higher elevation (400m higher than Paragominas, 600m higher than Belém) and further inland than the other locations. However, the $\delta^{18}O$ values for water in both the Carajás and Paragominas regions are lower than the long-term,

Table 1: $\delta^{18}O$ values of water samples from the region (all samples collected in June 1989).

sample	$\delta^{18}O(\permil)$
Carajás	
Seepage from Salobo mine adit	-5.4
Salobo River water	-5.3
Seepage into floor of N4E mine	-5.6
Paragominas	
Stream at Paragominas bauxite mine	-4.7
Seepage from roadcut near Paragominas	-4.6
Tocantins River	
Tucurui Dam water	-3.6

amount-weighted mean $\delta^{18}O$ values for precipitation at the coastal stations of Belém (-2.1), and Fortaleza (-2.4‰), established by Yurtsever and Gat [1981]. Salati et al. [1979] also reported a value of -3.7‰ for waters of the Tocantins River, compared with a value of -3.6‰ obtained for this study.

On the basis of the available data, we have assumed that modern waters have an average $\delta^{18}O$ value of -4.7 ± 0.5‰ in the Paragominas region, and -5.4 ± 0.5‰ in the Carajás region.

Mineral samples

The mineralogical and stable-isotope results from this study are presented in Table 2. The isotopic compositions of meteoric waters calculated to be in equilibrium with these minerals are plotted in Figure 2.

The $\delta^{18}O$ values for mineral samples from the Carajás region range from -6.5 to +19.8‰. The majority of kaolinite, gibbsite and microcrystalline iron oxide samples have $\delta^{18}O$ values in equilibrium with water approximately 1‰ lower than modern waters in the region. Coarse martite hematite and vein hematite samples have low $\delta^{18}O$ values between -3.3 and -6.5‰ while anatase and one microcrystalline iron oxide sample have high $\delta^{18}O$ values of +3.6 and +3.5‰ respectively.

Samples from the Paragominas region have $\delta^{18}O$ values ranging from +7.3 to +19.9‰. Kaolinite samples have a small range of values from, +18.7 to +19.9‰, and are in equilibrium with waters that have $\delta^{18}O$ values ~1‰ lower than modern waters. Gibbsite exhibits a wide range of $\delta^{18}O$ values, from +6.4 to +13.9‰, while samples of iron-oxide and anatase have values between +0.4 and +0.8‰.

DISCUSSION

Carajás region

N4E iron-ore deposit. Most of the oxygen-isotope data from the Carajás region were obtained from samples

collected from the N4E iron-ore deposit. Iron-oxide samples from this deposit can be divided into two groups: 1. coarsely crystalline (up to 1mm across) primary hematite plates and hematite pseudomorphs after magnetite (martite); and 2. microcrystalline hematite/goethite. In polished thin section, martite octahedra show a variety of degrees of alteration. In this paper, all alteration phases intermediate between magnetite and hematite are called kenomagnetite [Kellerud et al., 1969; Morris, 1983; Anand and Gilkes, 1984].

Martite octahedra in N4E-11 are predominantly kenomagnetite, with patches and rims of hematite, and only a thin coating of secondary microcrystalline goethite/hematite. Some of these crystals retain small magnetite cores. The XRD patterns exhibit comparatively broad peaks as is expected for a magnetite/kenomagnetite/hematite mixture. Electron microprobe analyses of the martite grains indicate consistently low levels of Al but appreciable concentrations of Si (~1%), and occasional traces of Ti, Zn and Mn. Foreign ions such as Ti and Zn stabilize the defect spinel structure of magnetite during its initial oxidation [Morris, 1983].

In the more weathered samples (N4E-2; N4E-4) magnetite cores are absent, and kenomagnetite in the martite octahedra is progressively replaced by fine secondary goethite/hematite. Some crystals are completely replaced by secondary iron oxide phases and are transluscent and reddish in colour in transmitted light. A similar phenomenon has been reported by Morris [1980], who found that kenomagnetite in the Hammersley banded iron deposits commonly transformed directly to goethite. Under the SEM, altered martite grains can be seen to have a thick coating of fine-grained secondary iron minerals (Plate 1A). When this coating is ultrasonically removed, the crystals appear to be covered with a honeycomb structure. The cause of the honeycomb structure is not known, but as the conversion of magnetite to hematite is accompanied by volume changes [Davis et al., 1968] it is possible that the structures are the result of volume changes accompanying the conversion (Plate 1B).

The second group of iron oxides consists of microcrystalline goethite/hematite. This material forms either a solid, and sometimes finely banded, cement (in canga), or an earthy matrix (in wet and soft hematite ore-types).

Because it is likely that the goethite and hematite mineral-water fractionation factors are sensitive to the degree of aluminium substitution in a crystal structure, an attempt was made to determine the Al-substitution in the microcrystalline iron minerals (Table 2). The d-spacings of the 111 plane in goethite and the 110 plane in hematite are dependent on the percentage of aluminium substituted in the mineral structure (Schwertman et al., 1979; Schulze, 1984). Goethite from deep in the supergene zone (N4E-11,

Table 2: $\delta^{18}O$ values and mineralogy of regolith samples from the Eastern Amazon region

sample	mineralogy[1]	method[2]	$\delta^{18}O_{impure}$[3]	$\delta^{18}O_{pure}$[4]	$\delta^{18}O_{H_2O}$[5]	comments
N4E Iron-ore Deposit						
N4E-1 coarse	Hm(D) Go(M) (0.16% H¶)	B	-2.3	-2.7 (Ma)	-:-	lateritic breccia ('canga') capping deposit; protore fragments cemented by microcrystalline Hm/Go/Gi (Al substitution: Go, 6.8%; Hm, 1.0%†)
N4E-1 fine	Go(D) Hm(M) (1.23% H¶)	B	-0.2	-0.2 (Feox)	-6.5	
N4E-2 coarse	Hm(D) Go(SD) (0.40% H¶)	B	-2.9	-4.8±0.5 (Ma)	-:-	friable 'wet hematite' ore; martite crystals loosely cemented in a Go/Hm matrix; ~20m depth
N4E-2 fine	Go(D) Hm(SD) An(Tr) (0.96% H¶)	B	-0.8	-0.8 (Feox)	-7.1	
N4E-3 fine	Go(D) Hm(M) An(Tr)	B	+3.5	+3.5 (Feox)	-2.8	microcrystalline Go-rich mud from a 1m wide vug at same locality as sample N4E-2; ~20m depth
N4E-3 HCl	An/Ru(D)	B	+3.6	+3.6 (An/Ru)	-3.2	
N4E-4 coarse	Hm(D) Go(M) (0.12% H¶)	B	-3.0	-3.3 (Ma)	-:-	transitional between 'wet' and 'soft' hematite ore, ~20m depth; (Al substitution: Go, 7.3%; Hm, 1.3%†)
N4E-4 fine	Go(D) Hm(SD) An(Tr) (1.01% H¶)	B	-0.4	-0.4 (Feox)	-6.6	
N4E-4 pseud.*	Hm(D) Go(M)	B	+0.3	+0.3 (Feox)	-6.0	
N4E-4 Km	Km(SD) Hm(SD) Go(M)	B	-0.6	-0.6 (Feox)	-6.8	
N4E-6 coarse	Hm(D)	B	-6.5	-6.5 (Ma)	-:-	'soft' hematite ore; loose martite octahedra and platy Hm with little microcrystalline Hm/Go cement; ~30m depth
N4E-8 coarse	Hm(D)	B	-6.5	-6.5 (Ma)	-:-	as for N4E-6; ~30m depth
N4E-10 coarse	Hm(D)	B	-5.0	-5.0 (Ma)	-:-	massive 'hard' hematite ore, ~30m depth
N4E-11 coarse	Hm(D)	B	-5.3	-5.3 (Ma)	-:-	'soft' hematite, Level 655, ~60m depth; loose martite octahedra with relict magnetite cores (Al-substitution Go, 1.8%; Hm, 0%†)
N4E-11 magnetite	Mg(SD) Km(SD)	B	-1.9	-1.9 (Mg)	-:-	
N-5 Bauxite Deposit						
N5-4 CDB	Gi(88%) An/Ru/Qz/Ka(12%)	C	+9.4	+9.7(Gi)	-6.2	light brown earthy bauxite; 45cm depth
N5-4 (residue)	An(D) Ru(M) Qz(M) Ka(M)	C	+6.9	-:-	-:-	
N5A-4 CDB	Gi(91%) An/Ru/Qz/Ka(9%)	C	+9.4	+9.7(Gi)	-6.2	light brown earthy bauxite with scattered red motiles, 1.5m depth
N5A-4 (residue)	An(D) Ru(M) Qz(M) Ka(M)	C	+6.1	-:-	-:-	
Azul Manganese Deposit						
Azul 3 untreated	mineralogy not determined	-	+1.9/+2.1			fine black indurated 'blocky' ore; ~4m depth
Azul 3 hydroxylamine	Cr(98%) Ka/Bo/An(2%)	D	+1.7	+1.3(Cr)	-:-	
Azul 3 residue	Ka/Bo/An(100%)	D	+14.4	-:-	-:-	

Table 2 (continued)

sample	mineralogy[1]	method[2]	δ¹⁸O$_{impure}$[3]	δ¹⁸O$_{pure}$[4]	δ¹⁸O$_{H_2O}$[5]	comments
Bahia Gold Deposit						
F174-20.5m <2μm	Ka (98%) An/Ru (2%)	A	+19.5	+19.7 (Ka)	-6.1	all samples from drillhole F-174, near the NW margin of the deposit.
F174-43.0m <2μm	Ka (97%) An/Ru/mica (3%)	A	+19.5	+19.8 (Ka)	-6.0	The hole passes through 14m of indurated pisolitic ironstone,
F174-63.6m <2μm	Ka (97%) An/Ru/mica (3%)	A	+19.5	+19.8 (Ka)	-6.0	followed by mottled and pallid zone kaolinitic clays and bottoms at
F174-81.4m <2μm	Ka (94%) An/Ru/mica (6%)	A	+18.2	+18.8 (Ka)	-7.0	121m in weathered basic intrusive rock. (sample depths given in
F174-81.4m residue	An/Ru/mica	A	+7.3	-.-	-.-	metres in the sample names)
F174-102.7m <2μm	Ka (95%) An/Ru/mica (5%)	A	+17.5	+18.1 (Ka)	-7.8	
F174-102.7m residue	An/Ru/mica	A	+5.9	-.-	-.-	
Paragominas Bauxite						
PARA-5A<2μm	Ka (95%) Qz/mica/An (5%)	A	+19.2	+19.7 (Ka)	-6.1	kaolinized Itapecuru Formation sediments beneath massive bauxite
PARA-5A residue	Qz(SD) mica(SD) An(M)	A	+9.8	-.-	-.-	layer at 15m depth, roadcut at Km-208 on Highway BR-010
PARA-5H<2μm	Ka (96%) mica/An (4%)	A	+19.5	+19.9 (Ka)	-5.9	lt. brown 'Belterra Clay' overlying bauxite, same locality, ~3m depth
PARA-6E pink raw	Gi(94%) Feox(1.5%) Ka/An(4.5%)	C	+7.7	+7.5 (Gi)	-8.4	massive, indurated pink bauxite from the Paragominas refractory
PARA-6E residue-1	Feox(17%) Ka/An(83%)	C	+12.5	+6.6±3.0 (Feox)	+0.4	bauxite mine; ~3m depth. The massive bauxite (PARA-6E pink)
PARA-6E residue-2	Ka(SD) An(SD)	C	+13.3	-.-	-.-	contains scattered 1cm diameter sub-vertical tubules infilled by fine
PARA-6E white raw	Gi(63%) Feox(2%) Ka/An/Ru(35%)	C	+14.4	+13.9 (Gi)	-2.1	white Gi/Ka-rich clay (PARA-6E white)
PARA-6E residue-1	Ka(66%) An/Ru(34%)	C	+15.6	+18.7±0.5 (Ka)	-7.1	
PARA-6E residue-2	An(D) Ru(M)	C	+7.3	+7.3 (An/Ru)	+0.8	
PARA-hp	Gi (85%) Ka(14%) An/Ru(1%)	C	+7.9	+6.4 (Gi)	-9.5	analysis from Bird et al. (1989)

[1] Hm = hematite; Go = goethite; Ma = martite; Km = kenomagnetite; Mg = magnetite; Ka = kaolinite; An = anatase; Ru = rutile; Gi = gibbsite; Qz = quartz; Bo = boehmite; Cr = cryptomelane; Feox = mixed iron oxides. Mineralogy by XRD, (D) = dominant; (SD) = sub-dominant; (M) = minor; (Tr) = trace. [2] see experimental methods section for a description of sample preparation. 3 δ¹⁸O of bulk sample in per mil (‰). [4] δ¹⁸O of pure mineral in per mil (‰), corrected for the presence of contaminating phases using the measured δ¹⁸O and oxygen yield of the contaminating phases. Where a mineral δ¹⁸O value did not require a correction for a contaminating phase the error was calculated assuming an error of ±0.25‰ for the δ¹⁸O values of the impure sample and contaminants and of ±2% in estimating the proportions of mineral and contaminants (assuming stoichiometric oxygen yields for each mineral). The combined error for corrected mineral δ¹⁸O values is ≤0.4‰ unless otherwise indicated. For calculations of 'pure' martite δ¹⁸O values, the hydrogen content of the sample (%H)was used to calculate the percentage of secondary goethite in the sample assuming a stoichiometric goethite has a hydrogen content. 5 δ¹⁸O water in equilibrium with mineral , calculated at 23°C (Carajás) or 25°C (Paragominas) using the following fractionation factors - kaolinite, 23°C = 1.0260, 25°C = 1.0256 (Land and Dutton, 1978); iron oxides, 23°C = 1.0063, 25°C = 1.0060 (Yapp, 1990a); gibbsite, 23°C and 25°C = 1.016 (Bird et al., 1990); anatase, 23°C = 1.0068, 25°C = 1.0065 (Zheng, pers. commun., as quoted in Bird et al., in press).†Al substitution (mole%) in microcrystalline secondary iron oxides, determined by XRD using the d-spacing of the 111 diffraction for goethite (Schulz, 1984) and the 110 diffraction for hematite (Schwertman et al., 1979). ‖weight % hydrogen (%H) determined by Carlo Erba gas chromatography.
* rhombohedral Hm/Go pseudomorphs of siderite (~1mm across) separated by hand-picking.

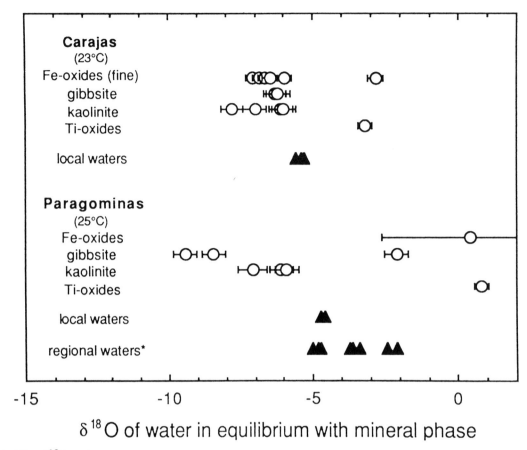

Fig. 2. The $\delta^{18}O$ values of modern waters, and palaeowaters calculated to be in equilibrium with regolith minerals in the Carajás and Paragominas regions. Errors are derived from errors in the estimates of pure mineral $\delta^{18}O$ values given in Table 2 (fractionation factors given in the footnote to Table 2). *Additional water $\delta^{18}O$ values for waters in the eastern Amazon region from Yurtsever and Gat, (1981) and Salati et al., (1979) and Table 1.

~60m depth - see Table 2) contains very little Al, reflecting the very Al-poor environment. In contrast, goethite in the upper parts of the supergene zone (N4E-4) and in the canga (N4E-1) contains around 7% Al. The higher degree of Al substitution is the result of the incorporation of Al released by weathering of small amounts of aluminosilicate minerals in the profile. Bulk Al_2O_3 analyses in the N4E deposit are likewise lowest in the weathered protore and reach maximum values in the canga [Melfi et al., 1988].

In addition to their distinctly different morphologies and chemistries, the microcrystalline (fine) and coarse iron-oxides in the N4E deposit also have distinctive ranges of $\delta^{18}O$ values (Table 2). Martites from the soft hematite (N4E-6, N4E-8, N4E-11), which show little development of secondary microcrystalline iron minerals, are characterized by very low $\delta^{18}O$ values ranging from -6.5 to -5.3‰, while martites from the wet hematite (N4E-4, N4E-2) and canga (N4E-1) exhibit higher $\delta^{18}O_{impure}$ values between -2.3 and -3.0‰. While all these values are

comparatively low, they are within the range that has been reported for hematites from other banded iron formations [Belevtsev et al., 1983; James and Clayton, 1962; Becker and Clayton, 1976].

The interpretation of the $\delta^{18}O$ data from the coarse iron oxides is complicated by two factors:

(a) The $\delta^{18}O$ values of hematite in the coarse iron oxides may have been affected both by weathering and by metamorphism, and may therefore reflect both processes. The similarity between $\delta^{18}O$ data for least weathered martites (N4E-6, N4E-8, N4E-11) and the $\delta^{18}O$ value of the 'hard' hematite (N4E-10), suggests that the isotopic composition of at least those samples reflects their formation in a metamorphic environment. Without information on the temperature of metamorphism, it is not possible to infer the nature of the metamorphic fluid responsible for the formation of the 'hard' hematite.

(b) Oxidation of magnetite to martite may proceed via diffusion of iron out of magnetite, through an essentially intact oxygen structure [e.g. Davis et al., 1968]. This raises

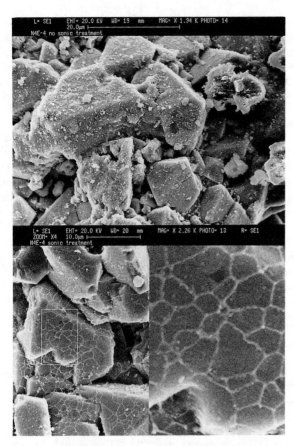

Plate 1.   Scanning electron micrographs of martite octahedra from the N4E iron ore deposit. (A) microcrystalline hematite/goethite coating martite in sample N4E-4. (B) Ultrasonically cleaned martite from sample N4E-4 (see text for discussion).

the possibility that martite may inherit part or all of the oxygen from the pre-existing magnetite. However, two observations suggest that the $^{18}O/^{16}O$ ratio of martite is reset during oxidation of magnetite. First, the $\delta^{18}O$ value of magnetite cores separated from N4E-11 martites is -1.9‰ (or possibly higher, as the sample may have contained martite impurities); this $\delta^{18}O$ value is substantially higher than martite in the same sample (-5.3‰, Table 2). Second, the $\delta^{18}O$ values of least altered martites are similar to that of hard hematite (N4E-10; -5.0‰), which is of metamorphic origin, and not derived from magnetite oxidation.

The higher $\delta^{18}O_{impure}$ values (up to -2.3‰) obtained for more highly weathered martites (N4E-4, N4E-2, N4E-1) indicate that a proportion of the martite hematite formed, or was isotopically reset, during weathering. A correction can be made for the presence of secondary goethite, and by implication secondary hematite (i.e. hematite not pseudomorphic after magnetite), based on the hydrogen content of the samples, the hematite/goethite ratio in the

fine fraction of the same sample (Table 2), and the isotopic composition of iron oxides in the fine fraction. This correction yields $\delta^{18}O_{pure}$ values (Table 2) for weathered martite hematite that are lower than $\delta^{18}O_{impure}$ values by 0.3 to 1.9 ‰. The uncertainty in this correction is up to ±0.5‰ (see Table 2), but even considering this error, the $\delta^{18}O$ values are consistently higher than those of purely metamorphic martites.

If martite hematite of weathering origin has a $\delta^{18}O$ of approximately -0.6‰ and metamorphic martite hematite, a value of -6.0‰, then the proportion of hematite in the weathered martite samples which has a weathering isotopic signature ranges from 20% (N4E-2) to 60% (N4E-1). This resetting of $\delta^{18}O$ values to those indicative of weathering could have occurred during the continued oxidation of magnetite cores during weathering, and/or during conversion of metamorphic kenomagnetite to hematite during weathering.

Kenomagnetite separated (by hand magnet) from N4E-4 has a $\delta^{18}O$ value of -0.6‰ similar to the microcrystalline iron oxides in the same sample. This similarity suggests that (i) the kenomagnetite formed during the weathering of relict magnetite cores, or (ii) kenomagnetite formed by oxidation of magnetite during metamorphism is susceptible to subsequent oxygen-isotope exchange as oxidation continues in the weathering environment.

With the exception of N4E-3 (discussed below), microcrystalline iron oxides from the N4E deposit exhibit a small range in $\delta^{18}O$ values, from -0.8 to +0.3‰. This suggests that the $\delta^{18}O$ value of the water in equilibrium with the minerals was around -6.6 ± 0.5‰, or approximately 1‰ lower than modern waters in the local area, but several per mil lower than some reported $\delta^{18}O$ values for precipitation in the eastern Amazon region (see Figure 2). The mineral-water fractionation factor for pure Al-free goethite was used in this calculation [Yapp, 1990], and as discussed above, the goethites analysed in this study contain around 7% Al. If the AlO(OH)-water fractionation factor for Al substituted into the FeO(OH) structure approximates that of boehmite [1.017; Bird et al., 1989], waters in equilibrium with the aluminous microcrystalline iron oxides might be up to 0.8‰ lower than calculated above. An adjustment of this sort has not been applied to the data in Table 2 or Figure 2. However, the example serves to illustrate that the $\delta^{18}O$ value of waters in equilibrium with the microcrystalline iron oxides may have been slightly lower than calculated (-6.6‰) but were unlikely to have been higher.

It is possible that some low-^{18}O fragments of the large martite grains were present in the fine-grained fraction of some samples; however the small range of values exhibited by the fine-grained iron-oxide samples suggests that the effect of such contamination was small.

The similarity between $\delta^{18}O$ values of microcrystalline iron oxides from the wet hematite (N4E-2) and from the canga (N4E-1), provides some useful genetic information. Hirato [1982] attributed the formation of the N4E canga to cementation of ore fragments by iron precipitated from solution as water migrates upward by capillary action during the dry season, and evaporates. Evaporation enriches the residual liquid in ^{18}O. Therefore, if Hirato is correct and evaporation was responsible for precipitation of the iron-oxide cements in the canga, their $\delta^{18}O$ values should be enriched in ^{18}O relative to the wet hematite, but this is not the case.

Evaporation may have played a role in the formation of goethite in N4E-3. This goethite (Al-substitution = ~5%), was collected from a large vug in the canga, as a yellowish friable mud and its $\delta^{18}O$ value suggests equilibrium with water approximately 4‰ enriched in ^{18}O relative to the water in equilibrium with the other fine-grained iron oxide samples, and 2.5‰ enriched relative to modern local groundwaters. This suggests that the goethite might have precipitated in the vug from evaporating, ponded groundwaters, although the $\delta^{18}O$ value of water calculated to be in equilibrium with the goethite is within the range of values observed in precipitation elsewhere in the eastern Amazon region.

Anatase from the same vug has a $\delta^{18}O$ value within 0.1‰ of that of the goethite, suggesting that the anatase-water fractionation may be similar to that of goethite at 23°C. This is consistent with data from titanium oxide mineral syntheses, although oxygen-isotope fractionation in the anatase-water system is complicated by the fact that ^{18}O-rich oxygen from sulphate ions in solution may be donated directly to the anatase structure [Bird et al., in press].

N-5, Bahia and Azul deposits. Two gibbsites from the N5 bauxite have identical $\delta^{18}O$ values of +9.7‰, corresponding to a water $\delta^{18}O$ of -6.2‰. This value overlaps the range of water compositions calculated from the N4E microcrystalline iron oxides, and is again lower than modern water $\delta^{18}O$ values. Kaolinites from the Bahia gold deposit were collected over a 100m section of core drilled through the kaolinitic regolith. The samples are in equilibrium with waters that have $\delta^{18}O$ values which overlap the field defined by the microcrystalline iron oxides and gibbsites.

The Bahia kaolinites that are closest to equilibrium with modern waters are highest in the regolith profile, while those deepest in the profile and closest to the weathering front have lower values. $^{40}Ar/^{39}Ar$ dating of potassium-bearing manganese minerals at Bahia has suggested a complex polygenetic history extending from the mid-Tertiary to the present [Vasconcelo, 1992]. Therefore, it is possible that the isotopic composition of kaolinite in the upper parts of the profile may have been reset during more recent weathering.

There are no published oxygen-isotope fractionation factors for the cryptomelane-water system at surficial temperatures, so it is not possible to plot the $\delta^{18}O$ value for water in equilibrium with cryptomelane ($K_{1-2}Mn_8O_{16}$) from the Azul mine on Figure 2. However, it is possible to estimate the cryptomelane-water oxygen-isotope fractionation factor, if we assume that the mineral is in equilibrium with water with a $\delta^{18}O$ value within the range of values measured for modern waters and calculated for palaeowaters in the region (-6.0 ± 0.9‰ 1s) at modern mean annual temperature.

The untreated sample of manganese ore from the deposit has a $\delta^{18}O$ value of +2.0‰. Treatment with hydroxylamine hydrochloride lowered this value by 0.3‰, presumably due to the dissolution of poorly crystalline manganese oxides in the sample. Pure cryptomelane from the sample has an estimated $\delta^{18}O$ value of +1.3 ± 0.3‰ (Table 2), implying an $a_{cryptomelane-water}$ of 1.0073 ± 0.009 at approximately 23°C. This value is comparable to a $\delta^{18}O$ value of +7.9‰ obtained for manganese oxides (predominantly todorokite) from a sea-floor manganese nodule forming at approximately 0°C from seawater with an approximate $\delta^{18}O$ value of 0‰ [Bar-Matthews and Matthews, 1990].

The Paragominas bauxite

The range of isotopic compositions calculated for waters in equilibrium with minerals from the Paragominas bauxite reveals a complex history (Table 2). Gibbsite separates from the deposit exhibit a range in $\delta^{18}O$ of 7.5‰. Fine powdery gibbsite which fills channels in the massive bauxite has a $\delta^{18}O$ value which is not in equilibrium with modern local waters, but within the range of modern rainfall $\delta^{18}O$ values reported for the eastern Amazon region (see Figure 2). The massive gibbsite, which forms the bulk of the deposit, has a very low $\delta^{18}O$ value of +7.4‰, consistent with an earlier analysis reported in Bird et al. [1989] of +6.4‰, from another outcrop of the bauxite. The $\delta^{18}O$ value of water in equilibrium with the massive gibbsite would have been approximately -9‰. This value is substantially lower than the modern value for water at this location. The existence of at least two generations of gibbsite with widely disparate $\delta^{18}O$ values is consistent with morphological and petrographic observations made by several authors for the existence of several periods of bauxitization [e.g. Hieronymus et al., 1990].

Kaolinite samples both from above and below the massive bauxite, and from channels within the bauxite have similar $\delta^{18}O$ values, ranging from +18.7 to +19.9‰. These kaolinites are, as in the Carajás region, in equilibrium with waters slightly depleted in ^{18}O relative to modern waters in the area. The concordance of kaolinite $\delta^{18}O$ values from the three major parts of the bauxite profile suggests that the kaolinitic profile existed prior to bauxitization. This is consistent with the model of formation for the Paragominas bauxite proposed by Kronberg et al. [1982], whereby bauxitization occurs in the sub-surface of a pre-existing profile by the focusing of transverse groundwater flow above a relatively impermeable zone.

Iron and titanium oxides in the massive bauxite have similar, and comparatively high, $\delta^{18}O$ values, although interpretation of these results is hampered by the large uncertainty in the case of the iron oxide sample ($\pm 3\text{‰}$) and by the complications in the interpretation of anatase $\delta^{18}O$ values as discussed above.

Palaeoclimatic implications

The $\delta^{18}O$ values for meteoric waters calculated from weathering minerals in the Carajás region have a comparatively small range of $-6.5 \pm 0.6\text{‰}$ (1s n=11) (excluding the N4E-3 samples because of possible evaporative modification of the fluid). This value is approximately 1‰ lower than for modern waters in the region ($-5.4 \pm 0.5\text{‰}$). Several competing effects may have caused the observed shift in the isotopic composition of waters in the region. If most of the samples formed prior to the major development of the polar ice-caps, then the 'pre-icecap' waters could be up to ~1‰ lower than modern waters reflecting the addition of low-^{18}O water now sequestered in the polar ice-caps [Shackleton, 1967; Savin and Yeh, 1981]. Alternatively, an increase in temperature of about 5°C would cause a decrease in mineral-water fractionation factors of around 0.6‰ for iron oxides and 1‰ for kaolinite. Temperature changes, either externally imposed by global changes, or locally imposed by uplift, may also have a small direct effect on the isotopic composition of rainfall. It is not possible to determine the relative importance of any of these mechanisms, but it would appear that the $\delta^{18}O$ data can be adequately explained without invoking any substantial changes in climate in the Carajás region throughout the Tertiary. The small range of values is remarkable given that weathering in the region has occurred episodically over at least the last 70Ma. It may be that this reflects the fact the the major periods of active weathering appear to occur during periods when the climate is comparatively warm and humid [Vasconcelos, 1992], and therefore conditions of formation are similar during each weathering event.

In contrast to the Carajás region, the extremely low $\delta^{18}O$ values obtained for the massive bauxite at Paragominas provides strong evidence for bauxitization under conditions that differed substantially from those of today. The 4-5‰ shift in the $\delta^{18}O$ value of the waters implied by the gibbsite $\delta^{18}O$ data is too large to be explained by analytical errors. It is also too large to have have resulted from changes in temperature, as the region has been in an equatorial position since the beginning of the Tertiary, and likewise, formation of the bauxite at a time when the polar ice-caps were significantly smaller could only reduce the difference by ~1‰.

Meteoric waters with low $\delta^{18}O$ values are common in parts of the tropics that have a monsoonal climate, or are strongly influenced by the Inter-Tropical Convergence Zone (ITCZ); this so-called monsoon, or amount, effect has been widely documented [Yurtsever and Gat 1981, Aharon, 1983; Bird, 1988]. Low $\delta^{18}O$ values in modern precipitation are infrequently reported (as low as -11‰) from the IAEA-WMO stations at Belém and Manaus, and Salati et al. [1979] speculated that they were result of yearly incursions of the ITCZ into the region. At present, the effect of the ITCZ is largely masked by the large amount of rainfall that is derived from evapo-transpired moisture in the basin.

The modern climate of the region is only weakly monsoonal in nature over the Amazon Basin. There is no seasonal reversal of winds at all over the adjacent ocean because the extent of the South American continent in the northern hemisphere is too small to produce a thermal low, and the North American continent too far away, to attract the ITCZ from its near-equatorial position [Nieuwolt, 1977].

The $\delta^{18}O$ values of waters in equilibrium with the massive bauxite (approximately -9‰) indicate that the region must once have experienced a strongly monsoonal climate. The extreme depletion in ^{18}O recorded by the gibbsites may partly result from the preferential formation of gibbsite during the most intense rainfall events (i.e. lowest $\delta^{18}O$) as discussed by Bird et al., [1990]; such behaviour also implies a strongly monsoonal climate.

The absence of a generation of low-$\delta^{18}O$ minerals in the Carajás region may indicate that the deposits at Carajás and Paragominas did not form at the same time. Alternatively, the ITCZ may not have moved as far south as the Carajás region, which consequently did not receive the low-^{18}O precipitation.

CONCLUSIONS

The physical and chemical separation techniques employed in this study have enabled the determination of the $\delta^{18}O$ values of a variety of regolith minerals, and a comparison of these values with those expected for equilibrium with modern conditions in both the Carajás and Paragominas areas of the eastern Amazon Basin. The data suggest that comparatively little change has occurred in the $\delta^{18}O$ composition of meteoric waters in the Carajás region, but that there has been a major shift in the isotopic composition of meteoric waters in the Paragominas region since the formation of the massive bauxite. The low-^{18}O signature associated with some gibbsites from the Paragominas bauxite is best explained by formation of the deposit in a monsoonal climatic regime.

Acknowledgements. We wish to thank the management and geologists of Docegeo, the exploration arm of Rio Doce Geologia e Mineração S/A, for providing access to, and logistic support for, fieldwork in the Carajás and Paragominas areas. We also want to thank P. Middlestead (U.W.O.) and J. Cali (A.N.U.) for laboratory assistance, and J-P Girard, S. M. Savin and an anonymous reviewer for thoughtful reviews of the manuscript.

REFERENCES

Aharon, P., Analysis of the anomalous $^{18}O/^{16}O$ and D/H isotope ratios in the tropical rainfall over the western Pacific Ocean, *Trans. Amer. Geophys. Union*, 64, 196, 1983.

Anand, R.R., and Gilkes, R.J., Mineralogical and chemical properties of weathered magnetite grains from lateritic saprolite, *J. Soil Sci.*, 35, 559-567, 1984.

Ayalon, A., and Longstaffe, F.J., Oxygen-isotope studies of diagenesis and porewater evolution in the Western Canada Sedimentary Basin: evidence from the Upper Cretaceous basal Belly River sandstone, Alberta, *J.Sed. Petrol.*, 58, 489-505, 1988.

Bar-Matthews, M., and Matthews, A., Chemical and stable-isotope fractionation in manganese oxide - phosphorite mineralization, Timna Valley, Israel, *Geol. Mag.*, 127, 1-12, 1990.

Barron, E.J., A warm equable Cretaceous: the nature of the problem, *Earth Sci. Rev.*, 19, 305-333, 1983.

Bird, M.I., Isotopically depleted rainfall and El Niño, *Nature*, 331, 489-490, 1988.

Bird, M.I., and Chivas, A.R., Oxygen-isotope dating of the Australian regolith, *Nature*, 331, 513-516, 1988.

Bird, M.I., and Chivas, A.R., Oxygen-isotope geochronology of the Australian regolith, *Geochim. Cosmochim. Acta*, 53, 3239-3256, 1989.

Bird, M.I., Chivas, A.R., and Andrew, A.S., A stable-isotope study of lateritic bauxites, *Geochim. Cosmochim. Acta*, 53, 1411-1420, 1989.

Bird, M.I., Chivas, A.R., and Andrew, A.S., Reply to comment by C.H-. Chen, K-K Liu and Shieh, Y.-N. on "A stable-isotope study of lateritic bauxites", *Geochim. Cosmochim. Acta*, 54, 1485-1486, 1990.

Bird, M.I., Longstaffe, F.J. and Fyfe, W.S., Oxygen-isotope fractionation in titanium oxide minerals at low temperature. *Geochim. Cosmochim. Acta*, in press.

Bird, M.I., Longstaffe, F.J., Fyfe, W.S. and Bildgen, P., Oxygen isotope systematics in a multiphase weathering system in Haiti. *Geochim. Cosmochim, Acta*, 56, 2831-2838, 1992.

Beauvais, A., Melfi, A., Nahon, D., and Trescases, J.J., Pétrologie du gisement lateritic manganésifere d'Azul (Brésil), *Mineral. Deposita*, 22, 124-134, 1987.

Becker, R.H., and Clayton, R.N., Oxygen-isotope study of a Precambrian banded iron-formation, Hammersley Range, Western Australia, *Geochim. Cosmochim. Acta*, 40, 1153-1165, 1976.

Belevetsev, Ya. N., Belevetsev, Ya. R., and Siroshtan, R.I., The Kirov Rog Basin, in *Iron-formations: Facts and problems*, edited by A.F. Trendall, and R.C. Morris, pp. 229-239, Elsevier, Amsterdam, 1983.

Bernadelli, A.L., Azul Manganese deposit, *Int. Symp. Arch. Early Prot. Evol. and Metall. (Excursion guide)*, Rio de Janeiro, 62-66, 1982.

Chao, T.T., Selective dissolution of manganese oxides from soils and sediments with acidified hydroxylamine hydrochloride, *Soil Sci. Soc. Amer. Proc.*, 36, 764-768, 1972.

Chen, C.-H., Liu, K.-K., and Shieh, Y.-N., Chemical and isotopic studies of bauxitization in the Tatun Volcanic area, northern Taiwan, *Chem. Geol.*, 68, 41-56, 1988.

Clayton, R.N., and Mayeda, T.K., The use of bromine pentafluoride in the extraction of oxygen in silicates for isotopic analysis, *Geochim. Cosmochim. Acta*, 27, 43-52, 1963.

Colinvaux, P.A., Ice-age Amazon revisited, *Nature*, 340, 188-189, 1989.

Davis, B.L., Rapp, G.Jr., and Walawender, M.L., Fabric and structural characteristics of the martization process, *Am. J. Sci.*, 266, 482-496, 1968.

Giral, S., Géochimie isotopique ($^{18}O/^{16}O$) de kaolinites de profils latéritiques d'Amazonie, *D.E.A. de Géologie*, Univ. d'Aix-Marseille III, pp. 43 unpublished, 1989.

Hirato, W.K., Serra Dos Carajás - Pará State: iron, manganese, copper and gold deposits, *Int. Symp. Arch. Early Prot. Evol. and Metall. (Excursion guide)*, Rio de Janeiro, 40-75, 1982.

Hieronymus, B., Kotschoubey, B., Boulegue, J., Benedetti, M., Godot, J.M., and Truckenbrodt, W., Aluminium behaviour in some alterites of eastern Amazonia (Brazil), *Proc. 2nd Int. Symp, Gochem. Earth's Surf. and Mineral Formation*, Aix-en-Provence, 74-77, 1990.

Hodges, S.C., and Zelazny, L.W., Determination of non-crystalline soil components by weight difference after selective dissolution, *Clays Clay Miner.*, 28, 35-42, 1980.

James, H.L., and Clayton, R.N., Oxygen-isotope fractionation in metamorphosed iron-formations of the Lake Superior region and in other iron rich rocks, in *Petrologic studies, a volume to honour A.F. Buddington*, edited by H.L. James, A.E.G. Engel, and B.F. Leonard, pp. 217-239, Geol. Soc. Amer., Denver, 1962.

Kampf, N., and Schwertmann, U., The 5M-NaOH concentration treatment for iron oxides in soils, *Clays Clay Miner.*, 10, 401-408, 1982.

Kellerud, G., Donnay, G., and Donnay, J.D.H., Omission solid solution in magnetite: kenotetrahedral magnetite, *Zeitscher Kristallographie*, 128, 1-17, 1969.

King, L., A geomorphologia do Brasil Oriental, *Rev. Bras. Geogr.*, 2, 147-265, 1956.

Kotschoubey, B., and Truckenbrodt, W., Evolução poligenética das bauxitas do distrito de Paragominas-Açailandia (Estados do Pará e Maranhão), *Rev. Bras. Geoc.*, 11, 193-202, 1981.

Kronberg, B.I., Fyfe, W.S., McKinnon, B.J., Couston, J.F., Stiliandifilho, B., and Nash, R.A., Model for bauxite formation: Paragominas (Brazil), *Chem. Geol.*, 35, 311-320, 1982.

Land, L.S., and Dutton, S.P., Cementation of a Pennsylvanian deltaic sandstone: isotopic data, *J. Sed. Petrol.*, 48, 1167-1176, 1978.

Mehra, O.P., and Jackson, M.L., Iron oxide removal from soils and clays by a dithionite - citrate system buffered by sodium bicarbonate, *Clays Clay Min.*, 7, 317-327, 1960.

Melfi, A.J., Trescases, J.J., Carvalho, A., Barros de Oliveira, S.M., Ribeiro Filho, E., and Laquintie-Formoso, M.L., The lateritic ore deposits of Brazil, *Sci. Géol., Bull. (Strasbourg)*, 41, 5-36, 1988.

Morris, R.C., A textural and mineralogical study of the relationship of iron ore to banded iron-formation in the Hamersley iron province of Western Australia, *Econ. Geol.*, 75, 184-209, 1980.

Morris, R.C., Supergene alteration of banded iron-formation, in *Iron-formations: Facts and problems*, edited by A.F. Trendall, and R.C. Morris, pp. 513-534, Elsevier, Amsterdam, 1983.

Nieuwolt, S., *Tropical Climatology*, Wiley, London, pp. 212, 1977.

O'Neil, J.R., Preservation of H, C and O isotopic ratios in the low-temperature environment, in *Stable-isotope geochemistry of low-temperature fluids*, edited by T.K. Kyser, pp. 85-128, Mineral. Assoc. Canada. Short Course, 1987.

Salati, E., Dall'Olio, A., Matsui, E., and Gat, J.R., Recycling of water in the Amazon Basin: An isotopic study, *Water Resour. Res.*, 15, 1250-1258, 1979.

Savin, S.M., and Yeh, H.-W., Stable-isotopes in ocean sediments, in *The Sea, Volume 7, The Oceanic Lithosphere*, edited by C. Emiliani, pp. 1521-1554, Wiley-Interscience, New York, 1981.

Savin, S.M., Abel, L., Barrera, E., Hodell, D., Kennett, J.P., Murphy, M., Keller, G., Killingley, J., and Vincent, E., The evolution of Miocene surface and near-surface marine temperatures: Oxygen isotopic evidence, *Geol. Soc. Amer. Mem., 163,* 49-82, 1985.

Schulze, D.G., The influence of aluminium on iron oxides: VII: Properties of Al-substituted goethites and estimation of Al from them, *Clays Clay Miner., 32,* 36-44, 1984.

Schwertmann, U., Fitzpatrick, R.W., Taylor, R.M., and Lewis, D.G., The influence of aluminium on iron oxides: II: Preparation and properties of Al-substituted hematites, *Clays Clay Miner., 27,* 105-112, 1979.

Shackleton, N., Oxygen-isotope analyses and Pleistocene palaeotemperatures re-assessed, *Nature, 215,* 15-17, 1967.

Shackleton, N., and Boersma, A., The climate of the Eocene ocean, *J. Geol. Soc. London, 138,* 153-157, 1981.

da Silva Alves, C.A., Depósito de bauxita do Plato N-5, *Resumos XXXV Congresso Brasiliero de Geologia.* Rio De Janeiro, 1984, 123-129, 1985.

Sombroek, W.G., *Amazon soils, a reconnaissance of the soils of the Brazilian Amazon region,* PUDOC Wageningen, The Netherlands, pp. 300, 1966.

Truckenbrodt, W., and Kotschoubey, B., Argila de Belterra - cobertura Tertiária das bauxitas Amazonicas, *Rev. Bras. Geoc., 11,* 203-208, 1981.

Vasconcelos, P.M., Timing and Rates of Evolution of Hydrochemical Systems in Semiarid and Humid Environments by Application of ^{40}K-^{40}Ar and Laser-Heating ^{40}Ar/^{39}Ar Dating of K-Bearing Weathering Product Minerals, Berkeley, pp. 239, unpublished.

Yapp, C.J., Oxygen-isotopes in iron (III) oxides 1. Mineral-water fractionation factors, *Chem. Geol., 85,* 329-335, 1990.

Yapp, C.J., Oxygen-isotopes in an oolitic ironstone and the determination of goethite $\delta^{18}O$ values by selective dissolution of impurities - the 5M NaOH method, *Geochim Cosmochim. Acta, 55,* 2627-2634, 1991.

Yeh, H.-W., D/H ratios and late stage dehydration of shales during burial, *Geochim.Cosmochim. Acta, 44,* 341-352, 1980.

Yurtsever, Y., and Gat, J.R., Atmospheric waters, in *Stable isotope hydrology: Deuterium and oxygen-18 in the water cycle,* edited by J.R. Gat and R. Gonfiantini, pp. 103-142,.Int. Atomic Energy Agency, Vienna, 1981.

M.I. Bird[*], F.J. Longstaffe, W.S. Fyfe, Department of Geology, University of Western Ontario, London Ontario, N6A 5B7, Canada.

B.I. Kronberg, Department of Geology, Lakehead University, Thunder Bay, Ontario, P7B 5E1, Canada.

A. Kishida, Rio Doce Geologia e Mineração S.A., Av. Pres. Wilson 231, 20030 Rio de Janeiro, RJ, Brazil.

[*]Present address: Research School of Earth Sciences, The Australian National University, GPO Box 4, Canberra ACT 2601, Australia.

Ancient climate from deuterium content of water in volcanic glass

IRVING FRIEDMAN, JIM GLEASON, and AUGUSTA WARDEN

U. S. Geological Survey, Denver, Colorado

Explosive eruptions of rhyolitic tephras may eject ash to great heights where it is distributed by winds over large areas. This ash, within a few thousand years after deposition, incorporates relatively large amounts of environmental water (up to 3.5 percent by weight) into its glass structure. This secondary hydration water is shown to retain its original deuterium concentration through time, and because the deuterium content of precipitation has been used for climate characterization, the hydration water, which is related to ancient precipitation, can be used as an indication of ancient climates. Samples of water extracted from dated volcanic ash from Iceland, New Zealand, and the western and central United States have been analyzed for deuterium, and the results of these analyses are used to reconstruct elements of climate ranging in age from 885 years to 2.1 million years before present. Based on the analysis of ash samples erupted between 13,000 and 6800 years before present, the climate in northern Nevada, Oregon, western Washington and in western and central Montana has not changed greatly from about 6000 y BP to present, and the climate in western Washington has remained constant from about 13,000 y BP to present. However eastern Washington and western Montana may have been about 2°-6° C cooler at the end of the Pleistocene and early Holocene than the present. The deuterium concentrations of surface waters in the central and western United States appear to have been similar at 0.6 Ma, 0.75 Ma, and 2.1 Ma, but to have differed from the present concentrations in portions of the study area.

INTRODUCTION

Concentrations of the heavy (mass=2) stable isotope of hydrogen (deuterium) vary greatly in natural water. These variations are caused by isotopic separation processes that occur when water changes its state from vapor, to liquid, to solid during the hydrologic cycle. The physical basis for these isotopic fractionations is understood, and has been used to relate deuterium variations in natural water to elements of climate [Dansgaard, 1964; Friedman et al., 1964]. The determination of the deuterium-hydrogen ratios (usually expressed in δD notation) of ancient-water samples allows comparisons to be made between ancient and present climates.

There have been many successful attempts to use the hydrogen isotopic compositions of geologic materials to elucidate continental paleoclimates. Some of these studies have utilized organic materials and others have used minerals formed at or near the earth's surface [Friedman, 1983]. In both cases, the δD of the material was related to that of the local meteoric water, with allowances being made for post-depositional isotopic alteration. This paper will describe the use in paleoclimate studies of another source of ancient water, namely the secondary water of hydration in rhyolitic volcanic glass.

When erupted, the glass in rhyolitic ash typically contains 0.1 to 0.3 weight percent water and has a great affinity for additional water. Immediately after eruption, water from the surface environment begins to diffuse into the glass. The hydration process ceases when the glass is saturated with about 3.5% water [Ross and Smith, 1955]. The δD of secondary water in volcanic glass can be used to calculate the δD of environmental water that was present during the first few thousand years after the deposition of the ash, this being the time necessary for the water to diffuse completely through and saturate the thin walls of the pumice and glass shards that constitute the ash. The diffusion of water into the glass follows the relation :

$$\text{Depth of hydration} = \text{constant (time)}^{0.5}$$

Measured diffusion rates are typically in the range of 1 to 10 micrometers squared per thousand years under surficial conditions and are a function of temperature, relative humidity, and the chemical composition of the glass [Friedman and

Climate Change in Continental Isotopic Records
Geophysical Monograph 78

Long, 1976]. We demonstrate later in this study that once the glass is completely hydrated (saturated), the water incorporated into the glass structure does not undergo further significant isotopic exchange with its environment. This is because the water is bound very strongly to the glass structure, and the driving force for further isotopic exchange is very small.

All other environmental factors being equal, the thickness of the walls of the glass shards determines the time span represented by the secondary water of hydration. Shards with wall thicknesses less than a few micrometers are completely saturated in a few thousand years, with most of the water representing the first 20% of the hydration time. Therefore, finer ash fractions, rather than coarser, are preferred for paleoclimate studies.

In most cases the small amount of magmatic water present (usually 0.1% to 0.3%) represents less than 10% of the water extracted from the glass for analysis, and it can be ignored in calculations of the δD of environmental water. However, in some cases this magmatic water represents a significant portion of the analyzed water and the analytical data should be corrected accordingly. This is especially true for samples of young glass (particularly those having thick walls) where the hydration has penetrated only a short distance into the glass.

To calculate the δD of the environmental water that was present during hydration of the glass, it is necessary to know the magnitude of the isotopic fractionation that occurs when water diffuses into the glass. This fractionation was first estimated by Friedman and Smith (1958) and recently determined more accurately by Friedman et al., (this volume) to be 1.0343. The delta value of the environmental water that diffused into the ash can be calculated using the following equation:

$$\delta D_{environmental\ water} = 1.0343\left[1000 + \delta D_{glass}\right] - 1000$$

The secondary water of hydration in samples of volcanic ash of known age from Iceland, New Zealand and the United States varying in age from 0.885 ka to 2.1 Ma was analyzed for deuterium, and the results show that these samples of ancient water have retained their original deuterium signature. Data derived from the analyses of samples of ash from the central and western United States and Canada will be used to elucidate paleoclimates of the region.

Experimental Procedure

Samples of volcanic ash were first cleaned by conventional mineral separation methods to remove contaminants including silt, clay, detrital rock fragments, and zeolites. They were then dried at 200°C in vacuo and heated to temperatures of about 1400°C in vacuo to release water and other volatiles, which were then passed over cupric oxide at 650°C. At the conclusion of this treatment, the water was converted to hydrogen gas by reaction with uranium metal at 850°C. The amount of hydrogen gas, measured volumetrically, was used to calculate the water content of the volcanic glass to an accuracy of ±5% of the amount present. The D/H ratio of the hydrogen was then analyzed using a Nier-type double-collecting 15 cm./60°-radius, magnetic-deflection mass spectrometer. Appropriate corrections to the data were made for the $^3H^+$ ion, as well as for other documented sources of error. The results are expressed in per mil [‰] δD notation.

$$\delta D = \left[\frac{R_{sample}}{R_{standard}} - 1\right] \times 1000$$

The standard is Vienna Standard Mean Ocean Water (V-SMOW). All results are normalized to values of 0‰ for V-SMOW, and -438‰ for Standard Light Antarctic Precipitation (SLAP), and are precise to ± 3‰ [2 sigma]. This estimate of error is based upon replicate water extraction and isotopic analysis of 5 different ash samples.

Retention of Isotopic Signature

In several cases where ash beds of different ages were found at the same locality (Table 1), nearly all samples gave different δD values. Thus, homogenization of hydrogen isotopes did not occur, suggestive that samples of ash buried at low (surficial) temperatures retain their δD signature through time.

Other evidence for the retention of isotopic signature through time comes from the δD values of secondary water of hydration in samples of Lava Creek ash (erupted 0.61 Ma) collected from sites throughout the western and central United States (Table 7, Figure 3). Although some of these samples contain water whose calculated δD values resembles that of the local surface water, most contain water whose δD differs by as much as 56‰ from the δD of present-day local surface water. If isotopic reequilibration had occurred, all of the samples should contain water whose δD values are similar to those of present-day local surface waters. It is difficult to prove that little or no change in δD has occurred in our samples through time, but the totality of our data strongly suggests that this is true.

TABLE 1. δD of Samples of Different Ages from the Same Locality

Locality	Age of Ash				
	2.1 Ma	0.75 Ma	0.61 Ma	11.2 Ky	6.8 Ky
Ventura, Calif.	-90	-86	-96		
Lake Tecopa, Calif.	-107	-108	-113		
Bear Canyon, Utah	--	-175	-178		
Onion Creek, Utah	--	-150	-170		
Red Rocks Pass, Idaho	--	-181	-188		
Marias Pass, Mont.	--	--	--	-176	-162
New Rockport Colony, Mont.	--	--	-174	-152	--
Townsend, Mont.	--	--	--	-170	-163
Cupper Creek, Oreg.	--	--	--	-156	-129
Creston Bog, Wash.	--	--	--	-155	-141
Big Meadow Lake, Wash.	--	--	--	-163	-168

ISOTOPIC COMPOSITION of MODERN and ANCIENT WATER

Before discussing the results from the individual sample localities, it is necessary to point out that many of these ash samples were hydrated in lakes and bogs in which they were deposited, and that the inflow for these water bodies may have originated at some distance, rather than locally. Differences between δD values of present-day water and ancient water at some sites may therefore be the result of changed drainage patterns rather than changes with time in the δD of the local precipitation.

The oceans are the ultimate source of water vapor that falls as precipitation. Changes in the δD of the oceans as deuterium-depleted ice is added and removed from the continents as a result of glaciation and deglaciation will result in corresponding changes in the δD of precipitation. The formation of the Antarctic ice sheet during the Miocene resulted in an increase in δD of the oceans of 10‰, and Pleistocene ice increased the δD value an additional 10‰. [Friedman and Hardcastle, 1988]

To compare the δD of ancient surface water, as calculated from the analysis of the water extracted from the ash, to that of present-day surface water, it is necessary to determine the δD of modern water from the localities where the ash samples were collected. For Iceland, we used data from Arneson (1975). For samples from the United States, the local δD was determined in most cases by collecting and analyzing water samples from streams whose drainage areas were close to the localities of interest [Gleason et al., unpublished data]. The assignment of local δD values to present-day groundwater ·is particularly difficult in Utah, where great variability exists in local climate, and where high and variable amounts of evaporation can occur. This may account for the greater variability between present-day δD and paleo δD in this area.

HOLOCENE and LATE PLEISTOCENE CLIMATE RECONSTRUCTIONS

Ash from the United States and Canada

Analytical data and information about occurrence of samples from localities in the United States and Canada are given in Tables 2 and 3 The estimated mean annual δD value of modern surface water at each of the localities where the samples were collected is also shown, as is the difference between the calculated δD and the measured δD of present-day surface water. Given the local variability of the deuterium concentration in surface water due to micro-climatic factors, and the yearly variations in the δD of precipitation, we believe that our estimates of the deuterium concentrations of local surface water are accurate to ± 5 ‰.

Mazama Ash. Mazama ash resulted from the 6.8ka eruption of Mount Mazama that formed Crater Lake in southwestern Oregon [Powers and Wilcox, 1964; Lemke et al., 1975; Bacon, 1983]. The locations of samples are shown in Figure 1 and the δD data are given in Table 2.

With the exception of the sample from Big Meadow Lake, all of the samples yielded δD values indicative of hydration by water with a δD value within ±10 that of present-day surface water at the sampling site. The suspiciously high water content of the Big Meadow Lake sample, 4.3%, suggests that some of this water may be contained in non-glass phases, perhaps clay minerals. Inasmuch as the remainder of the samples contain water that diffused into the glasses over a variable, but relatively short time period, we conclude that the climate for the period 6.8 to 5 ka in the states of Washington, Montana and Nevada was similar to the present.

Glacier Peak Ash. The eruption of Glacier Peak, in north-central Washington, was the source of the 11.2 ka Glacier Peak

TABLE 2. Samples of Mazama Ash (6.8 Ka)

Sample number	Map ref[‡].	State	Locality	Water wt. %	δD, ‰ V-SMOW			
					measured	calculated	present surface water	Diff.[¶] in ‰
3477-19	10	Montana	New Rockport Colony	2.8	-152	-123	-120	-3
3474-16	11	"	Marias Pass	2.9	-162	-133	-130	-3
3516-4	12	"	Townsend	1.7	-163	-134	-135	+1
3516-6	18	"	Alder	3.0	-175	-147	-145	-2
3477-21	23	Nevada	Willow Creek	2.7	-150	-121	-120	-1
3477-12	19	Washington	Great Western Lake	3.7	-146	-117	-120	+3
3477-13	20	"	Okanogan	2.6	-152	-123	-120	-3
3477-14	4	"	Creston Bog	2.0	-141	-112	-115	+3
3477-16	22	"	Bailey Lake	2.1	-148	-119	-120	+1
3477-17	21	"	Bonaparte Lake	3.1	-162	-133	-130	-3
3477-18	6	"	Marmes Rock Shelter	3.1	-138 .	-108	-110	+2
3477-15	7	"	Big Meadow Lake	4.3[§]	-168[§]	-139[§]	-120	-19[§]

[‡] Numbers refer to locations of samples shown in figure 1.

[¶] Difference in ‰ between calculated δD value and that of present-day surface water.

[§] These values are suspect--see discussion on p. 10.

TABLE 3. Samples of Late Pleistocene Ashes [Glacier Peak (11.2 ka); Mt. St. Helens J (10.6 - 11.6 ka); Whidby Is. (13 ka); Mt. St. Helens S (13.6 ka); Merritt (73 ka)]

Sample number	Map ref‡.	State or provence	Locality	Water wt. %	δD, ‰ V-SMOW			
					measured	calculated	present surface water	Diff.¶ in ‰
3516-5	12	Montana	Townsend	2.8	-170	-142	-135	-7
3516-3	13	"	Gibson Reservoir	2.5	-169	-140	-130	-10
3477-7	8	"	Kerr Dam	2.5	-171	-143	-130	-13
3477-8	5	"	Diversion Lake	3.2	-175	-147	-130	-17
3477-9	11	"	Marias Pass	2.8	-176	-148	-130	-18
3477-10	10	"	New Rockport Colony	2.7	-174	-146	-120	-26
3477-11	9	Oregon	Cupper Creek	3.1	-156	-127	-110	-17
3477-5	3	Washington	Trinity Mine	3.0	-152	-123	-120	-3
3477-5A	3	"	Trinity Mine	3.3	-153	-124	-120	-4
3464-11	3	"	Trinity Mine, G.P.-B	3.6	-131	-101	-120	+19
3464-12	3	"	Trinity Mine, G.P.-M	3.4	-153	-124	-120	-4
3464-13	3	"	Trinity Mine, G.P.-G	3.6	-149	-120	-120	0
3477-1	1	"	Natural Cave	2.2	-158	-129	-120	-9
3477-2	4	"	Creston Bog	2.0	-155	-126	-115	-11
3477-4	2	"	Brewster Flat	2.6	-159	-130	-120	-10
3477-6	6	"	Marmes Rock Shelter	2.7	-149	-120	-110	-10
3477-3	7	"	Big Meadow Lake	2.8	-163	-134	-120	-14
3464-25	14	"	Whidby Island	2.1	-134	-104	-45	-59
3464-15	15	"	Swift Creek, St Helens J	4.1	-120	-90	-90	0
3464-16	15	"	Swift Creek, St. Helens S	4.2	-123	-93	-90	-3
3464-14	16	"	Vantage, St. Helens S	3.5	-155	-126	-125	-1
3464-28	17	British Columbia	Merritt	4.3	-172	-144	-135§	-9

‡ Numbers refer to locations of samples shown in figure 2

¶ Difference in ‰ between calculated δD value and that of present-day surface water.

§Brown, 1970.

Fig. 1. Map showing locations of Mazama ash samples. The numbers adjacent to the sample localities refer to Table 2

ash [Powers and Wilcox, 1964; Wilcox, 1965; Lemke et al., 1975; Porter, 1978; Mehringer et al., 1984]. The sample sites are shown in Figure 2 and data are given in Table 3.

Both late-Wisconsin-age continental glaciers and alpine glaciers at high elevations were receding by the time of the Glacier Peak eruptions. Two of the three samples from Trinity Mine, the most westerly of our Washington samples, yielded δD values consistent with hydration by present-day surface water. The δD values of the remaining sample from this locality, Glacier Peak horizon B, indicates that it hydrated in water trhat was enriched in deuterium relative to present-day environmental water. It is possible that the δD of the surface water in which the various Glacier Peak ashes were deposited changed during the approximately 500-1000 years that elapsed between the deposition of horizon B and horizons M and G. The isotopic composition of samples from eastern Washington, and one from north-central Oregon, are indicative of water depleted in deuterium by 9 to 17‰ compared to present-day surface water in the region, indicating either that the average climate from 11.2 ka to about 9 ka was about 3°C colder than the present, or that the surface water in this region at that time was derived from melting mountain glaciers. This estimate of temperature change is not very precise because the δD of precipitation is determined by a number of factors besides temperature [Dansgaard, 1964;

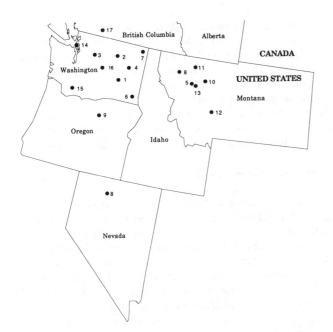

Figure 2. Map showing locations of 10 ka-13 ka ash samples. The numbers adjacent to the sample localities refer to Table 3

Friedman et al., 1964]. This uncertainty will also apply to paleoclimatic temperatures to be discussed later in this paper.

The δD values of samples from western Montana are also indicative of paleowaters depleted in deuterium compared to present-day surface water. The sample from Townsend, indicative of a 7‰ depletion is from an intermontane valley site, whereas the sample, from Gibson Reservoir was found in colluvium on a steep hillside. Both of these samples probably hydrated in water derived from local precipitation and indicate a paleoclimate about 2°C cooler than the present. The other four samples from western Montana were deposited in proglacial lakes. These samples have calculated δD values ranging from -143 to -148‰. Modern surface water from this region has δD values ranging from -120 to -130‰. Assuming that the lakes of late Wisconsin age contained meltwater of glacial ice that precipitated a few hundred to possibly a thousand years previously, these δD values, that are 13 to 26‰ lower than those of present-day water, indicate a late Pleistocene climate in which precipitation condensed at temperatures about 3°-6°C cooler than at present.

Prehistoric Mount St. Helens ash. An eruption of Mount St. Helens in 13.65 ±1.5 ka resulted in the deposition of ash layer S. An eruption dated at between about 10.6 and 11.6 ka yielded various layers of set J ashes (Table 3). The J layer collected at Swift Creek, near the volcano, gave a δD value for the hydrating water of -90‰, similar to that of present-day water. The S layer collected at the same site yielded a δD of -93‰. The S ash layer collected at Vantage, Wash. where the δD of present-day water is -125‰, yielded a calculated δD of -126‰ for the paleoenvironmental water. These results show that the δD value of surface waters in central and western

Washington was apparently the same as today during the interval from about 13 ka until the glass completed its hydration, 1 to 3 thousand years later.

Miscellaneous Late Pleistocene Ashes. Data on miscellaneous ash samples are given in Table 3. Pumice from late glacial drift collected on Whidbey Island in Puget Sound, Wash. (estimated age 13 ky), yielded a calculated δD of -104‰, which we take to be the value of late Pleistocene glacial melt-water in the Puget Sound area. Present-day water is estimated to have a δD of about -45‰. From this evidence we conclude not that the climate has changed greatly in *western* Washington State from 13 ky to the present, but that 13 ky glacial melt-water in the Puget Sound area originated from glaciers that formed at temperatures at least 15°C cooler than present temperatures in Puget Sound. The accumulation area for these glaciers may have been farther north, and/or at high elevations.

Precipitation in western Montana and eastern Washington from 11.2 ky to about 9 ky occurred at temperatures about 2°to 6° C cooler than the present, as revealed by the δD values for Glacier Peak ash. These are indicative of paleoenvironmental water with δD values about 10‰-25‰ lower than the δD of present-day surface water. Alternatively these low δD values may indicate that the hydration of these ash samples occurred in lakes fed by meltwater from ice of Pleistocene age.

Table 4 lists the calculated δD of the water from colocated samples of Glacier Peak (11.2 ky) and Mazama (6.8 ky) ash. The δD differences of 8 to 28‰ represent the change in the δD of surface water between these two time periods. Either the δD of precipitation from the 11.2 ky to 9 ky time period was lower than at present, or many of these ash samples hydrated in lakes fed by glacial meltwater depleted in δD.

Ash from Iceland

The data from samples of ash and perlite from Holocene and late Pleistocene eruptions of the samples discussed to this point are for volcanic ash that consists of glass shards that are the remnants of glass bubbles that have ruptured, or of pumice that has also suffered fracturing to yield thin-walled glass shards. This material is thin-usually less than 10 micrometers,

TABLE 4. Samples of Glacier Peak and Mazama Ash from the Same Locality

Sample State	Locality	δD, ‰ V-SMOW calculated Glacier Peak	Mazama	Difference per mil[¶]
Montana	Townsend	-142	-134	-8
"	Marias Pass	-148	-133	-15
"	New Rockport Colony	-146	-123	-23
Oregon	Cupper Creek	-127	-99	-28
Washington	Big Meadow Lake	-134	-139[§]	+5[§]
"	Creston Bog	-126	-112	-14

[¶] δD of Glacier Peak ash minus δD of Mazama ash.
[§] These values are suspect-see discussion on p. 10.

and the hydration-water quickly diffuses through these thin membranes. As a result, the small amount (typically less than 0.3%, and usually about 0.1%) of primary (magmatic) water originally present in the glass is diluted by the 3 to 4% of secondary hydration water, and no correction for the δD of primary water has been necessary in order to relate the measured δD value of to the δD of local environmental water. In contrast, it is neccessary to correct for magmatic water in samples of perlite, a material in which the secondary hydration proceded along cracks and fractures in fairly solid pieces of obsidian. This results in large volumes of the sample containing only magmatic water. These are separated by small volumes having secondary water that has diffused along cracks [Friedman et. al., 1966].

Two samples of 885y Hekla ash (Table 5), when corrected for magmatic water (an imprecise correction because of the lack of samples of unhydrated glass from this eruption), contain water that is about 10‰ depleted in deuterium as compared to present-day water. If uncorrected for magmatic water, the values for paleoenvironmental water would be lower by 20‰. Glacial meltwater, depleted in deuterium, might have contributed to the-now drained lake(s) into which these samples were deposited and in which much of the hydration took place. Alternately, the correction for magmatic water may be grossly in error, and the samples may contain present-day water.

Four samples, ranging in age from 6600 to 4000y, indicate paleoenvironmental waters that are depleted in deuterium by from 12 to 17‰ relative to modern water. Again, this might be due to the influence of glacial meltwater rather than changes in climate during the past 5000 years. The early post-glacial sample from Hráfntinnusker (3515-6) also shows a 15‰ depletion.

The δD of the sub-glacial perlite 3515-2 when corrected for magmatic water (assuming values of δD and concentration equal to that contained in the coexisting obsidian) indicates that this sample hydrated in an environment that contained glacial meltwater depleted in deuterium by about 39‰ compared to present-day environmental water. Deuterium depletions of this magnitude can be explained by changes in the temperature of condensation of precipitation, possibly caused by precipitation on the surface of glaciers that were elevated 1000 m or more above the present ground surface.

Kawakawa Ash from New Zealand

Data from samples of Kawakawa ash (20 ky) from localities in the Wairaki region of New Zealand are given in Table 6. M.K. Stewart [personal communication, 1988] estimated, on the basis of data derived from water samples collected from shallow wells, that the present-day δD value of surface water in the area where the Kawakawa ash samples were collected is -45 ±1‰. He also estimated, on the basis of $\delta^{18}O$ values of speleothems, that the δD of rainfall at 20ka would have been 3-5‰ more negative than at present. Our results suggest that the δD of surface water in the Wairaki region of New Zealand at 20 ka was -51‰, in good agreement with the estimate of Stewart.

LATE PLEISTOCENE and EARLIER CLIMATE RECONSTRUCTIONS

Late Pleistocene Canadian Ash

The analysis of ash collected at Merritt in British Columbia, Canada, and dated at 73 ky (Table 3), gave a calculated δD of –144‰ for the paleoenvironmental water. Present surface water in the area has a δD of about -135‰. Inasmuch as the

TABLE 5. Samples of Iceland Tephra

Sample number	Description, Locality	Age	Water wt. %	δD, ‰ V-SMOW			
				measured	calculated	present surface water	Diff.[‡] in ‰
3464-27	ash, Hekla #1 , Pjórsardalur	885 y BP	1.0	-119	-84[¶]	-75	-9
3515-10	ditto	885 y BP	0.6	-125	-87[¶]	-75	-12
3464-18	pumice, Hekla #4, Pjórsardalur	4,000 y BP	3.3	-122	-92	-75	-17
3464-20	ash,Hekla #5, Pjórsardalur	6,600 y BP	3.3	-117	-87	-75	-12
3515-9	pumice, Hráfntinnhrann	1,090 y BP	0.6	-109	-65[¶]	-75	+10
3515-11	lapilli, Sigalda	4,000 y BP	0.7	-118	-88	-75	-13
3515-12	ditto	4,000 y BP	1.8	-119	-89	-75	-14
3515-8	pumice, Sléttahrann	post-glacial	1.0	-109	-73[§]	-75	+2
3515-7	obsidian, Sléttahrann	ditto	0.12	-145			
3515-6	pumice, Hráfntinnusker	early post-glacial	1.7	-120	-90	-75	-15
3515-5	obsidian, Hráfntinnusker	ditto	0.12	-132			
3515-2	obsidian, Prestahnukur	ditto	0.25	-163			

[‡]Difference in ‰ between calculated δD value and that of present surface water.

[¶] Corrected for magmatic water, assuming 0.15% magmatic water with a δD of -147.

[§] Corrected for magmatic water, assuming values equal to coexisting obsidian.

TABLE 6. Samples of New Zealand Ash

Sample number	Locality	Water wt. %	δD,‰ V-SMOW			
			measured	calculated	present surface water	Diff.[‡] in ‰
3517-4	Oruani, Unit 6	2.9	-82	-51	-45	-5
3517-6	Broadlands, Unit 1	2.9	-82	-51	-45	-5

[‡]Difference in ‰ between calculated δD value and that of present surface water.

ash was deposited during a Pleistocene glacial maximum, the δD change of 10‰ can be accounted for by the change in the δD of the oceans during the glacial maxi,mum.

Lava Creek Ash

The Lava Creek ash originated from a large caldera-forming eruption in Yellowstone National Park, dated at 0.61 Ma, that deposited ash over a large portion of the western and central United States [Izett and Wilcox, 1982; Sarna-Wojcicki et al., 1984; Izett et al., 1988]. The data and locations of samples of Lava Creek ash are given in Table 7 and Figure 3.

The sample from Daniel Junction, Wyoming may have hydrated in water draining the nearby Wind River Range. Measured δD values of modern runoff are equal to that of paleoenvironmental water calculated from the δD value of the ash. The δD of the sample from Wascona Creek in Saskatchewan, Canada, indicates a paleoenvironmental water with a δD value of -165‰. This is much lower than the δD value of present-day surface water in the region, which ranges from -80 to -129‰, but it resembles the δD of water from subsurface aquifers (-149 to -164) McMonagle (1987). This suggests that subsurface water in this part of Saskatchewan is recharged by melted snow, while surface water, in addition to snow-melt, also includes summer precipitation enriched in deuterium. The δD from the ash sample implies that the climate in southern Saskatchewan 0.61 Ma was dominated by winter precipitation, in contrast to the present where 60% of precipitation occurs during the summer [Christiansen, 1965; Greer and Christiansen, 1963].

Samples of Lava Creek ash from eastern California, central Colorado, Wyoming, and Montana appear to have hydrated in

TABLE 7. Samples of Lava Creek Ash (0.61 Ma)

Sample number	Map ref.[‡]	State	Locality	Water wt. %	δD, ‰ V-SMOW			
					measured	calculated	present surface water	Diff.[¶] in ‰
3471-27	β	Arizona	Fort Thomas	4.1	-114	-84	-60	-24
3471-21	1	California	Lake Tecopa	1.8	-113	-83	-90	+7
3471-17	3	"	Xmas Can.(Searles)	2.1	-113	-83	-70	-13
3464-9	5[§]	"	Ventura	4.3	-95	-64	-40	-24
3474-20	5[§]	"	Ventura	4.2	-97	-66	-40	-26
3471-23	22	Colorado	Iliff	1.9	-146	-117	-115	-2
3471-19	16	"	Golden	2.5	-148	-119	-115	-4
3474-19	ε	"	Moffat County	3.9	-165	-136	-120	-16
3471-29	2	Idaho	Red Rocks Pass	3.8	-188	-160	-120	-40
3471-15	φ	Iowa	Little Sioux	1.7	-120	-90	-60	-30
3471-16	47	Kansas	Meade County	1.9	-118	-88	-40	-48
3474-1	4	Montana	Intake-7 1/2	3.1	-154	-125	-140	+15
3474-2	δ	Montana[†]	Lava Cr. "A"	3.0	-155	-126	-130	+4
3474-3	δ	"	Lava Cr. "B"	3.2	-160	-131	-130	-1
3471-28	11	Nebraska	Gilman Canyon	3.6	-121	-91	-50	-41
3471-18	3	South Dakota	Hartford	1.5	-155	-126	-70	-56
3471-26	1	Sask.-Can.	Wascona Creek	3.4	-193	-165	-120[#]	-45
3474-18	2	Texas	Austin	3.9	-89	-58	-30	-28
3471-20	γ	Utah	Arches N.M.	2.3	-151	-122	-110	-12
3464-21	3	"	Prom. Point	4.3	-160	-131	-120	-11
3471-14	4	"	Onion Creek	2.0	-170	-142	-110	-32
3471-22	α	"	Golden Spike	2.2	-176	-148	-125	-23
3471-25	9	"	Bear Canyon	2.4	-178	-150	-125	-25
3474-17	11	Wyoming	Wheatland	3.9	-145	-116	-118	+2
3471-24	12	"	Daniel Junction	1.5	-174	-146	-130	-16

[‡] Numbers are from Izett and Wilcox, 1982. Numbers and greek letters refer to Fig. 3.

[¶] Difference in ‰ between calculated δD value and that of present-day surface water.

[§] From Sama-Wojcicki et al., 1984.

[†] Silesia MT 7 1/2' quad., center NE 1/4 NE 1/4, sec. 22, T. 3 S. R 23 E.

[#]From McMonagle, 1987.

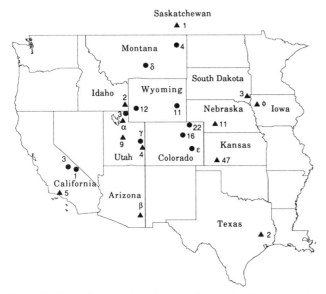

Figure 3. Map showing locations for Lava Creek ash samples. Circles represent samples whose calculated δD of their secondary water of hydration resembles those of present-day local surface water. Triangles represent samples whose calculated δD values show a deuterium-depletion of more than 20‰ compared to present-day local surface water.

water that is similar (± 20‰) in δD to present-day surface water. In contrast, samples from three other regions are more than 20‰ depleted in deuterium compared to present-day water. The first region consists of eastern Arizona and western California: the second region consists of Utah and eastern Idaho: the third region of South Dakota, Iowa, Kansas, Nebraska, Oklahoma and eastern Texas. The variability in the

difference in δD detween present-day and inferred paleowater from Utah may be related to the high, but variable, amounts of evaporation of surface water in this region.

The low δD values from these three regions may have been caused by one of several mechanisms. One of these is a change in the trajectories of storms that provide precipitation. If this were indeed the cause, it would suggest that a greater proportion of storms over these regions during deposition of Lava Creek ash originated over the central and northern Pacific Ocean rather than the Gulf of Mexico and the Gulf of California, as happens at present [Friedman et al., 1992].

Another possible cause of the low δD values for the third region is that drainage from the Rockies to the west may have been an important contributor to surface water in this region at 0.61 Ma. The low δD values that were found in samples from Idaho, Utah and Arizona could not have been caused by a change in the source of lake inflow, since at present the surface water in these areas originates in the mountains surrounding the lake basins. However the δD of present lakes in these areas is increased by evaporation. If at 0.61 Ma evaporation were greatly lower than at present, then the ancient lake water could have had a δD resembling that inferred from the analysis of the hydrated ash.

Bishop Tuff

The eruption that created the Bishop tuff in eastern California resulted in the formation of the Long Valley Caldera dated at 0.75 Ma, and resulted in the deposition of ash over a large area of the western United States [Izett et al., 1988]. The data for samples of Bishop tuff are given in Table 8 and plotted in Figure 4.

The pattern of δD values for the Bishop tuff is similar to that for the Lava Creek ash, suggesting that the climates of 0.75

TABLE 8. Samples of Bishop Tuff (0.75 Ma)

Sample number	Map ref.‡	State	Locality	Water wt. %	δD, ‰ V-SMOW			
					measured	calculated	present surface water	Diff.¶ in ‰
3471-8	8	California	Lake Tecopa	2.2	-108	-77	-90	+13
3464-8	43§	"	Ventura	3.0	-86	-55	-40	-15
3474-13	21	Colorado	Centerville	3.5	-139	-109	-110	+1
3471-9	24	"	San Luis Valley	2.5	-152	-123	-120	-3
3471-7	25	Idaho	Red Rocks Pass	2.4	-181	-153	-120	-33
3471-11	27	Nebraska	Bull Pasture	2.0	-113	-83	-50	-33
3474-14	32	Nevada	Calico Hills	3.8	-136	-106	-100	-6
3471-6	30	"	Jackass Flats	2.8	-141	-112	-100	-12
3471-13	34	New Mexico	Grama Siding	1.6	-110	-79	-50	-29
3474-10	36	Utah	Mineral Range	4.3	-154	-125	-120	-5
3474-11	41	"	Cove Fort	4.2	-155	-126	-120	-6
3474-9	38	"	Onion Creek	3.4	-150	-121	-110	-11
3471-5	39	"	Delta	2.4	-166	-137	-115	-22
3471-12	45	"	Bear Canyon	2.5	-175	-147	-125	-22
3471-10	46	Wyoming	Monolith Quarry	1.6	-162	-133	-135	+2

‡ From Izett et al., 1988.

¶ Difference in ‰ between calculated δD value and that of present-day surface water.

§ From Sarna-Wojcicki et al., 1984.

TABLE 9. Samples of Huckleberry Ridge Ash (2.1 Ma)

State number	Map ref.[‡]	State	Locality	Water wt. %	δD, ‰ V-SMOW			
					measured	calculated	present surface water	Diff.[¶] in ‰
3474-5	1	Arizona	California Wash	3.3	-91	-60	-60	0
3474-8	1	Calif.	Lake Tecopa	3.4	-107	-76	-90	+14
3464-10	5	"	Ventura	2.6	-101	-70	-40	-30
3474-6	44	Kansas	McPherson County	3.5	-105	-74	-40	-34
3471-4	52	"	Borchers	2.7	-126	-96	-40	-56
3471-3	5	Montana	Soda Springs	2.5	-190	-162	-140	-22
3474-4	1	Nevada	Pine Valley	3.2	-164	-135	-125	-10
3474-7	28	Texas	Snyder	3.7	-88	-57	-30	-7
3471-1	14	"	Finley	1.7	-109	-78	-60	-18
3471-2	1	Utah	Beaver	2.3	-185	-157	-120	-37

[‡] From Izett and Wilcox, 1982.
[¶] Difference in ‰ between calculated δD value and that of present-day surface water.

Ma and 0.61 Ma were similiar. Again, the samples from eastern California, western Nevada, and Colorado are within 20‰ of present-day waters. Some of the samples from Utah could have hydrated in lakes fed by runoff from the Wasatch Front that have δD values as low as -147‰. However a much lower evaporation than at present is required to explain the low δD found in these Bishop tuff samples. The sample from Red Rocks Pass, Idaho yielded a δD value indicative of environmental water appreciably lower in deuterium (-153‰) than any present-day groundwater in the region. Alternatively the presence of low δD water in the samples from Utah and Idaho might reflect a change in the deuterium content of regional precipitation, as described above for Lava Creek ash

samples. The environmental water in which samples from Nebraska and New Mexico hydrated have much lower in δD values than present groundwater, and the explanation for the Lava Creek ash samples from the this region may also apply here.

Huckleberry Ridge Ash

The Huckleberry Ridge ash was formed by an eruption in the Yellowstone Park region at 2.1 Ma [Izett and Wilcox, 1982]. The data for samples of Huckleberry Ridge ash are given in Table 9 and the localities are plotted in Figure 5.

The same pattern of δD values that we have seen for the Lava

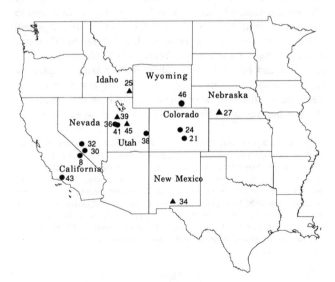

Figure 4. Map showing sample locations of Bishop Ash samples. Circles represent samples whose calculated δD of their secondary water of hydration resembles those of present-day local surface water. Triangles represent samples whose calculated δD values show a deuterium-depletion of more than 20‰ compared to present-day local surface water

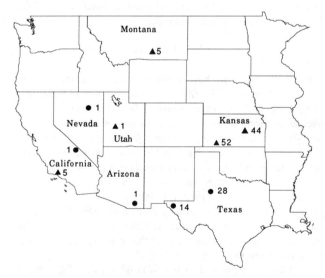

Figure 5. Map showing locations of Huckleberry Ridge ash samples. Circles represent samples whose calculated δD of their secondary water of hydration resembles those of present-day local surface water. Triangles represent samples whose calculated δD values show a deuterium-depletion of more than 20‰ compared to present-day local surface water

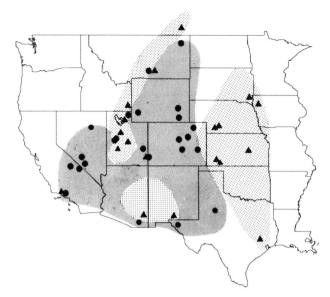

Figure 6. Map showing locations of Lava Creek, Bishop Tuff, and Huckleberry Ridge ash samples. The dark pattern shows areas of samples that have δD within ±20‰ of the present surface waters. The light pattern shows areas of samples that are depleted in δD by more than 20‰.

Creek ash and Bishop tuff applies to the Huckleberry Ridge ash samples. Samples from Arizona, Nevada, Texas and eastern California indicate a paleowater thatis similar to that expected for hydration by present-day water. All of the remaining 2.1 Ma ash samples hydrated in water that was depleted in deuterium by 22‰ to 56‰ compared to present surface water. The possible causes of this are the same as those discussed for the Lava Creek ash and Bishop tuff.

CONCLUSIONS

We believe that we have shown that the secondary water of hydration contained in rhyolitic volcanic ash represents a sample of ancient water whose δD value has remained unchanged from the time that the water was incorporated into the glass structure to the present. Using ash samples from dated eruptions (or ash that can be directly dated) allows the determination of the δD of environmental water of ancient, but known, age at the site of deposition of the ash.

Our results show that the δD of early post-glacial groundwater in south-central Iceland was about 15‰ lower than the present groundwater, and that sub-glacial samples from this region hydrated in water that was depleted by about 40‰ compared to the present.

A 20ka ash deposited in the Wairaki region of the North Island, New Zealand hydrated in water depleted in deuterium by 5‰ as compared to present-day environmental water, in good agreement with estimates based on other isotopic measurements.

Our data indicate that the climate in northern Nevada,

Oregon, western Washington, and western and central Montana has not changed greatly from about 6 ka to the present, and that the climate in western Washington has remained constant from about 13 ka to the present.

Our data also indicate that surface water from about 10 ka in eastern Washington and western Montana was depleted in deuterium compared to the present, suggesting a climate about 2° to 6° C cooler than the present.

Other isotopic climatic data derived from the δD of cellulose from trees that grew in the San Juan mountains of southern Colorado show that the δD of precipitation decreased from 9.6 ka to the present in southwestern Colorado. [Friedman et al., 1988]. Those data have been interpreted to indicate a reduction in monsoonal precipitation, derived from the Gulf of Mexico or Gulf of California, from 9.6 ka to the present. There is no contradiction between those data and the results of the present research, since there is little penetration of the present Gulf-derived monsoonal precipitation into northwestern Montana and eastern Washington.

Our results indicate that the climate in the central and western United States was the same at 2.1, 0.75, and 0.61 Ma. However, during these times the climate may have differed from that of the present in certain areas. The climate at these time periods was similar to the present in eastern California, Nevada, and in the central Rockies of Wyoming and Colorado. Surface paleowater in three other areas contained less deuterium than at present. These areas, shown in Figure 6, are: (1) eastern Arizona and New Mexico, (2) the mid-continent east of the 95th meridian, and (3) Utah and eastern Idaho. One possible cause of these δD differences is that a change occurred after 0.61 Ma in air-mass trajectories, resulting in an increased monsoonal flow of moisture northward from the Gulf of Mexico and Gulf of California to both the Great Basin and to the west-central United States. Changes in drainage and evaporation rates, particularly in Utah, New Mexico and Arizona, may also explain these δD differences. The collection and δD analyses of additional ash samples spanning the time interval 0.6 to 11 Ma should more closely constrain the time of this climatic event, as well as determine the climatic mechanisms involved in these δD changes.

Acknowledgements. We thank the many people who made this study possible by their contributions of samples. They include Glen Izett, Ray Wilcox, Andrei Sarna-Wojcicki, Marith Reheis, and Paul Carrara of the U.S. Geological Survey, Kristjan Saemundsson of the Science Institute of the University of Iceland, Diane Seward of the Department of Scientific and Industrial Research, New Zealand, Steven Porter of the Quaternary Research Institute of the University of Washington, Lester Davis of Montana State University, and Rick Hutchinson of the National Park Service. We also wish to acknowledge the contributions made by several reviewers, including Sam Savin and Paul Carrara.

REFERENCES

Arneson, B.,. Groundwater systems in Iceland traced by deuterium, *Science Institute, University of Iceland*, Reykjavik, 255 pp., 1975.

Bacon, C. R., Eruptive history of Mount Mazama and Crater Lake caldera, Cascade Range, U. S. A. *Journal of Volcanology and Geothermal Research*, **18**, 57-115, 1983.

Brown, R. M., Stevens, W. H., and Thurston, W. M., Deuterium content of Canadian waters-1, *AECL-2697*, Chalk River, Ontario, 21 pp.,1967.

Brown, R. M., Distribution of hydrogen isotopes in Canadian waters, p. 3-21, in *Isotope Hydrology 1970*, symposium proceedings, Vienna March 1970, International Atomic Energy, Vienna STI/PUB/255, 917 pp., 1970.

Christiansen, E.A.,. Geology and groundwater resources of the Kindersley area (72-N) Saskatchewan, *Saskatchewan Research Council, Geology Division, Report No.7*, 1965.

Dansgaard, W. Stable isotopes in precipitation, *Tellus,* **16**, 436-468, 1964.

Friedman, I., Paleoclimatic evidence from stable isotopes, in *Late Quaternary Environments of the United States*, H. E. Wright, ed., v. 1; The Late Pleistocene, S. C. Porter, ed., 1983, University of Minn. Press, Minn., 385-390, 1983.

Friedman, I., Carrara, P. and Gleason, J., Isotopic evidence for climate change in the San Juan Mountains, Colorado--9600 B.P. to present, *Quaternary Research* , **30**, 350-353, 1988.

Friedman, I., Gleason, J., Sheppard, R. A., and Gude, A. J. 3rd., Deuterium fractionation between water and silicic volcanic ash, *this volume*.

Friedman, I.and Hardcastle, K., Deuterium in interstitial water from deep sea cores, *Journal of Geophysical Research* **93**, 8249-8263, 1988.

Friedman, I. and Long, W., Hydration rate of obsidian, *Science*, **191**, 347-352, 1976.

Friedman, I., Redfield, A. C., Schoen, B., and Harris, J., The variation of the deuterium content of natural waters in the hydrologic cycle, *Reviews of Geophysics*, **2**, 177-224, 1964.

Friedman, I. and Smith, R. L., The deuterium content of water in some volcanic glass, *Geochimica et Cosmochimica Acta*, **15**, 218-228, 1958.

Friedman, I., Smith, R. L., Long, W. D., Hydration of natural glass and formation of perlite, *Geological Society of America Bulletin*, **77**, 323-328, 1966.

Friedman, I., Smith, G. I., Gleason, J. D., Warden, A., and Harris, J., Stable Isotope Composition of Waters in Southeastern California: Part 1, Modern Precipitation, *Journal of Geophysical Research*, **97**, 5795-5812, 1992.

Gleason, J., Warden, A., and Friedman, I., Deuterium in surface waters of the western United States, U.S. Geological Survey Professional Paper, manuscript in preparation.

Greer, J.E. and Christiansen, E.A., Geology and groundwater resources of Wynward area (72-P) Saskatchewan, *Saskatchewan Research Council, Geology Division, Report No. 3*, 1963.

Izett, G. A., Obradovich, J. D., and Mehnert, H. H., The Bishop Ash bed (Middle Pleistocene) and some older (Pliocene and Pleistocene) chemically and mineralogically similar ash beds in California, Nevada, and Utah, *U.S. Geological Survey Bulletin 1675*, 37 pp and map, 1988.

Izett, G. A. and Wilcox, R. A., Map showing localities and inferred distributions of the Huckleberry Ridge, Mesa Falls, and Lava Creek Ash beds (Pearlette Family Ash beds) of Pliocene and Pleistocene Age in the western United States and southern Canada, *U.S. Geological Survey*, Map I-1325, 1982.

Lemke, R. W., Mudge, M. R., Wilcox, R. E., and Powers, H. A., Geologic setting of the Glacier Peak and Mazama Ash-bed markers in west-central Montana. *U.S. Geological Survey Bulletin 1395-H*, 41 pp., 1975.

McMonagle, A. L., Stable isotope and chemical compositions of surface and subsurface waters in Saskatchewan. Thesis, Dept. Geology, Univ. of Saskatchewan, Saskatoon, Saskatchewan, Canada, 1987.

Mehringer, P. J.,Jr., Sheppard, J. C., and Foit, F. F., Jr., The age of the Glacier Peak tephra in west-central Montana. *Quaternary Research* **21**, 36-41, 1984.

Porter, S. C., Glacier Peak tephra in North Cascade Range, Washington: Stratigraphy, distribution, and relationship to late-glacial events. *Quaternary Research* **10**, 30-41, 1978.

Powers , H.A. and Wilcox, R. E., Volcanic ash from Mount Mazama (Crater Lake) and from Glacier Peak. *Science*, **144**, 1334-1336, 1964.

Ross, C. S., and Smith, R. L., Water and other volatiles in volcanic glasses, *American Mineralogist*, **40**, 1071-1089, 1955.

Sarna-Wojcicki, A. M., Bowman, H. R., Meyer, C. E., Russell, P. C., Woodward, M. J., McCoy, G., Rowe, J. J., Jr., Baedecker, P. A., Asaro, F., and Michael, H., Chemical analyses, correlations, and ages of Upper Pliocene and Pleistocene ash layers of east-central and southern California, *U.S. Geological Survey Professional Paper 1293*, 40 pp. and table, 1984.

Wilcox, R. E., Volcanic-ash chronology, in *The Quaternary of the United States*, edited by H. E. Wright, Jr. and D. G. Frey. Princeton University Press, Princeton, N. J., 807-818, 1965.

I. Friedman, J. Gleason and A. Warden, U.S. Geological Survey, Denver Federal Center, MS 963, Denver, Colorado 80225.

Deuterium Fractionation as Water Diffuses into Silicic Volcanic Ash

Irving Friedman, Jim Gleason, Richard A. Sheppard and Arthur J. Gude, 3rd

U. S. Geological Survey, Denver, Colorado

In order to use the isotopic composition of secondary water of hydration of volcanic glass as a measure of the isotopic composition of ancient environmental water, it is necessary to determine the fractionation of deuterium between water and hydrated glass (ash) that occurs when water diffuses into rhyolitic volcanic ash. Two methods gave a fractionation factor of $0.9668\pm.0005$ ($\alpha_{glass-water}$) for samples that have hydrated at surficial temperatures and atmospheric pressure. This value is in good agreement with an estimate of the fractionation factor reported by Friedman and Smith in 1958.

INTRODUCTION

For studies of climate change, it is important to secure samples of ancient materials that retain a record of paleoclimate. The deuterium content of natural water is related to both large-scale and local elements of climate [Friedman et al., 1964]. Thus a determination of deuterium content of samples of ancient water will allow comparisons to be made between ancient and modern climates [Friedman, et al., 1985].

Rhyolitic volcanic ash, as erupted, contains 0.1 to 0.3% water, but, immediately after eruption, it begins to react with environmental water, which is then incorporated into the glass structure. This process ceases when the glass is saturated with about 3.5 % water [Ross and Smith, 1955]. This secondary water of hydration does not undergo further isotopic exchange with its environment and represents a fractionated sample of ancient environmental water [Friedman et al., this volume]. In order to derive the actual deuterium content of the ancient environmental water present during hydration of the glass, it is necessary to know the fractionation factor $\alpha_{glass-water}$.

The first estimate of this fractionation was made by Friedman and Smith [1958], who measured the δD of perlite, a water-rich rhyolite glass formed by diffusion of environmental water into obsidian, and compared this to the δD of meteoric water at the localities where the perlite formed. They derived a fractionation factor of about 0.97 for this process, the glass being depleted in deuterium by approximately 30 per mil (‰) relative to the meteoric water.

We report on two determinations of this fractionation factor. The first was made by the extraction and isotopic analysis of water that had diffused into, and partially filled, small hollow volcanic glass spheres found in a seven million year old volcanic ash of the Miocene Chalk Hills Formation, southwestern Idaho. The second method of determining the fractionation factor at surficial temperature and pressure was by analyzing samples of the 20,000 year old Kawakawa ash that hydrated on shore in the North Island of New Zealand.

FRACTIONATION FACTOR DETERMINED from HOLLOW GLASS SPHERES

The first method of determining the fractionation factor at surficial temperature and pressure was by analyzing samples of ash from southwestern Idaho. This method utilized hollow glass spheres that had been deposited in a lake. These spheres are frozen bubbles formed by vesiculation of lava during a volcanic eruption of rhyolitic lava. They were apparently concentrated by flotation on a lake from a large amount of volcanic shards, which sank. The spheres were subsequently buried by later ash falls. They now form about 10 percent of a layer within the ash sequence. The spheres are 0.01 to 0.5 mm in diameter with wall thichness varying from 1 to 5 micrometers. Commonly, several spheres are attached to one another in a manner that indicates that they were originally part of a molten foam. Similar glass spheres were found in ejecta from the 1912 eruption of Katmai, where the spheres were observed floating on the sea [Perret , 1950]. Although the bubbles from the Katmai eruption appeared to be filled only with gas or vapor, those from the Idaho site contain a fluid phase as well. During many ash eruptions similar foam bubbles probably form, but the proportion of intact spheres relative to fractured bubbles is probably small and a specific, geologically rare, process is required to concentrate the spheres sufficiently so that they are observable.

Climate Change in Continental Isotopic Records
Geophysical Monograph 78

The spheres were generated during the eruption of the ash by the sudden exsolution of volatiles, mainly water. We postulate that as the glass spheres cooled, the water vapor that caused the spheres to form condensed and rapidly diffused into the glass walls of the spheres. We further postulate that some of this water was also taken up by salts present in the original vapor and formed a "brine" film on the inside surface of the bubble. With time, environmental water diffused into, and eventually completely saturated the glass [Friedman et al., 1966]. The water-saturated glass walls then acted as a semipermeable membrane, allowing water to diffuse into the sphere. This water diffusion was driven by the difference in osmotic pressure between the environmental water on the outside and the "brine" film inside the sphere. This osmotic-driven diffusion slowed down as the concentration of salts in solution inside the sphere was reduced by the dilution caused by the incoming environmental water, and it would have stopped when the osmotic pressures inside and outside the sphere were equal. The rate at which water entered the inside volume of the bubble therefore decreased exponentially with time. Apoproximately 10-20% of the space in the spheres is now filled with liquid. The presence of liquid-vapor filled inclusions in other rhyolitic glass samples has been named "superhydration" by Steen-MacIntyre [1975], who used the amount of water in the inclusions to date the eruption of the ash, a technique first suggested by Roedder and Smith [1964]. As water diffuses into silicic glass, the lighter molecule H_2O, rather than HDO is preferrentially bound to the glass, depleting this bound water in HDO. During the reverse process, when water leaves the glass to enter the hollow center of the glass sphere, the water entering the center will be enriched in the heavy molecule HDO. The result of this process is that the water in the hollow interior of the sphere will have the same isotopic composition as the water surrounding the sphere, but the water bound to the glass structure will be depleted in HDO. The fractionation factor that describes this relationship between the water in the glass and the water outside the glass is given by:

$$\alpha_{glass-water} = \frac{\left(\dfrac{HDO}{H_2O}\right)_{glass}}{\left(\dfrac{HDO}{H_2O}\right)_{water}}$$

Note that the negligible amount of magmatic water originally present, both as steam that expanded the bubbles, and as the approximately 0.1% water remaining in the glass after expansion of the glass during bubble formation, can be neglected in this discussion.

EXPERIMENTAL

After separation from the ash by flotation on water, the spheres in our study were dried at 50° C for 4 hours in a vacuum. They were then heated in a vacuum to 300° C, which resulted in rupturing of the spheres and the release of the formerly-fluid water. We found that very little additional water was released on continued heating from 300° C to ~800° C. The water was then converted to hydrogen gas by reaction with hot uranium metal, and the hydrogen was analyzed for deuterium on a Nier type 15 cm-60° magnetic sector mass spectrometer equipped to collect mass 2 and 3 (H_2 and HD) simultaneously.

At the conclusion of this phase of the experiment, the hydrated glass remaining after rupturing of the spheres was heated in vacuum to 1400° C to release the water contained in, and bound to the glass. This water was also converted to hydrogen and analyzed for deuterium as above. After appropriate mass spectometric corrections were made, the results are given in δD per mil units relative to Vienna Standard Mean Ocean Water (V-SMOW), where:

$$\delta D = \left[\frac{R_{sample}}{R_{standard}} - 1\right] \times 1000$$

$R_{standard}$=ratio of deuterium to hydrogen in V-SMOW, R_{sample}=ratio of deuterium to hydrogen in the sample. All analyses are precise to 2 per mil (2 sigma).

Other ash samples that did not contain hollow spheres were dried in vacuum at 250° and then heated to 1400° C in vacuum to extract water, which was then analyzed for δD as above.

The fractionation factor α was obtained from the relation:

$$\alpha_{glass-water} = \frac{1000 + \delta D_{glass}}{1000 + \delta D_{water}}$$

where δD_{water} is the δD of the liquid water in the interior of the sphere, and δD_{glass} is the δD of the water extracted from the molten glass.

The δD of the water inside the sphere is the integrated value (i.e. intergrated over the time during which the water diffused into the sphere) of the δD in the environmental (surface or ground) water. Using the data described above, the calculated fractionation factor will be in error if the isotopic composition of the meteoric water in the area where the ash sample was collected changed significantly and then remained constant for an appreciable time period, during the last 7 my. If this occurred, the liquid water in the sphere will be an integrated sample of meteoric water of different compositions, while the water bound to the glass will have an isotopic composition that is determined by the isotopic composition of meteoric water present during the last few thousand years.

The water in the interior of the bubbles from the Idaho samples has a δD of -147 per mil, whereas the water "dissolved" in the glass has a δD of -175 per mil. The fractionation factor calculated from these values is 0.9672. The value of environmental water that diffused into an ash sample can, therefore, be calculated by solving the following equation

$$\delta D_{environmental\ water} = \frac{\left(1000 + \delta D_{glass}\right)}{0.967} - 1000$$

It is interesting to note that the δD of present-day mereoric water at the collection site is -125 ±5 ‰, which implies that the δD of meteoric water in this region was more negative in the Miocene, when most of the diffusion of water into the hollow spheres occurred. A companion paper [Friedman et al., this volume] discusses the changes in δD of meteoric water with time.

FRACTIONATION FACTOR from ASH HYDRATED on LAND

The second method of determining the fractionation factor at surficial temperature and pressure was by analyzing samples of Kawakawa ash that hydrated on shore on the North Island of New Zealand. The water extracted from two samples of the ash had a δD of -82 per mil. The δD of present-day groundwater at this site is given by Stewart [written communication, 1988] as -45 ± 1‰. Stewart [1978] estimated, on the basis of published $\delta^{18}O$ data on New Zealand speleothems, that the δD of precipitation during the last glaciation in this location would have been 3-5‰ more negative than present precipitation. Using a value of -50‰ for surficial water that was present 20,000 years ago at this site yields a fractionation factor of 0.9663 that is in good agreement with that found from the glass spheres from Idaho.

CONCLUSION

The fractionation of deuterium that occurs when water at atmospheric pressure diffuses into rhyolitic glass has been determined by two methods to be 0.9668±.0005. This fractionation factor can be used to calculate the δD of ancient environmental water from the δD of secondary water of hydration extracted from volcanic ash samples. We recognize that rhyolite glass is not a phase that has strictly defined structure and composition; its chemistry and physical properties vary. However our experience suggests that the variation in these properties has only a small effect on the fractionation.

Acknowledgements. We thank W.G. Deuser, S.M. Savin, Paul Carrara, and an anonymous reviewer for their help in improving the manuscript.

REFERENCES

Friedman, I., Izett, G. A., and Gleason, J. D., Isotopic paleoclimate from hydrated volcanic ash (abstr.), *Geological Society of America, 98 Annual meeting Abstracts with Programs,* 587, 1985.

Friedman, I., Redfield, A. C., Schoen, B., and Harris, J., The variation of the deuterium content of natural waters in the hydrologic cycle, *Reviews of Geophysics,* 2, 177-224, 1964.

Friedman, I., and Smith, R. L., The deuterium content of water in some volcanic glass, *Geochimica et Cosmochimica Acta,* 1 4, 316-322, 1958.

Friedman, I., Smith, R. L., and Long, W. D.,. Hydration of natural glass and the formation of perlite, *Bulletin of the Geological Society of America,* 7 7, 323-328, 1966.

Friedman, I., Gleason, J., Warden, A., Ancient climate from the deuterium content of water in volcanic glass, *This volume.*

Perret, F. A., Volcanic ejectamenta, Volcanological Observations in *Carnegie Institution of Washington* Pub. 549, 115-116, 1950.

Roedder, E., and Smith, R. L., Liquid water in pumice vesicles, a crude but useful dating technique (abstract), *Geological Society of America Special Paper 82,* Abstracts for 1964, 164, 1964.

Ross, C. S., and Smith, R. L.,. Water and other volatiles in volcanic glasses, *American Mineralogist,* 4 0, 1071-1089, 1964.

Steen-McIntyre, V., Hydration and superhydration of tephra glass; A potential tool for estimating age of Holocene and Pleistocene ash beds, *Quaternary Studies* (ed. Suggate, R.P. & Cresswell, M.M., The Royal Soc. of New Zealand, Wellington), 271-278, 1975.

Stewart, M. K., Stable Isotopes in waters from the Wairakei geothermal area, New Zealand, *Department of Scientific and Industrial Reserarch Bulletin 220,* 113-119, 1978.

I. Friedman, J. Gleason, A.J. Gude, 3rd and R.A. Sheppard, U.S. Geological Survey, Denver Federal Center, MS 963, Denver, CO 80225.

Environmental Information from $^{13}C/^{12}C$ Ratios of Wood

STEVEN W. LEAVITT

Laboratory of Tree-Ring Research, University of Arizona
Tucson, Arizona

Wood is obtainable from living trees up to thousand of years old, and available as sub-fossils up to millions of years old. Current stable-carbon isotope fractionation models indicate that factors such as $\delta^{13}C$ of atmospheric CO_2, atmospheric CO_2 concentration, relative humidity, water stress and light may potentially influence plant $\delta^{13}C$. There are some circumstances, however, where a single variable (or two) may be dominant. $\delta^{13}C$ time series of pinyon pine display a trend of decreasing $\delta^{13}C$ apparently related to atmospheric $\delta^{13}C$ changes over the past two centuries, and possess high-frequency fluctuations related to soil moisture (drought) conditions. Radiocarbon-dated wood and plant material from the past 40,000 years also show changes in $\delta^{13}C$, most pronounced during the glacial to post-glacial transition. This may not simply reflect $\delta^{13}C$ of CO_2 because the Pleistocene-Holocene transition is characterized by major environmental changes related to CO_2 concentration, temperature, rainfall, etc. Seasonal variation of $\delta^{13}C$ within growth rings of trees appears to be strongly related to seasonal moisture stress. Additional research to quantify and calibrate will be required to fully access the environmental signal(s) contained in $\delta^{13}C$ of wood.

INTRODUCTION

The $\delta^{13}C$ of plant matter is depleted in ^{13}C relative to the isotopic composition of the atmospheric source, CO_2, which has a current $\delta^{13}C$ value of about -8‰ with respect to the PDB standard. C_3 plants, of which most trees are a subset, show a greater ^{12}C-enrichment than C_4 plants (which include many grasses), and fractionation models such as that of Farquhar et al. (1982) have been devised to explain the C_3 fractionation. The model is expressed in the following relationship:

$$\delta^{13}C_{plant} = \delta^{13}C_{air} - a - (b - a)C_i/C_{air} \qquad (1)$$

where a is a constant representing fractionation as CO_2 diffuses through the stomata ($\approx 4.4‰$), and b is a constant representing fractionation by the photosynthetic enzyme RuBP ($\approx 19‰$). The term C_i is the intercellular CO_2 concentration and C_{air} is the CO_2 concentration of the ambient air. When the ratio of C_i/C_{air} is high, then the plant shows the greatest departure from atmospheric $\delta^{13}C$ and the greatest ^{13}C depletion. When the ratio is very small, the difference between $\delta^{13}C$ of the air and the plant is minimum. Thus, not only is $\delta^{13}C_{air}$ important in determining $\delta^{13}C$ of the plant, but factors that influence C_i/C_{air} are also potentially very important. The rate of carbon fixation and the rate of stomatal conductance are the two primary factors

determining C_i/C_{air}, with high rates of conductance and low rates of fixation promoting high C_i/C_{air} (more negative $\delta^{13}C$). Environmental conditions that may influence these processes include light levels, soil moisture, relative humidity, and nutrient deficiencies. This formidable assortment of environmental variables potentially affecting $\delta^{13}C_{plant}$ suggests that not only would it be difficult to reconstruct changes in $\delta^{13}C_{air}$ from measurements on tree rings, but any attempt to reconstruct climate variables would be confounded by the multiple influences.

Before the full development of these plant stable-carbon isotope fractionation models, much empirical research had already been done on $^{13}C/^{12}C$ ratios in tree rings as a method of reconstructing global changes in $\delta^{13}C$ of atmospheric CO_2 (e.g., Farmer and Baxter, 1974; Stuiver, 1978). Other studies seemed to find climate signals in the $\delta^{13}C$ of tree rings (e.g., Libby et al., 1976; Harkness and Miller, 1980; Mazany et al., 1980). This paper illuminates some of the applications of $\delta^{13}C$ in tree rings which show promise despite the potential multiple influences.

METHODS

There is isotopic heterogeneity among trees, even at the same site, usually more in the sense of differences in absolute $\delta^{13}C$ among trees rather than differences among trends. This may be a consequence of the influence of subtle microenvironmental differences or it is perhaps related to genetic differences promoting slightly different isotopic fractionation under the same conditions. The range of $\delta^{13}C$ values among individual trees at a site is typically 2-3‰ and has been described in great detail

Climate Change in Continental Isotopic Records
Geophysical Monograph 78

elsewhere (Leavitt and Long, 1986; 1988a). In routine sampling, the pooling of tree rings from four trees, with four cores per tree, has been found to yield results representative of the site (Leavitt and Long, 1984).

Because of the great number of chemical constituents of wood, each with a different stable-carbon isotopic composition, whole wood is usually not used for analysis. Pectins and sugars tend to be isotopically heavier than bulk plant composition, whereas lignins and especially lipids tend to be isotopically lighter (Deines, 1980). Where cellulose and whole wood have been analyzed from the same samples, the cellulose is isotopically heavier (Leavitt and Long, 1982; 1991). Therefore, changes in the proportions of constituents, whether a consequence of physiological responses, or due to differential decomposition in the case of old wood, will cause changes in δ^{13}C of whole wood perhaps unrelated to any real environmental changes. Typically, either alpha-cellulose, the pure cellulose produced by polymerization of glucose, or holocellulose, the combination of alpha-cellulose, hemicelluloses (polymers of non-glucose sugars), and minor residual lignin, are isolated for isotopic studies involving wood. Conventional processing will first remove oils and resins with benzene (toluene now commonly substituted) and ethanol through soxhlet extraction, after which the samples may then be oxidized in a chlorinated solution. Sodium chlorite in a solution acidified with acetic acid is one method by which this "bleaching" step is accomplished to produce holocellulose (Green, 1963). The δ^{13}C results reported herein are with respect to the PDB standard.

δ^{13}C APPLICATIONS

There are several categories of study where δ^{13}C measurements have been employed with wood samples, usually to elucidate temporal changes in some environmental parameter. These are frequently accomplished with dated tree-ring series, but the use of complete wood samples to represent points in time over a very long time frame (tens of thousands of years) have also been used. Currently, perhaps the most fertile directions of isotopic inquiry involve reconstructions of drought from high-frequency δ^{13}C fluctuations, although the possibility that long-term changes in δ^{13}C reflect changes in δ^{13}C of atmospheric CO_2 and/or CO_2 concentration is intriguing and important.

Southwestern Pinyon Pine δ^{13}C Chronologies

One of the primary driving forces in ^{13}C/^{12}C tree-ring research over the past 20 years has been to develop a chronology of atmospheric δ^{13}C changes. Such a chronology would contain the input from fossil-fuels as well as the input from the biosphere, and by accounting for the fossil-fuel portion with calculations related to the fossil-fuel CO_2 dilution of atmospheric $^{14}CO_2$, the residual δ^{13}C trend should represent the activity of the biosphere as CO_2 source or sink (Stuiver, 1978). Several studies have addressed this question by combining tree-ring δ^{13}C time series from multiple trees to enhance the signal-to-noise ratio (Freyer and Belacy, 1983; Stuiver et al., 1984). Rather than accepting individual δ^{13}C chronologies at face value, efforts have also been

made to select and adjust in order to account for environmental effects that could mask the signal being sought. For example, Peng et al. (1983) examined most of the published tree-ring δ^{13}C chronologies and included in their master chronology only those derived from trees that were isolated and not water-stressed. Stuiver et al. (1984) developed chronologies from the Pacific coast of the Americas, and then attempted to correct for the variation in ratio of rates of carbon fixation to rates of stomatal conductance by means of ring area determinations.

Whereas the above studies usually represent a site with a single tree (and often a single radius), a network of 14 pinyon pine sites in the southwestern U.S. was sampled, and δ^{13}C chronologies were developed from four trees per site with four cores per tree by Leavitt and Long (1988b; 1989b). Figure 1 depicts the mean curve of the 56 trees for the period 1780 to present. It shows a long-term decline in δ^{13}C with short-term fluctuations superimposed. The fluctuations were not random, but rather a strong inverse relationship was found between the high-frequency fluctuations and changes in ring-width indices, so that each site curve could be normalized to unit ring index. Figure 1 also contains the mean curve after normalization, showing substantial damping of the high-frequency variation. Because of difference in absolute values among sites, the error bars would be even smaller than those depicted in Figure 1 if each site chronology were first normalized to its mean before the chronologies were averaged. The amount of decline from pre-Industrial to present is about 1.2-1.4‰, somewhat larger than that found by Stuiver et al. (1984) of ca. 1.1‰, but less than that found by Freyer and Belacy (1983) of 1.8-2.0‰. After removing the portion of the curve attributable to fossil-fuel inputs (assuming δ^{13}C changes modelled by Peng et al. (1983) from ^{14}C dilution), Leavitt and

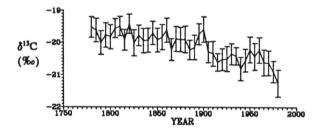

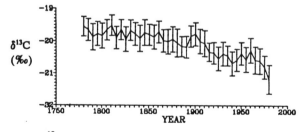

Fig. 1. δ^{13}C chronology from 56 pinyon pine trees in the American Southwest without correction for climate influence (upper) and with correction for climate influence (lower) using δ^{13}C-ring-width index relationships. Error bars are 2 standard errors of the mean. (After Leavitt and Long, 1988b; 1989b)

Long (1988b; 1989b) have interpreted the residual curve as indicating that the biosphere acted as a net CO_2 source until about 1950-1960, but since then has become a net sink.

It should be noted that Southern Hemisphere tree-ring $\delta^{13}C$ chronologies developed by Francey (1981) show a flat $\delta^{13}C$ trend from pre-Industrial to present, but there have been questions about the location of Francey's trees and associated impacts of respired CO_2 (Freyer, 1981). One of the major uncertainties about all the tree-ring reconstructions is the influence of increasing CO_2 concentration. Fractionation models indicate that increasing atmospheric CO_2 alone with no change in $\delta^{13}C_{air}$ could cause changes in C_i/C_{air} and hence in $\delta^{13}C_{plant}$. If the magnitude of this CO_2 effect could be quantified, then it could also be removed from the tree-ring $\delta^{13}C$ reconstructions to obtain a more accurate record of $\delta^{13}C_{air}$ changes. Ice core $\delta^{13}C_{air}$ reconstructions (e.g., Friedli et al., 1986) may provide a key to determining this CO_2 concentration effect on tree ring $\delta^{13}C$ chronologies, although ice core chronologies do not have the same time resolution as is possible from tree rings.

The high-frequency $\delta^{13}C$ fluctuations in the pinyon chronologies actually provide a wealth of information in their own right. The inverse correlation of the $\delta^{13}C$ with ring index indicates that there should be a correlation with climate, because ring widths of southwestern trees have already been found to relate to temperature and precipitation changes (Fritts, 1976). In the case of pinyon, $\delta^{13}C$ most strongly correlated with measures of drought (Leavitt and Long, 1989a). Drought indices are a function of temperature, precipitation and other factors, and provide a logical influence on plant carbon isotope fractionation. Namely, during drought stomata are likely to close and limit the pool of carbon available for photosynthesis, so that there is less discrimination against ^{13}C and hence higher $\delta^{13}C$ values. Under moist conditions, the opposite is true and discrimination is maximal with more negative $\delta^{13}C$. With this simple model Leavitt and Long (1989a) were able to use the relative heights of $\delta^{13}C$ maxima and minima at the 14 pinyon sites to map the spatial distribution of drought in the Southwest. Figure 2a depicts a map for the period 1930-34. Rather than plotting the raw $\delta^{13}C$ value, because of different absolute values among sites a spline curve was fitted to each chronology and a "DEL Index" was calculated as the ratios of the actual value to the corresponding spline value:

$$DEL\ Index = \left[\frac{\delta^{13}C_{measured}}{\delta^{13}C_{spline\ curve}} - 1 \right] \times 1000 \quad (2)$$

The more negative DEL Indices reflect stronger drought conditions, whereas the more positive DEL Indices indicate reduced moisture stress. The DEL Indices in Figure 2a indicate the strongest drought was located in the northwest portion of the network with the most abundant moisture in the southeast. For comparison, the actual Palmer Hydrological drought indices from weather records are plotted in Figure 2b. Actual average moisture for the period was lowest in the northern part of the

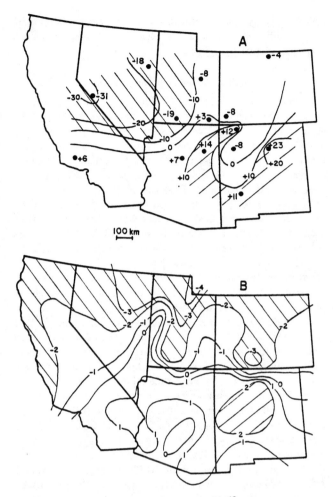

Fig. 2. Drought map reconstructed from pinyon $\delta^{13}C$ as DEL Indices for the pentad 1930-1934 (A) as compared to actual Palmer Hydrological Drought Indices (PHDI) average for 1931-1934 (B). Areas characterized as moist (DEL Indices > +10, and PHDI > 2) and dry (DEL Indices < -10, and PHDI < -2) are highlighted with diagonal lines. (After Leavitt and Long, 1989a)

area, and highest in the southeast. Limitations to agreement are probably at least partly related to the limited spatial distribution of the pinyon network, and the fact that the drought indices had really only been available to average for the period 1931-1934.

Plant physiologists had speculated on a possible link of plant $\delta^{13}C$ to moisture for a number of years before the pinyon chronologies were published. From the physiologist's viewpoint, the pathways for transpired water out of the plant and for CO_2 into the plant are both through stomata. Transpiration rates will largely be dependent on climate conditions and especially soil moisture availability. Therefore, $\delta^{13}C$ was seen as a measure of integrated water-use efficiency (Farquhar, et al., 1988), where high water-use efficiency is denoted by a high ratio of rates of carbon fixation to rates of water loss through transpiration. In fact, screening of agricultural crop strains for high water-use efficiency is being tested through selection based on relative $\delta^{13}C$ of the different strains under identical growing conditions (Farquhar and Richards, 1984).

$\delta^{13}C$ Changes in Plants Over Thousands of Years

Because of the sensitivity of plant $\delta^{13}C$ to factors such as $\delta^{13}C_{air}$ and water stress, some researchers have felt that in the absence of continuous tree-ring chronologies beyond ca. 9000 years B.P., the analysis of isolated fragments of old, radiocarbon-dated plant matter, preserved through special sedimentary circumstances, could provide a key to exploring past environmental variations. Krishnamurthy and Epstein (1990) obtained over 100 ^{14}C-dated wood specimens from around the world and determined the $\delta^{13}C$ of the cellulose component. They found a $\delta^{13}C$ peak in their record at ca. 18,000-20,000 years B.P. with a sharp decline of about 4‰ to the current interglacial period. The trend follows closely that of the CO_2 concentration changes reconstructed from ice core measurements, but Krishnamurthy and Epstein interpret the wood $\delta^{13}C$ changes as primarily a consequence of changes in $\delta^{13}C_{air}$.

Independently, Leavitt and Danzer (1991) tabulated the $\delta^{13}C$ of all categories of ^{14}C-dated plant matter reported in the journal *Radiocarbon*. Although the samples were treated by many radiocarbon laboratories and pretreatment methods were not usually listed, the database consisted of over 800 entries. A plot of the full set of data revealed a significant $\delta^{13}C$ decline of 0.8‰ from pre-10,000 years to post-10,000 years B.P. For the category of "wood" alone, a significant $\delta^{13}C$ decline of 3.0‰ was found (n=306). Because there were no angiosperms in the database older than 10,000 year B.P., there was some concern that the isotopically heavier pre-10,000 average could simply reflect the absence of angiosperms, which tend to be isotopically lighter than gymnosperms (Stuiver and Braziunas, 1987). However, when the subset of gymnosperms was examined alone, it showed a 2.8‰ decline as well. Rather than interpreting this as reflecting a change in $\delta^{13}C_{air}$, however, Leavitt and Danzer felt a change in C_i/C_{air} of 0.78 in full glacial to 0.87 in post-glacial was sufficient to produce the observed overall plant $\delta^{13}C$ changes without any changes in $\delta^{13}C_{air}$. Consistent with this, heavier $\delta^{13}C$ values in high elevation plants relative to those at lower elevation (Körner et al., 1988) suggested that lower CO_2 partial pressure (also found in full glacial time) may favor reduced C_i/C_a. Neither the "peat" category nor the "charcoal" category showed any significant $\delta^{13}C$ dislocation over this time period, and in general, all the data subsets tend to show a lot of scatter. However, for organic matter in marine sediment cores, Rau et al. (1991) reported a sharp $\delta^{13}C$ decline of ca. 2‰ from about 18,000 to 10,000 years B.P.

Additional work by Leavitt and Danzer (in press) with $\delta^{13}C$ of ^{14}C-dated plant matter from a single radiocarbon laboratory at the University of Arizona has reinforced some of the earlier findings. The database of over 1000 entries has the advantage that all samples were pretreated in the same manner, via acid-alkali-acid extraction. The full set of plant data shows a significant decline of 1.0‰ from pre- to post-10,000 years B.P. Although the

wood category alone (n=174) shows a non-significant decline of 0.2‰, categories of twigs (n=119) and leaves (n=19) exhibit significant $\delta^{13}C$ declines of 0.9 and 1.4‰, respectively. All categories of plant material from the genus *Juniperus* (n=153) also present a significant decline of 0.4‰.

Besides commonality of direction and magnitude of $\delta^{13}C$ change, another way of testing the signals to see if they are tracking the same real environmental changes is through examination and comparison of the timing of changes. Running averages for all plant material from the *Radiocarbon* data set and from the University of Arizona Radiocarbon Laboratory data set are contained in Figure 3 for the period 14,000 to 10,000 years B.P., a period of rapid transition from full glacial to post-glacial. Although the absolute values of the two data sets differ by 2-2.5‰, the maxima and minima correspond reasonably well. The biggest problem is in interpreting what variable(s) is(are) driving these apparent changes. Besides CO_2 concentration and $\delta^{13}C$ of atmospheric CO_2, variables such as relative humidity, rainfall distribution and amounts, cloudiness, temperature, and other factors likely changed from full glacial to present, and these factors may all influence plant carbon isotope fractionation. There is hope that at least the component related to $\delta^{13}C$ changes in atmospheric CO_2 may be determined from analysis of C_4 plant material, which may be a better monitor of that parameter than C_3 plants because fractionation in C_4 plants is effectively independent of C_i/C_a (Marino and McElroy, 1991). After that change is removed from the $\delta^{13}C$ chronologies from C_3 plants, then the residual will reflect the influence of these other environmental parameters. Active research is being pursued to develop such a long C_4-plant $\delta^{13}C$ chronology (Marino et al., 1992), as well as to obtain a long atmospheric $\delta^{13}C$ chronology directly from ice cores (Leuenberger, et al., 1992).

Seasonal $\delta^{13}C$ Changes Within Tree Rings

Another promising $\delta^{13}C$ measurement being made on wood is the intra-annual change from the beginning of the growing season

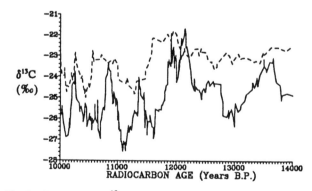

Fig. 3. Fluctuation in $\delta^{13}C$ of radiocarbon-dated plant matter during the period from 14,000 to 10,000 years B.P. These curves represent 11-point running averages of data tabulated from the journal *Radiocarbon* (solid) and from the database of materials dated only by the University of Arizona Radiocarbon Laboratory (dashed). (After Leavitt and Danzer, 1992)

to the end. Seasonal isotopic changes in plants were perhaps first reported by Lowden and Dyck (1974) who found changing whole-tissue $\delta^{13}C$ in leaves of grass and maple. Wilson and Grinsted (1977) subdivided tree rings of Monterey Pine from New Zealand into a number of segments. They found about 2‰ variation in both cellulose and lignin with the least negative $\delta^{13}C$ values in the middle of the growing season, and because of the abundant rainfall throughout the growing season, they attributed the changes to a positive correlation with temperature alone. Leavitt and Long (1982) found a strong correspondence between the changes in cellulose $\delta^{13}C$ in leaves of juniper and in the corresponding ring of the same year, and Leavitt and Long (1991) reported on the types of seasonal $\delta^{13}C$ signals observed in a number of different species from a number of different locations in North America. In the latter study, seasonal $\delta^{13}C$ results from a ponderosa pine whose microenvironment had been carefully monitored in previous research of Fritts (1976) show strongest correlations of $\delta^{13}C$ with drought.

In a detailed study of these seasonal $\delta^{13}C$ variations in modern trees (Leavitt, 1992), four maple trees and four white pine trees were sampled from each of two environments in southeastern Wisconsin, one in which the trees were growing very close together (closed canopy), and one in which they were more

widely spaced (open canopy). For each tree, equal-sized ring subdivisions were pooled from four sides of the tree to improve precision. Figure 4 shows some of the results from the individual trees of the closed sites for 1987 and 1988. These results reveal great consistency of patterns among trees of both species at the same site. The patterns in the trees from the nearby open sites are also quite similar. Additionally, a key to the environmental parameters that may be influencing these seasonal changes is revealed in the differences between 1987 and 1988. The 1987 pattern shows an increase of $\delta^{13}C$ at the beginning of the growing season to a maximum just prior to the end. Palmer Drought Severity Indices (PDSI) for this site declined from +3 (moderately wet) at the beginning of the growing season to -1.5 (mildly dry) approximately two thirds through the growing season, and increasing to +2 (moderately wet) at the end. In 1988, however, the upward $\delta^{13}C$ trend continues through the entire growing season. This pattern seems to confirm the pronounced 1988 drought conditions, in which the beginning of the growing season is characterized by PDSI of +3 (moderately wet), falling below -2 (moderately dry) about one-third through the season, and persisting below -3 and -4 (severely dry) for most of the remainder of the growing season. Thus, increasing $\delta^{13}C$ corresponds to increasing water stress, and decreasing $\delta^{13}C$ to

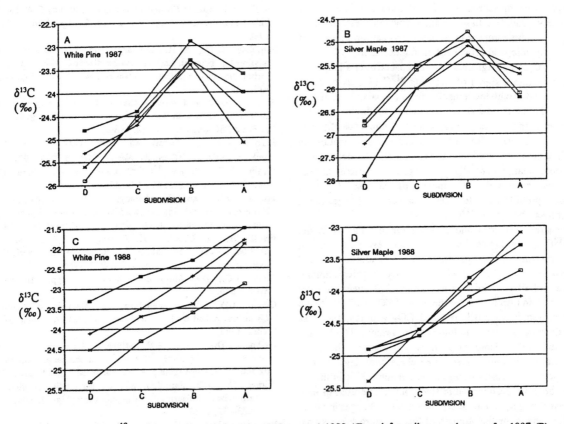

Fig. 4. Seasonal changes in $\delta^{13}C$ of four white pine trees for 1987 (A) and 1988 (C) and four silver maple trees for 1987 (B) and 1988 (D). Segment "D" is the inside of the ring and "A" is the outside. These trees were collected from closed-canopy stands in southeastern Wisconsin. (After Leavitt, 1992).

decreased water stress, the water stress probably being governed by both temperature and precipitation conditions. In 1986 (not shown), when moisture was high throughout the growing season, the seasonal $\delta^{13}C$ patterns are much flatter and there is less consistency among trees at each site than is seen for 1987 and 1988.

CONCLUSIONS

Interpretation of $\delta^{13}C$ measured in wood is fraught with pitfalls because of the complexity of carbon isotope fractionation models and multiplicity of factors which may influence the final difference between $\delta^{13}C$ of the air and that incorporated into the tree. However, there are several areas where $\delta^{13}C$ measurements in wood show promise with respect to atmospheric chemistry and climate.

Tree-ring $\delta^{13}C$ time series, at least for Northern Hemisphere trees, show a declining trend toward the present, consistent with expected changes in $\delta^{13}C$ of atmospheric CO_2. In principle, this chronology can be used to extract the history of activity of the global biosphere as net CO_2 source or sink. The most elaborate and carefully sampled of such chronologies, southwestern pinyon pine, indicates the biosphere had been a net CO_2 source until about 1950. However, there are some differences in the amount and timing of decline among different tree-ring $\delta^{13}C$ reconstructions, perhaps related to local environment and physiology. Direct measurements of $\delta^{13}C$ from CO_2 in ice cores may ultimately provide a baseline against which these tree-ring chronologies can be compared and deviations can be interpreted in terms of physiological influences on fractionation.

Individual ^{14}C-dated fragments of wood spaced over a long time period may contain a signal of changing $\delta^{13}C$ of CO_2 or other variables associated with the transition from glacial to post-glacial conditions. There are many environmental conditions that are likely to have changed over periods of tens of thousands of years, conditions which may also influence carbon fractionation in plants. Changing CO_2 concentrations alone may influence plant $\delta^{13}C$ without a change in $\delta^{13}C_{air\ CO2}$. Again, ice core $\delta^{13}C$ over this long time span and $\delta^{13}C$ analysis of sub-fossil C_4 plant material may provide a key to whether or not the changes found in wood reflect changes in $\delta^{13}C$ of CO_2.

Moisture stress provides one of the strongest signals in the pinyon $\delta^{13}C$ chronologies from the Southwest U.S. and in seasonal tree-ring $\delta^{13}C$ patterns from southwestern and midwestern trees. This signal is strongly coherent among individuals and sites. Moisture availability influences stomatal opening, which in turn influences the pool of CO_2 available to the carbon-fixing enzyme RuBP; reduced CO_2 will result in less discrimination against ^{13}C and higher $\delta^{13}C$ values. The linkage of water and carbon isotopes is already being used by plant physiologists (and now forest researchers working on soil moisture models) to assess water-use efficiency. This influence appears to be quite pronounced in trees and can be used to map year-to-year drought, and to determine seasonal changes in water status. The reason for the strength of this moisture stress signal may actually be its linkage to a number of driving variables. For example, moisture stress may be associated with low precipitation, high temperature, reduced cloudiness (high light levels), low relative humidity, and high evaporation.

Trees are not passive monitors, but rather interactive monitors of environment. Future research on $\delta^{13}C$ of wood will be directed toward improving the ratio of signal to noise, and to distinguishing among the influences of different environmental signals. In addition to help from records of ice cores and C_4 plants, other isotopes (e.g., δD and $\delta^{18}O$) may also prove valuable in this quest.

Acknowledgments. I thank Shelley Danzer and Teresa Newberry for help in much of the sample preparation and the isotopic analysis described herein. Samples were analyzed at the Laboratory of Isotope Geochemistry, Dept. of Geosciences, at The University of Arizona. This research was supported in part by NSF Grant #ATM-8906112 and #ATM-9096319.

REFERENCES

Deines, P., The isotopic composition of reduced organic carbon, in *Handbook of Environmental Isotope Geochemistry, Vol. 1*, edited by P. Fritz and J.Ch. Fontes, pp. 329-405, Elsevier, New York, 1980.

Farmer, J.G., and M.S. Baxter, Atmospheric carbon dioxide levels as indicated by the stable isotope record in wood, *Nature, 247*, 273-275, 1974.

Farquhar, G.D., K.T. Hubick, A.G. Condon, and R.A. Richards, Carbon isotope fractionation and plant water-use efficiency, in *Stable Isotopes in Ecological Research*, edited by P.W. Rundel, J.R. Ehleringer and K.A. Nagy, pp. 21-40, Springer-Verlag, New York, 1988.

Farquhar, G.D., M.H. O'Leary, and J.A. Baxter, On the relationship between carbon isotope discrimination and intercellular carbon dioxide concentration in leaves, *Aust. J. Plant Physiol., 9*, 121-137, 1982.

Farquhar, G.D. and R.A. Richards, Isotopic composition of plant carbon correlates with water-use efficiency of wheat genotypes, *Aust. J. Plant Physiol., 11*, 539-552, 1984.

Francey, R.J., Tasmanian tree rings belie suggested anthropogenic $^{13}C/^{12}C$ trends, *Nature, 190*, 232-235, 1981.

Freyer, H.D., Recent $^{13}C/^{12}C$ trends in atmospheric CO_2 and tree rings, *Nature, 293*, 679-680, 1981.

Freyer, H.D., and N. Belacy, $^{13}C/^{12}C$ records in Northern Hemisphere trees during the past 500 years- Anthropogenic impact and climatic superpositions, *J. Geophys. Res., 88*, 6844-6852, 1983.

Friedli, H., H. Lötscher, H. Oeschger, U. Siegenthaler, and B. Stauffer, Ice core record of the $^{13}C/^{12}C$ ratio of atmospheric CO_2 in the past two centuries, *Nature, 324*, 237-238, 1986.

Fritts, H.C., *Tree Rings and Climate*, 567 pp., Academic, New York, 1976.

Green, J.W., Wood cellulose, in *Methods of Carbohydrate Chemistry*, edited by R.L. Whistler, pp. 9-21, Academic Press, New York, 1963.

Harkness, D.D., and B.F. Miller, Possibility of climatically induced variations in the ^{13}C and ^{14}C enrichment patterns as recorded by 300-yr old Norwegian pine, *Radiocarbon, 22*, 291-298, 1980.

Körner, C., G.D. Farquhar, and Z. Roksandic, A global survey of carbon isotope discrimination in plants from high altitude, *Oecologia, 74*, 623-632, 1988.

Krishnamurthy, R.V., and S. Epstein, Glacial-interglacial excursion in the concentration of atmospheric CO_2: Effect in the $^{13}C/^{12}C$ ratio in wood cellulose, *Tellus, 42B*, 423-434, 1990.

Leavitt, S.W., Seasonal $^{13}C/^{12}C$ changes in tree rings: species and site coherence, and a possible drought influence, *Can. J. For. Res.*, *in press*, 1992.

Leavitt, S.W., and S.R. Danzer, Chronology from plant matter, *Nature*, *352*, 671, 1991.

Leavitt, S.W., and S.R. Danzer, $\delta^{13}C$ variations in C_3 plants over the past 50,000 years, *Radiocarbon, Proc. 14th Int. Radiocarbon Conf.*, *in press*, 1992.

Leavitt, S.W., and A. Long, Evidence for $^{13}C/^{12}C$ fractionation between tree leaves and wood, *Nature*, *298*, 742-744, 1982.

Leavitt, S.W., and A. Long, Sampling strategy for stable carbon analysis of tree rings in pine, *Nature*, *301*, 145-147, 1984.

Leavitt, S.W., and A. Long, Stable-carbon isotope variability in tree foliage and wood, *Ecology*, *67*, 1002-1010, 1986.

Leavitt, S.W., and A. Long, Intertree variability of $\delta^{13}C$ in tree rings, in *Stable Isotopes in Ecological Research*, edited by P.W. Rundel, J.R. Ehleringer and K.A. Nagy, pp. 21-40, Springer-Verlag, New York, 1988a.

Leavitt, S.W., and A. Long, Stable carbon isotope chronologies from trees in the southwestern United States, *Global Biogeochemical Cycles*, *2*, 189-198, 1988b.

Leavitt, S.W., and A. Long, Drought indicated in carbon-13/carbon-12 ratios of southwestern tree rings, *Water Resources Bull.*, *25*, 341-347, 1989a.

Leavitt, S.W., and A. Long, The atmospheric $\delta^{13}C$ record as derived from 56 pinyon trees at 14 sites in the southwestern United States, *Radiocarbon*, *31*, 469-474, 1989b.

Leavitt, S.W., and A. Long, Seasonal stable-carbon isotope variability in tree rings: possible paleoenvironmental signals, *Chemical Geology (Isotope Geos. Sect.)*, *87*, 59-70, 1991.

Leuenberger, M., U. Siegenthaler, and C.C. Langway, Carbon isotope composition of atmospheric CO_2 during the last ice age from an Antarctic ice core, *Nature*, *357*, 488-490, 1992.

Libby, L.M., L.F. Pandolfi, P.H. Payton, J. Marshall, B. Becker, and V. Giertz-Sienbenlist, Isotope tree thermometers, *Nature*, *261*, 284-288, 1976.

Lowden, J.A., and W. Dyck, Seasonal variations in the isotope ratios of carbon in maple leaves and other plants, *Can. J. Earth Sci.*, *11*, 79-88, 1974.

Marino, B.D., and M.B. McElroy, Isotopic composition of atmospheric CO_2 inferred from carbon in C4 plant cellulose, *Nature*, *349*, 127-131, 1991.

Marino, B.D., M.B. McElroy, R.J. Salawitch, and W.G. Spaulding, Glacial-to-interglacial variations in the carbon isotopic composition of atmospheric CO_2, *Nature*, *357*, 461-466, 1992.

Mazany, T., J.C. Lerman, and A. Long, Carbon-13 in tree-ring cellulose as an indicator of past climates, *Nature*, *287*, 432-435, 1980.

Peng, T.H., W.S. Broecker, H.D. Freyer, and S. Trumbore, A deconvolution of the tree ring based $\delta^{13}C$ record, *J. Geophys. Res.*, *88*, 3609-3620, 1983.

Rau, G.H., P.N. Froelich, T. Takahashi, and D.J. DesMarais, Does sedimentary organic $\delta^{13}C$ record variations in Quaternary ocean $[CO_2(aq)]$?, *Paleoceanography*, *6*, 335-347, 1991.

Stuiver, M., Atmospheric carbon dioxide and carbon reservoir changes, *Science*, *199*, 253-258, 1978.

Stuiver M., and T.F. Braziunas, Tree cellulose $^{13}C/^{12}C$ isotope ratios and climatic change, *Nature*, *328*, 58-60, 1987.

Stuiver, M., R.L. Burk, and P.D. Quay, $^{13}C/^{12}C$ ratios in tree rings and the transfer of biospheric carbon to the atmosphere, *J. Geophys. Res.*, *89*, 11731-11748, 1984.

Wilson, A.T., and M.J. Grinsted, $^{12}C/^{13}C$ in cellulose and lignin as palaeothermometers, *Nature*, *265*, 133-135, 1977.

S. W. Leavitt, Laboratory of Tree-Ring Research, University of Arizona, Tucson, AZ 85721.

Interpreting Past Climate From Stable Isotopes in Continental Organic Matter

Thomas W.D. Edwards

Department of Earth Sciences, Quaternary Sciences Institute, and Waterloo Centre for Groundwater Research,
University of Waterloo, Waterloo, Ontario, Canada

The isotopic composition of fossil organic material from continental deposits contains myriad information about past environmental conditions. Much effort has been devoted to study of $\delta^{18}O$ and δ^2H variations in organic matter, which depend strongly on isotopic effects occurring in the hydrologic cycle. Derivation of proxy temperature signals has been demonstrated in studies using δ^2H data from plant cellulose extracted from fossil wood, packrat middens, and peat, because plant tissues preserve isotopic signals inherited from the temperature-dependent signature of local meteoric water. More complex deconvolution of oxygen and hydrogen isotope data from terrestrial plant cellulose has also been undertaken using correlative or mechanistic models (calibrated through comparison of modern meteorological and isotopic data) that implicitly or explicitly consider the effects of secondary isotopic alteration of plant waters during evapotranspiration. Such models have been applied with notable success to isotopic studies of wood cellulose, which has proven to be a valuable source of quantitative proxy climate data, especially in the continuous time-series records offered by tree-ring sequences.

Promising results have also been obtained from isotopic study of oxygen in aquatic plant cellulose preserved in lake sediments, hydrogen in aquatic plant lipids, and oxygen and hydrogen in cellulose from peat deposits.

Introduction

Continental organic matter contains many potential paleoclimatic signals inherent in the isotopic composition of its constituent elements. Foremost among such isotopic signals are those preserved in the $^{18}O/^{16}O$ and $^2H/^1H$ ratios in plant tissues, which contain information about the isotopic composition of past environmental water associated with tissue synthesis. Understanding and documentation of the temperature- and moisture-dependent oxygen and hydrogen isotope fractionations that occur as water passes through the hydrologic cycle sets the stage for quantitative reconstruction of paleoclimate. Valuable information may be gleaned also from variations in $^{13}C/^{12}C$ ratio in plant matter, although interpretation in climatically relevant terms is commonly less straightforward.

The following discussions focus rather narrowly on quantitative interpretation of isotopic signals from the cellulose fraction of fossil plant matter, rather than attempting to review the broad field of paleoenvironmental research in which isotopic data from continental organic matter are routinely applied. This focus has been chosen for practical reasons (to limit the length of the discussions) and to highlight the rich store of quantitative information available from careful consideration of precisely measureable isotopic variations in organic matter. The article emphasizes the interpretation of isotopic data from wood cellulose, with less detailed consideration of isotopic data from lake sediments and peat.

Climate Change in Continental Isotopic Records
Geophysical Monograph 78
Copyright 1993 by the American Geophysical Union.

Wood

Hydrogen and oxygen isotope data from terrestrial plant matter contain signals originating from local meteoric water (which may provide proxy temperature), as well as moisture-dependent signals from enrichment of plant water by evapotranspiration prior to tissue synthesis [e.g. Ferhi and Létolle, 1977; Burk and Stuiver, 1981; Yapp and Epstein, 1982a]. Carbon isotope composition is fundamentally controlled by photosynthetic pathway, resulting in the well-known differentiation of C_3 and C_4 plants [Bender, 1971; Smith and Epstein, 1971]. Humidity-dependent variations occur in the carbon isotope composition of tissues from C_3 plants, related to water use efficiency [Francey and Farquhar, 1982; Farquhar et al., 1988], whereas strongly invariant carbon isotope effects may occur in C_4 plants [Marino and McElroy, 1991].

Following early attempts to use whole wood [e.g. Libby et al., 1976], studies have mainly been conducted using α-cellulose extracted from finely ground wood powder [Green, 1963]. Oxygen isotope analysis of the purified cellulose is usually undertaken on carbon dioxide gas produced by pyrolysis in a nickel or quartz tube [Thompson and Gray, 1977; Hardcastle and Friedman, 1974] or by mercuric chloride combustion [Rittenberg and Ponticorvo, 1956; Sternberg, 1989]. Hydrogen isotope determinations are conducted on cellulose that has been nitrated to eliminate exchangeable, non-carbon-bound hydrogen [DeNiro, 1981]; hydrogen gas for analysis is produced by reduction of combustion water over hot zinc or uranium [Epstein et al., 1976; Sternberg, 1989]. Carbon isotope ratios may be determined on carbon dioxide gas produced by combustion of either type of cellulose. Isotope ratios are expressed as "δ" values relative to SMOW (standard mean ocean water) for oxygen and hydrogen, and PDB (Pee Dee belemnite) for carbon, and have typical analytical uncertainties equal to or greater than $\pm$ 0.2, $\pm$ 2, and $\pm$ 0.1 ‰, respectively, depending on the method used and the natural inhomogeneity of the cellulose.

Oxygen, hydrogen, and carbon isotope data from terrestrial plant cellulose have been used as both qualitative and quantitative indicators of past environmental change. Qualitative interpretations may rely simply on correspondence between changes in isotopic composition and expected environmental change inferred from other sources of information. Examples include a study by Dubois and Ferguson [1985], which proposed that low cellulose δ^2H values in radiocarbon-dated pine from northern Scotland reflected "amount effects" on the isotopic composition of rainfall during Holocene wet phases. More recently Becker et al. [1991] attempted to identify post-Younger Dryas climatic amelioration in southern Germany on the basis of changing cellulose $\delta^{13}C$ and δ^2H values from tree-ring sequences, arguing that increasing $\delta^{13}C$ reflected declining humidity (increasing water use efficiency) and increasing δ^2H reflected a temperature-dependent rise in meteoric water δ^2H.

Such efforts often place useful qualitative constraints on paleoenvironmental interpretations, but commonly have limited potential to support independent reconstruction of climatically relevant parameters. Of more interest in the present context are methods for deriving quantitative information from isotopic variations in terrestrial plant cellulose. Reported studies fall into two principal (although not necessarily mutually exclusive) categories, depending on whether reconstructions are based on transfer functions describing empirical correlations between isotopic and climatic parameters (analogous to traditional dendroclimatological analysis) or on semi-empirical models that attempt to account mechanistically for fractionations occurring in the plants.

"Transfer Function" Approach

The transfer function approach improves upon qualitative analysis by establishing statistically significant correlations between cellulose isotopic data and pertinent environmental parameters for plants at modern sites, that may be used in turn to interpret cellulose data from fossil or subfossil deposits. An early example using hydrogen isotope data was reported by Yapp and Epstein [1977], who identified a strong linear relation between δ^2H of cellulose and associated environmental water at modern North American sites and applied it to infer late-glacial meteoric water composition in eastern United States from a suite of radiocarbon-dated fossil wood samples. In subsequent investigations these authors also documented strong relations between cellulose δ^2H and mean annual temperature in ring increments from selected trees [Yapp and Epstein, 1982b], reflecting direct transfer of temperature-dependent isotopic signals from local meteoric water to cellulose, as well as identifying weak humidity-dependent variations in the hydrogen-isotope separation between cellulose and water [Yapp and Epstein, 1982a]. Epstein and Krishnamurthy [1990] also used this approach for interpretation of hydrogen isotope data from tree-ring sequences obtained at 23 sites around the world to assess recent global temperature change.

Such simple transfer functions must be applied to paleoclimate reconstruction with caution, as their applicability is constrained nat-

urally by the range of climatic conditions sampled within their respective analogue data sets. The existence of humidity-dependent effects within the set of data used by Yapp and Epstein [1982a] provides a good case in point; valid estimation of past meteoric water isotopic composition using their observed linear relation between δ^2H of water and cellulose (which implicitly averages out the humidity effects) is inherently limited to sites that experienced the same style of secular climatic variation as that exhibited geographically by their modern data set.

More sophisticated multiple regression analysis of isotopic data from wood cellulose has also been undertaken to disentangle and quantify the simultaneous influence of different environmental factors. Of particular note are investigations by Ramesh et al. [1985, 1986] using ring sequences from silver fir trees growing in the Kashmir Valley, India. These studies verified the existence of statistically significant correlations between cellulose $\delta^{18}O$, δ^2H, $\delta^{13}C$, and various meteorological parameters (temperature, precipitation, humidity, cloudiness, etc), as well as demonstrating strong coherence in the isotopic variations between and within trees at the same site. Such observations strengthen the basis for using isotopic data from trees for paleoclimate analysis, as well as providing valuable independent information about the parameters that should be considered in mechanistic models. In turn, application of transfer functions based on the results of statistical analysis must be undertaken with due consideration of the mechanistic basis for the derived relations.

"Mechanistic" Approach

Efforts to understand how cellulose obtains its isotopic composition have fostered derivation of models to account for the isotopic effects occurring as water and carbon dioxide are assimilated by plants. Francey and Farquhar [1982; see also Farquhar et al., 1988] showed that carbon isotope variations in plants could be modelled in terms of assimilation rate of carbon dioxide and stomatal conductance, which are influenced by environmental factors such as relative humidity, water stress, and light levels. Unfortunately, owing to the complexity of possible interactions among environmental factors and strongly species-specific physiological characteristics, quantitative paleoclimate reconstruction using carbon isotope variations in wood cellulose remains rather elusive [see Stuiver and Braziunas, 1987], except perhaps in cases where dominance by an individual environmental factor can be assured [Leavitt, this volume].

Attempts to apply models that account for oxygen and hydrogen effects in plants have met with notable success, on the other hand, because of simpler environmental influences on oxygen and hydrogen assimilation into plant tissues. Burk and Stuiver [1981] showed that wood cellulose $\delta^{18}O$ could be modelled in terms of input (soil) water $\delta^{18}O$, humidity- and temperature-dependent isotopic enrichment arising from evapotranspiration of plant water, and biochemical fractionation effects occurring during cellulose synthesis. Subsequent studies [Edwards et al., 1985; Edwards and Fritz, 1986] suggested that cellulose $\delta^{18}O$ or δ^2H could be modelled adequately using only input water $\delta^{18}O$ or δ^2H and relative humidity as variables, and assuming fixed values for isotopic fractionations during plant water evapotranspiration and cellulose synthesis. Edwards and Fritz [1986] coupled independent equations of the form suggested by Yapp and Epstein [1982a] describing the oxygen and hydrogen isotopic relations between soil water and cellulose with a meteoric water line linking oxygen and hydrogen isotope content in the input water:

$$\delta^{18}O_{cellulose} = A\delta^{18}O_{water} + 1000(A - 1) \qquad (1)$$

$$\text{where } A = \alpha_n\alpha_e\alpha_k - \alpha_n(\alpha_e\alpha_k - 1)h$$

$$\delta^2H_{cellulose} = B\delta^2H_{water} + 1000(B - 1) \qquad (2)$$

$$\text{where } B = \beta_n\beta_e\beta_k - \beta_n(\beta_e\beta_k - 1)h$$

$$\delta^2H_{meteoric\ water} = C\delta^{18}O_{meteoric\ water} + D \qquad (3)$$

The terms α and β represent equilibrium (α_e and β_e) and kinetic (α_k and β_k) fractionations during plant water evapotranspiration, and biochemical (α_n and β_n) fractionations between cellulose and plant water, and h is photosynthetic humidity (the mean relative humidity during growth of the wood). Equations (1) and (2) are based implicitly on the Craig and Gordon [1965] model for steady-state evaporation from a terminal reservoir for which input exactly balances vapour loss, assuming isotopic equilibrium between ambient atmospheric vapour and input water, and (3) is the appropriate local meteoric water line. Solution of this set of equations for a pair of cellulose $\delta^{18}O$ and δ^2H values yields quantitative estimates of input water isotopic composition and photosynthetic humidity, which may be useful paleoclimate parameters. The model is shown graphically in Figure 1.

The Edwards and Fritz [1986] model was applied initially to interpret isotopic data from bulk fossil wood samples preserved in a kettle-fill sequence in southeastern Canada. The reconstructed postglacial climate history agreed closely with other independent sources of information, and was further reinforced by comparison with results of additional oxygen-isotope studies in lake sediments [Edwards and Fritz, 1988; Edwards and McAndrews, 1989].

In another application of the model Clague et al. [1992] compared present climate with that during the early Holocene period of elevated timberline at an alpine site in southwestern Canada, using samples of modern and fossil fir and pine. This study incorporated small modifications to the fractionation factors proposed by Edwards and Fritz [1986], adopted as a result of ongoing efforts to refine the characterization of the isotope effects occurring during plant water evapotranspiration [e.g. Buhay et al., 1991]. Edwards and Buhay [in press] also used the model in this form in a preliminary interpretation of recent climate history in southern Ontario (A.D. 1736-1968) from an elm ring-sequence, and presented an updated version of the postglacial climate reconstruction of Edwards and Fritz [1986] incorporating the small reductions in the magnitude of inferred shifts in water $\delta^{18}O$ and humidity resulting from use of altered fractionation factors.

Considerable scope for additional "fine-tuning" of the model exists to account for improved knowledge about plant physiology, such as possible species- or temperature-dependent variations in the biochemical fractionations, or to account for temperature-dependent variations in equilibrium exchange between plant water and atmospheric vapour or other factors affecting the evaporative-enrichment response of plant waters, etc. As noted by DeNiro et al.

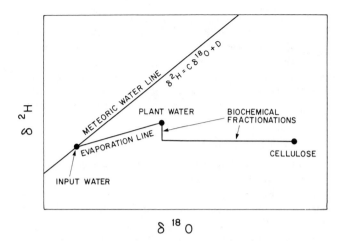

Fig. 1. Graphical representation of the Edwards and Fritz [1986] model. Potential paleoclimate signals exist in displacement along the meteoric water line (which may provide proxy temperature) and along the evaporation line (inverse to photosynthetic humidity). Note that the sensitivity of the model varies depending on the sense of climatic variation in meteoric water $\delta^{18}O$ versus h space, as the meteoric water and evaporation lines are not orthogonal.

[1988], and in spite of this inherent flexibility, such models continue to significantly understate the actual complexity of the system. Nevertheless, it is also apparent, even at the current state of development, that such models can effectively mimic the sum of the natural processes in order to derive useful paleoenvironmental information with a substantial degree of confidence. Ongoing development of the approach depends on incorporation of knowledge gained from studies of modern plants, especially better understanding of how isotopic signals are translocated within plants and the isotopic relations between plant water and cellulose [e.g. DeNiro and Cooper, 1989], kinetic isotope effects during evapotranspiration [e.g. Bariac et al., 1990; Flannagan et al., 1991a, b], and compartmentalization of water within plant tissues [e.g. Yakir et al., 1989].

The above-noted study by DeNiro and Cooper [1989] is worthy of special mention because it revealed particularly intriguing evidence from potato sprouts grown under controlled conditions that the oxygen isotope composition of cellulose precursors may be reset by equilibration with cell water prior to cellulose synthesis. If shown to be true for other plant cellulose, this would have potential to greatly simplify paleoclimate interpretation by eliminating the need to account for the effect of evapotranspiration on cellulose $\delta^{18}O$, and perhaps (by extension) cellulose δ^2H [DeNiro and Cooper, 1989]. Although many questions remain unanswered, the weight of empirical evidence presently favours the existence of significant moisture-dependent effects on both oxygen and hydrogen isotope ratios in wood cellulose [see discussions of Edwards, 1990, and DeNiro and Cooper, 1990]; however, ongoing studies may conceivably discover that other mechanisms could at times imitate such effects.

Another source of potential uncertainty in the application of such a model is the possibility of species-dependent effects, either in the biochemical reactions producing cellulose, or in the physical processes accompanying evapotranspiration. Although strong evidence of such complications is presently lacking, this problem can be circumvented effectively by using mono-specific sample suites or tree-ring sequences. Calibration of model parameters using recent ring sequences and instrumental meteorological data also permits indirect assessment of such effects to complement results from physiological study and growth experiments, as discussed below.

Interpretation of Model Output

Interpretation of input water isotopic composition, photosynthetic humidity, or other meteorological factors inferred from analysis of cellulose isotopic data (by whatever means) requires careful consideration of the conditions under which the plant lived. Trees using well-mixed groundwater having a relatively long subsurface residence time may be expected to incorporate isotopic signals representative of local weighted mean annual precipitation (and thus proxy temperature). In contrast, trees in more active hydrogeologic settings characterized by shorter groundwater residence times (such as talus slopes or other rapidly draining systems) may incorporate short-term seasonal fluctuations in isotopic composition that provide alternative information such as changing seasonal proportions [e.g. Yapp and Epstein, 1985; Clague et al., 1992] or rainfall amount [e.g. Lawrence and White, 1984].

Similarly, the relation between moisture conditions during the growth period, as indicated by photosynthetic humidity values (or, for example, precipitation inferred from transfer functions) and the local synoptic moisture regime may not be straightforward. Modelled humidity will be biased naturally toward periods of more rapid growth, which may be unrepresentative of average growth season conditions. The potential effects of such opportunism on the part of trees will be most marked in extreme climates, in which growth may be limited to short daily or periodic episodes, and least marked in temperate climates, in which favorable conditions persist over longer intervals. Even in temperate climates, isotopic composition of cellulose is likely to vary between early and late wood in a single annual ring [Epstein and Yapp, 1976; Luckman and Gray, 1990], in response to changing growth season conditions. As well, neighbouring trees at the same site could possibly sample different parts of the growing season because of differing periods of photosynthetic activity [Friedman et al., 1988].

Models like that of Edwards and Fritz [1986] also harbour potential for misrepresentation of past meteoric water $\delta^{18}O$ and humidity if water used by a tree did not lie on a known meteoric water line (i.e. if equation (3) is unknown or incorrect). A possible example of this occurs in a silver fir ring sequence reported by Ramesh et al. [1985]. Shown in Figure 2 is a comparison of oxygen and hydrogen isotope data and inferred input water $\delta^{18}O$ and photosynthetic humidity from two radii of a tree having a notably asymmetric cross-section. Ramesh et al. [1985] demonstrated that strong coherence existed between the isotopic signals from the normal and stunted radii, in spite of systematic differences in the δ^2H values. They attributed the uneven growth to internal physiological effects, and speculated that circumferential isotope variability reflected varying seasonal bias in wood around the trunk. These data were merged with δ^2H and $\delta^{13}C$ data from a third radius of the same tree

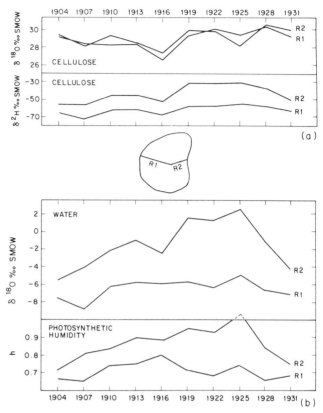

Fig. 2. a. Cellulose $\delta^{18}O$ and δ^2H profiles from a 30-year silver fir (*Abies pindrow*) ring-sequence [data from Ramesh et al., 1986]. The original ring-by-ring data are expressed here as three-year averages for simplicity.

b. Inferred meteoric water $\delta^{18}O$ and photosynthetic humidity, derived using the model of Edwards and Fritz [1986] with modified fractionations in equations (1) and (2) as suggested by Buhay et al. [1991; $\alpha_n = 1.0264$; $\alpha_e = 1.0095$; $\alpha_k = 1.0210$; $\beta_n = 0.9520$; $\beta_e = 1.0797$; $\beta_k = 1.0185$] and the Kashmir meteoric water line [Ramesh et al., 1986; $\delta^2H = 6.8\delta^{18}O + 2$] for equation (3). Note that both inferred parameters fall within expected ranges for R1, but not for R2.

and a radius from a neighbouring silver fir for the statistical comparison with growing season meteorological parameters mentioned previously [Ramesh et al., 1986].

As revealed in Figure 2, model-derived values for radius R1 (normal growth) fall within the ranges reported by the original authors, and thus provide plausible climate records, whereas values for radius R2 (stunted growth) range well beyond realistic values. The simplest explanation to account for this significant discrepancy is that water used by the R2 side of the tree did not fall on the Kashmir meteoric water line, and hence the model is not accounting correctly for isotopic effects during evapotranspiration. The isotopic enrichment of cellulose in R2 relative to R1 and the underestimation of isotopic enrichment during evapotranspiration (i.e. excessively high inferred water $\delta^{18}O$ and humidity) could occur if a portion of the cellulose on the R2 side of the tree had been derived from "pre-evaporated" soil water lying off the meteoric water line, perhaps as a result of shallower rooting on this side of the tree.

Moreover, the discrepancy between model-derived data from R1 and R2 varies over time, which may in itself constitute an additional climate signal; maximum divergence occurs during the latter half of the 30-year period represented by the ring sequence, which was slightly warmer and drier than the early part of the record [Ramesh et al., 1986; Table 1] and hence perhaps even more likely to foster isotopic differences between the radii by the postulated mechanism. It is tempting to speculate whether differential growth of the tree is not also a straightforward signal of recurrent moisture stress on the stunted side.

Climatic correlations derived by Ramesh et al. [1986] are broadly consistent with equivalent correlations inherent in both the Burk and Stuiver [1981] and Edwards and Fritz [1986] models. For example, Ramesh et al. [1986] found cellulose $\delta^{18}O$ to be most strongly influenced by relative humidity, with a regression coefficient of about -0.2 ‰ per 1% relative humidity, as opposed to -0.3 ‰ per 1% predicted for steady-state evaporation by both the Burk and Stuiver [1981] model and the Edwards and Fritz [1986] model, as configured here. Underestimation of sensitivity to humidity would be expected given the use of average growth season relative humidity, which is naturally biased by consistently high humidity at night, rather than average daytime relative humidity (i.e. photosynthetic humidity) considered in calibration of the models.

Cellulose δ^2H from the Kashmir site correlated most strongly (and negatively) with rainfall amount, which Ramesh et al. [1986] argued was an indirect confirmation that cellulose δ^2H reflected the δ^2H of associated meteoric water, in light of "amount effects" noted by previous authors for precipitation and cellulose [e.g. Dansgaard, 1964; Lawrence and White, 1984]. A positive correlation with temperature, of similar magnitude to that observed globally for precipitation, was taken as further indirect evidence of the isotopic link between cellulose and water. Their analysis did not reveal strong dependence on humidity, but could not discount a weak relation. These observations are consistent qualitatively with relations predicted by the modified Edwards and Fritz [1986] model, which suggests that variation in cellulose δ^2H is controlled primarily by changing water δ^2H, and secondarily by changing humidity (with a coefficient of about -1 ‰ per 1% change in photosynthetic humidity).

Carbon isotope variations in the Kashmir sample suite were strongly correlated with both humidity and cloudiness, in agreement with the expected influence of moisture and light levels as originally modelled by Francey and Farquhar [1982; see discussion of Ramesh et al., 1986].

ORGANIC LAKE SEDIMENTS AND PEAT

Lake Sediments

Carbon isotope analysis of organic matter in lake sediments has been used extensively to provide qualitative indications of paleolimnologic change, which may in turn provide indirect paleoclimatic information [e.g. see Stuiver, 1975; Håkannson, 1985; McKenzie, 1985]. Potential isotopic signals include productivity-driven enrichment of ^{13}C in dissolved inorganic carbon due to preferential uptake of ^{12}C by phytoplankton in the water column, which may lead to preservation of a proxy DIC $\delta^{13}C$ record in offshore or deep-water sediments. Carbon isotope variations related to chang-

ing proportions of allochtonous (e.g. terrestrial) and autochtonous organic matter can also provide qualitative information about local environmental change in nearshore sediments.

Studies have also been undertaken using the cellulose fraction of organic matter from lake sediments in order to resolve isotopic variations within a specific organic constituent. Cellulose is extracted from the fine fraction (< 149 μm) of the sediments using a technique similar to that used for purification of wood cellulose [Green, 1963], with additional washing and decanting or heavy liquid separation to remove detrital sediment particles [Edwards and McAndrews, 1989; Edwards and Elgood, 1992]. Investigations using cores from several Canadian lakes [Edwards and McAndrews, 1989; MacDonald et al., in prep; Edwards, unpublished data] suggest that cellulose in the fine fraction of offshore (profundal) sediments is commonly derived entirely from aquatic plants and algae, possibly because of rapid sedimentation and burial of phytoplankton in fecal pellets with little opportunity for oxidation [Schrader, 1971; Ferrante and Parker, 1977], compared to longer residence times of terrestrial organic matter in the oxic zone. Lake water $\delta^{18}O$ and DIC $\delta^{13}C$ histories from sediment cellulose have potential to provide qualitative paleohydrologic information analogous to that obtained from primary lacustrine carbonates [see Talbot, 1990]. Edwards and McAndrews [1989] and Edwards and Fritz [1988] showed that quantitative deconvolution of $\delta^{18}O$ profiles from sediment cellulose and carbonate (respectively) from lake sediment cores could be undertaken to reconstruct residual changes in lake water $\delta^{18}O$, independent of changing input water isotopic composition, using the regional meteoric water $\delta^{18}O$ history derived by Edwards and Fritz [1986].

Figure 3 illustrates an example of carbon and oxygen isotope stratigraphy for a sediment core from a small tundra lake in central Canada. A notable lack of correspondence exists between the whole organic and cellulose $\delta^{13}C$ profiles in this core, demonstrating that isotopic variations within undifferentiated organic material may mask more subtle $\delta^{13}C$ variations in aquatic cellulose. Cellulose $\delta^{18}O$ variations were apparently inherited from changing lake water $\delta^{18}O$, which was controlled primarily by varying evaporative enrichment of lake waters, signifying changing hydrologic balance related to advance and retreat of the Arctic Front (MacDonald et al., in prep.).

Considerably more climatically relevant information could be obtained if both oxygen and hydrogen isotope contents of lake water could be inferred from aquatic plant cellulose, since evaporative enrichment of lake waters relative to local meteoric water constitutes a possible climate signal, analogous to that for plant water enrichment in terrestrial cellulose [e.g. see Edwards and McAndrews, 1989]. Attempts to obtain useful hydrogen isotope data from aquatic plant cellulose have met with little success, apparently owing to substantial species-dependent variability in the biochemical fractionation effects between water and phytoplankton and aquatic plants [Stiller and Nissenbaum, 1980; Sternberg et al., 1984a, b; Sternberg, 1988; Edwards, unpublished data]. As shown by Sternberg [1988], however, consistent hydrogen isotope fractionation does occur between water and lipids in various aquatic plant species, permitting coupling of cellulose $\delta^{18}O$ with lipid δ^2H in a model that could be used to reconstruct $\delta^{18}O$ and δ^2H of associ-

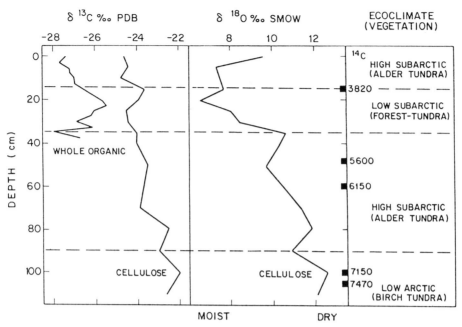

Fig. 3. Carbon and oxygen isotope data from a sediment core, Queen's Lake, Northwest Territories, Canada. Cellulose $\delta^{18}O$ mainly responded to varying evaporative enrichment, reflecting changing water balance. Covariance of cellulose $\delta^{18}O$ and $\delta^{13}C$ in the lower part of the core is consistent with closed-basin hydrology under dry climatic conditions [cf. Talbot 1990]. Covariance disappears in sediments deposited after about 6000 B.P. as open-basin conditions developed in response to climatic amelioration [MacDonald et al., in prep.]. Discrepancy between $\delta^{13}C$ of undifferentiated organic matter [Edwards, unpublished data] and cellulose $\delta^{13}C$ in the upper part of the core shows the influence of abundant non-cellulose organic material, probably of both aquatic and terrestrial origin.

ated water. Significant opportunities for paleoenvironmental reconstruction exist if further studies verify that aquatic plant lipids can be derived from sediments along with aquatic plant cellulose.

Peat

Plant matter preserved in peat deposits derives from a range of microenvironments, from terrestrial conditions inhabited by trees and emergent vegetation, subject to variable evaporative isotopic enrichment of oxygen and hydrogen in plant water, to fully saturated conditions that are essentially aquatic. Isotopic studies of peat include attempts to derive quantitative information from whole organic matter or cellulose [e.g. Schiegl, 1972; Dupont and Mook, 1987; Chatwin, 1981], relying on temperature-dependent isotopic signals inherited from local meteoric water, as well as more qualitative examinations of $\delta^{18}O$, $\delta^{2}H$, and $\delta^{13}C$ of the whole organic or cellulose fractions, to better understand environments of peat formation or to seek other paleoenvironmental signals [e.g. Brenningkmeijer et al., 1982; Dupont and Brenninkmeijer, 1984; Gouze et al., 1987; Aucour et al., this volume].

Possible species-dependent variations in biochemical isotope effects during cellulose synthesis could be especially significant for peat, because of the variety of plants (and photosynthetic pathways) that may be represented in a single sample. Varying carbon isotope effects may provide broad indications of changing climate, as reflected by changing ratio of C_3 to C_4 plants [see Aucour et al., this volume], but are unlikely to record more subtle environmental signals like those that appear to be preserved in tree cellulose. Biochemical fractionation effects for oxygen in peat plants (as in

other plants and algae) seem to be remarkably invariant, in contrast to the suspected variability in the hydrogen isotope effects [Sternberg, 1988]. Brenninkmeijer et al. [1982] concluded that covariant (but seemingly excessive) fluctuations in cellulose $\delta^{18}O$ and $\delta^{2}H$ from a core of an ombotrophic bog in The Netherlands reflected fundamental control by meteoric water isotopic composition ($\approx$ proxy temperature), overprinted by noise from varying biochemical effects on hydrogen, and evaporative enrichment effects on both hydrogen and oxygen from plant water transpiration. Dupont and Mook [1987] later derived tentative "relative deuterium temperature" records using hydrogen isotope data from several Dutch peat cores, by correcting for varying taxonomic composition of the peat.

If biochemical fractionation effects were similar for different plants in a particular peat assemblage, or specific plants could be selectively subsampled, then estimates of meteoric water isotopic composition could be derived from cellulose $\delta^{18}O$ and $\delta^{2}H$ data, using a model like that of Edwards and Fritz [1986] to filter out residual enrichment arising from evapotranspiration (which would indicate a spurious apparent humidity). Carrying out this exercise using the data reported by Brenninkmeijer et al. [1982], for example, yields highly reasonable values for $\delta^{18}O$ of water in the Engbertsdijksveen bog around the time of the Subboreal/Subatlantic transition (Figure 4). Reconstructed water $\delta^{18}O$ values are further supported by the observation that cellulose $\delta^{18}O$ and $\delta^{2}H$ obtained on two separates of a single sample (taken at 105 cm), which was chosen by the authors to illustrate species-dependent isotopic differences, produce identical estimates of water $\delta^{18}O$

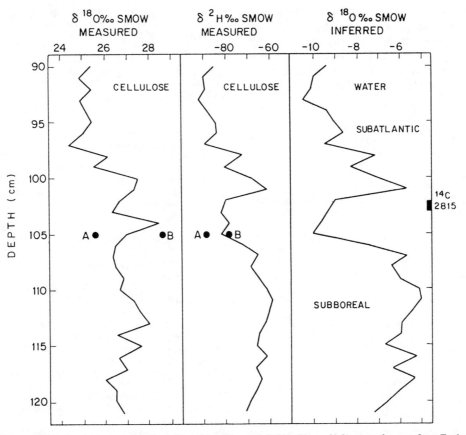

Fig. 4. Cellulose oxygen and hydrogen isotope data from Brenninkmeijer et al. [1982; Figure 2] for part of a core from Engbertsdijksveen bog, The Netherlands. The core interval represents about 350 years of accumulation, spanning the Subboreal-Subatlantic climatic deterioration. Bog water $\delta^{18}O$ was inferred using the Edwards and Fritz [1986] model, using the global meteoric water line of Craig [1961] as equation (3) and the fractionation factors listed in the caption of Figure 2. B and A refer to splits of the sample from 105 cm, containing and lacking *Eriophorum* plant matter, respectively. Both subsamples yield the same inferred input water $\delta^{18}O$ value, as would be expected if *Sphagnum* (the major constituent of the peat) and *Eriophorum* have similar biochemical isotope effects. The difference in the apparent discrepancies between $\delta^{18}O$ and δ^2H values for A and B illustrates the much larger relative influence that humidity-dependent effects have on cellulose $\delta^{18}O$ compared to cellulose δ^2H.

(while yielding, predictably, differing apparent humidities). Both of these observations are, in fact, consistent with isotopic data reported from plants in a modern Dutch bog [Brenningkmeijer et al., 1982; Figure 1]. Cellulose $\delta^{18}O$ and δ^2H data for various species of *Sphagnum* moss and other vascular plants that occur in the peat fall along a linear trend similar in slope to that of the evaporation line implicit in the Edwards and Fritz [1986] model as here configured. This suggests that biochemical effects may, after all, be rather similar in the major species within this taxonomic assemblage (although this exercise does not confirm that correct values for the biochemical fractionation factors are being used).

More conclusive information about possible species-dependent effects on cellulose δ^2H could be obtained by comparison with lipid δ^2H, as suggested by Sternberg [1988], and as discussed in the previous section.

CONCLUDING COMMENTS

Numerous isotopic signals exist in other types of continental organic matter, yet few offer potential for detailed quantitative time-series paleoclimate reconstruction equivalent to that obtainable from wood cellulose, and cellulose (and perhaps lipids) of sediments and peat. Promising exceptions may include isotopic data from plant remains in packrat middens [e.g. Long et al., 1990], mammal bone collagen [e.g. Cormie, 1991], insect chitin [Miller et al., 1988], or other substrates that may be preserved in fossil deposits.

Both qualitative and quantitative information about past environmental conditions may be obtained from isotopic study of continental organic matter. Although qualitative data may place useful constraints on interpretations based on other sources of information, the greatest value of isotopic studies is realized by the ability to make quantitative estimates of climatically relevant parameters independently of other data.

The most useful climatically relevant isotopic parameter is the isotopic composition of meteoric water. Aside from possibly providing paleotemperature, this parameter can afford a fundamental baseline at regional spatial scale for assessment of paleomoisture variations. Isotopic signals preserved in terrestrial cellulose provide one of the least ambiguous sources of information about changes in past meteoric water composition, which can in turn be used to deconvolute isotopic signals preserved in other materials.

Finally, it is clear that the existing models used to interpret such data remain first approximations that need to be continually tested and refined as new data become available from study of modern plants.

Acknowledgments. The author wishes to thank W.M. Buhay for extensive discussion of this topic, and the two anonymous reviewers for help in improving this paper.

REFERENCES

Aucour, A.-M., Hillaire-Marcel, C., and Bonnefille, R., A 30,000 year record of ¹³C and ¹⁸O changes in organic matter from an equatorial peat-bog, (this volume).

Bariac, T., Jusserand, C., Mariotti, A., Evolution spatio-temporelle de la composition isotopique de l'eau dans le continuum sol-plante-atmosphere, *Geochim. Cosmochim. Acta*, 54, 413-424, 1990.

Becker, B., Kromer, B., and Trimborn, P., A stable-isotope tree-ring timescale of the Late Glacial/Holocene boundary, *Nature*, 353, 647-649, 1991.

Bender, M.M., Variations in the ¹³C/¹²C ratios of plants in relation to the pathway of photosynthetic carbon dioxide fixation, *Phytochem.*, 1239-1244, 1971.

Brenningkmeijer, C.A.M., van Geel, B., and Mook, W.G., Variations in the D/H and ¹⁸O/¹⁶O ratios in cellulose extracted from a peat bog core, *Earth Planet. Sci. Lett.* 61, 283-290, 1982.

Buhay, W.M., Edwards, T.W.D., and Aravena, R., Fine-tuning the cellulose model for dendroclimatological interpretation, in *Proc. Int. Symp. Use Isot. Techn. Water Resour. Devel.*, International Atomic Energy Agency, Vienna, March 1991, IAEA-SM-319/10P, 122-123, 1991.

Burk, R.L., and Stuiver, M., Oxygen isotope ratios in tree cellulose reflect mean annual temperature and humidity, *Science*, 211, 1417-1419, 1981.

Chatwin, S.C., Holocene temperatures in the Mackenzie Valley determined by oxygen isotope analysis of peat cellulose, in *Proc. 4th Int. Conf. Permafrost*, 127-130, 1981.

Clague, J.J., Mathewes, R.W., Buhay, W.M., and Edwards, T.W.D., Early Holocene climate at Castle Peak, southern Coast Mountains, British Columbia, Canada, *Palaeogeogr., Palaeoclim., Palaeoecol.*, 95, 153-167, 1992.

Cooper, L.W., and DeNiro, M.J., Covariant stable oxygen and hydrogen isotope compositions of plant water: Species effects, *Ecol.*, 70, 1619-1628, 1989.

Cormie, A.B., Developing bone collagen stable hydrogen isotope analyses for paleoclimate research and enhancing interpretations with bone carbon, nitrogen and oxygen isotopes, *Ph.D. Thesis*, McMaster Univ., Hamilton, 343 pp., 1991.

Craig, H., Isotopic variations in meteoric waters, *Science* 133, 1702-1703, 1961.

Craig, H., and Gordon, L.I., Deuterium and oxygen-18 isotope variations in the ocean and marine atmosphere, in *Stable Isotopes in Oceanography Studies and Paleotemperatures*, Spoleto, July 26-27, 1965, Coniglio Nazionale delle Ricerche, Laboratorio di Geologia Nucleare, Pisa, 1-122, 1965.

Dansgaard, W., Stable isotopes in precipitation, *Tellus*, 16, 436-468, 1964.

DeNiro, M.J., The effects of different methods of preparing cellulose nitrate on the determination of the D/H ratios of non-exchangeable hydrogen of cellulose, *Earth Planet. Sci. Lett.*, 54, 177-185, 1981.

DeNiro, M.J., and Cooper, L.W., Post-photosynthetic modification of oxygen isotope ratios of carbohydrates in the potato: Implications for paleoclimatic reconstruction based on isotopic analysis of wood cellulose, *Geochim. Cosmochim. Acta*, 53, 2573-2580, 1989.

DeNiro, M.J., and Cooper, L.W., Water is lost from leaves and trunks of trees by fundamentally different mechanisms, *Geochim. Cosmochim. Acta*, 54, 1845-1846, 1990.

DeNiro, M.J., and Epstein, S., Relationship between oxygen isotope ratios of terrestrial plant cellulose, carbon dioxide, and water, *Science* 204, 51-53, 1979.

DeNiro, M.J., Sternberg, L.D., Marino, B.D., and Druzik, J.R., Relation between D/H ratios and ¹⁸O/¹⁶O ratios in cellulose from linen and maize – Implications for paleoclimatology and for sindonology, *Geochim. Cosmochim. Acta*, 52, 2189-2196, 1988.

Dubois, A.D., and Ferguson, D.K., The climatic history of pine in the Cairngorms based on radiocarbon dates and stable isotope analysis, with an account of the events leading up to its colonization, *Rev. Palaeobot. Palynol.*, 46, 55-80, 1985.

Dupont, L.M., and Brenningkmeijer, C.A.M., Paleobotanic and isotopic analysis of late Subboreal and early Subatlantic peat from Engbertsdijksveen VII, The Netherlands, *Rev. Paleobot. Palynol.*, 41, 241-271, 1984.

Dupont, L.M., and Mook, W.G., Palaeoclimate analysis of ²H/¹H ratios in peat sequences with variable plant composition, *Chem. Geol. (Isot. Geosci. Sect.)*, 66, 323-333, 1987.

Edwards, T.W.D., New contributions to isotope dendroclimatology from studies of plants, *Geochim. Cosmochim. Acta*, 54, 1843-1844, 1990.

Edwards, T.W.D., and Buhay, W.M., Isotope paleoclimatology of southern Ontario, in *Proc. Symp. Great Lakes Archaeol. Paleoecol.*, Quat. Sci. Inst., Univ. Waterloo, in press.

Edwards, T.W.D., and Elgood, R., Extraction of sediment cellulose for oxygen and carbon isotope analysis, *Tech. Proced. 28.0*, Environ. Isot. Lab., Univ. Waterloo, Waterloo, 9 pp., 1992.

Edwards, T.W.D., and Fritz, P., Assessing meteoric water composition and relative humidity from ¹⁸O and ²H in wood cellulose: paleoclimatic implications for southern Ontario, Canada, *Appl. Geochem.*, 1, 715-723, 1986.

Edwards, T.W.D., and Fritz, P., Stable isotope paleoclimate records for southern Ontario: comparison of results from marl and wood, *Can. J. Earth Sci.*, 25, 1397-1406, 1988.

Edwards, T.W.D., and McAndrews, J.H., Paleohydrology of a Canadian Shield lake inferred from ¹⁸O in sediment cellulose, *Can. J. Earth Sci.*, 26, 1850-1859, 1989.

Edwards, T.W.D., Aravena, R.O., Fritz, P., and Morgan, A.V., Interpreting paleoclimate from ¹⁸O and ²H in plant cellulose: comparison with evidence from fossil insects and relict permafrost in southwestern Ontario, *Can. J. Earth Sci.*, 22, 1720-1726, 1985.

Epstein, S., and Yapp, C., Climatic implications of the D/H ratio of hydrogen in C-H groups in tree cellulose. *Earth Planet. Sci. Lett.*, 30, 252-261, 1976.

Epstein, S., and Krishnamurthy, R.V., Environmental information in the isotopic record in trees, *Phil. Trans. R. Soc. Lond.*, A330, 427-439, 1990.

Epstein, S., Thompson, P., and Yapp, C.J., Oxygen and hydrogen isotopic ratios in plant cellulose, *Science*, 198, 1209-1215, 1977.

Epstein, S., Yapp, C.J., and Hall, J.H., The determination of the D/H ratio of non-exchangeable hydrogen in cellulose extracted from aquatic and land plants, *Earth Planet. Sci. Lett.*, 30, 241-251, 1976.

Farquhar, G.D., Hubick, K.T., Condon, A.G., and Richards, R.A., Carbon isotope fractionation and plant water-use efficiency, in *Stable Isotopes in Ecological Research*, edited by P.W. Rundel, J.R. Ehleringer, and K.A. Nagy, pp. 21-40, Springer-Verlag, New York, 1988.

Ferhi, A., and Létolle, R., Transpiration and evaporation as the principal factors in oxygen isotope variations of organic matter in land plants, *Physiol. Vég.*, 15, 363-370, 1977.

Ferrante, J.G., and Parker, J.I., Transport of diatom frustules by copepod faecal pellets to the sediments of Lake Michigan, *Limnol. Oceanogr.*, 22, 92-97, 1977.

Flannagan, L.B., Comstock, J.P., and Ehleringer, J.R., Comparison of modelled and observed environmental influences on the stable oxygen and hydrogen isotope composition of leaf water in *Phaseolus vugaris* L., *Plant Physiol.*, 96, 588-596, 1991a.

Flannagan, L.B., Bain, J.F., and Ehleringer, J.R., Stable oxygen and hydrogen isotope composition of leaf water in C₃ and C₄ plant species under field conditions, *Oecologia*, 88, 153-158, 1991b.

Friedman, I., Carrara, P., and Gleason, J., Isotopic evidence of Holocene climatic change in the San Juan Mountains, Colorado, *Quat. Res.*, 30, 350-353, 1988.

Francey, R.J., and Farquhar, G.D., An explanation of $^{13}C/^{12}C$ variations in tree rings, *Nature*, 297, 28-31, 1982.

Gouze, P., Ferhi, A., Fontes, J.-Ch., and Roche, M., Composition isotopique (^{18}O) de la matière organique des tourbières actuelles et holocènes en Bolivie: Résultats préliminaires et perspectives d'application en paléoclimatologie, *Géodynamique*, 2, 113-116, 1987.

Green, J.W., Wood cellulose, in *Methods in Carbohydrate Chemistry, Vol. II*, edited by R.L. Whistler, pp. 9-20, Academic Press, New York, 1963.

Håkansson, S., A review of various factors influencing the stable carbon isotope ratio of organic lake sediments by the change from glacial to postglacial environmental conditions, *Quat. Sci. Rev.*, 4, 135-146, 1985.

Hardcastle, K.G., and Friedman, I., A method for oxygen isotope analysis of organic matter, *Geophys. Res. Lett.*, 1, 165-167, 1974.

Lawrence, J.R., and White, J.W.C., Precipitation amounts during the growing season from the D/H ratios of Eastern White Pine, *Nature*, 311, 558-560, 1984.

Leavitt, S.W., Environmental information from $^{13}C/^{12}C$ ratios of wood, this volume.

Libby, L.M., Pandolfi, L.F., Payton, P.H., Marshall, J., Becker, B., and Giertz-Sienbenlist, V., Isotope tree thermometers, *Nature*, 261, 284-288, 1976.

Long, A., Warneke, L.A., Betancourt, J.L., and Thompson, R.S., Deuterium variations in plant cellulose from fossil packrat middens, in *Packrat Middens: The Last 40,000 Years of Biotic Change*, edited by J.L. Betancourt, T.R. van Devender, P.S. Martin, pp. 380-396, Univ. Arizona Press, Tucson, 1990.

Luckman, B.H., and Gray, J., Oxygen isotope ratios from tree rings containing compression wood, *Quat. Res.*, 33, 117-121, 1990.

Marino, B.D., and McElroy, M.B., Isotopic composition of atmospheric CO_2 inferred from carbon in C_4 plant cellulose, *Nature*, 349, 127-131, 1991.

MacDonald, G.M., Edwards, T.W.D., Moser, K.A., Pienitz, R., and Smol, J.P., Response of treeline vegetation and lakes to rapid climatic change caused by shifts in the Arctic Front, in prep.

McKenzie, J.A., Carbon isotopes and productivity in the lacustrine and marine environments, in *Chemical Processes in Lakes*, edited by W. Stumm, pp. 99-117, Wiley, Toronto, 1985.

Miller, R..F., Fritz, P., and Morgan, A.V. Climatic implications of D/H ratios in beetle chitin, *Palaeogeogr., Palaeoclim., Palaeoecol.*, 66, 277-288, 1988.

Ramesh, R., Bhattacharya, S.K., and Gopalan, K., Dendroclimatological implications of isotopic coherence in trees from Kashmir Valley, India, *Nature*, 317, 802-804, 1985.

Ramesh, R., Bhattacharya, S.K., and Gopalan, K., Climatic correlations in the stable isotope records of silver fir (*Abies pindrow*) trees from Kashmir, India, *Earth Planet. Sci. Lett.*, 79, 66-74, 1986.

Rittenberg, D., and Ponticorvo, L., A method for the determination of the ^{18}O concentration of the oxygen of organic compounds, *Int. J. Appl. Radiat. Isot.*, 1, 208-214, 1956.

Schiegl, W.G., Deuterium content of peat as a paleoclimatic recorder, *Science*, 175, 512-514, 1972.

Schrader, H.-J., Fecal pellets; role in sedimentation of pelagic diatoms, *Science*, 174, 55-57, 1971.

Smith, B.N., and Epstein, S., Two categories of $^{13}C/^{12}C$ ratios for higher plants, *Plant Physiol.*, 47, 380-384, 1971.

Sternberg, L., DeNiro, M.J., and Ting, I.P., Carbon, hydrogen, and oxygen isotope ratios of cellulose from plants having intermediary photosynthetic modes, *Plant Physiol.*, 74, 104-107, 1984a.

Sternberg, L.O., DeNiro, M.J., and Johnson, H.B., Isotope ratios of cellulose from plants having different photosynthetic pathways, *Plant Physiol.*, 74, 557-561, 1984b.

Sternberg, L.S.L., D/H ratios of environmental water recorded by D/H ratios of plant lipids, *Nature*, 333, 59-61, 1988.

Sternberg, L.S.L., Oxygen and hydrogen isotope measurements in plant cellulose analysis, in *Modern Methods of Plant Analysis. Vol. 10 Plant Fibers*, edited by H.F. Linokens and J.F. Jackson, pp. 89-99, Springer-Verlag, Heidelberg, 1989.

Stiller, M., and Nissenbaum, A., Variations of stable hydrogen isotopes in plankton from a freshwater lake, *Geochim. Cosmochim. Acta*, 44, 1099-1101, 1980.

Stuiver, M., Climate versus changes in ^{13}C content of the organic component of lake sediments during the Late Quaternary, *Quat. Res.*, 5, 251-262, 1975.

Stuiver, M., and Braziunas, T.F., Tree cellulose $^{13}C/^{12}C$ isotope ratios and climatic change, *Nature*, 328, 58-60, 1987.

Talbot, M.R., A review of the palaeohydrological interpretation of carbon and oxygen isotopic ratios in primary lacustrine carbonates, *Chem. Geol. (Isot. Geosci. Sect.)*, 80, 261-279, 1990.

Thompson, P., and Gray, J., Determination of $^{18}O/^{16}O$ ratio in compounds containing C, H, and O, *Int. J. Radiat. Isot.*, 28, 411-415, 1977.

Yakir, D., DeNiro, M.J., and Rundel, P.W., Isotopic inhomogeneity of leaf water: Evidence and implications for the use of isotopic signals transduced by plants, *Geochim. Cosmochim. Acta*, 53, 2769-2773, 1989.

Yapp, C.J., and Epstein, S., Climatic implications of meteoric water over North America (9,500-22,000 B.P.) as inferred from ancient wood cellulose C-H hydrogen, *Earth Planet. Sci. Lett.*, 34, 333-350, 1977.

Yapp, C.J., and Epstein, S., A re-examination of cellulose carbon-bound hydrogen ∂D measurements and some factors affecting plant-water D/H relationships, *Geochim. Cosmochim. Acta*, 46, 955-965, 1982a.

Yapp, C.J., and Epstein, S., Climatic significance of the hydrogen isotope ratios in tree cellulose, *Nature*, 297, 636-639, 1982b.

Yapp, C.J., and Epstein, S., Seasonal contributions to the climatic variations recorded in tree ring deuterium/hydrogen data, *J. Geophys. Res.*, 90, D2, 3747-3752, 1985.

T.W.D. Edwards, Department of Earth Sciences, University of Waterloo, Waterloo, Ontario, Canada N2L 3G1.

A 30,000 year Record of ^{13}C and ^{18}O Changes in Organic Matter from an Equatorial Peatbog

ANNE-MARIE AUCOUR[1] AND CLAUDE HILLAIRE-MARCEL

Centre de recherche en Géochimie isotopique et en Géochronologie (GEOTOP),
Université du Québec à Montreal

RAYMONDE BONNEFILLE

Laboratoire de Géologie du Quaternaire, CNRS, Marseille-Luminy

Coring of the Kashiru peatbog in Burundi (core Ka-2, 3° 28'S, 29°34'E, altitude: 2240 m) allowed the recovery of a 30,000 year palaeoclimatic record for Equatorial Africa. The δ^{13}C values of total organic matter (TOM) in core Ka-2 vary primarily as a function of C_3/C_4 plant inputs (essentially Gramineae and Cyperacae versus other plants). Maximum ^{13}C contents of TOM are observed between 30 and 15 ka, which suggests the dominance of C_4 plants in the local palaeovegetation during that period. Fossil plant remains are derived from emergent plants and their cellulose ^{18}O/^{16}O ratios are measured. They reflect (1) the isotopic composition of the meteoric waters used by the plants, (2) the evapotranspiration which is mostly controlled by the relative humidity of the air, and (3) the type of plant. The 30-12 ka and 3-0 ka intervals show high ^{18}O contents (δ^{18}O values of 24-27.5 ‰ and 26-28.5 ‰ versus SMOW respectively). The much lower δ^{18}O values (19-24‰) observed during the 7-3 ka interval indicate relatively high humidity and ^{18}O-depleted precipitation during that period. Combined with pollen data, the isotopic composition of organic matter in peatbogs provides additional information on vegetational mass budgets (C_3/C_4 plants) and on past water cycle in the equatorial atmosphere.

INTRODUCTION

The carbon and oxygen isotopic compositions of fossil organic matter from continental deposits have been related to the proportions of C_3 and C_4 plants in the palaeovegetation [Krishnamurthy et al., 1982; Guillet et al., 1988; Cerling et al., 1989; Hillaire-Marcel et al., 1989] and to palaeohydrological characteristics such as the isotopic composition of the source waters taken up by the plants and the relative humidity of the air [Burk and Stuiver, 1981; Edwards and Fritz, 1986]. The depletion in ^{13}C of organic matter in terrestrial plants relative to the ambient CO_2 is effectively controlled by the photosynthetic pathway with δ^{13}C values falling around -26‰ versus PDB for C_3 plants and around -13‰ for C_4 plants [Wickman, 1952; Craig, 1954; Bender, 1968; Smith and Epstein, 1971; Troughton, 1972]. The enrichment in ^{18}O of cellulose in the leaves of a terrestrial plant relative to the source water is expected to include (1) the biochemical enrichment of cellulose relative to the leaf water used in its synthesis [DeNiro and Epstein, 1981], and (2) the enrichment in ^{18}O of leaf water (relative to the source water), which has been related to the air relative humidity and to the isotopic composition of the local water vapour [Dongman et al., 1974; Allison et al., 1985] . The oxygen isotope composition of fossil leaf cellulose therefore constitutes a potential tracer of the palaeohumidity and of the isotopic composition of atmospheric palaeowaters, i.e., precipitation and atmospheric vapour. However, the isotopic composition of plant cellulose could also largely depend on plant type. Large differences in cellulose ^{18}O/^{16}O ratios have been observed among terrestrial plants having different photosynthetic pathways or physiognomies and growing within the same area [Sternberg and DeNiro, 1983; Sternberg et al., 1984a, 1984b].

The carbon and oxygen isotope study of organic material was undertaken in modern plants and in a core (Ka-2: Kashiru,

[1]also at Centre des Sciences de la Terre, Université Lyon 1, 43, boulevard du 11 novembre, 69622 Villeurbanne Cédex

Climate Change in Continental Isotopic Records
Geophysical Monograph 78

Burundi) collected in East Equatorial Africa. The $^{13}C/^{12}C$ ratios of total organic matter (TOM) in swamp plants is related to the photosynthetic pathways employed by the plants, among other things. Analysis of the core samples should indicate the changes in the C_3/C_4 composition of the local palaeovegetation during the last 30,000 years. The $^{18}O/^{16}O$ ratios of waters and plant cellulose from modern bog sites in Burundi document (1) the range in $\delta^{18}O$ values in plant cellulose at a given site in an equatorial bog, and (2) the present relationship between the $^{18}O/^{16}O$ ratios of plant cellulose, the $^{18}O/^{16}O$ ratios of the meteoric waters used by the plants, and the relative humidity of the air. $^{18}O/^{16}O$ ratios of fossil cellulose from core Ka-2 should provide proxy data on palaeohumidity and on the isotopic composition of palaeoprecipitation in Equatorial Africa during the last 30,000 years.

ENVIRONMENTAL SETTING

Kashiru peatbog (3°28'S, 29°34'E, altitude: 2240 m) occupies a small depression of 0.5 km² surface area within a 4 km² catchment area in the Burundi highlands (Figure 1). The bog is fed by rainfall and by the superficial aquifer from the weathered zone of the Precambrian basement. The mean daily air surface temperature (T) is estimated as 16°C and the mean relative humidity (h) as 0.8 on the basis of climatic data given by Bultot [1972]. Relative humidity measurements were made

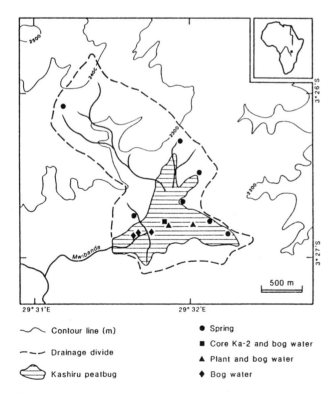

Fig. 1. Location map for Kahiru peatbog, water and plant samples, and core Ka-2.

Contour line (m)
Drainage divide
Kashiru peatbug
● Spring
■ Core Ka-2 and bog water
▲ Plant and bog water
◆ Bog water

in the early morning and in mid-afternoon [Griffiths, 1972] and their mean is therefore taken as the daytime humidity. At the climatic station of Kisozi, 15 km southeast of Kashiru, at 2150 m altitude, the mean annual precipitation is 1450 mm [Bultot, 1972]. No isotopic data on precipitation are available in Burundi. The International Atomic Energy Agency (IAEA) network has provided monthly precipitation $\delta^{18}O$ values in Equatorial Africa, for several of the years within the 1961-1974 period (Dar-es-Salaam: 6°53'S, 39°12'E, altitude 55 m, n = 9; Entebbe: 0°03'S, 32°27'E, altitude 1155 m, n = 8; Malange: 9°33'S, 16°22'E, altitude 1139 m, n = 6; Kinshasa: 4°22'S, 15°15'E, altitude 430 m, n = 5) [International Atomic Energy Agency, 1969, 1970, 1971, 1973, 1975, 1979, 1981]. The natural vegetation of the peatbog, preserved in 1982, was described as a *Xyris-Sphagnum* association with *Cyperus denudatus*, *Juncus* sp. [Bonnefille and Riollet,1988]. The bog was drained in 1983-1984 and partly exploited. At the time of this study, in September and October 1989, a part of the depression was cultivated. The other part was inundated, swampy and occupied by the natural bog vegetation. The water and plant samples were collected in the swampy part of the depression.

MATERIALS AND METHODS

Modern samples consist of herbarium specimens provided by the National Botanical Garden (Meise, Belgium) and the Museum of Natural History (Paris) for the ^{13}C study and of plants and waters collected at Kashiru in September and October 1989 for the ^{18}O study. Core Ka-2 was taken at Kashiru in 1989 (Figure 1). The lower part of the core (825-890 cm) consists of sandy and clayey peat. The peat is well preserved and brown in the upper 4 meters. Abundant charcoal was noticed between 350 and 400 cm. Subsampling consists of 5 cm-slices in the upper part (0 - 605 cm) and 10 cm-slices in the lower part (605 - 895 cm), taken at 20 cm intervals. The fresh peat samples were first boiled in a 5% NaOH solution, then washed on a 125 μm sieve in order to remove the fine particulate matter and humic substances. The carbonized plant remains (charcoal) and siliclastic fractions were removed under a binocular microscope. The non-carbonized plant fragments were subsequently boiled in a 4% HCl solution, washed in distilled water and dried. The following steps in sample processing were then applied to both modern and fossil plants (1) Soxhlet extraction with ether and then methanol,(2) chloriting treatment, and (3) alkaline extraction [Green, 1963]. The end product represents the purified alpha cellulose fraction.

Cellulose $^{18}O/^{16}O$ ratios were determined following pyrolysis with mercuric chloride [Epstein et al.,1977]. TOM, cellulose, and charcoal $^{13}C/^{12}C$ ratios were determined following combustion with cupric oxide [Stump and Frazer, 1973]. The isotopic composition of the water was determined by equilibration of water with CO_2 for oxygen [Epstein and Mayeda, 1953]. Isotope ratios are expressed as a δ value with

TABLE 1. [14]C Ages Obtained by Accelerator-Mass Spectrometrer Analysis on Plant Remains from Kashiru (Core Ka-2)

Depth (cm)	Age (ka)	Laboratory Number
195-197	4.450 ± 0.060	TO-2319
330-332	6.000 ± 0.060	TO-2320
420-422	12.160 ± 0.080	TO-2321
580-582	20.880 ± 0.140	TO-2322
835-840	30.220 ± 0.260	TO-2323

reference to SMOW for [18]O and to PDB for [13]C. The overall analytical uncertainty is 0.5‰ for cellulose $\delta^{18}O$ values, 0.1‰ for water $\delta^{18}O$ values and 0.1‰ for $\delta^{13}C$ values .

The chronology of core Ka-2 was established by five ^{14}C ages obtained by accelerator mass spectrometry (AMS) at the IsoTrace Laboratory of the University of Toronto on hand-picked plant remains.

ACCUMULATION RATES AND CHRONOLOGY

[14]C ages (Table 1) yield an average accumulation rate of 27 cm/kyr between 30 and 21 ka and of 18 cm/kyr between 21 and 12 ka. The accumulation rates is reduced between 12 and 6 ka (15 cm/kyr). In the upper part of the sequence, accumulation rate averages 87 cm/kyr between 6 and 4.5 ka and 44 cm/kyr from 4.5 ka to present. Accumulation rates found in this study are similar to the rates reported by Bonnefille et Riollet [1988] for core Ka-1, located in the vicinity of core Ka-2. Detailed chronology of several different cores is not yet available to confirm that the very low accumulation rates between 12 and 6 ka are due to changes in the drainage pattern within the swamp.

The ages referred to in the following paragraphs were interpolated from the five available AMS- [14]C ages (Table 1). However, it was assumed that the change in peat composition at 400 cm reflects a change in the accumulation rate and the ages between 332 and 400 cm were extrapolated from the accumulation rate of the 195-332 cm section of the core.

RESULTS

[13]C/[12]C Ratios of TOM in Recent Plants

Modern medium-altitude (1300-1700 m) and high-altitude (>1700 m) peatbog and swamp components from Rwanda and Burundi and from East Africa were investigated (Table 2). Among the investigated plants, C_4 plants are characterized by their high $\delta^{13}C$ values (-11.3 ± 0.7‰, n = 6) and C_3 plants by their low $\delta^{13}C$ values (-25.5 ± 2.3‰, n = 17). Both C_3 and C_4 species are found within the Gramineae and the Cyperaceae. The other components are exclusively C_3 plants (Table 2). These results are in agreement with studies of Gramineae

TABLE 2. $\delta^{13}C$ Values of Swamp and Bog Plants from East Africa.

	Family	Sampling site	$\delta^{13}C$ (‰)
Carex monostachya	Cyperaceae	Upper Hausberg Valley, 4297 m	-22.7
Carex runssoroensis	Cyperaceae	Parc National Albert, 3650 m,	-22.8
Cladium mariscus	Cyperaceae	Bujumbura, 780 m, Burundi	-26.6
Cladium mariscus	Cyperaceae	Bujumbura, 780 m, Burundi	-24.7
Cyperus denudatus	Cyperaceae	Bubanza, 900 m, Burundi	-28.0
Cyperus latifolius	Cyperaceae	Kashiru, 2240 m, Burundi	-10.6
Cyperus papyrus	Cyperaceae	Mosso ruijigi, Burundi	-11.0
Drypoteris gongylodes	Pteridophyta	Kigali, 1350 m, Rwanda	-27.4
Erica kingaensis	Ericaceae	Kashiru, 2240 m, Burundi	-25.9
Juncus effusus	Juncaceae	Bururi, 1750 m, Burundi	-28.5
Juncusoxycarpus	Juncaceae	Muramwya,2100 m,Burundi	-25.0
Miscanthidium erectum	Gramineae	Pretoria, South Africa	-11.9
Miscanthidium fuscescens	Gramineae	Misma Pori, South Africa	-12.0
Miscanthidium violaceum	Gramineae	Tanganyika	-11.9
Nymphea sp.	Nympheacae	Gitanga, 1800 m, Burundi	-24.7
Phragmites mauritianius	Gramineae	Gakungwe,780 m , Burundi	-26.7
Potamogeton sp.	Potamogetonaceae	Gitanga, 1800 m, Burundi	-25.9
Pycreus nigricans	Cyperaceae	2280 m, Rwanda	-10.3
Restio mahoni	Restionaceae	Mlanje mountains, South Africa	-22.2
Restio ocreatus	Restionaceae	Matjesfontein, South Africa	-22.6
Sphagnum sp.	Bryophyta	Kashiru, 2240 m, Burundi	-23.7
Syzygium cordatum	Myrtaceae	Mosso Ruijigi, Burundi	-27.1
Syzygium guineense	Myrtaceae	Kigeme, 2100 m, Burundi	-29.8

[Livingstone and Clayton, 1980] and Cyperaceae [Hesla et al.,1982] from East Africa. The Gramineae *Miscanthidium violaceum* is recognized as a C_4 species. This study confirms that the Cyperaceae *Cyperus latifolius*, *C. papyrus* and *Pycreus nigricans* are C_4 plants [Hesla et al., 1982].

$^{18}O/^{16}O$ Ratios of Modern Waters and Plant Celluloses

The mean $\delta^{18}O$ value of the spring waters, which feed the Kashiru bog, is -4.0 ± 0.2‰ (n = 7) (Table 3). At the sites of plant collection, the water $\delta^{18}O$ values are close to -3‰. The cellulose of the *Sphagnum* moss, which grows nearly submerged, shows an enrichment in ^{18}O relative to the environmental waters of 25‰. For emergent plants, $\delta^{18}O$ values in leaf cellulose range from 26 to 31‰, hence yielding an enrichment of 29 to 34‰ relative to the bog waters. Plants growing close to the water surface (*Polygonum pulchrum*, *Erica kingaensis*) have the lowest ^{18}O content and show an isotopic composition between those of mosses and of other emergent plants. A difference in $\delta^{18}O$ values of 3‰ is observed between the C_4 *Cyperus* (*C. latifolius*) and the C_3 *Cyperus* (*C. haspan*). The mean isotopic composition of the

TABLE 3. Description, Altitude and $\delta^{18}O$ Values of Waters and Plant Celluloses Collected in September and October 1989 at Kashiru

Description	Altitude (m)	$\delta^{18}O$(‰)
Spring waters	2240	-4.2
	2240	-3.6
	2260	-4.0
	2270	-4.1
	2270	-4.1
	2340	-4.1
	2245	-3.7
Bog surface waters	2240	-2.8
	2240	-3.0
	2240	-2.5
	2240	-2.0
	2240	-2.8
Channel water	2240	-2.7
Bog interstitial waters	2240	-3.6
	2240	-3.3
Outlet waters	2240	-3.6
	2240	-3.5
Plant celluloses:		
Cyperus latifolius (C_4 sedge)	2240	30.8
Cyperus haspan (C_3 sedge)	2240	27.8
Polygonum pulchrum (C_3 herbaceous plant)	2240	26.1
Erica kingaensis (C_3 shrub)	2240	26.5
Sphagnum sp (moss)	2240	21.9
Mimosa pigra (C_3 shrub)	2240	30.1
Pteridium aquilinum (fern)	2240	28.0
Eucalyptus (C_3 tree)	2240	29.4

sedges (29.3 ± 1.5‰, n = 2) compares well with those of the shrubs and trees (28.7 ± 1.9‰, n = 3) and of all the emergent plants (28.4 ± 1.6‰, n = 7).

$^{13}C/^{12}C$ Ratios of TOM, Charcoal and Celluloses(Core Ka-2)

Large variations in TOM $\delta^{13}C$ values from -28 to -15‰ (Figure 2) are observed throughout the core. High values (-17.3 ± 1.3‰, n = 18) characterized the 30-15 ka period. Lower values are observed before 30 ka (-22.4 ± 1.1‰, n = 2), between 15 and 12 ka (-22.4 ± 1.6‰, n = 5), between 7 and 4.5 ka (-25.7 ± 1.4‰, n = 9) and intermediate values from 4.5 ka to present (-21.9 ± 1.3‰, n = 10). Charcoal is present throughout the core and its abundance relative to the organic fraction > 125 μm is generally of 5-25%. Charcoal is generally depleted in ^{13}C relative to TOM. Differences in $\delta^{13}C$ greater than 2‰ between charcoal and TOM are observed in the 30-25 ka and 3-0 ka intervals with values of -23.5 ± 2.5‰ (n = 7) and of -24.8 ± 0.9‰ (n = 4) respectively. The cellulose recovery represents a small part of the dry sediment (< 2%). Cellulose is generally enriched in ^{13}C relative to TOM. The difference in $\delta^{13}C$ between cellulose and TOM is greater than 2‰ in the 30-15 ka and 3-0 ka intervals with cellulose $\delta^{13}C$ values of -13.4 ± 3.2‰ (n = 14) and -14.8 ± 3.0‰ (n = 7) respectively.

$^{18}O/^{16}O$ Ratios of Celluloses in Core Ka-2

The oxygen isotope composition (Figure 2) of the most recent cellulose (0-90 cm: $\delta^{18}O$ = 27.9 ± 0.7‰, n = 5) is close to the mean isotopic composition of modern emergent plants at Kashiru and is enriched in ^{18}O relative to the fossil cellulose below 90 cm. Low $\delta^{18}O$ values (19 to 24‰) are found from 7 to 3 ka and relatively high values from 30 to 12 ka (24 to 27.5‰) and from 3 ka to present (26 to 28.5‰)(Figure 2).

Plant remain composition in Core Ka-2

The peat remains (Figure 3) originate from emergent plants except in the 245 cm level, in which moss fragments (50-75%) are abundant. Dominance of monocotyledonous leaf remains, in the 30-12 ka and 3-0 ka intervals, coincides with relatively high cellulose $\delta^{13}C$ and $\delta^{18}O$ values, except in the 15-12 ka interval, which shows low $\delta^{13}C$ values (-20.8 ± 0.7‰, n = 5) and intermediate $\delta^{18}O$ values (24.7 ± 0.8‰, n = 5). Presence of other constituents (wood, fern and herbaceous remains) in the 7-3 ka interval corresponds to low $\delta^{18}O$ values.

DISCUSSION

$^{18}O/^{16}O$ Ratios of Modern Waters and Plant Celluloses

The $\delta^{18}O$ values of the springs feeding the Kashiru swamp are similar to the values reported for meteoric groundwaters in the

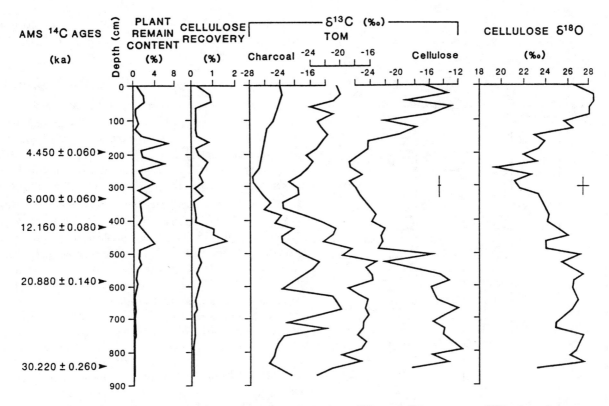

Fig. 2. Core Ka-2. Content in non-carbonized plant remains > 125 μm, cellulose recovery, $\delta^{13}C$ values of TOM, cellulose and charcoal, $\delta^{18}O$ values of cellulose.

East African highlands [Payne, 1970; Hillaire-Marcel and Casanova, 1987]. They are slightly more negative than the mean annual $\delta^{18}O$ values of the precipitation at the IAEA stations of Dar es Salaam (-2.7 ± 0.5°/oo) and of Entebbe (-3.0 ± 1.2°/oo) [International Atomic Energy Agency, 1969, 1970, 1971, 1973, 1975, 1979, 1981]. This effect may be attributed to preferential recharge in the rainy season as the mean $\delta^{18}O$ values of the rainy season precipitation are -3.1 ± 0.6°/oo at Dar es Salaam and -3.6 ± 1.7°/oo at Entebbe. At the sites of plant collection, the bog water collected at the end of the dry season are expected to show the highest yearly evaporative enrichment relative to the meteoric groundwater. The $\delta^{18}O$ values of the source water for the plants throughout the year (δ_i) can therefore be reasonably assumed to range from -3 to -4°/oo. These estimated values are similar to the mean annual $\delta^{18}O$ value of precipitation in East Equatorial Africa (δ_p ~ -3°/oo).

The enrichment in ^{18}O of the *Sphagnum* moss relative to the environmental water falls within the range (24 to 30°/oo) observed by DeNiro and Epstein [1981], and Fehri et al.[1983] for aquatic plants. It is smaller than the enrichment reported for temperate mosses [Brenninkmeijer et al., 1982]. This difference is probably due to the facts that the species analysed in this study is growing in a wet habitat and that high relative

humidity (about 80%) leads to saturation in the first several centimeters above the water surface in equatorial bogs.

Cellulose $^{18}O/^{16}O$ ratios are higher in emergent plants than in mosses. This is due to the transpirative enrichment of the leaf water in the emergent plants relative to the swamp water [e.g., Gonfiantini et al., 1965; Epstein et al., 1977]. The scatter in the $\delta^{18}O$ values of emergent plants from Kashiru is similar to that observed by Brenninkmeijer et al.[1982] in a temperate bog but much smaller than those reported by Sternberg and DeNiro [1983], and Sternberg et al. [1984a, 1984b] for other terrestrial sites. Swamp samples studied here show no significant difference in oxygen isotope ratios between shrubs and sedges and do not show that physiognomy significantly influences cellulose $^{18}O/^{16}O$ ratios as previously suggested by Sternberg and DeNiro [1983]. The difference between the cellulose $\delta^{18}O$ values from C_3 and C_4 sedges in this study (3°/oo) is smaller than that reported by Sternberg and DeNiro (1983) and Sternberg et al. [1984a, 1984b] at terrestrial sites (5 to 8°/oo). It has been found that, under humid and non-water stressed conditions, the difference in oxygen isotope ratios between C_3 and C_4 plants is eliminated [Sternberg et al., 1986]. Such conditions, which do prevail in equatorial swamps, probably attenuate the differences in ^{18}O contents between C_3 and C_4 plants, and between the different species.

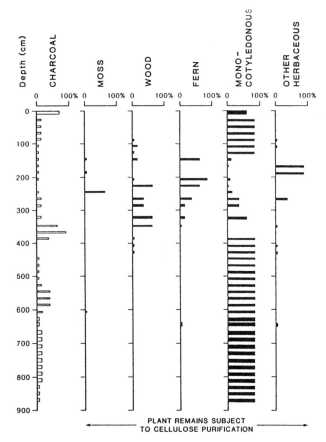

Fig. 3. Core Ka-2. Semi-quantitative estimation of abundances of charcoal and non- carbonized plant remains > 125 μm.

The theoretical $\delta^{18}O$ value of leaf cellulose in emergent plants ($\delta^{18}O_{leaf\ cel.}$) for the Kashiru site was estimated according to a model similar to that of Burk and Stuiver [1981] and Edwards et al.[1985]:

$$\delta^{18}O_{leaf\ cel.} = (1-f)(\varepsilon^* + \varepsilon_k(1-h) + h\,\delta_a + (1-h)\,\delta_i) + f\,\delta_i + \varepsilon \quad (1)$$

ε   net "biochemical" enrichment factor
ε*   water liquid-vapour equilibrium separation factor, function of temperature
ε_k water liquid-vapour kinetic enrichment factor
δ_a $\delta^{18}O$ value of the atmospheric water vapour
δ_{in} $\delta^{18}O$ value of the source water taken up by the plant
h   relative humidity of the air
f   fraction of the leaf water not subject to evaporation and having the same isotopic composition as the source water
Atmospheric water vapour is assumed in isotopic equilibrium with the rain at the air surface temperature. δ_i is taken equal to δ_p. The mean annual $\delta^{18}O$ leaf cellulose value was calculated because plant growth is continuous in the Burundi swamps. For $\varepsilon_k = 28°/oo$ and f = 0.2 [Allison et al., 1985], ε*(T=16°C) =

10.1 [Majoube, 1971], h = 0.8 and δ_p = -3°/oo, the calculated value, 28.5, yields a good approximation of the mean $\delta^{18}O$ value in emergent plant cellulose.

$^{13}C/^{12}C$ and $^{18}O/^{16}O$ Ratios of Organic Material in Core Ka-2

The cellulose analysed in core Ka-2 is derived from emergent plants except in the 245 cm level. The $\delta^{13}C$ value of TOM in emergent plant cellulose depends mostly on the photosynthetic pathway and to a lesser extent on the $\delta^{13}C$ value for ambient CO_2 and on the ratio of intracellular to ambient CO_2 concentrations [Farqhar et al., 1982]. In the core studied here, $\delta^{13}C$ values of cellulose cover the entire range between the C_3 anf the C_4 poles. Therefore, the variations in ^{13}C content of fossil cellulose can unambiguously be related to changes in the dominant photosynthetic pathway. The $\delta^{13}C$ values show that cellulose is predominantly derived from C_4 plants during the 30-15 ka and 3-0 ka period and from C_3 plants during the 15-12 ka and 7-4.5 ka period.

The variations in the ^{13}C content of TOM generally follow those of the cellulose, suggesting that the TOM $^{13}C/^{12}C$ ratio also responds to changes in the dominant photosynthetic pathway. From 15 to 12 ka and from 7 to 4.5 ka, the slight depletion in ^{13}C of TOM relative to cellulose is in agreement with measurements on recent plants [Deines, 1980; Benner et al., 1987]. The greater depletion observed during other periods (30-15 ka, 3-0 ka) may be due to a difference between the coarse plant remains (> 125 μm), from which cellulose was purified, and the total remains. Alternative explanations include (1) lighter isotopic composition for organic constituents in the sediment than in living plants, possibly indicating a bacterial source [Macko et al., 1991], and (2) preferential preservation during peat formation of plant compounds, such as lignin, which are depleted in ^{13}C relative to plant TOM [Benner et al., 1987]. The presence of charcoal, which shows lighter isotopic compositions than cellulose and TOM in the core, is also a factor. In some intervals (15-12 ka, 7-4.5 ka), the depletion of charcoal relative to TOM is of less than 2°/oo, which is consistent with the carbonization effect on the carbon isotopic composition of vegetal remains (DeNiro and Hastorf, 1985). During other periods (30-25 ka, 3-0 ka), the depletion is greater, which suggests a better conservation of C_3 (woody?) plants than C_4 sedges during swamp burning.

The comparison between the ^{18}O curve and plant remain composition shows a correspondence between ^{18}O content and the proportion of monocotyledonous fragments. However, the range of values measured in the core (245 cm excepted) from 22 to 28.5°/oo does not match that of modern emergent plants (26-31°/oo) and the ^{18}O shifts in the core are greater than the difference in $\delta^{18}O$ values between the modern monocotyledonous plants ($\delta^{18}O$ =29.3 ± 1.5°/oo, n = 2) and the emergent plants ($\delta^{18}O$ = 28.4 ± 1.6°/oo, n = 7). The minimum $\delta^{18}O$ value (19°/oo), at 245 cm depth, which coincides with abundant moss remains, is lower than the isotopic composition of modern *Sphagnum* moss (21.9‰). The 7-3 ka interval, in which wood remains are occasionally abundant, shows no relationship between the cellulose oxygen isotopic composition and the proportions of wood and leaf fragments. This contradicts the assumption of DeNiro and Cooper [1989] that wood should be depleted in ^{18}O relative to leaf.

The rather good correlation, for the cellulose fraction, between the $\delta^{13}C$ and the $\delta^{18}O$ ($r^2 = 0.645$, n = 36) indicates concomitant changes in the dominant photosynthetic pathway and ^{18}O content of plant cellulose. Two hypotheses can be put forward to explain this correlation: (1) the photosynthetic pathway directly controls the $^{18}O/^{16}O$ ratio of plant cellulose or (2) hydrological factors affect both the C_3/C_4 ratio of vegetation and the ^{18}O content of plant cellulose. In the latter hypothesis, the cellulose $^{18}O/^{16}O$ ratio can be used as a palaeohydrological indicator. Regarding the first hypothesis, the difference in $\delta^{18}O$ values between C_3 and C_4 sedges, which is observed in modern swamps, (3‰) is smaller than the shifts in ^{18}O content measured in the core. As discussed for modern plants, the difference in cellulose $\delta^{18}O$ values between C_3 and C_4 plants should be minimized under humid and non-water stressed conditions, such as those prevailing in the modern swamps of Burundi [Sternberg et al., 1986]. It is unlikely that past conditions were drier than at present, as indicated by the general depletion in ^{18}O content of the peat cellulose below 90 cm relative to the most recent cellulose (0-90 cm). These evidences suggest that the shift in

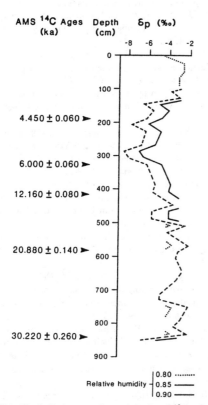

Fig. 4. Core Ka-2. Estimates of precipitation $\delta^{18}O$ values (δ_p).

photosynthetic pathway cannot account for all the ^{18}O variations in the core and that the cellulose $^{18}O/^{16}O$ ratio can potentially record palaeohydrological changes. First estimates of h and δ_p during the last 30,000 years can be obtained by using equation (1) and a constraint given by the trend of decreasing δ_p values with increasing air humidity as observed today in Equatorial Africa [Gonfiantini, 1985; Aucour, 1992]. Modern conditions are taken as h = 0.8 and δ_p = -3‰ and are identical to the estimated values for the most recent sediments. The depletion in ^{18}O of fossil celluloses relative to the most recent samples indicates an increase in h and a decrease in δ_p in the last 30,000 years. Therefore, h is taken to increase from 0.8 (modern value) to 0.85 and 0.9 and corresponding δ_p values (calculated from equation (1)) are presented when they are lower than -3‰ (modern value) (Figure 4). The δ_p estimate of -3.5‰ (for h = 0.85) during the 30-15 ka interval compares well with results from Atmospheric General Circulation Models (AGCMs) experiments at 18 ka [Jousseaume and Jouzel, 1987]. The δ_p estimates are -5‰ (for h = 0.85) during the 15-12 ka period, and -7‰ (for h = 0.85) or -5.5‰ (for h = 0.9) during the 7-4.5 ka period. They decrease to -3.5‰ (for h = 0.85) at ca 3 ka.

Comparison with the Pollen Record from Kashiru

Pollen data from the Kashiru site (core Ka-1) have yielded a detailed record of vegetational [Bonnefille, 1987; Bonnefille and Riollet, 1988] and climatic [Bonnefille et al.,1990] changes in the Burundi mountains. The pollen record indicates (1) grassland extension and a cool and dry climate between 30 and 13 ka, (2) forest development, warmer and wetter conditions after 7 ka (3) a decrease of the arboreal components, probably reflecting a trend towards drier conditions after 4.5 ka.

The extension of grassland, the reduction of the forest and the dry conditions inferred from the pollen study correspond, for the 30-15 ka and 3-0 ka periods, to isotopic evidences for preeminence of C_4 photosynthetic pathway in palaeovegetation, relatively low h and high δ_p values. High proportions of C_4 plants are consistent with the observation that C_4 grass species are favoured by dry environments [e.g., Livingstone and Clayton, 1980]. Forest development and humid conditions assessed by the pollen record from 7 to 4.5 ka coincide to isotopic evidences for preeminence of C_3 photosynthetic pathway in palaeovegetation, relatively high h and low δ_p values. However, the change in dominant photosynthetic pathway inferred from the isotopic data between 15 and 12 ka is not recorded in the pollen data. Isotopic data show slightly higher h values for the 30-12 ka period than those today whereas the pollen data indicate lower rainfall [Bonnefille et al., 1990]. The two interpretations are not in conflict since the temperature decrease of 4 ± 2°C between 30 and 13 ka [Bonnefille et al., 1990] implies a reduced water vapour content of the atmosphere for h values sligthly higher than they are today.

CONCLUSION

The ^{13}C content of the organic fraction from intertropical peats depends on the ratio of C_3 to C_4 plants in the accumulated remains from local bog vegetation. It reflects important changes in the C_3 and C_4 biomass proportions in the organic matter during the last 30,000 years.

The ^{18}O studies yield information on atmospheric water cycle. Simple calculations, which take into account the cellulose-water biochemical fractionation and the evaporative enrichment in ^{18}O of water in emergent leaves, yield a rather good estimate of the mean cellulose $\delta^{18}O$ values in modern emergent plant leaves and in recent sediments at Kashiru. The scatter in $\delta^{18}O$ values of emergent plants in the modern swamp is smaller than that reported at other terrestrial sites, probably because of the humid and non-water stressed conditions prevailing in equatorial swamps which minimize the variability in leaf water evaporation.

In core Ka-2, cellulose of plant remains shows significant shifts in $\delta^{18}O$ values. These shifts are greater than the range in $\delta^{18}O$ values observed in modern emergent plants. The cellulose ^{18}O profile is therefore expected to record the relative humidity (h) and the $\delta^{18}O$ value of precipitation (δ_p).

The combination of the ^{13}C and ^{18}O records indicates that several episodes have marked the palaeovegetation and palaeoclimatic history:

(1) the 30-15 ka period was characterized by high proportions of C_4 plants, relative humidity (0.85) slightly higher than modern humidity, and precipitation slightly depleted in ^{18}O (by 0-0.5°/oo) relative to modern precipitation,

(2) the 15-12 ka period was marked by higher proportions of C_3 plants, humidity (0.85) slightly higher than modern humidity and precipitation slightly depleted in ^{18}O (by 0-2°/oo) relative to modern precipitation,

(3) the 7 to 4.5 ka period was characterized by high proportions of C_3 plants, high humidity (0.85-0.9), and precipitation depleted in ^{18}O content by 2-4°/oo relative to modern precipitation,

(4) The period from 4.5 ka to present saw increased proportions of C_4 plants, decrease in humidity, and increase in precipitation ^{18}O content towards modern conditions.

The record is insufficient to ascertain conditions between 12 and 7 ka because of the low sedimentation rate.

Acknowledgments Comments from J.-C. Mareschal, S. Macko, J.C. Fontes, D. Holländer and T.E. Edwards helped to clarify several aspects of the manuscript. Field studies were supported by grants from the French C.N.R.S. (to R.B.) and analytical studies by grants from NSERC-Canada (to C. H.-M.). Government of Canada Award to AMA, the help of Appollinaire Yengayenge and the support of Burundian Authorities (University of Bujumbura, Department of Education) are acknowledged. We also thank Caroline Guilmette for laboratory assistance, Michelle Laithier for the illustrations, and Claude Jaupart at Institut de Physique du Globe, Paris, for permiting access of computer facilities during the redaction of this paper.

REFERENCES

Allison, G. B., Gat, J. R., Leaney, F. W. D., The relationship between deuterium and oxygen 18 delta values in leaf water, *Chemical Geology*, *58*, 145-156, 1985.

Aucour, A.M., Composition isotopique ($^{18}O/^{16}O, ^{13}C/^{12}C$) de la matière organique de sédiments de tourbières du Burundi (0-40 000 ans). Relation avec les changements climatiques, PhD thesis, 139 pp.,Université du Québec à Montreal, Canada, 1992.

Bender, M.M., Mass spectrometric studies of carbon-13 variations in corn and other grasses, *Radiocarbon* , *10* , 468-472, 1968.

Benner, R., Fogel, M.L., Sprague, E.K., Hdson, R.E., Depletion in ^{13}C of lignin and its implications for stable carbon isotope studies, *Nature*, *329*, 708-710, 1987.

Bonnefille, R., Evolution climatique et forestière au Burundi au cours des quarante derniers milliers d'années, *C. R. Acad. Sci. Paris*, *305* (*série II*) , 1021-1026, 1987.

Bonnefille, R., Riollet, G., The Kashiru pollen sequence (Burundi). Palaeoclimatic implications for the last 40 000 yr BP in tropical Africa , *Quaternary Research*, *30*, 19-35, 1988.

Bonnefille, R., Roeland, J.C., Guiot, J., Temperature and rainfall estimates for the past 40,000 years in Equatorial Africa, *Nature*, *346*, 347-349.

Brenninkmeijer, C. A. M., Van Geel, B., Mook W. G., Variations in the D/H and $^{18}O/^{16}O$ ratios in cellulose extracted from a peat bog core, *Earth and Planetary Science Letters*, *61*, 283-290 1982.

Bultot, F., Rwanda and Burundi, in *Climates of Africa* edited by J.F. Griffiths, pp. 349-366, Elsevier, Amsterdam, 1972.

Burk, R. L., Stuiver, M., Oxygen isotope ratios in trees reflect mean annual temperature and humidity, *Science*, *211*, 1,417-1,419, 1981.

Cerling, T.E., Quade, J., Wang, Y., Bowman, J. R., Carbon isotopes in soils and palaeosols as ecology and palaeoecology indicators, *Nature*, *341.*, 138-139, 1989.

Coleman, M.L., Sheppard, T.J., Durham, J.J., Rouse, J.E., Moore, G.R., Reduction of water with zinc for Hydrogen Isotope Analysis, *Analytical Chemistry*, *54* , 993-995, 1982.

Craig, H., Carbon -13 in plants and the relationship between carbon- 13 and carbon-14 in nature, *Journal of Geology*, *62*, 115-149, 1954.

Craig, H., Gordon, L.I. , Deuterium and oxygen -18 variations in the ocean and the marine atmosphere, in *Stable isotopes in oceanographic studies and paleotemperatures* edited by E. Tongiorgi, pp .9-130, Consiglio Nazionale delle Richerche, Roma, 1965.

Deines, P., The isotopic composition of reduced organic carbon, in *Handbook of Environmental Isotopes Geochemistry*, vol. 1, *The terrestrial environment* A edited by P. Fritz and J.C. Fontes, pp. 9-130, Elsevier, Amsterdam, 1980.

DeNiro, M. J., Epstein, S., Isotopic composition of cellulose from aquatic organisms, *Geochimica et Cosmochimica Acta* , *45* , 1,885-1,894, 1981.

DeNiro, M. J., Hastorf, F.C.A., Alteration of $^{15}N/^{14}N$ and $^{13}C/^{12}C$ ratios of plant matter during the initial stages of diagenesis: Studies utilizing archaelogical specimens from Peru, *Geochimica et Cosmochimica Acta* , *49* , 97-115, 1985.

DeNiro, M. J., Cooper, L. W., Postphotosynthetic modification of oxygen isotope ratios of carbohydrates in the potato: Implications for paleoclimatic reconstruction based upon isotopic analysis of wood cellulose, *Geochimica et Cosmochimica Acta* , *53* , 2,573-2,580, 1989.

Dongman, G., Nurnberg, H.W., Forstel, H., Wagener, K., On the enrichment of $H_2^{18}O$ in the leaves of transpiring plants, Radiation and Environmental Biophysics, 11, 41-52, 1974.

Edwards, T. W. D., Fritz, P., Assessing meteric water isotopic composition and relative humidity from ^{18}O and ^2H in wood cellulose: paleoclimatic implications for southern Ontario, Canada., Applied Geochemistry, 1, 715-723, 1986.

Epstein, S., Mayeda, T. , Variation of ^{18}O content of water from natural sources, Geochimica et Cosmochimica Acta, 42, 213-224, 1953.

Epstein, S., Thompson, P. , Yapp, C.J., Oxygen and hydrogen isotope ratios in plant cellulose, Science, 198, 1,209-1,215, 1977.

Farqhar, GD., O'Leary, M.H., Berry, J.A., On the relationship between carbon isotope discrimination and the intercellular carbon dioxyde concentration in leaves, Australian Journal of Plant Physiology, 9, 121-137, 1982.

Fehri, A., Bariac, T., Letolle, R., $^{18}O/^{16}O$ ratios in leaf water and cellulose of aquatic and terrestrial plants, in Panel proceedings series I.A.E.A., Palaeoclimates and palaeowaters: a collection of environmental isotope studies, pp. 85-99, I.A.E.A., Vienna, 1983.

Gonfiantini, R., On the isotopic composition of precipitation in tropical stations, Acta Amazonica, 15(1-2), 121-139,1985.

Gonfiantini, R., Gratziu, S., Tongiorgi, E., Oxygen isotopic composition of water in leaves, in Proceedings series I.A.E.A., Isotopes and Radiation in Soil-Plant nutrition Studies , pp. 405-410, I.A.E.A., Vienna, 1965.

Green, J.W., Wood cellulose, Methods in carbohydrate Chemistry (III) edited by R. L. Whistler and J. W. Green, pp. 9-21, Academic Press, New York, 1963.

Griffiths, J.F., Climates of Africa, 604 pp., Elsevier, Amsterdam, 1972.

Grootes, P. M., Stuiver, M., Thompson, L. G., Mosley-Thompson, E., Oxygen isotope changes in tropical ice, Quelccaya, Peru, Journal of Geophysical Research, 94, 1,187-1,194, 1989.

Guillet, B., Faivre, P., Mariotti, A., Khobzi, J., The ^{14}C dates and $^{13}C/^{12}C$ ratios of soil organic matter as a mean of studying the past vegetations in intertropical regions: examples from Colombia (South America), Palaeogeography, Palaeoclimatology, Palaeoecology, 65, 51-58, 1988.

Hesla, B.I., Tieszen, L.L., Imbamba, S.K., A systematic survey of C_3 and C_4 photosynthesis in the Cyperaceae of Kenya, East Africa, Photosynthetica, 16, 196-205, 1982.

Hillaire-Marcel, C., Aucour , A.M., Bonnefille, R., Riollet, G. , Vincens, A., Williamson, D., ^{13}C/palynological evidence of differential residence times of organic carbon prior to its sedimentation in East african rift lakes and peat bogs, Quaternary Science Reviews, 8, 207-210 1989.

Hillaire-Marcel, C., Casanova, J., Isotopic hydrology and paleohydrology of the Magadi (Kenya)-Natron (Tanzania) Basin during the late Quaternary, Palaeogeography, Palaeoclimatology, Palaeoecology, 58, 155-181, 1987.

International Atomic Energy Agency, Environmental isotope data No. 1. World survey of isotope concentration in precipitation (1953-1963), Technical Reports Series No 96, I.A.E.A., Vienna, 1969.

International Atomic Energy Agency, Environmental isotope data No. 2. World survey of isotope concentration in precipitation (1964-1965), Technical Reports Series No 117, I.A.E.A., Vienna, 1970.

International Atomic Energy Agency, Environmental isotope data No. 3. World survey of isotope concentration in precipitation (1966-1967), Technical Reports Series No 129 , I.A.E.A., Vienna,1971.

International Atomic Energy Agency, Environmental isotope data No. 4. World survey of isotope concentration in precipitation (1968-1969), Technical Reports Series No 147, I.A.E.A., Vienna, 1973.

International Atomic Energy Agency, Environmental isotope data No. 5. World survey of isotope concentration in precipitation (1970-1971), Technical Reports Series No 165, I.A.E.A., Vienna, 1975.

International Atomic Energy Agency, Environmental isotope data No. 6. World survey of isotope concentration in precipitation (1972-1975), Technical Reports Series No 192, I.A.E.A., Vienna, 1979.

International Atomic Energy Agency, Statistical treatment of environmental isotope data in precipitation, Technical Reports Series No 206, I.A.E.A., Vienna,1981.

Jousseaume, S., Jouzel, J., Simulation of paleoclimatic tracers using atmospheric general circulation models, in Abrupt climatic changes: evidences and implications edited by W. H. Berger et L. D. Labeyrie, pp. 369-381, Reidel, Dordrecht, 1987.

Krishnamurthy, R.V., DeNiro, M.J., Pant, R.K., Isotope evidence for Pleistocene climatic changes in Kashmir, India, Nature, 298, 640-641, 1982.

Livingstone, D.A., Clayton , W.D., An atitudinal cline in tropical african grass floras and its paleoecological significance, Quaternary Research, 13, 392-402, 1980.

Macko, S.A., Engel, M.H., Hartley, G., Hatche, P., Helleur, R., Jackman, P., Silfer, J.A., Isotopic compositions of individual carbohydrates as indicators of early diagenesis of organic matter in peat, Chemcal Geology, 93, 147-161, 1991.

Majoube, M., Fractionnement en oxygène 18 entre l'eau et sa vapeur, J. Chim. Phys., 68 (10), 1,423-1,436, 1971.

Payne, B.R., Water balance of Lake Chala and its relation to groundwater from tritium and stable isotope data, Journal of Hydrology, 11, 47-58, 1970.

Smith, B. N., Epstein, S., Two categories of $^{13}C/^{12}C$ ratios for higher plants, Plant Physiology, 47, 380-384, 1971.

Sternberg, L., DeNiro, M. J., Isotopic composition of cellulose from C_3, C_4 and CAM plants growing near one another, Science, 220, 947-949, 1983.

Sternberg, L., DeNiro, M.J., Johnson, H.B., Isotope ratios of cellulose from plants having different photosynthetic pathways, Plant Physiology, 74, 557-561, 1984 a.

Sternberg, L., DeNiro, M.J., Ting, I. P., Carbon, hydrogen and oxygen isotope ratios of cellulose from plants having intermediary photosynthetic modes, Plant Physiology, 74, 104-107, 1984 b.

Sternberg, L., DeNiro M. J., Sloan, M. E., Black, C.C. Jr, Compensation point and isotopic characteristics of C_3/C_4 intermediates and hybrids in Panicum, Plant Physiology, 80, 242-245, 1986.

Stump, R.K., Frazer , J.W., Simultaneous determination of carbon, hydrogen, and nitrogen in organic compounds, Report UCID-16198, Univ. Calif., Livermore, 7 pp., 1973.

Trewartha, G.T., Earth' s problem climates, 371 pp., University of Wisconsin Press, Madison, 1981.

Troughton, J.H., Carbon isotope fractionation in plants, in Proceeding of the eight conference of radiocarbon dating, vol. 2, pp. 39-57, Royal Society of New Zealand, Wellington, 1972.

Wickman, F. E., Variations in the relative abundances of the carbon isotopes in plants, Geochimica et Cosmochimica Acta, 2, 243-254, 1952.

A.M.Aucour and C. Hillaire-Marcel, Centre de Recherche en Géochimie Isotopique et en Géochronologie (GEOTOP), Université du Québec à Montreal, B.P. 8888, Succ. A, H3C 3P8 Montreal, Québec, Canada.

R. Bonnefille, Laboratoire de Géologie du Quaternaire, C.N.R.S., Luminy, Case 907, 13288 Marseille Cédex 9, France.

Tree-Ring ^{14}C as a Possible Indicator of Climate Change

JOHN L. JIRIKOWIC, ROBERT M. KALIN AND OWEN K. DAVIS

Department of Geosciences, University of Arizona, Tucson

Analyses of the tree-ring "Calibration" Δ^{14}C data set shows intermittent high ^{14}C anomalies. During these anomalies, the time series nature of the ^{14}C data set changes markedly. Such non-stationarity suggests ^{14}C variation results from a dynamic, non-linear set of processes. The latest ^{14}C anomaly occurs during the past millennium and coincides with the historic profound solar activity minima. To test the hypothesis of solar modulation of global climate and cosmogenic isotope anomalies, we sought evidence for brief climatic events coincident with major anomalies in a rapid deposition site which had not been previously reported. Using detailed pollen analysis and precise ^{14}C dating, we have studied climatic change during the Homeric-Greek (2830-2550, 2360-21600 cal BP) and Noachan (4880-4660 cal BP) anomalies, at Mission Cross Bog, Elko Co., Nevada. Through "wiggle-matching" with the ^{14}C calibration curve, we could date the precise interval in the sediment record which corresponds to the Homeric and Greek ^{14}C production anomalies. We located the exact age of the Greek anomaly only, correspondng to a wet period in the pollen diagram. We also discovered two wet periods that do not match any cosmogenic isotope anomalies. Hence, the presence of brief climatic episodes cannot be used to date the sequence. By dating and analyzing pollen during the Homeric and Noachan anomalies, we may confirm or refute the climate-cosmogenic isotope anomaly association hypothesis.

INTRODUCTION

The connection between cosmogenic isotope anomalies and climate first was postulated by de Vries [1958] and was further developed by Suess [1965], Stuiver [1965], Damon [1968], Eddy [1977], and Raisbeck et al. [1990]. de Vries [1958] noted the correspondence between reduced ^{14}C production, colder global climate, and reduced solar activity during the Maunder (sunspot) Minimum, *AD* 1645-1715. Lamb [1977, Appendix] and Pittock and Shapiro [1982] note the many failures to use very short sunspot cycles (22 and 11-yr) to forecast climate. Willson and Hudson's [1991] measurements show total solar irradiance varied by $\simeq 0.2\%$ during Solar Cycle 21. Several lines of evidence suggest solar irradiance decreased by 0.24% during the Maunder Minimum [Lean et al., 1992] and global temperature was $\simeq 1°$C less than today [Grove, 1988]. During the Maunder and other Little Ice Age sunspot minima (the Dalton *AD* 1805-1835, Spörer *AD* 1400-1510, and Wolf *AD* 1290-1350), the atmospheric Δ^{14}C exceeded background values by $\simeq 10$ $^0/_{00}$. Eddy [1977] initiated research into the possible climate implications of longer solar activity variations represented by Δ^{14}C anomalies. This study describes the apparent nature of these anomalies and investigates their reliability as indicators of climate change.

Previous studies of the tree-ring Δ^{14}C activity records show both global uniformity and an extremely complex time series description [see Damon and Sonett, 1992, for review]. Previous studies based upon spectral analysis have described these periodic characteristics and suggested that such methods have likely reached the apogee of their usefulness [see for example, Stuiver and Braziunas, 1990; Damon and Jirikowic, 1992]. Ordinary spectral methods assume stationarity, uniform mean and variation over the length of the time series being analyzed. As intimated above, we believe the time series to be non-stationary. The pronounced Δ^{14}C maxima, caused by solar activity minima, vary much more strongly than the rest of the tree-ring Δ^{14}C time series. Because presently observed solar irradiance variations over a single solar activity cycle seem far too small to produce global climate response, we present examples of analyses emphasizing non-stationarity. Such profound changes in solar activity have far greater likelihood of forcing a coherent global climate signal reconstructable from individual paleoclimatic records.

A coherent climatic signal should be strongest during episodes of high Δ^{14}C periodicity. Following our previous nomenclature [see Davis et al., 1992] these episodes are (with

Climate Change in Continental Isotopic Records
Geophysical Monograph 78

increasing age) the (A) Spörer-Maunder-Wolf, (B) Homeric-Greek, (C) Sumarian-Noachan, and (D) Jericho intervals. The paired names reflect the presence of at least two major anomalies in each interval. Stuiver and Brazinas [1988] have noted consistent features between such pairs.

Schmidt and Gruhle [1988] have presented a compelling connection between cool-wet climate and increased ^{14}C production between 750-200 BC using climatic reconstructions and ^{14}C dating from the same tree-ring series. In western North America, we have shown the general correspondence between major cosmogenic isotope anomalies and cold episodes at San Joaquin Marsh, Orange County, California [Davis, 1992; Davis et al., 1992]; and in Montezuma Well, Yavapai County, Arizona [Davis and Shafer, in press]. The Greek (2450-2250 BP, all years calendar unless otherwise noted) and Homeric (2830-2550 BP) Δ^{14}C anomalies associate with major cooling events in western North America [Davis, 1992; Davis et al., 1992], and northern Europe [Schmidt and Gruhle 1988].

To further test the association of climatic events and cosmogenic anomalies, we propose to locate cosmogenic isotope anomaly-correlative climatic events that have not been previously assigned. The underlying working hypothesis is that decreased solar activity results in increased production of cosmogenic isotopes, and causes global cooling. Therefore, climatic events should be detectable in sediments correlative with the time of cosmogenic anomalies in many regions of the world, given a estimated lag in radiocarbon cycle response [Seigenthaler et al., 1980]. Because solar forcing represents only one of the many climate forcings, a single paleoclimatic reconstruction cannot be considered conclusive. Site selection and proxies of volcanic, dynamic and other climate forcings must be considered when interpreting a putative solar forcing. The Mission Cross Bog study presented here represents only one example of the type of paleoclimate investigations necessary to present convincing evidence of solar forcing.

TIME SERIES ANALYSES OF TREE-RING ^{14}C

High-precision ^{14}C measurements of dendrochronology-dated tree-rings shows atmospheric ^{14}C activity varied ±5% during the Holocene. Data from spatially separated sample sites show some residual offset even after correction for isotopic fractionation and standardization (reported as Δ^{14}C $^{0}/_{00}$, Stuiver and Pollach [1979]). For example, Δ^{14}C from single annual-rings during the historic period from sites on the Olympic Peninsula in the Pacific Northwest (Stuiver and Quay [1982]) have an average -2.5 $^{0}/_{00}$ offset from a site near Tucson (Damon et al., [1989]; upper panel, Fig. 1). Although the measurement error averages 1.5 $^{0}/_{00}$, the consistency of this offset suggests the proximity to the North Pacific, a region where "old" carbon upwells with deep ocean waters and is released into the atmosphere, causes the Washington data to be negatively offset. Lerman et al. [1970] interpreted a similar causative mechanism for spatial Δ^{14}C variations.

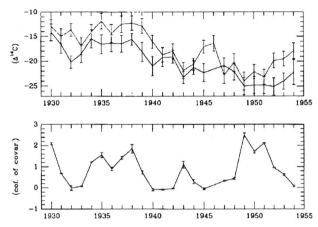

Fig. 1. Δ^{14}C variations from tree-ring cellulose sampled annually during this century. The upper panel shows measurements by the University of Arizona Radiocarbon Lab from trees growing in the American Southwest (solid line; from Damon et al., [1989]) and measurements from the University of Washington Lab from trees in the Pacific Northwest (dotted line; from Stuiver and Quay, [1981]). Note the offset due to the spatial proximity to the North Pacific Ocean, a source of "old" carbon from deep waters. The lower panel shows the coefficient of covariation, at statistical measurement of the short-term similarity between the two Δ^{14}C series. The coefficient of covariation is most sensitive to relative changes in slope. Generally, the value is high with notable short-term departures. The measurement error propagated through the coefficient of covariation calculation (error bars) show little significant difference between the two data sets except the anomalous periods.

The coefficient of covariation measures the short-term statistical correspondence between two data sets. For these two data sets, the coefficient of covariation remains high except during three distinct events, *AD* 1932-1933, 1940-1942 and 1945-1948, when the two data sets' slopes diverge (lower panel, Fig. 1). Covariate spectral analysis shows little evidence of distortion over periods between $\simeq$ 10 and 2 years (see Fig. 2). For near-biennial periods, the gain spectrum suggests the Washington data varies more strongly than the Arizona data. This may confirm the Washington data set's greater sensitivity to oceanic influences (or it may be due to greater random error in these measurements). The middle panel of Fig. 2 shows the coherency-squared spectrum. Coherency-squared estimates the Fourier transform of the squared cross-correlation function. For periods between 10 and 2 years, the coherency-squared value remains above the 95% confidence line suggesting the data sets share most variance within this time scale. Coherency-squared dips below the 95% confidence line for periods longer than 10-years most likely because the unreliability of spectral estimates for longer periods approaching the length of the data set. For the shorter periods of high coherence, the phase lag spectrum (lowermost panel, Fig. 2) remains near zero. This implies little phase distortion results from this spatial effect. The spatial consistency of Δ^{14}C variations allows high-precision data sets from separate localities to be combined

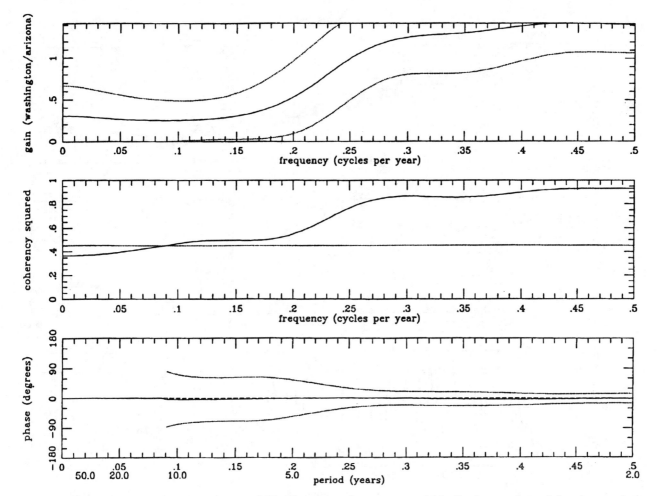

Fig. 2. Cospectral analysis between the Arizona and Washington annually-sampled data using a windowed discrete Fourier transform. The solid lines represent the median and the dotted lines the 95% confidence interval. The uppermost panel shows the gain (a ratio of spectral power). The gain lays near unity between 2- and 10-year periods. This suggests both series have the same amount of variance between these time scales (although the Washington data may have slightly stronger short-term variations due to slightly higher measurement error). The middle panel shows coherency squared (the Fourier transform of the cross-correlation function). Greater than 95% coherency squared suggests a high probability of a shared signal. The lowermost panel shows the phase lag spectrum. It remains within 95% probability zero for all periods below 10 years. The poor correspondence of the annual $\Delta^{14}C$ data sets long of 10-year periods may be most likely due to the poor performance of estimated spectral methods with periods that approach the length of the data sets being analyzed.

into a global $\Delta^{14}C$ data set reflecting atmospheric $\Delta^{14}C$ through the Holocene.

Overlaying several bidecadally-sampled tree-ring $\Delta^{14}C$ data sets displays little evidence of distortion due to the spatial distribution of the sample sites (upper panel, Fig. 3). Spatial consistency and high-precision permitted combining these well-dated data sets into the "Calibration Data" [Stuiver and Kra, 1986] (middle panel, Fig. 3). Based upon theoretical cosmogenic isotope production and radiocarbon models, the calibration $\Delta^{14}C$ data set may be considered a record of ^{14}C production through the Holocene [Stuiver et al., 1991].

After removing a $\simeq 11$ ky year-trend caused by changing ^{14}C production with changing geomagnetic field strength

[Damon, 1988], several anomalous high-$\Delta^{14}C$ intervals remain their amplitude largely unaffected by the geomagnetic dipole field strength [Damon, 1988]. Schmidt and Gruhle [1988] have studied one high-$\Delta^{14}C$ anomaly in particular. Using correlation between $\Delta^{14}C$ and tree-ring index chronologies in central Europe, Schmidt and Gruhle [1988] proposed that an interval between 750 and 550 BC witnessed both rapid changes in $\Delta^{14}C$ (the "Hallstattzeit Disaster" or Homeric, after Landscheidt [1987] and regional climate deterioration. Additionally, Schmidt and Gruhle cross-correlated $\Delta^{14}C$ and tree-ring indices from this 200-year interval forward through the present millennium. This analysis showed evidence that the Homeric anomaly recurs. Using a similar cross-correlation between $\Delta^{14}C$ during the

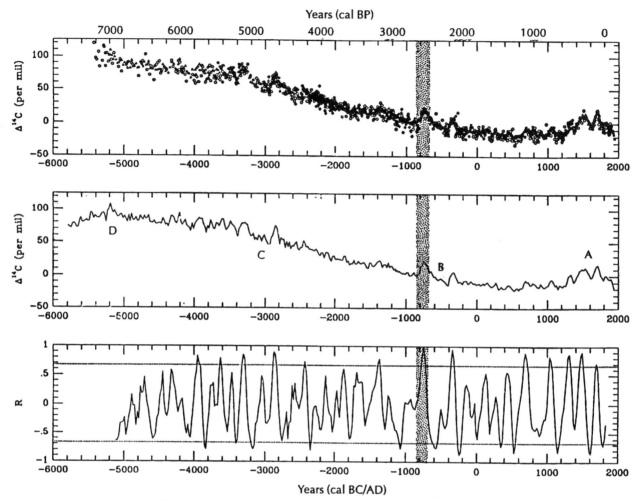

Fig. 3. The bidecadal Δ^{14}C data sets point-by-point in the uppermost panel were combined into the calibration Δ^{14}C data set. Although these data were measured by different laboratories from samples from spatial separate sites, the scatter remains low until 5000 BC. The middle panel shows the combined calibration Δ^{14}C data set. The lowermost panel shows the Pearson correlation coefficient at zero lag between the Δ^{14}C data during the Homeric ("Hallstattzeit Disaster") interval (shaded) and an interval of the calibration Δ^{14}C data set of the same length centered upon that date. The dotted lines represent the 95% confidence interval calculated using Fisher's Z-transform.

Homeric and the rest of the Δ^{14}C archive, we note several recurrences during the length of the Δ^{14}C data set (see lower panel, Fig. 3). Times of high correlation correspond to high Δ^{14}C anomalies. Similar ^{14}C anomalies were first described by de Vries [1958] and have been termed "Suess Wiggles". We prefer to label the non-anthropogenic anomalies "de Vries Effects" since de Vries [1858] proposed that these anomalies occur due to climate-induced changes in the radiocarbon cycle. The de Vries effects have coincided with historical solar activity minima (such as the Maunder Minimum and the "Bronze Age" (Homeric) Minimum, Eddy [1976]). The Maunder solar activity minimum groups with other historic solar activity minima (the Dalton, Spörer and Wolf minima) into a prolonged interval of anomalously high Δ^{14}C labeled "A" in Fig. 3. Other high Δ^{14}C anoma-

lies also tend to group. For example, the Homeric Δ^{14}C anomaly follows the "Greek" Δ^{14}C anomaly centered near 350 BC (also after Landscheidt [1987]). We have labeled this anomalous region "B". The earlier "Noachan" and "Sumarian" anomalies fall within our "C" region and the earliest named anomaly, the "Jericho", lays within our "D" region. We have tentatively identified an "E" region near the very beginning of the tree-ring Δ^{14}C calibration data set. Because the paucity of well-dated tree-ring sites and coincidence with the Younger Dryas neoglacial event, the confident demarcation of this Δ^{14}C anomaly awaits further analyses. Notably, these groups of anomalous Δ^{14}C events (we shall henceforth call these groups of anomalous Δ^{14}C events "Hallstattzeit" episodes after Schmidt and Gruhle [1988]) have a somewhat predictable occurrence every 2000 to 2500-

years. Although spectral analyses cannot confidently estimate periodicities near one fourth of the total length of the data set, many workers have cited evidence for a periodicity between 2000 and 2500-years (see Damon and Sonett [1992], for review).

Lingenfelter and Ramaty [1970] predicted the association of solar activity minima with $\Delta^{14}C$ maxima due to variable heliomagnetic shielding of the Earth. Such a production effect should be strongly influenced by the strength of the geomagnetic dipole field. As noted above, however, these $\Delta^{14}C$ anomalies show a marked independence from the strength of the geomagnetic dipole field strength. This result strongly suggests these anomalies result from some terrestrial dynamic response, perhaps forced by bolometric or spectral solar variations. Summarizing, global $\Delta^{14}C$ anomalies recur throughout the Holocene, coincide with some regional climate indicators, coincide with solar activity minima during the past millennium, group into Hallstattzeit episodes and may also include a response to a solar-forced change in the radiocarbon cycle. The analyses below will attempt to describe the nature of the $\Delta^{14}C$ anomalies and their implications for continental paleoclimate studies.

The recurrence the $\Delta^{14}C$ anomalies can be shown with spectral analysis. Two recent reviews by Damon and Sonett [1992] and Thomson [1990] confirm a persistent spectral peak $\simeq$210-years. A second spectral peak $\simeq$80-years has been indicated by analyses of $\Delta^{14}C$ and solar activity data, but appears more ephemeral. Thomson's [1990] analysis

shows an indistinct broad feature in his analysis of a lower precision $\Delta^{14}C$ data set. This peak has been strongly indicated in other spectral analyses [Damon and Sonett, 1992; Damon and Jirikowic, 1992; Stuiver and Braziunas, 1990].

The spectral analysis methods used by these previous authors rely to some extent upon an assumption of stationarity, the time invariance of statistical characteristics over at least the length of the time series. The strongly variant character of the $\Delta^{14}C$ anomalies implies this assumption may not be inviolate. Statistical analyses similar to the interval correlations described above confirmed that even after removing the $\simeq$11 ky geomagnetic effect upon ^{14}C production, the $\Delta^{14}C$ variance increased dramatically during the Hallstattzeit episodes independently of the geomagnetic dipole field strength. Adapting the spectral methods described by Priestley [1988] and used by Berger et al. [1990], we spectrally analyzed intervals of the high-precision calibration $\Delta^{14}C$. By incrementing the central point of the interval though the entire length of the $\Delta^{14}C$ data set, we can produce a two-dimensional spectral power function of period and time. The interval length determines the range in time and period investigated. Fig. 4 shows one such function represented as a contour map.

For example, in Fig. 4 we began by spectrally analyzing a 640-year interval centered 300 years after the earliest samples of the published $\Delta^{14}C$ series (7120 BC). We then performed a spectral analysis between periods of 640 to 40 years (limited by interval length and the Nyquist period).

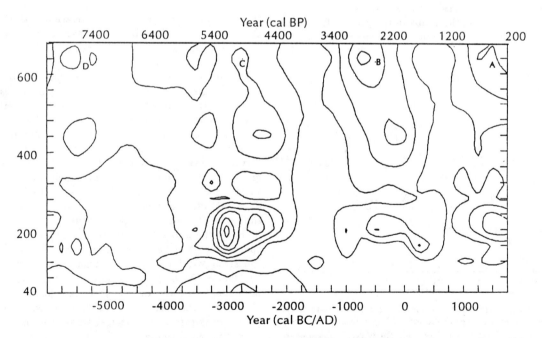

Fig. 4. The evolutive two-dimensional spectral function for the calibration $\Delta^{14}C$ data set calculated as Z-scores such that each contour represents a standard deviation from the mean of a population of 100 red-noise spectra with the same statistical characteristics as the calibration $\Delta^{14}C$ data set. Note that the periodicity (nonrandomness) varies strongly with time maximizing during intervals similar to the Homeric. We have called these intervals "Hallstattzeit" episodes.

For the particular spectral function shown in Fig. 4, we used the Discrete Fourier Spectral estimate (decimation-in-time, windowed to reduce distortions such as sync function side-lobes, chosen to give the optimum resolution without bias, see Thomson's [1990] discussion for an example of some mathematical conditioning needed in spectral analyses). Conventional estimate error analyses such as the chi-squared estimate require stationarity. Instead, we constructed a population of 100 red-noise spectra with the same statistical characteristics (moments, autoregression, etc.) to establish a two-dimensional distribution function of random power spectral density. Based upon this distribution function, the estimated power spectral density of the interval is represented as a Z-score multiple of the this distribution's standard deviation. In essence, each contour interval represents a multiple of the standard deviation of the distribution functions. Thus, the Z-value of two describes a period where the estimated power spectral density has a $\simeq$ 95% probability of nonrandomness. After moving the window central point incrementally by 160 years, the process begins anew producing another probability spectral estimate. After the final spectral analysis on an interval centered at AD 1790, we used the kriging algorithm to interpolate the probability spectral estimates into a two-dimensional probability spectral function. Contours represent Z-score values of the power compared to the randomized power distribution. Similar analyses with other spectral estimation methods (including Maximum Entropy, Maximum Likelihood and Multi-Taper Methods) yielded similar results confirming robustness to the method used to estimate the power series.

As shown in Fig. 4, the probability spectral function's most notable features occur during the recurring intervals

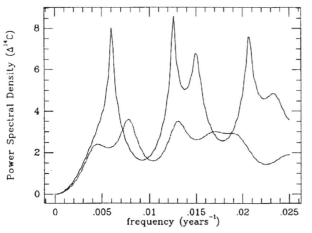

Fig. 5. Fourier transform functions of the combined strongly-periodic (solid line) and weakly-periodic (dotted line) intervals to the calibration Δ^{14}C data set. Note that the y-axis is power spectral density. Although the combined weakly-periodic transform function has four-times the frequency resolution as the strongly-periodic transform function, the spectral peaks appear much less distinct. This dispersion of spectral power may be due to frequency modulation or the response of a dissipative system.

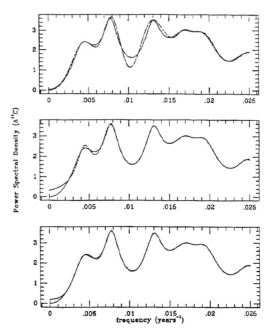

Fig. 6. Peak-shape fits (dot-dash line) to the low variability interval Fourier transform function (solid line) of the calibration Δ^{14}C data set using nonlinear least squares. Upper panel shows the Gaussian-shape fits, the middle panel show the Lorentzian-shape fits and the lower panel shows a Gaussian-shape fit to the $\simeq$ 220-year period peak and Lorentzian-shape to the shorter-period peaks (see Table 1).

that demonstrated high cross-correlation with the Homeric Δ^{14}C anomaly. These intervals have again been labeled A, B, C, D. Well-resolved $\simeq$ 210-year period domes and a more poorly resolved ridges lengthening from about a 360-year period to the 640-year period upper limit show contiguous regions of greater than a 67% probability of nonrandomness.

The clear distinction between the strongly-periodic Hallstattzeit episodes and the almost spectrally featureless regions between these intervals confirms non-stationarity. After combining the strongly-periodic and weakly-periodic intervals into two more stationary data sets, we used the Discrete Fourier Transform to more finely analyze the time-variant nature of the Δ^{14}C data set. Taking care to reduce the affect of aliasing and sample biasing so that the shapes of the spectral features suffer small distortion, Fig. 5 shows the two filtered transform functions. Note these spectra have not been transformed into Z-scores. Although the strongly-period intervals' transform function has only one fourth the frequency resolution of the weakly-periodic intervals' transform, the spectral peaks have a much sharper shape. For example, the $\simeq$ 80-year period feature present in the weakly-periodic spectral transform function has distinctly rounded shape.

Hypothesizing that this spectral peak represents response to a periodic forcing, the shape may be modeled as a Gaussian or as a Lorentzian function. The choice of model function has a fundamental impact upon interpretation. A Gaussian shape strongly implies that the $\simeq 80$-year period varies about a central value. Such a quasi-periodicity would result if the period varies with some longer-term variation such as the non-stationarity observed in Fig. 4. A Lorentzian shape implies that the data represents the response of a dissipative dynamic system to a periodic forcing. In this example, the first forcing function would likely be the 88-year Gleissberg solar activity cycle, the second the 210-year Suess cycle [Damon and Sonett, 1992]. The radiocarbon cycle may be the responsive dynamic system. Fig. 6 displays a nonlinear least-squares fit of Gaussian and Lorentzian shapes to the weakly-periodic spectral transform function (upper and middle panel, respectively). The Lorentzian-shape closely fits all peaks save the longest period. With a Gaussian-shape peak fit for this longest peak and Lorentzian-shapes for all the others, the nonlinear least-squares fit models all but 0.2% of the weakly-periodic spectral transform functions variance (lower panel Fig. 6). The poor resolution of the strongly-periodic portion spectral transform prohibited similar analyses on this data set.

Table 1 summarizes the result of the nonlinear least-square modeling of the weakly-periodic portion's spectral transform. Statistical comparison between the model peak shapes suggests that the Lorentzian shape appear to describe the $\simeq 80$-year spectral feature better than the Gaussian shape (based upon the standard error of the half-width, the most sensitive model parameter). This implies that the shape of the $\simeq 80$-year spectral peak may result from the radiocarbon system dissipating the Gleissberg solar activity forcing ^{14}C production. In contrast, the $\simeq 200$-year spectral feature may be better modeled by the Gaussian shape peak. An interpretation consistent with this result would be that the $\simeq 200$-year variation is modulated by the $\simeq 2100$-year recurrence shown in Figs. 3 and 4. This interpretation would be consistent with results by Damon and Sonett [1992]. For a complete quantitative verification of these interpretations, a conventional "goodness-of-fit" parameter such a second order norm may not be appropriate because the data has been transformed and filtered. To use such a measure, a weighting kernal function would have to be developed. Until such work is complete, comparisons between the standard error of the most sensitive model parameter, half-width, shown in Table 1, must suffice for now.

As a test of these interpretations, the detrended calibration $\Delta^{14}C$ data set was narrowly-filtered around 210 and 88-year periods. The choice of 210- and 88-year periods during the discussions below results from practical considerations. After testing a range of central periods between 220 and 200 years and 90 to 70 years, the choice of 210- and 88-year periods minimized side-lobe leakage from the time-domain filters employed below. The results shown in Fig. 7 show that the amplitude of an 88-year period varies gradually over the entire length of the data set. The largest amplitudes occur near 4000 BC, toward the end of the geomagnetic dipole field strength minimum during the Holocene. The lowest amplitudes occur near AD 0, when the geomagnetic dipole field strength peaked. In contrast, the 210-year period shows

TABLE 1. Least Squares Fit to the Discrete Fourier Power Spectrum of the Calibration $\Delta^{14}C$ Time Series

Period	Type	Amplitude	Halfwidth	Std-Error
227	Gaussian	2.38	0.00123	0.00002
226	Lorentzian	1.90	0.00122	0.00005
125	Gaussian	3.47	0.00109	0.00005
129	Lorentzian	3.01	0.00142	0.00004
77	Gaussian	3.45	0.00152	0.00011
77	Lorentzian	2.62	0.00142	0.00009
60	Gaussian	1.07	0.00083	0.00034
59	Lorentzian	1.43	0.00165	0.00010
53	Gaussian	2.88	0.00195	0.00071
52	Lorentzian	2.02	0.00202	0.00027
40	Gaussian	1.86	0.00188	0.00130
40	Lorentzian	1.49	0.00212	0.00033

little response to the geomagnetic dipole field strength. Instead, the amplitude appears more strongly affected by the Homeric-like anomalous intervals (marked A, B, C and D in the lower panel). This supports and extends the interpretation above. Summarizing, it appears that the Homeric-like events recurring every 1500 to 2500 years modulate the frequency and amplitude of the century-scale $\Delta^{14}C$ variations.

Evidence of a phase shift during the Homeric-like episodes would strengthen the modulation interpretation. Investigation of a phase-shift followed a similar path to the frequency and amplitude modulation analyses above. Narrow-band fil-

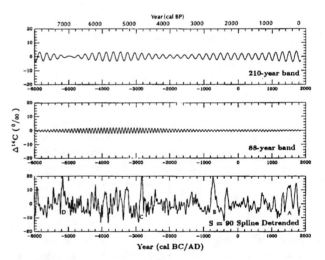

Fig. 7. Amplitude modulation of the 210-year and 88-year periodicities in the spline detrended calibration $\Delta^{14}C$ data set. The cubic-spline detrended calibration $\Delta^{14}C$ data set (stiffness=90) shown in the lowermost panel was narrow-banded filtered centered near 210-year (upper panel) and 88-year (middle panel) periods. The 88-year band shows modulation by the $\simeq 11$ ky geomagnetic dipole field strength maximum variation when the field strength was lower during the Fifth Millennium BC and minimum variation during the First Millennium AD when the field strength was higher. Note the 210-year period band shows significant modulation by the recurrences of Hallstattzeit episodes and little evident modulation by the geomagnetic dipole field strength.

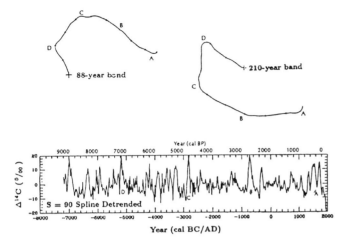

Fig. 8. Phase modulation of the 210-year and 88-year periodicities in the spline detrended calibration Δ^{14}C data set prepared similarly as in Fig. 7. Phase for each time increment is represented as a unit-length vector in polar coordinates. The angular coordinate of each vector represents the relative phase (zero-phase arbitrarily set). Beginning at the cross-hairs at 7000 BC, the vectors are connected head-to-tail for each successive increment in time (see Attolini *et al.*, [1984]). Ticks cross the trace every 914-years. Note that both the 88-year and 210-year periodicities change phase continuously during the Hallstattzeit episodes and during the latest portion of the calibration Δ^{14}C data set due to the dumping of "old" carbon into the atmosphere due to fossil fuel burning.

tering the calibration Δ^{14}C data set around the 210- and 88-year periods, the unit-length vector in polar coordinate space represents the phase as an angle over a window. The vectors for each window are connected head-to-tail to develop a trace of the evolution of phase with time [Attolini et al., 1984]. Two such traces ("cyclograms" or "hodographs" in the literature) projected onto polar coordinate space in the upper panel of Fig. 8 originate at 7120 BC and are ticked every 914 years. Phase shifts, represented as bends in the cyclogram trace, occur during the Hallstattzeit episodes. This surprising result confirms and confounds. These results confirm that the Hallstattzeit episodes modulate the $\simeq$ 220-year period variation. They also show a relationship between the $\simeq$ 80-year Gleissberg solar activity cycle and the Hallstattzeit episodes not suggested in the earlier analyses. One possible clue to interpreting this result occurs at the very end of the trace. Both traces curve sharply at the very end of the calibration Δ^{14}C data set beginning near *AD* 1800. As seen in the lower panel of Fig. 8, the Δ^{14}C values dip sharply at this time. This dip described by Suess [1968] results from the burning of ^{14}C-poor fossil fuels. This change in the radiocarbon cycle has dramatically shifted the phase in both the 210- and 88-year period variations. A change in the radiocarbon cycle would be a consistent interpretation for the Hallstattzeit episodes. Unlike the Suess

effect, however, this phase shift appears strongly period-dependent. Diffusion carbon cycle models have predicted a period-dependent phase response to ^{14}C production forcing [Seigenthaler et al., 1980]. Thus, the analyses presented constrains the interpretation of the Hallstattzeit episodes. They strongly suggest that the radiocarbon cycle responds dynamically to forcing, either externally due to changes in solar activity, or internally due to redistribution of carbon resulting from climatic forcing. Fortunately, researchers (beginning with Lorenz [1964]) have made much progress during the past thirty years in describing such complex geophysical systems.

Thorough methods to find the parameters of a chaotic dynamical systems require much larger data sets than the calibration Δ^{14}C data set. Undaunted, we have already reconneered using the nonlinear least squares model fit of Gaus-

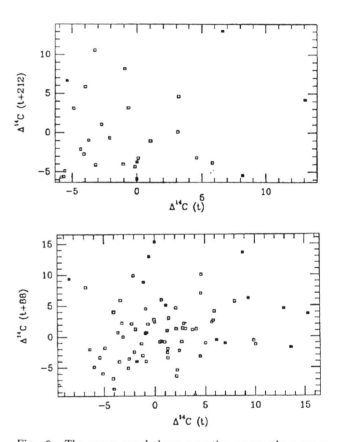

Fig. 9. The upper panel shows a section across phase space developed using a method described by Packard *et al.* [1980] using a sampling period of 210-years. Solid squares show the Hallstattzeit episodes, open squares show the weakly-period intervals. Although the visible evidence for two separate attractors is ambiguous, statistical tests suggest some distinction between the two distributions (see Table 2). The lower section across phase space developed similarly as the upper panel but for a sampling period of 88-years.

sian and Lorentzian wave shapes. The preliminary Gaussian model fit suggested mild frequency modulation of the $\simeq$ 220-year period. In a phase portrait, closed-loops represent such a frequency modulation. The Lorentzian model fit to the $\simeq$ 80-year period implies dissipation would create an attractor in phase space. Emboldened, we may even hypothesize that the two markedly distinctive strongly-periodic and weakly-period represent two states of ^{14}C response similar to behaviors observed in nonlinear harmonic oscillators. Using methods described by Packard et al. [1980], we have initiated studies investigating the radiocarbon cycle as a chaotic dynamical system. Preliminary results are shown in Fig. 9 as sections through phase space. Such sections are constructed by choosing the appropriate sampling frequency, usually a quite heuristic exercise. Our earlier investigations strongly suggest the utility of the 210- and 88-year periods, however, so they become consistent choices. Sampling the calibration Δ^{14}C data set at these periods may show the presence of multiple attractors. The strongly-periodic Hallstattzeit episodes (closed squares) do not appear to have a distinct distribution separate from the weakly-periodic intervals (open squares) for either the 210-or the 88-year sampling periods. Statistical analyses suggest some probability that the strongly-period and weakly-period intervals do have different distributions, however (see Table 2). Yet these results remain ambiguous.

More data would reduce uncertainty significantly. Increasing the length of the calibration Δ^{14}C data set may soon become limited by the scarcity of dendrochronology-dated wood. Even if enough precisely-dated sample material could be found, the transition from a glacial climate may prove too great a non-stationarity for the analyses above to be usefully employed. Another approach would sample Δ^{14}C and other data set on pentadal or annual time-scales. Such an approach has proven successful for sunspot indices [Morfill, et al., 1991; Blinov and Kremliovskij, in press] and less successful for Δ^{14}C [Blinov and Kremliovskij, in press]. The persistence of the radiocarbon cycle may reduce the amount of useful information obtained by sampling Δ^{14}C finer. A third approach would be to include related data sets. Schmidt and Gruhle's [1988] work strongly suggested a regional continental climate connection. Investigation of paleoclimate records with respect to the two state system hypothesis developed from the Δ^{14}C analyses above may confirm or confound.

TABLE 2: Statistical Measures of Significant Difference between High-variability and Low-variability Episodes in the Calibration Δ^{14}C Time Series

Sampling	F-Ratio	95%	t-score	95%
212 years	5.22	$\geq$ 2.42	1.72	$\geq$ 1.70
88 years	5.21	$\geq$ 1.84	3.14	$\geq$ 1.67

COINCIDENT ANOMALOUS ^{14}C AND CLIMATIC EVENTS

As an example study site, we have chosen Mission Cross Bog, Elko Co., Nevada [Thompson, 1984], because its sedimentation rate is fairly rapid (18 m in 8000 yr) and its ^{14}C dating suggested sedimentological changes at or around the time of large ^{14}C production anomalies. At 4260±150 yr BP the sediments change from lake mud (gyttja and diatomite) to peat, and Abies (fir) pollen percentages increase dramatically (Fig. 10). Our approach was to select sedimentary intervals that should be contemporaneous with radiocarbon anomalies, based on the available dates. We collected a series of short-interval, 10 cm, ^{14}C samples around each hypothesized ^{14}C production anomaly, and sampled for pollen analysis at 4 cm intervals.

Although Quaternary palynology typically focuses upon long-term (millennium-scale) vegetation change, this is a matter of emphasis rather than a physical limitation of the technique of pollen analysis. The typical sampling frequency is too low to record century-scale fluctuations, such as the 100-200 yr events we seek to compare to isotopic anomalies. However, the rate of vegetation change that can be studied is limited only by the rate of sediment deposition. In basins with annually-laminated sediments, pollen changes can be studied on an annual [MacDonald et al., 1991] or seasonal [Tippett, 1964] scale. Our 4-cm sampling density should census changes at 5-20 year intervals.

For this preliminary study, we report the results of the radiocarbon and pollen analysis of two intervals (Fig. 10). Each was selected based sedimentary evidence of environmental change, and proximity to an available ^{14}C date. The first interval (620-688 cm) was selected to match the Homeric-Greek Hallstattzeit episode. The sediments span a layer of humified peat around the date of 2470±100 (A-3701). We postulated that the humified layer, indicative of desiccation, might lie between the Homeric and Greek anomalies. The second interval (1270-1300 cm) was selected to match Noachan anomaly (4900-4700 yr). The sediments are between dates of 3720±120 yr BP (A-3702) and 4260±150 (A-3703), and they span a thin diatomite layer (called "clay" in Thompson [1984]) that we correlated with the Noachan anomaly.

The results of the radiocarbon dating and pollen analysis are shown in Fig. 10. The lower interval is dated 4538-4473 cal yr BP based on the calibrated age of 4065±60 for 1300 cm, and the average calibrated age of 4050±40, 4000±50, and 3940±40 for 1270 cm. Therefore, the pollen samples collected in lower interval are 300 years younger than the Noachan anomaly.

Radiocarbon dating such brief sediment intervals is difficult. But, due to the effects of the radiocarbon anomalies, precise calibrated time ranges for organic sediments can be determined using a technique called "wiggle-matching" [Van Geel and Mook, 1989]. A series of four additional ^{14}C samples was collected from between 620-820 cm in the core (Fig. 10). Precise ($\simeq$35 yr) radiocarbon dates of each of these

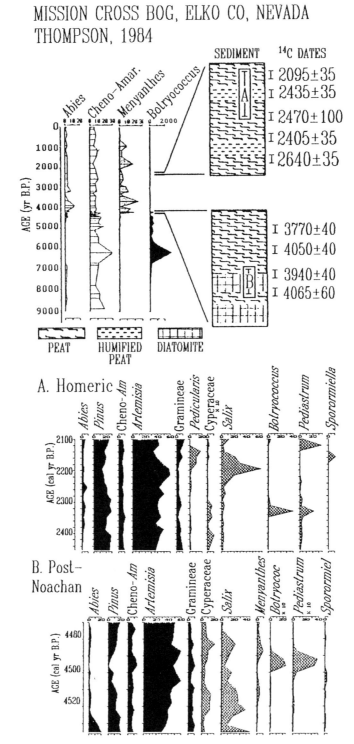

Fig. 10. Data used to test anomaly-climate association. Above, percentage pollen diagram for Mission Cross Bog is from Thompson's [1984] dissertation plotted *versus* ¹⁴C age. Sediment columns at right show stratigraphy for the intervals sampled (A. Homeric-Greek, B. Noachan lower), with ¹⁴C dates at right. Below, percentage pollen diagrams for the two sediment intervals are plotted versus calibrated ¹⁴C age.

YEAR BEFORE 1950

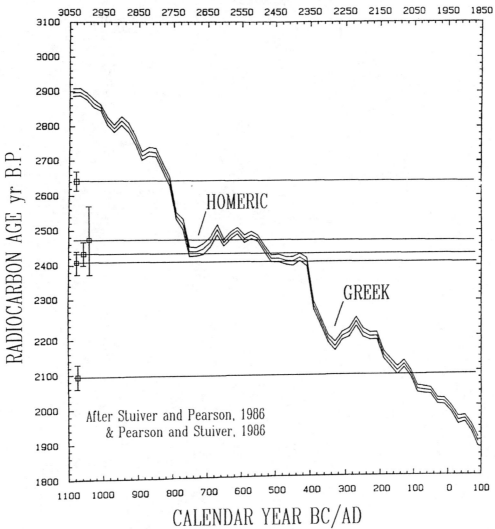

Fig. 11. The radiocarbon date-calendar year calibration curve (after Stuiver and Pearson [1986]; Pearson and Stuiver [1986]) showing "Wiggle-matched" Mission Cross Bog sediment dates and the Homeric and Greek ^{14}C production anomalies. Horizontal lines show radiocarbon dates on organic sediments. Error bars show the standard deviation of the ^{14}C measurements.

samples are plotted against the ^{14}C calibration curve calendar dates (Fig. 11). Note that the ^{14}C production anomalies have a different appearance in Fig. 11 then when represented by Δ^{14}C. The high production (increasing Δ^{14}C) stage of the ^{14}C production anomaly is evident as regions of steep negative slope. Near zero slope regions show the low production (decreasing Δ^{14}C) stage of the ^{14}C anomaly. Due to the large ^{14}C production anomalies during this period, there is a high (> 95%) degree of confidence that the pollen samples collected between 620–680 cm span the Greek anomaly and younger sediments, but misses the Homeric anomaly entirely.

In estimating sediment age, we considered superposition (i.e. oldest on the bottom) and sedimentation rates. The sedimentation rates derived from the calibrate dates are consistent with the averages based on Thompson's [1984] dates Homeric-Greek 4.18 yr cm^{-1} [Thompson, 1984] versus 4.82 yr cm^{-1} (this report); Noachan 2.17 yr cm^{-1} [Thompson, 1984] versus 1.54 yr cm^{-1} (this report).

The pollen diagrams for the two intervals record rapid oscillations in the aquatic environment that accompanies the sedimentary changes observed in these periods (Fig. 10). In the upper interval, Salix (willow) exceeds 20% from 632–644 cm (2157–2215 cal yr BP), suggesting expansion of

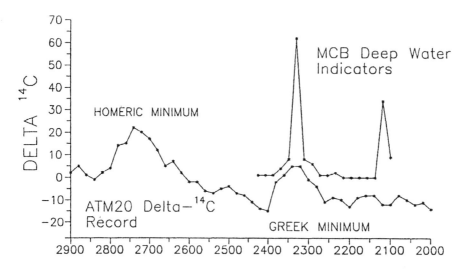

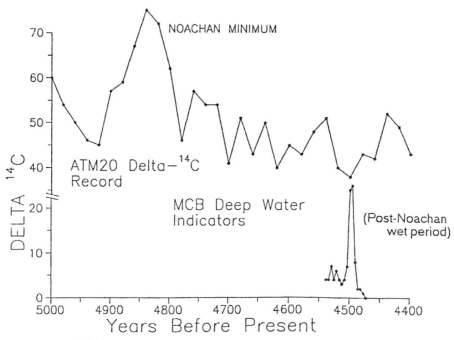

Fig. 12. Comparison of the $\Delta^{14}C$ record with moisture summary on pollen analysis of Mission Cross Bog. The deep water indicators are the sum of percentages of planktonic algae *Pediastrum* and *Botryococcus*, shown as square-root of percent in the lower graph.

woody vegetation due to drying of the bog surface. This sediment interval is narrower than, but within the humified layer shown in Fig. 10. The *Salix* peak is accompanied by increased *Artemisia* (sagebrush) percentages and decreased *Pinus* (pine) percentages, also indicative of aridity. Although it is only 50 years long, the simultaneous changes of the upland and aquatic environments argue for climatic change.

From 664-672 cm (2310-2348 cal yr BP), increased *Pediastrum* and *Botryococcus* (planktonic green algae) reach

maxima of 22% and 40%, respectively, suggesting a brief standing-water event that coincides with the Greek isotopic anomaly (Fig. 12). Concurrently, *Pinus* and *Abies* increase slightly, and *Artemisia* decreases. Again, the simultaneous changes of the upland and aquatic environments suggest climatic change. However, a second standing-water event (624 cm, 2119 cal yr BP, *Pediastrum* = 33%), does not correspond to a ¹⁴C production anomaly (Fig. 12).

In the lower interval, *Pediastrum* and *Botryococcus* percentages increase to 387% and 247% (1278-1284 cm, 4490-

4503 cal yr BP) in the diatomite layer, corroborating the indications of increased water depth. The percentages of bog-surface plants (*Cyperaceae, Salix,* and *Menyanthes*) are higher above and below the interval, in the peat bracketing the diatomite (Fig. 10). During this interval *Pinus* and *Abies* decline to < 1%, possibly suggesting a suppression of flowering for several years. However, despite this event's apparent severity, its calibrated age is 300 years after the Noachan isotope anomaly (Fig. 12).

CONCLUSIONS

We are encouraged that our close-interval sampling has recorded vegetation changes on decadal time scales. The changes seen in the pollen percentages (Fig. 10) match and corroborate those inferred from sediment stratigraphy. The 632-644 cm and 1278-1284 cm intervals both record environmental events of < 100 yr duration, in more than 2 samples, and the events are reflected in both aquatic and upland pollen types. We conclude that pollen analysis of Mission Cross Bog can detect rapid environmental changes on the scale of the climatic oscillations comparable to cosmogenic isotope anomalies. We also are encouraged that the precise radiocarbon dates on organic sediments allow the calibration of sediment samples to calendar time, and increase our confidence of vegetation changes on decadal time-scales.

An unsupportive result suggests that we have detected at least one brief event that does not correlate with cosmogenic isotope anomalies. The high-water events (increased *Pediastrum* and *Botryococcus* %) at 624 and 1278-1284 cm do not correspond to cosmogenic isotope anomalies. The underlying hypothesis of solar modulation of both cosmogenic isotope anomalies and climate does not exclude other climate-forcing events. Thus, the presence of these "non-anomaly" events does not falsify the hypothesis of solar modulation. However, it does show that brief climatic events in the pollen record do not necessarily correlate only with cosmogenic isotope anomalies, and such events cannot be solely used to date the profile. Our selection of a potentially-correlative interval at 1278-1284 cm, based entirely on the sediment stratigraphy failed to match the age of a major cosmogenic isotope anomaly.

We cannot conclusively support the solar modulation hypothesis as a causal mechanism for both ^{14}C production anomalies and rapid climate changes evident in the pollen record at Mission Cross Bog. However, the Greek anomaly, for which we have pollen samples, matched a wet period to a ^{14}C production anomaly. Further evidence supporting or refuting the solar modulation hypothesis could be gained by locating and analyzing pollen from the sediments correlative to the Homeric and Noachan anomalies.

Our next step will be to locate segments of the core dating to the Homeric and Noachan anomalies and collect close-interval pollen samples in those segments. Based on the existing dates, the Homeric interval should be at ≃700-725 cm, the Noachan at ≃1350-1375 cm. We already have located an earlier interval (7291-7175 BP) that dates to a [14]C

anomaly noted by Stuiver and Braziunas [1988]. Pollen from this interval should suggest cold-wet climate.

Our preliminary investigations have convinced us that rapid climatic events can be detected by pollen analysis of Mission Cross Bog. Although the Greek anomaly is matched by a wet period, we have neither supported nor refuted the hypothesis of solar modulation of global climate. Detailed study of sediments confidently dated to the Homeric, Noachan, and earlier anomalies are needed.

The time series analysis suggests rapid climate change may appear in tree-ring ^{14}C data sets. Transitions from weakly-period to a strongly-periodic variations occur over comparatively short (> 200-years) time-scales. Intervals like the Greek-Homeric Hallstattzeit episode show increased variability. This increased variability may be represented in climate records as well. The case for a strong physical link (such as solar forcing of climate change) between ^{14}C and climate anomalies has yet to be made, however. We believe the utility beyond dating for ^{14}C becomes apparent, however, as an indicator of solar and geomagnetic variation and changes in the radiocarbon cycle that may associate with other archives of continental climate change.

Acknowledgements. We benefitted from advice from Professors Charles P. Sonett and Paul E. Damon of the University of Arizona. We are grateful to Robert S. Thompson for preliminary analysis of Mission Cross Bog sediments. His pollen analysis, and our own studies will be published later. The comments of two anonymous reviewers were essential in clarifying and improving this paper. This work was partially supported by NSF Grants BNS-8902140, SES-9009974, ATM-8919535 and the State of Arizona.

REFERENCES

Attolini, M. R., S. Cecchini, and M. Galli, Time series variation analysis with Fourier vector amplitudes, *Il Nove Cimento, 7,* 245-253, 1984.

Berger, A., J. L. Miélice, and I. van der Mersch, Evolutive spectral analysis of sunspot data over the past 300 years, *Transactions of the Royal Society of London, A, 330,* 529-541, 1990.

Blinov, A. V., and M. N. Kremliovskij, Reflection of solar activity dynamics in radionuclide data, *Radiocarbon,* in press.

Damon, P. E., Radiocarbon and Climate (A comment on a paper by H. Suess), *Meteorological Monographs, 8,* 151-154, 1968.

Damon, P. E., J. C. Lerman, and A. Long, Temporal fluctuations of atmospheric ^{14}C causal factors and implications, *Annual Review of Earth and Planetary Science, 6,* 457-494, 1978.

Damon, P. E., Production and decay of radiocarbon and its modulation by geomagnetic field-solar activity changes with possible implications for global environment, in *Secular, Solar, and Geomagnetic Variations in the Last 10,000 Years,* edited by F. R. Stephenson and A. W. Wolfdale, Kluwer Academic Press, Dordrecht, 267-285, 1988.

Damon, P. E., S. Cheng, and T. W. Linick, Fine and hyperfine structure in the spectrum of secular variations of atmospheric ^{14}C, *Radiocarbon, 31,* 704-718, 1989.

Damon, P. E ., and C. P. Sonett, Solar and terrestrial components of the atmospheric ^{14}C variation spectrum, in *The Sun in Time,* edited by C. P. Sonett, M. S. Giampapa, and M. S. Matthews, University of Arizona Press, Tucson, 360-388, 1992.

Davis, O. K., Rapid climatic change in coastal southern California inferred from Pollen Analysis of San Joaquin Marsh, *Quaternary Research, 37,* 89-100, 1990.

Davis, O. K., J. L. Jirikowic, and R. M. Kalin, The Radiocarbon record of solar variability and Holocene climatic change in coastal southern California, in *Proceedings 8th PACLIM Workshop*, Technical Report Interagency Ecological Studies Program Sacramento-San Joaquin Estuary, 1992.

Davis, O. K., and D. S. Shafer, An early-Holocene maximum for the Arizona monsoon recorded at Montezuma Well, central Arizona, *Palaeogeography Palaeoclimatology Palaeoecology*, in press.

de Vries, H. L., Variation in concentration of radiocarbon with time and location on Earth, *Koninkl Nederlandse Akad Wetenschappen, Proc. Ser.B, 61*, 94-102, 1958.

Eddy, J. A., The Maunder Minimum, *Science, 192*, 1189-1202, 1976.

Eddy, J. A., Climate and the changing Sun, *Climate Change, 1*, 173-190, 1977. Grove, J. M., The Little Ice Age, *Methuen*, London, 1988.

Lamb, H. H., Climate History and the Future, *Princeton University Press*, Princeton, 1977.

Landscheidt, T., Long-term forecasts of solar cycles and climate change, Climate History, Periodicity and Predictability, edited by M. R. Rampino, J. E. Sanders, W. S. Newman and L. K. Königsson, *Van Nostrand Reinhold Co*, New York, 421-455, 1987.

Lean, J., A. Skumanich and O. White, Estimating the Sun's radiative output during the Maunder Minimum, *Geophysical Research Letters, 19*, 1591-1594, 1992.

Lerman, J. C., W. G. Mook, and J. C. Vogel, ¹⁴C in tree rings from different localities, in *Radiocarbon Variations and Absolute Chronology*, edited by I. U. Olsson, Almqvist & Wiksell, Stockholm, 275-301, 1970.

Lorenz, E. N., The problem of deducing the climate from governing equations, *Tellus, 16*, 1-16, 1964.

MacDonald, G. M., C. P. S. Larsen, J. M. Szeicz, and K. A. Moser, The reconstruction of Boreal forest fire history from lake sediments A comparison of charcoal, pollen, sedimentological, and geochemical indices, *Quaternary Science Reviews, 10*, 53-72, 1991.

Morfill, G. E., H. Scheingraber, and C. P. Sonett, Sunspot number variations stochastic or chaotic, in *The Sun in Time*, edited by C. P. Sonett, M. S. Giampapa, and M. S. Matthews, University of Arizona Press, Tucson, 30-58, 1991.

Packard, N. H., J. P. Crutchfield, J. D. Farmer, and R. S. Shaw, Geometry from a time series, *Physical Review Letters, 45*, 712-716, 1980.

Pearson, G. W., and M. Stuiver, High-precision calibration of the radiocarbon time scale, 500-2500 BC, *Radiocarbon, 28 (2B)*, 839-862, 1986.

Pittock, A. B., and N. G. Shapiro, Assessment of evidence of the effect of solar variations on weather and climate, in *Solar Variability, Weather, and Climate*, National Academy Press, Washington, 1982.

Priestley, M. B., Non-linear and Non-stationary Time Series Analysis, *Academic Press*, San Diego, 237 pp., 1988.

Raisbeck, G. M., F. Yiou, J. Jouzel, and J. R. Petit, 10Be and 2H in polar ice cores as a probe of the solar variability's influence on climate, *Transactions of the Royal Society of London, A, 330*, 463-470, 1990.

Schmidt, B., and W. Gruhle, Radiokohlenstoffgehalt und dendrochronologie, *Naturwissenshaftliche Rundschau, 5*, 177-182, 1988.

Seigenthaler, U., M. Heimann, and H. Oeschger, ¹⁴C variations caused by changes in the global carbon cycle, *Radiocarbon, 22*, 177-191, 1980.

Stuiver, M., Carbon-14 content of 18th and 19th century wood Variations correlated with sunspot activity, *Science, 149*, 533-537, 1965.

Stuiver, M., and H. A. Pollach, Reporting of ¹⁴C data, *Radiocarbon, 19*, 355-363, 1977.

Stuiver, M., and P. D. Quay, Atmospheric ¹⁴C changes resulting from fossil fuel CO2 release and cosmic ray variability, *Earth and Planetary Science Letters, 53*, 349-362, 1981.

Stuiver, M., and R. S. Kra (Eds.), Calibration Issue, *Radiocarbon, 28 (2B)*, 805-1030, 1986.

Stuiver, M., and G. W. Pearson, High-precision calibration of the radiocarbon time scale, *AD* 1950-500 BC, *Radiocarbon, 28 (2B)*, 805-838, 1986.

Stuiver, M., and T. F. Braziunas, The solar component of the atmospheric ¹⁴C record, in *Secular, Solar, and Geomagnetic Variations in the Last 10,000 Years*, edited by F. R. Stephenson and A. W. Wolfdale, Kluwer Academic Press, Dordrecht, 245-266, 1988.

Stuiver, M., T. F. Braziunas, B. Becker, and B. Kromer, Climatic, solar, oceanic, and geomagnetic influences on late-glacial and Holocene atmospheric ¹⁴C/¹²C change, *Quaternary Research, 35*, 1-24, 1991.

Suess, H. E., Climatic changes, solar activity and the cosmic-ray production rate of natural radiocarbon, *Meteorological Monographs, 8*, 146-150, 1968.

Tippett, R., An investigation into the nature of the layering of deep-water sediments in two eastern Ontario lakes, *Canadian Journal of Botany, 42*, 1693-1704, 1964.

Thomson, D. J., Time series analysis of Holocene climatic data, *Transactions of the Royal Society of London, A, 330*, 601-616, 1990.

Van Geel, B., and W. G. Mook, High-resolution ¹⁴C dating of organic deposits using natural atmospheric ¹⁴C variations, *Radiocarbon, 31*, 151-155, 1989.

Willson, R. C., and H. S. Hudson, The Sun's luminosity over a complete solar cycle, *Nature, 351*, 42-44, 1991.

O.K. Davis, J.L. Jirikowic and R.M. Kalin, Department of Geosciences, University of Arizona, Tucson, Arizona.

Hydrogen Isotopic Exchange and Stable Isotope Ratios in Cellulose, Wood, Chitin, and Amino Compounds

ARNDT SCHIMMELMANN

University of California at San Diego, Scripps Institution of Oceanography, La Jolla, CA 92093-0215, U.S.A.

RANDALL F. MILLER

New Brunswick Museum, 277 Douglas Avenue, Saint John, New Brunswick, Canada E2K 1E5

STEVEN W. LEAVITT

University of Arizona, Laboratory of Tree-Ring Research, Building #58, Tucson, AZ 85721, U.S.A.

Hydrogen bound to oxygen and nitrogen in organic compounds may exchange isotopically with ambient water hydrogen. This exchangeable hydrogen is also called 'active', 'mobile', or 'labile'. Most chemically complex compounds cannot be derivatized (for example, nitrated) to entirely exclude the influence of exchangeable hydrogen from the measurement of the stable isotope ratio of non-exchangeable, carbon-bound hydrogen. This interference can be overcome by independently equilibrating aliquots of organic sample with water vapors of different hydrogen isotopic compositions, followed by determinations of the bulk D/H ratios and mass-balance calculations. Our survey of the hydrogen isotopic exchangeability and chemical stability of chitin, wood, cellulose, and collagen at equilibration temperatures between 104 and 163°C proves the applicability of the isotope equilibration method for macromolecular substrates. Isotopic time-series of non-derivatized chitin, cellulose, and whole wood demonstrate their potential for paleoenvironmental isotopic studies. Sample size requirements of the isotope equilibration method are typically much smaller than for methods involving prior chemical derivatization.

INTRODUCTION

Stable hydrogen isotope ratios of modern, archaeological, and fossil organic substrates bear potentially valuable geochemical, environmental, dietary, climatic, and archaeological information [Buchardt and Fritz, 1980; Estep and Dabrowski, 1980; Connan et al., 1992; Smith et al., 1983; Schimmelmann and DeNiro, 1986a; Schimmelmann et al., 1986, 1987; Miller et al., 1988; Friedman et al., 1988; Miller, 1991; articles in this volume]. The problem of adsorbed moisture and uncontrolled isotopic exchange between organic hydrogen bound to oxygen (O-H) and nitrogen (N-H) with ambient water hydrogen limits the usefulness of measuring total D/H ratios in most organic compounds. Known exceptions are hydrocarbons and lipids [Schoell, 1984a, b; Sternberg, 1988], nitrated cellulose [Epstein et al., 1976; articles in this volume], and chemical derivatives of chitin [Schimmelmann and DeNiro, 1986a; Miller et al., 1988] all of which contain little or no exchangeable

hydrogen. Carbon-bound hydrogen (C-H) is non-exchangeable and thus isotopically conservative, as shown for cellulose for temperatures up to 250°C [Marino and DeNiro, 1987]. Below 170°C the isotopic integrity of C-H is preserved even over geologic time [Schoell, 1981]. Only high activation energies provided by exposure to radiation [Dahl et al., 1988] and reactions involving radicals (for example, via 'cracking') facilitate isotopic exchange between C-H and ambient water hydrogen [Hoering, 1975; Schoell, 1984a]. Chemical strategies have been developed for wood and insect exoskeletons to limit the amount of exchangeable hydrogen and to determine bulk δD values of extracted cellulose nitrate and chitin derivatives for paleoclimatic reconstruction. For collagen, lignin, and other chemically more complex substrates there are currently no derivatization techniques available to chemically eliminate exchangeable hydrogen. Our isotopic equilibration approach takes a first step to facilitate paleoclimatic exploitation of these compounds.

Isotopic control of exchangeable hydrogen via equilibration with water vapor of known isotopic composition allows the determination of the concentration and stable isotopic composition of the non-exchangeable hydrogen in organic matter (for underlying theory and equations, see Schimmelmann [1991]). This method is applicable to organic compounds that are chemically stable in water vapor at a chosen equilibration temperature regardless of the ratio of C-H versus O-H and N-H. This study establishes the applicability of the equilibration method for non-derivatized chitin, whole wood, non-nitrated cellulose, and collagen. Limitations become apparent from our preliminary data for monomeric amino compounds.

Preliminary tests of the technique using the well-known cellulose/cellulose-nitrate system [Schimmelmann, 1991] are continued here with holocellulose and whole wood to evaluate the isotopic response of lignin in wood. Without questioning the value and the high precision of the conventional cellulose nitrate method, we point out preparative limitations in the degree of nitration of cellulose and in the accuracy of the nitration method.

MATERIALS AND METHODS

Chitin was chemically isolated from whole or fragmented arthropod exoskeletons, minimizing the amounts of calcareous, protein, lipid, and other non-amino-sugar components of chitin from arthropods [Miller, 1984; Schimmelmann and DeNiro, 1986b]. The same chitin isolates had served in the form of derivatized substrates for isotopic determination of C-H hydrogen in earlier studies [Schimmelmann and DeNiro, 1986a; Miller et al., 1988]. Additional fossil insect chitins with ages between 10.5 and 12 ka were recovered as fragments from a buried organic-rich sequence near Lismore, Nova Scotia [Stea and Mott, 1989; Miller and Morgan, 1991]. Quaternary insect fossils, particularly beetles, are useful indicators of past climatic and environmental conditions. Beetle fragments are separated from organic-rich sediments using a kerosene floatation technique [Morgan and Morgan, 1980].

Whole wood samples represent annual growth increments from a 1970-1977 tree-ring sequence of oil pine (*Pinus tabulaeformis*) collected in 1991 by associates at the Xian Laboratory of Loess and Quaternary Geology in a state forest north of Xian, China. After grinding in a Wiley mill to sub-mm size, holocellulose (a mixture of α-cellulose, hemicelluloses and trace amounts of lignin) was extracted from splits of the annual samples [Green, 1963].

The isotopic response of bulk hydrogen to thermal stress was evaluated by equilibration experiments at different temperatures, between 104 and 163°C, for 20 hours each. Substrates tested were chitin (from lobster, *Homarus americanus*), cellulose (non-arboreal, provided by L. Sternberg, prepared according to Sternberg [1989]), whole ground wood, and collagen (type I, insoluble from bovine achilles tendon; Sigma Chemical Co.).

Six monomeric amino compounds were equilibrated in water vapor at 114°C over 20 hours: DL-α-amino-n-butyric acid (ABA, Sigma Chemical Co.), DL-phenylalanine (PA, Calbiochem), glycine amide hydrochloride (GA·HCl, Calbiochem), α-glucosamine hydrochloride (GlcN·HCl, Eastman), thiosemicarbazide (TSC, Sigma Chemical Co.), and semicarbazide hydrochloride (SC·HCl, Sigma Chemical Co.).

Open-ended quartz equilibration ampules [Schimmelmann, 1991] were loaded with samples weighing between 1 to 12 mg, depending on sample availability and hydrogen content. For the

isotopic equilibration we used distilled San Diego tap water (δD = -87 per mil) and distilled tap water mixed with D_2O, yielding a δD value of +704 per mil. The procedures for equilibration with water vapors, the subsequent combustion of organic hydrogen to H_2O, the reduction to H_2, and the isotopic determination of hydrogen stable isotope ratios are described elsewhere [Schimmelmann, 1991]. Isotopic results are reported in $δD_{SMOW}$ notation in per mil referring to Standard Mean Ocean Water (SMOW) isotope standard. The overall precision of the measurements, given as standard deviation of the mean calculated for six analyses of aliquots of a single cotton sample, is ± 1.5 per mil.

RESULTS AND DISCUSSION

Isotopic Response of Macromolecular Compounds to Thermal Stress During Isotopic Equilibration

Aliquots of chitin, cellulose, ground wood, and collagen were equilibrated with two isotopically different water vapors at temperatures between 104 and 163°C, for 20 hours each (Figure 1). Consistent yields and elemental ratios confirm the chemical stability of the substrates up to about 150°C. At higher temperatures, the substrates' color became light brown and the hydrogen yields diminished, presumably due to diagenetic polycondensation and elimination of functional groups.

The isotopic exchangeability, expressed in % exchangeable hydrogen of total hydrogen, increases in chitin from 15.8% at 114°C to 17.2% at 157°C, but shows no trend in cellulose, averaging 26.7 ± 0.2% over the same temperature range. At 104°C, however, the same cellulose has an exchangeability of only 23.1% [Schimmelmann, 1991]. In theory, 30% of the hydrogen in cellulose $[C_6H_7O_2(OH)_3]_n$ is O-H and thus potentially exchangeable, but in practice cellulose is a fibrous, closely-knit agglomerate of biopolymers in which not all O-H are physically accessible [Urbanski, 1965; Young and Rowell, 1986]. The chitin structure offers even more reduced sterical accessibility of its functional groups [Muzzarelli, 1977]. The few data on hydrogen exchangeability in collagen indicate 16.7% at about 110°C, with an estimated increase to 17.7% at 150°C. Whole wood at 120°C has 13.6% exchangeable hydrogen, with an increase to 14.7% at about 150°C.

Our results on the hydrogen isotopic exchangeability indicate that with increasing temperature more hydrogen is subject to isotopic exchange. At room temperature, hydrogen isotopic exchange between water and exchangeable hydrogen in low-molecular organic compounds occurs within seconds to a few hours in liquid phase [Worley and Garteiz, 1969; Carlson et al., 1972; MacCarthy and Bowman, 1981; and others] and several orders of magnitude faster in gas phase [Guarini et al., 1992]. The rate-controlling factor for isotopic exchange in larger molecules at temperatures above 100°C for 20 hours is therefore predominantly the permeability, or sterical accessibility of functional groups within the substrate's structure. The exchange rate varies depending on chemical composition and structure. Isotopic equilibration requires stringent control of temperature to achieve reproducibility. It is important to realize that for all substrates one needs to establish a temperature dependence and time regime for isotopic equilibration

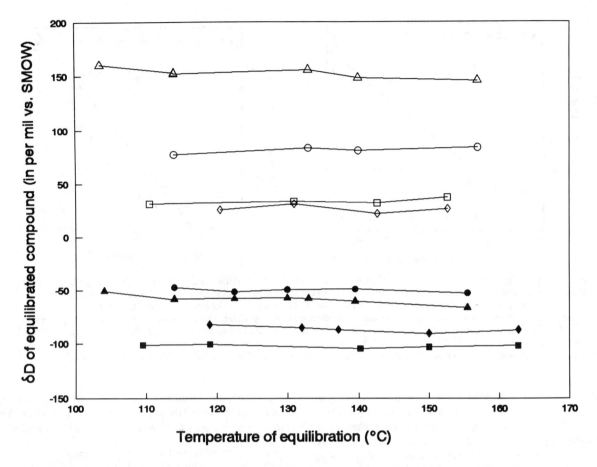

Fig. 1. Bulk hydrogen stable hydrogen isotope ratios from cellulose (triangles), chitin (circles), whole wood (diamonds), and collagen (squares) that were equilibrated for 20 hours at different temperatures, with two isotopically different water vapors (solid symbols: -87 per mil; open symbols: +704 per mil).

that achieves reproducible exchangeability while not compromising the chemical stability.

The δD values in Figure 1 show the expected trends due to thermal changes in the isotope fractionation effect ε between exchangeable organic hydrogen and water vapor [Schimmelmann, 1991]. The overall mean isotope fractionation effect ε of a chemical substrate with more than one type of O-H and/or N-H is the weighted average of the fractionation effects of individual functional groups. With increasing temperature ε decreases [Bigeleisen and Ishida, 1975a, b]. In cellulose, where there is no N-H and no additional hydrogen becomes exchangeable between 114 and 157°C, the mean isotopic decrease for bulk hydrogen of 0.19 per mil per degree temperature increase translates into 0.7 per mil $°C^{-1}$ of the exchangeable hydrogen. Thus, even for cellulose, isotopic reproducibility mandates isothermal conditions for each set of experiments, or that the isotopic results are corrected for changes in the fractionation effect ε. The chemical composition (i.e., the relative abundances of types of O-H and N-H) also influences ε, although the variance should be of minor importance within a given compound class, for example collagen.

Isotopic Equilibration of Chitin

Ideally, chitin is a biopolymer consisting of N-acetyl-D-glucosamine units, but natural chitin is partly deacetylated. Chitin is hardened by 'sclerotization', i.e. chemical cross-linking with non-amino-sugar compounds such as peptides and pigments [Muzzarelli, 1977]. Exchangeable hydrogen in chitin is therefore not limited to O-H and N-H in (N-acetyl)-D-glucosamine. Two different chemical approaches were developed for the exclusion of exchangeable hydrogen. Nitration of hydroxyl-groups in chitin decreases the amount of exchangeable O-H hydrogen. Miller et al. [1988] used the bulk δD values in 'chitin nitrate' for correlation with isotopic precipitation patterns and thus for paleoclimatic reconstruction. This straightforward and convenient method requires 5 to 10 mg of sample. For paleoclimatology on fossil chitin this is still a large sample, when one considers the enormous effort to concentrate sub-mm chitin fragments from soil or peat. Also, the bulk δD values of nitrated chitin are still affected by N-H hydrogen and some non-nitrated O-H, including the O-H of firmly adsorbed water. The other method excludes all exchangeable hydrogen by polycondensation of chitose (prepared from

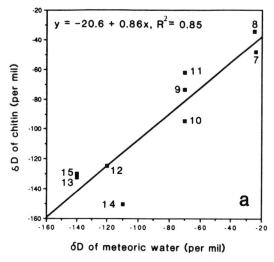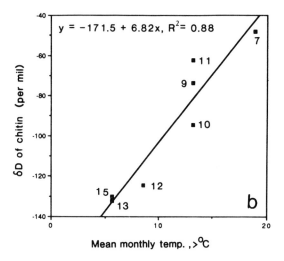

Fig. 2. δD values of modern North American beetle chitin equilibrated with -87 per mil water for 20 hours at 130°C versus (a) δD values of local meteoric water based on IAEA and UW samples, and (b) versus mean local temperature for months when the average temperature was greater than 0°C (for underlying statistics, see Miller et al. [1988]). Chitin sources are identified by numbers: 7 = scarab beetle, cf. Euetheola sp., Kentwood, Louisiana; 8 = water beetle, Hydrophilidae, Venice, Florida; 9 = June beetle, Phyllophaga, Waterloo, Ontario; 10 = ground beetle, Stenolophus comma, Waterloo, Ontario; 11 = grapevine beetle, Pelidnota punctata, Waterloo, Ontario; 12 = ground beetles, Carabidae, Churchill, Montana; 13 = ground beetles, Carabidae, Whatever Lake, Northwest Territories, Canada; 14 = ground beetle, Carabidae, Skagway, Alaska; 15 = undetermined, Whatever Lake, Northwest Territories, Canada.

hydrolyzed chitin) to form furanes [Schimmelmann and DeNiro, 1986a]. The δD values resulting from this accurate, yet labor-intensive method represent only four out of ten non-exchangeable C-H hydrogen positions in D-glucosamine. The sample size requirement of approximately 100 mg of chitin as starting material for one isotopic measurement is prohibitive for paleoclimatic applications. In contrast, the isotopic equilibration method used in this study is capable of measuring δD values in 1 mg of chitin. Chitin hydrogen isotope ratios from different analytical methods are not directly compatible, because the various methods focus on distinctly different hydrogen pools.

A survey of the hydrogen exchangeability of 13 modern chitin isolates at 130°C over 20 hours yielded a mean hydrogen exchangeability of 15.3 ± 2.9% (n = 13). This value is significantly below the theoretical 23.1% for 'ideal' chitin, poly-N-acetyl D-glucosamine. In addition to our results from the thermal stress study, this underscores the chitin's resistance to hydrogen isotopic exchange.

δD values of chitin isolates from modern North American beetles that were equilibrated with water vapor of -87 per mil are plotted against average annual δD values of local meteoric water determined from nearby sites (Figure 2a) and against mean local temperature for months when the average temperature was greater than 0°C (Figure 2b) (for underlying statistics, see Miller et al. [1988]). The R² values of 0.85 and 0.88, respectively, suggest that the isotopic equilibration of chitin is a useful approach for the reconstruction of isotopic precipitation patterns and temperature.

A preliminary time-series of δD values from equilibrated fossil beetle chitin from Lismore (Nova Scotia) is shown in Figure 3. Although we are cautious with the interpretation of these δD data from bulk undetermined Coleoptera fragments, we note that the

isotopic trends are tantalizing and not inconsistent with temperature trends seen in some late glacial sequences in Europe (i.e., a climatic warming prior to 12,200 BP, an Older Dryas cold interval

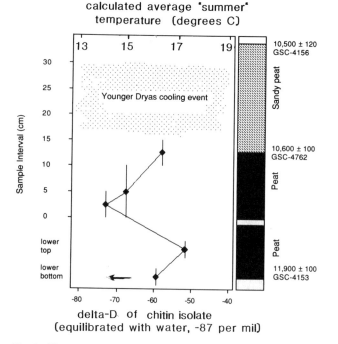

Fig. 3. Hydrogen isotopic ratios of fossil chitins from Lismore (Nova Scotia) after isotopic equilibration with water vapor of -87 per mil at 134°C over 20 hours. Stratigraphy adapted from Miller and Morgan [1991]. Paleotemperatures were calculated based on the regression shown in Figure 2b.

12,200-12,050 BP, followed by Allerød warming 12,000-10,700 BP; Wright [1989]). The δD data also agree well with palynological evidence suggesting a pre-Younger Dryas climatic cooling in eastern Canada between 11,220 and 11,010 BP [Levesque, 1992]. Numerous biological factors remain to be considered [Miller et al., 1988] but clearly the technique merits investigation.

Isotopic Time-Series of Whole Wood and Holocellulose

The two major components of wood are cellulose and lignin. Stable isotope ratios in chemically complex whole wood are considered ill-defined and unsuitable to serve as reliable paleoclimatic proxies, although some data have been published [Libby and Pandolfi, 1977]. We are not endorsing the use of wood as a substitute for cellulose. Our experiments described here are intended to evaluate the influence of lignin as a major component in wood, in comparison with the associated holocellulose.

Aliquots of annual ground whole wood and its extracted holocellulose from a 1970-1977 tree-ring sequence of *Pinus tabulaeformis* were equilibrated with water vapors of -87 and +704 per mil at 130°C for 20 hours. The resulting bulk hydrogen stable isotope ratios (shown in Figure 4) translate into an isotope

exchangeability of 21.9 ± 1.7% (n = 8) for wood-derived holocellulose. Interestingly, this arboreal holocellulose exhibits about 5% less exchangeability than the non-woody cellulose described earlier. The reason may lie in biological structural differences, as well as in the different methods employed in the extraction of the celluloses. For whole wood we determined a relatively low exchangeability of 17.4 ± 1.0% (n = 8). In comparison to lignin, the second main component of wood, which has a phenolic structure with few O-H positions [Hedges and Mann, 1979, and references therein], the carbohydrate cellulose is rich in O-H, expressing itself in a larger exchangeability. Whole wood hydrogen is more depleted in deuterium than holocellulose hydrogen, therefore lignin should have even more negative δD values than whole wood. Results of mass-balance calculations of δD values for non-exchangeable hydrogen in holocellulose are indicated by lines in Figure 4, for three different fractionation effects ε = 70, 76, and 80 (equations for calculation are given by Schimmelmann [1991]). The offset among these lines is about 1-3 per mil. The correct value for ε at 130°C can be determined experimentally (see above), eliminating this source of uncertainty.

The ratio of holocellulose versus lignin (C/L) in a tree-ring, which is determined by species-specific differences [Panshin and de Zeeuw, 1980] and environmental factors, therefore influences

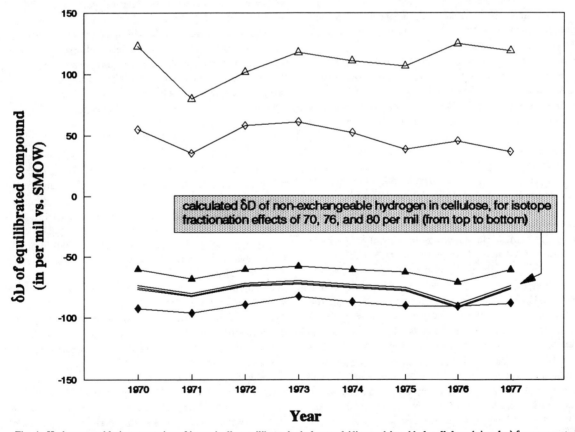

Fig. 4. Hydrogen stable isotope ratios of isotopically equilibrated whole wood (diamonds) and holocellulose (triangles) from separated annual growth rings of *Pinus tabulaeformis*. Isotopic equilibrations took place with water vapors of -87 per mil (solid symbols) and +704 per mil (open symbols) at 130°C over 20 hours. Lines without symbols indicate δD values of non-exchangeable hydrogen in holocellulose, calculated for various fractionation effects ε.

the hydrogen exchangeability and the hydrogen isotopic composition of equilibrated, bulk hydrogen. Inter-annual differences in C/L may thus be partly responsible for non-parallelity of the holocellulose and wood time-series in Figure 4. The time-series of holocellulose and whole wood (equilibrated with water vapor of -87 per mil) correlate with R = 0.64.

The similar patterns expressed by time-series in Figure 4 suggest that, under conditions of isotopic equilibration, the hydrogen isotopic signature of C-H in lignin is isotopically conservative. This raises the possibility that δD values of C-H in lignin may be exploited as paleoclimatic proxy, after proper isotopic evaluation of biological and environmental constraints of its biosynthesis and post-mortem diagenesis. Our results encourage further hydrogen isotopic studies on lignin, for example utilizing tree remnants submerged in bogs where cellulose may have been degraded, whereas the more refractory lignin is still preserved [Hedges et al., 1985]. Kohara [1956] describes buried 150,000 yr old wood that is reduced nearly completely to lignin.

Accuracy of δD Data in Cellulose and Cellulose Nitrate

The isotopic equilibration method for cellulose takes into account all O-H hydrogen that is isotopically exchangeable at a given temperature. Furthermore, the fractionation effect ε for cellulose can be determined experimentally, using a nitrated portion of the cellulose for comparative mass-balance calculations (ε = 80 per mil at 114°C, Schimmelmann [1991]), contributing significantly to the accuracy of the equilibration method.

The nitration of cellulose is undoubtedly an economic and highly successful means of controlling O-H in cellulose. We are not suggesting to abandon nitration in favor of isotopic equilibration. However, there appears to be a widespread overly optimistic and complacent view on the accuracy of the method. The nitration of cellulose in acid baths is unable to achieve a nitrogen content of 14.15% N for theoretical, fully nitrated cellulose, because the structure of cellulose sets practical limits for the access of nitric acid to functional O-H groups [Urbanski, 1965, and numerous references therein; DeNiro, 1981]. We argue that water with a smaller molecular size than that of nitric acid is able to access and exchange isotopically with some of the non-nitrated O-H in cellulose nitrate. Multiple determinations of the hydrogen exchangeability in a batch of properly nitrated cellulose (by Sternberg, via method of Sternberg [1989]) yielded a remaining hydrogen exchangeability of 3.9% (at 114°C over 20 hours, Schimmelmann [1991]). Some problems associated with the nitration of cellulose, namely that hydrogen stable isotope data show systematic differences among various nitration techniques, and that celluloses from woody plants require special precaution to avoid 'contamination' [DeNiro, 1981], may be due to insufficient elimination of O-H. We recognize that the nitration of cellulose in a given lab with a fairly constant isotopic composition of its reagents and ambient moisture may produce δD data for cellulose nitrate with high precision (defined as closeness of agreement between randomly selected individual measurements). To test the accuracy (defined as the closeness of agreement between an observed value and an accepted reference value), we suggest to

perform inter-laboratory comparisons using a common batch of cellulose.

Monomeric Amino Compounds

The chemical integrity of four monomeric amino compounds ABA, PA, GA·HCl, and GlcN·HCl over 20 hours of equilibration in water vapor at 114°C was monitored by measuring the carbon stable isotope ratios as well as the C/N elemental ratios of unequilibrated and of equilibrated aliquots. ABA is stable and the experimentally determined hydrogen exchangeability of 34.6% is close to the theoretical value of 33.3% (3 out of 9 hydrogen atoms are O-H or N-H). PA with a theoretical hydrogen exchangeability of 27.3% shows an experimental exchangeability of only 8.5%, but the carbon isotopic and C/N elemental ratios are conservative. Apparently PA forms amide bonds via intermolecular condensation reactions at 114°C, thus eliminating significant amounts of exchangeable hydrogen. TSC, SC·HCl, and GA·HCl prove to be chemically unstable at 114°C, resulting in changes of the carbon isotopic and C/N elemental compositions. The experimentally determined hydrogen exchangeability of 81.8% in GA·HCl is more than 10% above the theoretical level and may indicate chemical conversion. The hydrochloride of the deacetylated monomer of the biopolymer chitin, GlcN·HCl, is also chemically unstable at 114°C and yields irreproducible data. The general thermal instability of aminosugar monomers and their derivatives [Muzzarelli, 1977, and references therein] and that of other tested monomeric amino compounds contrasts sharply with the excellent stability of chitin and collagen (Figure 1). Isotopic equilibration of chemically reactive, monomeric amino compounds appears to be feasible only at temperatures below 100°C. Such equilibration would need to be performed under reduced pressure to avoid the formation of liquid water and phase transition isotope fractionation.

CONCLUSIONS

Most tested monomeric amino compounds are chemically unstable during the isotopic equilibration at 114°C. In contrast, the macromolecular biochemical compounds chitin and collagen, as well as cellulose and whole wood, are stable, at least up to 150°C, and permit reproducible determination of hydrogen stable isotope ratios. A similar isotopic response of holocellulose and whole wood to isotopic equilibration suggests that lignin isolates themselves may contain analytically accessible, isotopically conservative hydrogen.

Paleodietary, paleoclimatic, and paleoenvironmental stable isotope studies can utilize the hydrogen isotopic composition of chemically complex substrates, as demonstrated here for chitin, non-nitrated cellulose, whole wood, and collagen. The isotope equilibration method minimizes chemical preparation of samples. Subsequently, the sample size requirement is typically smaller than for methods which require prior chemical derivatization. For chitin, whole wood and cellulose we found that one milligram is sufficient for one equilibration, yielding 5 to 25 μmol of hydrogen. This is extremely important for analysis of paleontological materials which yield small samples. Further use of the isotopic equilibration method might be extended to collagen from fossil bones, dentin from teeth, humic compounds and phyllosilicates.

Acknowledgments. The tree-ring material was collected through a cooperative field excursion by the Laboratory of Tree-Ring Research (with Malcolm Hughes) and the Xian Laboratory of Loess and Quaternary Geology (An Zhiseng, Liu Rongmo, Sun Xiang and Liu Yu). Gracie Gutierrez processed the tree rings to holocellulose under NSF Grant No. ATM-9143581. Additional cellulose was provided by Leonel Sternberg. Ralph Stea and Bob Mott aided the collection of fossil insect samples. Two anonymous reviewers provided valuable suggestions. A. Schimmelmann acknowledges suppport from Scripps Industrial Associates, from a Biomedical Research Support Grant, from the Department of Energy research grant number DE-FG03-92ER14245, and from the Universitywide Energy Research Group account #6-506290.

REFERENCES

Bigeleisen, J., and T. Ishida, Isotope chemistry and molecular structure. Total deuterium isotope effects, *J. Phys. Chem.*, *62*, 80-88, 1975a.

Bigeleisen, J., and T. Ishida, Erratum: "Total deuterium isotope effects", *J. Phys. Chem.*, *63*, 1702, 1975b.

Buchardt, B., and P. Fritz, Environmental isotopes as environmental and climatological indicators, in *Handbook of Environmental Isotope Geochemistry*, vol. 1, edited by P. Fritz and J. C. Fontes, pp. 473-504, Elsevier, Amsterdam, 1980.

Carlson, G. L., W. G. Fateley, and F. F. Bentley, A simple deuteration technique for the identification of OH and NH vibrations in the far infrared region, *Spectrochim. Acta*, *28A*, 177-179, 1972.

Connan, J., A. Nissenbaum, and D. Dessort, Molecular archaeology: Export of Dead Sea asphalt to Canaan and Egypt in the Chalcolithic-Early Bronze Age (4th-3rd millennium BC), *Geochim. Cosmochim. Acta*, *56*, 2743-2759, 1992.

Dahl, J., R. Hallberg, and I. R. Kaplan, The effects of radioactive decay of uranium on elemental and isotopic ratios of Alum Shale kerogen, *Appl. Geochem.*, *3*, 583-589, 1988.

DeNiro, M. J., The effects of different methods of preparing cellulose nitrate on the determination of the D/H ratios of non-exchangeable hydrogen of cellulose, *Earth Planet. Sci. Lett.*, *54*, 177-185, 1981.

DeNiro, M. J., and S. Epstein, Hydrogen isotope ratios of mouse tissues are influenced by a variety of factors other than diet, *Science*, *214*, 1374-1375, 1981.

Epstein, S., C. J. Yapp, and J. H. Hall, The determination of the D/H ratios of non-exchangeable hydrogen in cellulose extracted from aquatic and land plants, *Earth Planet. Sci. Lett.*, *30*, 241-251, 1976.

Estep, M. F., and H. Dabrowski, Tracing food webs with stable hydrogen isotopes, *Science*, *209*, 1537-1538, 1980.

Friedman, I., P. Carrara, and J. Gleason, Isotopic evidence of Holocene climatic change in the San Juan Mountains, Colorado, *Quat. Res.*, *30*, 350-353, 1988.

Green, J. W., Wood cellulose, in *Methods of Carbohydrate Chemistry*, edited by R. L. Whistler, pp. 9-21, Academic Press, San Diego, 1963.

Guarini, A., G. Guglielmetti, and N. Andriollo, Labile hydrogen counting in biomolecules using deuterated reagents in desorption chemical ionization and fast atom bombardment mass spectrometry, *Anal. Chem.*, *64*, 204-210, 1992.

Hedges, J. I., and D. C. Mann, The characterization of plant tissues by their lignin oxidation products, *Geochim. Cosmochim. Acta*, *43*, 1803-1807, 1979.

Hedges, J. I., G. L. Cowie, J. R. Ertel, R. J. Barbour, and P. G. Hatcher, Degradation of carbohydrates and lignins in buried woods, *Geochim. Cosmochim. Acta*, *49*, 701-711, 1985.

Hoering, T. C., The biogeochemistry of the stable hydrogen isotopes, *Annual Report Director Geophysical Laboratory, Carnegie Institution Yearbook 1974-1975*, Carnegie Institution, Geophysical Laboratory, Washington D.C., 598-604, 1975.

Kohara, J., Permanence of wood, *Mokuzai Gakkaishi*, *19*, 191-195, and *20*, 195-200, 1956.

Levesque, A. J., A Younger Dryas precursor in maritime Canada

(abstract), *Geol. Assoc. Can./ Mineral. Assoc. Can., Wolfville '92*, *17*, A65, 1992.

Libby, L. M., and L. J. Pandolfi, Climate periods in tree, ice and tides, *Nature*, *266*, 415-417, 1977.

MacCarthy, P., and S. J. Bowman, Mulls of deuterated solid samples for infrared spectrometry, *Anal. Chem.*, *53*, 1151-1152, 1981.

Marino, B. D., and M. J. DeNiro, Isotopic analysis of archaeobotanicals to reconstruct past climates: effects of activities associated with food preparation on carbon, hydrogen and oxygen isotope ratios of plant cellulose, *J. Archaeol. Sci.*, *14*, 537-548, 1987.

Miller, R. F., Stable isotopes of carbon and hydrogen in the exoskeleton of insects: Developing a tool for paleoclimatic research, Ph.D. thesis, Univ. Waterloo, Canada, 1984.

Miller, R. F., Chitin paleoecology, *Biochem. Syst. Ecol.*, *19*, 401-411, 1991.

Miller R. F., and A. V. Morgan, Late-Glacial Coleoptera fauna from Lismore, Nova Scotia, *Atlantic Geol.*, *27*, 193-197, 1991.

Miller, R. F., P. Fritz, and A. V. Morgan, Climatic implications of D/H ratios in beetle chitin, *Palaeogeogr., Palaeoclim., Palaeoecol.*, *66*, 277-288, 1988.

Muzzarelli, R. A. A., *Chitin*, 309 pp., Pergamon Press, Oxford, 1977.

Panshin, A. J., and C. de Zeeuw, *Textbook of Wood Technology*, 722 pp., McGraw Hill, New York, 1980.

Schimmelmann, A., Determination of the concentration and stable isotopic composition of non-exchangeable hydrogen in organic matter, *Anal. Chem.*, *63*, 2456-2459, 1991.

Schimmelmann, A., and M. J. DeNiro, Stable isotopic studies on chitin III. The $^{18}O/^{16}O$ and D/H ratios in arthropod chitin, *Geochim. Cosmochim. Acta*, *50*, 1485-1496, 1986a.

Schimmelmann, A., and M. J. DeNiro, Stable isotopic studies on chitin. Measurements on chitin/chitosan isolates and D-glucosamine hydrochloride from chitin, in *Chitin in Nature and Technology*, edited by R. A. A. Muzzarelli, C. Jeuniaux and G. W. Gooday, , pp. 357-364, Plenum, New York, 1986b.

Schimmelmann, A., M. J. DeNiro, M. Poulicek, M.-F. Voss-Foucart, G. Goffinet, and C. Jeuniaux, Isotopic composition of chitin from arthropods recovered in archaeological contexts as palaeoenvironmental indicators, *J. Archaeol. Sci.*, *13*, 553-566, 1986.

Schimmelmann, A., P. Lavens, and P. Sorgeloos, Carbon, nitrogen, oxygen and hydrogen stable isotope ratios in *Artemia* from different geographical origin, in *Artemia Research and its Applications, Vol. 1, Morphology, Genetics, Strain characterization, Toxicology*, edited by P. Sorgeloos, D. A. Bengtson, W. Decleir, and E. Jaspers, pp. 167-172, Universa Press, Wetteren, Belgium, 1987.

Schoell, M., D/H-Isotopenverhältnisse in organischen Substanzen, Erdölen und Erdgasen, *Rep. T 81-204*, 79 pp., Bundesanstalt für Geowissenschaften und Rohstoffe, Hannover, Germany, 1981.

Schoell, M., Wasserstoff und Kohlenstoff Isotope in Organischen Substanzen, Erdölen und Erdgasen, *Geol. Jb.*, *D67*, Schweizerbart, Hannover, 1-161, 1984a.

Schoell, M., Recent advances in petroleum isotope geochemistry, *Org. Geochem.*, *6*, 645-663, 1984b.

Smith, J. W., D. Rigby, P. W. Schmidt, and D. A. Clark, D/H ratios of coals and the paleoaltitude of their deposition, *Nature*, *302*, 322-323, 1983.

Stea, R. R., and R. J. Mott, Deglaciation environments and evidence for glaciers of Younger Dryas age in Nova Scotia, Canada, *Boreas*, *18*, 169-187, 1989.

Sternberg, L., D/H ratios of environmental water recorded by D/H ratios of plant lipids, *Nature*, *333*, 59-61, 1988.

Sternberg, L., Oxygen and hydrogen isotope measurements in plant cellulose analysis, in *Modern Methods of Plant Analysis, New Series, Vol. 10, Plant Fibers*, edited by H. F. Linskens and J. F. Jackson, pp. 89-99, Springer, Berlin, 1989.

Urbanski, T., *Chemistry and Technology of Explosives*, vol. 2, 517 pp., Macmillan, New York, 1965.

Worley, J. D., and D. A. Garteiz, Generation of deuterated materials in a nonpolar solvent, *J. Chem. Education*, *46*, 608-610, 1969.

Wright, H. E., The amphi-Atlantic distribution of the Younger Dryas paleoclimatic oscillation, *Quat. Sci. Rev.*, *8*, 295-306, 1989.

Young, R. A., and R. M. Rowell, *Cellulose: Structure, Modification, and Hydrolysis*, 379 pp., Wiley, New York, 1986.

A. Schimmelmann, Scripps Institution of Oceanography, La Jolla, CA 92093-0215.

Randall F. Miller, New Brunswick Museum, 277 Douglas Avenue, Saint John, New Brunswick, Canada E2K 1E5.

Steven W. Leavitt, University of Arizona, Laboratory of Tree-Ring Research, Building #58, Tucson, AZ 85721.